Representations of Reductive Groups

Proceedings of Symposia in
PURE MATHEMATICS

Volume 101

Representations of Reductive Groups

Conference in honor of Joseph Bernstein
Representation Theory & Algebraic Geometry
June 11–16, 2017
Weizmann Institute of Science, Rehovot, Israel
and The Hebrew University of Jerusalem, Israel

Avraham Aizenbud
Dmitry Gourevitch
David Kazhdan
Erez M. Lapid
Editors

AMERICAN MATHEMATICAL SOCIETY
Providence, Rhode Island

2010 *Mathematics Subject Classification.* Primary 11F27, 11F70, 20C08, 20C33, 20G05, 20G20, 20G25, 22E47.

Library of Congress Cataloging-in-Publication Data

Names: Conference on Representation Theory and Algebraic Geometry in honor of Joseph Bernstein (2017 : Jerusalem, Israel) | Aizenbud, Avraham, 1983- editor. | Gourevitch, Dmitry, 1981- editor. | Kazhdan, D. A. (David A.), 1946- editor. | Lapid, Erez, 1971- editor. | Bernstein, Joseph, 1945- honoree.

Title: Representations of reductive groups : Conference on Representation Theory and Algebraic Geometry in honor of Joseph Bernstein, June 11-16, 2017, The Weizmann Institute of Science & The Hebrew University of Jerusalem, Jerusalem, Israel / Avraham Aizenbud, Dmitry Gourevitch, David Kazhdan, Erez M. Lapid, editors.

Description: Providence, Rhode Island : American Mathematical Society, [2019] | Series: Proceedings of symposia in pure mathematics ; Volume 101 | Includes bibliographical references.

Identifiers: LCCN 2018041948 | ISBN 9781470442842 (alk. paper)

Subjects: LCSH: Representations of groups–Congresses. | Representations of algebras–Congresses.

Classification: LCC QA176 .C665 2019 | DDC 512/.22–dc23

LC record available at https://lccn.loc.gov/2018041948

DOI: http://dx.doi.org/10.1090/pspum/101

---

10 9 8 7 6 5 4 3 2 1 24 23 22 21 20 19

To Joseph Bernstein, in gratitude

Contents

Preface

The conference "Representation Theory and Algebraic Geometry" took place on June 11-16, 2017 at the Weizmann Institute of Science and the Hebrew University of Jerusalem. The conference honored Joseph Bernstein upon his 72nd birthday and his recent retirement from Tel Aviv University.

Representation theory in its broadest sense is one of the areas that Joseph Bernstein has fundamentally transformed through his seminal contributions, together with many of his collaborators. An in-depth account of the Mathematics of Joseph Bernstein appeared recently in the special 2016 issue of Selecta Mathematica, which also contains some of the topics not covered by these proceedings. Suffice it to say that Bernstein has had a towering influence on Mathematics, in particular in Israel, through his writings, guidance of students, communication with colleagues and generously sharing his ideas.

The papers in this volume are mostly authored by the conference speakers and their collaborators, although the contents do not necessarily match exactly with the lectures themselves. While the underlying theme is representation theory, the papers are quite diverse. The lion's share of the papers concerns different aspects of representations of p-adic groups and Hecke algebras, including in characteristic p. Even here, the papers differ significantly in methodology and emphasis and represent the full spectrum of techniques from the algebraic to the analytic. Other topics, such as representations of finite groups, real groups and supergroups, analytic aspects of automorphic forms and invariant theory are also covered. This diversity reflects the wide interests of Joseph Bernstein in representation theory and beyond.

We would like to thank all the speakers of the conference, most of whom contributed to this volume. We are also very grateful to the many referees for this volume whose assiduous and selfless work was absolutely crucial.

The conference was the pinnacle of a special trimester on representation theory of reductive groups over local fields and applications to automorphic forms, which took place at the Weizmann Institute in the spring of 2017. Altogether there were over 150 participants including more than 100 foreign visitors. We take this opportunity to thank them all, both lecturers and students who made this event so successful.

The conference and the trimester were made possible by the generous support of the Chorafas Institute for Scientific Exchange, the Arthur and Rochelle Belfer Institute, the Israel Science Foundation and the European Research Council. We thank these organizations wholeheartedly.

Last but not least, we thank the administrative staff of the Weizmann Institute, especially at the faculty of Mathematics and Computer Science, as well as that of the Institute of Mathematics of the Hebrew University of Jerusalem, who worked diligently towards the success of the conference.

It is a great pleasure to dedicate this volume to Joseph Bernstein.

The editors

Proceedings of Symposia in Pure Mathematics
Volume **101**, 2019
https://doi.org/10.1090/pspum/101/01788

Character values and Hochschild homology

Roman Bezrukavnikov and David Kazhdan

Abstract. We present a conjecture (and a proof for $G = SL(2)$) generalizing a result of J. Arthur which expresses a character value of a cuspidal representation of a p-adic group as a weighted orbital integral of its matrix coefficient. It also generalizes a conjecture by the second author proved by Schneider-Stuhler and (independently) the first author. The latter statement expresses an elliptic character value as an orbital integral of a pseudo-matrix coefficient defined via the Chern character map taking value in zeroth Hochschild homology of the Hecke algebra. The present conjecture generalizes the construction of pseudo-matrix coefficient using compactly supported Hochschild homology, as well as a modification of the category of smooth representations, the so called compactified category of smooth G-modules. This newly defined "compactified pseudo-matrix coefficient" lies in a certain space $\mathcal{K}$ on which the weighted orbital integral is a conjugation invariant linear functional, our conjecture states that evaluating a weighted orbital integral on the compactified pseudo-matrix coefficient one recovers the corresponding character value of the representation.

We also discuss the properties of the averaging map from $\mathcal{K}$ to the space of invariant distributions, partly building on works of Waldspurger and Beuzart-Plessis.

Contents

1. Introduction

Let G be a reductive group over a local nonArchimedean field F.

The goal of the article is to present an algebraic expression for a character of an admissible representation of G on a compact element.

The statement is presented as a conjecture (see Conjecture 4.3) for a general reductive group, it is proved in the paper for $G = SL(2)$. We also describe a modification of the category $Sm = Sm(G)$ of finitely generated smooth representations, the so called *compactified category* of smooth G-modules, which plays a key role in our algebraic description of character values and may have an independent interest.

To describe the context for these constructions recall a conjecture of [**14**] proved in [**19**] and [**7**].

Let $\mathcal{H} = \cup_K \mathcal{H}_K$ be the Hecke algebra of locally constant compactly supported $\mathbb{C}$-valued measures. Thus $Sm(G)$ is identified with the category of finitely generated nondegenerate $\mathcal{H}$ modules [**10**].

Let $C(\mathcal{H}) = \mathcal{H}/[\mathcal{H},\mathcal{H}] = \mathcal{H}_G = HH_0(\mathcal{H}) = HH_0(Sm)$ be the cocenter of $\mathcal{H}$. Here $\mathcal{H}_G$ denotes coinvariants with respect to the conjugation action, while HH_* stands for Hochschild homology, and its second appearance refers to the notion of Hochschild homology of an abelian category.

Since $\mathcal{H}$ is Noetherian and has finite homological dimension, there is a well defined Chern character (also called the Hattori-Stallings or Dennis trace) map $ch : K^0(Sm) \to C(\mathcal{H})$ (we will abbreviate $ch([M])$ to $ch(M)$). It has been conjectured in [**14**] and proven in [**19**], [**7**] that for an *elliptic* regular semisimple element $g \in G$ and an admissible representation ρ we have

$$\chi_\rho(g) = O_g(ch(\rho)), \tag{1}$$

where O_g denotes the orbital integral. Here we use that $O_g : \mathcal{H} \to \mathbb{C}$ being conjugation invariant factors through $C(\mathcal{H})$.

If ρ is a cuspidal irreducible representation then (assuming that G has a compact center) a matrix coefficient $m_\rho \in \mathcal{H}$ is a representative of the class $ch(\rho) \in C(\mathcal{H})$. Thus in this case (1) reduces to an earlier result of Arthur [**2**]. However, the latter applies also to nonelliptic regular semisimple elements: for such an element g and a cuspidal irreducible representation ρ Arthur has proved that

$$\chi_\rho(g) = WO_g(m_\rho), \tag{2}$$

where WO_g denotes the *weighted orbital integral.* Our Conjecture 4.3 provides a generalization of (1) to all regular semisimple compact elements g, which for a cuspidal representation ρ reduces to (2).

The first step in this direction is a generalization of the map $ch : K^0(Sm) \to C(\mathcal{H})$. For our present purposes we need to modify both the source and the target of this map. We replace the target $C(\mathcal{H}) = \mathcal{H}_G$ by $\mathcal{K}_G$ where $\mathcal{K} \subset \mathcal{H}$ is a subspace invariant under the conjugation action of G, the so called space of "weightless" functions.[1] Definition and some properties of $\mathcal{K}$ are discussed in section 2. The key

[1]This adjective reflects the fact that weighted orbital integrals restricted to this space are independent of the choices involved in choosing the weight function on an orbit. This space has appeared in the literature (see [**6**], [**24**] and references therein) where it was called the space of

property is that $WO_g|_{\mathcal{K}}$ is a G-invariant functional for any regular semisimple element $g \in G$; furthermore, there is a well defined averaging map Av from $\mathcal{K}$ to the space of invariant generalized functions on G and for $\phi \in \mathcal{K}$ the value of $WO_g(\phi)$ coincides with $Av(\phi)(g)$ (the latter is well defined since $Av(\phi)$ is in fact a locally constant function on the set of regular elements in G). We also provide a conjecture with a proof for $PGL(2, F)$, $char(F) = 0$ describing the image of the averaging map from $\mathcal{K}$ to the space of invariant distributions.

To describe the source of the map generalizing ch we need some new ingredients. One of them is the so called *compactified category* of smooth (finitely generated) representations $\overline{Sm}$.

The abelian category $\overline{Sm}$ is defined in section 3. Recall that according to Bernstein [**9**], Sm can be identified with the category of coherent sheaves of modules over a certain sheaf of algebras over a scheme which is an infinite union of affine algebraic varieties, the spectrum $\mathfrak{Z}$ of the Bernstein center of G. The category $\overline{Sm}$ can be described as the category of coherent sheaves of modules over a certain coherent sheaf of algebras over a (componentwise) compactification of $\mathfrak{Z}$. An admissible module ρ can also be viewed as an object in $\overline{Sm}$, so we can apply the Chern character to the class of ρ obtaining $\bar{ch}(\rho) \in HH_0(\overline{Sm})$.

We also need another invariant of the category Sm, namely the *compactly supported* Hochschild homology $HH_*^c(Sm)$ which is the derived global sections with compact support in the sense of [**12**] of localized Hochschild homology $R\underline{Hom}_{\mathcal{H}\otimes\mathcal{H}^{op}}(\mathcal{H}, \mathcal{H})$.

We have natural maps $HH_*^c(Sm) \to HH_*(\overline{Sm}) \to HH_*(Sm)$; for an admissible module ρ we have its compactly supported Chern character $ch^c(\rho) \in HH_0^c(Sm)$, so that $\bar{ch}(\rho)$ and $ch(\rho)$ equal the images of $ch^c(\rho)$ under the corresponding maps.

The first statement in the main conjecture (a theorem for $SL(2)$) provides a natural isomorphism

$$\mathcal{K}_G^c \cong Im(HH_0^c(Sm) \to HH_0(\overline{Sm})),$$

where $\mathcal{K}^c \subset \mathcal{K}$ is the subspace of measures supported on compact elements. By the previous paragraph, $\bar{ch}(\rho)$ belongs to that image, thus we obtain a homological construction of an element in $\mathcal{K}_G^c$ from an admissible representation, the so called "compactified pseudo-matrix coefficient" of the representation.

The second main statement (proved for $SL(2)$) asserts that for a compact regular element g and an admissible representation ρ we have $WO_g(\bar{ch}(\rho)) = \chi_\rho(g)$. Notice that for a noncompact regular element g the value of $\chi_\rho(g)$ coincides with a character value of the Jacquet functor applied to ρ [**11**], in particular it vanishes for a cuspidal module.

We view Conjecture 4.3 as an algebraic statement underlying some aspects of Arthur's local trace formula [**1**], while equality (1) underlies the elliptic part of the local trace formula (see also Remark 3.16 below). We plan to develop this theme in a future publication.

---

strongly cuspidal function. We refrain from using this terminology since we use the term "cuspidal function" in the sense of [**14**] where it refers to a function acting by zero in any parabolically induced representation, thus in our terminology $\mathcal{K}$ contains the space of cuspidal functions. The term "cuspidal function" was used in a different sense in [**6**], [**24**] etc., so that $\mathcal{K}$ is contained in the set of cuspidal functions in the sense of *loc. cit.*

Acknowledgements. We thank Joseph Bernstein, Vladimir Drinfeld, Dmitry Kaledin and Dmitry Vaintrob for many useful discussions over the years. In particular, the definition of the compactified category was conceived as a result of discussions with Kaledin and it took its present form partly due to discussions with Drinfeld.

We also thank Raphaël Beuzart-Plessis, Dan Ciubotaru, Xuhua He, Ju-Lee Kim, Eric Opdam and Jean-Loup Waldspurger for helpful correspondence.

R.B. was partly supported by NSF grant DMS-1601953, D.K. acknowledges that this project received funding from ERC under grant agreement No. 669655, their collaboration was partly supported by the US-Israel Binational Science Foundation.

2. Weightless functions and invariant distributions

Let $\mathcal{H} = \mathcal{H}(G)$ be the Hecke algebra of compactly supported locally constant measures on G, the convolution product on $\mathcal{H}(G)$ will be denoted by $*$. For an open subsemigroup $S \subset G$ we let $\mathcal{H}(S) \subset \mathcal{H}(G)$ denote the subalgebra of measures supported on S. We denote by $\mathcal{D}$ the space of generalized functions on G (that is the space of linear functionals on $\mathcal{H}(G)$) and by $\mathcal{D}^G \subset \mathcal{D}$ the subspace of invariant generalized functions. Let $\mathcal{H}_{cusp}$, $\mathcal{D}^G_{cusp}$ be the cuspidal part of $\mathcal{H}$, $\mathcal{D}^G$, i.e. $\mathcal{H}_{cusp}$ consists of functions acting by zero in any parabolically induced representation and $\mathcal{D}^G_{cusp}$ consists of distributions vanishing on the orthogonal complement of $\mathcal{H}_{cusp}$ (cf. footnote 1 above). Until the end of the section we assume for simplicity of notation that G has compact center.

Then *averaging* with respect to conjugations yields a well defined map $H_0(G, \mathcal{H}_{cusp}) \to \mathcal{D}^G_{cusp}$ given by $f \mapsto \int\limits_G \frac{{}^g f}{dg} dg$ for a Haar measure dg on G.

In this section we define a larger subspace $\mathcal{K}$ in $\mathcal{H}$ on which the averaging map is still well defined and conjecture that the map τ defines an embedding $H_0(G, \mathcal{K}) \hookrightarrow \mathcal{D}^G$. Moreover, we propose a conjectural description of the image of τ. We prove this conjecture in the case when $G = PGL(2)$.

2.1. The conjecture. We start with some notation. Let $\mathcal{O} \subset F$ be the ring of integers and π be a generator of the maximal ideal $\mathfrak{m}$ of $\mathcal{O}$. Let $val : F^\star \to \mathbb{Z}$ be the valuation such that $val(\pi) = 1$. We define $\|x\| = q^{-val(x)}, x \in F^\star$ where $q = |\mathcal{O}/\mathfrak{m}|$.

For any smooth F-variety X we denote by $\mathcal{S}(X)$ the space of locally constant measures on X with compact support. In the case when X is a homogeneous G-variety with a G-invariant measure dx the map $f \to f/dx$ identifies $\mathcal{S}(X)$ with the space of compactly supported locally constant functions on X. In this case we will not distinguish between functions and measures on X. Also, we will work with spaces $Y_P = (G/U_P \times G/U_P)/L$ for a parabolic subgroup $P = LU_P \subset G$. In this case $\mathcal{S}(Y_P)$ will denote the space of integral kernels of operators $\mathcal{S}(G/U_P) \to \mathcal{S}(G/U_P)$, i.e. locally constant compactly supported sections of the G-equivariant locally constant sheaf $pr_1^*(\mu)$, where $pr_1 : Y_P \to G/P$ is the first projection.

In particular, we fix a Haar measure dg on G and identify the space $\mathcal{S}(G) = \mathcal{H}(G)$ with the space of compactly supported locally constant functions on G. Then Δ is identified with the space of distributions, we also get the L^2 pairing $\langle \ , \ \rangle$ on $\mathcal{H}(G)$.

Let $\mathfrak{g}$ be the Lie algebra of G. We will assume that $\mathfrak{g}$ has a finite number of nilpotent conjugacy classes and that there exists a G-equivariant F-analytic bijection ϕ between a neighbourhood of 0 in $\mathfrak{g}$ and a neighbourhood of e in G. We also assume that for a semisimple element $s \in G$ the Lie algebra $\mathfrak{z}$ of its centralizer $Z_G(s)$ admits a $Z_G(s)$-invariant complement. These assumptions are well known to hold if $char(F) = 0$ or if $char(F) > N$ for some N depending on the rank of G, see [**15**, §1.8] for more precise information.

DEFINITION 2.1.
- We denote by $\mathcal{G}_{G,e}$ the space of germs of Ad-invariant distributions near 0 on $\mathfrak{g}$ which are restrictions of linear combinations of the Fourier transforms of invariant measures on a nilpotent orbits. Using the bijection ϕ we consider $\mathcal{G}_{G,e}$ as a space of germs of Ad-invariant distributions on G near the identity.
- For a semisimple element $s \in G$ we denote by $\mathcal{G}_{Z_G(s),s}$ the space of germs of distributions on $Z_G(s)$ at s obtained from the space $\mathcal{G}_{Z_G(s),e}$ by the shift by s.
- Let $s \in G$ be a semisimple element, $X_s = G/Z_G(s), r : G \to X_s$ be the natural projection and $\gamma : X_s \to G$ be a continuous section. We denote by dz a G-invariant measure on X_s.
- We denote by $\tilde{\kappa}: Z_G(s) \times X_s \to G$ the map given by $(z, x) \to \gamma(x)sz(\gamma(x))^{-1}$.

Let $\mathfrak{z} \subset \mathfrak{g}$ be the Lie algebra of $Z_G(s)$. By assumption there exists a complementary $\mathfrak{z}$-invariant subspace $W \subset \mathfrak{g}$ and the map $\kappa_0 : \mathfrak{z} \oplus W \to \mathfrak{g}, (z, w) \to z + Ad(s)w$ is a bijection. Therefore there exists an open neighborhood $R \subset Z_G(s)$ of e such that the restriction κ of $\tilde{\kappa}$ on $R \times X_s$ is an open embedding.

DEFINITION 2.2. For any $\bar{\psi} \in \mathcal{G}_{Z_G(s),s}$ we choose a representative $\tilde{\psi} \in \mathcal{D}^{Z_G(s)}(Z_G(s))$ of $\bar{\psi}$.

- For a function $f \in \mathcal{S}(G)$ and $z \in R \subset X_s$ we define $f_z \in \mathcal{S}(Z_G(s))$ by $f_z := f(\kappa(z, x))$.
- We define a function $\bar{f}$ on R by $\bar{f}(z) := \tilde{\psi}(f_z)$.
- We define a distribution $\psi(\tilde{\psi})$ on G by $\psi(f) := \int_{X_s} \bar{f}(z)$.
- We denote by $[\psi(\tilde{\psi})]$ the germ of the distribution ψ at s.

It is clear that for any two choices $\tilde{\psi}, \tilde{\psi}'$ of representatives of $\bar{\psi}$ the difference $\psi(\tilde{\psi}) - \psi(\tilde{\psi}')$ vanishes on a G-invariant open neighborhood of s. Therefore the germ $[\bar{\psi}]$ does not depend on a choice of a representative of $\bar{\psi}$.

DEFINITION 2.3.
- We denote by $\mathcal{G}_s$ the space of germs at s of Ad-invariant distributions of the form $[\psi], \psi \in \mathcal{G}_{Z_G(s),s}$.
- We denote by $\mathcal{E} \subset \mathcal{D}^G$ the subspace of distributions α such that
  a) there exists a compact subset C in G such that $supp(\alpha) \subset G(C)$
  and
  b) for any semisimple $s \in G$ the germ of α at s belongs to $\mathcal{G}_s$.

REMARK 2.4. If $char(F) = 0$ then $\mathcal{E}$ admits an equivalent description as the space of invariant distributions α satisfying the following requirements:

a) there exists a compact subset C in G such that $supp(\alpha) \subset G(C)$;

b) there exists a compact open subgroup $K \subset G$ such that for every element z in the Bernstein center satisfying $\delta_K * z = 0$ we have $\alpha * z = 0$.

Equivalence of the two definitions of $\mathcal{E}$ follows from[2] [**13**, Theorem 16.2].

DEFINITION 2.5.
- We define the space $\mathcal{K}(G)$ of *weightless* functions as the subspace in $\mathcal{S}(G)$ of functions f such that $\int_{u\in U_Q} f(lu)du = 0, l \in L$ for all proper parabolic subgroups $Q = LU_Q \subset G$.
- For a closed conjugation invariant subset X of G we define the space $\mathcal{K}(X) = \mathcal{K}_X \subset \mathcal{S}(X)$ as the subspace of functions f such that

$$\int_{u\in U_Q} f(lu)du = 0,$$

for all proper parabolic subgroups $Q = LU_Q \subset G$ and $l \in L$ such that $lU_Q \subset X$.

REMARK 2.6. For $f \in \mathcal{H}$ and a parabolic $P = LU_P \subset G$ let $A_P(f) \in \mathcal{S}(Y_P)$ denote its orishperic transform, i.e. the integral kernel of the action of f on $\mathcal{S}(G/U_P)$. Let $\Delta_{Y_P} \subset Y_P$ be the preimage of diagonal under the projection $Y_P \to (G/P)^2$. Then $\Delta_{Y_P} \cong (G/U_P \times L)/L$, where L acts on the first factor by right translations and on the second one by conjugation. It is easy to see that for $f \in \mathcal{H}$ we have $f \in \mathcal{K}$ iff for any parabolic subgroup $P \subsetneq G$ we have $A_P(f)|_{\Delta_{Y_P}} = 0$.

The following result is due to J.-L. Waldspurger (see [**24**, Lemma 9]).

PROPOSITION 2.7. *For $f \in \mathcal{H}(G)$ the following are equivalent:*
a) $f \in \mathcal{K}$.
b) For any $h \in \mathcal{H}(G)$ the function $g \mapsto \langle {}^g f, h\rangle$ has compact support.

For any $f \in \mathcal{S}(G)$ we define distribution $\hat{f}$ by:

$$\langle \hat{f}, h\rangle := \int_{g\in G} \langle {}^g f, h\rangle dg.$$

For future reference we mention the following.

LEMMA 2.8. *For $f \in \mathcal{K}$ the distribution $\hat{f}|_{G^{rs}}$ (where G^{rs} is the open set of regular semisimple elements) is a locally constant function. For $g \in G^{rs}$ we have*

$$\hat{f}(g) = WO_g(f),$$

where WO_g denotes the weighted orbital integral.

Proof follows from the definition and basic properties of the weighted orbital integral, see e.g. [**3**, §I.11]. □

The group G acts on $\mathcal{K}$ by conjugation. It is clear that the map $f \to \hat{f}$ factors through a map $\tau : H_0(G, \mathcal{K}) \to \mathcal{D}^G$. For any $f \in \mathcal{K}$ we denote by $[f]$ it image in $H_0(G, \mathcal{K})$.

CONJECTURE 2.9. *a) $\hat{f} \in \mathcal{E}$ for $f \in \mathcal{K}$.*
b) The map τ defines an isomorphism between $H_0(G, \mathcal{K})$ and $\mathcal{E}$.
c) $\dim(H_0(G, \mathcal{K}(\Omega))) = 1$ *for any regular semisimple conjugacy class $\Omega \subset G$.*

REMARK 2.10. One can check that $\hat{f}$ satisfies the conditions of Remark 2.4, thus if $char(F) = 0$ then part (a) of the conjecture follows from Harish-Chandra's Theorem [**13**, Theorem 16.2]; see [**23**, Corollary 5.9], [**6**, Proposition 5.6.1] for details.

[2]We thank Raphaël Beuzart-Plessis for pointing this out to us.

REMARK 2.11. It is clear that part c) follows from a) and b).

REMARK 2.12. If the centralizer of an element $g \in \Omega$ is an anisotropic (compact) torus then statement (c) clearly follows from uniqueness (up to scaling) of a Haar measure on G. In the case when that centralizer has split rank one the statement is checked in the next subsection.

2.2. Almost elliptic orbits. To simplify the wording we assume in this subsection that the center of G is compact. A regular semisimple element $g \in G$ will be called almost elliptic if the split rank of its centralizer is at most one. We now prove Conjecture 2.9(c) in the case when Ω consists of almost elliptic elements.

Fix $g \in \Omega$ and let T be the centralizer of G, thus $\Omega \cong G/T$. In view of Remark 2.12 it suffices to consider the case when the split rank of T equals one; we also assume without loss of generality that G is almost simple.

There are exactly two parabolic subgroups P, $P' \subsetneq G$ containing T. Let U, U' be their unipotent radicals.

Consider the complex

$$0 \to \mathcal{K}_\Omega \to \mathcal{S}(G/T) \to \mathcal{S}(G/TU) \oplus \mathcal{S}(G/TU') \to \mathbb{C} \to 0; \tag{3}$$

here $\mathcal{S}$ stands for the space of locally constant compactly supported measures as before, the third arrow send ϕ to $(pr_*(\phi), pr'_*(\phi))$, where $pr : G/T \to G/TU$, $pr' : G/T \to G/TU'$ are the projections and the fourth arrow sends (ϕ, ϕ') to $\int \phi - \int \phi'$.

LEMMA 2.13. *The complex* (3) *is exact.*

PROOF. Exactness at all the terms except for $\mathcal{S}(G/TU) \oplus \mathcal{S}(G/TU')$ is clear. Suppose that $(\psi, \psi') : \mathcal{S}(G/TU) \oplus \mathcal{S}(G/TU') \to \mathbb{C}$ is a linear functional vanishing on the image of $\mathcal{S}(G/T)$. Let $\tilde{\psi} : \mathcal{S}(G) \to \mathbb{C}$ be the composition of the direct image map $\mathcal{S}(G) \to \mathcal{S}(G/TU)$ with ψ. Then ψ is a right TU invariant generalized function on G. On the other hand, $-\psi$ is equal to the composition of the direct image map $\mathcal{S}(G) \to \mathcal{S}(G/TU')$ with ψ', which shows that $\tilde{\psi}$ is also right TU' invariant. Since G is assumed to be almost simple, U and U' together generate G, thus we see that $\tilde{\psi}$ is right G invariant. It follows that $\tilde{\psi}$ is proportional to the functional $\phi \mapsto \int \phi$, hence the functional (ψ, ψ') factors through the differential in (3), which yields exactness of (3). □

We can now finish the proof of Conjecture 2.9(c) in the present case. Breaking (3) into short exact sequences we get

$$0 \to \mathcal{K}_\Omega \to \mathcal{S}(G/T) \to M \to 0,$$

$$0 \to M \to \mathcal{S}(G/TU) \oplus \mathcal{S}(G/TU') \to \mathbb{C} \to 0.$$

Considering the corresponding long exact sequences on homology we see that it suffices to check that the map $\mathbb{C} = H_0(G, \mathcal{S}(G/T)) \to H_0(G, M)$ is nonzero while the map $H_1(G, \mathcal{S}(G/T)) \to H_1(G, M)$ has one-dimensional cokernel. The former statement is clear since the composition $H_0(G, \mathcal{S}(G/T)) \to H_0(G, M) \to H_0(G, \mathcal{S}(G/TU))$ is nonzero. To check the latter recall that $H_i(G, \mathbb{C}) = 0$ for $i > 0$ since the resolution of $\mathbb{C}$ provided by the simplicial complex for computation of homology of the Bruhat-Tits building $\mathfrak{B}$ shows that $H_i(G, \mathbb{C}) \cong H_i(\mathfrak{B}/G)$, while $\mathfrak{B}/G$ is a product of simplices. Thus

$$H_1(G, M) \cong H_1(G, \mathcal{S}(G/TU) \oplus \mathcal{S}(G/TU')) \cong H_1(T, \mathbb{C}) \oplus H_1(T, \mathbb{C}).$$

Since $\mathcal{S}(G/TU) \cong H_1(T, \mathbb{C})$ we see that

$$CoKer\,(H_1(G, \mathcal{S}(G/T)) \to H_1(G, M)) \cong H_1(T, \mathbb{C}),$$

which is one-dimensional. □

2.3. The case of $PGL(2)$. To simplify the argument we assume in this subsection that $char(F) = 0$.

THEOREM 2.14. *Conjecture 2.9 is true for $G = PGL(2)$.*

The rest of the subsection is devoted to the proof of the Theorem.

CLAIM 2.15. We have $\mathcal{K}_{cusp} \subset \mathcal{K}$.

PROPOSITION 2.16. *The map $\tau : H_0(G, \mathcal{K}) \to \mathcal{D}^G$ is an embedding.*

PROOF. We need more notation.

DEFINITION 2.17.
- For $\epsilon \geq 0$ we define $G^\epsilon = \{g \in G || p(g)| \leq \epsilon\}$, where $p(g) = \frac{tr^2(\tilde{g})}{det(\tilde{g})} - 4$; here $\tilde{g} \in GL(2, F)$ is a representative of g.
- We let G_s, G_e, N, $\bar{N}$ denote, respectively, the sets of regular semisimple split, regular semisimple elliptic, regular unipotent and all unipotent elements.
- We set

$$\mathcal{K}_u = \mathcal{K}(\bar{N}) := \{f \in \mathcal{S}(\bar{N}) \mid \int\limits_{u \in U} f(u)du = 0 \ \forall B = TU \subset G\},$$

$$\mathcal{K}_0 := \{f \in \mathcal{K}_u | f(e) = 0\},$$

where B runs over the set of Borel subgroups in G.
- For $f \in \mathcal{K}$ we denote by $\kappa(f) \in \mathcal{K}_u$ the restriction of f to $\bar{N}$ and by $[\kappa(f)]$ the image of $\kappa(f)$ in $H_0(G, \mathcal{K}_u)$.

We start with the following geometric statement.

LEMMA 2.18. *Let $f \in \mathcal{S}(G)$ be such that $f|_{\bar{N}} \in \mathcal{K}_u$. Then there exists $\epsilon > 0$ such that $f|_{G^\epsilon} \in \mathcal{K}$.*

PROOF. Recall notations of Remark 2.6. We have $\Delta_{Y_B} = G/B \times T$, where T is the (abstract) Cartan subgroup of G. It is easy to see that condition $f|_{\bar{N}} \in \mathcal{K}_u$ is equivalent to vanishing of the restriction of $A_B(f)$ to $G/B \times \{1\} \subset \Delta_{Y_B}$. Also, condition $f|_{G^\epsilon} \in \mathcal{K}$ is equivalent to vanishing of $A_B(f)$ on $G/B \times T_\epsilon \subset \Delta_{Y_B}$, where $T_\epsilon = G_\epsilon \cap T$ (here we abuse notation by identifying the abstract Cartan subgroup T with an arbitrarily chosen Cartan subgroup). Since $A_B(f)$ is locally constant for $f \in \mathcal{H}$, the statement follows from compactness of G/B. □

COROLLARY 2.19. *For any $f \in \mathcal{K}$ such that $[\kappa(f)] = 0$ there exists $f' \in \mathcal{K}$ with the same image in $H_0(G, \mathcal{K})$ and such that $f'|_{\bar{N}} = 0$.*

PROOF. Since $[f|_{\bar{N}}] = 0$ we can write the restriction of f to $\bar{N}$ as a finite sum $\sum_i (\tilde{f}_i^{g_i} - \tilde{f}_i)$, $\tilde{f}_i \in \mathcal{K}'_0, g_i \in G$. As follows from the Lemma 2.18 we can choose $f_i \in \mathcal{K}$ such that $\tilde{f}_i = f_i|_{\bar{N}}$. Then the function $f' := f - \sum_i (f_i^{g_i} - f_i)$ satisfies the conditions of Corollary. □

PROPOSITION 2.20. *The space $H_0(G, \mathcal{K}_u)$ is two dimensional.*

PROOF. We first show that $\dim(H_0(G, \mathcal{K}_0)) = 1$.

Let $\mathcal{B}$ be the variety of Borel subgroups, $p : N \to \mathcal{B}$ the map which associates to $u \in N$ the Borel subgroup containing u. By definition we have an exact sequence

$$0 \to \mathcal{K}_0 \to \mathcal{S}(N) \to \mathcal{S}(\mathcal{B}) \to 0$$

and therefore an exact sequence

$$H_1(G, \mathcal{S}(N)) \to H_1(G, \mathcal{S}(\mathcal{B})) \to H_0(G, \mathcal{K}_0) \to H_0(G, \mathcal{S}(N)) \to H_0(G, \mathcal{S}(\mathcal{B})).$$

LEMMA 2.21. $H_1(G, \mathcal{S}(N)) = 0$.

PROOF. Fix a Borel subgroup $B = TU$. We can write U as a union of open compact subgroup $U_1 \subset ...U_n \subset ....$ Therefore $\mathcal{S}(N) = \mathcal{S}(G/U)$ is the direct limit of $\mathcal{S}(G/U_n)$. Since the functor $M \mapsto H_1(G, M)$ commutes with direct limits it is sufficient to show that $H_1(G, \mathcal{S}(G/U_n)) = \{0\}$. Since $U_n \subset G$ is a compact subgroup the space $H_1(G, \mathcal{S}(G/U_n))$ is a direct summand of $\mathcal{S}(G)$. Since $H_1(G, \mathcal{S}(G)) = 0$ the Lemma is proven. □

Since G acts transitively on N and on $\mathcal{B}$ we have $H_0(G, \mathcal{S}(N)) \widetilde{\longrightarrow} H_0(G, \mathcal{S}(\mathcal{B})) = \mathbb{C}$. Since $H_1(G, \mathcal{S}(N)) = 0$ we see that the map $H_1(G, \mathcal{S}(\mathcal{B})) \to H_0(G, \mathcal{K}_0)$ is an isomorphism. Since $\mathcal{B} = G/B$ we have:

$$H_1(G, \mathcal{S}(\mathcal{B})) = H_1(B, \mathbb{C}) = H_1(T, \mathbb{C}) = \mathbb{C}.$$

So $\dim(H_0(G, \mathcal{K}_0)) = 1$.

To conclude the argument, recall the short exact sequence $0 \to \mathcal{K}_0 \to \mathcal{K}_u \to \mathbb{C} \to 0$.

Since $H_1(G, \mathbb{C}) = 0$ we have an exact sequence:

$$0 \to H_0(G, \mathcal{K}_0) \to H_0(G, \mathcal{K}_u) \to \mathbb{C} \to 0$$

So $\dim(H_0(G, \mathcal{K}_u) = 2$. □

Let $r : \mathcal{K}_{cusp} \to \mathcal{K}_u$ be the restriction and $[r] : \mathcal{K}_{cusp} \to H_0(G, \mathcal{K}_u)$ be the composition of r and projection $\mathcal{K}_u \to H_0(G, \mathcal{K}_u)$.

LEMMA 2.22. *The map* $[r]$ *is onto.*

PROOF. Recall the map $\tau : H_0(G, \mathcal{K}) \to \mathcal{D}^G$, $[f] \to \hat{f}$. Let $\bar{\mathcal{D}}^G$ be the space of germs of invariant distributions at e and $\bar{\tau} : H_0(G, \mathcal{K}) \to \bar{\mathcal{D}}^G$ be the composition of τ with the restriction map.

Corollary 2.19 implies that the map $\bar{\tau}$ vanishes on the kernel of the map $H_0(G, \mathcal{K}) \to H_0(G, \mathcal{K}_u)$. Thus it suffices to show that $\bar{\tau}|_{\mathcal{K}_{cusp}}$ has rank at least two, i.e. that there exist irreducible cuspidal representations ρ_1, ρ_2, such that their characters restricted to any G-invariant open neighborhood of identity are not proportional. This is easily done by inspecting the character tables, see e.g. [**20**, §2.6]. □

COROLLARY 2.23. *a) For any* $f \in \mathcal{K}$ *there exists* $f_{cusp} \in \mathcal{K}_{cusp}$ *such that* $[\kappa(f)] = [\kappa(f_{cusp'})]$.

b) For any $f_0 \in \mathcal{K}_u$ *there exists* $f \in \mathcal{K}_{cusp}$ *such that* $[f_0] = [\kappa(f)]$. □

Let $s \in G$ be a regular split semisimple element and $\Omega \subset G$ be the conjugacy class of s.

PROPOSITION 2.24. $\dim(H_0(G, \mathcal{K}_\Omega)) = 1$.

PROOF. Let $T = Z_G(t)$ be the split torus and $B, B' \subset G$ be Borel subgroups containing T. Since $\Omega = G/T$ we have maps $r : \Omega \to G/B$ and $r' : \Omega \to G/B'$ and therefore morphisms $r_\star : \mathcal{S}(\Omega) \to \mathcal{S}(G/B)$ and $r'_\star : \mathcal{S}(\Omega) \to \mathcal{S}(G/B')$.

As a special case of Lemma 2.13 we get:

LEMMA 2.25. *The sequence*

$$0 \to \mathcal{K}_\Omega \to \mathcal{S}(\Omega) \to \mathcal{S}(G/B) \oplus \mathcal{S}(G/B') \to \mathbb{C} \to 0, \tag{4}$$

where the last map l is given by $(\nu, \nu') \mapsto \int \nu - \int \nu'$, is exact. □

Let $L := ker(l)$. We have an exact sequence

$$0 \to L \to \mathcal{S}(G/B) \oplus \mathcal{S}(G/B') \to \mathbb{C} \to 0.$$

Using that G has homological dimension one, we get that the corresponding long exact sequence of homology contains the following fragment:

$$0 \to H_1(G, L) \to H_1(B, \mathbb{C}) \oplus H_1(B', \mathbb{C}) \to H_1(G, \mathbb{C}).$$

It is easy to see that

$$\dim(H_1(B, \mathbb{C})) = \dim(H_1(B', \mathbb{C})) = \dim(H_1(T, \mathbb{C})) = 1$$

Since the quotient $G/[G, G]$ is finite we see that $H_1(G, \mathbb{C}) = 0$. Therefore $\dim(H_1(G, L)) = 2$.

On the other hand we have an exact sequence

$$0 \to \mathcal{K}_\Omega \to \mathcal{S}(\Omega) \to L \to 0$$

and therefore an exact sequence

$$H_1(G, \mathcal{S}(\Omega)) \to H_1(G, L) \to H_0(G, \mathcal{K}_\Omega) \to H_0(G, \mathcal{S}(\Omega)).$$

Since $H_1(G, \mathcal{S}(\Omega)) = H_1(T, \mathbb{C})$ we see $\dim(H_1(G, \mathcal{S}(\Omega)) = 1$.

LEMMA 2.26. *The map $a : H_1(G, \mathcal{S}(\Omega)) \to H_1(G, L)$ is an embedding.*

PROOF. It is sufficient to show that the map $H_1(G, \mathcal{S}(\Omega)) \to H_1(G, \mathcal{S}(G/B))$ induced by the composition $p_\star \circ a : L \to \mathcal{S}(G/B)$ is an embedding.

Since $H_1(G, \mathcal{S}(\Omega)) = H_1(T, \mathbb{C})$, $H_1(G, \mathcal{S}(G/B)) = H_1(B, \mathbb{C})$ and $H_{>0}(U, \mathbb{C}) = 0$, we see that this map is an isomorphism. □

We can now finish the proof of Proposition 2.24. Since G acts transitively on G/B the map $\mu \to \int_{G/B} \mu$ defines an isomorphism $H_0(G, \mathcal{S}(G/B)) \to \mathbb{C}$. On the other hand, since $\int_\Omega \nu = 0$ for any $\nu \in \mathcal{K}_\Omega$ the map $H_0(G, \mathcal{K}_\Omega) \to H_0(G, \mathcal{S}(\Omega))$ equals zero. So $\dim(H_0(G, \mathcal{K}_\Omega)) = 1$. □

Recall that $G_s \subset G$ is the subset of regular split semisimple elements, let $\mathcal{K}_s \subset \mathcal{K}$ be the subspace of functions in $\mathcal{K}$ supported on G_s. We fix a Cartan subgroup T, the Weyl group $W = \mathbb{Z}/2\mathbb{Z}$ acts on T and on G/T in the usual way. Then the map

$$(T - \{e\}) \times G/T \to G_s, (s, g) \to gsg^{-1}$$

induces an isomorphism

$$\mathcal{S}(G_s) \to (\mathcal{S}(T - \{e\}) \otimes \mathcal{S}(G/T))^W. \tag{5}$$

COROLLARY 2.27. *a) For $f \in \mathcal{K}_s$ the distribution $\hat{f}$ is locally constant on G_s.*
b) The map $f \mapsto (t \mapsto \hat{f}(t))$ induces an isomorphism

$$H_0(G, \mathcal{K}_s) \to \mathcal{S}(T - \{e\})^W.$$

PROOF. The isomorphism (5) is clearly compatible with the averaging map $f \mapsto \hat{f}$, which implies a). Likewise, restriction to an orbit is compatible with averaging, thus in view of (5) it suffices to show that for a fixed orbit $\Omega \subset G_s$ the map $f \mapsto \hat{f}(t)$, $t \in \Omega$ induces an isomorphism $H_0(G, \mathcal{K}(\Omega)) \to \mathbb{C}$. By Proposition 2.24 it suffices to see that this map is nonzero. This follows, for example, from the fact that the character of a cuspidal representation does not necessarily vanish on G_s, while the character of an irreducible cuspidal representation is obtained by averaging from its matrix coefficient. □

Now we can finish the proof of Proposition 2.16. Let $f \in \mathcal{K}$ be such that $\hat{f} = 0$. It follows from Corollary 2.19 and Lemma 2.18 that we can can find f' with the same image in $H_0(G, \mathcal{K})$ such that $f' = f_s + f_e$ where f_s is supported on regular split semisimple elements and f_e on regular elliptic elements. The condition $\hat{f} = 0$ implies that $\hat{f'} = 0$, hence $\hat{f_s} = 0$ and $\hat{f_e} = 0$. It is easy to see that the condition $\hat{f_e} = 0$ implies that $[f_e] = 0$. So we may assume that the support of f is contained in the subset $G_s \subset G$ of regular split semisimple elements. Now Proposition 2.16 follows from Corollary 2.27. □

Let $\tilde{\mathcal{D}}_e$ be the space of germs of distributions at e and $\mathcal{D}_e \subset \tilde{\mathcal{D}}_e$ be the subspaces spanned by germs of characters of irreducible representations.

LEMMA 2.28. *The space $\mathcal{D}_e$ is 2-dimensional. It is spanned by germs of characters of irreducible cuspidal representations.*

PROOF. The second statement is a special case of a theorem of Harish-Chandra [**13**]. The first one also follows from *loc. cit.*, as it shown there that more generally the space $\mathcal{D}_e$ has a basis indexed by unipotent orbits. □

Recall that $\mathcal{E} \subset \mathcal{D}$ is the subspace of distributions α satisfying the following three conditions:

a) There exists a compact subset C in G such that $supp(\alpha) \subset C^G$.
b) The restriction of α on $G - \{e\}$ is given by a locally constant function.
c) The germ of α at e belongs to $\mathcal{D}_e$.

LEMMA 2.29. $\tau(\mathcal{K}) \subset \mathcal{E}$.

PROOF. Fix $f \in \mathcal{K}$. It is clear that the distribution $\hat{f}$ satisfies condition a).

To prove that $\hat{f}$ satisfies condition b) we have to show that for any semisimple element $s \in G - \{e\}$ there exists an open neighborhood $R \subset G$ of s such that the restriction $\hat{f}|_R$ is a constant. If s is split then this follows from Corollary 2.27(a), if s is elliptic the proof is similar.

To prove that $\hat{f}$ satisfies condition c) we observe that Corollary 2.23 implies existence of $f_{cusp} \in \mathcal{K}_{cusp}$ such that $[\kappa(f)] = [\kappa(f_{cusp})]$.

It is easy to see that when f is a matrix coefficient of an irreducible cuspidal representation ρ then $\hat{f}$ is proportional to the character of ρ. Thus condition c) is satisfied by $\alpha_{cusp} = \widehat{f_{cusp}}$. However, by Lemma 2.18 and Corollary 2.19 the germs of α and α_{cusp} at e coincide. □

PROPOSITION 2.30. $\tau(\mathcal{K}) = \mathcal{E}$.

PROOF. It remains to show that every $\alpha \in \mathcal{E}$ is in the image of τ. Lemma 2.28 shows that there exists $\beta \in \mathcal{E}$ which is a linear combination of characters of cuspidal representations such $\alpha - \beta$ vanishes on an open neighborhood of e. Thus we have $\alpha - \beta = \alpha_s + \alpha_e$, where α_s is supported on G_s, while α_e is supported on G_e.

Now α_s is in the image of τ by Corollary 2.27, while α_e is in the image of τ by a similar argument. Also, β is in the image of τ since the character of an irreducible cuspidal representation ρ equals $\hat{f}$ where f is a matrix coefficient of ρ.

Proposition 2.30 and therefore Theorem 2.14 are proven. □

3. The compactified category of smooth modules

3.1. Definition of the compactified category. For a parabolic $P = LU$ let $L^0 \subset L$ be the subgroup generated by compact subgroups; thus L^0 is the kernel of the unramified characters of L. Set $\check{\Lambda}_P = L/L^0$.

Let Λ_P be the group of F-rational characters of L and Λ_P^+ be the subset of P-dominant weights, i.e. weights which are (non-strictly) dominant with respect to any (not necessarily F-rational) Borel subgroup $B \subset P$. We have a nondegenerate pairing between the lattices $\check{\Lambda}_P$ and Λ_P given by:

$$\langle xL^0, \lambda \rangle = val_F(\lambda(x)).$$

Let $\check{\Lambda}_P^+$ be the subsemigroup defined by:

$$\check{\Lambda}_P^+ = \{x \in \check{\Lambda}_P \mid \langle x, \lambda \rangle \geq 0 \ \ \forall \lambda \in \Lambda_P^+\},$$

and let $L_P^+ \subset L$ be the preimage of $\check{\Lambda}_P^+$ under the projection $L \to \check{\Lambda}_P$.

For a pair of parabolics $P \supset Q$ let $L_Q^{P+} \subset L_Q$ denote the image of $L_P^+ \cap Q$ in $L_Q = Q/U_Q$. It is easy to see that $L_Q^{P+} \supset L_Q^+$.

For an open submonoid $M \subset G$ we let $Sm(M)$ denote the category of nondegenerate finitely generated $\mathcal{H}(M)$-modules; this is easily seen to be equivalent to the category of finitely generated smooth M-modules.

For parabolic subgroups $P \supset Q$ we have the "Jacquet" functor $J_Q^P : Sm(L_P^+) \to Sm(L_Q^{P+})$, $M \mapsto M_{\overline{U_Q}}$, where $\overline{U_Q}$ is image of U_Q in $L_P = P/U_P$.

To simplify the wording in the following definition we fix a minimal parabolic P_0, then by a standard parabolic we mean a subgroup P containing P_0.

DEFINITION 3.1. The *compactified category* of smooth G-modules $\overline{Sm}^{qc} = \overline{Sm}^{qc}(G)$ is the category whose object is a collection (M_P) indexed by standard parabolic subgroups $P = L_P U_P$, where M_P is a smooth module over L_P^+, together with isomorphisms

$$J_Q^P(M_P) \cong \mathcal{H}(L_Q^{P+}) \otimes_{\mathcal{H}(L_Q^+)} M_Q \tag{6}$$

fixed for every pair of standard parabolic subgroups $P = L_P U_P \supset Q = L_Q U_Q$; here $\mathcal{H}$ denotes the algebra of locally constant compactly supported distributions. The isomorphisms are required to satisfy the associativity identity for each triple of parabolics $P_1 \supset P_2 \supset P_3$.

An object in the compactified category is called coherent if the module M_P is finitely generated for all P.

We let $\overline{Sm} = \overline{Sm}(G) \subset \overline{Sm}^{qc}$ denote the full subcategory of coherent objects.

It is easy to see that $\overline{Sm}(G)$ is an abelian category, the functor sending $(M_P) \in \overline{Sm}$ to M_G identifies $Sm(G)$ with a Serre quotient of $\overline{Sm}$.

We also have an adjoint functor $Sm(G) \to \overline{Sm}^{qc}$. This functor sends admissible modules but not general finitely generated modules to $\overline{Sm}$.

EXAMPLE 3.2. Let $G = SL(2)$. In this case the category $\overline{Sm}$ admits the following more direct description. A component of the spectrum $\mathfrak{Z}$ of Bernstein center is in this case either a point or an affine curve, thus $\mathfrak{Z}$ admits a canonical (componentwise) compactification $\overline{\mathfrak{Z}}$. Notice that $\partial\mathfrak{Z} = \overline{\mathfrak{Z}} \setminus \mathfrak{Z}$ is identified with the set $(O^\times)^*$ of characters of $T_0 = O^\times$. Let $T^+ = O \setminus \{0\} \subset F^\times = T$, thus $T^+ \cong O^\times \times \mathbb{Z}_{\geq 0}$. Set $\tilde{\mathfrak{Z}}^+ = Spec(\mathcal{H}(T^+)) \cong (O^\times)^* \times \mathbb{A}^1$. Notice that we have a natural map $\pi : \tilde{\mathfrak{Z}}^+ \to \overline{\mathfrak{Z}}$ inducing an isomorphism $(O^\times)^* \times \{0\} \to \partial\mathfrak{Z}$. Moreover, π is etale at $(O^\times)^* \times \{0\}$.

The full Hecke algebra $\mathcal{H}$ defines a quasicoherent sheaf of algebras on $\mathfrak{Z}$, we now describe its extension to a quasicoherent sheaf of algebras $\overline{\mathcal{H}}$ on $\overline{\mathfrak{Z}}$. The latter depends on the choice of a maximal open compact subgroup $K_0 = SL(2, O)$. Fixing this choice we set $\tilde{\mathcal{H}}^+ = End_{T^+}(\mathcal{S}(G/U^+))$, where $(G/U)^+ = O^2 \setminus \{0\} \subset F^2 \setminus \{0\} = G/U$. We also let $\tilde{\mathcal{H}} = End_T(\mathcal{S}(G/U))$. It is clear that $\tilde{\mathcal{H}}^+$ defines a quasicoherent sheaf on $\tilde{\mathfrak{Z}}^+$ whose restriction to the open subset $\tilde{\mathfrak{Z}} := \tilde{\mathfrak{Z}}^+ \setminus (O^\times)^* \times \{0\}$ is the quasicoherent sheaf defined by $\tilde{\mathcal{H}}$.

The action of $\mathcal{H}$ on $\mathcal{S}(G/U)$ defines a homomorphism $\pi^*(\mathcal{H}) \to \tilde{\mathcal{H}}$ which is an isomorphism on a Zariski neighborhood of $\partial\tilde{\mathfrak{Z}}^+ = \tilde{\mathfrak{Z}}^+ \setminus \tilde{\mathfrak{Z}}$. Thus we get a well defined quasicoherent sheaf of algebras $\overline{\mathcal{H}}$ on $\overline{\mathfrak{Z}}$ such that $\overline{\mathcal{H}}|_{\mathfrak{Z}} = \mathcal{H}$ and the induced map $\pi^*(\overline{\mathcal{H}})|_{\tilde{\mathfrak{Z}}} \to \tilde{\mathcal{H}}$ extends to a map $\pi^*(\overline{\mathcal{H}}) \to \tilde{\mathcal{H}}^+$ which is an isomorphism on a neighborhood of $\partial\tilde{\mathfrak{Z}}^+$.

It is clear that K_0^2 acts on $\overline{\mathcal{H}}$ and for an open subgroup $K \subset K_0$ the subsheaf $\overline{\mathcal{H}}_K$ of K^2 invariants is a coherent sheaf of algebras.

We leave it to the reader to show that although $\overline{\mathcal{H}}$ depends on an auxiliary choice, different choices lead to algebras which are canonically Morita equivalent. Thus we can consider the category of sheaves of nondegenerate $\overline{\mathcal{H}}$-modules which can be checked to be canonically equivalent to $\overline{Sm}$. If the subgroup $K \subset K_0$ is nice in the sense of [**9**] then for every component X of $\overline{\mathfrak{Z}}$ either the coherent sheaf of algebras $\overline{\mathcal{H}}_K|_X$ is zero or the corresponding summand in $\overline{Sm}$ (respectively, $\overline{Sm}^{qc}$) is canonically equivalent to the category of coherent (respectively, quasicoherent) sheaves of $\overline{\mathcal{H}}_K|_X$-modules.

3.2. Compactified center and a spectral description of the compactified category. Let $Z = Z(G)$ be the Bernstein center of G and $\mathfrak{Z} = Spec(Z)$ be its spectrum. By the main result of [**9**] (the set of closed points of) $\mathfrak{Z}$ is in bijection with the set $Cusp(G)$ of *cuspidal data*, i.e. the set of G-conjugacy classes of pairs (L, ρ), where $L \subset G$ is a Levi subgroup and ρ is a cuspidal irreducible representation of L.

3.2.1. *Compactified center.* Let $\overline{\mathfrak{Z}}$ denote its compactification described as follows. We have a canonical isomorphism $\mathfrak{Z} = \tilde{\mathfrak{Z}}/W$ where W is the Weyl group, and $\tilde{\mathfrak{Z}}$ parametrizes pairs (L, ρ) where L is a Levi subgroup containing a fixed maximally split Cartan T and ρ is a cuspidal representation of L. The complex torus ${}^LT = \mathcal{X}(L)$ acts on the union $\tilde{\mathfrak{Z}}_L$ of components corresponding to a given Levi subgroup $L \supset T$; here $\mathcal{X}(L)$ stands for the group of unramified characters of L acting on the set of representations by twisting. Notice that ${}^LT$ is a torus with

$X^*({}^LT) = L/L^0$. The action is transitive on each component and the stabilizer of each point is finite. The space $X^*({}^LT)_{\mathbb{R}} = X_*(Z(L))_{\mathbb{R}}$ (where X_* stands for the lattice of F-rational cocharacters) contains hyperplanes corresponding to the roots of $Z(L)$ in $\mathfrak{g}$; the fan formed by these hyperplanes defines an equivariant compactification $\overline{{}^LT}$ of ${}^LT$. We set

$$\overline{\mathfrak{Z}} = \tilde{\overline{\mathfrak{Z}}}/W, \qquad \tilde{\overline{\mathfrak{Z}}} = \bigcup_L \overline{{}^LT} \times^{{}^LT} \tilde{\mathfrak{Z}}_L,$$

where the right hand side makes sense because the action of W on $\tilde{\mathfrak{Z}}$ extends to the compactification, here we use the notation $X \times^H Y = (X \times Y)/H$. Notice that every component of $\tilde{\mathfrak{Z}}_L$ is of the form ${}^LT/A$ for a finite subgroup $A \subset^L T$, thus the corresponding component of $\overline{{}^LT} \times^{{}^LT} \tilde{\mathfrak{Z}}_L$ is identified with $\overline{{}^LT}/A$.

For a parabolic $P = LU$ let $Z^0(L) \subset Z(L)$, $Z^+(L) \subset Z(L)$ be the subalgebras consisting of distributions supported on L^0 and L_P^+ respectively, set also $\mathfrak{Z}^0(L) = Spec(Z^0(L))$, $\mathfrak{Z}^+(L) = Spec(Z^+(L))$.

It is clear that

$$\mathfrak{Z}^0(L) = \mathfrak{Z}(L)/\mathcal{X}(L), \tag{7}$$

where $\mathcal{X}(L)$ is the group of unramified characters of L.

PROPOSITION 3.3. *a) $\overline{\mathfrak{Z}}$ admits a canonical stratification indexed by conjugacy classes of parabolic subgroups, where the stratum $\overline{\mathfrak{Z}}_P$ corresponding to the class of a parabolic P is identified with $\mathfrak{Z}^0(L)$.*

b) The embedding $\mathfrak{Z}^0(L) \to \overline{\mathfrak{Z}}$ canonically extends to a map $\mathfrak{Z}^+(L) \to \overline{\mathfrak{Z}}$ which is etale on a Zariski neighborhood of $\overline{\mathfrak{Z}}_P \cong \mathfrak{Z}^0(L)$.

Given two parabolics $P \subset Q$ we have a canonical map $c_P^Q : \mathfrak{Z}^+(L_P) \to \mathfrak{Z}^+(L_Q)$ which is compatible with maps to $\overline{\mathfrak{Z}}$.

Moreover, for three parabolics $P_1 \subset P_2 \subset P_3$ we have

$$c_{P_1}^{P_3} = c_{P_2}^{P_3} c_{P_1}^{P_2}.$$

PROOF. Let $\tilde{\overline{\mathfrak{Z}}}_L = \overline{{}^LT} \times^{{}^LT} \tilde{\mathfrak{Z}}_L$.

It is a standard fact that ${}^LT$-orbits in $\overline{{}^LT}$ are in bijection with parabolic subgroups containing L, so that the orbit $\overline{{}^LT}_Q$ corresponding to a parabolic $Q = MU_Q$ is identified with $\mathcal{X}(L)/\mathcal{X}(M)$. The stratification of $\overline{{}^LT}$ by ${}^LT$-orbits induces a stratification on $\tilde{\overline{\mathfrak{Z}}}_L$, the stratum corresponding to a parabolic Q will be denoted by $\tilde{\overline{\mathfrak{Z}}}_L\{Q\}$.

Fix a conjugacy class $\mathbf{P}$ of parabolic subgroups and set $\tilde{\overline{\mathfrak{Z}}}_L(\mathbf{P}) = \bigcup_{Q \in \mathbf{P}} \tilde{\overline{\mathfrak{Z}}}_L\{Q\}$. Let $\tilde{\overline{\mathfrak{Z}}}_{\mathbf{P}} = \bigcup_L \tilde{\overline{\mathfrak{Z}}}_L(\mathbf{P})$.

It is clear that $(\tilde{\overline{\mathfrak{Z}}}_{\mathbf{P}})$ is a stratification of $\tilde{\overline{\mathfrak{Z}}}$ and each stratum is W-invariant. Thus $\overline{\mathfrak{Z}}_{\mathbf{P}} := \tilde{\overline{\mathfrak{Z}}}_{\mathbf{P}}/W$ are strata of a stratification of $\overline{\mathfrak{Z}}$.

The map $Q \mapsto \tilde{\overline{\mathfrak{Z}}}_L\{Q\}$ is easily seen to be W-equivariant, it follows that for a parabolic $P = LU \in \mathbf{P}$ we have

$$\overline{\mathfrak{Z}}_{\mathbf{P}} \cong \bigcup_{M, T \subset M \subset L} \tilde{\overline{\mathfrak{Z}}}_M\{P\}/W_L.$$

The above isomorphism $\overline{{}^LT_Q}\cong\mathcal{X}(L)/\mathcal{X}(M)$ shows that $\tilde{\bar{\mathfrak{z}}}_M\{P\}\cong\bar{\mathfrak{z}}_M(L)/\mathcal{X}(L)$. Passing to the union over M and taking quotient by the action of W_L (which commutes with the action of $\mathcal{X}(L)$) we get $\bar{\mathfrak{z}}_{\mathbf{P}}\cong\mathfrak{z}(L)/\mathcal{X}(L)$ which yields (a) in view of (7).

To check (b) observe that for parabolic subgroups $Q=MU_Q\supset P=LU_P\supset T$ the cone $\mathbb{R}^{\geq 0}\Lambda_Q^+$ belongs to the fan defining the toric variety $\overline{{}^LT}$. Let $V_L\{Q\}$ be the corresponding affine open subset in $\overline{{}^LT}$ and $\mathcal{V}_L\{Q\}$ be the corresponding open affine in $\tilde{\bar{\mathfrak{z}}}_L$. Thus $\mathcal{V}_L\{Q\}$ is a Zariski open neighborhood of $\tilde{\bar{\mathfrak{z}}}_L\{Q\}$.

It is easy to see that $\mathcal{V}_L\{Q\}$ is W_M invariant and $\mathcal{V}_L\{Q\}/W_M\cong\mathfrak{z}^+(L)$. Since $\bar{\mathfrak{z}}=\tilde{\bar{\mathfrak{z}}}/W$, claim (b) follows from the fact that the stabilizer of any point $x\in\tilde{\bar{\mathfrak{z}}}\{Q\}$ is contained in W_M.

c) follows by inspection. □

In order to relate $\overline{Sm}$ to $\bar{\mathfrak{z}}$ we will need the following general concept. Let X be an algebraic variety. By a *quasicoherent enrichment* of a category $\mathcal{C}$ over X we will mean assigning to objects $M,\ N\in\mathcal{C}$ an object $\underline{\mathrm{Hom}}(M,N)\in QCoh(X)$ together with an isomorphism $Hom(M,N)=\Gamma(\underline{\mathrm{Hom}}(M,N))$ and maps $\underline{\mathrm{Hom}}(M_1,M_2)\otimes_{\mathcal{O}_X}\underline{\mathrm{Hom}}(M_2,M_3)\to\underline{\mathrm{Hom}}(M_1,M_3)$ satisfying the associativity constraint and compatible with the composition of morphisms in $\mathcal{C}$. If the quasicoherent sheaf $\underline{\mathrm{Hom}}(M,N)$ is actually coherent for all $M,N\in\mathcal{C}$ we say that the enrichment is coherent.

PROPOSITION 3.4. *The category $\overline{Sm}$ (respectively, $\overline{Sm}^{qc}(G)$) admits a natural lifts to a category coherent (respectively, quasicoherent) enrichment over $\bar{\mathfrak{z}}$.*

COROLLARY 3.5. *The categories $\overline{Sm}$, $\overline{Sm}^{qc}$ split as a direct sum indexed by components of $\bar{\mathfrak{z}}$.* □

Before proceeding to prove the Proposition we state a general elementary Lemma.

LEMMA 3.6. *Let $X=\coprod X_i$ be a scheme with a fixed stratification (i.e. the closure of X_i coincides with $\coprod_{j\leq i}X_j$ for some partial order $\leq$ on the set I of strata). Set $U_i=\coprod_{j\geq i}X_j$, this is an open subset of X. Suppose that for each i we are given a map $u_i:Y_i\to U_i$, such that*

i) $X_i\times_X Y_i\widetilde{\longrightarrow}X_i$

ii) u_i is etale over a Zariski neighborhood of X_i.

iii) For $j\leq i$ set $Y_{ji}=Y_j\times_X U_i$. Then the map $Y_{ji}\to U_i$ factors through a map $u_{ji}:Y_{ji}\to Y_i$. Moreover, for $k<j<i$ the map $Y_{ki}\to U_j\supset U_i$ factors through a map $u_{kji}:Y_{ki}\to Y_{ji}$.

Let $Y_d=\coprod_{i_1<\cdots<i_d}Y_{i_1i_d}$. Then the diagram $Y_3\substack{\rightarrow\\\rightarrow\\\rightarrow}Y_2\rightrightarrows Y_1\to X$ satisfies descent for quasicoherent sheaves, i.e. $QCoh(X)$ is equivalent to the category of quasicoherent sheaves on Y_1 with isomorphisms of the two pull-backs to Y_2 whose pull-backs to Y_3 satisfy the natural compatibility.

REMARK 3.7. To fix ideas let us first prove the Lemma for sheaves in the analytic topology assuming we work over the base field $\mathbb{C}$. Then it is easy to see that we can find an open subset $Y_i^o\subset Y_i$ for each i so that Y_i^o maps isomorphically to a neighborhood of X_i in X. Moreover, we can arrange it so that the images of Y_i^o and Y_j^o have a nonempty intersection only if $i\leq j$ or $j\leq i$. Replacing Y_i by Y_i^o does not affect the category of gluing data; however, (Y_i^o) is just an open covering of X, so the claim is clear.

3.2.2. *Proof of Lemma* 3.6. Let X_0 be a closed stratum. Running an inductive argument, we can assume the theorem is known for the stratified space $X \setminus X_0$. Then we are reduced to proving the claim in the situation when the stratification consists of two strata $X = X_0 \coprod X_1$. Replacing Y_0 by its open subset containing X_0 clearly does not affect the category of descent data, so we can assume without loss of generality that $Y_1 \to X$ is etale. By a standard argument the claim reduces to exactness of the complex of sheaves on X:

$$0 \to \mathcal{O} \to (Y_0 \to X)_*(\mathcal{O}) \oplus (Y_1 \to X)_*(\mathcal{O}) \to (Y_{01} \to X)_*(\mathcal{O}) \to 0.$$

The complex is clearly exact over X_1, so it is enough to show that local cohomology of this complex with support on X_1 vanishes. This reduces to showing that $\alpha^!(\mathcal{O}) \widetilde{\longrightarrow} \beta^!(\mathcal{O})$, where $\beta : X_0 \to X$, $\alpha_1 : X_0 \to Y_0$. This follows from conditions (i), (ii). □

3.2.3. *Proof of Proposition* 3.4. Proposition 3.3 implies that $X = \overline{\mathfrak{Z}}$ with the stratification of Proposition 3.3(a), and $Y_i = \mathfrak{Z}(L_i^+)$ satisfy the conditions of Lemma 3.6. It is easy to see from the definition of $\overline{Sm}^{qc}$ that the collection of quasi-coherent sheaves $Hom(M_P, N_P)$ provides gluing data described in Lemma 3.6. Also, for $M, N \in \overline{Sm}$ the module $Hom(M_P, N_P)$ is finitely generated, so the resulting quasicoherent sheaf is in fact coherent. □

3.3. The spectral description of $\overline{Sm}$. Recall that for every component $X \subset \mathfrak{Z}$ the choice of a sufficiently small nice (in the sense of [**9**]) open compact subgroup $K \subset G$ defines a coherent sheaf of algebras $\mathcal{A} = \mathcal{A}_X(K)$ on X with an equivalence between $\mathcal{A}_X(K)$ modules and the corresponding summand in smooth G-modules.

Let $\mathfrak{B}$ be the (reductive) Bruhat-Tits building of G. We fix a special vertex $x \in \mathfrak{B}$ and let $K_x \subset G$ denote the corresponding maximal compact subgroup.

We also fix a maximally split Cartan subgroup T, such that the corresponding apartment $A_T \subset \mathfrak{B}$ contains x. Thus A_T is an affine space with underlying vector space $V = X_*(T) \otimes \mathbb{R}$. Let $V_+ \subset A_T$ be a Weyl cone with vertex at x and B the corresponding minimal parabolic subgroup.

It follows from the Iwasawa decomposition (see e.g. [**22**, §3.3.2]) that we have a natural bijection

$$P' \backslash G / K_x \cong \check{\Lambda}_P, \tag{8}$$

where P' denotes the commutator group. Let $G^+(x, P)$ be the union of cosets corresponding to elements in $\check{\Lambda}_P^+ \subset \check{\Lambda}_P$.

Let $(U_P \backslash G)^+(x)$ be the image of $G^+(x, P)$ in $U_P \backslash G$ and $\mathcal{S}^+(U_P \backslash G)(x) \subset \mathcal{S}(U_P \backslash G)$ be the subspace of functions whose support is contained in $(U_P \backslash G)^+(x)$.

PROPOSITION 3.8. *a) The left action of* L *on* $\mathcal{S}(U_P \backslash G)$ *restrict to an* L_P^+ *action on* $\mathcal{S}^+(U_P \backslash G)(x)$.

b) Let $K \subset K_x$ *be an open subgroup. The* $\mathcal{H}(L_P^+)$*-module* $\mathcal{S}^+(U_P \backslash G)(x)^K$ *is finitely generated and projective.*

c) There exists a unique object $\mathcal{P}(x)^K \in \overline{Sm}$ *such that* $\mathcal{P}(x)_P^K = \mathcal{S}^+(U_P \backslash G)(x)^K$ *while the isomorphism* (6) *for* $P = G$ *comes from the natural arrow* $\mathcal{S}^+(U_Q \backslash G)(x)^K \to \mathcal{S}(U_Q \backslash G)^K = J_Q^G(\mathcal{S}(G)^K)$.

d) For every open subgroup $K \subset K_x$ *the object* $\mathcal{P}(x)^K$ *is locally projective, i.e. the functor* $\underline{\mathrm{Hom}}(\mathcal{P}(x)^K, \quad)$ *is exact.*

e) For every component X there exists an open subgroup $S \subset K_x$ such that for any open subgroup $K \subset S$ the object $\mathcal{P}_x^K$ is local generator of the corresponding summand $\overline{Sm}_X \subset \overline{Sm}$. The latter property means that $\underline{\mathrm{Hom}}(\mathcal{P}_\sigma(x)^K, \ )$ is conservative i.e. it $\underline{\mathrm{Hom}}(\mathcal{P}_\sigma(x)^K, \mathcal{M}) \neq 0$ for $0 \neq \mathcal{M} \in \overline{Sm}_X$.

PROOF. a) Bijection (8) intertwines the left action of L on $P'\backslash G/K_x$ with the action of $\check{\Lambda}_P = L/L^0$ on itself by translations, this implies part a). The space $\mathcal{S}^+(U_P\backslash G)(x)^K$ splits as a direct sum indexed by $P\backslash G/K$, each summand is isomorphic to the space $\mathcal{S}(L_P^+)^{K_L}$ for some open compact subgroup $K_L \subset K$, this implies b).

Notice that given modules M_P and isomorphisms (6) for $P = G$ as in Definition 3.1, the rest of the isomorphisms (6) satisfying the requirements of the Definition are defined uniquely (if they exist) provided that each module M_P is torsion free as a module over $Z^+(L_P)$. This implies uniqueness in c). Existence follows from the fact that $G^+(x, P) \subset G^+(x, Q)$ for parabolic subgroups $P \subset Q$.

Statement d) follows from b).

Finally e) follows from the corresponding statement about $\mathcal{H}_K$ established in **[9]**. □

Set $\overline{\mathcal{A}}_X = \overline{\mathcal{A}}_X(x, K) := \underline{\mathrm{Hom}}(\mathcal{P}(x)^K, \mathcal{P}(x)^K)|_X$, this is a sheaf of algebras on X.

COROLLARY 3.9. *Given a component $X \subset \overline{\mathfrak{Z}}$ for any small enough open subgroup $K \subset K_x$ we have a canonical equivalence between the category of (quasi)coherent sheaves of $\overline{\mathcal{A}}_X(x, K)$-modules and the summand in $\overline{Sm}$ (respectively, $\overline{Sm}^{qc}$) corresponding to X.*

3.4. Compactified category and filtered modules. For a simple root α let P_α be the corresponding maximal proper parabolic and $\overline{\mathfrak{Z}}_\alpha$ be the closure of the corresponding stratum in $\overline{\mathfrak{Z}}$. It is easy to see that $\overline{\mathfrak{Z}}_\alpha$ is a divisor. For a weight λ set $D_\lambda = \sum_\alpha \langle \check{\alpha}, \lambda \rangle [\overline{\mathfrak{Z}}_\alpha]$ where the sum runs over the set of simple roots, let also $\mathcal{O}(\lambda) = \mathcal{O}_{\overline{\mathfrak{Z}}}(D_\lambda)$ and $\mathcal{F}(\lambda) = \mathcal{F} \otimes_{\mathcal{O}(\overline{\mathfrak{Z}})} \mathcal{O}(\lambda)$.

A coherent sheaf $\mathcal{F}$ on $\overline{\mathfrak{Z}}$ is determined by the graded module $\oplus_\lambda \Gamma(\mathcal{F}(\lambda))$, over the homogeneous coordinate ring $\oplus \Gamma(\mathcal{O}(\lambda))$.

If $\mathcal{F}$ is torsion free then the natural map $\mathcal{F}(\mu) \to \mathcal{F}(\lambda + \mu)$, $\lambda \in \Lambda^+$ is injective and $\bigcup_\lambda \mathcal{F}(\lambda) = j_* j^*(\mathcal{F})$ where $j : \mathfrak{Z} \to \overline{\mathfrak{Z}}$ is the embedding. Thus the category of torsion free coherent sheaves $Coh_{tf}(\overline{\mathfrak{Z}})$ admits a full embedding into the category of $\mathcal{O}(\mathfrak{Z})$-modules equipped with a filtration indexed by Λ^+ compatible with the natural filtration on the ring $\mathcal{O}(\mathfrak{Z})$: to a sheaf $\mathcal{F}$ it assigns the module $\Gamma(j^*(F))$ with tho filtration by the subspaces $\Gamma(\mathcal{F}(\lambda))$

Applying it to the sheaf of rings $\overline{\mathcal{A}}_X(x, K)$ we get a filtration on the Hecke algebra $F^{spec}_{\leq \lambda}(\mathcal{H}_K)$.

Thus we obtain the following:

PROPOSITION 3.10. *We have a full embedding from the category $\overline{Sm}_K^{tf}$ of torsion free objects in $\overline{Sm}_K$ to the category of modules over $\mathcal{H}_K$ equipped with a filtration compatible with the filtration F^{spec} on $\mathcal{H}_K$.*

It is clear that the associated graded of the filtered module in the image of such an embedding is finitely generated; recall that a filtration with this property is called a *good filtration.*

We also have the left adjoint functor Loc from the category of $\mathcal{H}_K$ modules with a good filtration to $\overline{Sm}_K$.

3.4.1. *A geometric description of the filtration.* We now provide a more explicit description of the filtration $F^{spec}_{\leq\lambda}(\mathcal{H}_K)$.

Recall that the two sided cosets of K_x in G are indexed by X^+_B (see e.g. [**22**, §3.3.3]), for $\lambda \in X^+_B$ let G_λ denote the corresponding coset. Let $F^{geom}_{\leq\lambda}(\mathcal{H}_K)$ be the space of functions whose support is contained in $\bigcup\limits_{\mu\leq\lambda} G_\mu$.

PROPOSITION 3.11. *For every open compact $K \subset K_x$ there exists $\lambda_0 \in X^+_B$ such that:*

$$F^{spec}_{\leq\lambda}(\mathcal{H}_K) = \{h \in \mathcal{H}_K \mid \forall \mu \in \lambda_0 + X^+_B : \quad h * F^{geom}_{\leq\mu}(\mathcal{H}_K) \subset F^{geom}_{\leq\mu+\lambda}(\mathcal{H}_K).\} \tag{9}$$

Proof It follows from the definition that

$$h \in F^{spec}_{\leq\lambda}(\mathcal{H}_K) \iff \forall P, \mu \ : \ \mathcal{S}((U_P\backslash G)_{\leq\mu}) * h \subset \mathcal{S}((U_P\backslash G)_{\leq\bar{\lambda}_P+\mu}), \tag{10}$$

where $\bar{\lambda}_P$ is the image of λ in $\check{\Lambda}_P$ and $(U_P\backslash G)_{\leq\mu} = \bigcup\limits_{\nu\leq\mu} (U_P\backslash G)_\nu$ for the standard partial order $\leq$ on $\check{\Lambda}_P$; here $(U_P\backslash G)_\nu$ is the image of the coset corresponding to ν under bijection (8).) Let $X_P = (G/U_P \times G/U_{P^-})/L$. We have $K_x\backslash X_P/K_x \cong \check{\Lambda}_P$, let $(X_P)_\mu$ denote the two-sided coset corresponding to $\mu \in \check{\Lambda}_P$. Then the condition in the right hand side of (10) is equivalent to:

$$\forall P, \mu \ : \ h * \mathcal{S}((X_P)_{\leq\mu}) \subset \mathcal{S}((X_P)_{\leq\lambda+\mu}), \tag{11}$$

where $(X_P)_{\leq\mu} = \bigcup\limits_{\nu\leq\mu} (X_P)_\nu$. this is clear by considering the projection $X \to G/P^-$ with fiber G/U_P. Assume now that h lies in the set in the right hand side of (9). Applying the map B_I of [**8**, Definition 5.3] and using [**8**, Lemma 5.5], we see that h also satisfies (11). This shows that the right hand side of (9) is contained in the left hand side. We proceed to check the opposite inclusion. The Rees ring $\oplus_\lambda F^{spec}_{\leq\lambda}(\mathcal{H}_K)$ is finite over its center which is a finitely generated commutative ring, hence the Rees ring is finitely generated. Thus it suffices to check that for a finite set of generators $h_i \in F^{spec}_{\leq\lambda_i}(\mathcal{H}_K)$ we have:

$$h_i * F^{geom}_{\leq\mu}(\mathcal{H}_K) \subset F^{geom}_{\leq\mu+\lambda_i}(\mathcal{H}_K) \ \ \forall \mu \in \lambda_0 + \Lambda^+$$

for some $\lambda_0 \in \Lambda^+$. Existence of such a λ_0 for a given h_i follows from [**8**, Lemma 5.5]. Since we consider a finite set of h_i, there exists λ_0 which satisfies the requirement for all h_i. □)

REMARK 3.12. It easily follows from the construction that $gr(F^{spec})$ is a Noetherian ring.

Notice that the geometric filtration on $\mathcal{H}$ is also compatible with the algebra structure; however, its associated graded is neither Noetherian, nor finitely generated in general. This is closely related to the fact that the intertwining operator acting on the space of functions over $\underline{(G/U)}(O/\pi^n O)$ is not an isomorphism for $n > 1$ (here $\underline{(G/U)}$ is a scheme over O coming from the O-group scheme with generic fiber G corresponding to $x \in \mathfrak{B}$).

EXAMPLE 3.13. In view of Proposition 3.11 the geometric filtration F^{geom} on $\mathcal{H}$ makes it into a filtered module over the filtered algebra $(\mathcal{H}, F^{spec})$. Applying the functor *Loc* to that filtered module we get an object in $\overline{Sm}$. We denote it by $\overline{\mathcal{H}}'$ and call it the *intertwining* object in $\overline{Sm}$.

LEMMA 3.14. *Assume that* $G = SL(2)$.
The intertwining object $\overline{\mathcal{H}}'$ *is equivalently described by* $\overline{\mathcal{H}}'_G = \mathcal{H}$,

$$\overline{\mathcal{H}}'_B = \{f \in \mathcal{S}(G/U) \mid supp(I^{-1}(f)) \subset (G/U_-)^+\},$$

where I^{-1} *denotes the inverse* intertwining operator *taking values in functions of bounded support.*[3]

PROOF. Without loss of generality we can assume that x is the standard vertex, so that $K_x = SL(2, O)$, we will write K_0 instead of K_x.

We need to check that for $h \in \mathcal{H}_K$ and large $\lambda \in \Lambda$ we have:

$$h \in F^{geom}_{\leq \lambda} \iff supp(I_P^{-1} A_P(h)) \subset (X_P)_{\leq \lambda},$$

where $X_{\leq \lambda}$ is the union of two-sided K_0 cosets corresponding to weights $\mu \leq \lambda$ and $A : \mathcal{H} \to \mathcal{S}(Y_P)$. By [**8**, Theorem 7.6] $I^{-1}A = B^\star$ (notations of *loc. cit.*). It follows from [**8**, Lemma 5.5] that for large λ we have

$$supp(h) \subset G_{\leq \lambda} \iff supp(B^\star(f)) \subset X_{\leq \lambda}.$$

The claim follows. □

REMARK 3.15. A similar statement can also be checked for an arbitrary reductive group G based on a generalization of [**8**, Theorem 7.6] to an arbitrary parabolic subgroup.

REMARK 3.16. We expect the local trace formula [**1**] to be closely related to the computation of Chern character of $\overline{\mathcal{H}}'$ taking values in the appropriate Hochschild homology group.

4. Hochschild homology and character values

Recall the notion of *Hochschild homology* of an abelian category [**18**], [**16**]. For a coherent sheaf of algebras A over a quasi-projective algebraic variety X over a field we have $HH_*(X; A) \cong HH_*(A - mod)$, where $A - mod$ is the abelian category of coherent sheaves of A-modules and $HH_*(X; A)$ is defined as the derived global section of a naturally defined object $R\underline{Hom}_{A\otimes A^{op}}(A, A)$ in the derived category of sheaves on X, here the isomorphism is shown in [**17**].

One can also define the *compactly supported* Hochschild homology $HH^c_*(X; A)$ as the derived global sections with compact support in the sense of [**12**] of $R\underline{Hom}_{A\otimes A^{op}}(A, A)$. If X is projective then we also have

$$HH^c_*(X; A) \cong HH_*(X; A) \cong HH_*(A - mod),$$

where the first isomorphism is clear since $R\Gamma = R\Gamma_c$ for a projective scheme. Also, for an open subscheme $U \subset X$ we have a natural push forward map $HH^c_*(U; A|_X) \to HH^c_*(X, A)$.

Assume that the sheaf of algebras A (locally) has finite homological dimension. Then we have the *Chern character* map $ch : K^0(A - mod) \to HH_0(A - mod)$, see e.g. [**17**, §4.2] for the definition (it is called the Euler class map in *loc. cit.*).

[3]Here by a bounded set in $G/U_- = F^2 \setminus \{0\}$ we mean a subset with compact closure in F^2.

EXAMPLE 4.1. For future reference we spell out this general construction in some simple special cases. We leave the proofs of these standard facts to the interested reader.

(1) Assume X is affine, so that $A - mod$ is the category of finitely generated modules over a Noetherian ring which we also denote by A. Then $HH_*(X;A) = HH_*(A)$ is computed by the bar complex of A, in particular $HH_0(X;A) = HH_0(A) = A/[A,A]$. A finitely generated projective module M is isomorphic to $A^{\oplus n}e$ for an idempotent $e \in Mat_n(A)$ for some n. Then $ch(M) = \sum(e_{ii}) \mod [A,A]$.
(2) Assume that $X = U_1 \cup U_2$ for affine open subsets U_1, U_2. Let $A_i = \Gamma(U_i, A)$, $i = 1, 2$ and $A_{12} = \Gamma(U_1 \cap U_2, A)$. Then $HH_*(X;A)$ is computed by the complex
$$Cone\,(Bar(A_1) \oplus Bar(A_2) \to Bar(A_{12}))\,[-1].$$
For a locally projective module coherent sheaf M of A-modules we can find integers n, m and idempotents $e^1 \in Mat_n(A_1)$, $e^2 \in Mat_m(A_2)$ together with isomorphisms $\Gamma(U_1, M) \cong A_1^{\oplus n}e^1$, $\Gamma(U_2, M) \cong A_2^{\oplus m}e^2$ and matrices $a \in Mat_{nm}(A_{12})$, $b \in Mat_{mn}(A_{21})$ which induce the inverse isomorphisms between $A_{12}^{\oplus n}e^1$ and $A_{12}^{\oplus m}e^2$ coming from the identification of both with $\Gamma(U_1 \cap U_2, M)$. In this setting $ch(M)$ is represented by the cocycle $(\sum e^1_{ii}, \sum e^2_{jj}, \sum a_{ij} \otimes b_{ji}) \in A_1 \oplus A_2 \oplus A_{12}^{\otimes 2}$.
(3) Let $U \subset X$ be an affine open subset such that its complement Z is also affine. Let $A_1 = \Gamma(U, A)$, and let A_1, A_{12} be the algebras of sections of A on the formal neighborhood and the punctured formal neighborhood of Z respectively. Then the statements of part (b) continue to hold *mutatis mutandis*.

DEFINITION 4.2. In the above setting, for an object $\mathcal{M} \in A - mod$ with *proper* support we have the Chern character $ch^c(\mathcal{M}) \in HH_0^c(X;A)$ as follows. Fix a proper subscheme $Z \subset X$ containing support of cohomology of $\mathcal{M}$. Then [**17**, §5.7] yields $ch_Z(\mathcal{M}) \in R\Gamma_Z(X, R\underline{Hom}_{A\otimes A^{op}}(A,A))$. We define $ch^c(\mathcal{M})$ to be the image of $ch_Z(\mathcal{M})$ under the canonical map $R\Gamma_Z(X,F) \to R\Gamma_c(X,F)$, $F = R\underline{Hom}_{A\otimes A^{op}}(A,A)$.

It is immediate to check that $ch^c(\mathcal{M})$ is independent of the auxiliary choice of Z.

In particular, we have Chern character map $\bar{ch} : K^0(\overline{Sm}) \to HH_0(\overline{Sm})$ and $ch^c : K^0(Adm) \to HH_0^c(Sm)$, where $Adm \subset Sm$ is the subcategory of admissible modules.

Let $\mathcal{K}^c \subset \mathcal{K}$, $\mathcal{K}^{nc} \subset \mathcal{K}$ be the subspace of measures supported on compact (respectively, noncompact) elements.

CONJECTURE 4.3. *a) We have a canonical isomorphism:*
$\mathcal{K}^c_G \cong Im(HH_0^c(Sm) \to HH_0(\overline{Sm}))$.

b) For $\rho \in Adm$ and $g \in G$ we have $WO_g(\bar{ch}(\rho)) = \chi_\rho(g)$ if g is compact regular semisimple. (Here we use that $\bar{ch}(\rho)$ is the image of $ch^c(\rho)$ thus it belongs to $Im(HH_0^c(Sm) \to HH_0(\overline{Sm}))$ which we identify with $\mathcal{K}^c_G$ by a).

The proof of Conjecture for $G = SL(2)$ is presented in the next section; we plan to present the proof in the general case in a later publication.

5. $SL(2)$ calculations

THEOREM 5.1. *Conjecture 4.3 holds for $G = SL(2)$.*

The rest of the section is devoted to the proof of the Theorem. From now on set $G = SL(2)$.

5.1. Explicit complexes for Hochschild homology. Applying the general construction of section 3.3 with $\sigma = \{x\}$ where x is the vertex with stabilizer $G(O)$ we arrive at a sheaf of algebras $\mathcal{A}$ on $\overline{\mathfrak{Z}}$ such that a direct summand in $\overline{Sm}$ is canonically identified with the category of coherent sheaves of $\mathcal{A}$-modules. It is easily seen to coincide with the sheaf $\overline{\mathcal{H}}_K$ introduced in Example 3.2. We keep notations of that Example.

We also let $\hat{\mathcal{H}}_K^+$ denote its sections on the formal neighborhood $\widehat{\partial\mathfrak{Z}}$ of $\partial\mathfrak{Z} = \overline{\mathfrak{Z}} \setminus \mathfrak{Z}$ and $\hat{\mathcal{H}}_K$ the sections on the punctured formal neighborhood of $\partial\mathfrak{Z}$. We set $\hat{\mathcal{H}}^+ = \bigcup\limits_K \hat{\mathcal{H}}_K^+$ and $\hat{\mathcal{H}} = \bigcup\limits_K \hat{\mathcal{H}}_K$. We also let $X = (G/U \times G/U^-)/T$ be the set of rank one matrices in $Mat_2(F)$ and $X_+ = X \cap Mat_2(O)$.

Then we have $\hat{\mathcal{H}}_K^+ = End_{T^+}(\mathcal{S}(G/U)_+^K) \otimes_{\mathbb{C}[t]} \mathbb{C}[[t]]$, $\hat{\mathcal{H}}_K = \hat{\mathcal{H}}_+^K \otimes_{\mathbb{C}[[t]]} \mathbb{C}((t))$. We have $\hat{\mathcal{H}}_K = End_{\hat{\mathcal{H}}(T)}(\mathcal{S}_b^K(G/U))$, where $\mathcal{S}_b(G/U)^K \cong \mathcal{S}(G/U) \otimes_{\mathbb{C}[t,t^{-1}]} \mathbb{C}((t))$ is the space of functions on G/U with bounded support.

Also notice that $\hat{\mathcal{H}}$ (though not $\hat{\mathcal{H}}_+$) carries a $G \times G$ action.

LEMMA 5.2. *a) $HH_*(\overline{Sm})$ is computed by the complex*

$$Bar(\overline{Sm}) := cone[Bar(\mathcal{H}) \oplus Bar(\hat{\mathcal{H}}^+) \to Bar(\hat{\mathcal{H}})][-1]. \tag{12}$$

b) $HH_*^c(Sm)$ is computed by the complex

$$Bar^c(Sm) := cone[Bar(\mathcal{H}) \to Bar(\hat{\mathcal{H}})][-1]. \tag{13}$$

c) $HH_*^c(Sm)$ is canonically isomorphic to derived G coinvariants in the two-term complex

$$\mathcal{H} \to \hat{\mathcal{H}} \tag{14}$$

(placed in degrees 0,1).

PROOF. For a complex $\mathcal{F}$ of coherent sheaves on $\overline{\mathfrak{Z}}$ the complexes

$$cone[\Gamma(\mathfrak{Z}, \mathcal{F}) \oplus \hat{\Gamma}^+(\mathcal{F}) \to \hat{\Gamma}(\mathcal{F})][-1],$$

$$cone[\Gamma(\mathfrak{Z}, \mathcal{F}) \to \hat{\Gamma}(\mathcal{F})][-1]$$

compute, respectively, $R\Gamma(\mathcal{F})$ and $R\Gamma_c(\mathfrak{Z}, \mathcal{F})$, where $\hat{\Gamma}^+(\mathcal{F})$ and $\hat{\Gamma}(\mathcal{F})$ denote, respectively, sections of $\mathcal{F}$ on the formal neighborhood and on the punctured formal neighborhood of $\partial\mathfrak{Z}$. Applying this to a complex representing $R\underline{Hom}_{\overline{\mathcal{H}}_K \otimes \overline{\mathcal{H}}_K^{op}}(\overline{\mathcal{H}}_K, \overline{\mathcal{H}}_K)$ and observing that we have canonical quasi-isomorphisms $\hat{\Gamma}^+(\mathcal{F}) \to Bar(\hat{\mathcal{H}}^+)$, $\hat{\Gamma}(\mathcal{F}) \to Bar(\hat{\mathcal{H}})$ we get statements (a,b).

Statement (c) follows from isomorphisms $HH_*(M) \cong H_*(G, M)$, $HH_*^{\hat{\mathcal{H}}}(N) \cong HH_*^{\mathcal{H}}(N)$ for an $\mathcal{H}$-bimodule M and an $\hat{\mathcal{H}}$-bimodule N. □

5.2. Calculation of HH_0. In this subsection we prove part (a) of the Theorem.

PROPOSITION 5.3. *a) We have a short exact sequence:*

$$0 \to \mathcal{K}_G \to HH_0^c(Sm) \to \widehat{\mathcal{S}(T)}/\mathcal{S}(T) \to 0. \tag{15}$$

b) We have a natural isomorphism $\mathcal{K}_G^c \cong Im(HH_0^c(Sm) \to HH_0(\overline{Sm}))$.

We start the proof with the following

LEMMA 5.4. *G acts trivially on the cokernel of the map $H \to \mathcal{S}(Y_\Delta)$.*

Proof follows from Lemma 2.25. □

COROLLARY 5.5. *We have:*
$H_1(G, CoKer(H \to \mathcal{S}(Y_\Delta))) = 0.$
$H_1(G, CoKer(H \to \widehat{\mathcal{S}(Y_\Delta)})) = \widehat{\mathcal{S}(T)}/\mathcal{S}(T),$
where $\widehat{\mathcal{S}(T)}$ is the Tate completion of $\mathcal{S}(T)$ (functions on the torus).

Proof of Proposition 5.3. By Lemma 5.2 (c), $HH_*^c(Sm)$ is identified with the derived G-coinvariants in the complex $\mathcal{H} \to \widehat{\mathcal{S}(Y)}$ which is clearly isomorphic to (14). We have a short exact sequence

$$0 \to \mathcal{S}(Y_0) \to \mathcal{S}(Y) \to \mathcal{S}(Y_\Delta) \to 0,$$

where $Y_0 = Y \setminus Y_\Delta$ is the open set of non-colinear pairs of vectors. Thus the action of G on Y_0 is almost free, i.e. the stabilizer of every point in $\{\pm 1\}$, and the orbits of G on Y_0 are indexed by the finite set $F^\times/(F^\times)^2$. This shows that $H_1(G, \mathcal{S}(Y_0)) = 0$ and $H_0(G, \mathcal{S}(Y_0))$ is finite dimensional. Denoting as above $\widehat{\mathcal{S}(Y)} = \mathcal{S}(Y) \otimes_{\mathbb{C}[\varpi,\varpi^{-1}]} \mathbb{C}((\varpi))$ and similarly for $\widehat{\mathcal{S}(Y_0)}$ we see $H_i(G, \mathcal{S}(Y_0)) = 0$ for all i. Thus derived G coinvariants of the complex (14) are canonically isomorphic to derived G-coinvariants of the quotient complex $\mathfrak{C} = \mathcal{H} \to \widehat{\mathcal{S}(Y_\Delta)}$. Notice that $H^0(\mathfrak{C}) = \mathcal{K}$, $H^1(\mathfrak{C}) = CoKer(\mathcal{H} \to \widehat{\mathcal{S}(Y_\Delta)})$ and $\mathfrak{C}$ is quasi-isomorphic to the complex with zero differential since the category of G-modules has homological dimension one. Thus part(a) of the Proposition follows from Corollary 5.5.

Lemma 5.2 shows that $Bar^c(Sm)$ is canonically quasiisomorphic to $\mathbb{C} \overset{L}{\otimes}_G \mathfrak{C}$. One the other hand, we have a short exact sequence of complexes

$$0 \to Bar^c(Sm) \to Bar(\overline{Sm}) \to Bar(\widehat{H(T^+)}) \to 0,$$

which yields a long exact sequence on cohomology:

$$\cdots \to HH_{i+1}(\widehat{H(T^+)}) \to HH_i^c(Sm) \to HH_i(\overline{Sm}) \to \cdots$$

We are interested in $i = 0$. We have $HH_1(\widehat{H(T^+)}) \cong \widehat{H(T^+)}$. To finish the proof we need another

LEMMA 5.6. *a) Let us identify*

$$HH_1(\widehat{H(T)}) = H_1(G, \widehat{\mathcal{S}(Y_\Delta)}) = \widehat{\mathcal{S}(T)}.$$

Then the image of the natural map $HH_1(\widehat{H(T^+)}) \to HH_1(\widehat{H(T)})$ coincides with $\varpi\widehat{\mathcal{S}(T^+)}$, i.e. with the space of functions supported on ϖT^+, where $\varpi \in F$ is a uniformizer.

b) In view of (15)*, part (a) yields a map* $\varpi\widehat{\mathcal{S}(T^+)}\cap\mathcal{S}(T)\to\mathcal{K}_G\subset HH_0^c(Sm)$. *This map induces an isomorphism onto* $(\mathcal{K}_{nc})_G$ *where* $\mathcal{K}_{nc}\subset\mathcal{K}$ *is the space of measures supported on noncompact elements.*

Proof of the Lemma. By Hochschild-Kostant-Rosenberg, we can identify Hochschild homology with forms. It is easy to see that elements in $\mathcal{S}(T^0)\subset\widehat{\mathcal{S}(T)}$ (where T^0 is the maximal compact subgroup in the torus T) locally on each component of Bernstein center are proportional to $\frac{dz}{z}$ for a local coordinate coming from a global coordinate on $\mathbb{C}^\times$. On the other hand, the image of $HH_1(\widehat{H(T^+)})$ equals $\mathbb{C}[[z]]dz$, this proves (a).

To prove (b) observe that the map in question can be described as follows.

We have a short exact sequence of G-modules

$$0\to\mathcal{K}\to\mathcal{H}\to I\to 0,$$

where $I=Im(\mathcal{H}\to\mathcal{S}(Y_\Delta))$ which yields the Bockstein homomorphism $\phi: H_1(G,I)\to H_0(G,\mathcal{K})$. By Corollary 5.5 we have $H_1(G,I)\cong H_1(G,\mathcal{S}(Y_\Delta))=H_1(G,\mu_{\mathbb{P}^1})\otimes\mathcal{S}(T)=\mathcal{S}(T)$. In view of (a) it suffices to check that $\phi:\mathcal{S}(\varpi T^+)\widetilde{\longrightarrow}H_0(G,\mathcal{K}^{nc})$.

We have a direct sum decomposition $\mathcal{H}=\mathcal{H}^c\oplus\mathcal{H}^{nc}$ compatible with the decomposition $\mathcal{S}(Y_\Delta)=\mathcal{S}(Y_\Delta^c)\oplus\mathcal{S}(Y_\Delta^{nc})$, where $Y_\Delta^c=\mathbb{P}^1\times O^\times$ and $Y_\Delta^{nc}=Y_\Delta\setminus Y_\Delta^{nc}$. Let us further decompose $Y_\Delta^{nc}=Y_\Delta^+\cup Y_\Delta^-$, where $Y_\Delta^+=\mathbb{P}^1\times(\varpi O\setminus\{0\})$, $Y_\Delta^-=\mathbb{P}^1\times(F\setminus O)$.

The image of ϕ clearly coincides with the image of $H_1(G,\mathcal{S}(Y_\Delta^+))$ under the map coming from the class in $Ext^1(\mathcal{S}(Y_\Delta^+),\mathcal{K})$ induced from the above short exact sequence. It is clear that it is contained in $H_0(G,\mathcal{K}^{nc})$. It coincides with $H_0(G,\mathcal{K}^{nc})$ for the following reason. Consider a regular orbit $O=G/T$, then the image $\mathcal{K}_O$ of $\mathcal{K}$ in $\mathcal{S}(O)$ is the kernel of the pushforward map $\mathcal{S}(G/T)\to\mu_{G/B}\oplus\mu_{G/B^-}$. Then $(\mathcal{K}_O)_G$ is one dimensional and the Bockstein map $H_1(G,\mu_{G/B})\to H_0(G,\mathcal{K}_O)$ is nonzero. □

The Lemma clearly implies statement (b) of Proposition. □

5.3. Cocycles for Chern character and Euler characteristic. Let now M be a locally projective object of $\overline{Sm}$.

Thus there exist idempotents $e\in Mat_n(\mathcal{H})$, $e_+\in Mat_m(\hat{\mathcal{H}}_+)$ and $a,b\in Mat_{n,m}(\hat{\mathcal{H}})$, such that $ab=e$, $ba=e_+$ with isomorphisms $M|_{\mathfrak{Z}}\cong\mathcal{H}^n e$, $M_{\widehat{\partial\mathfrak{Z}}}\cong\hat{\mathcal{H}}_+^m e_+$, so that a, b induces the two inverse isomorphisms between the restrictions of $\mathcal{H}^n e$ and $\hat{\mathcal{H}}_+e_+$ to the punctured formal neighborhood of $\partial\mathfrak{Z}$ coming from the identification with the restriction of M.

LEMMA 5.7. *In the above notation,* $\bar{ch}(M)$ *is represented by the cocycle for* $Bar(\overline{Sm})$ $c=(c_{\mathcal{H}},c_+,c)$*, where*
$c=\sum\limits_i e_{ii}\in\mathcal{H}$, $c_+=\sum_j(e_+)_{jj}\in\hat{\mathcal{H}}_+$, $\hat{c}=\sum\limits_{i,j}a_{ij}\otimes b_{ji}\in\hat{\mathcal{H}}\otimes\hat{\mathcal{H}}$.

PROOF. This is a special case of Example 4.1(c). □

To state the next result we need the following notation. Fix $g\in G(O)$ normalizing K thus g acts by conjugation on the sheaf of algebras $\mathcal{A}$ and on global sections of an object in $\mathcal{A}-mod$.

We introduce a zero-cochain τ_g for the dual complex $Bar(\overline{Sm})^*$, $\tau_g=(\tau_{\mathcal{H}}(g),\tau_+(g),\hat{\tau}(g))$, where $\tau_{\mathcal{H}}(g)=WO_g\in\mathcal{H}^*$ (the weighted orbital integral), $\tau_+(g)=0\in\hat{\mathcal{H}}_+^*$ and $\hat{\tau}(g):\alpha\otimes\beta\mapsto Tr([\alpha,\Pi]\beta\circ g,\widehat{\mathcal{S}(X)})$, where $\alpha,\beta\in\hat{\mathcal{H}}$ act on

$\widehat{\mathcal{S}(X)}$ by left multiplication, Π is the operator of multiplication by the characteristic function of X_+ acting on $\widehat{\mathcal{S}(X)}$ and g is acting on the right.

PROPOSITION 5.8. *a) The cochain τ_g is a cocycle. It satisfies:*

$$Tr(g, RHom(\overline{\mathcal{H}}', \mathcal{M})) = \langle \tau_g, \bar{ch}(M) \rangle,$$

for $M \in \overline{Sm}$, where $\overline{\mathcal{H}}'$ is the intertwining bimodule introduced at the end of section 3.

b) For $\phi \in \mathcal{K}_G^c = Im(HH_0^c(Sm) \to HH_0(\overline{Sm}))$ we have:

$$\langle \phi, \tau_g \rangle = WO_g(\phi).$$

c) The image of τ_g in $HH_0^c(Sm)$ is independent of the auxiliary choices in the definition of τ_g.

PROOF. Part (c) clearly follows from (b).

We deduce (b) from Lemma 3.14. Let us present the map $\mathcal{K}_G^c \to HH_0(Bar^c(Sm))$ by an explicit cocycle. Fix $f \in \mathcal{K} \cap \mathcal{H}_K$. Consider the decomposition of $\mathcal{S}(G/U)^K$ as a direct sum of spaces of functions on preimages of K orbits in G/B. Then the action of f on this space has zero block diagonal components with respect to this decomposition. Also, the image of f in $\hat{\mathcal{H}} = \widehat{\mathcal{S}(Y)}$ is contained in $\mathcal{S}(Y_0)$, while $\widehat{\mathcal{S}(Y_0)}_G = 0$ by the proof of Proposition 5.3. Thus $f = \sum h_i - {}^{g_i} h_i$, $g_i \in G$, $h_i \in \widehat{\mathcal{S}(Y_0)}$. It is easy to check that the image of $\bar{f}$ in $HH_0(Bar^c(Sm))$ is represented by $(f, \sum g_i \delta_K \otimes h_i g^{-1}) \in \mathcal{H} \oplus \hat{\mathcal{H}} \otimes \hat{\mathcal{H}}$.

It remains to check that for h_i as above we have

$$Tr([g_i, \Pi] h_i g_i^{-1} \circ g, \widehat{\mathcal{S}(X)}) = 0. \tag{16}$$

We claim that the following stronger statement holds: the operator $[g_i, \Pi] h_i g_i^{-1}$ acting on $\mathcal{S}(G/U)$ has zero diagonal blocks with respect to the above block decomposition. To see this pick an open compact subgroup K' such that $h_i \in \mathcal{H}_{{}^{g_i^{-1}}K'}$ for all i. Then g_i^{-1} sends a summand of the decomposition corresponding to a K'-orbit $U \subset G/B$ to a summand corresponding to the ${}^{g_i^{-1}}K'$ orbits $g_i^{-1}(U)$, the endomorphism h_i acts by an operator with zero diagonal blocks with respect to the decomposition corresponding to ${}^{g_i^{-1}}K'$ orbits, while $[g_i, \Pi]$ sends the summand corresponding to an orbit $g_i^{-1}(U)$ to the summand corresponding to the orbit U (this is clear since Π preserves all the direct sum decompositions above). Equality (16) and hence part (b) of the Proposition follows.

The proof of part (a) occupies the next subsection.

5.4. Tate cocycle and weighted orbital integral.

5.4.1. *The Tate cocycle.* A general reference for the following material is [**5**] or [**4**], especially [**4**], §4.2.13, p 142; §7.13.18, p 344, developing the theme started in [**21**].

Let V be a Tate vector space, then the Lie algebra $End(V)$ of continuous endomorphism of V has a canonical central extension, which we will denote by $\widetilde{End}(V)$.

Let $End^b(V) \subset End(V)$ denote the subspace of endomorphisms with bounded image, and $End^d(V) \subset End(V)$ be the subspace of endomorphisms having open kernel (here b stands for bounded, and d for discrete). Then $End^b(V)$, $End^d(V)$ are two-sided ideals in the associative algebra $End(V)$, and hence ideals in the

Lie algebra $End(V)$. Further, the trace functional is defined on the intersection $End^{bd} = End^b(V) \cap End^d(V)$, $tr : End^{bd} \to k$; it satisfies

$$(17) \qquad \begin{aligned} tr([E, E^{bd}]) &= 0; \\ tr([E^b, E^d]) &= 0 \end{aligned}$$

for $E^b \in End^b(V)$, $E^d \in End^d(V)$, $E^{bd} \in End^{bd}(V)$, $E \in End(V)$.

The central extension $\widetilde{End}(V) \to End(V)$ is specified by the requirement that it is trivialized on the ideals $End^b(V)$, $End^d(V)$, i.e. the embeddings $End^b(V) \hookrightarrow End(V)$, $End^d(V) \hookrightarrow End(V)$ are lifted to (fixed) homomorphisms $s_b : End^b(V) \hookrightarrow \widetilde{End}(V)$, $s_d : End^d(V) \hookrightarrow \widetilde{End}(V)$, where s_b, s_d intertwine the adjoint action of $End(V)$; and for $E^{bd} \in End^{bd}(V)$ we have

$$(18) \qquad s_b(E^{bd}) - s_d(E^{bd}) = tr(E^{bd}) \cdot c,$$

where $c \in \widetilde{End}(V)$ is the generator of the kernel of the projection $\widetilde{End}(V) \to End(V)$.

Suppose now that a decomposition of V

$$(19) \qquad V = V^+ \oplus V^-$$

into a sum of a bounded open and a discrete subspaces is given, and let Π denote the projection to the bounded open V^+ along the discrete V^-. Then for any $E \in End(V)$ in the right-hand side of

$$E = E \cdot \Pi + E \cdot (Id - \Pi)$$

the first summand lies in $End^b(V)$, and the second one in $End^d(V)$. Thus we get a splitting $s = s_\Pi : End(V) \to \widetilde{End}(V)$,

$$(20) \qquad s_\Pi(E) = s_b(E \cdot \Pi) + s_d(E \cdot (Id - \Pi)).$$

It is also easy to see that for E preserving V_- (respectively, V_-) the element $s_\Pi(E)$ is independent of the choice of the complement V_- (respectively, V_+). Thus we get a canonical splitting s_{V^+} (respectively, s_{V^-}) of the central extension on the subalgebras $End_{V_+}(V)$, $End_{V^-}(V)$ consisting of endomorphisms preserving V_+ (respectively, V_-).

Denote by $C(E_1, E_2)$ the corresponding 2-cocycle of $End(V)$, i.e. a bi-linear functional, such that

$$[s(E_1), s(E_2)] = s([E_1, E_2]) + C(E_1, E_2) \cdot c.$$

Then we have

$$(21) \qquad C(E_1, E_2) = Tr(E_1 \circ \Pi \circ E_2 \circ (Id - \Pi) - E_2 \circ \Pi \circ E_1 \circ (Id - \Pi)).$$

Finally, suppose that another discrete cobounded space $W \subset V$ is fixed. The splitting s_W of the central extension on $End_W(V)$ yields a linear functional $\sigma_W = s_W - s_\Pi|_{End_W(V)}$ on $End_W(V)$.

EXAMPLE 5.9. We have

$$(22) \qquad \sigma_W(Id) = \dim(V^+ \cap W) - \operatorname{co}\dim_V(V^+ + W).$$

Also, suppose that F is an automorphism of V such that either $F(V^+) \subset V^+$, or $V^+ \subset F(V^+)$; set $d_{V^+}(F) = -\dim(V^+/F(V^+))$ in the former and $d_{V_+}(F) = \dim(F(V^+)/V^+)$ in the latter case. Then

$$(23) \qquad C(F, F^{-1}) = d_{V^+}(F).$$

Both equalities follow directly from the definitions.

Consider now the complex

$$Cone\,(Bar(End_{V^+}(V)) \oplus Bar(End_W(V)) \to Bar(End(V)))\,[-1] \tag{24}$$

and define a zero-cochain for the dual complex $\epsilon = (0, \sigma_W, C)$.

It is easy to check that the zero-cochain ϵ is in fact a cocycle whose cohomology class does not depend on the choice of V^- for a fixed V^+.

5.4.2. *Weighted orbital integral and Tate cocycle.* We now apply this in the following example. Consider the Tate space

$$V = \hat{\mathcal{H}} \cong \widehat{\mathcal{S}(Y)} \cong \widehat{\mathcal{S}(X)} \tag{25}$$

where X was defined in the second paragraph of §5.1, the first isomorphism was discussed above and the second one is induced by the inverse intertwining operator. Let W be the image of $\mathcal{H}$ in the space (25) and $\widehat{\mathcal{S}(X)}^+$ be the space of functions supported on X_+ while $\widehat{\mathcal{S}(X)}^-$ is the space of functions supported on the complement of X_+.

The group K_0 acts on all these spaces. Fix a representation ρ of K_0 and set $V_\rho = Hom_{K_0}(\rho, V)$, $W_\rho = Hom_{K_0}(\rho, \mathcal{H})$ and $V_\rho^\pm = Hom_{K_0}(\rho, \widehat{\mathcal{S}(X)}^\pm)$.

The complex (12) maps naturally to the complex (24) constructed from $V = V_\rho$ $V^\pm = V_\rho^\pm$ and $W = W_\rho$, let a_ρ denote that map.

LEMMA 5.10. *We have* $a_\rho^*(\epsilon) = \int_{K_0} \tau_g Tr(g, \rho) dg$.

PROOF. Equality of components in $(\hat{\mathcal{H}}^{\otimes 2})^*$ follows by inspection.

It remains to check equality of components in $\mathcal{H}^*$. Recall (see e.g. [**3**, §I.11]) that $WO_g(f) = \phi(0)$, where ϕ is a linear function on dominant weights such that for large λ we have

$$\phi(\lambda) = Tr(f \circ \Pi_\lambda \circ g, \mathcal{H}).$$

Here f acts on $\mathcal{H}$ by convolution on the left, g acts by right translation and Π_λ is the characteristic function of $G_{\leq\lambda}$ which is the union of two sided $K_0 = G(O)$ cosets corresponding to $\mu \leq \lambda$.

It follows that for a locally constant function ψ on K_0 supported on regular semisimple elements there exists an affine linear function $\phi_\psi(\lambda)$, such that

$$\phi_\psi(\lambda) = Tr(f \circ \Pi_\lambda \circ \psi, \mathcal{H}) \qquad \text{for } \lambda \gg 0,$$

$$\int_{K_0} WO_g(f)\psi(g)dg = \phi_\psi(0),$$

where ψ acts by convolution on the right.

Comparing this with the definition of Tate cocycle ϵ and of the intertwining bimodule $\overline{\mathcal{H}}'$ (see Example 3.13) and using Lemma 3.14, we see that for ψ as above:

$$\int_{K_0} WO_g(f)\psi(g)dg = \sigma_{\mathcal{H}}(f \otimes \psi),$$

where $f \otimes \psi$ acts on the Tate space (25) via its natural $\mathcal{H}$-bimodule structure. Both sides of the last equality are continuous in ψ with respect to the L^1 norm: this is clear for the right hand side and it follows from Theorem 2.14 for the left hand side. Thus validity of the equality for ψ supported on regular semisimple elements implies its validity for all ψ. □

5.4.3. *Sheaves of algebras on curves.* We now apply the above construction in the following setting. Let C be a smooth[4] complete curve with a finite collection of points $\mathbf{x} = \{x_1, \ldots, x_n\}$. Let $\hat{C}_{\mathbf{x}}$, $\hat{C}^0_{\mathbf{x}}$ be the formal neighborhood and the punctured formal neighborhood of $\mathbf{x}$ respectively. Let $\mathcal{V}$ be a torsion free coherent sheaf on C, and set $V = \Gamma(\hat{C}^0_{\mathbf{x}}, \mathcal{V})$, $W = \Gamma(C \setminus \mathbf{x}, \mathcal{V})$, $V^+ = \Gamma(\hat{C}_{\mathbf{x}}, \mathcal{V})$. Fixing formal coordinates z_i at x_i and trivializations of $\mathcal{V}$ on the formal neighborhood of x_i we get an isomorphism $V \cong \mathbb{C}((z))^N$ sending V_+ to $\mathbb{C}[[z]]^N$, thus we get a splitting $V = V^+ \oplus V^-$ where $V^- \cong z^{-1}\mathbb{C}[z^{-1}]^N$.

Let $\mathcal{A}$ be a torsion free coherent sheaf of algebras on C with a right action on V; we assume also that $\mathcal{A}$ has finite homological dimension (i.e. that the algebra of sections of $\mathcal{A}$ on an affine open set has this property).

Let $A = \Gamma(C \setminus \mathbf{x}, \mathcal{A})$, $\hat{A}^+ = \Gamma(\hat{C}_{\mathbf{x}}, \mathcal{A})$ and $\hat{A} = \Gamma(\hat{C}^0_{\mathbf{x}}, \mathcal{A})$. Then the complex

$$Cone\left(Bar(A) \oplus Bar(\hat{A}^+) \to Bar(\hat{A})\right)[-1] \tag{26}$$

computes Hochschild homology of the category $\mathcal{A} - mod$. On the other hand, it maps naturally to the complex (24), let α denote this map.

LEMMA 5.11. *Suppose that $\mathcal{V}$ is a locally projective sheaf of right $\mathcal{A}$-modules. Then for a coherent sheaf $\mathcal{M}$ of (left) $\mathcal{A}$-modules we have:*

$$\langle ch(\mathcal{M}), \alpha^*(\epsilon)\rangle = \chi(\mathcal{V} \otimes_{\mathcal{A}} \mathcal{M}), \tag{27}$$

where χ denotes Euler characteristic.

PROOF. Both sides of (27) do not change if we replace $\mathcal{A}$ by $\underline{End}_{\mathcal{O}_C}(\mathcal{V})^{op}$ and $\mathcal{M}$ by $\underline{End}_{\mathcal{O}_C}(\mathcal{V})^{op} \otimes_{\mathcal{A}} \mathcal{M}$. Thus we can assume without loss of generality that $\mathcal{A} = \underline{End}_{\mathcal{O}_C}(\mathcal{V})^{op}$. In that case we have a canonical equivalence of categories $Coh(C) \cong \mathcal{A} - mod$, $\mathcal{F} \mapsto \mathcal{F} \otimes_{\mathcal{O}_C} \mathcal{V}^*$.

Furthermore, the operator $[\mathcal{F}] \mapsto [\mathcal{F} \otimes \mathcal{V}]$ induces an automorphism of the rational Grothendieck group $K^0(Coh(C)) \otimes \mathbb{Q}$, thus it suffices to prove (27) for $\mathcal{A} = \underline{End}_{\mathcal{O}_C}(\mathcal{V})^{op}$, $\mathcal{M} = \mathcal{F} \otimes \mathcal{V}^*$, where $\mathcal{F} = \mathcal{V} \otimes \mathcal{F}'$, thus $\mathcal{F} = \mathcal{F}' \otimes_{\mathcal{O}_C} \mathcal{A}$. Now, both sides of (27) do not change if we replace $\mathcal{A}$ by $\mathcal{O}_C$ and $\mathcal{M}$ by $\mathcal{F}'$; the locally projective module $\mathcal{V}$ is replaced by the corresponding coherent sheaf $\mathcal{V}|_{\mathcal{O}_C}$. Thus we have reduced to the case $\mathcal{A} = \mathcal{O}_C$.

It is easy to see that both sides of (27) are additive on short exact sequences as a function of both $\mathcal{V}$ and $\mathcal{F}$. A locally free sheaf on a curve admits a filtration whose associated graded is a sum of line bundles; also, since C is smooth the Grothendieck group $K^0(Coh(C))$ is generated by line bundles. Thus we can assume without loss of generality that both $\mathcal{M}$ and $\mathcal{V}$ are line bundles.

If $\mathcal{M}$ is a trivial line bundle then $ch(\mathcal{M})$ is represented by the cocycle $(1, 1, 1\otimes 1)$, so (27) follows from (22). Notice also that both sides of (27) for a fixed $\mathcal{V}$ depend only on the degree of the line bundle $\mathcal{M}$ (here we assume without loss of generality that the curve C is irreducible): this is clear for the right hand side, while for the left hand side it follows from the fact that a regular function on an abelian variety is constant. Thus it suffices to consider the case when $\mathcal{M} = \mathcal{O}(nx)$, $x \in \mathbf{x}$. In this case $ch(\mathcal{M})$ is represented by the cocycle $(1, 1, f \otimes f^{-1})$ where f is a function on the formal punctured neighborhood of $\mathbf{x}$ having order n at x and order 0 at other points. In this case (27) follows from (22) and (23). □

---

[4]This assumption is likely unnecessary but it allows one to simplify the statements and the proofs.

COROLLARY 5.12. *a) Suppose that $\mathcal{V} \cong \mathbf{e}\mathcal{A}$ is a direct summand in a free rank one module for an idempotent $\mathbf{e} \in \Gamma(C, \mathcal{A})$. Then for a coherent sheaf $\mathcal{M}$ of $\mathcal{A}$-modules we have:*

$$\langle ch(\mathcal{M}), \alpha^*(\epsilon)\rangle = \chi(\mathbf{e}\mathcal{M}),$$

where χ denotes the Euler characteristic.

b) Suppose that $\mathcal{M}$ has finite support contained in $C \setminus \mathbf{x}$. Then the equality in (a) holds under a weaker assumption that $\mathcal{V}|_{C\setminus \mathbf{x}} \cong \mathcal{A}\mathbf{e}|_{C\setminus \mathbf{x}}$.

5.4.4. *Proof of Proposition 5.8(a).* It suffices to check that we get an equality upon averaging against a character of any representation ψ of $K_0 = G(O)$.

In view of Corollary 5.12(b) it suffices to check that the cochain $\int_{K_0} \tau_g Tr(g, \psi) dg$ is obtained as in Corollary 5.12 for an appropriate choice of trivialization for $\mathcal{V}$ on the formal neighborhood of $\mathbf{x}$, where $C \subset \overline{\mathfrak{Z}}$ is the union of one dimensional components containing representations with nonzero K-invariant vectors, $\mathbf{x} = (\overline{\mathfrak{Z}} \backslash \mathfrak{Z}) \cap \mathbf{C}$, $\mathcal{A} = \overline{\mathcal{H}}$, z_i are the natural local coordinates and $\mathcal{V} = Hom_{K_0}(\psi, \overline{\mathcal{H}}')$. Applying Lemma 5.10 we reduce to showing that $a^*_\psi(\epsilon)$ has the required form.

It follows from Lemma 3.14 that the space V_+ of sections of $\overline{\mathcal{H}}'$ on the formal neighborhood of $\overline{\mathfrak{Z}} \setminus \mathfrak{Z}$ is identified with $\widehat{\mathcal{S}(X_+)}^{K\times K}$ (notation introduced in the second paragraph of §(5.1)); moreover, we can choose a trivialization of the sheaf $\overline{\mathcal{H}}$ on the formal neighborhood of $\overline{\mathfrak{Z}} \backslash \mathfrak{Z}$ so that constant sections correspond to functions supported on the set $X(O) = X_+ \setminus (tX_+)$. Then the space V_- is identified with $\mathcal{S}(X \setminus X_+)^{K\times K}$. The desired equality now follows by inspecting the definitions. □

5.5. Proof of part (b) of Theorem 5.1. For $g \in K_0$ the formula follows from Proposition 5.8. The general case of a compact element follows by conjugating with an element of $GL(2, F)$. □

References

[1] J. Arthur, *A local trace formula*, Inst. Hautes Études Sci. Publ. Math. **73** (1991), 5–96. MR1114210

[2] J. Arthur, *The characters of supercuspidal representations as weighted orbital integrals*, Proc. Indian Acad. Sci. Math. Sci. **97** (1987), no. 1-3, 3–19 (1988), DOI 10.1007/BF02837809. MR983600

[3] J. Arthur, *An introduction to the trace formula*, Harmonic analysis, the trace formula, and Shimura varieties, Clay Math. Proc., vol. 4, Amer. Math. Soc., Providence, RI, 2005, pp. 1–263. MR2192011

[4] A. Beilinson, V. Drinfeld, *Quantization of Hitchin's Integrable System and Hecke Eigensheaves,* unpublished manuscript available at: http://www.math.uchicago.edu/~mitya/langlands/hitchin/BD-hitchin.pdf

[5] A. Beilinson, B. Feigin, B. Mazur, *Notes on Conformal Field Theory,* unpublished manuscript available at: http://www.math.stonybrook.edu/~kirillov/manuscripts.html

[6] R. Beuzart-Plessis, *A local trace formula for the Gan-Gross-Prasad conjecture for unitary groups: the archimedean case,* preprint arXiv:1506.01452.

[7] R. Bezrukavnikov, *Homological properties of representations of p-adic groups related to geometry of the group at infinity,* Ph.D. thesis, preprint arXiv:math/0406223.

[8] R. Bezrukavnikov and D. Kazhdan, *Geometry of second adjointness for p-adic groups*, Represent. Theory **19** (2015), 299–332, DOI 10.1090/ert/471. With an appendix by Yakov Varshavsky, Bezrukavnikov and Kazhdan. MR3430373

[9] J. N. Bernstein, *Le "centre" de Bernstein* (French), Representations of reductive groups over a local field, Travaux en Cours, Hermann, Paris, 1984, pp. 1–32. Edited by P. Deligne. MR771671

[10] J. N. Bernstein and A. V. Zelevinsky, *Representations of the group $GL(n,F)$, where F is a local non-Archimedean field* (Russian), Uspehi Mat. Nauk **31** (1976), no. 3(189), 5–70. MR0425030

[11] W. Casselman, *Characters and Jacquet modules*, Math. Ann. **230** (1977), no. 2, 101–105, DOI 10.1007/BF01370657. MR0492083

[12] R. Hartshorne, *Residues and duality*, Lecture notes of a seminar on the work of A. Grothendieck, given at Harvard 1963/64. With an appendix by P. Deligne. Lecture Notes in Mathematics, No. 20, Springer-Verlag, Berlin-New York, 1966. MR0222093

[13] Harish-Chandra, *Admissible invariant distributions on reductive p-adic groups*, University Lecture Series, vol. 16, American Mathematical Society, Providence, RI, 1999. With a preface and notes by Stephen DeBacker and Paul J. Sally, Jr. MR1702257

[14] D. Kazhdan, *Cuspidal geometry of p-adic groups*, J. Analyse Math. **47** (1986), 1–36, DOI 10.1007/BF02792530. MR874042

[15] D. Kazhdan and Y. Varshavsky, *Endoscopic decomposition of certain depth zero representations*, Studies in Lie theory, Progr. Math., vol. 243, Birkhäuser Boston, Boston, MA, 2006, pp. 223–301, DOI 10.1007/0-8176-4478-4_10. MR2214251

[16] B. Keller, *On the cyclic homology of exact categories*, J. Pure Appl. Algebra **136** (1999), no. 1, 1–56, DOI 10.1016/S0022-4049(97)00152-7. MR1667558

[17] B. Keller, *On the cyclic homology of ringed spaces and schemes*, Doc. Math. **3** (1998), 231–259. MR1647519

[18] R. McCarthy, *The cyclic homology of an exact category*, J. Pure Appl. Algebra **93** (1994), no. 3, 251–296, DOI 10.1016/0022-4049(94)90091-4. MR1275967

[19] P. Schneider and U. Stuhler, *Representation theory and sheaves on the Bruhat-Tits building*, Inst. Hautes Études Sci. Publ. Math. **85** (1997), 97–191. MR1471867

[20] A. J. Silberger, PGL_2 *over the p-adics: its representations, spherical functions, and Fourier analysis*, Lecture Notes in Mathematics, Vol. 166, Springer-Verlag, Berlin-New York, 1970. MR0285673

[21] J. Tate, *Residues of differentials on curves*, Ann. Sci. École Norm. Sup. (4) **1** (1968), 149–159. MR0227171

[22] J. Tits, *Reductive groups over local fields*, Automorphic forms, representations and L-functions (Proc. Sympos. Pure Math., Oregon State Univ., Corvallis, Ore., 1977), Proc. Sympos. Pure Math., XXXIII, Amer. Math. Soc., Providence, R.I., 1979, pp. 29–69. MR546588

[23] J.-L. Waldspurger, *Une formule intégrale reliée à la conjecture locale de Gross-Prasad* (French, with English summary), Compos. Math. **146** (2010), no. 5, 1180–1290, DOI 10.1112/S0010437X10004744. MR2684300

[24] J.-L. Waldspurger, *Caractères de représentations de niveau 0*, arXiv 1602.01440.

Department of Mathematics, Massachusetts Institute of Technology, Cambridge, Massachusetts 02139

Department of Mathematics, Hebrew University of Jerusalem, 9190401 Jerusalem, Israel

Proceedings of Symposia in Pure Mathematics
Volume **101**, 2019
https://doi.org/10.1090/pspum/101/01789

Schwartz space of parabolic basic affine space and asymptotic Hecke algebras

Alexander Braverman and David Kazhdan

ABSTRACT. Let F be a local non-archimedean field and G be the group of F-points of a split connected reductive group over F. In arXiv:1704.03019 we define an algebra $\mathcal{J}(G)$ of functions on G which contains the Hecke algebra $\mathcal{H}(G)$ and is contained in the Harish-Chandra Schwartz algebra $\mathcal{C}(G)$. We consider $\mathcal{J}(G)$ as an algebraic analog the algebra $\mathcal{C}(G)$.

Given a parabolic subgroup P of G with a Levi subgroup M and the unipotent radical U_P we write $X_P := G/U_P$. Let $\mathcal{S}_c(X_P)$ be the space of locally constant functions on X_P with compact support and $\mathcal{S}_{cusp,c}(X_P) \subset \mathcal{S}_c(X_P)$ be subspace of functions whose right shifts span a cuspidal representation of M.

In this paper we study two versions of the Schwartz space of X_P. The first is $\mathcal{S}(X_P) := \mathcal{J}(\mathcal{S}_c(X_P))$ and the 2nd is the space spanned by functions of the form $\Phi_{Q,P}(\phi)$ where Q is another parabolic with the same Levi subgroup, $\phi \in \mathcal{S}_c(X_Q)$ and $\Phi_{Q,P}$ is a normalized intertwining operator from $L^2(X_Q)$ to $L^2(X_P)$. We formulate a series of conjectures about these spaces; for example, we conjecture that $\mathcal{S}'(X_P) \subset \mathcal{S}(X_P)$ and that this embedding is an isomorphism on the M-cuspidal part. We give a proof of some of our conjectures (cf. Theorem 1.9).

1. Introduction and statement of the results

Let F be a local field, G be a connected split reductive group over F and $\mathcal{M}(G)$ the category of complex smooth representation of G. The analysis of the category $\mathcal{M}(G)$ is done in two steps:

the first is to define the notion of *elementary* representations and and (partially or conjecturally) *classify* these representations

the second is to *describe* the category $\mathcal{M}(G)$ in terms of these *elementary* representations.

Harish-Chandra considers *square-integrable* representations of Levi subgroups as elementary while Bernstein considers only *cuspidal* ones as elementary.

The approach of Harish-Chandra which is based on the analysis of the Schwartz algebra $\mathcal{C}(G)$ of smooth rapidly decreasing functions on G. His approach works for all local fields and also provides a complete reduction of the description of the K-theory of the category $\mathcal{M}(G)$ to a description of *elementary* representations.

The first author is partially supported by NSERC. The second author acknowledges that this project received funding from ERC under grant agreement No. 669655.

On the other hand the approach of Harish-Chandra is not algebraic and therefore not compatible with the categorical Langlands program.

The approach of Bernstein is based on the analysis of algebra $\mathcal{H}(G)$ of locally constant compactly supported functions on G. This approach works only for non-archimedean local fields and in the second step is much more difficult to implement. For example, the only elementary objects relevant for a description of the subcategory $\mathcal{M}^I(G) \subset \mathcal{M}(G)$ of representations generated by the subspace of Iwahori-fixed vectors are unramified characters of the maximal split torus $T \subset G$. On the other hand an important advantage of the Bernstein's analysis is it algebraic nature. In particular there exists a natural categorification of Hecke algebras $\mathcal{H}(G)$ by appropriate monoidal categories of sheaves on the group $G(F)$ considered as an ind-pro variety over the residue field k (at least in the case when $F = k((t))$).

From now we consider only the case when the field F is non-archimedean. In [**4**] we introduced an algebraic analogue $\mathcal{J}(G) \subset \mathcal{C}(G)$ of the Harish-Chandra algebra $\mathcal{C}(G)$ containing $\mathcal{H}(G)$. The algebra $\mathcal{J}(G)$ is finitely generated as a $\mathcal{H}(G)$-module and it indecomposable direct summands are parametrized by irreducible square-integrable representations of Levi subgroups of G. To simplify notations we omit the term *algebraic* and say that $\mathcal{J}(G)$ is *the Schwartz algebra* of functions on G. One of results of this paper is an explicit description of the part of $\mathcal{J}(G)$ corresponding to cuspidal representations of maximal Levi subgroups.

In addition to an *algebraic enlargement* of the algebra $\mathcal{H}(G) = \mathcal{S}(G)$ we consider *algebraic enlargements* of spaces $\mathcal{S}(G/U)$ where U is a radical of a parabolic $P = LU$ and $\mathcal{S}(G/U)$ is the space of locally constant compactly supported functions on G/U. Actually we propose two such enlargements- the one defines as $\mathcal{J}(G)\ calS(G/U) \subset L^2(G/U)$ and the other the subspace of $L^2(G/U)$ spanned by images of normalized intertwining operators. We say that these spaces of functions on G/U also *the Schwartz functions on G/U*.

We show that these definitions coincide on the cuspidal component for maximal parabolic subgroups P and discuss the conjecture that the same is true without the conditions of maximality. In the case when P is the Borel subgroup these definitions coincide with one which appears in [**3**].

1.1. Notation. Let F be a local field non-archimedean with ring of integers $\mathcal{O}$; we shall fix a generator κ of the maximal ideal of $\mathcal{O}$. Typically, we shall denote algebraic varieties over F by boldface letters (e.g. $\mathbf{G}, \mathbf{X}$ etc.) and the corresponding sets of F-points – by the corresponding ordinary letters (i.e. G, X etc.).

with a Borel subgroup $\mathbf{B}$, unipotent radical $\mathbf{U}$ be the unipotent radical of $\mathbf{B}$ and $\mathbf{T} := \mathbf{B}/\mathbf{U}$ be the Cartan torus. We fix an imbedding of $\mathbf{T}$ into $\mathbf{B}$.

Let Λ be the lattice of cocharacters of $\mathbf{T}$ and $\Lambda^\vee$ be the lattice of characters of $\mathbf{T}$. We fix a Haar measure dg on G and we denote by $\mathcal{H}(G)$ the Hecke algebra of locally constant compactly supported functions on G. A choice of a Haar measure defines a structure of a locally unital associative algebra on $\mathcal{H}(G)$. As well-known the category $\mathcal{M}(G)$ of smooth G-modules is equivalent to the category of non-degenerate $\mathcal{H}(G)$-modules.

1.2. Functions on X_P. Let P be a parabolic subgroup of G with a Levi subgroup M and unipotent radical U_P. Let $X_P = G/U_P$. This space has a natural $G \times M$ action. Therefore the space $\mathcal{S}_c(X_P)$ of locally constant compactly supported functions on X_P becomes a $G \times M$ module; for convenience we are going to twist the M action by the square root of the absolute value of the determinant of the M-action on the Lie algebra $\mathfrak{u}_P$ of U_P.

It is easy to see that X_P possesses unique (up to constant) G-invariant measure, hence we can talk about $L^2(X_P)$. It has a natural unitary action of $G \times M$ (the action of M is unitary if we twist it by $|\delta_P|^{1/2}$ where $\delta_P : \mathbf{M} \to \mathbb{G}_m$ is the determinant of the action of $\mathbf{M}$ on the Lie algebra of $\mathbf{U_P}$).

1.3. The algebra $\mathcal{J}(G)$ and the space $\mathcal{S}(X_P)$. In [**4**] we have defined certain algebra $\mathcal{J}(G)$ of functions on G which contains the Hecke algebra $\mathcal{H}(G)$ which can be thought of as an algebraic version of the Harish-Chandra Schwartz space $\mathcal{C}(G)$ (the definition will be recalled in Section 3); in particular, $\mathcal{J}(G)$ is a smooth $G \times G$ -module. It is explained in *loc. cit.* that $\mathcal{J}(G)$ acts on $L^2(X_P)$ for any P, so we can set

$$\mathcal{S}(X_P) = \mathcal{J}(G) \cdot \mathcal{S}_c(X_P),$$

where $\mathcal{S}_c(X_P)$ stands for the space of locally constant functions with compact support on X_P. The space $\mathcal{S}(X_P)$ is a smooth $G \times M$-module.

1.4. Intertwining operators – L^2-version. The following result is essentially Theorem 2.1 of [**1**]:

THEOREM 1.5. *Let P and Q be two associate parabolics, i.e. two parabolics with the same Levi subgroup M. Then there exists a $G \times M$-equivariant unitary isomorphism $\Phi_{P,Q} : L^2(X_P) \widetilde{\to} L^2(X_Q)$. These isomorphisms satisfy the following properties:*

(1) $\Phi_{P,P} = \mathrm{id}$
(2) *For 3 parabolic subgroups P, Q, R with the same Levi subgroup M we have* $\Phi_{Q,R} \circ \Phi_{P,Q} = \Phi_{P,R}$.

Note that (1) and (2) together imply that $\Phi_{Q,P} \circ \Phi_{P,Q} = \mathrm{id}$.
Warning. The operator $\Phi_{P,Q}$ is not canonical - it depends on various choices. In what follows we are going to choose some operators $\Phi_{P,Q}$ satisfying the above requirements.

1.6. Another version of a space of functions on X_P. For a parabolic subgroup P of G with chosen Levi subgroup M let $\mathrm{Ass}(P)$ denote the set of all parabolics Q containing M as a Levi subgroup. We now define another version of the $\mathcal{S}'(X_P)$ of the Schwartz space of functions on X_P by setting

$$\mathcal{S}'(X_P) = \sum_{Q \in \mathrm{Ass}(P)} \Phi_{Q,P}(\mathcal{S}_c(X_Q)).$$

REMARK 1.7. It is easy to see that the space $\mathcal{S}'(X_P)$ does not depend on a choice of intertwining operators $\Phi_{Q,P}$.

The expected relationship between the two versions of the Schwartz space is described by the following conjecture:

CONJECTURE 1.8. (1) $\mathcal{S}'(X_P) \subset \mathcal{S}(X_P)$.
(2) $\mathcal{S}'(X_P)_{\mathrm{cusp}} = \mathcal{S}(X_P)_{\mathrm{cusp}}$. *Here by $\mathcal{S}(X_P)_{\mathrm{cusp}}$ (resp. $\mathcal{S}'(X_P)_{\mathrm{cusp}}$) we denote the M-cuspidal part of $\mathcal{S}(X_P)_{\mathrm{cusp}}$ (resp. of $\mathcal{S}'(X_P)_{\mathrm{cusp}}$).*
(3) *The operator $\Phi_{P,Q}$ defines an isomorphism between $\mathcal{S}(X_P)$ and $\mathcal{S}(X_Q)$.*

(4) *The right action of $\mathcal{J}(M)$ on $L^2(X_P)$ preserves $\mathcal{S}(X_P)$ (thus making $\mathcal{S}(X_P)$ into a $(\mathcal{J}(G), \mathcal{J}(M))$-bimodule (It is easy to see that the Schwartz algebra $\mathcal{C}(G)$ acts on $L^2(X_P)$ on the right; hence $L^2(X_P)$ is also a right module over $\mathcal{J}(M)$).*

We denote by $\mathcal{S}(X_P)_{\text{cusp}}$ (resp. $\mathcal{S}'(X_P)_{\text{cusp}}$) the M-cuspidal part of $\mathcal{S}(X_P)_{\text{cusp}}$ (resp. of $\mathcal{S}'(X_P)_{\text{cusp}}$).

Below is the main result of this paper:

THEOREM 1.9. (1) *Let P be a maximal parabolic of G. Then we have $\mathcal{S}'(X_P)_{\text{cusp}} = \mathcal{S}(X_P)_{\text{cusp}}$.*

(2) *Assume that $\mathcal{S}'(X_P)_{\text{cusp}} \subset \mathcal{S}(X_P)_{\text{cusp}}$. Then*

(a) $\mathcal{S}'(X_P)_{\text{cusp}} = \mathcal{S}(X_P)_{\text{cusp}}$

(b) The operator $\Phi_{P,Q}$ defines an isomorphism between $\mathcal{S}(X_P)_{\text{cusp}}$ and $\mathcal{S}(X_Q)_{\text{cusp}}$.

The rest of the paper is devoted to the proof of Theorem 1.9.

1.10. Examples. Let us look at two extreme cases. First, consider the case $P = G$. In this case $\text{Ass}(P)$ consists of 1 element and therefore $\mathcal{S}'(X_P) = \mathcal{H}(G)$; similarly, $\mathcal{S}(X_P) = \mathcal{J}(G)$. Conjecture 1.8 is obvious in this case. Also, this example shows that $\mathcal{S}'(X_P) \neq \mathcal{S}(X_P)$ in general since $\mathcal{H}(G) \neq \mathcal{J}(G)$.

Second, let us consider the case when P is a Borel subgroup B of G. Then M is a maximal split torus T of G and we have $\mathcal{H}(T) = \mathcal{J}(T)$, so assertion (4) is obvious. On the other hand, any representation of T is cuspidal, so we have $\mathcal{S}(X_B)_{\text{cusp}} = \mathcal{S}(X_B)$ and $\mathcal{S}'(X_B)_{\text{cusp}} = \mathcal{S}'(X_B)$. Hence assertion (2) says in this case that $\mathcal{S}'(X_B) = \mathcal{S}(X_B)$. This statement was conjectured in [**4**] (note that $\mathcal{S}'(X_B)$ is exactly the Schwartz space of [**3**]), where we proved this conjecture for the I-invariant part of both spaces (here I stands for an Iwahori subgroup of G).

2. Tempered representations and Harish-Chandra algebra

DEFINITION 2.1. For any smooth representation (π, V) of G and smooth vectors $v \in V, v^\star \in V^\vee$ we define the function $m_{v,v^\star}$ on G

$$m_{v,v^\star}(g) := v^\star(\pi(g)v), v \in V$$

and say that $m_{v,v^\star}$ is a *matrix coefficient* of the representation π.

The algebra $\mathcal{H}(G)$ can be embedded into the Harish-Chandra Schwartz algebra $\mathcal{C}(G)$. The algebra $\mathcal{C}(G)$ does not act on all smooth representations of G but only on tempered ones. Let G be a locally compact group. In this subsection we only consider unitary representations $(\pi, V, (,))$ of G.

DEFINITION 2.2. (1) We say that a representation $(\pi, V, (,))$ is in the closure of a representation $(r, W, (,))$ if for any vectors $v_1, \dots, v_n \in V$, a compact C in G and $\epsilon > 0$ there exist vectors $w_1, \dots w_n \in W$ such that

$$|(\pi(g)v_i, v_j) - (r(g)v_i, v_j)| \leq \epsilon \text{ for all } i, j, 1 \leq i, j \leq n \text{ and } g \in C.$$

(2) A representation (π, V) of G is *tempered* if it is in the closure of the regular representation of G.

(3) We denote by $\mathcal{M}_t(G)$ the category of tempered representations.

(4) We denote by $\hat{G}_t$ the set of tempered irreducible representations. As follows Schur's lemma we can consider $\hat{G}_t$ as a subset of the set $\hat{G}$ of irreducible Banach representations of G.

The following results are proven in [**5**] and [**9**].

CLAIM 2.3. (1) A unitary irreducible representation $(\pi, V, (,))$ of a reductive group G over a non-archimedean local field is tempered iff matrix coefficients of π belong to $L^{2+\epsilon}(G/Z(G))$ for any $\epsilon > 0$.

(2) For any tempered representation V of G the action of the algebra $\mathcal{H}(G)$ on V extends to a continuous action of the Harish-Chandra algebra $\mathcal{C}(G)$.

(3) Let P be a parabolic subgroup of G with a Levi group M and σ be a tempered irreducible representation of M. Then the unitary induced representation $\pi_\sigma = ind_P^G(\sigma)$ is tempered.

(4) For a generic unitary character $\chi : M \to S^1$ the representation $\pi_{\sigma\otimes\chi}$ (which is tempered) is irreducible.

(5) Any representation which is a Hilbert space integral of tempered representations is tempered.

Let us now discuss the Harish-Chandra algebra $\mathcal{C}(G)$.

DEFINITION 2.4. (1) Recall that for any $g \in G$, there exists a unique dominant coweight $\lambda(g)$ of $\mathbf{T}$ such that $g \in \mathbf{G}(\mathcal{O})\lambda(g)(\kappa)\mathbf{G}(\mathcal{O})$. We define a function Δ on G by $\Delta(g) := q^{\langle\lambda,\rho\rangle}$.

(2) We say that a function $f : G \to \mathbb{C}$ is a *Schwartz function* if

(a) There exists an open compact subgroup K of G such that f is two-sided K-invariant.

(b) For any polynomial function $p : G \to F$ and $n > 0$, there exists a constant $C = C_{p,n} \in \mathbb{R}_{>0}$ such that

$$\Delta(g)|f(g)| \leq C \ln^{-n}(1 + |p(g)|)$$

for all $g \in G$.

We denote by $\mathcal{C}(G)$ the space of Schwartz functions.

Obviously we have an inclusion $\mathcal{H}(G) \hookrightarrow \mathcal{C}(G)$ and $\mathcal{H}(G) \subset \mathcal{C}(G)$ is dense in the natural topology of $\mathcal{C}(G)$.

The following statements are well known (see for example [**9**]).

CLAIM 2.5. (1) $\mathcal{C}(G)$ has an algebra structure with respect to convolution.

(2) Any tempered representation of $\mathcal{H}(G)$ extends to a continuous representation of $\mathcal{C}(G)$.

LEMMA 2.6. *The natural unitary representation of $G \times M$ on the smooth part of $L^2(X_P)$ is tempered.*

PROOF. Since the right action of M on X_P is free we can write the space $L^2(X_P)$ as a Hilbert space integral

$$L^2(X_P) = \int_{\rho\in\hat{M}_t} \pi(\rho) \otimes \rho \, d\mu_M$$

where μ_M is the Plancherel measure on $\hat{M}_t$ and $\pi(\rho)$ is a unitary representation of G. Since the representation of $G \times M$ on $L^2(X_P)$ has a simple spectrum it is easy to see that $\pi(\rho) = i_{GP}(\rho)$. Now Lemma 2.6 follows from Claim 2.5. □

COROLLARY 2.7. *The natural representation of $\mathcal{H}(G)$ on the smooth part of $L^2(X_P)$ extends to a representation of $\mathcal{C}(G)$.*

3. Paley-Wiener theorems and the definition of the algebra $\mathcal{J}(G)$

3.1. The Paley-Wiener theorem for $\mathcal{H}(G)$. Let P be a parabolic subgroup of G with a Levi group M. The set $\Psi(M)$ of unramified characters of M is equal to $\Lambda_M^\vee \otimes \mathbb{C}^\times$ where $\Lambda_M^\vee \subset \Lambda^\vee$ is the subgroup of characters of $\mathbf{T}$ trivial on $\mathbf{T} \cap [\mathbf{M}, \mathbf{M}]$. So $\Psi(M)$ has a structure of a complex algebraic variety; the algebra of polynomial functions on X_M is equal to $\mathbb{C}[\Lambda_M]$ where Λ_M is the lattice dual to $\Lambda_M^\vee$. We denote by $\Psi_t(M) \subset \Psi(M)$ the subset of unitary characters.

For any $(\sigma, V) \in \mathcal{M}(M)$ we denote by $i_{GP}(\sigma)$ the corresponding unitarily induced object of $\mathcal{M}(G)$. As a representation of $G(\mathcal{O})$ this representation is equal to $\mathrm{ind}_{P(\mathcal{O})}^{G(\mathcal{O})}(\sigma)$. So for any unramified character $\chi : M \to \mathbb{C}^*$ the space V_χ of the representation $i_{GP}(\sigma \otimes \chi)$ is isomorphic to the space V_σ of the representation $i_{GP}(\sigma)$ and is independent on a choice of χ. Since X_M has a structure of an algebraic variety over $\mathbb{C}$ it make sense to say that a family $\eta_\chi \in \mathrm{End}(V_\chi), \chi \in \Psi(M)$ is regular or smooth.

We denote by $Forg : \mathcal{M}(G) \to Vect$ the forgetful functor, by $\widetilde{\mathcal{E}(G)} = \{e(\pi)\}$ the ring of endomorphisms of $Forg$ and define $\mathcal{E}(G) \subset \widetilde{\mathcal{E}(G)}$ as the subring of endomorphisms η_π such that

1) For any Levi subgroup M of G and $\sigma \in Ob(\mathcal{M}(M))$, the endomorphisms $\eta_{i_{GP}(\sigma \otimes \chi)}$ are regular functions of χ.

2) There exists an open compact subgroup K of G such that η_π is $K \times K$-invariant for every π.

By definition, we have a homomorphism

$$PW : \mathcal{H}(G) \to \mathcal{E}(G), \quad f \mapsto \pi(f).$$

The following is usually called "the matrix Paley-Wiener theorem" (cf. [**2**], Theorem 25):

THEOREM 3.2. *The map PW is an isomorphism.*

3.3. The Paley-Wiener $\mathcal{C}(G)$. As follows from the Claim 2.3 the representations $\pi_{\sigma \otimes \chi}$ of G belong to $\mathcal{M}_t(G)$. for any tempered representation σ of M and a unitary character χ of M.

Let $\mathcal{E}_t(G)$ be the subring of endomorphisms $\{\eta\}$ of the forgetful functor $Forgt : \mathcal{M}_t(G) \to Vect$ such that

(1) The function $\chi \to \eta(\pi_{\sigma \otimes \chi})$ is a smooth function of $\chi \in \Psi_t(M)$ for any Levi subgroup M of G and $\sigma \in \hat{M}_t$.
(2) There exists an some open compact subgroup K of G such that η is $K \times K$-invariant.

The following version of the matrix Paley-Wiener theorem is contained in the last section of [**10**].

CLAIM 3.4. The map $f \to \pi(f)$ defines an isomorphism between algebras $\mathcal{C}(G)$ and $\mathcal{E}_t(G)$.

3.5. The definition of the algebra $\mathcal{J}(G)$. Let P be a parabolic subgroup with Levi group M. We say that an unramified character $\chi : M \to \mathbb{C}^*$ is (non-strictly) positive if for any coroot α of $\mathbf{G}$, such that the corresponding root subgroup lies in the unipotent radical $\mathbf{U_P}$ of $\mathbf{P}$ (which in particular defines a homomorphism $\alpha : F^* \to Z(M)$), we have $|\chi(\alpha(x))| \geq 1$ for $|x| \geq 1$.

Let $\mathcal{E}_{\mathcal{J}}(G)$ be ring of collections $\{\eta_\pi \in \mathrm{End}_{\mathbb{C}}(V)|$ for tempered irreducible $(\pi, V)\}$ which extend to a rational function $E_{i_{GP}(\sigma\otimes\chi)} \in \mathrm{End}_{\mathbb{C}}(\sigma \otimes \chi)$ for every tempered irreducible representation σ of M and which are

a) regular on the set of characters χ such that χ^{-1} is (non-strictly) positive.

b) $K \times K$-invariant for some open compact subgroup K of G.

As follows from the definition, we have an embedding $\mathcal{E}_{\mathcal{J}}(G) \to \mathcal{E}_t(G)$.

DEFINITION 3.6. We define $\mathcal{J}(G)$ to be the preimage of $\mathcal{E}_{\mathcal{J}}(G)$ in $\mathcal{C}(G)$. Note that we have natural embeddings $\mathcal{H}(G) \subset \mathcal{J}(G) \subset \mathcal{C}(G)$.

Next, let us explain certain direct sum decompositions of algebras $\mathcal{C}(G)$ and $\mathcal{J}(G)$.

DEFINITION 3.7. (1) Let $(M, \sigma), (M', \sigma')$ be a pair of square-integrable irreducible representations of Levi subgroups of G. We write $(M, \sigma) \sim (M', \sigma')$ if there exists an element $g \in G$ and a unramified character $\chi \in \Psi(M)$ such that $M' = M^g$ and $(\sigma')^g$ is equivalent to $\sigma \otimes \chi$.

(2) We denote by R the set of equivalence classes of such representations (M, σ).

(3) For any $r = (M, \sigma)$ we denote by $\mathcal{M}_r(G) \subset \mathcal{M}_t(G)$ the subcategory of representations in the closure of $\bigcup_{\chi\in\Psi(M)} i_{GP}(\sigma \otimes \chi)$ where $P = MU_P$ is a parabolic subgroup and by $\mathcal{C}_r(G) \subset \mathcal{C}(G)$ the corresponding subalgebra.

(4) For any $r \in R$ we define $\mathcal{J}_r(G) := \mathcal{C}_r(G) \cap \mathcal{J}(G)$

The following statement is contained in [**10**].

CLAIM 3.8. (1) The subcategory $\mathcal{M}_r(G)$ does depends neither on a representative (M, σ) of r nor on the choice of a parabolic P.

(2) $\mathcal{M}_t(G) = \bigoplus_{r\in R} \mathcal{M}_r(G)$

COROLLARY 3.9. $\mathcal{J}(G) = \bigoplus_{r\in R} \mathcal{J}_r(G)$

4. Intertwining operators

Until the end of this section we fix $r \in R$ and choose a representative (M, σ) of r and a parabolic subgroup $P = MU_P \subset G$. We write $X_P := G/U_P$ and denote by dx a G-invariant measure on X_P.

For any character $\chi \in \Psi(M)$ we define $(\pi_P(\chi), V_{P,\chi}) := ind_P^G \sigma_\chi$. The restriction of the representation $\pi_P(\chi)$ to the subgroup $G(\mathcal{O})$ does not depend on χ (and is equal to $ind_{P\cap G(\mathcal{O})}^{G(\mathcal{O})} \sigma_{M(\mathcal{O})}$. We denote this space by V_P. Of course the image of $V_{P,\chi}$ in the space of M-valued functions on X_P depends on χ. For any $f \in \mathcal{V}_P, \chi \in \Psi(M)$ we denote by f_χ the corresponding function of X_P.

DEFINITION 4.1. (1) We denote $\mathbb{C}(X_P)$ the space of smooth complex-valued functions on X_P and $\mathcal{S}_c(X_P) \subset \mathcal{S}(X_P)$ be the subspace of compactly supported functions.

(2) For any pair $P = MU_P, Q = MU_Q$ of associated parabolic subgroups we denote by $I_{P,Q} : \mathcal{S}_C(X_P) \to \mathbb{C}(X_Q)$ the *geometric (or non-normalized) intertwining operator* given by

$$I_{P,Q}(f)(g) = \int_{U_Q} f(gu)du$$

The following statement is contained in Section 2 of [**1**] (Theorem 2.7); it is a slightly stronger version of Theorem 1.5 .

CLAIM 4.2. (1) There exists a non-empty open subset $\Psi_+(M) \subset \Psi(M)$ such that the integral $I_{P,Q}(f) = \int_{U_Q} f(gu)du$ is absolutely convergent for all $f \in \mathcal{V}_{P,\chi}, \chi \in \Psi_+(M)$.

(2) For any $f \in \mathcal{V}_P$ the function $I_{P,Q}(f_\chi) \in \mathcal{V}_Q$ is a rational function of χ.

(3) The map $I_{P,Q} : \mathcal{V}_{P,\chi} \to \mathcal{V}_{Q,\chi}$ is G- covariant.

(4) There exist rational $\mathbb{C}$-valued functions $r_{P,Q}(\chi)$ such that the operators $\Phi_{P,Q} := r_{P,Q} I_{P,Q}$ satisfy the following:
   (a) $\Phi_{P,Q}$ is regular on the set of non-strongly elements with respect to P.
   (b) $\Phi_{P,Q} \circ \Phi_{Q,P} = Id$
   (c) $\Phi_{P,Q}(\chi)$ is unitary for unitary characters χ.
   (d) For any three associate parabolic subgroups $P_i = MU_i, 1 \leq i \leq 3$ we have $\Phi_{P_1,P_2} \circ \Phi_{P_2,P_3} = \Phi_{P_1,P_3}$.

4.3. Functions on X_P. We define two subspace of $L^2(X_P)$

DEFINITION 4.4. (1) $\mathcal{S}(X_P) := \mathcal{J}(G) \cdot \mathcal{S}_c(X_P)$

(2) $\mathcal{S}'(X_P) := \sum_Q \Phi_{P,Q}(\mathcal{S}_c(X_Q))$
where the sum is over the set of parabolic subgroup $Q = MU_Q$ of G associated with P.

REMARK 4.5. Operators $\Phi_{P,Q}$ are not canonical but it is easy to see that any two choices differ by the multiplication by an regular invertible function on $\Psi(M)$ and therefore the space $\mathcal{S}'(X_P)$ is well defined.

PROPOSITION 4.6. *There exists a surjective (but not necessarily injective) morphism of $G \times M$- modules $\mathcal{J}(G)_{U_P} \to \mathcal{S}(X_P)$.*

PROOF. We define a map $\alpha_P : \mathcal{C}(G) \to L^2(X_P)$ by

$$\alpha(f)(g) := \int_{U_P} f(gu)du$$

where the absolute convergence of the integral follows from Theorem 4.4.3 in [**7**]. It is clear that α_P factorizes through the map $\zeta_P : \mathcal{C}(G)_{U_P} \to L^2(X_P)$. It is clearly sufficient to prove the following statement. □

LEMMA 4.7. (1) $\alpha_P(\mathcal{J}(G)) \subset \mathcal{S}(X_P)$

(2) *α_P defines a surjection from $\mathcal{J}(P)_{U_P}$ onto $\mathcal{S}(X_P)$.*

PROOF. Since the map α commutes with the action of the algebra $\mathcal{J}(G)$ and $\mathcal{J}(G)\mathcal{H}(G) = \mathcal{J}(G)$ it is sufficient to see that $\alpha_P(\mathcal{H}(G)) = \mathcal{S}_c(X_P)$. But the last claim is obviously true. □

Let us explain why ζ is not necessarily injective. Let $G = SL(2, F)$ and P be a Borel subgroup of G; thus X_P can be naturally identified with $F^2 \backslash \{0\}$. Let St denote the Steinberg representation of G. Then $\mathcal{J}(G)$ contains a direct summand isomorphic to $\mathrm{End}_f(\mathrm{St})$ where End_f stands for endomorphisms of finite rank. It is easy to see that any homomorphism of G-modules from St to $L^2(F^2 \backslash \{0\})$ is equal to 0. Hence the above subalgebra must act by 0 on $L^2(F^2 \backslash \{0\})$. On the other hand, $\dim \mathrm{St}_{U_P} = 1$ hence $\mathrm{St} \otimes \mathrm{St}_{U_P} \simeq \mathrm{St}$ is a non-zero subspace of $\mathcal{J}(G)_{U_P}$ which lies in the kernel of ζ.

REMARK 4.8. The [3] contains a wrong statement on the injectivity of ζ.

We now pass to the proof of Theorem 1.9. First we are going to discuss the explicit form of normalized intertwining operators for maximal parabolics.

5. Intertwining operators in the cuspidal corank 1 case

In this section we assume that $M \subset G$ be a Levi subgroup of semi-simple corank 1. Then there exist two parabolic subgroups $P = MU_P$ and $P_- = MU_-$ containing M. We would like to give an explicit description of the normalized intertwining operator Φ_{P,P_-}.

We denote by $M_+ \subset M$ the subset of elements m such that the map $u \to mum^{-1}$ contracts U_P to e. To simplify notations we assume that G is semisimple.

REMARK 5.1. If G is reductive we have to fix a character of the center of G.

We can identify $\mathbb{C}^\star$ with $\Psi(M), z \to \chi_z$ in such a way that $|\chi_z(m)| < 1$ for $|z| \leq 1, m \in M_+$.

Let σ be an irreducible unitary cuspidal representation of M. We write $(\pi_z, \mathcal{V}_z)$ for the representation of G on $i_{GP}(\sigma \otimes \chi_z)$. The following statement is contained in [2].

We denote by $W_\sigma \subset N_G(M)/M$ the subgroup of elements x such that $\sigma^x = \sigma \otimes \chi$ for some $\chi \in \Psi(M)$. Since $|N_G(M)/M| \leq 2$ we see that either $W_\sigma = \{e\}$ or $W_\sigma = \{S_2\}$.

CLAIM 5.2.
(1) If $W_\sigma = \{e\}$ then representations $\mathcal{V}_z$ are irreducible for all $z \in \mathbb{C}^\star$.
(2) If $|z| \neq 1$ and the representation $(\pi_z, \mathcal{V}_z)$ is reducible then $\mathcal{V}_z$ has unique irreducible G-submodule $\overline{\mathcal{W}}_z \subset \mathcal{V}_z$ and the quotient representation $\overline{\mathcal{W}}_z := \mathcal{V}_z/\overline{\mathcal{W}}_z$ is irreducible.
(3) If $|z| = 1$ then either π_z is irreducible or is the direct sum of two tempered irreducible representations $\mathcal{V}_z = \mathcal{V}_z^+ \oplus \mathcal{V}_z^-$.

We consider now the case when $W_\sigma = \{S_2\}$. We shall write $I(z) : \mathcal{V}_{P,\chi} \to \mathcal{V}_{Q,\chi}$ instead of I_{P,χ_z}. The following statement is proven in [8].

CLAIM 5.3. Either the geometric intertwining operator $I(z)$ is regular and invertible for all $z \in \mathbb{C}^\star$ of there exist $z_0 \in \mathbb{C}^\star, |z_0| = 1$ such that I has a first order pole at z_0.

In the first case we have $\Phi = I$ satisfies conditions of Claim 4.2.

We consider now the case when I has a pole. Clearly, we can assume without loss of generality that $z_0 = 1$. The next statement is also contained in [8].

CLAIM 5.4.
(1) $I(z)$ has a first order pole at $z = 1$ and $(z-1)I(z)(1) = a \cdot \mathrm{Id}$, where $a \in \mathbb{C}^\times$.
(2) There exists $c > 1$ such that operators $I^{\pm 1}(z)$ are regular and invertible for $z \notin \{1, c^{\pm 1}\}$.
(3) I is regular at c, $\ker(I(c)) = \mathcal{W}_c$ and $I(c)$ defines an isomorphism $\overline{\mathcal{W}}_c \to \mathcal{W}_{c^{-1}}$.

(4) I^{-1} is regular at $c^{-1}, \ker(I(c^{-1})) = \mathcal{W}_{c^{-1}}$ and $I(c^{-1})$ defines an isomorphism $\overline{\mathcal{W}}_{c^{-1}} \to \mathcal{W}_c$.

Now we define $\Phi(z) = I(z)\frac{(1-z)}{(1-cz^{-1})}$.

It is now clear that Φ can be considered as a normalized intertwining operator Φ_{P,P_-}; if we define a similar operator $\Phi_{P_-,P}$ then these operators satisfy the conditions of Claim 4.2.

To formulate the next statement we need to slightly change the point of view. Namely, we would like to modify Φ so that it becomes an operators from $\mathcal{V}_{P,z}$ to $\mathcal{V}_{P,z^{-1}}$. For this let us choose an element $n \in N_G(M)$ so that n^2 belongs to the center of M and $nPn^{-1} = P_-$. Multiplying σ by some element of $\Psi(M)$ we can assume that $\mathrm{Ad}(n)(\sigma) \simeq \sigma$. Then the right multiplication by n defines an isomorphism between $\mathcal{S}_c(X_P)_\sigma$ and $\mathcal{S}_c(X_{P_-})_\sigma$ which commutes with G and commutes with M up to the action of $\mathrm{Ad}(n)$. Hence, for every $z \in \mathbb{C}^*$ it defines an isomorphism between $\mathcal{V}_{P,z^{-1}}$ and $\mathcal{V}_{P_-,z}$. Composing Φ with the inverse of this isomorphism we get a (rational) isomorphism $\widetilde{\Phi}(z)$ between $\mathcal{V}_{P,z}$ and $\mathcal{V}_{P,z^{-1}}$.

COROLLARY 5.5. (1) *$\mathcal{S}'$ is the space of regular functions $f : \mathbb{C}^\star - \{c\} \to V$ such that the function $(z-c)f(z)$ is regular at $z = c$, $((z-c)f(z))(c) \in \mathcal{W}_c$ and such that $f(z)$ is K-invariant for some open compact subgroup K of G.*

(2) *$\mathcal{J}_\sigma$ is the space of regular functions $h : \mathbb{C}^\star - \{c\} \to \mathrm{End}\,(V)$ such that*
   (a) *the function $(z-c)h(z)$ is regular at $z = c$*
   (b) *$((z-c)h(z))(c) \in \mathrm{Hom}(\overline{\mathcal{W}}_c, \mathcal{W}_c)$*
   (c) *$f(z)$ is two-sided K-invariant for some open compact subgroup K of G*
   (d) *$h(z^{-1}) = \widetilde{\Phi}(z)h(z)\widetilde{\Phi}^{-1}(z))$.*

PROOF. The first assertion follows immediately from the definition of $\mathcal{S}'$ and the construction of $\Phi(z)$. Let us prove the 2nd assertion. First of all, it is clear that any h satisfying (a)-(d) belongs to $\mathcal{J}_\sigma$. On the other hand, let h be any element of $\mathcal{J}_\sigma$. Then by definition it defines a rational function $h : \mathbb{C}^\star - \{c\} \to \mathrm{End}\,(V)$ which satisfies (c) and (d) and which does not have poles when $|z| \leq 1$. Since $\widetilde{\Phi}(z)$ is an isomorphism for $z \neq c^{\pm 1}, 1$ it follows that h can have a pole only at c. Now (d) and Claim 5.4 imply conditions (a) and (b). □

6. Proof of Theorem 1.9

The assertion (1) of Theorem 1.9 is equivalent to the following

LEMMA 6.1. *Let G, P, σ be as in the previous section. Let $\mathcal{S} = \mathcal{S}(X_P)_\sigma, \mathcal{S}' = \mathcal{S}'(X_P)_\sigma$. Then $\mathcal{S}' = \mathcal{S}$.*

PROOF. It is clear from Corollary 5.5 that for any $j \in \mathcal{J}_\sigma$ and $f \in \mathcal{S}_c(X_P)_\sigma$ we have $j(f) \in \mathcal{S}'$. It also clear that the quotient $\mathcal{S}'/\mathcal{S}_c(X_P)_\sigma$ is isomorphic to $\mathcal{W}_c$ as a representation of $\mathcal{H}$. So to prove the equality $\mathcal{S}' = \mathcal{S}$ it is sufficient to find $j \in \mathcal{J}_\sigma$ which is not regular at c. To find such j choose a function r in Corollary 5.5 which satisfies the first two conditions of Corollary, has a pole at c and has a second order zero at c^{-1}. Now take $h(z) = r(z) + \Phi(z)r(z)\Phi^{-1}(z)$.

□

We now pass to the 2nd assertion of Theorem 1.9.

6.2. In this subsection we show that the following statement

CONJECTURE 6.3. $\mathcal{S}'(X_P)_{\text{cusp}} \subset \mathcal{S}(X_P)_{\text{cusp}}$.

implies the validity of the part (2) of Theorem 1.9.

First we show that this Conjecture implies the validity of the $2(b)$. For this it is enough to show that for every $\phi \in \mathcal{S}(X_Q)$ we have $\Phi_{Q,P}(\phi) \in \mathcal{S}(X_P)$. By definition of $\mathcal{S}(X_Q)$ we have

$$\phi = j \cdot f$$

where $f \in \mathcal{S}_c(X_Q)$ and $j \in \mathcal{J}(G)$. Then we have

$$\Phi_{Q,P}(\phi) = j \cdot \Phi_{Q,P}(f) \in \mathcal{J}(G) \cdot \mathcal{S}'(X_P) \subset \mathcal{J}(G) \cdot \mathcal{S}(X_P) = \mathcal{S}(X_P).$$

So, to complete the proof it is enough to show that the inclusion $\mathcal{S}'(X_P)_{\text{cusp}} \subset \mathcal{S}(X_P)_{\text{cusp}}$ implies that $\mathcal{S}'(X_P)_{\text{cusp}} = \mathcal{S}(X_P)_{\text{cusp}}$. Let us as before choose a unitary cuspidal representation σ of M and let $\mathcal{S}(X_P)_\sigma$, $\mathcal{S}'(X_P)_\sigma$ be the corresponding direct summands of $\mathcal{S}(X_P)_{\text{cusp}}$, $\mathcal{S}'(X_P)_{\text{cusp}}$. These are modules over $\mathbb{C}[\Psi(M)]$ and it is clear that if we take K-invariant vectors for some open compact subgroup K of G, these modules become finitely generated. Hence it is enough to show that the embedding $\mathcal{S}'(X_P)_{\text{cusp}} \subset \mathcal{S}(X_P)_{\text{cusp}}$ is surjective in the formal neighbourhood of every $\chi \in \Psi(M)$.

Let us first assume that χ is non-strictly negative with respect to P. Then by definition we have $\mathcal{S}\widehat{(X_P)_{\sigma,\chi}} = \mathcal{S}_c\widehat{(X_P)_{\sigma,\chi}}$ where $\mathcal{S}\widehat{(X_P)_{\sigma,\chi}}$ and $\mathcal{S}_c\widehat{(X_P)_{\sigma,\chi}}$ denote the formal completions of the corresponding spaces at χ. Since $\mathcal{S}_c(X_P) \subset \mathcal{S}'(X_P)$ the desired surjectivity follows. Let now χ be arbitrary. Then there exists an associate parabolic Q such that χ is non-strictly negative with respect to Q. Since we have $\mathcal{S}'(X_P) = \mathcal{S}'(X_Q)$ and $\mathcal{S}(X_P) = \mathcal{S}(X_Q)$, it follows from the above argument that the map $\mathcal{S}'\widehat{(X_P)_{\sigma,\chi}} \to \mathcal{S}\widehat{(X_P)_{\sigma,\chi}}$ is surjective for every χ.

7. Some further questions

7.1. The $\mathcal{J}$-version of the Jacquet functor. Let us assume the 4th assertion of Conjecture 1.8. Then we can define a functor

$$r^{\mathcal{J}}_{GP} : \text{Right } \mathcal{J}(G)\text{-modules} \to \text{Right } \mathcal{J}(M)\text{-modules}$$

by setting

$$r^{\mathcal{J}}_{GP}(\pi) = \pi \underset{\mathcal{J}(G)}{\otimes} \mathcal{S}(X_P). \tag{7.1}$$

It would be interesting to investigate exactness properties of this functor and compute it in some examples. Note that manifestly we have $r^{\mathcal{J}}_{GP}(\mathcal{J}(G)) = \mathcal{S}(X_P)$. Also note that Conjecture 1.8 implies that at least non canonically the functor $r^{\mathcal{J}}_{GP}$ depends only on M and not on P (the non-canonicity comes from the fact that the operators $\Phi_{P,Q}$ are not canonically defined).

7.2. The spherical part. Let us define the spherical part of $\mathcal{S}(X_P)$ by setting

$$\mathcal{S}_{\text{sph}}(X_P) = \mathcal{S}(X_P)^{\mathbf{G}(\mathcal{O}) \times \mathbf{M}(\mathcal{O})}. \tag{7.2}$$

We would like to describe this space explicitly. For this note that set-theoretically we have

$$\mathbf{G}(\mathcal{O})\backslash X_P/\mathbf{M}(\mathcal{O}) = \mathbf{M}(\mathcal{O})\backslash M/\mathbf{M}(\mathcal{O}).$$

Hence elements of $\mathcal{S}_{\mathrm{sph}}(X_P)$ can be thought of as $\mathbf{M}(\mathcal{O})\times\mathbf{M}(\mathcal{O})$-invariant functions on M. Recall that the Satake isomorphism (for M) says that the spherical Hecke algebra $\mathcal{H}_{\mathrm{sph}}(M)$ consisting of compactly supported $\mathbf{M}(\mathcal{O})\times\mathbf{M}(\mathcal{O})$-invariant functions on M is isomorphic to the complexified Grothendieck ring of the category of finite-dimensional representations of the Langlands dual group $M^\vee$ (considered as a group over $\mathbb{C}$). We shall denote the corresponding map $K_0(\mathrm{Rep}(M^\vee)\to\mathcal{H}_{\mathrm{sph}}(M)$ by Sat_M.

Let $G^\vee$ denote the Langlands dual group of G and let $P^\vee$ be the corresponding parabolic subgroup of $G^\vee$ with unipotent radical $U_{P^\vee}$. Let $\mathfrak{u}_{\mathfrak{p}^\vee}$ denote the Lie algebra of $U_{P^\vee}$; it has a natural action of $M^\vee$.

Now let

$$f_P=\sum_{i=0}^{\infty}\mathrm{Sat}_M([\mathrm{Sym}^i(\mathfrak{u}_{\mathfrak{p}^\vee})]). \tag{7.3}$$

Using the Satake correspondence we can regard f_P as a $\mathbf{G}(\mathcal{O})\times\mathbf{M}(\mathcal{O})$-invariant function on X_P.

CONJECTURE 7.3. *$\mathcal{S}_{\mathrm{sph}}(X_P)$ is a free right $\mathcal{H}_{\mathrm{sph}}(M)$-module generated by f_P.*

In the case when P is a Borel subgroup this conjecture is proved in [**3**].

7.4. Iwahori part and K-theory. Let $\mathcal{H}_{\mathrm{aff}}(\mathbf{G})$ denote the affine Hecke algebra of G. This is an algebra over $\mathbb{C}[v,v^{-1}]$; its specialization at $v=q^{1/2}$ is isomorphic to the Iwahori-Hecke algebra $\mathcal{H}(G,I)$ of G.

Let $\mathcal{N}_{G^\vee}$ (resp. $\mathcal{N}_{M^\vee}$) denote the nilpotent cone in the Lie algebra of $G^\vee$ (resp. in the Lie algebra of $M^\vee$). Let also $\mathcal{B}_{G^\vee},\mathcal{B}_{M^\vee}$ denote the corresponding flag varieties. The cotangent bundle $T^*\mathcal{B}_{G^\vee}$ maps naturally to $\mathcal{N}_{G^\vee}$. Thus we can define

$$\mathbf{St}_{G^\vee,M^\vee}=T^*\mathcal{B}_{G^\vee}\underset{\mathcal{N}_{G^\vee}}{\times}T^*\mathcal{B}_{M^\vee}.$$

This variety is acted on by the group $G^\vee\times\mathbb{C}^\times$ (where the second factor acts on $\mathcal{N}_{G^\vee}$ by the formula $t(x)=t^2x$ where $t\in\mathbb{C}^\times$ and $x\in\mathcal{N}_{G^\vee}$. Thus we can consider the complexified equivariant K-theory $K_{M^\vee\times\mathbb{C}^\times}(\mathbf{St}_{G^\vee,M^\vee})$. This is a vector space over $\mathbb{C}[v,v^{-1}]=K_{\mathbb{C}^\times}(pt)$. It is easy to see (generalizing the standard construction of [**6**]) that it has a structure of $\mathcal{H}_{\mathrm{aff}}(\mathbf{G})\otimes\mathcal{H}_{\mathrm{aff}}(\mathbf{M})$.

CONJECTURE 7.5. *The specialization of $K_{M^\vee\times\mathbb{C}^\times}(\mathbf{St}_{G^\vee,M^\vee})$ at $v=q^{1/2}$ is isomorphic to $\mathcal{S}'(X_P)^I$.*

It would be interesting to extend this Conjecture to $\mathcal{S}(X_P)$ instead of $\mathcal{S}'(X_P)$.

References

[1] J. Arthur, *Intertwining operators and residues. I. Weighted characters*, J. Funct. Anal. **84** (1989), no. 1, 19–84, DOI 10.1016/0022-1236(89)90110-9. MR999488

[2] J. Bernstein, *Draft of : Representations of p-adic groups*, Fall 1992. Lectures by Joseph Bernstein, Written by Karl E. Rumelhart.

[3] A. Braverman and D. Kazhdan, *On the Schwartz space of the basic affine space*, Selecta Math. (N.S.) **5** (1999), no. 1, 1–28, DOI 10.1007/s000290050041. MR1694894

[4] A. Braverman and D. Kazhdan, *Remarks on the asymptotic Hecke algebra*, arXiv:1704.03019.

[5] M. Cowling, U. Haagerup, and R. Howe, *Almost L^2 matrix coefficients*, J. Reine Angew. Math. **387** (1988), 97–110. MR946351

[6] D. Kazhdan and G. Lusztig, *Proof of the Deligne-Langlands conjecture for Hecke algebras*, Invent. Math. **87** (1987), no. 1, 153–215, DOI 10.1007/BF01389157. MR862716

[7] A. J. Silberger, *Introduction to harmonic analysis on reductive p-adic groups*, Mathematical Notes, vol. 23, Princeton University Press, Princeton, N.J.; University of Tokyo Press, Tokyo, 1979. Based on lectures by Harish-Chandra at the Institute for Advanced Study, 1971–1973. MR544991

[8] A. J. Silberger, *Special representations of reductive p-adic groups are not integrable*, Ann. of Math. (2) **111** (1980), no. 3, 571–587, DOI 10.2307/1971110. MR577138

[9] P. Schneider and E.-W. Zink, *The algebraic theory of tempered representations of p-adic groups. I. Parabolic induction and restriction*, J. Inst. Math. Jussieu **6** (2007), no. 4, 639–688, DOI 10.1017/S1474748007000047. MR2337311

[10] P. Schneider and E.-W. Zink, *The algebraic theory of tempered representations of p-adic groups. II. Projective generators*, Geom. Funct. Anal. **17** (2008), no. 6, 2018–2065, DOI 10.1007/s00039-007-0647-2. MR2399091

Department of Mathematics, University of Toronto, Toronto, Ontario, Canada M5S 2E4 – and – Perimeter Institute for Theoretical Physics, Waterloo, Ontario, Canada N2L 2Y5 – and – Skolkovo Institute of Science and Technology, Moscow, Russia 121205

Department of Mathematics, Hebrew University of Jerusalem, 9190401 Jerusalem, Israel

Proceedings of Symposia in Pure Mathematics
Volume **101**, 2019
https://doi.org/10.1090/pspum/101/01790

Explicit local Jacquet-Langlands correspondence: The non-dyadic wild case

Colin J. Bushnell and Guy Henniart

Abstract. Let F be a non-Archimedean locally compact field of residual characteristic p with $p \neq 2$. Let n be a power of p and let G be an inner form of the general linear group $\mathrm{GL}_n(F)$. We give a transparent parametrization of the irreducible, totally ramified, cuspidal representations of G of parametric degree n. We show that the parametrization is respected by the Jacquet-Langlands correspondence, relative to any other inner form. This expresses the Jacquet-Langlands correspondence for such representations within a single, compact formula.

Introduction

1. Let F be a non-Archimedean local field of residual characteristic p. Let $n \geqslant 1$ and let G be an inner form of the general linear group $\mathrm{GL}_n(F)$. In other words, there is a central simple F-algebra A, of dimension n^2, such that $G = A^\times$. Let $\mathcal{A}^\square(G)$ be the set of equivalence classes of *essentially square-integrable,* smooth, complex representations of G. Let G' be another inner form of $\mathrm{GL}_n(F)$. We study the canonical bijection

$$T_G^{G'} : \mathcal{A}^\square(G) \xrightarrow{\approx} \mathcal{A}^\square(G')$$

provided by the Jacquet-Langlands correspondence [**10**], [**1**]. We make a narrow, but significant, contribution to the analysis of the correspondence in explicit terms.

Let $\pi \in \mathcal{A}^\square(G)$ and let $d(\pi)$ be the *parametric degree* of π, in the sense of [**6**]. Thus $d(\pi)$ is a positive integer dividing n. If $d(\pi) = n$, then π is cuspidal. The converse holds if G is the split group $\mathrm{GL}_n(F)$ but not in general. The parametric degree is preserved by the Jacquet-Langlands correspondence.

Here we concentrate on the case where π is of parametric degree n and *totally wildly ramified.* This means that n is a power of p and, if $\chi \neq 1$ is an unramified character of $F^\times$, the twist $\chi\pi$ of π is not equivalent to π. When $p \neq 2$, such representations admit a transparent description that we use to describe the correspondence via a compact explicit formula. The case $p = 2$ has sufficiently many distinctive features to merit a separate treatment that we defer for the time being.

With this result to hand, the way is open to follow the framework of [**6**] and [**8**] (but without complications arising from the transfer factors of automorphic

---

2010 *Mathematics Subject Classification.* Primary 22E50; Secondary 11S37.
Key words and phrases. Local Jacquet-Langlands correspondence, cuspidal representation, simple character, endo-class.

induction [**13**]) to an explicit description of the Jacquet-Langlands correspondence for representations $\pi \in \mathcal{A}^\square(G)$ with $d(\pi) = n$. The more difficult case is that where $d(\pi) < n$ while π is cuspidal. With the newly available Endo-class Transfer Theorem of [**22**] and [**11**] recalled below, that general case is substantially less mysterious than hitherto. However, it seems unlikely that one will resolve the question finally without the detail of the complementary special case treated here. The one fully known case of [**24**], [**7**] indicates the prospect of further subtlety.

2. For the convenience of the reader, we interpolate an outline summary of recent developments in the broader context. Again, G is an inner form of $\mathrm{GL}_n(F)$, but we impose no restriction on n or p for the time being.

The papers [**3, 17, 18, 19, 20, 21**] of Sécherre and collaborators contain a complete description of the representations $\pi \in \mathcal{A}^\square(G)$ in terms of *simple characters* and *simple types*. Allowing for a few novel features and a higher level of technical intricacy, it is parallel to the split case $G = \mathrm{GL}_n(F)$ of [**9**]. In particular, any $\pi \in \mathcal{A}^\square(G)$ contains a simple type and hence a simple character θ_π. The representations π that contain a given simple type are classified via a scheme following that of the split case [**19**].

Simple characters, as a class, have a fundamental naturality property. Working at first in the split case of [**9**], let θ be a simple character in $G = \mathrm{GL}_n(F)$. Thus θ is attached to a hereditary $\mathfrak{o}_F$-order in the matrix algebra $\mathrm{M}_n(F)$. If $\mathfrak{a}'$ is a hereditary order in $\mathrm{M}_{n'}(F)$ then, subject to minor combinatorial constraints, one can construct from θ a simple character θ' in $\mathrm{GL}_{n'}(F)$, attached to $\mathfrak{a}'$. We refer to θ' as a "transfer" of θ. If, for $i = 1, 2$, we are given a simple character θ_i in $G_i = \mathrm{GL}_{n_i}(F)$, one can always find an integer n_3 and a hereditary order $\mathfrak{a}_3$ in $A_3 = \mathrm{M}_{n_3}(F)$ that admits a transfer θ_i' of θ_i, $i = 1, 2$. One knows from [**4**] that if, for some choice of datum $(A_3, \mathfrak{a}_3)$, the transfers θ_i' intertwine (and so are conjugate) in $A_3^\times$, then the same is true for all such choices. When this holds, one says that θ_1 is *endo-equivalent to* θ_2. Endo-equivalence is an equivalence relation on the class of all simple characters in all groups $\mathrm{GL}_n(F)$, $n \geqslant 1$. The set of endo-equivalence classes (endo-classes for short) is an arithmetic object of considerable interest: see section 6 of [**8**] for an overview.

The achievement of [**3**] is an extension of this relation to the class of all simple characters in all inner forms of all $\mathrm{GL}_n(F)$. Every endo-equivalence class, in this extended context, contains a simple character in some split group $\mathrm{GL}_n(F)$, $n \geqslant 1$. On such characters, the two notions of endo-equivalence are the same. For general G and $\pi \in \mathcal{A}^\square(G)$, the endo-class of a simple character θ_π contained in π is uniquely determined by π. One cannot avoid asking how this fundamental invariant behaves with respect to the Jacquet-Langlands correspondence.

THEOREM (Endo-class Transfer Theorem [**22**], [**11**]). *For $i = 1, 2$, let G_i be an inner form of* $\mathrm{GL}_n(F)$. *Let $\pi_i \in \mathcal{A}^\square(G_i)$ and let θ_i be a simple character contained in π_i. If $\pi_2 = T_{G_1}^{G_2}(\pi_1)$, then θ_1 is endo-equivalent to θ_2.*

The proof of this result takes an unexpected form. Let ℓ be a prime number different from p. In a series of papers including [**14**] and [**15**], Mínguez and Sécherre develop a theory of ℓ-modular representations of the inner forms G of $\mathrm{GL}_n(F)$ and of reduction, modulo ℓ, of representations in characteristic zero. In [**16**], they show that reduction modulo ℓ is compatible with the Jacquet-Langlands correspondence. Using these results, for varying ℓ, Sécherre and Stevens show that the Endo-class

Transfer Theorem holds in general *provided it holds when* $d(\pi_i) = n$ *and* π_1 *is totally ramified* [**22**], that is, $\chi\pi_1 \not\cong \pi_1$ for any non-trivial unramified character χ of $F^\times$. By combining properties of certain simple characters, relative to unramified base field extension, and trace comparisons of a sort familiar from [**6**] or [**8**], Dotto [**11**] reduces the problem to the split groups and so despatches the outstanding special case. That method also yields the relation between the simple types contained in corresponding representations π_i of parametric degree n.

We do not rely here on the Endo-class Transfer Theorem. In the case to hand, it emerges as a corollary of the main result.

3. We return to the main theme. From now on, G is an inner form of $\mathrm{GL}_n(F)$, where $n = p^r$, $r \geqslant 1$. Beyond the very first stages, we assume $p \neq 2$. A representation $\pi \in \mathcal{A}^\square(G)$ is *totally wildly ramified* if $\pi \not\cong \chi\pi$ for any unramified character $\chi \neq 1$ of $F^\times$. Let $\mathcal{A}^{\text{m-wr}}(G)$ be the set of totally wildly ramified representations $\pi \in \mathcal{A}^\square(G)$ such that $d(\pi) = n$. Let $G = A^\times$, where A is a central simple F-algebra of dimension n^2.

We start with a simple stratum $[\mathfrak{a}, l, 0, \beta]$ in A, in which the field extension $E = F[\beta]/F$ is totally ramified of degree n. (This determines $\mathfrak{a}$ up to conjugation, so it may be treated as fixed.) The stratum gives rise to a pair $H^1(\beta, \mathfrak{a}) \subset J^1(\beta, \mathfrak{a})$ of open subgroups of $U^1_\mathfrak{a}$ and, modulo the choice of a suitable character of F, a finite set $\mathcal{C}(\mathfrak{a}, \beta)$ of *simple characters* of the group $H^1(\beta, \mathfrak{a})$. The G-normalizer of any $\theta \in \mathcal{C}(\mathfrak{a}, \beta)$ is the group $\mathrm{J} = \mathrm{J}(\beta, \mathfrak{a}) = E^\times J^1(\beta, \mathfrak{a})$. If $\pi \in \mathcal{A}^{\text{m-wr}}(G)$, one may choose β so that π contains some $\theta \in \mathcal{C}(\mathfrak{a}, \beta)$, and π determines θ up to G-conjugation. The natural representation Λ of J on the θ-isotypic subspace of π is irreducible and $\pi \cong c\text{-}\mathrm{Ind}_\mathrm{J}^G\, \Lambda$.

The representation Λ is not well adapted to our purposes, so we follow [**5**] and adopt a variant strategy. First, there is a canonical "Lagrangian" subgroup $I^1(\beta, \mathfrak{a})$ lying between $H^1(\beta, \mathfrak{a})$ and $J^1(\beta, \mathfrak{a})$ (section 1). It is normalized by $E^\times$, and we set $\mathrm{I} = \mathrm{I}(\beta, \mathfrak{a}) = E^\times I^1(\beta, \mathfrak{a})$. For the given θ, one may choose the element β so that θ admits extension to a character of I (2.1, 2.2): it is at this point we require the restriction $p \neq 2$. Any $\pi \in \mathcal{A}^{\text{m-wr}}(G)$ contains some such character λ, unique up to conjugation. Moreover, $\pi = c\text{-}\mathrm{Ind}_\mathrm{I}^G\, \lambda$ (2.3).

The one-dimensional parameters (I, λ) behave transparently with respect to finite, unramified base field extension. They can be transferred, via such an extension, to any inner form G' of G. This transfer process, set out in section 3, specializes to the standard transfer of simple characters as in [**3, 17**], but it also suggests a *parametric transfer* of representations between $\mathcal{A}^{\text{m-wr}}(G)$ and $\mathcal{A}^{\text{m-wr}}(G')$, say $\pi \mapsto \pi'$. At first, the parametric transfer is not well-defined, as it apparently depends on a fairly random choice.

There is a second approach (in section 4). There, one extends the base field to the completion $\widetilde{F}$ of the maximal unramified extension of F. The parametric transfer process applies equally over $\widetilde{F}$ but, because of completeness, it is achieved via a conjugation in the group of $\widetilde{F}$-points of G. This version of the parametric transfer is equally ill-defined but is equivalent to the first one (4.5 Proposition). However, the fact that it is given by a conjugation over the complete field $\widetilde{F}$ enables comparison of the characters of π and π'. In section 5, we show that these characters agree at enough points to ensure that π, π' are related by the Jacquet-Langlands correspondence. The conclusion in section 6 is that *the parametric transfer, in either version, is the Jacquet-Langlands correspondence.* In particular, the parametric transfer process is well-defined and implies the endo-class transfer theorem in this case.

4. This parametric transfer is stronger than endo-class transfer but is not sufficiently explicit for our purposes. Take $G = A^\times$ and θ as before. Generalizing an idea of [**5**], we write down an explicit character ${}_I\theta$ of $I^1(\beta, \mathfrak{a})$ extending θ (2.4). This character depends on the choice of β, but is otherwise canonical. If ξ is a character of $E^\times$, agreeing with θ on $U^1_E = E^\times \cap H^1(\beta, \mathfrak{a})$, we may use it to extend ${}_I\theta$ to a character $\xi \odot_\beta \theta$ of $\mathrm{I}(\beta, \mathfrak{a})$. The first point is that, if $\pi \in \mathcal{A}^{\text{m-wr}}(G)$ contains θ, there is a *unique* choice of ξ such that π contains $\xi \odot_\beta \theta$ (2.4). The representation π is then induced by $\xi \odot_\beta \theta$, so we write $\pi = \pi_G(\xi \odot_\beta \theta)$.

The second point is that, if $G' = A'^\times$ is an inner form of G, the parametric transfer process starts from an F-embedding $\iota : E \to A'$. The embedding ι is used, as in section 3, to transfer θ to a simple character $\iota\theta$ in G', attached to a simple stratum $[\mathfrak{a}', l, 0, \iota(\beta)]$ in A'. It obviously transfers ξ to a character $\iota\xi$ of $\iota(E)^\times$ allowing us to form the extended simple character $\iota\xi \odot_{\iota(\beta)} \iota\theta$ in G'. We so obtain the formula

$$\boxed{T^{G'}_G\ \pi_G(\xi \odot_\beta \theta) = \pi_{G'}(\iota\xi \odot_{\iota(\beta)} \iota\theta)}$$

of 6.2 Theorem.

5. This paper up-dates and supersedes the relevant parts of our earlier work [**5**]. That concerned only the relation between $\mathrm{GL}_n(F)$ and $\mathrm{GL}_1(D)$, where D is a central F-division algebra of dimension n^2: not enough of the general machinery of [**3, 17, 18, 19, 20, 21**] was available at that time, so [**5**] could only rely on [**2**]. However, a lot of the effort in [**5**] is centred on $\mathrm{GL}_n(F)$, and is available to ease our task here. On the other hand, the step from $\mathrm{GL}_1(D)$ to a general inner form requires some effort and re-organization of the detail into a more efficient and flexible form.

1. The Lagrangian subgroup

Let A be a central simple F-algebra of dimension n^2, $n = p^r$, and set $G = A^\times$. Let $\mathcal{A}^{\text{m-wr}}(G)$ denote the set of equivalence classes of irreducible, smooth, complex representations of G that are cuspidal, totally ramified and of parametric degree n. In this and the following section, we describe the elements π of $\mathcal{A}^{\text{m-wr}}(G)$ as representations induced from a canonical family of *characters* of open, compact modulo centre, subgroups of G.

We recall something of the *simple characters in* A, following a simplified version of the foundational account of [**17**]: since we deal only with a very special case, the more elaborate technical structures of [**17**] are not needed here. From there, we develop a modified version of the method of [**5**]. In this section, we allow the possibility $p = 2$.

1.1. Let $\mathfrak{a}$ be a *minimal hereditary* $\mathfrak{o}_F$*-order* in A. Thus $\mathfrak{a}$ is a principal order: if $\mathfrak{p}$ is the Jacobson radical of $\mathfrak{a}$, there exists $\varPi \in G$ such that $\mathfrak{p} = \varPi\mathfrak{a} = \mathfrak{a}\varPi$. Any two minimal hereditary orders in A are G-conjugate.

We use the concept of *simple stratum in* A, following [**17**].

NOTATION. Let $\mathcal{S}^{\mathrm{wr}}(\mathfrak{a})$ be the set of elements β of G satisfying the following conditions.

(1) There is an integer $l>0$ such that the quadruple $[\mathfrak{a},l,0,\beta]$ is a simple stratum in A.
(2) The field extension $F[\beta]/F$ is of degree n.

Since $\mathfrak{a}$ is minimal and n is a power of p, these conditions imply that $F[\beta]/F$ is *totally wildly ramified.* The order $\mathfrak{a}$ is stable under conjugation by $F[\beta]^{\times}$ — one says that $\mathfrak{a}$ is $F[\beta]$-*pure* — and $\mathfrak{a}$ is the unique hereditary order in A with this property. The integer l is given by $\beta^{-1}\mathfrak{a}=\mathfrak{p}^{l}$.

PROPOSITION. *Let $\beta\in\mathcal{S}^{\mathrm{wr}}(\mathfrak{a})$ and write $E=F[\beta]$. Let B be a central simple F-algebra of dimension n^2, let $\iota:E\to B$ be an F-embedding, and let $\mathfrak{b}$ be an ιE-pure hereditary $\mathfrak{o}_F$-order in B. The stratum $[\mathfrak{b},l,0,\iota\beta]$ is then simple, the order $\mathfrak{b}$ is minimal and $\iota\beta\in\mathcal{S}^{\mathrm{wr}}(\mathfrak{b})$.*

PROOF. See Proposition 2.25 of [**17**]. □

1.2. Let $\beta\in\mathcal{S}^{\mathrm{wr}}(\mathfrak{a})$. Following [**17**], the simple stratum defined by $\mathfrak{a}$ and β gives rise to a pair of $\mathfrak{o}_F$-orders in A,

$$\mathfrak{H}(\beta,\mathfrak{a})\subset\mathfrak{J}(\beta,\mathfrak{a})\subset\mathfrak{a},$$

and families of open subgroups

$$\begin{aligned}H^k(\beta,\mathfrak{a})&=1+\mathfrak{H}(\beta,\mathfrak{a})\cap\mathfrak{p}^k,\\ J^k(\beta,\mathfrak{a})&=1+\mathfrak{J}(\beta,\mathfrak{a})\cap\mathfrak{p}^k,\end{aligned}\qquad k\geqslant 1,$$

of the principal unit group $U^1_{\mathfrak{a}}=1+\mathfrak{p}$.

Fix a character ψ^F of F that is trivial on $\mathfrak{p}_F$ but not trivial on $\mathfrak{o}_F$: one says that ψ^F *is of level one.* As in [**17**], use ψ^F to define the set $\mathcal{C}(\mathfrak{a},\beta,\psi^F)$ of *simple characters* of $H^1(\beta,\mathfrak{a})$.

Write $\psi^A=\psi^F\circ\mathrm{tr}_A$, where $\mathrm{tr}_A:A\to F$ is the reduced trace map. For $\alpha\in A$, define a function ψ^A_α by

$$\psi^A_\alpha(x)=\psi^A(\alpha(x-1)),\quad x\in A. \tag{1.2.1}$$

1.3. We recall from [**17**] *passim* the behaviour of these structures relative to unramified base field extension.

Let K/F be a finite, *unramified* field extension. The K-algebra $A_K=A\otimes_F K$ is central simple of dimension n^2. Set $G_K=A_K^{\times}$. The ring $\mathfrak{a}_K=\mathfrak{a}\otimes_{\mathfrak{o}_F}\mathfrak{o}_K$ is a minimal hereditary $\mathfrak{o}_K$-order in A_K, with Jacobson radical $\mathfrak{p}_K=\mathfrak{p}\otimes_{\mathfrak{o}_F}\mathfrak{o}_K$. We habitually identify A with the subring $A\otimes 1$ of A_K.

PROPOSITION. *Let K/F be a finite unramified field extension and let $\beta\in\mathcal{S}^{\mathrm{wr}}(\mathfrak{a})$.*

(1) *The element $\beta\otimes 1$ of G_K lies in $\mathcal{S}^{\mathrm{wr}}(\mathfrak{a}_K)$ and*

$$\begin{aligned}\mathfrak{H}(\beta\otimes 1,\mathfrak{a}_K)&=\mathfrak{H}(\beta,\mathfrak{a})\otimes_{\mathfrak{o}_F}\mathfrak{o}_K,\\ H^k(\beta,\mathfrak{a})&=H^k(\beta\otimes 1,\mathfrak{a}_K)\cap G,\quad k\geqslant 1.\end{aligned}$$

Similarly for the J-groups.

(2) *Let* ψ^K *be a character of* K*, of level one, such that* $\psi^K \mid F = \psi^F$*. If* $\theta \in \mathcal{C}(\mathfrak{a}_K, \beta \otimes 1, \psi^K)$*, the character* $\theta^F = \theta \mid H^1(\beta, \mathfrak{a})$ *lies in* $\mathcal{C}(\mathfrak{a}, \beta, \psi^F)$*. The restriction map*

$$\begin{aligned} \mathcal{C}(\mathfrak{a}_K, \beta \otimes 1, \psi^K) &\longrightarrow \mathcal{C}(\mathfrak{a}, \beta, \psi^F), \\ \theta &\longmapsto \theta^F, \end{aligned}$$

is surjective.

Proof. If the degree $[K{:}F]$ is divisible by n, then $A_K \cong \mathrm{M}_n(F)$ and all assertions follow directly from the definitions in [**17**], particularly 3.3. The general case then follows by transitivity. □

From now on, we follow convention and write $\beta = \beta \otimes 1 \in A_K$.

1.4. Denote by $\mathrm{K}_\mathfrak{a}$ the group of $g \in G$ for which $g\mathfrak{a}g^{-1} = \mathfrak{a}$. Equivalently, $\mathrm{K}_\mathfrak{a}$ is the G-normalizer of $U_\mathfrak{a} = \mathfrak{a}^\times$. It is generated by $U_\mathfrak{a}$ and any element $\varPi$ such that $\varPi\mathfrak{a} = \mathfrak{p}$.

Let $\beta \in \mathcal{S}^{\mathrm{wr}}(\mathfrak{a})$ and let $\theta \in \mathcal{C}(\mathfrak{a}, \beta, \psi^F)$. The G-normalizer of θ is the group

$$\mathrm{J}(\beta, \mathfrak{a}) = F[\beta]^\times J^1(\beta, \mathfrak{a}).$$

In particular, $\mathrm{J}(\beta, \mathfrak{a})$ is an open subgroup of $\mathrm{K}_\mathfrak{a}$ that does not depend on the choice of $\theta \in \mathcal{C}(\mathfrak{a}, \beta, \psi^F)$. An element g of G intertwines the character θ if and only if $g \in \mathrm{J}(\beta, \mathfrak{a})$. (For these facts, see [**17**] Théorème 3.50.)

The point of the section is to construct a canonical subgroup $I^1(\beta, \mathfrak{a})$ of G, lying between $J^1(\beta, \mathfrak{a})$ and $H^1(\beta, \mathfrak{a})$. The group

$$\mathcal{J}^1(\beta, \mathfrak{a}) = J^1(\beta, \mathfrak{a})/H^1(\beta, \mathfrak{a}) \cong \mathfrak{J}^1(\beta, \mathfrak{a})/\mathfrak{H}^1(\beta, \mathfrak{a})$$

is a vector space over the finite residue field $\mathrm{k}_F = \mathrm{k}_{F[\beta]}$. In particular, it is a vector space over the field $\mathbb{F}_p$ of p elements. Let $\theta \in \mathcal{C}(\mathfrak{a}, \beta, \psi^F)$. Using the commutator convention $[x, y] = x^{-1}y^{-1}xy$, the pairing

$$(x, y) \longmapsto \theta([x, y]), \quad x, y \in J^1(\beta, \mathfrak{a}), \tag{1.4.1}$$

induces an $\mathbb{F}_p$-bilinear form on $\mathcal{J}^1(\beta, \mathfrak{a})$. This form is nondegenerate and alternating [**17**] Théorème 3.52.

Lemma. *The pairing* (1.4.1) *satisfies*

$$\theta([1{+}x, 1{+}y]) = \psi^A_\beta(1{+}xy{-}yx),$$

for $x, y \in \mathfrak{J}^1(\beta, \mathfrak{a})$ *and* $\theta \in \mathcal{C}(\mathfrak{a}, \beta, \psi^F)$*.*

Proof. When $A \cong \mathrm{M}_n(F)$, the result is 6.1 Proposition of [**5**]. We reduce the general case to that one. Let K/F be a finite unramified extension such that $A_K \cong \mathrm{M}_n(K)$. By 1.3 Proposition, there exists $\theta_K \in \mathcal{C}(\mathfrak{a}_K, \beta, \psi^K)$ such that $\theta = \theta_K \mid H^1(\beta, \mathfrak{a})$. For $x, y \in \mathfrak{J}^1(\beta, \mathfrak{a}) \subset \mathfrak{J}^1(\beta, \mathfrak{a}_K)$, we have

$$\theta([1{+}x, 1{+}y]) = \theta_K([1{+}x, 1{+}y]) = \psi^{A_K}_\beta(1{+}xy{-}yx)$$

loc. cit. On the other hand, $\psi^A_\beta = \psi^{A_K}_\beta \mid A$ so

$$\theta([1{+}x, 1{+}y]) = \psi^A_\beta(1{+}xy{-}yx),$$

as required. □

The pairing (1.4.1) is thus independent of the choice of $\theta \in \mathcal{C}(\mathfrak{a}, \beta, \psi^F)$: we name it h_β, or h_β^F when we need to specify the base field.

Let $k \geqslant 1$ be an integer. Define $\mathcal{J}^k = \mathcal{J}^k(\beta, \mathfrak{a})$ as the image of $J^k(\beta, \mathfrak{a})$ in $\mathcal{J}^1(\beta, \mathfrak{a})$. By Théorème 3.52 of [**17**], the pairing h_β is nondegenerate on $\mathcal{J}^k(\beta, \mathfrak{a})$, $k \geqslant 1$. So, for each $k \geqslant 1$, there is a unique subspace $\mathcal{U}^k(\beta, \mathfrak{a})$ of $\mathcal{J}^1(\beta, \mathfrak{a})$ such that

$$\mathcal{J}^k(\beta, \mathfrak{a}) = \mathcal{U}^k(\beta, \mathfrak{a}) \perp \mathcal{J}^{k+1}(\beta, \mathfrak{a}),$$

the sum being orthogonal with respect to the alternating form h_β. One has $\mathcal{J}^k(\beta, \mathfrak{a}) \neq \mathcal{J}^{k+1}(\beta, \mathfrak{a})$ if and only if $2k$ is a jump of the stratum $[\mathfrak{a}, l, 0, \beta]$, so $\mathcal{U}^k(\beta, \mathfrak{a}) = 0$ for all but finitely many k. The definition ensures that the form h_β is nondegenerate on $\mathcal{U}^k(\beta, \mathfrak{a})$. It follows that $\mathcal{J}^k(\beta, \mathfrak{a})$ can be expressed as an *orthogonal* sum

$$\mathcal{J}^k(\beta, \mathfrak{a}) = \sum_{i \geqslant k} \mathcal{U}^i(\beta, \mathfrak{a}),$$

in which only finitely many terms are nonzero.

PROPOSITION. *Let $\beta \in \mathcal{S}^{\mathrm{wr}}(\mathfrak{a})$ and set $E = F[\beta]$. There exists a unique $\mathfrak{o}_F$-lattice $\mathfrak{I}^1(\beta, \mathfrak{a})$ with the following properties:*

(1) $\mathfrak{H}^1(\beta, \mathfrak{a}) \subset \mathfrak{I}^1(\beta, \mathfrak{a}) \subset \mathfrak{J}^1(\beta, \mathfrak{a})$;
(2) *$\mathfrak{I}^1(\beta, \mathfrak{a})$ is stable under conjugation by $\mathrm{J}(\beta, \mathfrak{a})$;*
(3) *the image $\mathcal{I}^1(\beta, \mathfrak{a})$ of $\mathfrak{I}^1(\beta, \mathfrak{a})$ in the alternating space $\mathcal{J}^1(\beta, \mathfrak{a})$ is a maximal totally isotropic subspace that is the sum of its intersections with the subspaces $\mathcal{U}^k(\beta, \mathfrak{a})$, $k \geqslant 1$.*

The lattice $\mathfrak{I}^1(\beta, \mathfrak{a})$ has the following additional properties.

(4) *If $\beta' \in \mathcal{S}^{\mathrm{wr}}(\mathfrak{a})$ and $\mathcal{C}(\mathfrak{a}, \beta', \psi^F) = \mathcal{C}(\mathfrak{a}, \beta, \psi^F)$, then $\mathfrak{I}^1(\beta', \mathfrak{a}) = \mathfrak{I}^1(\beta, \mathfrak{a})$.*
(5) *If K/F is a finite unramified extension, then*

$$\mathfrak{I}^1(\beta, \mathfrak{a}_K) = \mathfrak{I}^1(\beta, \mathfrak{a}) \otimes_{\mathfrak{o}_F} \mathfrak{o}_K. \tag{1.4.2}$$

PROOF. In the case of $G \cong \mathrm{GL}_n(F)$, the result is 6.4 Proposition of [**5**]. To deal with the first assertion in the general case, it is enough to show that there is a unique $\mathrm{J}(\beta, \mathfrak{a})$-stable subspace $\mathcal{I}^1(\beta, \mathfrak{a})$ of $\mathcal{J}^1(\beta, \mathfrak{a})$ satisfying condition (3).

Let K/F be a finite unramified extension such that $A_K \cong \mathrm{M}_n(K)$. The group $\mathcal{J}^1(\beta, \mathfrak{a})$ is a k_F-vector space. Likewise, $\mathcal{J}^1(\beta, \mathfrak{a}_K)$ is a k_K-vector space and 1.3 Proposition implies

$$\mathcal{J}^k(\beta, \mathfrak{a}_K) = \mathcal{J}^k(\beta, \mathfrak{a}) \otimes_{\mathrm{k}_F} \mathrm{k}_K, \quad k \geqslant 1.$$

Equally, if $\Gamma = \mathrm{Gal}(K/F)$ then

$$\mathcal{J}^k(\beta, \mathfrak{a}) = \mathcal{J}^k(\beta, \mathfrak{a}_K)^\Gamma, \quad k \geqslant 1.$$

Indeed, $\mathcal{V} \mapsto \mathcal{V}^\Gamma$ is a bijection between the set of Γ-stable k_K-subspaces $\mathcal{V}$ of $\mathcal{J}^1(\beta, \mathfrak{a}_K)$ and the set of k_F-subspaces of $\mathcal{J}^1(\beta, \mathfrak{a})$, the inverse being $\mathcal{W} \mapsto \mathcal{W} \otimes \mathrm{k}_K$.

Remark also that, by the preceding lemma and the choice of ψ^K, the pairing h_β^F is the restriction of h_β^K to $\mathcal{J}^1(\beta, \mathfrak{a})$.

We first prove that $\mathcal{U}^k(\beta, \mathfrak{a}_K)$ is Γ-invariant. For $\gamma \in \Gamma$, let ${}^\gamma\psi^K$ be the character $x \mapsto \psi^K(x^\gamma)$, $x \in K$. There is a unique $t_\gamma \in U_K$ such that

${}^{\gamma}\psi^K(x) = \psi^K(t_\gamma x)$, $x \in K$. If $j \in \mathcal{J}^{1+k}(\beta, \mathfrak{a}_K)$ and $u \in \mathcal{U}^k(\beta, \mathfrak{a}_K)$, then

$$\begin{aligned} h^K_\beta(j, u^\gamma) &= \psi^K(\mathrm{tr}_{A_K}(\beta(ju^\gamma - u^\gamma j))) \\ &= {}^{\gamma}\psi^K(\mathrm{tr}_{A_K}(\beta(j^{\gamma^{-1}}u - uj^{\gamma^{-1}}))) \\ &= h^K_\beta(t_\gamma j^{\gamma^{-1}}, u) = 1, \end{aligned}$$

since $\mathcal{J}^{1+k}(\beta, \mathfrak{a}_K)$ is Γ-invariant. Thus $u^\gamma \in \mathcal{U}^k(\beta, \mathfrak{a}_K)$, as desired. It follows that

$$\mathcal{J}^k(\beta, \mathfrak{a}) = \mathcal{J}^{1+k}(\beta, \mathfrak{a}) \perp \mathcal{U}^k(\beta, \mathfrak{a}_K)^\Gamma.$$

The summands here are h^K_β-orthogonal, so they are h^F_β-orthogonal whence

$$\mathcal{U}^k(\beta, \mathfrak{a}_K)^\Gamma = \mathcal{U}^k(\beta, \mathfrak{a}).$$

The uniqueness property of $\mathfrak{I}^1(\beta, \mathfrak{a}_K)$ implies that $\mathcal{I}^1(\beta, \mathfrak{a}_K)$ is Γ-stable. Consider the subspace $\mathcal{I}^1(\beta, \mathfrak{a}_K)^\Gamma$ of $\mathcal{J}^1(\beta, \mathfrak{a})$. It is the sum of its intersections with the spaces $\mathcal{U}^k(\beta, \mathfrak{a})$ and is totally isotropic. Comparing dimensions, it is a maximal totally isotropic subspace of $\mathcal{J}^1(\beta, \mathfrak{a})$. As $\mathcal{I}^1(\beta, \mathfrak{a}_K)$ is stable under conjugation by $\mathrm{J}(\beta, \mathfrak{a}_K)$, so $\mathcal{I}^1(\beta, \mathfrak{a}_K)^\Gamma$ is stable under conjugation by $\mathrm{J}(\beta, \mathfrak{a}_K) \cap G = \mathrm{J}(\beta, \mathfrak{a})$. Thus $\mathcal{I}^1(\beta, \mathfrak{a}_K)^\Gamma$ has all the properties demanded of $\mathcal{I}^1(\beta, \mathfrak{a})$.

It remains to show that these properties determine $\mathcal{I}^1(\beta, \mathfrak{a}) = \mathcal{I}^1(\beta, \mathfrak{a}_K)^\Gamma$ uniquely. Let $\mathcal{I}_0$ be a maximal totally isotropic subspace of $\mathcal{J}^1(\beta, \mathfrak{a})$ satisfying the required conditions. The subspace $\mathcal{I}_0 \otimes \mathrm{k}_K$ of $\mathcal{J}^1(\beta, \mathfrak{a}_K)$ then has the necessary intersection property in (3). We show it is totally isotropic. Suppose the contrary. There then exist $x, y \in \mathcal{I}_0$ and a root of unity ζ in K such that $h^K_\beta(1{+}x, 1{+}\zeta y) \neq 1$. That is,

$$\begin{aligned} h^K_\beta(1{+}x, 1{+}\zeta y) &= \psi^K(\mathrm{tr}_{A_K}(\beta\zeta(xy{-}yx))) \\ &= \psi^K(\zeta \mathrm{tr}_{A_K}(\beta(xy{-}yx))) \neq 1. \end{aligned}$$

Therefore $\mathrm{tr}_{A_K}(\beta(xy{-}yx)) = \mathrm{tr}_A(\beta(xy{-}yx))$ does not lie in $\mathfrak{p}_F$. By additive duality in F and the choice of ψ^F, there exists $\zeta_0 \in \mathfrak{o}_F$ such that

$$\psi^F(\zeta_0 \mathrm{tr}_A(\beta(xy{-}yx))) = h^F_\beta(1{+}\zeta_0 x, 1{+}y) \neq 1.$$

That is, the space $\mathcal{I}_0$ is not totally isotropic. This contradiction implies that $\mathcal{I}_0 \otimes \mathrm{k}_K$ is totally isotropic. It surely has properties (2) and (3), so it equals $\mathcal{I}^1(\beta, \mathfrak{a}_K)$. Therefore $\mathcal{I}_0 = \mathcal{I}^1(\beta, \mathfrak{a}_K)^\Gamma$, as required.

If we fix β for the moment, the construction of $\mathfrak{I}^1(\beta, \mathfrak{a})$ has been done entirely in terms of a randomly chosen $\theta \in \mathcal{C}(\mathfrak{a}, \beta, \psi^F)$. If θ also lies in $\mathcal{C}(\mathfrak{a}, \beta', \psi^F)$, that is, if $\mathcal{C}(\mathfrak{a}, \beta, \psi^F) = \mathcal{C}(\mathfrak{a}, \beta', \psi^F)$, the uniqueness property of the first part implies $\mathfrak{I}^1(\beta', \mathfrak{a}) = \mathfrak{I}^1(\beta, \mathfrak{a})$, as required for (4). Part (5) has already been done. □

Write

(1.4.3) $$I^1(\beta, \mathfrak{a}) = 1 + \mathfrak{I}^1(\beta, \mathfrak{a}).$$

Corollary. *Let $f : A \to A'$ be an isomorphism of F-algebras. If $\mathfrak{a}' = f(\mathfrak{a})$, then $\mathcal{S}^{\mathrm{wr}}(\mathfrak{a}') = f(\mathcal{S}^{\mathrm{wr}}(\mathfrak{a}))$ and $I^1(f(\beta), \mathfrak{a}') = f(I^1(\beta, \mathfrak{a}))$, $\beta \in \mathcal{S}^{\mathrm{wr}}(\mathfrak{a})$.*

Proof. The first statement is 1.1 Proposition and the second follows from the uniqueness property of $\mathfrak{I}^1(\beta, \mathfrak{a})$. □

The group $I^1(\beta, \mathfrak{a})$ is the *Lagrangian subgroup* of the section title and [**5**].

2. Extensions of simple characters

The notation is carried over from 1.4. We admit the case $p = 2$ as far as the end of 2.1.

2.1. By definition, the group $I^1(\beta, \mathfrak{a})/H^1(\beta, \mathfrak{a})$ is a totally isotropic subspace of $J^1(\beta, \mathfrak{a})/H^1(\beta, \mathfrak{a})$, so a simple character $\theta \in \mathcal{C}(\mathfrak{a}, \beta, \psi^F)$ admits extension to a character ξ of the group $I^1(\beta, \mathfrak{a})$. The group $J^1(\beta, \mathfrak{a})/I^1(\beta, \mathfrak{a})$ acts on the set of such extensions by conjugation. Since $I^1(\beta, \mathfrak{a})/H^1(\beta, \mathfrak{a})$ is a maximal totally isotropic subspace of $J^1(\beta, \mathfrak{a})/H^1(\beta, \mathfrak{a})$, this set of extensions is a principal homogeneous space over $J^1(\beta, \mathfrak{a})/I^1(\beta, \mathfrak{a})$.

LEMMA. *Let ξ be a character of $I^1(\beta, \mathfrak{a})$ such that $\theta = \xi \mid H^1(\beta, \mathfrak{a})$ lies in $\mathcal{C}(\mathfrak{a}, \beta, \psi^F)$. An element g of G intertwines ξ if and only if $g \in \mathrm{J}(\beta, \mathfrak{a})$ and $\xi^g = \xi$.*

PROOF. If g intertwines ξ, it surely intertwines θ and so lies in $\mathrm{J}(\beta, \mathfrak{a})$. In particular, g normalizes the character θ and also the group $I^1(\beta, \mathfrak{a})$. The lemma follows. □

In other words, the G-intertwining of ξ is the $\mathrm{J}(\beta, \mathfrak{a})$-normalizer of ξ. Our aim is to control this normalizer. To this end, we introduce a finer version of the set $\mathcal{C}(\mathfrak{a}, \beta, \psi^F)$ of simple characters attached to $\beta \in \mathcal{S}^{\mathrm{wr}}(\mathfrak{a})$, following section 8.1 of [**5**].

DEFINITION. Let $\beta \in \mathcal{S}^{\mathrm{wr}}(\mathfrak{a})$ and define the positive integer l by $\beta\mathfrak{a} = \mathfrak{p}^{-l}$. Let $\theta \in \mathcal{C}(\mathfrak{a}, \beta, \psi^F)$. Say that θ is *adapted to* β if the following conditions hold.

(1) If l is even and a jump of β, then $\theta \mid H^{l/2}(\beta, \mathfrak{a}) = \psi^A_\beta$.
(2) If $2k$ is a jump of β, such that $0 < 2k < l$, and if $[\mathfrak{a}, l, 2k, \gamma]$ is a simple stratum equivalent to $[\mathfrak{a}, l, 2k, \beta]$, there exists $\phi \in \mathcal{C}(\mathfrak{a}, \gamma, \psi^F)$ such that

$$\theta \mid H^k(\beta, \mathfrak{a}) = \phi\, \psi^A_{\beta-\gamma} \mid H^k(\beta, \mathfrak{a}). \tag{2.1.1}$$

Let a-$\mathcal{C}(\mathfrak{a}, \beta, \psi^F)$ be the set of $\theta \in \mathcal{C}(\mathfrak{a}, \beta, \psi^F)$ that are adapted to β.

As in [**5**] 8.1, this definition does not depend on the choice of γ in part (2). We review that argument. As a matter of notation, if $[\mathfrak{a}, q, 0, \alpha]$ is a simple stratum and $t \geqslant 0$ is an integer, let $\mathcal{C}(\mathfrak{a}, t, \alpha)$ be the set of characters of $H^{1+t}(\alpha, \mathfrak{a})$ of the form $\theta \mid H^{1+t}(\alpha, \mathfrak{a})$, for some $\theta \in \mathcal{C}(\mathfrak{a}, \alpha)$. In part (2) of the definition, the restriction map $\mathcal{C}(\mathfrak{a}, k{-}1, \gamma) \to \mathcal{C}(\mathfrak{a}, k, \gamma)$ is bijective, as follows from [**5**] 8.1. Also,

$$\begin{aligned} J^{1+k}(\beta, \mathfrak{a}) &= J^{1+k}(\gamma, \mathfrak{a}), \\ H^{1+k}(\gamma, \mathfrak{a}) &= H^{1+k}(\beta, \mathfrak{a}), \\ J^k(\beta, \mathfrak{a}) &= H^k(\gamma, \mathfrak{a}) J^{1+k}(\beta, \mathfrak{a}). \end{aligned} \tag{2.1.2}$$

Let $[\mathfrak{a}, m, 2k, \gamma']$ be a simple stratum equivalent to $[\mathfrak{a}, m, 2k, \beta]$. We then have $H^k(\gamma', \mathfrak{a}) = H^k(\gamma, \mathfrak{a})$ and a bijection

$$\begin{aligned} \mathcal{C}(\mathfrak{a}, k{-}1, \gamma) &\longrightarrow \mathcal{C}(\mathfrak{a}, k{-}1, \gamma'), \\ \theta &\longmapsto \theta\, \psi^A_{\gamma'-\gamma}. \end{aligned}$$

The asserted independence of γ follows directly.

PROPOSITION. *Let $\beta \in \mathcal{S}^{\mathrm{wr}}(\mathfrak{a})$.*

(1) *There exists $\theta \in \mathcal{C}(\mathfrak{a}, \beta, \psi^F)$ that is adapted to β: the set a-$\mathcal{C}(\mathfrak{a}, \beta, \psi^F)$ is not empty.*

(2) *If* $\vartheta \in \mathcal{C}(\mathfrak{a},\beta,\psi^F)$, *there exists* $\beta' \in \mathcal{S}^{\mathrm{wr}}(\mathfrak{a})$ *such that* $\vartheta \in a\text{-}\mathcal{C}(\mathfrak{a},\beta',\psi^F)$. *In particular,* $\mathcal{C}(\mathfrak{a},\beta',\psi^F) = \mathcal{C}(\mathfrak{a},\beta,\psi^F)$ *and* $I^1(\beta',\mathfrak{a}) = I^1(\beta,\mathfrak{a})$.

PROOF. If A is either a full matrix algebra or a division algebra, the result is proved in [**5**] 8.1. The general case is identical and there is no need to repeat the details. □

2.2. *From this point on, we assume that* $p \neq 2$*:* the immediate reason for this restriction is given in the remark at the end of the sub-section.

DEFINITION. Let $\beta \in \mathcal{S}^{\mathrm{wr}}(\mathfrak{a})$ and write $E = F[\beta]$.

(1) Set

$$\mathrm{I}(\beta,\mathfrak{a}) = E^\times I^1(\beta,\mathfrak{a}). \tag{2.2.1}$$

(2) Define $\mathcal{D}(\mathfrak{a},\beta,\psi^F)$ to be the set of characters λ of the group $\mathrm{I}(\beta,\mathfrak{a})$ such that $\lambda \,\big|\, H^1(\beta,\mathfrak{a}) \in a\text{-}\mathcal{C}(\mathfrak{a},\beta,\psi^F)$.

(3) Set

$$\mathcal{D}(\mathfrak{a},\psi^F) = \bigcup_{\beta\in\mathcal{S}^{\mathrm{wr}}(\mathfrak{a})} \mathcal{D}(\mathfrak{a},\beta,\psi^F). \tag{2.2.2}$$

PROPOSITION. *Let* $\beta \in \mathcal{S}^{\mathrm{wr}}(\mathfrak{a})$ *and write* $E = F[\beta]$.

(1) *The restriction map* $\mathcal{D}(\mathfrak{a},\beta,\psi^F) \to a\text{-}\mathcal{C}(\mathfrak{a},\beta,\psi^F)$ *is surjective.*

(2) *Let* K/F *be a finite unramified extension. If* $\lambda \in \mathcal{D}(\mathfrak{a}_K,\beta,\psi^K)$, *the character*

$$\lambda^F = \lambda \,\big|\, \mathrm{I}(\beta,\mathfrak{a}) \tag{2.2.3}$$

lies in $\mathcal{D}(\mathfrak{a},\beta,\psi^F)$. *The map*

$$\begin{aligned} \mathcal{D}(\mathfrak{a}_K,\beta,\psi^K) &\longrightarrow \mathcal{D}(\mathfrak{a},\beta,\psi^F),\\ \lambda &\longmapsto \lambda^F, \end{aligned}$$

is surjective.

(3) *If* $f : A \to A'$ *is an isomorphism of* F*-algebras, then*

$$f(\mathrm{I}(\beta,\mathfrak{a})) = \mathrm{I}(f(\beta), f(\mathfrak{a}))$$

and f *induces a bijection*

$$\begin{aligned} \mathcal{D}(\mathfrak{a},\beta,\psi^F) &\longrightarrow \mathcal{D}(f(\mathfrak{a}), f(\beta), \psi^F),\\ \lambda &\longmapsto \lambda\circ f^{-1}. \end{aligned}$$

PROOF. Suppose first that $A \cong \mathrm{M}_n(F)$. Let $\theta \in a\text{-}\mathcal{C}(\mathfrak{a},\beta,\psi^F)$. As in 8.4 Proposition of [**5**], there is a character ξ of $I^1(\beta,\mathfrak{a})$ that extends θ and is stable under conjugation by $E^\times$. The character ξ then extends to a character λ of $\mathrm{I}(\beta,\mathfrak{a})$, whence follows (1) in this case.

In the general case, let K/F be unramified of degree divisible by n, so that $A_K \cong \mathrm{M}_n(K)$.

LEMMA 1. *The restriction map* $\mathcal{C}(\mathfrak{a}_K,\beta,\psi^K) \to \mathcal{C}(\mathfrak{a},\beta,\psi^F)$ *induces a surjection* $a\text{-}\mathcal{C}(\mathfrak{a}_K,\beta,\psi^K) \to a\text{-}\mathcal{C}(\mathfrak{a},\beta,\psi^F)$.

PROOF. If $\vartheta \in a\text{-}\mathcal{C}(\mathfrak{a}_K, \beta, \psi^K)$, then 2.1 Definition implies that $\vartheta \mid H^1(\beta, \mathfrak{a})$ lies in $a\text{-}\mathcal{C}(\mathfrak{a}, \beta, \psi^F)$. So, we take $\theta \in a\text{-}\mathcal{C}(\mathfrak{a}, \beta, \psi^F)$ and construct a character $\theta_K \in a\text{-}\mathcal{C}(\mathfrak{a}_K, \beta, \psi^K)$ that extends θ.

Let $t \geqslant 0$ be the least integer for which there exists $\vartheta \in \mathcal{C}(\mathfrak{a}_K, \beta, \psi^K)$ agreeing with θ on $H^{1+t}(\beta, \mathfrak{a})$ and satisfying the conditions of the definition relative to any even jump $2k$, with $k > t$.

Recall that l is defined by $\mathfrak{p}^l = \beta^{-1}\mathfrak{a}$, where $\mathfrak{p}$ is the Jacobson radical of $\mathfrak{a}$. If l is even and a jump of β, the definition yields $t < l/2$. If l is the only even jump, we get $t = 0$ since there are no further restrictions to be observed. In all other cases, we still have $t < l/2$. If $t < k$ for any even jump $2k$, we again get $t = 0$. Otherwise, let $2k$ be the greatest even jump such that $k \leqslant t$. We may adjust our choice of ϑ so that it agrees with θ on $H^{1+k}(\beta, \mathfrak{a})$ without affecting the conditions already imposed in our hypothesis. That is, we can assume that $t = k$ where $2k$ is an even jump. Take a simple stratum $[\mathfrak{a}, l, 2k, \gamma]$ equivalent to $[\mathfrak{a}, l, m, \beta]$. On $H^k(\beta, \mathfrak{a})$, θ takes the form $\phi\psi^A_{\beta-\gamma}$, for some $\phi \in \mathcal{C}(\mathfrak{a}, \gamma, \psi^F)$. The standard construction of simple characters implies that, on $H^{1+k}(\beta, \mathfrak{a}_K)$, the character ϑ takes the form $\varphi\psi^{A_K}_{\beta-\gamma}$, for some $\varphi \in \mathcal{C}(\mathfrak{a}_K, \gamma, \psi^K)$. Surely φ agrees with ϕ on $H^{1+k}(\beta, \mathfrak{a})$. However, 8.1 Proposition of [**5**] implies that φ agrees with ϕ on $H^k(\beta, \mathfrak{a})$. We could therefore have chosen our original ϑ to agree with θ on the larger group $H^k(\beta, \mathfrak{a})$, contrary to our definition of t. So, in all cases, $t = 0$ and the lemma is proven. □

NOTE. Lemma 1 and its proof remain valid when $p = 2$.

Continuing with the proof of part (1) of the proposition, let $\theta \in a\text{-}\mathcal{C}(\mathfrak{a}, \beta, \psi^F)$. Take $\theta_K \in a\text{-}\mathcal{C}(\mathfrak{a}_K, \beta, \psi^K)$ such that $\theta = \theta_K \mid H^1(\beta, \mathfrak{a})$. Since $A \cong \mathrm{M}_n(K)$, we know that θ_K admits extension to a character λ_K of $\mathrm{I}(\beta, \mathfrak{a}_K)$. The restriction $\lambda_K^F = \lambda_K \mid \mathrm{I}(\beta, \mathfrak{a})$ provides an extension of θ, and part (1) of the proposition is proven.

We next have to show that any extension λ of θ to $\mathrm{I}(\beta, \mathfrak{a})$ arises in this way. The character $\lambda \mid I^1(\beta, \mathfrak{a})$ differs from $\lambda_K \mid I^1(\beta, \mathfrak{a})$ by a character ϕ of $I^1(\beta, \mathfrak{a})/H^1(\beta, \mathfrak{a})$ that is stable under conjugation by a prime element ϖ of E.

LEMMA 2. *The character ϕ extends to a character ϕ_K of $I^1(\beta, \mathfrak{a}_K)/H^1(\beta, \mathfrak{a}_K)$ stable under conjugation by ϖ.*

PROOF. Identify $I^1(\beta, \mathfrak{a})/H^1(\beta, \mathfrak{a})$ with the k_F-space $\mathfrak{J}^1(\beta, \mathfrak{a})/\mathfrak{H}^1(\beta, \mathfrak{a})$. When viewed as a character of this group, ϕ is trivial on the image of the map $A_\varpi : x \mapsto \varpi x \varpi^{-1} - x$. As $I^1(\beta, \mathfrak{a}_K)/H^1(\beta, \mathfrak{a}_K) = I^1(\beta, \mathfrak{a})/H^1(\beta, \mathfrak{a}) \otimes_{\mathrm{k}_F} \mathrm{k}_K$, the result follows straightaway. □

Following Lemma 2, we could have chosen λ_K to agree with λ on $I^1(\beta, \mathfrak{a})$. In other words, $\lambda_K \mid \mathrm{I}(\beta, \mathfrak{a}) = \tau \otimes \lambda$, for a character τ of $\mathrm{I}(\beta, \mathfrak{a})/I^1(\beta, \mathfrak{a}) = E^\times/U^1_E$. Surely τ is the restriction of a character τ_K of $KE^\times/U^1_{KE}$. Replacing λ_K by $\tau_K^{-1} \otimes \lambda_K$, we get $\lambda_K \mid \mathrm{I}(\beta, \mathfrak{a}) = \lambda$. This proves (2) in the case where $[K{:}F]$ is divisible by n, and the general case follows by transitivity.

In part (3), the map f carries $a\text{-}\mathcal{C}(\mathfrak{a}, \beta, \psi^F)$ bijectively to $a\text{-}\mathcal{C}(f(\mathfrak{a}), f(\beta), \psi^F)$, as follows directly from the definition. □

REMARK. The definition here of the set $\mathcal{D}(\mathfrak{a}, \beta, \psi^F)$ is different from, and more inclusive than, the one used in [**5**]. We have found it more convenient.

When $p = 2$, both approaches fail: there are examples where $\mathcal{D}(\mathfrak{a}, \beta, \psi^F)$ is empty ([**5**] 8.3 Remark).

2.3. We relate the sets $\mathcal{D}(\mathfrak{a}, \beta, \psi^F)$, $\mathcal{D}(\mathfrak{a}, \psi^F)$ to representations of G. Observe that, as a consequence of part (3) of 2.2 Proposition, the group $\mathrm{K}_{\mathfrak{a}}$ acts on the set $\mathcal{D}(\mathfrak{a}, \psi^F)$ by conjugation.

THEOREM. *Let $\mathfrak{a}$ be a minimal hereditary $\mathfrak{o}_F$-order in A, and let $\beta \in \mathcal{S}^{\mathrm{wr}}(\mathfrak{a})$.*

(1) *If $\lambda \in \mathcal{D}(\mathfrak{a}, \beta, \psi^F)$, the induced representation*

$$\pi_G(\lambda) = c\text{-}\mathrm{Ind}_{\mathrm{I}(\beta,\mathfrak{a})}^G \lambda$$

is irreducible and cuspidal. Its equivalence class lies in $\mathcal{A}^{\text{m-wr}}(G)$.

(2) *The map*

$$\lambda \longmapsto \pi_G(\lambda), \quad \lambda \in \mathcal{D}(\mathfrak{a}, \psi^F),$$

induces a canonical bijection

$$\pi_G : \mathrm{K}_{\mathfrak{a}} \backslash \mathcal{D}(\mathfrak{a}, \psi^F) \xrightarrow{\approx} \mathcal{A}^{\text{m-wr}}(G). \tag{2.3.1}$$

PROOF. Let $\lambda \in \mathcal{D}(\mathfrak{a}, \beta, \psi^F)$, and put $\vartheta = \lambda \,|\, I^1(\beta, \mathfrak{a})$, $\theta = \lambda \,|\, H^1(\beta, \mathfrak{a})$. There is a unique irreducible representation η of $J^1(\beta, \mathfrak{a})$ containing θ [**18**] 2.2. By definition, $I^1(\beta, \mathfrak{a})/H^1(\beta, \mathfrak{a})$ is a maximal isotropic subspace of the alternating space $J^1(\beta, \mathfrak{a})/H^1(\beta, \mathfrak{a})$, so η is induced by any character of $I^1(\beta, \mathfrak{a})$ extending θ. If Λ is the representation of $\mathrm{J}(\beta, \mathfrak{a})$ induced by λ, the Mackey restriction formula shows that $\Lambda \,|\, J^1(\beta, \mathfrak{a})$ is the irreducible representation η. Therefore Λ is irreducible. Any $g \in G$ that intertwines θ lies in $\mathrm{J}(\beta, \mathfrak{a})$, so

$$\pi_G(\lambda) = c\text{-}\mathrm{Ind}_{\mathrm{I}(\beta,\mathfrak{a})}^G \lambda = c\text{-}\mathrm{Ind}_{\mathrm{J}(\beta,\mathfrak{a})}^G \Lambda \tag{2.3.2}$$

is irreducible and cuspidal. Since it contains the simple character $\theta \in \mathcal{C}(\mathfrak{a}, \beta, \psi^F)$ and the field extension $F[\beta]/F$ is totally ramified of degree n, the representation $\pi_G(\lambda)$ is totally ramified of parametric degree n. That is, $\pi_G(\lambda) \in \mathcal{A}^{\text{m-wr}}(G)$.

Conversely, let $\pi \in \mathcal{A}^{\text{m-wr}}(G)$. As in [**20**], there exists an extended maximal simple type Λ in G, inducing π. Since $\pi \in \mathcal{A}^{\text{m-wr}}(G)$, Λ is a representation of a group $\mathrm{J}(\beta, \mathfrak{a})$, for some $\beta \in \mathcal{S}^{\mathrm{wr}}(\mathfrak{a})$, that contains a simple character $\theta \in \mathcal{C}(\mathfrak{a}, \beta, \psi^F)$. By 2.1 Proposition, we may assume $\theta \in a\text{-}\mathcal{C}(\mathfrak{a}, \beta, \psi^F)$. Let ϑ be a character of $I^1(\beta, \mathfrak{a})$, extending θ. Any two choices of ϑ are $J^1(\beta, \mathfrak{a})$-conjugate, so ϑ occurs in Λ. By part (1) of 2.2 Proposition, we may take ϑ to be $F[\beta]^\times$-stable, so there exists $\lambda \in \mathcal{D}(\mathfrak{a}, \beta, \psi^F)$ extending ϑ and occurring in Λ. The representation of $\mathrm{J}(\beta, \mathfrak{a})$ induced by λ is then Λ, giving $\pi = \pi_G(\lambda)$, as desired.

The group $\mathrm{K}_{\mathfrak{a}}$ acts on the set $\mathcal{D}(\mathfrak{a}, \psi^F)$ by conjugation. For $\lambda \in \mathcal{D}(\mathfrak{a}, \psi^F)$, the equivalence class of $\pi_G(\lambda)$ depends only on the $\mathrm{K}_{\mathfrak{a}}$-orbit of λ, so $\lambda \mapsto \pi_G(\lambda)$ induces a surjective map $\mathrm{K}_{\mathfrak{a}} \backslash \mathcal{D}(\mathfrak{a}, \psi^F) \to \mathcal{A}^{\text{m-wr}}(G)$. To prove it is injective, take $\beta_i \in \mathcal{S}^{\mathrm{wr}}(\mathfrak{a})$ and $\lambda_i \in \mathcal{D}(\mathfrak{a}, \beta_i, \psi^F)$, $i = 1, 2$, and suppose that $\pi_G(\lambda_1) = \pi_G(\lambda_2) = \pi$, say. The simple characters $\theta_i = \lambda_i \,|\, H^1(\beta_i, \mathfrak{a})$ intertwine in G. They are therefore $\mathrm{K}_{\mathfrak{a}}$-conjugate [**20**] Theorem 6.1. We may assume they are equal, say $\theta_1 = \theta_2 = \theta$, implying that the sets $\mathcal{C}(\mathfrak{a}, \beta_i, \psi^F)$ are the same. In particular, $I^1(\beta_1, \mathfrak{a}) = I^1(\beta_2, \mathfrak{a})$ (1.4 Proposition) and the same holds for the H^1, J^1 and J groups. After applying a $J^1(\beta_i, \mathfrak{a})$-conjugation, we can assume that the λ_i agree on $I^1(\beta_i, \mathfrak{a})$ and intertwine in G. This intertwining is implemented by an element x which intertwines θ and so lies in $\mathrm{J}(\beta_i, \mathfrak{a})$. The element x normalizes $I^1(\beta_i, \mathfrak{a})$ and fixes the character $\lambda_i \,|\, I^1(\beta_i, \mathfrak{a})$. Therefore $x \in \mathrm{I}(\beta_i, \mathfrak{a})$ and x fixes λ_i. Thus $\lambda_1 = \lambda_2$, as required. □

2.4. Let $\beta \in \mathcal{S}^{\mathrm{wr}}(\mathfrak{a})$, let $\theta \in a\text{-}\mathcal{C}(\mathfrak{a},\beta,\psi^F)$ and write $E = F[\beta]$. Following [**5**] 8.3, 8.4, we write down an $E^\times$-invariant character ${}_I\theta$ of $I^1(\beta,\mathfrak{a})$. The definition of ${}_I\theta$ does depend on the choice of β, subject to $\theta \in a\text{-}\mathcal{C}(\mathfrak{a},\beta,\psi^F)$, but is otherwise canonical.

Let $\beta^{-1}\mathfrak{a} = \mathfrak{p}^l$, where $\mathfrak{p}$ is the Jacobson radical of $\mathfrak{a}$. If β has no even jumps, that is, if $H^1(\beta,\mathfrak{a}) = J^1(\beta,\mathfrak{a})$, then $I^1(\beta,\mathfrak{a}) = H^1(\beta,\mathfrak{a})$ and we set ${}_I\theta = \theta$. This is certainly $E^\times$-invariant. Otherwise, let $2s_1 < 2s_2 < \cdots < 2s_t$ be the even jumps of β. Reverting to the notation of 1.4, let $\mathcal{I}$ be the image of $I^1(\beta,\mathfrak{a})$ in $J^1(\beta,\mathfrak{a})/H^1(\beta,\mathfrak{a})$. Let $[\mathfrak{a}, m, 2s_k, \gamma]$ be a simple stratum equivalent to $[\mathfrak{a}, m, 2s_k, \beta]$. The inclusion of $H^{s_k}(\gamma,\mathfrak{a})$ in $J^{s_k}(\beta,\mathfrak{a})$ induces an isomorphism of $H^{s_k}(\gamma,\mathfrak{a})/H^{1+s_k}(\gamma,\mathfrak{a})$ with $J^{s_k}(\beta,\mathfrak{a})/J^{1+s_k}(\beta,\mathfrak{a})$ (2.1.2). Let I_k be the inverse image, in $H^{s_k}(\gamma,\mathfrak{a})$, of $\mathcal{I} \cap \mathcal{U}^k(\beta,\mathfrak{a})$, $1 \leqslant k \leqslant t$. We then have

$$I^1(\beta,\mathfrak{a}) = H^1(\beta,\mathfrak{a})\, I_1\, I_2\, \dots\, I_t, \tag{2.4.1}$$

with all factors in the product commuting modulo the kernel of θ. If $2s_t = l$, we define

$$ {}_I\theta(1{+}x) = \psi^A_\beta(1{+}x - x^2/2), \quad 1{+}x \in I_t. \tag{2.4.2}$$

Otherwise, let $2s_k < l$ and choose a simple stratum $[\mathfrak{a}, l, 2s_k, \gamma_k]$ equivalent to $[\mathfrak{a}, l, 2s_k, \beta]$. By 2.1 Definition, $\theta \,\big|\, H^{s_k}(\beta,\mathfrak{a}) = \vartheta\psi^A_{\beta-\gamma_k} \,\big|\, H^{s_k}(\beta,\mathfrak{a})$, for some $\vartheta \in \mathcal{C}(\mathfrak{a},\gamma_k,\psi^F)$. We set

$$ {}_I\theta(1{+}x) = \vartheta(1{+}x)\, \psi^A_{\beta-\gamma_k}(1{+}x{-}x^2/2), \quad 1{+}x \in I_k. \tag{2.4.3}$$

Following [**5**] 8.1–3, the product formula (2.4.1) and the expressions (2.4.2), (2.4.3) define ${}_I\theta$ as a character of $I^1(\beta,\mathfrak{a})$. As such, it is stable under conjugation by $E^\times$. It depends on the chosen element β such that $\theta \in a\text{-}\mathcal{C}(\mathfrak{a},\beta,\psi^F)$. It does not depend on the choices of γ_k (*cf.* 2.1).

To go a step further, note that $E^\times \cap H^1(\beta,\mathfrak{a}) = E^\times \cap I^1(\beta,\mathfrak{a}) = U^1_E$. Let ξ be a character of $E^\times$ agreeing with θ on $U^1_E = E^\times \cap H^1(\beta,\mathfrak{a})$. The formula

$$\xi \odot_\beta \theta : ux \longmapsto \xi(u)\, {}_I\theta(x), \quad u \in E^\times,\ x \in I^1(\beta,\mathfrak{a}), \tag{2.4.4}$$

defines $\xi \odot_\beta \theta$ as a character of $\mathrm{I}(\beta,\mathfrak{a})$. Surely, $\xi \odot_\beta \theta \in \mathcal{D}(\mathfrak{a},\beta,\psi^F)$. It is necessary to refer to β in the $\odot$-notation, since everything depends on the choice of β for which $\theta \in a\text{-}\mathcal{C}(\mathfrak{a},\beta,\psi^F)$.

PROPOSITION. *Let $\pi \in \mathcal{A}^{\text{m-wr}}(G)$ contain the character $\theta \in a\text{-}\mathcal{C}(\mathfrak{a},\beta,\psi^F)$. Set $E = F[\beta]$.*

(1) *The representation π contains the character ${}_I\theta$ of $I^1(\beta,\mathfrak{a})$.*
(2) *There is a unique character ξ of $E^\times$, agreeing with θ on U^1_E, such that π contains $\xi \odot_\beta \theta$ and, consequently, $\pi = \pi_G(\xi \odot_\beta \theta)$.*
(3) *The map $\xi \mapsto \pi_G(\xi \odot_\beta \theta)$ is a bijection between the set of characters ξ of $E^\times$, that agree with θ on U^1_E, and the set of elements of $\mathcal{A}^{\text{m-wr}}(G)$ that contain θ.*

PROOF. Surely π contains a character ϕ of $I^1(\beta,\mathfrak{a})$ extending θ. Since the space $I^1(\beta,\mathfrak{a})/H^1(\beta,\mathfrak{a})$ is a maximal totally isotropic subspace of $J^1(\beta,\mathfrak{a})/H^1(\beta,\mathfrak{a})$, the character ϕ is $J^1(\beta,\mathfrak{a})$-conjugate to ${}_I\theta$, which therefore occurs in π. If η is the unique irreducible representation of $J^1(\beta,\mathfrak{a})$ that contains θ, then η occurs in π with multiplicity one. So, there is a unique character $\lambda \in \mathcal{D}(\mathfrak{a},\beta,\psi^F)$ that occurs in π and extends ${}_I\theta$. Surely there exists a unique character ξ, of the required form, such that $\lambda = \xi \odot_\beta \theta$. □

2.5. This construction behaves properly with respect to unramified base field extension.

PROPOSITION. *Let $\beta \in \mathcal{S}^{\mathrm{wr}}(\mathfrak{a})$, let $\theta \in a\text{-}\mathcal{C}(\mathfrak{a}, \beta, \psi^F)$ and let ξ be a character of $E^\times = F[\beta]^\times$ agreeing with θ on U^1_E. Let K/F be a finite unramified extension and let $\theta_K \in a\text{-}\mathcal{C}(\mathfrak{a}_K, \beta, \psi^K)$ agree with θ on $H^1(\beta, \mathfrak{a})$. If ξ_K is a character of $KE^\times$ agreeing with θ_K on U^1_{KE} and with ξ on $E^\times$, then*

$$\begin{aligned} {}_I\theta_K \,\big|\, I^1(\beta, \mathfrak{a}) &= {}_I\theta, \\ \xi_K \odot_\beta \theta_K \,\big|\, \mathrm{I}(\beta, \mathfrak{a}) &= \xi \odot_\beta \theta. \end{aligned}$$

PROOF. Immediate. □

3. Change of group and endo-classes

Let A and B be central simple F-algebras of dimension n^2, where n is a power of p. Let $\mathfrak{a}$ and $\mathfrak{b}$ be minimal hereditary $\mathfrak{o}_F$-orders in A and B respectively. Write $G = A^\times$ and $H = B^\times$. Other notation is carried over from sections 1 and 2.

If K/F is an unramified extension of finite degree divisible by n, the K-algebras A_K and B_K are both isomorphic to $\mathrm{M}_n(K)$. In this section, we use such isomorphisms to exploit the naturality properties of the sets $\mathcal{D}(\mathfrak{a}, \beta, \psi^F)$ laid out in 2.2 Proposition. In the case to hand, these properties restrict to a transfer of simple characters between G and H that preserves endo-classes.

More to the point, we obtain a process, called *parametric transfer*, for moving representations between $\mathcal{A}^{\text{m-wr}}(G)$ and $\mathcal{A}^{\text{m-wr}}(H)$. By the end of the section, it still depends on one choice made in the construction.

3.1. We start with a basic formal result.

PROPOSITION. *Let $\beta \in \mathcal{S}^{\mathrm{wr}}(\mathfrak{a})$ and write $E = F[\beta]$. Let K/F be a finite unramified extension of degree divisible by n.*

(1) *There exists an F-embedding $\iota : E \to B$ such that $\iota(E^\times) \subset \mathrm{K}_\mathfrak{b}$ and $\iota(\beta) \in \mathcal{S}^{\mathrm{wr}}(\mathfrak{b})$. An F-embedding $\iota' : E \to B$ has the same property if and only if $\iota' = \mathrm{Ad}\, x \circ \iota$, for some $x \in \mathrm{K}_\mathfrak{b}$.*

(2) *The map ι extends to an isomorphism $\iota_K : A_K \to B_K$ of K-algebras such that $\iota_K(\mathfrak{a}_K) = \mathfrak{b}_K$. Any such extension ι_K has the property*

$$\begin{aligned} \iota_K(I^1(\beta, \mathfrak{a}_K)) &= I^1(\iota(\beta), \mathfrak{b}_K), \\ \iota_K(\mathrm{I}(\beta, \mathfrak{a}_K)) &= \mathrm{I}(\iota(\beta), \mathfrak{b}_K), \end{aligned} \tag{3.1.1}$$

and induces a bijection

$$\begin{aligned} \mathcal{D}(\mathfrak{a}_K, \beta, \psi^K) &\longrightarrow \mathcal{D}(\mathfrak{b}_K, \iota(\beta), \psi^K), \\ \lambda &\longmapsto \lambda \circ \iota_K^{-1}, \end{aligned} \tag{3.1.2}$$

(3) *If $\iota'_K : A_K \to B_K$ also extends ι and has the property $\iota'_K(\mathfrak{a}_K) = \mathfrak{b}_K$, there exists $y \in KE^\times$ such that $\iota'_K = \iota_K \circ \mathrm{Ad}\, y$.*

PROOF. There surely exists an F-embedding $\iota : E \to B$. Since E/F is totally ramified of degree n, there is a *unique* minimal hereditary $\mathfrak{o}_F$-order $\mathfrak{b}_1$ in B such that $\iota(E^\times) \subset \mathrm{K}_{\mathfrak{b}_1}$, as in 1.1. Replacing ι by $\mathrm{Ad}\, x \circ \iota$, for some $x \in H$, we may take

$\mathfrak{b}_1 = \mathfrak{b}$. That is, ι satisfies the first assertion of (1) and the second follows from 1.1 Proposition. The third assertion of (1) follows from the uniqueness of $\mathfrak{b}$.

We extend ι, by linearity, to a K-embedding ι_K of the field $K \otimes_F E = KE = K[\beta]$ in B_K. Since $\iota(\beta) \in \mathcal{S}^{\mathrm{wr}}(\mathfrak{b})$, we have $\iota_K(\beta) \in \mathcal{S}^{\mathrm{wr}}(\mathfrak{b}_K)$ (*cf.* 1.3). Since $A_K \cong B_K$, the Skolem–Noether theorem implies that ι_K extends to an isomorphism $A_K \to B_K$ of K-algebras. The image $\iota_K(\mathfrak{a}_K)$ is a hereditary $\mathfrak{o}_K$-order stable under conjugation by $\iota_K(K[\beta])^\times = K[\iota(\beta)]^\times$, so $\iota_K(\mathfrak{a}_K) = \mathfrak{b}_K$. The map ι_K carries $H^1(\beta, \mathfrak{a}_K)$ to $H^1(\iota(\beta), \mathfrak{b}_K)$ and similarly for the J^1-groups. Further, the map $\theta \mapsto \theta \circ \iota_K^{-1}$ is a bijection

$$\text{(3.1.3)} \qquad \mathcal{C}(\mathfrak{a}_K, \beta, \psi^K) \longrightarrow \mathcal{C}(\mathfrak{b}_K, \iota(\beta), \psi^K).$$

The property (3.1.3) follows from 2.2 Proposition (3). Moreover, the map (3.1.1) restricts to a bijection

$$a\text{-}\mathcal{C}(\mathfrak{a}_K, \beta, \psi^K) \longrightarrow a\text{-}\mathcal{C}(\mathfrak{b}_K, \iota(\beta), \psi^K),$$

by 2.2 Lemma 1. Thus (3.1.2) is a bijection as required.

Finally, if ι'_K is another extension as in (3), then $\iota'_K = \iota_K \circ \operatorname{Ad} z$, for some $z \in G$. It also agrees with ι_K on KE, while the field KE is its own centralizer in A_K. □

Take an F-embedding $\iota : E \to B$ and an extension $\iota_K : A_K \to B_K$, as in the proposition. Composing the bijection $\mathcal{D}(\mathfrak{a}_K, \beta, \psi^K) \to \mathcal{D}(\mathfrak{b}_K, \iota(\beta), \psi^K)$ of (3.1.2) with the restriction map $\mathcal{D}(\mathfrak{b}_K, \iota(\beta), \psi^K) \to \mathcal{D}(\mathfrak{b}, \iota(\beta), \psi^F)$ of 2.2 Proposition, we get a surjective map

$$\text{(3.1.4)} \qquad \begin{aligned} \mathcal{D}(\mathfrak{a}_K, \beta, \psi^K) &\longrightarrow \mathcal{D}(\mathfrak{b}, \iota(\beta), \psi^F), \\ (\mathrm{I}, \kappa) &\longmapsto (\iota_K(\mathrm{I}) \cap H, \iota\kappa^F), \end{aligned}$$

where we abbreviate $\mathrm{I} = \mathrm{I}(\beta, \mathfrak{a}_K)$, so that $\iota_K(\mathrm{I}) \cap H = \mathrm{I}(\iota(\beta), \mathfrak{b})$, and

$$\iota\kappa^F = \kappa \circ \iota_K^{-1} \mid \iota_K(\mathrm{I}) \cap H.$$

We form the representation $\pi_H(\iota\kappa^F) = c\text{-}\mathrm{Ind}_{\iota_K(\mathrm{I})\cap H}^H\, \iota\kappa^F \in \mathcal{A}^{\text{m-wr}}(H)$, as in 2.3.

Corollary. *Let ι, $\iota' : E \to B$ be embeddings as in the proposition. If $\kappa \in \mathcal{D}(\mathfrak{a}_K, \beta, \psi^K)$, then $\pi_H(\iota\kappa^F) \cong \pi_H(\iota'\kappa^F)$.*

Proof. Part (3) of the proposition shows that $\pi_H(\iota\kappa^F)$ depends only on ι, not on the choice of extension ι_K, while (1) shows it is independent of ι. □

3.2. The procedure of 3.1 is essentially independent of the choice of K/F. For, if L/K is a finite unramified extension, we have a commutative diagram

$$\begin{array}{ccc} \mathcal{D}(\mathfrak{a}_L, \beta, \psi^L) & \xrightarrow{\approx} & \mathcal{D}(\mathfrak{b}_L, \iota(\beta), \psi^L) \\ \downarrow & & \downarrow \\ \mathcal{D}(\mathfrak{a}_K, \beta, \psi^K) & \xrightarrow{\approx} & \mathcal{D}(\mathfrak{b}_K, \iota(\beta), \psi^K) \end{array}$$

in which the vertical arrows are the surjective restriction maps.

3.3. Take $\theta \in a\text{-}\mathcal{C}(\mathfrak{a}, \beta, \psi^F)$. It is the restriction of some $\theta_K \in a\text{-}\mathcal{C}(\mathfrak{a}_K, \beta, \psi^K)$ (2.2 Lemma 1). We use an embedding ι, satisfying the conditions of 3.1 Proposition, to form

$$\theta_K \circ \iota_K^{-1} \in a\text{-}\mathcal{C}(\mathfrak{b}_K, \iota\beta, \psi^K).$$

We set

$$\iota\theta_K^F = \theta_K \circ \iota_K^{-1} \mid H^1(\iota\beta, \mathfrak{b}). \tag{3.3.1}$$

Thus $\iota\theta_K^F \in a\text{-}\mathcal{C}(\mathfrak{b}, \iota\beta, \psi^F)$. Using the language of [**3**], we have:

PROPOSITION. *The simple characters θ, $\iota\theta_K^F$ are endo-equivalent.*

PROOF. This is Theorem 1.13 plus Remark 6.9 of [**3**]. □

3.4. Let $\pi \in \mathcal{A}^{\text{m-wr}}(G)$. As in 2.3, there is a character $\lambda \in \mathcal{D}(\mathfrak{a}, \psi^F)$ that induces π, and this λ is unique up to $\mathrm{K}_{\mathfrak{a}}$-conjugation. We may choose $\beta \in \mathcal{S}^{\text{wr}}(\mathfrak{a})$ so that $\lambda \in \mathcal{D}(\mathfrak{a}, \beta, \psi^F)$. Following the procedure of 3.1, we set $E = F[\beta]$ and choose

(1) an F-embedding $\iota : E \to B$ such that $\iota(E)^\times \subset \mathrm{K}_{\mathfrak{b}}$,
(2) a finite unramified field extension K/F of degree divisible by n,
(3) a K-isomorphism $\iota_K : A_K \to B_K$ extending ι, and
(4) a character $\lambda_K \in \mathcal{D}(\mathfrak{a}_K, \beta, \psi^K)$ extending λ.

Having made these choices, we get a representation $\pi' = \pi_H(\iota\lambda_K^F) \in \mathcal{A}^{\text{m-wr}}(H)$. Following (3.1) Corollary, π' actually depends only on the choice of λ^K in (4). We make no effort at this stage to eliminate that dependence. We say that a representation $\pi' \in \mathcal{A}^{\text{m-wr}}(H)$, obtained from $\pi \in \mathcal{A}^{\text{m-wr}}(G)$ by such a choice, is a *parametric transfer of π*.

Observe that, if we have a third algebra C, a representation $\pi'' \in \mathcal{A}^{\text{m-wr}}(C^\times)$ that is a parametric transfer of π' (relative to the same β) is also a parametric transfer of π.

3.5. We look back at the constructions of 2.4, 2.5. Thus we take $\beta \in \mathcal{S}^{\text{wr}}(\mathfrak{a})$, $\theta \in a\text{-}\mathcal{C}(\mathfrak{a}, \beta, \psi^F)$ and set $E = F[\beta]$. Let ξ be a character of $E^\times$ agreeing with θ on U_E^1. We form the character $\xi \odot_\beta \theta \in \mathcal{D}(\mathfrak{a}, \beta, \psi^F)$ as in (2.4.4). Let $\theta_K \in a\text{-}\mathcal{C}(\mathfrak{a}_K, \beta, \psi^K)$ agree with θ on $H^1(\beta, \mathfrak{a})$. By 2.5 Proposition, ${}_I\theta = {}_I\theta_K \mid H^1(\beta, \mathfrak{a})$. So, if ξ_K is a character of $KE^\times$ that agrees with θ_K on U_{KE}^1 and with ξ on $E^\times$, we have $\xi_K \odot_\beta \theta_K \in \mathcal{D}(\mathfrak{a}_K, \beta, \psi^K)$ and

$$\xi_K \odot_\beta \theta_K \mid \mathrm{I}(\beta, \mathfrak{a}) = \xi \odot_\beta \theta. \tag{3.5.1}$$

Chasing through the definitions, we find:

PROPOSITION. *If $\lambda = \xi \odot_\beta \theta$ and $\lambda_K = \xi_K \odot_\beta \theta_K$, then*

$$\iota\lambda_K^F = (\xi \circ \iota^{-1}) \odot_{\iota\beta} \iota\theta_K^F.$$

4. Transfer via completion

We analyze more deeply the embeddings ι of 3.1 using a technique of passing to a limit, as suggested by the diagram in 3.2.

Apart from results in 4.5 concerning the character sets $\mathcal{D}$, everything in this section holds equally when $p = 2$. The notation follows on from the preceding sections but, from 4.3 onwards, it is convenient to choose our "base point" A to be the matrix algebra $\mathrm{M}_n(F)$. That entails no loss of generality.

4.1. We need some new notation.

Let F_∞/F be a maximal unramified extension of F. Thus F_∞ is the union of all finite unramified extensions K/F inside some algebraic closure of F. The discrete valuation υ_F on F extends to a discrete valuation $F_\infty^\times \to \mathbb{Z}$, also denoted υ_F. The associated discrete valuation ring is $\mathfrak{o}_\infty = \bigcup \mathfrak{o}_K$, with K ranging as before. The maximal ideal of $\mathfrak{o}_\infty$ is $\varpi_F\mathfrak{o}_\infty$, where ϖ_F is a prime element of F. The residue field $\mathrm{k}_\infty = \mathfrak{o}_\infty/\mathfrak{p}_\infty$ is an algebraic closure of the residue field k_F of F.

Let $\widetilde{F}$ be the completion of F_∞ with respect to υ_F. Thus υ_F extends to a discrete valuation on $\widetilde{F}$. The associated discrete valuation ring is the closure of $\mathfrak{o}_\infty$ in $\widetilde{F}$: we denote it by $\tilde{\mathfrak{o}}$. This has maximal ideal $\varpi_F\tilde{\mathfrak{o}}$. The residue field $\widetilde{\mathrm{k}} = \tilde{\mathfrak{o}}/\varpi_F\tilde{\mathfrak{o}}$ is equal to k_∞. The group $\widetilde{F}^\times$ is generated by ϖ_F, the group $\tilde{\boldsymbol{\mu}}$ of roots of unity in F_∞ of order relatively prime to p, and the principal unit group $1+\varpi_F\tilde{\mathfrak{o}}$.

Let $\Omega = \mathrm{Gal}(F_\infty/F)$. Thus Ω is procyclic and canonically isomorphic to $\mathrm{Gal}(\widetilde{\mathrm{k}}/\mathrm{k}_F)$. It is topologically generated by the arithmetic Frobenius σ_F, that acts on $\tilde{\boldsymbol{\mu}}$ as $\zeta \mapsto \zeta^q$, where $q = |\mathrm{k}_F|$. Every element of Ω extends uniquely to a continuous F-automorphism of $\widetilde{F}$, and Ω is so identified with the group of continuous F-automorphisms of $\widetilde{F}$. If K/F is a finite unramified extension and $\Omega_K = \mathrm{Gal}(F_\infty/K)$, the set of Ω_K-fixed points in $\widetilde{F}$ is again K.

4.2. We return to the situation of section 2. Thus $n = p^r$, for an integer $r \geqslant 1$, and A is a central simple F-algebra of dimension n^2. Let $\mathfrak{a}$ be a minimal hereditary $\mathfrak{o}_F$-order in A. Let $\beta \in \mathcal{S}^{\mathrm{wr}}(\mathfrak{a})$ and set

$$\mathrm{I}_\infty(\beta,\mathfrak{a}) = \bigcup_{K/F} \mathrm{I}(\beta,\mathfrak{a}_K).$$

Here K/F ranges over the finite sub-extensions of F_∞/F and the union is taken in $\bigcup_K A_K = A \otimes F_\infty$. Let $\tilde{\mathrm{I}}(\beta,\mathfrak{a})$ be the closure of $\mathrm{I}_\infty(\beta,\mathfrak{a})$ in the topological group $G_{\widetilde{F}} = \big(A \otimes_F \widetilde{F}\big)^\times$.

Proposition. *Let $\beta \in \mathcal{S}^{\mathrm{wr}}(\mathfrak{a})$ and define*

$$\widetilde{\mathfrak{J}}^1(\beta,\mathfrak{a}) = \mathfrak{J}^1(\beta,\mathfrak{a}) \otimes_{\mathfrak{o}_F} \tilde{\mathfrak{o}}, \quad \tilde{I}^1(\beta,\mathfrak{a}) = 1 + \widetilde{\mathfrak{J}}^1(\beta,\mathfrak{a}).$$

We then have $\tilde{\mathrm{I}}(\beta,\mathfrak{a}) = \widetilde{F}[\beta]^\times \tilde{I}^1(\beta,\mathfrak{a})$. If K/F is a finite sub-extension of $\widetilde{F}/F$, then

$$I^1(\beta,\mathfrak{a}_K) = \tilde{I}^1(\beta,\mathfrak{a}) \cap G_K,$$
$$\mathrm{I}(\beta,\mathfrak{a}_K) = \tilde{\mathrm{I}}(\beta,\mathfrak{a}) \cap G_K.$$

Proof. The proof is immediate. □

Definition. Choose, once for all, a character $\tilde{\psi}$ of $\widetilde{F}$, of level one, such that $\tilde{\psi}\,|\,F = \psi^F$. If K/F is a finite extension contained in $\widetilde{F}$, set $\psi^K = \tilde{\psi}\,|\,K$.

For $\beta \in \mathcal{S}^{\mathrm{wr}}(\mathfrak{a})$, define

$$\widetilde{\mathcal{D}}(\mathfrak{a},\beta,\tilde{\psi}) = \varprojlim_{K/F} \mathcal{D}(\mathfrak{a}_K,\beta,\psi^K), \tag{4.2.1}$$

the limit being taken with respect to the canonical restriction maps. The elements of $\widetilde{\mathcal{D}}(\mathfrak{a},\beta,\tilde{\psi})$ are characters of the group $\mathrm{I}_\infty(\beta,\mathfrak{a})$. Each such character extends uniquely to a continuous character of $\tilde{\mathrm{I}}(\beta,\mathfrak{a})$, so we regard $\widetilde{\mathcal{D}}(\mathfrak{a},\beta,\tilde{\psi})$ as a set of characters of the group $\tilde{\mathrm{I}}(\beta,\mathfrak{a})$.

Let K/F be a finite extension inside $\widetilde{F}$. By 2.2 Proposition and the definition (4.2.1), the restriction map

$$\begin{aligned}\widetilde{\mathcal{D}}(\mathfrak{a},\beta,\tilde{\psi}) &\longrightarrow \mathcal{D}(\mathfrak{a}_K,\beta,\psi^K),\\ \tilde{\lambda} &\longmapsto \tilde{\lambda}^K = \tilde{\lambda}\,\big|\,\mathrm{I}(\beta,\mathfrak{a}_K),\end{aligned} \tag{4.2.2}$$

is surjective.

4.3. From now on, the following notation will be standard.

NOTATION. Set $A = \mathrm{M}_n(F)$ and fix a prime element ϖ_F of F. Let $\mathfrak{a}$ be the standard minimal hereditary $\mathfrak{o}_F$-order in A, consisting of all $a \in \mathrm{M}_n(\mathfrak{o}_F)$ that are upper triangular when reduced modulo $\mathfrak{p}_F$. The order $\mathfrak{a}$ has a standard prime element $\varPi$, such that $\varPi^n = \varpi_F I_n$. Specifically, all entries x_{ij} of the matrix $\varPi$ are zero except $x_{i,i+1} = 1$, $1 \leqslant i \leqslant n-1$, and $x_{n1} = \varpi_F$. The Jacobson radical of $\mathfrak{a}$ is then $\mathfrak{p} = \varPi\mathfrak{a} = \mathfrak{a}\varPi$.

Set $\widetilde{A} = A \otimes_F \widetilde{F}$ and $\tilde{\mathfrak{a}} = \mathfrak{a} \otimes_{\mathfrak{o}_F} \tilde{\mathfrak{o}}$. The group $\varOmega$ acts on $\widetilde{A}$ via the first tensor factor, the F-algebra $\widetilde{A}^{\varOmega}$ of $\varOmega$-fixed points being A. Let τ be a topological generator of the pro-cyclic group $\varOmega$. Thus $\tau = \sigma_F^z$, for some $z \in \widehat{\mathbb{Z}}^\times$, where $\widehat{\mathbb{Z}}$ is the profinite completion of $\mathbb{Z}$ and σ_F is the arithmetic Frobenius.

PROPOSITION. *Let m be positive divisor of n, say $n = md$. The F-algebra $B = \widetilde{A}^{\tau\varPi^m}$ of $\mathrm{Ad}\,\tau\varPi^m$-fixed points in $\widetilde{A}$ is a central simple F-algebra of dimension n^2 and Hasse invariant*

$$\mathrm{inv}_F\, B = -d^{-1}z + \widehat{\mathbb{Z}} \in \mathbb{Q}/\mathbb{Z}.$$

In particular, $B \cong \mathrm{M}_m(D)$, for a central F-division algebra D of dimension d^2.

PROOF. Let Δ be the algebra of diagonal matrices in A and $\widetilde{\Delta}$ that in $\widetilde{A}$. The vector space A is then the direct sum of the spaces $\Delta\varPi^i$, $0 \leqslant i < n$, and likewise for $\widetilde{A}$. In $\widetilde{A}$, each of the spaces $\widetilde{\Delta}\varPi^i$ is stable under both $\mathrm{Ad}\,\tau$ and $\mathrm{Ad}\,\varPi$. An elementary argument shows that

$$\dim_F \big(\widetilde{\Delta}\varPi^i\big)^{\tau\varPi^m} = n, \quad 0 \leqslant i < n.$$

It follows that B has F-dimension n^2 and that the canonical map $\widetilde{F} \otimes_F B \to \widetilde{A}$ is an isomorphism of $\widetilde{F}$-algebras. Consequently, B is a central simple F-algebra of dimension n^2.

We deal first with the special case $m = 1$. The F-algebra $L = \widetilde{\Delta}^{\tau\varPi}$ is then an unramified field extension of F, of degree n. The automorphism $\mathrm{Ad}\,\varPi$ stabilizes L, where it acts as the Galois automorphism induced by $\tau^{-1} = \sigma_F^{-z}$. That is, $\sigma_F^{-z}\,\big|\,L = \sigma_F^{-z_0}\,\big|\,L$, for an integer z_0 uniquely determined modulo n. In other words, $z_0 + n\widehat{\mathbb{Z}} = z + n\widehat{\mathbb{Z}}$. The algebra B is thus the classical cyclic division algebra of Hasse invariant $-z_0/n$ (mod $\mathbb{Z}$): see the Appendix to section 1 in [**23**].

In the general case, let $e_i \in \Delta$ be the diagonal idempotent matrix with 1 in the i-th place. In particular, e_i is indecomposable and $\mathbf{e} = \{e_i : 1 \leqslant i \leqslant n\}$ is a complete set of orthogonal, indecomposable idempotents in A or in $\widetilde{A}$. Viewing the e_i as indexed by the elements of $\mathbb{Z}/n\mathbb{Z}$, the automorphism $\mathrm{Ad}\,\tau$ fixes each e_i, while $\mathrm{Ad}\,\varPi$ maps e_i to e_{i-1}.

Since $\varPi^n = \varpi_F$, each orbit of $\mathrm{Ad}\,\varPi^m$ on the set $\mathbf{e}$ has d elements, where $n = md$, and there are m distinct orbits. For each such orbit O, let e_O be the sum of its elements. Thus $e_O \in B = \widetilde{A}^{\tau\varPi^m}$ and e_O is an idempotent in B.

The F-algebra $e_O B e_O$ has e_O as unit element. Moreover, $\widetilde{F}\otimes e_O B e_O = e_O \widetilde{A} e_O$ and $e_O \widetilde{A} e_O \cong \mathrm{M}_d(\widetilde{F})$. Consequently, $e_O B e_O$ is a central simple F-algebra of dimension d^2.

On the other hand, the ring $e_O \mathfrak{a} e_O$ is the standard minimal hereditary order in $e_O A e_O = \mathrm{M}_d(F)$ and its standard prime element is $\Pi_O = e_O \Pi^m e_O$. It satisfies $\Pi_O^d = \varpi_F e_O$ and $v_F(\det_{e_o B e_o} \Pi_O) = 1$. Moreover, $e_O B e_O = (e_O \widetilde{A} e_O)^{\tau \Pi_O}$. The special case above shows that $e_O B e_O$ is a central F-division algebra D of dimension d^2 with the required Hasse invariant.

The space $\widetilde{A}$ is the direct sum of its subspaces $e_O \widetilde{A} e_{O'}$, as O and O' range independently over the set $\langle \mathrm{Ad}\, \Pi^m \rangle \backslash \mathbf{e}$. It follows that B is the direct sum of the spaces $e_O B e_{O'}$ and $\widetilde{F} \otimes e_O B e_{O'} = e_O \widetilde{A} e_{O'}$. The right B-module $e_O B$ is an $(e_O B e_O, B)$-bimodule in the obvious way. As left module over $D = e_O B e_O$, it has dimension m, and so the right action of B on $e_O B$ induces a non-trivial algebra homomorphism $B \to \mathrm{M}_m(D)$. This is the desired isomorphism $B \to \mathrm{M}_m(D)$. □

REMARK.

(1) Since $\Pi^n = \varpi_F$ and τ commutes with Π, the operators $\mathrm{Ad}\,(\tau \Pi^m)^n$, $\mathrm{Ad}\, \tau^n$ are the same. The field $K = \widetilde{F}^{\tau^n}$ of τ^n-fixed points in $\widetilde{F}$ is of degree n over F. In the notation of the proposition, $\widetilde{A}^{\tau^n} = A_K = B_K = \mathrm{M}_n(K)$. Moreover, the set $\mathfrak{b}_K = \tilde{\mathfrak{a}} \cap B_K = \mathfrak{a}_K$ is a minimal hereditary $\mathfrak{o}_K$-order in B_K.

(2) We likewise have $\tilde{\mathfrak{a}}^\tau = \mathfrak{a}$. As $B = \widetilde{A}^{\tau \Pi^m}$, so $\mathfrak{b} = \tilde{\mathfrak{a}}^{\tau \Pi^m} = \mathfrak{b}_K \cap B$ is a minimal hereditary $\mathfrak{o}_F$-order in B.

(3) If C is a central simple F-algebra of dimension n^2, we may choose the positive divisor m of n and the topological generator τ of Ω so that $C \cong \widetilde{A}^{\tau \Pi^m}$.

4.4. We write $\widetilde{U} = \tilde{\mathfrak{a}}^\times$ and work in the group $\Omega \widetilde{G} = \Omega \ltimes \widetilde{G}$, where $\widetilde{G} = \widetilde{A}^\times$.

PROPOSITION. *Let $u \in \widetilde{U}$. There exists $y \in \widetilde{U}$ such that $u \tau \Pi^m = y \tau \Pi^m y^{-1}$.*

PROOF. We start with an elementary and familiar observation.

COHOMOLOGICAL LEMMA.

(1) *If $x \in \widetilde{\mathrm{k}}^\times$, there exists $y \in \widetilde{\mathrm{k}}^\times$ such that $x = y^\tau y^{-1}$.*

(2) *If $x \in \widetilde{\mathrm{k}}$, there exists $y \in \widetilde{\mathrm{k}}$ such that $x = y^\tau - y$.*

PROOF. In either part, the element x lies in some finite field k/k_F. In the first part, an elementary argument gives a finite extension ℓ/k such that $\mathrm{N}_{\ell/\mathrm{k}_F}(x) = 1$. If $\Gamma = \mathrm{Gal}(\ell/\mathrm{k}_F)$, the triviality of the Tate cohomology group $\widehat{\mathrm{H}}^{-1}(\Gamma, \ell^\times)$ (Hilbert 90) gives the result. In part (2), we choose ℓ so that $\mathrm{Tr}_{\ell/\mathrm{k}_F}(x) = 0$. That $\widehat{\mathrm{H}}^{-1}(\Gamma, \ell) = 1$ implies the result. □

The group $\tilde{\mathfrak{p}} = \Pi \tilde{\mathfrak{a}}$ is the Jacobson radical of $\tilde{\mathfrak{a}}$. Recall that $\tilde{\boldsymbol{\mu}}$ denotes the group of roots of unity in F_∞ of order prime to p. Reduction modulo $\varpi_F \tilde{\mathfrak{o}}$ induces an isomorphism $\tilde{\boldsymbol{\mu}} \to \widetilde{\mathrm{k}}^\times$. Set $\widetilde{U}^k = 1 + \tilde{\mathfrak{p}}^k$, $k \geqslant 1$. Thus $\widetilde{U}$ decomposes as a semi-direct product

$$\widetilde{U} = \tilde{\boldsymbol{\mu}}^n \ltimes \widetilde{U}^1.$$

The groups $\tilde{\boldsymbol{\mu}}^n = \tilde{\boldsymbol{\mu}} \times \tilde{\boldsymbol{\mu}} \times \cdots \times \tilde{\boldsymbol{\mu}}$ (with n factors) and $\widetilde{U}^1$ are both stable under conjugation by τ and Π^m separately.

LEMMA 1. *Let $u \in \tilde{\boldsymbol{\mu}}^n$. There exists $y \in \tilde{\boldsymbol{\mu}}^n$ such that $u\tau\Pi^m = y\tau\Pi^m y^{-1}$.*

PROOF. The Galois automorphism τ acts on each factor $\tilde{\boldsymbol{\mu}}$ in the natural way while $\operatorname{Ad}\Pi$ permutes them. Write $u = (u_1, \dots, u_n)$ and likewise for y. We have

$$\tau\Pi^m y\Pi^{-m}\tau^{-1} = (\tau(y_{1+m}), \tau(y_{2+m}), \dots, \tau(y_{n+m})),$$

all subscripts being read modulo n. So, if we set $d = n/m$, we have to solve m independent systems of equations in $z_i \in \tilde{\boldsymbol{\mu}}$, of the form

$$v_i = z_i/\tau(z_{i+1}), \quad 1 \leqslant i \leqslant d, \tag{4.4.1}$$

for given $v_i \in \tilde{\boldsymbol{\mu}}$. The Cohomological Lemma gives an element $z_1 \in \tilde{\boldsymbol{\mu}}$ such that

$$z_1/\tau^d(z_1) = \prod_{i=1}^{d-1} \tau^{i-1}(v_i).$$

We solve for z_j, $2 \leqslant j \leqslant d$, directly from (4.4.1). □

Given $u \in \widetilde{U}$, Lemma 1 shows there exists $y_0 \in \widetilde{U}$ such that $y_0^{-1}u\tau\Pi^m y_0 = u_1\tau\Pi^m$, for some $u_1 \in \widetilde{U}^1$. We now set $\widetilde{U}^k = 1+\tilde{\mathfrak{p}}^k$, $k \geqslant 1$, and proceed iteratively.

LEMMA 2. *For an integer $k \geqslant 1$, let $u_k \in \widetilde{U}^k$. There exists $y_k \in \widetilde{U}^k$ such that*

$$y_k^{-1}u_k\tau\Pi^m y_k = u_{k+1}\tau\Pi^m,$$

for some $u_{k+1} \in \widetilde{U}^{k+1}$.

PROOF. Let $\mathfrak{q}$ be the Jacobson radical of $\mathfrak{b}$. We have $\tilde{\mathfrak{a}} = \tilde{\mathfrak{o}}\mathfrak{b}$ and $\tilde{\mathfrak{p}} = \tilde{\mathfrak{o}}\mathfrak{q}$. We use the isomorphism

$$\widetilde{U}^k/\widetilde{U}^{k+1} \cong \tilde{\mathfrak{p}}^k/\tilde{\mathfrak{p}}^{k+1} \cong \mathfrak{q}^k/\mathfrak{q}^{k+1} \otimes_{\Bbbk_F} \widetilde{\Bbbk}.$$

The automorphism $\tau\Pi^m$ acts trivially on the first tensor factor, and as τ on the second. Write $u_k = 1+x$, where $x \in \tilde{\mathfrak{p}}^k$ takes the form

$$x+\tilde{\mathfrak{p}}^{k+1} = \sum_i q_i \otimes \zeta_i,$$

for a finite number of terms with $q_i \in \mathfrak{q}^k/\mathfrak{q}^{k+1}$ and $\zeta_i \in \widetilde{\Bbbk}$. The Cohomological Lemma gives $z_i \in \widetilde{\Bbbk}$ such that $\zeta_i \equiv z_i - \tau z_i\tau^{-1} \pmod{\tilde{\mathfrak{p}}^{k+1}}$. Setting

$$z+\tilde{\mathfrak{p}}^{k+1} = \sum_i q_i \otimes z_i,$$

the element $y_k = 1+z$ has the required property. □

Using the same notation, set $Y_k = y_0y_1\dots y_k$, $k \geqslant 0$. The completeness of $\widetilde{F}$ ensures that the sequence $\{Y_k\}$ converges to the desired element y of $\widetilde{U}$. □

4.5. Take an algebra $B = \widetilde{A}^{\tau\Pi^m}$, as in 4.3 Proposition, along with the minimal hereditary $\mathfrak{o}_F$-order $\mathfrak{b} = \tilde{\mathfrak{a}}^{\tau\Pi^m}$ in B. Set $H = B^\times$.

PROPOSITION. *Let $\beta \in \mathcal{S}^{\mathrm{wr}}(\mathfrak{a})$, and write $E = F[\beta]$.*

1. *There exists $\gamma \in E$, of valuation m, and $y \in \widetilde{U}$ satisfying $y^{-1}E^\times y \subset \mathrm{K}_{\mathfrak{b}} \subset H$ and $\tau\Pi^m = y^{-1}\tau\gamma y$.*
2. *The element y of (1) satisfies*

$$y^{-1}\tilde{\mathrm{I}}(\beta,\mathfrak{a})y = \tilde{\mathrm{I}}(y^{-1}\beta y,\mathfrak{b}) \quad \text{and}$$
$$\tilde{\lambda}\circ \operatorname{Ad} y \in \widetilde{\mathcal{D}}(\mathfrak{b}, y^{-1}\beta y, \tilde{\psi}).$$

(3) *Let* $\iota : E \to B$ *be an* F*-embedding such that* $\iota(E^\times) \subset \mathrm{K}_{\mathfrak{b}}$*, let* K/F *be unramified of finite degree divisible by* n *and extend* ι *to a* K*-isomorphism* $\iota_K : A_K \to B_K$*. Let* $\tilde\lambda \in \widetilde{\mathcal{D}}(\mathfrak{a}, \beta, \tilde\psi)$ *and write*

$$\tilde\lambda^K = \tilde\lambda \,|\, \mathrm{I}(\beta, \mathfrak{a}_K),$$
$$(\iota\tilde\lambda^K)^F = \big(\tilde\lambda^K \circ \iota_K^{-1}\big) \,|\, \mathrm{I}(\iota(\beta), \mathfrak{b}).$$

The character $(\iota\tilde\lambda^K)^F$ *lies in* $\mathcal{D}(\mathfrak{b}, \iota(\beta), \psi^F)$ *and*

$$c\text{-}\mathrm{Ind}_{\mathrm{I}(y^{-1}\beta y,\mathfrak{b})}^{H}\big(\tilde\lambda \circ \mathrm{Ad}\, y \,|\, \mathrm{I}(y^{-1}\beta y, \mathfrak{b})\big) \cong c\text{-}\mathrm{Ind}_{\mathrm{I}(\iota(\beta),\mathfrak{b})}^{H}\big((\iota\tilde\lambda^K)^F\big).$$

PROOF. Let ϖ be a prime element of E and set $\gamma = \varpi^m$. There is a unit $u \in \widetilde{U}$ such that $\gamma = u\Pi^m$. By 4.4 Proposition, there exists $y \in \widetilde{U}$ such that $y^{-1}\tau\gamma y = \tau\Pi^m$. Since $\varpi \in A$, it commutes with τ. Therefore $y^{-1}\varpi y$ commutes with $y^{-1}\tau\gamma y = \tau\Pi^m$. That is, $y^{-1}\varpi y \in H$, whence $y^{-1}Ey \subset B$. On the other hand, the group $y^{-1}E^\times y$ normalizes $\tilde{\mathfrak{a}}$, so it also normalizes $\tilde{\mathfrak{a}} \cap B = \mathfrak{b}$. That is, $y^{-1}E^\times y \subset \mathrm{K}_{\mathfrak{b}}$.

In (2), we have $A_K = B_K$ and $\mathfrak{a}_K = \mathfrak{b}_K$. Abbreviating $E_y = y^{-1}Ey$, we have $E_y^\times \subset \mathrm{K}_{\mathfrak{b}}$, so $KE_y^\times \subset \mathrm{K}_{\mathfrak{b}_K} = \mathrm{K}_{\mathfrak{a}_K}$. By 3.1 Proposition (2), there exists $x \in \mathrm{K}_{\mathfrak{a}_K}$ so that $x^{-1}gx = y^{-1}gy$, for all $g \in KE^\times$. In the language of 3.1, ι is the embedding $\mathrm{Ad}\, y^{-1} : E \to B$ and $\mathrm{Ad}\, x^{-1}$ is the extension ι_K of ι to a K-isomorphism $A_K \to B_K = A_K$. The first assertion in (2) now follows from 3.1 Proposition on passing to the limit over K. The second assertion of (2) follows the same course.

In (3), we use 3.1 Proposition again: ι extends to a K-automorphism ι_K of A_K, stabilizing $\mathfrak{a}_K$. This has the form $\iota_K = \mathrm{Ad}\, y_0$, for some $y_0 \in U_{\mathfrak{a}_K}$. So, by definition, $\pi_H((\iota\tilde\lambda^K)^F)$ is induced by $\tilde\lambda \circ \mathrm{Ad}\, y_0 \,|\, \tilde{\mathrm{I}}(y_0^{-1}\beta y_0, \mathfrak{b}) \cap H$. But, as in 3.1 Proposition, $\mathrm{Ad}\, y = \mathrm{Ad}\, zy_0x$, for some $x \in H$ and $x \in \widetilde{F}[\beta]^\times$. The factor $\mathrm{Ad}\, z$ has no effect on the inducing datum, while $\mathrm{Ad}\, x$ does not change the equivalence class of the induced representation. □

5. The basic character relation

We prove our pivotal result. We use the notation of 4.3 along with a central simple F-algebra B of dimension n^2, realized in the form $B = \widetilde{A}^{\tau\Pi^m}$ as in 4.5. We set $H = B^\times$ and $\mathfrak{b} = \tilde{\mathfrak{a}} \cap B$, again as in 4.5. Recall that, in this scheme, $A = \mathrm{M}_n(F)$ and $G = \mathrm{GL}_n(F)$. We assume throughout that $p \neq 2$.

5.1. We evaluate characters of representations of G and H at a certain class of elements as follows.

DEFINITION. Let H^{wr} be the set of elements h of H satisfying the following conditions:

(1) $v_F(\det_B(h))$ is not divisible by p and
(2) there exists a minimal hereditary $\mathfrak{o}_F$-order $\mathfrak{b}_1$ in B such that $h \in \mathrm{K}_{\mathfrak{b}_1}$.

LEMMA 1. *Let* $h \in H^{\mathrm{wr}}$ *and let* $\mathfrak{b}_1$ *be a minimal hereditary* $\mathfrak{o}_F$*-order in* B *such that* $h \in \mathrm{K}_{\mathfrak{b}_1}$*. The algebra* $L = F[h]$ *is a field, totally ramified of degree* n *over* F*, and* $L^\times \subset \mathrm{K}_{\mathfrak{b}_1}$*.*

PROOF. If ϖ is a prime element of F, we can find integers such that $h' = \varpi^a h^b$ satisfies $v_F(\det_B(h')) = 1$. It follows that $h' \in H^{\mathrm{wr}}$ and $F[h'] = L$. Also, $h' \in \mathrm{K}_{\mathfrak{b}_1}$. In other words, we may assume that $\det_B(h) = 1$. Thus h is a prime element of the hereditary order $\mathfrak{b}_1$, in the sense that $h\mathfrak{b}_1$ is the Jacobson radical of $\mathfrak{b}_1$. The reduced characteristic polynomial of any such element is Eisenstein. It follows that the algebra $L = F[h]$ is a totally ramified field extension of F of degree n and $\mathfrak{o}_L = \mathfrak{o}_F[h]$. Thus $U_L \subset U_{\mathfrak{b}_1}$ and $L^\times \subset \mathrm{K}_{\mathfrak{b}_1}$, as required. □

In particular, the reduced characteristic polynomial $\mathrm{ch}_B(t;h) \in F[t]$ of h is irreducible over F. Let $H^{\mathrm{wr}}_{\mathrm{reg}}$ be the set of $h \in H^{\mathrm{wr}}$ for which $\mathrm{ch}_B(t;h)$ is also separable. Thus an element of H^{wr} is "elliptic quasi-regular", in the sense of [**4**] A.2, while any $h \in H^{\mathrm{wr}}_{\mathrm{reg}}$ is elliptic regular in the customary sense. The sets H^{wr}, $H^{\mathrm{wr}}_{\mathrm{reg}}$ are stable under conjugation by H

The sets G^{wr}, $G^{\mathrm{wr}}_{\mathrm{reg}}$ are defined in the same way.

LEMMA 2. *Let $g \in G^{\mathrm{wr}}$. There is a unique H-conjugacy class of elements $h \in H^{\mathrm{wr}}$ such that $\mathrm{ch}_B(t;h) = \mathrm{ch}_A(t;g)$. Equality of reduced characteristic polynomials induces canonical bijections*

$$\begin{aligned} \mathrm{Ad}\,H\backslash H^{\mathrm{wr}} &\longrightarrow \mathrm{Ad}\,G\backslash G^{\mathrm{wr}}, \\ \mathrm{Ad}\,H\backslash H^{\mathrm{wr}}_{\mathrm{reg}} &\longrightarrow \mathrm{Ad}\,G\backslash G^{\mathrm{wr}}_{\mathrm{reg}}. \end{aligned} \tag{5.1.1}$$

The proof is immediate. The lemma clearly remains valid on replacing $G = \mathrm{GL}_n(F)$ by an inner form. We refer to the bijections (5.1.1), and their inverses, as *association.*

5.2. We use the following additional notation.

NOTATION.

(1) Let $\beta \in \mathcal{S}^{\mathrm{wr}}(\mathfrak{a})$, let $\lambda \in \mathcal{D}(\mathfrak{a},\beta,\psi^F)$ and write $\pi_G = \pi_G(\lambda)$ for the representation of G induced by λ, as in 2.3.

(2) Let $\tilde\lambda \in \widetilde{\mathcal{D}}(\mathfrak{a},\beta,\tilde\psi)$ satisfy $\tilde\lambda \,|\, \mathrm{I}(\beta,\mathfrak{a}) = \lambda$. Let $y \in \widetilde{U}$ satisfy 4.5 Proposition. Let π_H be the representation of H induced by the character $\tilde\lambda \circ \mathrm{Ad}\,y \,|\, \mathrm{I}(y^{-1}\beta y,\mathfrak{b})$.

THEOREM. *If $g \in G^{\mathrm{wr}}$ is associate to $h \in H^{\mathrm{wr}}$, then*

$$\mathrm{tr}\,\pi_G(g) = \mathrm{tr}\,\pi_H(h). \tag{5.2.1}$$

The proof occupies the rest of the section.

5.3. We work first with the group H. With notation as in 5.2, write

$$\begin{aligned} \tilde\kappa &= \tilde\lambda \circ \mathrm{Ad}\,y \in \widetilde{\mathcal{D}}(\mathfrak{b}, y^{-1}\beta y, \tilde\psi), \\ \kappa &= \tilde\kappa \,\big|\, \mathrm{I}(y^{-1}\beta y,\mathfrak{b}). \end{aligned}$$

Thus $\kappa \in \mathcal{D}(\mathfrak{b}, y^{-1}\beta y, \psi^F)$ and π_H is induced by κ. Abbreviate $\mathrm{I}(y^{-1}\beta y,\mathfrak{b}) = I_H$ and $\widetilde{\mathrm{I}}(y^{-1}\beta y,\mathfrak{b}) = \tilde I_H$.

Let $h \in H^{\mathrm{wr}}$. The Mackey induction formula gives

$$\mathrm{tr}\,\pi_H(h) = \sum_{\substack{x\in H/I_H,\\ x^{-1}hx\in I_H}} \kappa(x^{-1}hx). \tag{5.3.1}$$

The condition $x^{-1}hx \in I_H$ is equivalent to $hxI_H = xI_H$, that is, to xI_H being a fixed point for the natural left translation action of h on H/I_H. We may therefore re-write this character expansion in the form

$$\mathrm{tr}\,\pi_H(h) = \sum_{x\in(H/I_H)^h} \kappa(x^{-1}hx). \tag{5.3.2}$$

REMARK. Since h is elliptic quasi-regular, the argument of [**6**] 1.2 Lemma applies to show that the expansion (5.3.2) has only finitely many terms. Consequently, there are no convergence issues to be considered.

The natural conjugation action of $\tau\Pi^m$ on $\widetilde{G}$ stabilizes $\tilde{I}_H$, so $\tau\Pi^m$ acts on the coset space $\widetilde{G}/\tilde{I}_H$ by conjugation. We write $\big(\widetilde{G}/\tilde{I}_H\big)^{\tau\Pi^m}$ for the set of fixed points. On the other hand, $\tilde{I}_H^{\tau\Pi^m} = \tilde{I}_H \cap H = I_H$.

LEMMA. *The canonical map* $H/I_H \to \big(\widetilde{G}/\tilde{I}_H\big)^{\tau\Pi^m}$ *is a bijection.*

PROOF. The map in question is surely injective. For $x \in \widetilde{G}$, the coset $x\tilde{I}_H$ is $\tau\Pi^m$-fixed if and only if $\tau\Pi^m x = xj\tau\Pi^m$, for some $j \in \tilde{I}_H$. Any such element j must lie in $\tilde{I}_H^0 = \widetilde{U} \cap \tilde{I}_H$. There exists $k \in \tilde{I}_H^0$ such that $j\tau\Pi^m = k\tau\Pi^m k^{-1}$: this is proved is the same way as 4.4 Proposition but is easier since $\tilde{I}_H^0 = \tilde{\boldsymbol{\mu}} \times \tilde{I}_H^1$. For this element k, we have

$$\tau\Pi^m x = xk\tau\Pi^m k^{-1}, \quad \text{or} \quad \tau\Pi^m xk = xk\tau\Pi^m,$$

giving $xk \in H$, as desired. □

The obvious inclusion of $\widetilde{G}$ in $\langle\tau\Pi^m\rangle \ltimes \widetilde{G}$ induces a bijection

$$\widetilde{G}/\tilde{I}_H = \langle\tau\Pi^m\rangle \ltimes \widetilde{G}/\langle\tau\Pi^m\rangle \ltimes \tilde{I}_H.$$

We use it to extend the translation action of $\widetilde{G}$ on $\widetilde{G}/\tilde{I}_H$ to one of $\langle\tau\Pi^m\rangle \ltimes \widetilde{G}$. The set of $\tau\Pi^m$-fixed points in $\widetilde{G}/\tilde{I}_H$ for this action is H/I_H (by the lemma), so we may re-write (5.3.2) as

$$\operatorname{tr}\pi_H(h) = \sum_{x\in(\widetilde{G}/\tilde{I}_H)^{\langle h,\tau\Pi^m\rangle}} \tilde{\kappa}(x^{-1}hx), \quad h \in H^{\mathrm{wr}}. \tag{5.3.3}$$

5.4. We apply the same argument to the representation π_G. Abbreviating $\tilde{I} = \tilde{\mathrm{I}}(\beta, \mathfrak{a})$, we find

$$\operatorname{tr}\pi_G(g) = \sum_{x\in(\widetilde{G}/\tilde{I})^{\langle g,\tau\rangle}} \tilde{\lambda}(x^{-1}gx), \quad g \in G^{\mathrm{wr}}. \tag{5.4.1}$$

When evaluating this finite sum, there is no loss entailed in assuming $g \in \mathrm{K}_{\mathfrak{a}}$.

LEMMA 1. *Let* $g \in \mathrm{K}_{\mathfrak{a}} \cap G^{\mathrm{wr}}$ *and let* $g_0 \in F[g]$ *have valuation* $v_{F[g]}(g_0) = m$. *There exists* $t \in \widetilde{U}$ *such that* $t^{-1}F[g]^\times t \subset \mathrm{K}_{\mathfrak{b}}$ *and* $t^{-1}g_0\tau t = \tau\Pi^m$.

The proof is identical to that of 4.5 Proposition (1), so we say no more of it. The element $h = t^{-1}gt$ lies in $\mathrm{K}_{\mathfrak{b}} \cap H^{\mathrm{wr}}$ and is associate to g. We therefore compare (5.4.1) with (5.3.2) evaluated at this element h.

We relate the index sets $(\widetilde{G}/\tilde{I})^{\langle\tau,g\rangle}$, $(\widetilde{G}/\tilde{I}_H)^{\langle\tau\Pi^m,h\rangle}$. To do this, we view $\Omega \ltimes \widetilde{G}$ as acting, by left translation, on $\widetilde{G}/\tilde{I}$ and $\langle\tau\Pi^m\rangle \ltimes \widetilde{G}$ likewise on $\widetilde{G}/\tilde{I}_H$.

LEMMA 2. *The map*

$$\begin{aligned} \Phi : \widetilde{G}/\tilde{I} &\longrightarrow \widetilde{G}/\tilde{I}_H, \\ x\tilde{I} &\longmapsto t^{-1}xy\tilde{I}_H, \end{aligned}$$

induces a bijection

$$\big(\widetilde{G}/\tilde{I}\big)^{\langle\tau,g\rangle} \xrightarrow{\ \approx\ } \big(\widetilde{G}/\tilde{I}_H\big)^{\langle\tau\Pi^m,h\rangle}.$$

PROOF. The defining properties of $y \in \widetilde{U}$ (*cf.* 5.2 Notation) are

$$y^{-1}E^\times y \subset \mathrm{K}_{\mathfrak{b}} \quad \text{and} \quad \tau\Pi^m = y^{-1}\tau\gamma y,$$

for a certain element γ of E as in 4.5 Proposition. We also have $y^{-1}E^\times y \subset \mathrm{I}(y^{-1}\beta y, \mathfrak{b}) = I_H \subset \tilde{I}_H$.

Likewise, $t \in \widetilde{U}$ satisfies

$$t^{-1}F[g]^\times t \subset \mathrm{K}_{\mathfrak{b}} \quad \text{and} \quad t^{-1}g_0\tau t = \tau\Pi^m.$$

We have set $h = t^{-1}gt$.

Immediately, the map Φ is a bijection $\widetilde{G}/\tilde{I} \to \widetilde{G}/\tilde{I}_H$. Let $x \in \widetilde{G}$ and suppose that $x\tilde{I}$ is fixed by τ and g. By the obvious analogue of 5.3 Lemma, we may assume $x \in G$. Since $gxI = xI$, the element x conjugates g into I. The algebra $F[g]$ is a field and g in minimal over F, in the sense of [**9**] (1.4.14), whence it follows readily that x conjugates $F[g]^\times$ into I. Thus

$$h\Phi(x\tilde{I}) = ht^{-1}xy\tilde{I}_H = t^{-1}gxy\tilde{I}_H = \Phi(gx\tilde{I}) = \Phi(x\tilde{I}),$$

as desired.

Now consider

$$\begin{aligned}\tau\Pi^m\Phi(x\tilde{I}) &= \tau\Pi^m t^{-1}xy\tilde{I}_H = t^{-1}g_0\tau xy\tilde{I}_H \\ &= t^{-1}g_0x\tau y\tilde{I}_H = t^{-1}g_0x\gamma^{-1}y\tilde{I}_H\tau\Pi^m.\end{aligned}$$

Since $\gamma \in E$, we have $y^{-1}\gamma^{-1} \in I_H$, whence

$$\tau\Pi^m\Phi(x\tilde{I}) = t^{-1}g_0xy\tilde{I}_H\tau\Pi^m = \Phi(g_0x\tilde{I})\tau\Pi^m = \Phi(x\tilde{I})\tau\Pi^m.$$

Thus $\Phi(x\tilde{I})$ is fixed by $\tau\Pi^m$ and h, as required. The argument is reversible and the lemma is proven. □

5.5. We prove 5.2 Theorem. Let $xI \in (G/I)^g$. The bijection Φ of 5.4 Lemma 2 gives a coset $x'I_H = t^{-1}xyj(x)I_H \in (H/I_H)^h$, for some $j(x) \in \tilde{I}_H$ uniquely determined modulo I_H. The contribution to (5.3.2) from the coset $x'I_H$ is

$$\begin{aligned}\kappa(x'^{-1}hx') &= \tilde{\kappa}(x'^{-1}hx') \\ &= \tilde{\kappa}(j(x)^{-1}y^{-1}x^{-1}tht^{-1}xyj(x)).\end{aligned}$$

Since $\tilde{\kappa}$ is a character of $\tilde{I}_H$, it is invariant under conjugation by $j(x)$. Recalling that $\tilde{\kappa} = \tilde{\lambda} \circ \mathrm{Ad}\, y$, this expression reduces to

$$\kappa(x'^{-1}hx') = \tilde{\lambda}(x^{-1}tht^{-1}x) = \lambda(x^{-1}gx).$$

Lemma 2 now implies $\operatorname{tr}\pi_G(g) = \operatorname{tr}\pi_H(h)$, as required. □

6. Consequences

We derive from 5.2 Theorem the main results of the paper.

6.1. Let G, H be inner forms of $\mathrm{GL}_n(F)$, and let

$$T_G^H : \mathcal{A}^{\text{m-wr}}(G) \longrightarrow \mathcal{A}^{\text{m-wr}}(H)$$

denote the bijection induced by the Jacquet-Langlands correspondence.

THEOREM. *Let G and H be inner forms of $\mathrm{GL}_n(F)$. Let $\pi \in \mathcal{A}^{\text{m-wr}}(G)$ and let $\pi' \in \mathcal{A}^{\text{m-wr}}(H)$ be a parametric transfer of π.*

(1) *The representations π, π' are related by*

$$\pi' = T_G^H(\pi).$$

(2) *If $\rho \in \mathcal{A}^{\text{m-wr}}(H)$ satisfies*

$$\operatorname{tr}\rho(h) = \operatorname{tr}\pi(g),$$

for all $g \in G_{\text{reg}}^{\text{wr}}$ with associate $h \in H_{\text{reg}}^{\text{wr}}$, then $\rho = T_G^H(\pi) = \pi'$.

Consequently, the equivalence class of a parametric transfer π' of π depends only on that of π, and not on the choices made in the definition in 3.4. The theorem implies the endo-class transfer theorem in this special case.

COROLLARY. *Let $\pi \in \mathcal{A}^{\text{m-wr}}(G)$ and let $\rho = T_G^H(\pi) \in \mathcal{A}^{\text{m-wr}}(H)$. If θ_π, θ_ρ are simple characters contained in π, ρ respectively, then θ_π, θ_ρ are endo-equivalent.*

The proofs are in 6.4, following some preparatory material in 6.3.

6.2. Before proceeding to the proofs, we return to the discussions leading to 3.5 in order to write the Jacquet-Langlands correspondence in explicit form. Re-setting the notation, let A and B be central simple F-algebra of dimension n^2, let $\mathfrak{a}$ and $\mathfrak{b}$ be minimal hereditary $\mathfrak{o}_F$-orders in A and B respectively and write $G = A^\times$, $H = B^\times$.

Let $\beta \in \mathcal{S}^{\text{wr}}(\mathfrak{a})$, write $E = F[\beta]$ and let $\theta \in a\text{-}\mathcal{C}(\mathfrak{a}, \beta, \psi^F)$. Let ξ be a character of $E^\times$ such that $\xi \mid U_E^1 = \theta \mid U_E^1$. Let $\iota : E \to B$ be an F-embedding such that $\iota(E^\times) \subset \mathrm{K}_{\mathfrak{b}}$, extended to an isomorphism $\iota_K : A_K \to B_K$. Define $\iota\theta_K^F \in a\text{-}\mathcal{C}(\mathfrak{b}, \iota(\beta), \psi^F)$ as in (3.3.1).

THEOREM. *Abbreviating $\tau = \iota\theta_K^F \in a\text{-}\mathcal{C}(\mathfrak{b}, \iota(\beta), \psi^F)$, we have*

$$T_G^H\, \pi_G(\xi \odot_\beta \theta) = \pi_H((\xi \circ \iota^{-1}) \odot_{\iota\beta} \tau). \tag{6.2.1}$$

PROOF. The definitions ensure that $\pi_H((\xi\circ\iota^{-1})\odot_{\iota\beta}\tau)$ is a parametric transfer of $\pi_G(\xi \odot_\beta \theta)$ (3.5 Proposition), so the result follows from 6.1 Theorem. □

6.3. In this sub-section, the algebra B will be *a division algebra*. We need some special properties of the characters of irreducible smooth representations of $H = B^\times$.

LEMMA. *Let $\pi \in \mathcal{A}^{\text{m-wr}}(H)$ and let ρ be an irreducible smooth representation of H. The following conditions are equivalent.*

(1) $\operatorname{tr}\pi(g) = \operatorname{tr}\rho(g)$ *for all* $g \in H_{\text{reg}}^{\text{wr}}$.
(2) $\operatorname{tr}\pi(g) = \operatorname{tr}\rho(g)$ *for all* $g \in H^{\text{wr}}$.
(3) $\rho \cong \pi$.

PROOF. The characters $\operatorname{tr}\pi$, $\operatorname{tr}\rho$ are locally constant functions on H. The set $H^{\mathrm{wr}}_{\mathrm{reg}}$ is dense in H^{wr}. Thus (1) is equivalent to (2) and surely (3) implies (1).

Let X be the group of unramified characters χ of $F^\times$ such that $\chi^n = 1$. Let $h \in H^{\mathrm{wr}}$, $\upsilon_F(\det_B(h)) = 1$. We form the function

$$\begin{aligned}\Phi_\pi(g) &= n^{-1}\sum_{\chi\in X}\chi(\det_B(gh^{-1}))\operatorname{tr}\pi(g)\\ &= n^{-1}\sum_{\chi\in X}\chi(\det_B(h^{-1}))\operatorname{tr}\chi\pi(g), \quad g\in H.\end{aligned}$$

Define Φ_ρ similarly. Both functions Φ_π, Φ_ρ are supported in the set of $g \in H$ with $\upsilon_F(\det_B(g)) \equiv 1 \pmod n$. This set is contained in H^{wr} so (2) implies $\Phi_\pi(g) = \Phi_\rho(g)$ for all $g \in H$. The set of characters of irreducible smooth representations of H, viewed as functions on H, is linearly independent. We conclude that $\rho = \chi\pi$, for some $\chi \in X$ with $\chi(\det_B(h)) = 1$. That is, $\chi = 1$ so $\rho \cong \pi$, as required for (3). □

6.4. We prove 6.1 Theorem. Since the Jacquet-Langlands correspondence and parametric transfer are transitive, it is enough to prove the theorem and the corollary under the assumption $G = \mathrm{GL}_n(F)$.

Initially take $H = B^\times$, for a division algebra B. Let $\pi \in \mathcal{A}^{\mathrm{m\text{-}wr}}(G)$ and let $\rho \in \mathcal{A}^{\mathrm{m\text{-}wr}}(H)$ be a parametric transfer of π. Set $\rho' = T^H_G(\pi)$. Let $g \in G^{\mathrm{wr}}_{\mathrm{reg}}$ and let $h \in H^{\mathrm{wr}}_{\mathrm{reg}}$ be associate to g. Since $n = p^r$, $p \neq 2$, we have $\operatorname{tr}\rho'(h) = \operatorname{tr}\pi(g)$ by definition. However, 5.2 Theorem and 4.5 Proposition (3) together give $\operatorname{tr}\pi(g) = \operatorname{tr}\rho(h)$ so 6.3 Lemma implies $\rho = \rho'$. This yields an intermediate conclusion:

LEMMA. *Let $\pi_1, \pi_2 \in \mathcal{A}^{\mathrm{m\text{-}wr}}(G)$. If $\operatorname{tr}\pi_1(g) = \operatorname{tr}\pi_2(g)$ for all $g \in G^{\mathrm{wr}}_{\mathrm{reg}}$, then $\pi_1 = \pi_2$.*

We pass to the case where H is arbitrary. Let $\pi \in \mathcal{A}^{\mathrm{m\text{-}wr}}(G)$, let $\rho \in \mathcal{A}^{\mathrm{m\text{-}wr}}(H)$ be a parametric transfer of π and let π' be the unique element of $\mathcal{A}^{\mathrm{m\text{-}wr}}(G)$ for which $\rho = T^H_G\,\pi'$. Let $g \in G^{\mathrm{wr}}_{\mathrm{reg}}$ and let $h \in H^{\mathrm{wr}}_{\mathrm{reg}}$ be associate to g. By 5.2 Theorem and 4.5 Proposition (3) again,

$$\operatorname{tr}\pi(g) = \operatorname{tr}\rho(h) = \operatorname{tr}\pi'(g).$$

The relation $\operatorname{tr}\pi(g) = \operatorname{tr}\pi'(g)$ holds for all $g \in G^{\mathrm{wr}}_{\mathrm{reg}}$, and the lemma implies $\pi = \pi'$. This completes the proof of the theorem. □

The corollary now follows from 3.3 Proposition. □

REMARK. We could have argued here in terms of elliptic quasi-regular elements $g \in G^{\mathrm{wr}}$. However, elliptic regular elements suffice to give the result and the extra precision can be useful in the context of linear independence of characters.

CORRECTION. On this subject, 3.1 Corollary 3 of [**5**] is wrong for rather trivial reasons. For a counterexample, take $\pi \in \mathcal{A}^{\mathrm{m\text{-}wr}}(G)$ and consider the set of representations $\chi\pi$, as χ ranges over all unramified characters of $F^\times$ such that $\chi^n = 1$. The set of characters $\operatorname{tr}\chi\pi$ is then linearly dependent on both $G^{\mathrm{wr}}_{\mathrm{reg}}$ and G^{wr}. This is essentially the only counterexample. The error has no effect on either [**5**] or this paper.

References

[1] A. I. Badulescu, *Correspondance de Jacquet-Langlands pour les corps locaux de caractéristique non nulle* (French, with English and French summaries), Ann. Sci. École Norm. Sup. (4) **35** (2002), no. 5, 695–747, DOI 10.1016/S0012-9593(02)01106-0. MR1951441

[2] P. Broussous, *Extension du formalisme de Bushnell et Kutzko au cas d'une algèbre à division* (French), Proc. London Math. Soc. (3) **77** (1998), no. 2, 292–326, DOI 10.1112/S0024611598000471. MR1635145

[3] P. Broussous, V. Sécherre, and S. Stevens, *Smooth representations of* $\mathrm{GL}_m(D)$ *V: Endo-classes*, Doc. Math. **17** (2012), 23–77. MR2889743

[4] C. J. Bushnell and G. Henniart, *Local tame lifting for* $\mathrm{GL}(N)$. *I. Simple characters*, Inst. Hautes Études Sci. Publ. Math. **83** (1996), 105–233. MR1423022

[5] C. J. Bushnell and G. Henniart, *Local tame lifting for* $\mathrm{GL}(n)$ *III: explicit base change and Jacquet-Langlands correspondence.* J. reine angew. Math. **508** (2005), 39-100.

[6] C. J. Bushnell and G. Henniart, *The essentially tame Jacquet-Langlands correspondence for inner forms of* $\mathrm{GL}(n)$, Pure Appl. Math. Q. **7** (2011), no. 3, Special Issue: In honor of Jacques Tits, 469–538, DOI 10.4310/PAMQ.2011.v7.n3.a2. MR2848585

[7] C. J. Bushnell and G. Henniart, *Explicit functorial correspondences for level zero representations of p-adic linear groups*, J. Number Theory **131** (2011), no. 2, 309–331, DOI 10.1016/j.jnt.2010.09.003. MR2736858

[8] C. J. Bushnell and G. Henniart, *To an effective local Langlands Correspondence.* Memoirs Amer. Math. Soc. **231** (2014), no. 1087, iv+88.

[9] C. J. Bushnell and P. C. Kutzko, *The admissible dual of* $\mathrm{GL}(N)$ *via compact open subgroups*, Annals of Mathematics Studies, vol. 129, Princeton University Press, Princeton, NJ, 1993. MR1204652

[10] P. Deligne, D. Kazhdan, and M.-F. Vignéras, *Représentations des algèbres centrales simples p-adiques* (French), Representations of reductive groups over a local field, Travaux en Cours, Hermann, Paris, 1984, pp. 33–117. MR771672

[11] A. Dotto, *The inertial Jacquet-Langlands correspondence.* arXiv:1707.00635.

[12] M. Grabitz, *Zur Konstruktion einfacher Charaktere und der Fortsetzungen ihrer Heisenbergdarstellungen für lokale zentral-einfache Algebren.* Thesis, Humboldt University Berlin, 2000.

[13] G. Henniart and R. Herb, *Automorphic induction for* $\mathrm{GL}(n)$ *(over local non-Archimedean fields)*, Duke Math. J. **78** (1995), no. 1, 131–192, DOI 10.1215/S0012-7094-95-07807-7. MR1328755

[14] A. Mínguez and V. Sécherre, *Représentations lisses modulo* ℓ *de* $\mathrm{GL}_m(D)$ (French, with English and French summaries), Duke Math. J. **163** (2014), no. 4, 795–887, DOI 10.1215/00127094-2430025. MR3178433

[15] A. Mínguez and V. Sécherre, *Types modulo* ℓ *pour les formes intérieures de* GL_n *sur un corps local non archimédien* (French, with English summary), Proc. Lond. Math. Soc. (3) **109** (2014), no. 4, 823–891, DOI 10.1112/plms/pdu020. With an appendix by Vincent Sécherre et Shaun Stevens. MR3273486

[16] A. Mínguez and V. Sécherre, *Correspondance de Jacquet-Langlands locale et congruences modulo* ℓ (French, with English summary), Invent. Math. **208** (2017), no. 2, 553–631, DOI 10.1007/s00222-016-0696-y. MR3639599

[17] V. Sécherre, *Représentations lisses de* $\mathrm{GL}(m, D)$. *I. Caractères simples* (French, with English and French summaries), Bull. Soc. Math. France **132** (2004), no. 3, 327–396. MR2081220

[18] V. Sécherre, *Représentations lisses de* $\mathrm{GL}(m, D)$. *II. β-extensions* (French, with English summary), Compos. Math. **141** (2005), no. 6, 1531–1550, DOI 10.1112/S0010437X05001429. MR2188448

[19] V. Sécherre, *Représentations lisses de* $\mathrm{GL}_m(D)$. *III. Types simples* (French, with English and French summaries), Ann. Sci. École Norm. Sup. (4) **38** (2005), no. 6, 951–977, DOI 10.1016/j.ansens.2005.10.003. MR2216835

[20] V. Sécherre and S. Stevens, *Représentations lisses de* $\mathrm{GL}_m(D)$. *IV. Représentations supercuspidales* (French, with English and French summaries), J. Inst. Math. Jussieu **7** (2008), no. 3, 527–574, DOI 10.1017/S1474748008000078. MR2427423

[21] V. Sécherre and S. Stevens, *Smooth representations of GL(m,D), VI : semisimple types.* Int. Math. Res. Not. IMRN **13** (2012), 2994–3039.

[22] V. Sécherre and S. Stevens, *Towards an explicit local Jacquet-Langlands correspondence beyond the cuspidal case.* arXiv:1611.04317v1.

[23] J-P. Serre, *Local class field theory.* Algebraic Number Theory (J.W.S. Cassels and A. Fröhlich, eds.), Academic Press, London, 1967.

[24] A. J. Silberger and E.-W. Zink, *An explicit matching theorem for level zero discrete series of unit groups of* $\mathfrak{p}$*-adic simple algebras*, J. Reine Angew. Math. **585** (2005), 173–235, DOI 10.1515/crll.2005.2005.585.173. MR2164626

Department of Mathematics, King's College London, Strand, London WC2R 2LS, United Kingdom

Email address: colin.bushnell@kcl.ac.uk

Laboratoire de Mathématiques d'Orsay, UnivṖaris-Sud, CNRS, Université Paris-Saclay, 91405 Orsay, France

Email address: Guy.Henniart@math.u-psud.fr

Proceedings of Symposia in Pure Mathematics
Volume **101**, 2019
https://doi.org/10.1090/pspum/101/01791

On the Casselman-Jacquet functor

T.-H. Chen, D. Gaitsgory, and A. Yom Din

Dedicated to J. Bernstein

Abstract. We study the Casselman-Jacquet functor J, viewed as a functor from the (derived) category of $(\mathfrak{g}, K)$-modules to the (derived) category of $(\mathfrak{g}, N^-)$-modules, N^- is the negative maximal unipotent. We give a functorial definition of J as a certain right adjoint functor, and identify it as a composition of two averaging functors $\mathrm{Av}_!^{N^-} \circ \mathrm{Av}_*^N$. We show that it is also isomorphic to the composition $\mathrm{Av}_*^{N^-} \circ \mathrm{Av}_!^N$. Our key tool is the *pseudo-identity* functor that acts on the (derived) category of (twisted) D-modules on an algebraic stack.

Contents

Introduction

0.1. The Casselman-Jacquet functor.

0.1.1. Let G be a real reductive algebraic group, $\mathfrak{g}$ the complexification of $\mathrm{Lie}(G)$, K the complexification of a maximal compact subgroup in $G(\mathbb{R})$, and $\mathfrak{n}, \mathfrak{n}^- \subset \mathfrak{g}$ the complexifications of the Lie algebras of the unipotent radicals of opposite minimal parabolics in G. Let $(\mathfrak{g}, K)$-mod denote the corresponding category of $(\mathfrak{g}, K)$-modules. Recall that a Harish-Chandra module is a $(\mathfrak{g}, K)$-module that is of finite length (equivalently, finitely generated and acted on locally finitely by the center of $U(\mathfrak{g})$).

In his work on representations of real reductive groups, W.A. Casselman introduced a remarkable functor on the category of Harish-Chandra modules: it is defined by the formula

$$\mathcal{M} \mapsto \widehat{J}(\mathcal{M}) := \underset{\underset{k}{\longleftarrow}}{\lim}\, \mathcal{M}/\mathfrak{n}^k \cdot M, \tag{0.1}$$

where $\mathfrak{n}$ is the unipotent radical of a minimal parabolic.

A key property of the functor $\widehat{J}$ is that it is *exact* and *conservative*; this provided a new tool for the study of the category of Harish-Chandra modules, leading to an array of powerful results.

0.1.2. The functor (0.1) has an algebraic cousin, denoted J, and defined as follows.

Pick a cocharacter $\mathbb{G}_m \to G$ that is dominant and regular in the split Cartan, and let $J(\mathcal{M})$ be the subset of $\widehat{J}(\mathcal{M})$ on which $\mathbb{A}^1 = \mathrm{Lie}(\mathbb{G}_m)$ acts locally finitely, i.e., the direct sum of generalized eigenspaces with respect to the generator $t \in \mathbb{A}^1$:

$$J(\mathcal{M}) \simeq \underset{\lambda}{\oplus} J(\mathcal{M})_\lambda := \underset{\lambda}{\oplus} \widehat{J}(\mathcal{M})_\lambda.$$

One shows that the entire $\widehat{J}(\mathcal{M})$ can be recovered as

$$\underset{\lambda}{\Pi} J(\mathcal{M})_\lambda,$$

so the information contained in J is more or less equivalent to that possessed by $\widehat{J}$. In particular, the functor J is also exact.

We will refer to J as the *Casselman-Jacquet* functor.

0.1.3. An important feature of the functor J, and one relevant to this paper, is that it can be extracted from $\widehat{J}$ using the Lie algebra $\mathfrak{n}^-$ (the unipotent radical of the opposite parabolic) rather than the split Cartan.

Namely, an elementary argument shows that $J(\mathcal{M})$ can be identified with the subset of vectors in $\widehat{J}(\mathcal{M})$ on which $\mathfrak{n}^-$ acts locally nilpotently.

Thus, we can think of J as a functor

$$(\mathfrak{g}, K)\text{-mod}_\chi \to (\mathfrak{g}, M_K \cdot N^-)\text{-mod}_\chi,$$

where our notations are as follows:

- $(\mathfrak{g}, K)$-mod denotes the abelian category of $(\mathfrak{g}, K)$-modules (K is the algebraic group corresponding to the maximal compact);
- $(\mathfrak{g}, M_K \cdot N^-)$-mod denotes the abelian category of $(\mathfrak{g}, M_K \cdot N^-)$-modules ($N^-$ and M_K are the algebraic groups corresponding to the (opposite) maximal unipotent and compact part of the Levi, respectively);
- The subscript χ indicates that we are considering categories of modules with a fixed central character χ.

0.1.4. *Our goals.* The primary goals of the present paper are as follows:

–Extend the definitions of the functors $\widehat{J}$ and J from the abelian categories to the corresponding derived categories $\mathfrak{g}\text{-mod}_\chi^K$, $\mathfrak{g}\text{-mod}_\chi^{M_K \cdot N^-}$, etc., (and in particular, explain their functorial meaning);

–Express the functor J as a double-averaging functor, and thus reprove the corresponding result from the paper [**CY**], where it was obtained by interpreting J via nearby cycles using [**ENV**];

–Record a conjecture that states that the functor J is (up to some twist) the *right* adjoint of the functor of *averaging* with respect to K:

$$\text{Av}_*^{K/M_K} : \mathfrak{g}\text{-mod}_\chi^{M_K \cdot N^-} \to \mathfrak{g}\text{-mod}_\chi^K,$$

and explain that this is analogous to Bernstein's "2nd adjointness theorem" for $\mathfrak{p}$-adic groups.

0.1.5. In the course of realizing these goals we will encounter another operation of interest: Drinfeld's pseudo-identity functor on the category of $M_K \cdot N$-equivariant (twisted) D-modules on the flag variety X.

This functor will be used in the proofs of the main results, and as such, may seem to be not more than a trick. However, in the sequel to this paper it will be explained that this functor plays a conceptual role at the categorical level.

0.2. Functorial interpretation of the Casselman-Jacquet functor.

0.2.1. We first give a functorial interpretation of the (derived version of the) functor $\widehat{J}$.

Namely, in Sects. 2.2 and 2.3, we show that (the derived) version of this functor identifies with the composition

$$\mathfrak{g}\text{-mod}_\chi \overset{\text{Av}_*^N}{\longrightarrow} \mathfrak{g}\text{-mod}_\chi^N \overset{(\text{Av}_*^N)^R}{\longrightarrow} \mathfrak{g}\text{-mod}_\chi.$$

Here Av_*^N is the functor of $*$-averaging with respect to N, i.e., the right adjoint to the forgetful functor

$$\textbf{oblv}_N : \mathfrak{g}\text{-mod}_\chi^N \to \mathfrak{g}\text{-mod}_\chi, \tag{0.2}$$

and $(\text{Av}_*^N)^R$ is the (a priori, discontinuous) *right* adjoint of Av_*^N.

In fact, we show this in a rather general situation when instead of $\mathfrak{g}\text{-mod}_\chi$ we consider the category A-mod, where A is an associative algebra, equipped with a *Harish-Chandra structure* with respect to N (see Sect. 1.1.2 for what this means).

0.2.2. Next, we consider the functor J, in the general setting of a category $\mathcal{C}$ equipped with an action of G (for example, for $\mathcal{C} = A$-mod for an associative algebra A, equipped with a Harish-Chandra structure with respect to all of G), see Sect. 1.1.1 for what this means.

We define the (derived version of the) functor J as the composite

$$J := \mathrm{Av}_*^{N^-} \circ (\mathrm{Av}_*^N)^R \circ \mathrm{Av}_*^N : \mathcal{C}^{M_K} \to \mathcal{C}^{M_K \cdot N^-}.$$

Assume that the following property is satisfied (which is the case for $\mathcal{C} = \mathfrak{g}\text{-mod}_\chi$ or $\mathcal{C}$ being the category $\text{D-mod}_\lambda(X)$ of twisted D-modules on the flag variety):

(*) *The "long intertwining functor"*

$$\Upsilon := \mathrm{Av}_*^N \circ \mathbf{oblv}_{N^-} : \mathcal{C}^{M_K \cdot N^-} \to \mathcal{C}^{M_K \cdot N}, \tag{0.3}$$

given by forgetting N^--equivariance and then averaging with respect to N, is an equivalence.

In this case we show (see Proposition 2.1.4 and its variant in Sect. 4.1.3) that we have a canonical isomorphism of functors

$$J \simeq \mathrm{Av}_!^{N^-} \circ \mathrm{Av}_*^N. \tag{0.4}$$

In the above formula, $\mathrm{Av}_!^{N^-}$ is the !-averaging functor with respect to N^-, i.e., the *left* adjoint to (0.2) (with N replaced by N^-).

The isomorphism (0.4) had been initially obtained in [**CY**]; we will comment on that in Sect. 0.2.4 below.

0.2.3. Finally, in the particular case of $\mathcal{C} = \mathfrak{g}\text{-mod}_\chi$ we will show (see Theorem 2.4.2 and its variant in Sect. 4.1.3) that J is canonically isomorphic to its Verdier dual functor

$$J \simeq \mathrm{Av}_*^{N^-} \circ \mathrm{Av}_!^N, \tag{0.5}$$

when applied to objects in $\mathfrak{g}\text{-mod}_\chi^{M_K}$ whose cohomologies are finitely generated over $\mathfrak{n}$ (or are direct limits of such).

We will deduce (0.5) from a similar statement for $\mathcal{C} = \text{D-mod}_\lambda(X)$ (see Theorem 2.4.3 and its variant in Sect. 4.1.3), where the isomorphism in question holds for objects on $\text{D-mod}_\lambda(X)^{M_K}$ that are ULA with respect to the projection $X \to N\backslash X$ (see Sect. 3.2.1 for what this means).

The isomorphism (0.5) implies that the functor J is t-exact when applied to $\mathfrak{g}\text{-mod}_\chi^K$. In particular, it shows that the procedure in Sect. 0.1.3 *does not omit higher cohomologies* (in principle, the functor of taking $\mathfrak{n}^-$-locally nilpotent vectors should be derived).

0.2.4. The isomorphism (0.5), applied to objects from $\mathfrak{g}\text{-mod}_\chi^K$, had been obtained in [**CY**] as a combination of the following two results:

–One is the main theorem of [**ENV**] that shows that under the localization equivalence (for this one assumes that χ is regular), the functor J (defined as in Sect. 0.1.3) corresponds to a certain nearby cycles functor

$$\Psi : \text{D-mod}_\lambda(X)^K \to \text{D-mod}_\lambda(X)^{M_K \cdot N^-}.$$

This was done by explicitly analyzing the V-filtration on the corresponding D-module.

–The other is the key result from the paper [**CY**] itself, which establishes an isomorphism

$$\Psi \simeq \mathrm{Av}_*^{N^-} \circ \mathrm{Av}_!^N,$$

again on $\text{D-mod}_\lambda(X)^K$;

The isomorphism (0.4) was then deduced from (0.5) using the Verdier self-duality property of the nearby cycles functor Ψ, which implies that

$$\mathrm{Av}_*^{N^-} \circ \mathrm{Av}_!^N \simeq \Psi \simeq \mathrm{Av}_!^{N^-} \circ \mathrm{Av}_*^N.$$

So, the present paper gives another, in a sense more direct proof of (0.4) and (0.5), which does not appeal to the nearby cycles functor (however, we do *not* to imply that the latter is irrelevant: see Sect. 0.4.3).

0.3. The pseudo-identity functor and the ULA condition.

0.3.1. The pseudo-identity functor is a certain canonical endofunctor of the category of (twisted) D-modules on any algebraic stack, denoted

$$\text{Ps-Id}_{\mathcal{Y}} : \text{D-mod}_\lambda(\mathcal{Y}) \to \text{D-mod}_\lambda(\mathcal{Y}),$$

see Sect. 3.1. Its definition was suggested by V. Drinfeld and was recorded in [**Ga1**].

This functor is *uninteresting* (equals to the identity functor up to a shift) when $\mathcal{Y}$ is a smooth separated scheme, but has some very interesting properties on certain algebraic stacks that appear in geometric representation theory, see, e.g., [**Ga3**].

0.3.2. In this paper we apply this functor to the stack $\mathcal{Y}$ equal to $H\backslash X$, where X a proper scheme acted on by an algebraic group H (in our applications, X will be the flag variety of G and $H = M_K \cdot N$).

We prove the following result (see Theorem 3.2.6):

THEOREM 0.3.3. *Let f denote the projection $X \to H\backslash X$. Then the functors*

$$f_![2\dim(H)] \text{ and } \text{Ps-Id}_{H\backslash X} \circ f_*[2\dim(X)]$$

are canonically isomorphic when evaluated on objects of $\text{D-mod}_\lambda(X)$ *that are ULA with respect to f.*

In other words, this theorem says that the functor $\text{Ps-Id}_{H\backslash X}$ intertwines the !- and *- direct images along f.

0.3.4. From Theorem 0.3.3 we deduce:

COROLLARY 0.3.5. *For X being the flag variety of G, the functor* $\text{Ps-Id}_{M_K\cdot N\backslash X}$ *induces a self-equivalence of* $\text{D-mod}_\lambda(M_K \cdot N\backslash X)$*; moreover, this self-equivalence is canonically isomorphic (up to a shift) to the composition*

$$\text{D-mod}_\lambda(M_K \cdot N\backslash X) \xrightarrow{\Upsilon^{-1}} \text{D-mod}_\lambda(M_K \cdot N^-\backslash X) \xrightarrow{(\Upsilon^-)^{-1}} \text{D-mod}_\lambda(M_K \cdot N\backslash X).$$

where Υ is the long intertwining functor of (0.3), *and Υ^- is the analogous functor where the roles of N and N^- are swapped.*

0.3.6. The application of the functor $\text{Ps-Id}_{M_K\cdot N\backslash X}$ in this paper is the following one:

Combining Corollary 0.3.5 and Theorem 0.3.3 we obtain an isomorphism of functors

$$\mathrm{Av}_*^{N^-} \circ \mathrm{Av}_!^N \text{ and } \mathrm{Av}_!^{N^-} \circ \mathrm{Av}_*^N : \text{D-mod}_\lambda(M_K\backslash X) \rightrightarrows \text{D-mod}_\lambda(M_K \cdot N^-\backslash X), \tag{0.6}$$

on the subcategory objects of $\text{D-mod}_\lambda(M_K\backslash X)$ that are ULA with respect to the projection $M_K\backslash X \to M_K \cdot N\backslash X$ (or are direct limits of such).

0.3.7. Finally, let us comment on the relationship between the isomorphism of functors (0.6) and the isomorphism

$$\text{Av}_*^{N^-} \circ \text{Av}_!^N \text{ and } \text{Av}_!^{N^-} \circ \text{Av}_*^N : \mathfrak{g}\text{-mod}_\chi^{M_K} \rightrightarrows \mathfrak{g}\text{-mod}_\chi^{M_K \cdot N^-}, \tag{0.7}$$

on the subcategory consisting of modules which are finitely generated over $\mathfrak{n}$ (or are direct limits of such).

The point is that the isomorphisms (0.6) and (0.7) are logically equivalent, using the following observation (Proposition 2.4.5):

The functors

$$\text{Loc} : \mathfrak{g}\text{-mod}_\chi \rightleftarrows \text{D-mod}_\lambda(X) : \Gamma(X, -)$$

map the corresponding subcategories to one-another.

0.4. The "2nd adjointness" conjecture.

0.4.1. Let us consider the "principal series" functor in the context of $(\mathfrak{g}, K)$-modules. We stipulate this to be the functor

$$\mathfrak{g}\text{-mod}_\chi^{M_K \cdot N^-} \xrightarrow{\mathbf{oblv}_{N^-}} \mathfrak{g}\text{-mod}_\chi^{M_K} \xrightarrow{\text{Av}_*^{K/M_K}} \mathfrak{g}\text{-mod}_\chi^K.$$

Tautologically, this functor is the *right* adjoint of the functor

$$\mathfrak{g}\text{-mod}_\chi^K \xrightarrow{\mathbf{oblv}_*^{K/M_K}} \mathfrak{g}\text{-mod}_\chi^{M_K} \xrightarrow{\text{Av}_!^{N^-}} \mathfrak{g}\text{-mod}_\chi^{M_K \cdot N^-}.$$

A priori, it is not clear that $\text{Av}_*^{K/M_K} \circ \mathbf{oblv}_{N^-}$ itself should admit a *right* adjoint given by a nice formula. However, based on the analogy with Bernstein's 2nd adjointness theorem (see Sect. 4.4) we propose the following conjecture:

CONJECTURE 0.4.2. *The functor $J \circ \mathbf{oblv}_{K/M_K}$ (up to a shift) provides a right adjoint to the principal series functor $\text{Av}_*^{K/M_K} \circ \mathbf{oblv}_{N^-}$.*

In the sequel to this paper further evidence towards the validity of Conjecture 0.4.2 will be provided, and the logical equivalence between Conjecture 0.4.2 and [**Yo**, Conjectures 9.1.4, 9.1.6] will be explained.

In addition to Conjecture 0.4.2, we make a similar conjecture when the category $\mathfrak{g}\text{-mod}_\chi$ is replaced by $\text{D-mod}_\lambda(X)$.

0.4.3. At the moment, it is not clear to the authors how to write down either the unit or the counit for the conjectural adjunction between $J \circ \mathbf{oblv}_{K/M_K}$ or $\text{Av}_*^{K/M_K} \circ \mathbf{oblv}_{N^-}$, either in the context of $\mathfrak{g}\text{-mod}_\chi$ or in that of $\text{D-mod}_\lambda(X)$.

The following, however, seems very tempting:

In the paper [**BK**] it is explained that the in the context of $\mathfrak{p}$-adic groups, Bernstein's 2nd adjointness can be obtained by analyzing the *wonderful degeneration* of G, i.e., the geometry of the *wonderful compactification* near the stratum of the boundary corresponding to the given parabolic.

Now, as was mentioned above, one of the main results of [**CY**] says that the functor J for $\text{D-mod}_\lambda(X)$ is isomorphic to the nearby cycles functor along the same wonderful degeneration.

So it would be very nice to adapt the ideas of [**BK**] to prove Conjecture 0.4.2. However, so far, we do not know how to carry this out.

0.5. Organization of the paper.

0.5.1. The main body of the paper starts with Sect. 1 where we recall (but also reprove and supply proofs that we could not find in the literature) the following topics:

–The notion of action of an algebraic group on a (DG) category; the associated notions of equivariance and the *- and !-averaging functors;

–The Beilinson-Bernstein localization theory;

–Translation functors;

–The long intertwining functor between N- and N^--equivariant categories (either for $\mathfrak{g}$-modules, or D-modules on the flag variety).

The reader may consider skipping this section on the first pass, and return to it when necessary.

0.5.2. In Sect. 2 we initiate the study of the Casselman-Jacquet functor. However, in order to simplify the exposition, in this section instead of working with a real reductive group (or the corresponding symmetric pair), we work in a completely algebraic situation.

I.e., in this section we take N to be a maximal unipotent subgroup in a reductive group G, and consider the Casselman-Jacquet functor J as a functor

$$\mathfrak{g}\text{-mod}_\chi \to \mathfrak{g}\text{-mod}_\chi^N.$$

(Analogous results in the case of symmetric pairs require very minor modifications, which will be explained in Sect. 4.1.3).

–We define the functor J (in the context of a category $\mathcal{C}$ acted on by G) as the composition

$$\mathrm{Av}_*^{N^-} \circ (\mathrm{Av}_*^N)^R \circ \mathrm{Av}_*^N : \mathcal{C} \to \mathcal{C}^{N^-}.$$

–We show that for $\mathcal{C} = A$-mod (for an associative algebra A equipped with a Harish–Chandra structure with respect to G), the functor

$$(\mathrm{Av}_*^N)^R \circ \mathrm{Av}_*^N : A\text{-mod} \to A\text{-mod}$$

is given by $\mathfrak{n}$-adic completion.

–We show that if the functor

$$\Upsilon := \mathrm{Av}_*^N \circ \mathbf{oblv}_{N^-} : \mathcal{C}^{N^-} \to \mathcal{C}^N$$

is an equivalence, then J identifies canonically with

$$\mathrm{Av}_!^{N^-} \circ \mathrm{Av}_*^N.$$

–We state that for $\mathcal{C} = \text{D-mod}_\lambda(X)$, the functor J is canonically isomorphic to $\mathrm{Av}_*^{N^-} \circ \mathrm{Av}_!^N$, when evaluated on objects that are ULA with respect to $X \to N\backslash X$.

–From here we deduce the corresponding isomorphism for $\mathfrak{g}\text{-mod}_\chi$ (on objects that are finitely generated with respect to $\mathfrak{n}$).

–Finally, we show the equivalence between the ULA and $\mathfrak{n}$-f.g. conditions under the localization functor

$$\mathrm{Loc} : \mathfrak{g}\text{-mod}_\chi \to \text{D-mod}_\lambda(X).$$

0.5.3. In Sect. 3, our ostensible goal is to prove the isomorphism

$$\mathrm{Av}_!^{N^-} \circ \mathrm{Av}_*^N \simeq \mathrm{Av}_*^{N^-} \circ \mathrm{Av}_!^N \tag{0.8}$$

on objects of $\text{D-mod}_\lambda(X)$ that are ULA with respect to $X \to N\backslash X$.

In order to do this we introduce the pseudo-identity functor $\text{Ps-Id}_{\mathcal{Y}}$, which is an endofunctor on the category of twisted D-modules on an algebraic stack $\mathcal{Y}$.

We deduce (0.8) from a key geometric result, Theorem 3.2.6.

0.5.4. In Sect. 4 we adapt the results of the preceding sections to the context of a symmetric pair, and thereby deduce the results announced earlier in the introduction.

Finally, we state our version of the 2nd adjointness conjecture and explain the analogy with the corresponding assertion (which is a theorem of J. Bernstein) in the case of $\mathfrak{p}$-adic groups.

0.6. Conventions and notation.

0.6.1. Throughout the paper we will be working over an algebraically closed field k of characteristic 0, and we let G be a connected reductive group over k. Throughout the paper, X will denote the flag variety of G.

In Sects. 1-3 we let N be the unipotent radical of a Borel subgroup of G, and by N^- the unipotent radical of an opposite Borel.

In Sect. 4 we will change the context, and assume that G is endowed with involution θ; we let $K := G^\theta$. Let P be a minimal parabolic compatible with θ; in particular $P^- := \theta(P)$ is an opposite parabolic. For the duration of Sect. 4, we let N be the unipotent radical of P and N^- the unipotent radical of P^-.

0.6.2. This paper will make a (mild) use of higher algebra, in that we will be working with DG categories rather than with triangulated categories (the reluctant reader can avoid this, see Sect. 0.7). See [**DrGa1**, Sect. 0.6] for a concise summary of the theory of DG categories.

Unless specified otherwise, our DG categories will be assumed *cocomplete*, i.e., contain infinite direct sums. Similarly, unless specified otherwise, functors between DG categories will be assumed *continuous*, i.e., preserving infinite direct sums.

We denote by Vect the DG category of chain complexes of vector spaces.

For a DG category $\mathcal{C}$ and $\mathbf{c}_0, \mathbf{c}_1 \in \mathcal{C}$ we will denote by $\mathcal{H}om_{\mathcal{C}}(\mathbf{c}_0, \mathbf{c}_1) \in \text{Vect}$ the Hom complex between them.

For a DG category $\mathcal{C}$ we will denote by $\mathcal{C}^c$ the full (but not cocomplete) subcategory consisting of compact objects.

0.6.3. If $\mathcal{C}$ is endowed with a t-structure, we will denote by $\mathcal{C}^{\leq 0}$ (resp., $\mathcal{C}^{\geq 0}$) the subcategory of connective (resp., coconnective) objects, and by $\mathcal{C}^\heartsuit = \mathcal{C}^{\leq 0} \cap \mathcal{C}^{\geq 0}$ its heart.

We will say that a functor between DG categories $\mathcal{C}_1$ and $\mathcal{C}_2$, each endowed with a t-structure, is left t-exact (resp., right t-exact, t-exact) if it sends $\mathcal{C}_1^{\geq 0}$ to $\mathcal{C}_2^{\geq 0}$ (resp., $\mathcal{C}_1^{\leq 0}$ to $\mathcal{C}_2^{\leq 0}$, both of the above).

0.6.4. For an associative algebra A we will denote by A-mod the corresponding DG category of A-modules (and *not* the abelian category). The same applies to $\mathfrak{g}$-mod for a Lie algebra $\mathfrak{g}$.

For a smooth scheme Y, equipped with a twisting λ, we let $\text{D-mod}_\lambda(Y)$ denote the DG category of twisted D-modules on Y.

For an algebraic group H, we denote by $\text{Rep}(H)$ the DG category of H-representations.

0.7. How to get rid of DG categories?

0.7.1. Unlike its sequel, in this paper we can make do by working with triangulated categories, rather than derived ones.

In general, the necessity to use DG categories arises for two reasons:

–We perform operations on DG categories (e.g., tensor two DG categories over a monoidal DG category acting on them).

–We take limits/colimits in a given DG category.

Both operations are actually present in this paper, but the general procedures can be replaced by *ad hoc* constructions.

0.7.2. We will be working with the notion of DG category acted on by an algebraic group H; if $\mathcal{C}$ is such a category, we will be considering the corresponding category $\mathcal{C}^H$ of H-equivariant objects, equipped with the forgetful functor $\mathbf{oblv}_H : \mathcal{C}^H \to \mathcal{C}$. The passage

$$\mathcal{C} \rightsquigarrow (\mathcal{C}^H, \mathbf{oblv}_H)$$

cannot be intrinsically defined within the world of triangulated categories, and that is why we need DG categories.

However, in our particular situation, H will be unipotent, and $\mathcal{C}^H$ can be defined as the *full subcategory* of $\mathcal{C}$, consisting of H-invariant objects. This does make sense at the triangulated level, where we regard $\mathcal{C}$ as a triangulated category equipped with the action of the monoidal triangulated category D-mod(H).

0.7.3. In Sect. 4 the DG categories $\mathcal{C}$ that we consider will themselves arise in the form $\mathcal{C} = \mathcal{C}_0^H$ for a *non-unipotent* H. However, this will only occur in the following examples:

(a) $\mathcal{C}_0$ is the category of (twisted) D-modules on a scheme Y acted on by H;

(b) $\mathcal{C}_0$ is the category $\mathfrak{g}\text{-mod}_\chi$, where $\mathfrak{g}$ is a Lie algebra and χ is its central character, and $(\mathfrak{g}, H)$ is a Harish-Chandra pair.

In both these examples, there are several ways to define the corresponding category $\mathcal{C} = \mathcal{C}_0^H$ "by hand".

Note, however, that, typically, in neither of these cases will $\mathcal{C}$ be the derived category of the heart of its natural t-structure.

0.7.4. Finally, the only limits and colimits procedures that we consider will be indexed by filtered sets (in fact, by $\mathbb{N}$), and they will consist of objects inside the heart of a t-structure. So the limit/colimit objects will stay in the heart.

0.8. Acknowledgements. The second and the third authors would like to thank their teacher J. Bernstein for many illuminating discussions related to representations of real reductive groups and Harish-Chandra modules; the current paper is essentially an outcome of these conversations. The third author would like to thank Sam Raskin for very useful conversations on higher categories. The first author would like to thank the Max Planck Institute for Mathematics for support, hospitality, and a nice research environment.

The research of D.G. has been supported by NSF grant DMS-1063470. The research of T.H.C. was partially supported by NSF grant DMS-1702337.

1. Recollections

In this section we recall some facts and constructions pertaining to the notion of action of a group on a DG category, to the Beilinson-Bernstein localization theory, translations functors, and the long intertwining functor.

1.1. Groups acting on categories: a reminder.

1.1.1. In this paper we will extensively use the notion of (strong) action of an algebraic group H on a DG category $\mathcal{C}$; for the definition see [**Ga2**, Sect. 10.2] (in the terminology of *loc.cit.*, these are categories acted on by H_{dR}).

One of the possible definitions is that such a data is equivalent to that of co-action of the co-monoidal DG category D-mod(H) on $\mathcal{C}$, where the co-monoidal structure on D-mod(H) is given by !-pullback with respect to the product operation on H:

$$\mathcal{C} \overset{\text{co-act}}{\longrightarrow} \text{D-mod}(H) \otimes \mathcal{C}. \tag{1.1}$$

We can regard D-mod(H) also as a *monoidal* category, with respect to the operation of *convolution*, i.e., *-pushforward with respect to the product operation on H. If H acts on $\mathcal{C}$, we obtain also a monoidal action of D-mod(H) on $\mathcal{C}$ by the formula

$$\text{D-mod}(H) \otimes \mathcal{C} \overset{\text{Id}\boxtimes\text{co-act}}{\longrightarrow} \text{D-mod}(H) \otimes \text{D-mod}(H) \otimes \mathcal{C} \overset{\overset{!}{\otimes}\boxtimes\text{Id}}{\longrightarrow}$$
$$\text{D-mod}(H) \otimes \mathcal{C} \overset{p_*\otimes\text{Id}}{\longrightarrow} \text{Vect} \otimes \mathcal{C} \simeq \mathcal{C},$$

where p_* denotes the pushforward functor $\text{D-mod}(H) \to \text{D-mod}(\text{pt}) = \text{Vect}$.

We denote the corresponding monoidal operation by

$$\mathcal{F} \in \text{D-mod}(H),\ \mathbf{c} \in \mathcal{C} \mapsto \mathcal{F} \star \mathbf{c}.$$

1.1.2. Here are some examples of groups acting on categories that we will use:

(i) Let H act on a scheme/algebraic stack Y. Then H acts on D-mod(Y).

(i') Suppose that Y is equipped with a twisting λ (see [**GR**, Sect. 6] for what this means) that is H-equivariant (the latter means that the twisting descends to one on the quotient stack $H\backslash Y$). Then H acts on the category $\text{D-mod}_\lambda(Y)$.

(ii) H acts on the category $\mathfrak{h}$-mod of modules over its own Lie algebra.

(ii') Let χ be the character of the center $Z(\mathfrak{h}) = U(\mathfrak{h})^{\text{Ad}_H} \subset Z(U(\mathfrak{h}))$. Then H acts on the category $\mathfrak{h}\text{-mod}_\chi$, the latter being the category of $\mathfrak{h}$-modules on which $Z(\mathfrak{h})$ acts via χ.

(iii) Let A be a (classical) associative algebra, equipped with a Harish-Chandra structure with respect to H. I.e., we are given an action of H on A by automorphisms, and a map of Lie algebras $\phi : \mathfrak{h} \to A$ such that

- ϕ is H-equivariant;
- The adjoint action of $\mathfrak{h}$ on A (coming from ϕ) equals the derivative of the given H-action on A.

Then A-mod is acted on by H. This example contains examples (ii) and (ii') (and also (i) and (i') for Y affine) as particular cases.

An example of this situation is when $A = U(\mathfrak{g})$, where $(\mathfrak{g}, H)$ is a Harish-Chandra pair, or A is the quotient of $U(\mathfrak{g})$ by a central character.

1.1.3. If $\mathcal{C}$ is acted on by H, there is a well-defined category $\mathcal{C}^H$ of H-equivariant objects in $\mathcal{C}$, equipped with a pair of adjoint functors

$$\mathbf{oblv}_H : \mathcal{C}^H \rightleftarrows \mathcal{C} : \text{Av}_*^H.$$

One way to define $\mathcal{C}^H$ is as the totalization of the co-Bar co-simplicial category

$$\mathcal{C} \rightrightarrows \text{D-mod}(H) \otimes \mathcal{C} ...$$

associated with the co-action of D-mod(H) on $\mathcal{C}$. Under this identification, $\mathbf{oblv}_H$ is given by evaluation on 0-simplices.

Equivalently, we can define $\mathcal{C}^H$ as the category of co-modules over the co-monad

$$\mathbf{oblv}_H \circ \mathrm{Av}_*^H := k_H \star -$$

acting on $\mathcal{C}$, where $k_H \in$ D-mod(H) is the *constant sheaf* D-module.

1.1.4. Note that the functor $\mathbf{oblv}_H$ is not necessarily fully faithful. In fact, the composition

$$\mathrm{Av}_*^H \circ \mathbf{oblv}_H : \mathcal{C}^H \to \mathcal{C}^H$$

is given by tensor product with $\mathrm{C}^*_{\mathrm{dR}}(H)$ (de Rham cochains on H), where the unit of the adjunction $\mathrm{Id} \to \mathrm{Av}_*^H \circ \mathbf{oblv}_H$ corresponds to the canonical map $k \to \mathrm{C}^*_{\mathrm{dR}}(H)$.

The above implies, among the rest, that $\mathbf{oblv}_H$ *is* fully faithful if H is unipotent.

1.1.5. The functor $\mathbf{oblv}_H$ does not necessarily admit a left adjoint. Its partially defined left adjoint[1] will be denoted by $\mathrm{Av}_!^H$. Concretely, the means that $\mathrm{Av}_!^H$ is defined on the full subcategory of $\mathcal{C}$ consisting of objects c for which the functor

$$\mathcal{C}^H \to \mathrm{Vect}, \quad c' \mapsto \mathcal{H}om_{\mathcal{C}}(c, \mathbf{oblv}_H(c'))$$

is co-representable.

1.1.6. Given two subgroups $H_1 \subset H_2$ we will denote by

$$\mathbf{oblv}_{H_2/H_1} : \mathcal{C}^{H_2} \rightleftarrows \mathcal{C}^{H_1} : \mathrm{Av}_*^{H_2/H_1}$$

the corresponding adjoint pair of functors.

Similarly, will denote by $\mathrm{Av}_!^{H_2/H_1}$ the partially defined left adjoint to $\mathbf{oblv}_{H_2/H_1}$.

1.1.7. When $\mathcal{C} =$ D-mod$_\lambda(Y)$ we have a canonical identification

$$\mathcal{C}^H = \text{D-mod}_\lambda(H\backslash Y),$$

where Av_*^H is given by the $*$-pushforward functor

$$f_* : \text{D-mod}_\lambda(Y) \to \text{D-mod}_\lambda(H\backslash Y),$$

and hence $\mathbf{oblv}_H$ is given by the $*$-pullback functor f^*. Note that the functor f^* is well-defined on all D-modules (and not just holonomic ones) because the morphism f is smooth.

Let us be in Example (iii) above with $A = U(\mathfrak{g})$, where $(\mathfrak{g}, H)$ is a Harish-Chandra pair. Then the corresponding category $\mathfrak{g}$-mod^H is by definition the (derived) category of $(\mathfrak{g}, H)$-modules. For $\mathfrak{g} = \mathfrak{h}$ we have

$$\mathfrak{g}\text{-mod}^H = \mathrm{Rep}(H),$$

the category of H representations.

1.1.8. Suppose that $\mathcal{C}$ is equipped with a t-structure, so that the co-action functor (1.1) is t-exact, where D-mod(H) is taken with respect to the *left* D-module t-structure.

Then the co-monad

$$\mathbf{c} \mapsto k_H \star \mathbf{c}$$

is left t-exact. This implies that the category $\mathcal{C}^H$ carries a t-structure, uniquely characterized by the property that the forgetful functor $\mathbf{oblv}_H$ is t-exact.

---

[1]For the terminology of partially defined left adjoints etc. see, for example, appendix A of [**DrGa2**].

In this case, the functor Av_*^H, being the right adjoint of a t-exact functor, is left t-exact.

We will need the following technical assertion:

LEMMA 1.1.9. *Assume that the t-structure on* $\mathcal{C}$ *is left-complete, i.e., for an object* $\mathbf{c} \in \mathcal{C}$*, the map*

$$\mathbf{c} \to \lim_n \tau^{\geq -n}(\mathbf{c})$$

is an isomorphism. Then:

(a) *The t-structure on* $\mathcal{C}^H$ *is also left-complete and for* $\mathbf{c} \in \mathbf{C}$*, the natural map*

$$\mathrm{Av}_*^H(\mathbf{c}) \to \lim_n \mathrm{Av}_*^H(\tau^{\geq -n}(\mathbf{c}))$$

is an isomorphism.

(b) *If for an object* $\mathbf{c} \in \mathbf{C}$*, the partially defined functor* $\mathrm{Av}_!^H$ *is defined on every* $\tau^{\geq -n}(\mathbf{c})$*, then it is defined on* $\mathbf{c}$ *itself, and the natural map*

$$\mathrm{Av}_!^H(\mathbf{c}) \to \lim_n \mathrm{Av}_!^H(\tau^{\geq -n}(\mathbf{c}))$$

is an isomorphism.

(b') *If the t-structure on* $\mathcal{C}$ *is compatible with filtered colimits, and for an object* $\mathbf{c} \in \mathbf{C}$*, the partially defined functor* $\mathrm{Av}_!^H$ *is defined on every* $H^n(\mathbf{c})$*, then it is defined on* $\mathbf{c}$ *itself.*

1.2. Localization theory.

1.2.1. Let G be an algebraic group acting on a smooth variety X, equipped with a G-equivariant twisting λ. Let A be an associative algebra equipped with a Harish-Chandra structure with respect to G, and let us be given a map

$$A \to \Gamma(X, \mathrm{D}_\lambda), \tag{1.2}$$

as associative algebras equipped with Harish-Chandra structures with respect to G.

Then the functor

$$\Gamma(X, -) : \text{D-mod}_\lambda(X) \to \mathrm{Vect}$$

naturally factors as

$$\text{D-mod}_\lambda(X) \to \Gamma(X, \mathrm{D}_\lambda)\text{-mod} \to A\text{-mod} \to \mathrm{Vect}\,.$$

By a slight abuse of notation, we will denote the resulting functor

$$\text{D-mod}_\lambda(X) \to A\text{-mod}$$

by the same symbol Γ. It admits a left adjoint, denoted Loc. Both these functors are compatible with the G-actions.

Note that the functor Loc is fully faithful if and only if the map (1.2) is an isomorphism.

1.2.2. The example of this situation of interest for us is, of course, when G is reductive and X is the flag variety of G. In this case G-equivariant twistings on X are in bijection with elements of $\mathfrak{t}^*$ (the dual vector space of the abstract Cartan $\mathfrak{t}$), where we take the ρ-shift into account.

Given $\lambda \in \mathfrak{t}^*$, we take

$$A = U(\mathfrak{g})_\chi := U(\mathfrak{g}) \underset{Z(\mathfrak{g})}{\otimes} k,$$

where $Z(\mathfrak{g}) \to k$ is the homomorphism χ corresponding to λ via the Harish-Chandra map $Z(\mathfrak{g}) \to \mathrm{Sym}(\mathfrak{t})$.

It is a theorem of Kostant that in this case the corresponding map

$$U(\mathfrak{g})_\chi \to \Gamma(X, \mathrm{D}_\lambda)$$

is an isomorphism, i.e., the functor Loc is fully faithful.

The following is the Localization Theorem of [**BB1**] (amplified by [**BB2**]):

THEOREM 1.2.3. *Consider* $\mathrm{D\text{-}mod}_\lambda(X)$ *as equipped with the* left *D-module t-structure.*

(a) *Let λ be such that $\check{\alpha}(\lambda) \neq 0$ for any coroot $\check{\alpha}$ of G (we call such λ "regular"). Then the functors Γ and* Loc *are mutually inverse equivalences.*

(b) *Let λ be such that $\check{\alpha}(\lambda) \notin \mathbb{Z}^{<0}$ for any coroot $\check{\alpha}$ of G (we call such λ "dominant"). Then The functor Γ is t-exact.*

(c) *Let λ be regular and dominant. Then the functors Γ and* Loc *are mutually inverse equivalences, compatible with the t-structures.*

1.2.4. *A warning.* When χ is *irregular*, the algebra $U(\mathfrak{g})_\chi$ has an infinite cohomological dimension. This implies, in particular, that the natural t-structure on $\mathfrak{g}\text{-mod}_\chi$ does *not* descend to one on $\mathfrak{g}\text{-mod}_\chi^c$.

Let $\mathfrak{g}\text{-mod}_\chi^{\mathrm{f.g.}} \subset \mathfrak{g}\text{-mod}_\chi$ be the full subcategory consisting of objects with finitely many non-vanishing cohomology groups, and with each cohomology finitely generated as a $U(\mathfrak{g})$-module. We have

$$\mathfrak{g}\text{-mod}_\chi^c \subset \mathfrak{g}\text{-mod}_\chi^{\mathrm{f.g.}},$$

but this inclusion is *not* an equality (the latter would be equivalent to $U(\mathfrak{g})_\chi$ having a finite cohomological dimension).

The functor $\Gamma(X, -)$ sends compact objects in $\mathrm{D\text{-}mod}_\lambda(X)$ to $\mathfrak{g}\text{-mod}_\chi^{\mathrm{f.g.}}$, but not necessarily to $\mathfrak{g}\text{-mod}_\chi^c$.

A related fact is that in this case, the functor Loc has an unbounded cohomological amplitude on the left (it is right t-exact, being the left adjoint of a left t-exact functor $\Gamma(X, -)$).

1.2.5. In what follows we will need the following observation:

PROPOSITION 1.2.6. *Let $H_1 \subset H_2$ be a pair of subgroups of G, and let $\mathcal{M}$ be an object of $\mathfrak{g}\text{-mod}_\chi^{H_1}$. Assume that the functor $\mathrm{Av}_!^{H_2/H_1}$ is defined on* $\mathrm{Loc}(\mathcal{M})$. *Then $\mathrm{Av}_!^{H_2/H_1}$ is defined on $\mathcal{M}$ and we have*

$$\mathrm{Loc}(\mathrm{Av}_!^{H_2/H_1}(\mathcal{M})) \simeq \mathrm{Av}_!^{H_2/H_1}(\mathrm{Loc}(\mathcal{M})) \text{ and } \mathrm{Av}_!^{H_2/H_1}(\mathcal{M}) \simeq \Gamma \circ \mathrm{Av}_!^{H_2/H_1} \circ \mathrm{Loc}(\mathcal{M}).$$

We will prove the following abstract version of Proposition 1.2.6. Let

$$\begin{array}{ccc} \mathcal{C}'_2 & \xrightarrow{i_2} & \mathcal{C}_2 \\ {\scriptstyle G'}\downarrow & & \downarrow{\scriptstyle G} \\ \mathcal{C}'_1 & \xrightarrow{i_1} & \mathcal{C}_1 \end{array}$$

be a commutative diagram of categories, and assume that the diagram

$$\begin{array}{ccc} \mathcal{C}_2' & \xrightarrow{i_2} & \mathcal{C}_2 \\ {\scriptstyle F'}\uparrow & & \uparrow{\scriptstyle F} \\ \mathcal{C}_1' & \xrightarrow{i_1} & \mathcal{C}_1, \end{array}$$

where F (resp., F') is the left adjoint of G (resp., G'), also commutes.

Let $\mathbf{c}_1 \in \mathcal{C}_1$ be an object, and assume that the partially defined left adjoint $(i_2)^L$ is defined on $F(\mathbf{c}_1)$.

PROPOSITION 1.2.7. *Assume that the functors F and F' are fully faithful. Then the partially defined left adjoint $(i_1)^L$ is defined on $\mathbf{c}_1$ and $F'((i_1)^L(\mathbf{c}_1)) \simeq (i_2)^L(F(\mathbf{c}_1))$.*

PROOF. Denote $\mathbf{c}_2 := F(\mathbf{c}_1)$ and $\mathbf{c}_2' := (i_2)^L(\mathbf{c}_2)$. We claim that the map

$$F' \circ G'(\mathbf{c}_2') \to \mathbf{c}_2' \tag{1.3}$$

is an isomorphism. (If this is this case, then it is easy to see that the object $\mathbf{c}_1' := G(\mathbf{c}_2')$ is the value of $(i_1)^L$ on $\mathbf{c}_1$).

Let $\widetilde{\mathbf{c}}_2'$ denote the cone of (1.3). Then $G'(\widetilde{\mathbf{c}}_2') = 0$. We claim that this implies that $\widetilde{\mathbf{c}}_2' = 0$, which is equivalent to the map $\mathbf{c}_2' \to \widetilde{\mathbf{c}}_2'$ being zero.

Indeed, for *any* $\widetilde{\mathbf{c}}_2' \in \mathcal{C}_2'$, we have

$$\mathcal{H}om_{\mathcal{C}_2'}(\mathbf{c}_2', \widetilde{\mathbf{c}}_2') \simeq \mathcal{H}om_{\mathcal{C}_2}(\mathbf{c}_2, i_2(\widetilde{\mathbf{c}}_2')) \simeq \mathcal{H}om_{\mathcal{C}_1}(\mathbf{c}_1, G \circ i_2(\widetilde{\mathbf{c}}_2')) \simeq \mathcal{H}om_{\mathcal{C}_1}(\mathbf{c}_1, i_1 \circ G'(\widetilde{\mathbf{c}}_2')).$$

Hence, if $G'(\widetilde{\mathbf{c}}_2') = 0$, then $\mathcal{H}om_{\mathcal{C}_2'}(\mathbf{c}_2', \widetilde{\mathbf{c}}_2') = 0$, as desired. □

1.3. Translation functors. In this subsection we take G to be a reductive group and X its flag variety.

1.3.1. Let λ be a dominant weight and let μ be a dominant integral weight. Set $\lambda' = \lambda + \mu$, and let χ and χ' be the corresponding characters of $Z(\mathfrak{g})$.

Let V^μ be the irreducible G-module with highest weight λ.

1.3.2. Let us view $U(\mathfrak{g})_{\chi'} \otimes (V^\mu)^*$ as a $(U(\mathfrak{g}), U(\mathfrak{g})_{\chi'})$-bimodule, where $U(\mathfrak{g})_{\chi'}$ acts on the $U(\mathfrak{g})_{\chi'}$-factor in $U(\mathfrak{g})_{\chi'} \otimes (V^\mu)^*$ by right multiplication and $U(\mathfrak{g})$ acts diagonally, with the action on the $U(\mathfrak{g})_{\chi'}$-factor being that by left multiplication.

It is easy to show that when viewed as a $U(\mathfrak{g})$-module, $U(\mathfrak{g})_{\chi'} \otimes (V^\mu)^*$ splits as a direct sum

$$(U(\mathfrak{g})_{\chi'} \otimes (V^\mu)^*)_\chi \oplus (U(\mathfrak{g})_{\chi'} \otimes (V^\mu)^*)_{\neq\chi},$$

where the set-theoretic support of $(U(\mathfrak{g})_{\chi'} \otimes (V^\mu)^*)_\chi$ over $\mathrm{Spec}(Z(\mathfrak{g}))$ is $\{\chi\}$ and the set-theoretic support of $(U(\mathfrak{g})_{\chi'} \otimes (V^\mu)^*)_{\neq\chi}$ over $\mathrm{Spec}(Z(\mathfrak{g}))$ is finite and disjoint from χ.

This decomposition automatically respects the right $U(\mathfrak{g})_{\chi'}$-action.

The key observation is that the assumptions on λ and λ' imply that $(U(\mathfrak{g})_{\chi'} \otimes (V^\mu)^*)_\chi$ is *scheme-theoretically* supported at $\chi \in \mathrm{Spec}(Z(\mathfrak{g}))$, i.e., $(U(\mathfrak{g})_{\chi'} \otimes (V^\mu)^*)_\chi$ is well-defined as an object of $(\mathfrak{g}\text{-mod}_\chi)^\heartsuit$.

1.3.3. We define the functor

$$T_{\chi'\to\chi} : \mathfrak{g}\text{-mod}_{\chi'} \to \mathfrak{g}\text{-mod}_{\chi}$$

to be given by

$$\mathcal{M} \mapsto (U(\mathfrak{g})_{\chi'} \otimes (V^{\mu})^*)_{\chi} \underset{U(\mathfrak{g})_{\chi'}}{\otimes} \mathcal{M}.$$

This functor is t-exact, since $(U(\mathfrak{g})_{\chi'} \otimes (V^{\mu})^*)_{\chi}$ is projective as a right $U(\mathfrak{g})_{\chi'}$-module (being a direct summand of such). At the level of abelian categories it is described as follows: for $\mathcal{M} \in \mathfrak{g}\text{-mod}_{\chi'}$, the tensor product $\mathcal{M} \otimes (V^{\mu})^*$ splits as a direct sum

$$(\mathcal{M} \otimes (V^{\mu})^*)_{\chi} \oplus (\mathcal{M} \otimes (V^{\mu})^*)_{\neq\chi},$$

where $(\mathcal{M} \otimes (V^{\mu})^*)_{\chi} \in \mathfrak{g}\text{-mod}_{\chi}$ and the set-theoretic support of $(\mathcal{M} \otimes (V^{\mu})^*)_{\neq\chi}$ over $\mathrm{Spec}(Z(\mathfrak{g}))$ is finite and disjoint from χ. Then

$$T_{\chi'\to\chi}(\mathcal{M}) = (\mathcal{M} \otimes (V^{\mu})^*)_{\chi}. \tag{1.4}$$

REMARK 1.3.4. At the level of derived categories, we had to define the functor $T_{\chi'\to\chi}$ using the bimodule $(U(\mathfrak{g})_{\chi'} \otimes (V^{\mu})^*)_{\chi}$, rather than by the formula (1.4), because for an object of $\mathfrak{g}$-mod, to belong to $\mathfrak{g}\text{-mod}_{\chi}$ is not a property, but extra structure.

1.3.5. The basic property of the translation functor $T_{\chi\to\chi'}$ is that it makes the following diagram commute:

$$\begin{array}{ccc} \text{D-mod}_{\lambda'}(X) & \xrightarrow{-\otimes\mathcal{O}(-\mu)} & \text{D-mod}_{\lambda}(X) \\ {\scriptstyle\Gamma}\downarrow & & \downarrow{\scriptstyle\Gamma} \\ \mathfrak{g}\text{-mod}_{\chi'} & \xrightarrow{T_{\chi'\to\chi}} & \mathfrak{g}\text{-mod}_{\chi}, \end{array}$$

where the G-equivariant line bundle $\mathcal{O}(\mu)$ on X is normalized so that it is ample and $\Gamma(X, \mathcal{O}(\mu)) \simeq V^{\mu}$.

1.3.6. We let

$$T_{\chi\to\chi'} : \mathfrak{g}\text{-mod}_{\chi} \to \mathfrak{g}\text{-mod}_{\chi'}$$

be the left adjoint functor of $T_{\chi'\to\chi}$.

Tautologically, it makes the following diagram commute:

$$\begin{array}{ccc} \text{D-mod}_{\lambda}(X) & \xrightarrow{-\otimes\mathcal{O}(\mu)} & \text{D-mod}_{\lambda'}(X) \\ {\scriptstyle\mathrm{Loc}}\uparrow & & \uparrow{\scriptstyle\mathrm{Loc}} \\ \mathfrak{g}\text{-mod}_{\chi} & \xrightarrow{T_{\chi\to\chi'}} & \mathfrak{g}\text{-mod}_{\chi'}. \end{array}$$

Consider the object

$$T_{\chi\to\chi'}(U(\mathfrak{g})_{\chi}) \in \mathfrak{g}\text{-mod}_{\chi'}.$$

The functor $T_{\chi\to\chi'}$ is given by

$$\mathcal{M} \mapsto T_{\chi\to\chi'}(U(\mathfrak{g})_{\chi}) \underset{U(\mathfrak{g})_{\chi}}{\otimes} \mathcal{M},$$

where we regard $T_{\chi\to\chi'}(U(\mathfrak{g})_{\chi})$ as a $(U(\mathfrak{g})_{\chi'}, U(\mathfrak{g})_{\chi})$-bimodule.

For what follows, we will need to describe the above object $T_{\chi\to\chi'}(U(\mathfrak{g})_{\chi})$ more explicitly.

1.3.7. Consider the tensor product

$$U(\mathfrak{g})_\chi \otimes V^\mu$$

as a $(U(\mathfrak{g}), U(\mathfrak{g})_\chi)$-bimodule as in Sect. 1.3.2, and consider the corresponding decomposition

$$(U(\mathfrak{g})_\chi \otimes V^\mu)_{\chi'} \oplus (U(\mathfrak{g})_\chi \otimes V^\mu)_{\neq\chi'}$$

according to set-theoretic support over $\operatorname{Spec}(Z(\mathfrak{g}))$.

It is no longer true that $(U(\mathfrak{g})_\chi \otimes V^\mu)_{\chi'}$ is scheme-theoretically supported at $\chi' \in \operatorname{Spec}(Z(\mathfrak{g}))$. Let $(U(\mathfrak{g})_\chi \otimes V^\mu)'_{\chi'}$ be the maximal quotient of $(U(\mathfrak{g})_\chi \otimes V^\mu)_{\chi'}$ with scheme-theoretic support at χ', i.e., the maximal quotient on which $U(\mathfrak{g})$ acts via $U(\mathfrak{g})_{\chi'}$.

PROPOSITION 1.3.8. *We have a canonical isomorphism of* $(U(\mathfrak{g})_{\chi'}, U(\mathfrak{g})_\chi)$-*bimodules*

$$T_{\chi\to\chi'}(U(\mathfrak{g})_\chi) \simeq (U(\mathfrak{g})_\chi \otimes V^\mu)'_{\chi'};$$

in particular, $T_{\chi\to\chi'}(U(\mathfrak{g})_\chi)$ *lies in* $\mathfrak{g}\text{-mod}^\heartsuit_{\chi'}$.

PROOF. Since $T_{\chi\to\chi'}$ is the left adjoint of a t-exact functor and since $U(\mathfrak{g})_\chi \in \mathfrak{g}\text{-mod}^\heartsuit_\chi$ is projective, we obtain that $T_{\chi\to\chi'}(U(\mathfrak{g})_\chi)$ is a projective object in $\mathfrak{g}\text{-mod}^\heartsuit_{\chi'}$.

The functor

$$\mathcal{M} \mapsto H^0\left(T_{\chi\to\chi'}(U(\mathfrak{g})_\chi) \underset{U(\mathfrak{g})_\chi}{\otimes} \mathcal{M}\right), \quad \mathfrak{g}\text{-mod}^\heartsuit_\chi \to \mathfrak{g}\text{-mod}^\heartsuit_{\chi'}$$

provides a left adjoint to the functor

$$T_{\chi'\to\chi} : \mathfrak{g}\text{-mod}^\heartsuit_{\chi'} \to \mathfrak{g}\text{-mod}^\heartsuit_\chi.$$

However, it is easy to see that the above left adjoint is given by

$$\mathcal{M} \mapsto (\mathcal{M} \otimes V^\mu)'_{\chi'},$$

where:

- $\mathcal{M} \otimes V^\mu \simeq (\mathcal{M} \otimes V^\mu)_{\chi'} \oplus (\mathcal{M} \otimes V^\mu)_{\neq\chi'}$ is the decomposition of $\mathcal{M} \otimes V^\mu$ according to set-theoretic support over $\operatorname{Spec}(Z(\mathfrak{g}))$;
- $(\mathcal{M} \otimes V^\mu)'_{\chi'}$ is the maximal quotient of $(\mathcal{M} \otimes V^\mu)_{\chi'}$ with scheme-theoretic support at $\chi' \in \operatorname{Spec}(Z(\mathfrak{g}))$.

Hence, we obtain

$$T_{\chi\to\chi'}(U(\mathfrak{g})_\chi) \simeq H^0\left(T_{\chi\to\chi'}(U(\mathfrak{g})_\chi)\right) \simeq (U(\mathfrak{g})_\chi \otimes V^\mu)'_{\chi'},$$

as desired. □

From here, and using Noetherianness, we obtain:

COROLLARY 1.3.9. *The object* $T_{\chi\to\chi'}(U(\mathfrak{g})_\chi)$, *viewed as a* $(U(\mathfrak{g}), U(\mathfrak{g})_\chi)$-*bimodule admits a resolution with terms of the form* $U(\mathfrak{g})_\chi \otimes V$, *where* V *is a finite-dimensional representation of* G.

1.4. The long intertwining operator.

1.4.1. In this subsection we take G to be a reductive group and X its flag variety. For a given $\lambda \in \mathfrak{t}^*$, we consider the corresponding categories $\text{D-mod}_\lambda(X)^N$ and $\text{D-mod}_\lambda(X)^{N^-}$.

The goal of this subsection is to supply a (possibly new) proof of the following well-known statement (see also [**CY**, Theorem 5.2] for a similar statement in a more general setting of the Matsuki correspondence):

PROPOSITION 1.4.2. *The partially defined functor* $\text{Av}_!^{N^-}$ *is defined on the essential image of* $\mathbf{oblv}_N$, *and the composition* $\text{Av}_!^{N^-} \circ \mathbf{oblv}_N$ *provides an inverse to the functor*

$$\Upsilon := \text{Av}_*^N \circ \mathbf{oblv}_{N^-}, \quad \text{D-mod}_\lambda(X)^{N^-} \to \text{D-mod}_\lambda(X)^N.$$

As a consequence we will now deduce:

PROPOSITION 1.4.3. *Consider the action of G on the category $\mathfrak{g}\text{-mod}_\chi$ for a given central character χ. Then the partially defined functor* $\text{Av}_!^{N^-}$ *is defined on the essential image of*

$$\mathbf{oblv}_N : \mathfrak{g}\text{-mod}_\chi^N \to \mathfrak{g}\text{-mod}_\chi,$$

and the composition $\text{Av}_!^{N^-} \circ \mathbf{oblv}_N$ *provides an inverse to the functor*

$$\Upsilon := \text{Av}_*^N \circ \mathbf{oblv}_{N^-}, \quad \mathfrak{g}\text{-mod}_\chi^{N^-} \to \mathfrak{g}\text{-mod}_\chi^N.$$

PROOF OF PROPOSITION 1.4.3. Choose a twisting λ that gives rise to χ via the Harish-Chandra homomorphism. The fact that $\text{Av}_!^{N^-}$ is defined on $\mathfrak{g}\text{-mod}_\chi^N$ follows from the corresponding fact for $\text{D-mod}_\lambda(X)^N$ using Proposition 1.2.6. Moreover, for $\mathcal{M} \in \mathfrak{g}\text{-mod}_\chi^N$, we have:

$$\text{Loc} \circ \text{Av}_!^{N^-} \circ \mathbf{oblv}_N(\mathcal{M}) \simeq \text{Av}_!^{N^-} \circ \text{Loc} \circ \mathbf{oblv}_N(\mathcal{M})$$

Further, we have

$$\begin{aligned}(\text{Av}_*^N \circ \mathbf{oblv}_{N^-}) \circ (\text{Av}_!^{N^-} \circ \mathbf{oblv}_N) \simeq \Gamma \circ \text{Loc} \circ (\text{Av}_*^N \circ \mathbf{oblv}_{N^-}) \circ (\text{Av}_!^{N^-} \circ \mathbf{oblv}_N) \simeq \\ \Gamma \circ (\text{Av}_*^N \circ \mathbf{oblv}_{N^-}) \circ (\text{Loc} \circ \text{Av}_!^{N^-} \circ \mathbf{oblv}_N) \simeq \\ \Gamma \circ (\text{Av}_*^N \circ \mathbf{oblv}_{N^-}) \circ (\text{Av}_!^{N^-} \circ \text{Loc} \circ \mathbf{oblv}_N) \simeq \\ \simeq \Gamma \circ (\text{Av}_*^N \circ \mathbf{oblv}_{N^-}) \circ (\text{Av}_!^{N^-} \circ \mathbf{oblv}_N) \circ \text{Loc} \simeq \Gamma \circ \text{Loc} \simeq \text{Id},\end{aligned}$$

where the fifth isomorphism follows from Proposition 1.4.2.

Similarly,

$$\begin{aligned}(\text{Av}_!^{N^-} \circ \mathbf{oblv}_N) \circ (\text{Av}_*^N \circ \mathbf{oblv}_{N^-}) \simeq \Gamma \circ \text{Loc} \circ (\text{Av}_!^{N^-} \circ \mathbf{oblv}_N) \circ (\text{Av}_*^N \circ \mathbf{oblv}_{N^-}) \simeq \\ \simeq \Gamma \circ \text{Av}_!^{N^-} \circ \text{Loc} \circ \mathbf{oblv}_N \circ (\text{Av}_*^N \circ \mathbf{oblv}_{N^-}) \simeq \\ \Gamma \circ (\text{Av}_!^{N^-} \circ \mathbf{oblv}_N) \circ (\text{Av}_*^N \circ \mathbf{oblv}_{N^-}) \circ \text{Loc} \simeq \\ \simeq \Gamma \circ \text{Loc} \simeq \text{Id}.\end{aligned}$$

□

1.4.4. The rest of this subsection is devoted to the proof of Proposition 1.4.2. The key idea is to use the theorem of T. Braden's (see, e.g., [**DrGa2**]).

Choose a regular dominant coweight $\mathbb{G}_m \to T$, and consider the resulting $\mathbb{G}_m$-action on X. Note that the orbits of N (resp., N^-) are $\mathbb{G}_m$-invariant. In particular, every N- or N^--equivariant D-module on X is $\mathbb{G}_m$-monodromic.

Moreover, the disjoint union of N-orbits, to be denoted X^+ (resp., the disjoint union of N^--orbits, to be denoted X^-) is the attracting (resp., repelling) locus for the above action. Denote $X^0 = X^{\mathbb{G}_m}$ and denote by

$$i^+ : X^0 \rightleftarrows X^+ : q^+, \quad p^+ : X^+ \to X$$

and

$$i^- : X^0 \rightleftarrows X^- : q^-, \quad p^- : X^- \to X$$

the corresponding maps.

We will consider the functor of *hyperbolic restriction* from the full subcategory

$$\text{D-mod}_\lambda(X)^{\text{mon}} \subset \text{D-mod}_\lambda(X)$$

consisting of $\mathbb{G}_m$-monodromic objects to $\text{D-mod}_\lambda(X^0)$. Braden's theorem insures that this functor is well-defined:

$$(i^-)^! \circ (p^-)^* \simeq (q^-)_! \circ (p^-)^* =: \mathsf{H}^- \underset{\sim}{\overset{\Phi}{\leftarrow}} \mathsf{H}^+ := (q^+)_* \circ (p^+)^! \simeq (i^+)^* \circ (p^+)^!,$$

where the isomorphisms on the sides are known as the *contraction principle.*

1.4.5. The fact that $\text{Av}_!^{N^-}$ is defined on the essential image of $\mathbf{oblv}_N$ follows from the holonomicity. Here is, however, an alternative argument:

We claim that $\text{Av}_!^{N^-}$ is defined on all of $\text{D-mod}_\lambda(X)^{\text{mon}}$. Indeed, for an object $\mathcal{F} \in \text{D-mod}_\lambda(X)$, the object $\text{Av}_!^{N^-}(\mathcal{F})$ is defined if

$$(p^-)^* \circ \text{Av}_!^{N^-}(\mathcal{F}) \in \text{D-mod}_\lambda(X^-)^{N^-}$$

is defined, and that is if and only if

$$(i^-)^! \circ \mathbf{oblv}_{N^-} \circ (p^-)^* \circ \text{Av}_!^{N^-}(\mathcal{F}) \in \text{D-mod}_\lambda(X^0)$$

is defined.

We rewrite

$$(1.5) \quad (i^-)^! \circ \mathbf{oblv}_{N^-} \circ (p^-)^* \circ \text{Av}_!^{N^-}(\mathcal{F}) \simeq (q^-)_! \circ \mathbf{oblv}_{N^-} \circ (p^-)^* \circ \text{Av}_!^{N^-}(\mathcal{F}) \simeq$$
$$\simeq (q^-)_! \circ \mathbf{oblv}_{N^-} \circ \text{Av}_!^{N^-} \circ (p^-)^*(\mathcal{F}) \simeq (q^-)_! \circ (p^-)^*(\mathcal{F}) \simeq \mathsf{H}^-(\mathcal{F}),$$

where the first isomorphism is given by the contraction principle, and the third isomorphism is due to the fact that the projection q^- is N^--invariant.

1.4.6. From (1.5) we obtain a canonical isomorphism

$$(1.6) \qquad \mathsf{H}^- \circ \mathbf{oblv}_{N^-} \circ \text{Av}_!^{N^-}(\mathcal{F}) \simeq \mathsf{H}^-(\mathcal{F}), \quad \mathcal{F} \in \text{D-mod}_\lambda(X)^{\text{mon}}.$$

Similarly, we have:

$$(i^+)^* \circ \mathbf{oblv}_N \circ (p^+)^! \circ \text{Av}_*^N(\mathcal{F}) \simeq (q^+)_* \circ \mathbf{oblv}_N \circ (p^+)^! \circ \text{Av}_*^N(\mathcal{F}) \simeq$$
$$\simeq (q^+)_* \circ \mathbf{oblv}_N \circ \text{Av}_*^N \circ (p^+)^!(\mathcal{F}) \simeq (q^+)_* \circ (p^+)^!(\mathcal{F}) \simeq \mathsf{H}^+(\mathcal{F}),$$

i.e.,

$$(1.7) \qquad \mathsf{H}^+ \circ \mathbf{oblv}_N \circ \text{Av}_*^N(\mathcal{F}) \simeq \mathsf{H}^+(\mathcal{F}), \quad \mathcal{F} \in \text{D-mod}_\lambda(X)^{\text{mon}}.$$

1.4.7. A priori, the functor $\mathrm{Av}_!^{N^-} \circ \mathbf{oblv}_N$ is the left adjoint of the functor $\mathrm{Av}_*^N \circ \mathbf{oblv}_{N^-}$. Let us prove that the unit of the adjunction

$$\mathcal{F} \to (\mathrm{Av}_*^N \circ \mathbf{oblv}_{N^-}) \circ (\mathrm{Av}_!^{N^-} \circ \mathbf{oblv}_N)(\mathcal{F}) \tag{1.8}$$

is an isomorphism.

To show that (1.8) is an isomorphism, it is enough to show that it induces an isomorphism after applying the functor $\mathsf{H}^+ \circ \mathbf{oblv}_N$. The resulting map is

$$\mathsf{H}^+ \circ \mathbf{oblv}_N(\mathcal{F}) \to (\mathsf{H}^+ \circ \mathbf{oblv}_N) \circ (\mathrm{Av}_*^N \circ \mathbf{oblv}_{N^-}) \circ (\mathrm{Av}_!^{N^-} \circ \mathbf{oblv}_N)(\mathcal{F}) \overset{(1.7)}{\simeq}$$
$$\simeq \mathsf{H}^+ \circ \mathbf{oblv}_{N^-} \circ (\mathrm{Av}_!^{N^-} \circ \mathbf{oblv}_N)(\mathcal{F}) \overset{\Phi}{\to} \mathsf{H}^- \circ \mathbf{oblv}_{N^-} \circ (\mathrm{Av}_!^{N^-} \circ \mathbf{oblv}_N)(\mathcal{F}) \overset{(1.6)}{\simeq}$$
$$\simeq \mathsf{H}^- \circ \mathbf{oblv}_N(\mathcal{F}).$$

By unwinding the definition of the natural transformation $\Phi : \mathsf{H}^+ \to \mathsf{H}^-$ (see [**DrGa2**, Equation (3.5)]), we obtain that the above map is induced by the map

$$\mathsf{H}^+ \circ \mathbf{oblv}_N(\mathcal{F}) \overset{\Phi}{\to} \mathsf{H}^- \circ \mathbf{oblv}_N(\mathcal{F}),$$

and hence is an isomorphism by Braden's theorem.

1.4.8. It remains to show that the functor $\mathrm{Av}_*^N \circ \mathbf{oblv}_{N^-}$ is conservative. But this follows from the fact that its composition with $\mathsf{H}^+ \circ \mathbf{oblv}_N$ is conservative:

$$\mathsf{H}^+ \circ \mathbf{oblv}_N \circ \mathrm{Av}_*^N \circ \mathbf{oblv}_{N^-} \overset{(1.7)}{\simeq} \mathsf{H}^+ \circ \mathbf{oblv}_{N^-} \overset{\Phi}{\simeq} \mathsf{H}^- \circ \mathbf{oblv}_{N^-},$$

while the latter is evidently conservative on $\text{D-mod}_\lambda(X)^{N^-}$.

2. Casselman-Jacquet functor as averaging

In this section we introduce the main character of this paper – the Casselman-Jacquet functor, and state the main results.

Throughout this section G is a reductive group, X is its flag variety. We let N (resp., N^-) be the unipotent radical of a Borel subgroup (resp., an opposite Borel) in G.

2.1. Casselman-Jacquet functor in the abstract setting.

2.1.1. Let $\mathcal{C}$ be a category acted on by N. Consider the (fully faithful) forgetful functor

$$\mathbf{oblv}_N : \mathcal{C}^N \to \mathcal{C}.$$

As was mentioned in Sect. 1.1.3, it admits a right adjoint functor Av_*^N, so that we have an adjoint pair $(\mathbf{oblv}_N, \mathrm{Av}_*^N)$.

We will now consider the (usually, *discontinuous*) right adjoint of Av_*^N, denoted

$$(\mathrm{Av}_*^N)^R : \mathcal{C}^N \to \mathcal{C}.$$

We will also consider the monad $(\mathrm{Av}_*^N)^R \circ \mathrm{Av}_*^N$ acting on $\mathcal{C}$.

REMARK 2.1.2. As we shall see in Sect. 2.3, the monad $(\mathrm{Av}_*^N)^R \circ \mathrm{Av}_*^N$ is not as bizarre as one could initially think: in the case when $\mathcal{C}$ is the category of modules over an associative algebra, equipped with a Harish-Chandra structure with respect to N, the functor $(\mathrm{Av}_*^N)^R \circ \mathrm{Av}_*^N$ is that of $\mathfrak{n}$-adic completion.

2.1.3. Suppose now that the N-action on $\mathcal{C}$ comes by restriction from a G-action. We define the functor

$$J : \mathcal{C} \to \mathcal{C}^{N^-}$$

to be the composition

$$\mathrm{Av}_*^{N^-} \circ (\mathrm{Av}_*^N)^R \circ \mathrm{Av}_*^N.$$

By construction, the functor J is the right adjoint of the functor

$$\mathbf{oblv}_N \circ \mathrm{Av}_*^N \circ \mathbf{oblv}_{N^-} : \mathcal{C}^{N^-} \to \mathcal{C}.$$

From here we obtain:

PROPOSITION 2.1.4. *Let $\mathcal{C}$ be such that the functor*

$$\Upsilon := \mathrm{Av}_*^N \circ \mathbf{oblv}_{N^-} : \mathcal{C}^{N^-} \to \mathcal{C}^N$$

is an equivalence. Then the functor J identifies with

$$\mathrm{Av}_!^{N^-} \circ \mathbf{oblv}_N \circ \mathrm{Av}_*^N.$$

PROOF. If Υ is an equivalence, then the functor $\mathrm{Av}_!^{N^-}$ is defined on the essential image of $\mathbf{oblv}_N$, and

$$\mathrm{Av}_!^{N^-} \circ \mathbf{oblv}_N : \mathcal{C}^N \to \mathcal{C}^{N^-},$$

which is the (a priori partially defined) left adjoint of Υ, is well-defined and is the inverse of Υ.

In particular, $\mathrm{Av}_!^{N^-} \circ \mathbf{oblv}_N$ is *right* adjoint to Υ, so

$$\mathrm{Av}_*^{N^-} \circ (\mathrm{Av}_*^N)^R \simeq \Upsilon^R \simeq \mathrm{Av}_!^{N^-} \circ \mathbf{oblv}_N.$$

Composing, we obtain

$$J = \mathrm{Av}_*^{N^-} \circ (\mathrm{Av}_*^N)^R \circ \mathrm{Av}_*^N \simeq \mathrm{Av}_!^{N^-} \circ \mathbf{oblv}_N \circ \mathrm{Av}_*^N,$$

as desired. □

REMARK 2.1.5. According to Sect. 1.4, the functor Υ is an equivalence in the following cases of interest:

$$\mathcal{C} = \text{D-mod}_\lambda(X) \text{ and } \mathcal{C} = \mathfrak{g}\text{-mod}_\chi.$$

2.2. Casselman-Jacquet functor as completion. In this subsection N can be any unipotent group.

2.2.1. In this subsection we will consider the particular case of $\mathcal{C} = \mathfrak{n}$-mod, equipped with the natural N-action. Note that in this case $\mathcal{C}^N \simeq \mathrm{Rep}(N)$.

We will see that the endofunctor $(\mathrm{Av}_*^N)^R \circ \mathrm{Av}_*^N$ of $\mathfrak{n}$-mod, can be described explicitly as the functor of $\mathfrak{n}$-adic completion.

2.2.2. Consider the inverse family of $\mathfrak{n}$-bimodules indexed by natural numbers

$$i \mapsto U(\mathfrak{n})^i := U(\mathfrak{n})/\mathfrak{n}^i \cdot U(\mathfrak{n}),$$

and the corresponding family of endofunctors of $\mathfrak{n}$-mod:

$$\mathcal{M} \mapsto U(\mathfrak{n})^i \underset{U(\mathfrak{n})}{\otimes} \mathcal{M} =: \mathcal{M}/\mathfrak{n}^i \cdot \mathcal{M}.$$

We claim:

PROPOSITION 2.2.3. *The functor $(\mathrm{Av}_*^N)^R \circ \mathrm{Av}_*^N$ identifies canonically with*

$$\mathcal{M} \mapsto \lim_i \mathcal{M}/\mathfrak{n}^i \cdot \mathcal{M}.$$

2.2.4. *Example.* Let $N = \mathbb{G}_a$. We identify the category $\mathfrak{n}$-mod with $k[t]$-mod, and

$$\operatorname{Rep}(N) \subset \mathfrak{n}\text{-mod}$$

with

$$k[t]\text{-mod}_{\{0\}} \subset k[t]\text{-mod},$$

the full subcategory consisting of modules, on whose cohomologies t acts locally nilpotently.

In this case, the assertion of the proposition is well-known: the functor in question is that of t-adic completion.

2.2.5. The rest of this subsection is devoted to the proof of Proposition 2.2.3.

Let $\mathfrak{n}\text{-mod}^{\text{non-nilp}} \subset \mathfrak{n}$-mod be the full subcategory equal to the *right orthogonal* of $\operatorname{Rep}(N)$. The assertion of the proposition amounts to the following two statements:

(a) For $\mathcal{M} \in \mathfrak{n}\text{-mod}^{\text{non-nilp}}$, the limit $\lim\limits_{i} \mathcal{M}/\mathfrak{n}^i \cdot \mathcal{M}$ is zero.

(b) For any $\mathcal{M} \in \operatorname{Rep}(N)$, the cofiber of the canonical map

$$\mathcal{M} \to \lim_{i} \mathcal{M}/\mathfrak{n}^i \cdot \mathcal{M}$$

belongs to $\mathfrak{n}\text{-mod}^{\text{non-nilp}}$.

Let $\mathbf{triv} \in \operatorname{Rep}(N) \subset \mathfrak{n}$-mod denote the trivial representation. First we notice:

LEMMA 2.2.6. $\mathcal{M} \in \mathfrak{n}\text{-mod}^{\text{non-nilp}} \Leftrightarrow \mathbf{triv} \underset{U(\mathfrak{n})}{\otimes} \mathcal{M} = 0.$

PROOF. The category $\operatorname{Rep}(N)$ is compactly generated by $\mathbf{triv}$. Hence,

$$\mathcal{M} \in \mathfrak{n}\text{-mod}^{\text{non-nilp}} \Leftrightarrow \mathcal{H}om_{\mathfrak{n}\text{-mod}}(\mathbf{triv}, \mathcal{M}) = 0 \Leftrightarrow \mathrm{C}^\bullet(\mathfrak{n}, \mathcal{M}) = 0.$$

Since, $\mathrm{C}^\bullet(\mathfrak{n}, -)$ is isomorphic to, up to a cohomological shift, to $\mathrm{C}_\bullet(\mathfrak{n}, -)$, we obtain that

$$\mathcal{M} \in \mathfrak{n}\text{-mod}^{\text{non-nilp}} \Leftrightarrow \mathrm{C}_\bullet(\mathfrak{n}, \mathcal{M}) = 0 \Leftrightarrow \mathbf{triv} \underset{U(\mathfrak{n})}{\otimes} \mathcal{M} = 0.$$

□

PROOF OF *(a)*. Since each $U(\mathfrak{n})^i$ admits a filtration with subquotients isomorphic to $\mathbf{triv}$, from the Lemma, we obtain that if $\mathcal{M} \in \mathfrak{n}\text{-mod}^{\text{non-nilp}}$, then all the terms in $\lim\limits_{i} \mathcal{M}/\mathfrak{n}^i \cdot \mathcal{M}$ are zero. □

PROOF OF *(b)*. We need to show that for any $\mathcal{M} \in \operatorname{Rep}(N)$, the map

$$\mathbf{triv} \underset{U(\mathfrak{n})}{\otimes} \mathcal{M} \to \mathbf{triv} \underset{U(\mathfrak{n})}{\otimes} \left(\lim_{i} U(\mathfrak{n})^i \underset{U(\mathfrak{n})}{\otimes} \mathcal{M} \right)$$

is an isomorphism. We know that the functor

$$\mathbf{triv} \underset{U(\mathfrak{n})}{\otimes} - \simeq \mathrm{C}_\bullet(\mathfrak{n}, -)$$

commutes with limits (again, because it is isomorphic up to a shift to $\mathrm{C}^\bullet(\mathfrak{n}, -)$), so we need to show that the map

$$\mathbf{triv} \underset{U(\mathfrak{n})}{\otimes} \mathcal{M} \to \lim_{i} \left(\mathbf{triv} \underset{U(\mathfrak{n})}{\otimes} U(\mathfrak{n})^i \underset{U(\mathfrak{n})}{\otimes} \mathcal{M} \right)$$

is an isomorphism. In other words, we need to show that

$$\lim_i \left(\text{coFib}(\mathbf{triv} \to \mathbf{triv} \underset{U(\mathfrak{n})}{\otimes} U(\mathfrak{n})^i) \underset{U(\mathfrak{n})}{\otimes} \mathcal{M}\right)$$

is zero.

Since $\text{Rep}(N)$ is *co-generated* by R_N, it suffices to show that

$$\lim_i \left(\text{coFib}(\mathbf{triv} \to \mathbf{triv} \underset{U(\mathfrak{n})}{\otimes} U(\mathfrak{n})^i) \underset{U(\mathfrak{n})}{\otimes} R_N\right)$$

is zero.

Now, since R_N admits a *finite* resolution with terms of the form $U(\mathfrak{n}) \otimes V$ (where $V \in \text{Vect}$), it suffices to show that

$$\lim_i \left(\text{coFib}(\mathbf{triv} \to \mathbf{triv} \underset{U(\mathfrak{n})}{\otimes} U(\mathfrak{n})^i) \underset{U(\mathfrak{n})}{\otimes} U(\mathfrak{n}) \otimes V\right) \simeq \lim_i \left(\text{coFib}(k \to \mathbf{triv} \underset{U(\mathfrak{n})}{\otimes} U(\mathfrak{n})^i) \otimes V\right)$$

is zero.

We will show that the inverse system (in Vect)

$$i \mapsto \text{coFib}(k \to \mathbf{triv} \underset{U(\mathfrak{n})}{\otimes} U(\mathfrak{n})^i)$$

is *null*, i.e., for every i there exists $i' > i$, such that the map

$$\text{coFib}(k \to \mathbf{triv} \underset{U(\mathfrak{n})}{\otimes} U(\mathfrak{n})^{i'}) \to \text{coFib}(k \to \mathbf{triv} \underset{U(\mathfrak{n})}{\otimes} U(\mathfrak{n})^i)$$

is zero.

The objects of Vect involved are compact (i.e., have finitely many cohomologies and each cohomology is finite-dimensional). Hence, we can dualize, and the assertion becomes equivalent to the fact that the direct system

$$i \mapsto \text{Fib}\left(\text{C}^\bullet(\mathfrak{n}, (U(\mathfrak{n})^i)^*) \to k\right)$$

is null. Again, by compactness, the latter is equivalent to the fact that

$$\underset{i}{\text{colim}}\ \text{Fib}\left(\text{C}^\bullet(\mathfrak{n}, (U(\mathfrak{n})^i)^*) \to k\right) = 0,$$

i.e., that the map

$$\underset{i}{\text{colim}}\ \text{C}^\bullet(\mathfrak{n}, (U(\mathfrak{n})^i)^*) \to k$$

is an isomorphism. Since $\text{C}^\bullet(\mathfrak{n}, -)$ commutes with colimits (being isomorphic up to a shift to $\text{C}_\bullet(\mathfrak{n}, -)$), this is equivalent to the map

$$\text{C}^\bullet\left(\mathfrak{n}, (\underset{i}{\text{colim}}\,(U(\mathfrak{n})^i)^*\right) \to k \tag{2.1}$$

being an isomorphism.

We now notice that the pairing

$$f, u \mapsto u(f)(1), \quad f \in R_N,\ u \in U(\mathfrak{n})$$

defines an isomorphism

$$R_N \simeq \underset{i}{\text{colim}}\,(U(\mathfrak{n})^i)^*$$

as $\mathfrak{n}$-modules.

Under this isomorphism, the above map (2.1) identifies with

$$\mathrm{C}^\bullet(\mathfrak{n}, R_N) \to R_N \overset{f\mapsto f(1)}{\longrightarrow} k,$$

which is evidently an isomorphism. □

□[Proposition 2.2.3]

2.3. The Casselman-Jacquet functor for A-modules.

2.3.1. Let A be an associative algebra equipped with a Harish-Chandra structure with respect to N (see Sect. 1.1.2 for what this means). In particular, A-mod is acted on by N, and we have the restriction functor

$$A\text{-mod} \to \mathfrak{n}\text{-mod},$$

and its left adjoint $\mathfrak{n}\text{-mod} \to A\text{-mod}$, both compatible with N-actions.

We have commutative diagrams

$$\begin{array}{ccccc} A\text{-mod}^N & \xrightarrow{\mathbf{oblv}_N} & A\text{-mod} & \xrightarrow{\mathrm{Av}_*^N} & A\text{-mod}^N \\ \downarrow & & \downarrow & & \downarrow \\ \mathrm{Rep}(N) & \xrightarrow{\mathbf{oblv}_N} & \mathfrak{n}\text{-mod} & \xrightarrow{\mathrm{Av}_*^N} & \mathrm{Rep}(N) \end{array}$$

and

$$\begin{array}{ccccc} A\text{-mod}^N & \xrightarrow{\mathbf{oblv}_N} & A\text{-mod} & \xrightarrow{\mathrm{Av}_*^N} & A\text{-mod}^N \\ \uparrow & & \uparrow & & \uparrow \\ \mathrm{Rep}(N) & \xrightarrow{\mathbf{oblv}_N} & \mathfrak{n}\text{-mod} & \xrightarrow{\mathrm{Av}_*^N} & \mathrm{Rep}(N). \end{array}$$

By passing to right adjoints in the second diagram, we obtain that the following diagram commutes as well:

$$\begin{array}{ccccc} A\text{-mod} & \xrightarrow{\mathrm{Av}_*^N} & A\text{-mod}^N & \xrightarrow{(\mathrm{Av}_*^N)^R} & A\text{-mod} \\ \downarrow & & \downarrow & & \downarrow \\ \mathfrak{n}\text{-mod} & \xrightarrow{\mathrm{Av}_*^N} & \mathrm{Rep}(N) & \xrightarrow{(\mathrm{Av}_*^N)^R} & \mathfrak{n}\text{-mod}. \end{array} \tag{2.2}$$

2.3.2. From the commutation of (2.2) and Proposition 2.2.3 we obtain:

COROLLARY 2.3.3. *The endofunctor* $(\mathrm{Av}_*^N)^R \circ \mathrm{Av}_*^N$ *on* A-mod *is right t-exact.*

REMARK 2.3.4. By unwinding the constructions one can show that for $\mathcal{M} \in A\text{-mod}^\heartsuit$, the action of A on

$$H^0((\mathrm{Av}_*^N)^R \circ \mathrm{Av}_*^N(\mathcal{M})) \simeq \lim_k H^0(\mathcal{M}/\mathfrak{n}^k \cdot \mathcal{M})$$

is given by the following formula:

For an element $a \in A$, let k be an integer so that

$$\mathrm{ad}_{\xi_1} \circ ... \circ \mathrm{ad}_{\xi_k}(a) = 0 \quad \forall \xi_1, ..., \xi_k \in \mathfrak{n}.$$

Then the action of a is well-defined as a map

$$H^0(\mathcal{M}/\mathfrak{n}^{k'+k} \cdot \mathcal{M}) \to H^0(\mathcal{M}/\mathfrak{n}^{k'} \cdot \mathcal{M}),$$

thereby giving a map on the inverse limit.

2.3.5. Let A be left-Noetherian. Let

$$A\text{-mod}^{\mathfrak{n}\text{-f.g.}} \subset A\text{-mod} \tag{2.3}$$

be the full subcategory consisting of modules that map to compact (i.e., finitely generated) objects under the forgetful functor $A\text{-mod} \to \mathfrak{n}\text{-mod}$. Note that $A\text{-mod}^{\mathfrak{n}\text{-f.g.}}$ is *not necessarily* contained in $A\text{-mod}^c$ (unless A has a finite cohomological dimension).

Let

$$\text{Ind}(A\text{-mod}^{\mathfrak{n}\text{-f.g.}})$$

be the ind-completion of $A\text{-mod}^{\mathfrak{n}\text{-f.g.}}$. Ind-extending the tautological embedding

$$A\text{-mod}^{\mathfrak{n}\text{-f.g.}} \hookrightarrow A\text{-mod},$$

we obtain a functor

$$\text{Ind}(A\text{-mod}^{\mathfrak{n}\text{-f.g.}}) \to A\text{-mod}. \tag{2.4}$$

Ind-extending the t-structure on $A\text{-mod}^{\mathfrak{n}\text{-f.g.}}$, we obtain one on $\text{Ind}(A\text{-mod}^{\mathfrak{n}\text{-f.g.}})$. The functor (2.4) is t-exact, but not in general fully faithful. However, as in [**Ga4**, Proposition 2.3.3], one shows that the functors

$$\text{Ind}(A\text{-mod}^{\mathfrak{n}\text{-f.g.}})^{\geq -n} \to A\text{-mod}^{\geq -n} \tag{2.5}$$

are fully faithful for every n.

Let $\text{Ind}^\wedge(A\text{-mod}^{\mathfrak{n}\text{-f.g.}})$ denote the left-completion of $\text{Ind}(A\text{-mod}^{\mathfrak{n}\text{-f.g.}})$ in its t-structure. Since A-mod is left-complete in its t-structure, the functor (2.4) extends to a functor

$$\text{Ind}^\wedge(A\text{-mod}^{\mathfrak{n}\text{-f.g.}}) \to A\text{-mod}. \tag{2.6}$$

The functor (2.6) is fully faithful, since the functors (2.5) have this property. Its essential image consists of objects of A-mod, whose cohomologies are filtered colimits of objects from

$$A\text{-mod}^{\heartsuit,\mathfrak{n}\text{-f.g.}} = A\text{-mod}^\heartsuit \cap A\text{-mod}^{\mathfrak{n}\text{-f.g.}}.$$

REMARK 2.3.6. Suppose that A has a finite cohomological dimension, in which case the functor (2.4) is fully faithful, and hence so is the functor

$$\text{Ind}(A\text{-mod}^{\mathfrak{n}\text{-f.g.}}) \to \text{Ind}^\wedge(A\text{-mod}^{\mathfrak{n}\text{-f.g.}}).$$

However, it is not clear to the authors whether the latter is an equivalence.

2.3.7. Assume now that A has a Harish-Chandra structure with respect to G. We claim:

PROPOSITION 2.3.8.

(a) *The functor* $J : A\text{-mod} \to A\text{-mod}^{N^-}$ *is left t-exact when restricted to* $A\text{-mod}^{\mathfrak{n}\text{-f.g.}}$.

(b) *Assume that the corresponding functor* Υ *is an equivalence. Then the functor* J *is left t-exact when restricted to* $\text{Ind}^\wedge(A\text{-mod}^{\mathfrak{n}\text{-f.g.}})$.

PROOF. First off, point (b) reduces to point (a) as follows: if Υ is an equivalence, then by Proposition 2.1.4, the functor J is commutes with colimits. Then we use Lemma 1.1.9.

We now prove point (a).

The functor $\mathrm{Av}_*^{N^-}$, being the right adjoint of the t-exact functor $\mathbf{oblv}_{N^-}$, is left t-exact (on all of A-mod). Hence, it suffices to show that the functor $(\mathrm{Av}_*^N)^R \circ \mathrm{Av}_*^N$ is left t-exact when restricted to $A\text{-mod}^{\mathfrak{n}\text{-f.g.}}$. We will show that it is in fact t-exact.

Using the commutation of (2.2), it suffices to show that the functor $(\mathrm{Av}_*^N)^R \circ \mathrm{Av}_*^N$ is t-exact when restricted to $\mathfrak{n}\text{-mod}^{\mathrm{f.g.}}$. Using Proposition 2.2.3, it suffices to show that the functor

$$\mathcal{M} \mapsto \lim_k \mathcal{M}/\mathfrak{n}^k \cdot \mathcal{M}$$

sends $\mathcal{M} \in \mathfrak{n}\text{-mod}^{\mathrm{f.g.},\heartsuit}$ to an object in $\mathrm{Vect}^\heartsuit$. However, the latter is known: this is Casselman's generalization of the Artin-Rees lemma. □

2.3.9. Note that the functor

$$\mathcal{M} \mapsto H^0(\mathrm{Av}_*^{N^-}(\mathcal{M})), \quad A\text{-mod}^\heartsuit \to (A\text{-mod}^{N^-})^\heartsuit$$

is that of sending $\mathcal{M}$ to its submodule consisting of elements that are locally nilpotent with respect to the action of $\mathfrak{n}^-$.

Hence, from Proposition 2.3.8 we obtain:

COROLLARY 2.3.10.
(a) *Under the assumptions of Proposition* 2.3.8*(a), the functor*

$$\mathcal{M} \mapsto H^0(J(\mathcal{M})), \quad (A\text{-mod}^{\mathfrak{n}\text{-f.g.}})^\heartsuit \to (A\text{-mod}^{N^-})^\heartsuit$$

sends $\mathcal{M}$ *to the submodule of* $\lim\limits_k \mathcal{M}/\mathfrak{n}^k \cdot \mathcal{M}$ *consisting of elements that are locally nilpotent with respect to the action of* $\mathfrak{n}^-$.

(b) *Under the assumptions of Proposition* 2.3.8*(b), ditto for the category* $(\mathrm{Ind}^\wedge(A\text{-mod}^{\mathfrak{n}\text{-f.g.}}))^\heartsuit$.

2.4. The Casselman-Jacquet functor for $\mathfrak{g}$-modules.

2.4.1. The goal of this subsection is to prove the following:

THEOREM 2.4.2. *Consider* G *acting on the category* $\mathfrak{g}\text{-mod}_\chi$.
(a) *There is a canonically defined natural transformation of functors*

$$\mathrm{Av}_*^{N^-} \circ \mathbf{oblv}_N \circ \mathrm{Av}_!^N \to \mathrm{Av}_!^{N^-} \circ \mathbf{oblv}_N \circ \mathrm{Av}_*^N \simeq J, \tag{2.7}$$

where the LHS is a partially defined functor.

(b) *The functor* $\mathrm{Av}_!^N$ *is defined and the above natural transformation is an isomorphism, when evaluated on* $\mathrm{Ind}^\wedge(\mathfrak{g}\text{-mod}_\chi^{\mathfrak{n}\text{-f.g.}})$.

We will deduce Theorem 2.4.2 from the following result that will be proved in Sect. 3.4.5.

Let $\mathrm{D}\text{-mod}_\lambda(X)^{\mathrm{ULA}} \subset \mathrm{D}\text{-mod}_\lambda(X)$ be the full subcategory of objects that are ULA with respect to the projection $X \to N\backslash X$, see Sect. 3.2.1 for what this means. Consider the corresponding subcategory

$$\mathrm{Ind}^\wedge(\mathrm{D}\text{-mod}_\lambda(X)^{\mathrm{ULA}}) \subset \mathrm{D}\text{-mod}_\lambda(X),$$

see Sect. 3.2.3.

We claim:

THEOREM 2.4.3. *Consider G acting on the category* D-mod$_\lambda(X)$.

(a) *There is a canonically defined natural transformation of functors*

$$\mathrm{Av}_*^{N^-} \circ \mathbf{oblv}_N \circ \mathrm{Av}_!^N \to \mathrm{Av}_!^{N^-} \circ \mathbf{oblv}_N \circ \mathrm{Av}_*^N \simeq J,$$

where the LHS is a partially defined functor.

(b) *The functor* $\mathrm{Av}_!^N$ *is defined and the above natural transformation is an isomorphism, on objects from* $\mathrm{Ind}^\wedge(\text{D-mod}_\lambda(X)^{\mathrm{ULA}})$.

2.4.4. *Proof of Theorem* 2.4.2. Consider the adjoint functors

$$\mathrm{Loc} : \mathfrak{g}\text{-mod}_\chi \rightleftarrows \text{D-mod}_\lambda(X) : \Gamma. \tag{2.8}$$

By Proposition 1.2.6, the functor $\mathrm{Av}_!^{N^-} \circ \mathbf{oblv}_N \circ \mathrm{Av}_*^N$ on $\mathfrak{g}\text{-mod}_\chi$ is defined and identifies with

$$\Gamma \circ \mathrm{Av}_!^{N^-} \circ \mathbf{oblv}_N \circ \mathrm{Av}_*^N \circ \mathrm{Loc}\,.$$

Similarly, the functor $\mathrm{Av}_*^{N^-} \circ \mathbf{oblv}_N \circ \mathrm{Av}_!^N$, viewed as taking values in $\mathrm{Pro}(\mathfrak{g}\text{-mod}_\chi^{N^-})$, identifies with

$$\Gamma \circ \mathrm{Av}_*^{N^-} \circ \mathbf{oblv}_N \circ \mathrm{Av}_!^N \circ \mathrm{Loc}\,.$$

Hence, point (a) of Theorem 2.4.2 follows from point (a) of Theorem 2.4.3.

For point (b), we will use Proposition 1.2.6 and the following assertion proved in Sect. 2.5:

PROPOSITION 2.4.5.

(a) *The functor* Loc *sends objects in* $\mathrm{Ind}^\wedge(\mathfrak{g}\text{-mod}_\chi^{\mathfrak{n}\text{-f.g.}})$ *to objects in* $\mathrm{Ind}^\wedge(\text{D-mod}_\lambda(X)^{\mathrm{ULA}})$.

(b) *The functor* Γ *sends*

$$\text{D-mod}_\lambda(X)^{\mathrm{ULA}} \to \mathfrak{g}\text{-mod}_\chi^{\mathfrak{n}\text{-f.g.}}$$

and

$$\mathrm{Ind}^\wedge(\text{D-mod}_\lambda(X)^{\mathrm{ULA}}) \to \mathrm{Ind}^\wedge(\mathfrak{g}\text{-mod}_\chi^{\mathfrak{n}\text{-f.g.}}).$$

Now, the assertion of Theorem 2.4.2(b) follows from that of Theorem 2.4.3(b) and Proposition 1.2.6.

2.5. ULA vs finite-generation. In this subsection we prove Proposition 2.4.5.

2.5.1. Let $\mathfrak{n} \otimes \mathcal{O}_X$ be the Lie algebroid on X corresponding to the $\mathfrak{n}$-action on X. Let $'\mathrm{D}$ be its universal enveloping D-algebra. We have a commutative diagram

$$\begin{array}{ccc} \text{D-mod}_\lambda(X) & \longrightarrow & '\text{D-mod}(X) \\ {\scriptstyle \Gamma(X,-)}\downarrow & & \downarrow{\scriptstyle \Gamma(X,-)} \\ \mathfrak{g}\text{-mod}_\chi & \longrightarrow & \mathfrak{n}\text{-mod}, \end{array} \tag{2.9}$$

where the vertical arrows are taken by taking global sections, and the horizontal arrows are given by restriction.

An object $\mathcal{M} \in \text{D-mod}_\lambda(X)$ is ULA with respect to $X \to N\backslash X$ if and only if $X|_{'\mathrm{D}}$ is finitely generated.

2.5.2. To prove Proposition 2.4.5(b) we have to show that the right vertical arrow in (2.9) sends finitely generated objects to finitely generated objects.

The latter is enough to do at the associated graded level, and the assertion follows from the fact that X is proper.

2.5.3. Let us prove Proposition 2.4.5(a). It suffices to show that for $\mathcal{M} \in \mathfrak{g}\text{-mod}_\chi^\heartsuit \cap \mathfrak{g}\text{-mod}_\chi^{\mathfrak{n}\text{-f.g.}}$, the object $\text{Loc}(\mathcal{M})|_{'\text{D}}$ has finitely generated cohomologies.

We first consider the case when λ is dominant and *regular*, in which case the functor Loc is t-exact.

Let

$$\text{Loc}_\mathfrak{n} : \mathfrak{n}\text{-mod} \to ' \text{D-mod}(X)$$

be the functor left adjoint to

$$\Gamma : '\text{D-mod}(X) \to \mathfrak{n}\text{-mod}.$$

From (2.9), we obtain a natural transformation

$$\text{Loc}_\mathfrak{n}(\mathcal{M}|_\mathfrak{n}) \to \text{Loc}(\mathcal{M})|_{'\text{D}}.$$

Moreover, for $\mathcal{M} \in \mathfrak{g}\text{-mod}_\chi^\heartsuit$, the above map is *surjective* at the level of H^0. Since $\text{Loc}(\mathcal{M}) \in \text{D-mod}_\lambda(X)^\heartsuit$, this proves the required assertion.

2.5.4. We now consider the case of a general dominant λ. Let μ be a dominant integral weight such that $\lambda' = \lambda + \mu$ is regular. Let χ' be the corresponding character of $Z(\mathfrak{g})$. Consider the translation functor

$$T_{\chi \to \chi'} : \mathfrak{g}\text{-mod}_\chi \to \mathfrak{g}\text{-mod}_{\chi'},$$

and the commutative diagram

$$\begin{array}{ccc} \text{D-mod}_\lambda(X) & \xrightarrow{-\otimes \mathcal{O}(\mu)} & \text{D-mod}_{\lambda'}(X) \\ {\scriptstyle \text{Loc}}\uparrow & & {\scriptstyle \text{Loc}}\uparrow \\ \mathfrak{g}\text{-mod}_\chi & \xrightarrow{T_{\chi \to \chi'}} & \mathfrak{g}\text{-mod}_{\chi'}, \end{array}$$

see Sect. 1.3.6.

It is clear that an object $\mathcal{F} \in \text{D-mod}_\lambda(X)$ is ULA with respect to $X \to N\backslash X$ if and only if $\mathcal{F} \otimes \mathcal{O}(\mu) \in \text{D-mod}_{\lambda'}(X)$ has the same property.

Furthermore, from the description of the functor $T_{\chi \to \chi'}$ in Corollary 1.3.9 it is clear that it sends objects in $\mathfrak{g}\text{-mod}_\chi^{\mathfrak{n}\text{-f.g.}}$ to objects in $\mathfrak{g}\text{-mod}_{\chi'}$ whose cohomologies are in $\mathfrak{g}\text{-mod}_{\chi'}^{\mathfrak{n}\text{-f.g.}}$. Hence, the assertion of Proposition 2.4.5(a) for λ follows from that for λ'.

REMARK 2.5.5. The above prove shows not only that $\text{Loc}(\mathcal{M}) \in \text{Ind}^\wedge(\text{D-mod}_\lambda(X)^{\text{ULA}})$, but that it is actually in $\text{Ind}(\text{D-mod}_\lambda(X)^{\text{ULA}})$.

Indeed, Corollary 1.3.9 implies that $T_{\chi \to \chi'}$ sends objects in $\mathfrak{g}\text{-mod}_\chi^{\mathfrak{n}\text{-f.g.}}$ to objects that can be expressed as filtered colimits of objects in $\mathfrak{g}\text{-mod}_{\chi'}^{\mathfrak{n}\text{-f.g.}}$.

3. The pseudo-identity functor

The goal of this section is to prove Theorem 2.4.3. In the process of doing so we will introduce the *pseudo-identity functor*, which will also be the main character of the sequel to this paper.

3.1. The pseudo-identity functor: recollections.

3.1.1. Let $\mathcal{Y}$ be a quasi-compact algebraic stack with an affine diagonal, and let us be given a twisting λ on $\mathcal{Y}$. Let $-\lambda$ denote the opposite twisting.

We identify
$$\text{D-mod}_\lambda(\mathcal{Y})^\vee \simeq \text{D-mod}_{-\lambda}(\mathcal{Y})$$
via Verdier duality. This allows to identify the category
$$\text{D-mod}_{-\lambda,\lambda}(\mathcal{Y} \times \mathcal{Y})$$
with that of continuous endofunctors on $\text{D-mod}_\lambda(\mathcal{Y})$. Explicitly,
$$\mathcal{Q} \in \text{D-mod}_{-\lambda,\lambda}(\mathcal{Y} \times \mathcal{Y}) \mapsto F_\mathcal{Q}, \quad F_\mathcal{Q}(\mathcal{F}) := (p_2)_*(\mathcal{Q} \overset{!}{\otimes} (p_1)^!(\mathcal{F})),$$
where p_1, p_2 are the two projections $\mathcal{Y} \times \mathcal{Y} \rightrightarrows \mathcal{Y}$, and where for a morphism f we denote by f_* the *renormalized* direct image functor (see [**DrGa1**, Sect. 9.3]).

Under this identification, the identity functor corresponds to
$$(\Delta_\mathcal{Y})_*(\omega_\mathcal{Y}) \in \text{D-mod}_{-\lambda,\lambda}(\mathcal{Y} \times \mathcal{Y}),$$
where we note that the functor
$$\Delta_* : \text{D-mod}(\mathcal{Y}) \to \text{D-mod}_{-\lambda,\lambda}(\mathcal{Y} \times \mathcal{Y})$$
is well-defined because the pullback of the $(-\lambda, \lambda)$-twisting along the diagonal map is canonically trivialized.

3.1.2. The pseudo-identity functor
$$\text{Ps-Id}_\mathcal{Y} : \text{D-mod}_\lambda(\mathcal{Y}) \to \text{D-mod}_\lambda(\mathcal{Y})$$
is the functor corresponding to the object
$$(\Delta_\mathcal{Y})_!(k_\mathcal{Y}) \in \text{D-mod}_{-\lambda,\lambda}(\mathcal{Y} \times \mathcal{Y}),$$
where $k_\mathcal{Y}$ is the "constant sheaf" on $\mathcal{Y}$, i.e., the D-module Verdier dual to $\omega_\mathcal{Y}$.

3.1.3. A stack $\mathcal{Y}$ equipped with a twisting λ is said to be *miraculous* if the endofunctor $\text{Ps-Id}_\mathcal{Y}$ is an equivalence.

The following will be proved in the sequel to this paper (however, we do not use this result here):

THEOREM 3.1.4. *Suppose that $\mathcal{Y}$ has a finite number of isomorphism classes of k-points. Then $\mathcal{Y}$ is miraculous.*

3.2. Pseudo-identity, averaging and the ULA property.

3.2.1. Let $f : \mathcal{Z} \to \mathcal{Y}$ be a smooth morphism between smooth algebraic stacks. Let us recall what it means for an object $\mathcal{F} \in \text{D-mod}_\lambda(\mathcal{Z})$ to be ULA with respect to f, see, e.g. [**Ga1**, Sect. 3.4].

The property of being ULA is local in the smooth topology on the source and the target, so we can assume that $\mathcal{Z} = Z$ and $\mathcal{Y} = Y$ are schemes.

In this case, the sheaf of rings D of differential operators on Z contains a subsheaf of rings, denoted $'\text{D}$, consisting of differential operators *vertical* with respect to f (i.e., these are those differential operators that commute with functions pulled-back from Y). Locally, $'\text{D}$ is generated by functions and vertical fields that are parallel to the fibers of f.

We shall say that $\mathcal{M} \in \text{D-mod}(Z)$ is ULA with respect to f if it is finitely generated when considered as a $'\text{D}$-module (i.e. it has finitely many non-zero cohomologies, and its cohomologies are locally finitely generated over $'\text{D}$).

REMARK 3.2.2. The above definition of ULA can be thought of as the D-module version of the one given in [**BG**, Sect. 5] in the setting of étale sheaves.

3.2.3. Let $\text{D-mod}_\lambda(\mathcal{Z})^{\text{ULA}} \subset \text{D-mod}_\lambda(\mathcal{Z})$ be the full subcategory that consists of objects that are ULA with respect to f.

If $\mathcal{Z}$ is a scheme, then $\text{D-mod}_\lambda(\mathcal{Z})$ has finite cohomological dimension, and we have

$$\text{D-mod}_\lambda(\mathcal{Z})^{\text{ULA}} \subset \text{D-mod}_\lambda(\mathcal{Z})^c.$$

As in Sect. 2.3.5 we define the corresponding functor

$$\text{Ind}(\text{D-mod}_\lambda(\mathcal{Z})^{\text{ULA}}) \to \text{D-mod}_\lambda(\mathcal{Z})$$

and a fully faithful embedding

$$\text{Ind}^\wedge(\text{D-mod}_\lambda(\mathcal{Z})^{\text{ULA}}) \hookrightarrow \text{D-mod}_\lambda(\mathcal{Z}),$$

whose essential image consists of objects whose cohomologies are filtered colimits of objects from

$$\text{D-mod}_\lambda(\mathcal{Z})^{\heartsuit,\text{ULA}} := \text{D-mod}_\lambda(\mathcal{Z})^{\heartsuit} \cap \text{D-mod}_\lambda(\mathcal{Z})^{\text{ULA}}.$$

REMARK 3.2.4. We note that as in Remark 2.3.6 it is not clear to the authors whether, when $\mathcal{Z}$ is a scheme, the inclusion $\text{Ind}(\text{D-mod}_\lambda(\mathcal{Z})^{\text{ULA}}) \subset \text{Ind}^\wedge(\text{D-mod}_\lambda(\mathcal{Z})^{\text{ULA}})$ is actually an equivalence.

3.2.5. We take X to be a smooth proper scheme acted on by a group H, and we take $\mathcal{Z} = X$ and $\mathcal{Y} = H\backslash X$.

We claim:

THEOREM 3.2.6. *Let λ be a H-equivariant twisting on X.*

(a) *There exists a canonically defined natural transformation*

$$\text{Av}_!^H \to \text{Ps-Id}_{H\backslash X} \circ \text{Av}_*^H[2\dim(X)], \tag{3.1}$$

(where the left-hand side is a partially defined functor).

(b) *The map* (3.1) *is an isomorphism when evaluated on objects from* $\text{Ind}^\wedge(\text{D-mod}_\lambda(\mathcal{Z})^{\text{ULA}})$.

3.2.7. *Proof of Theorem* 3.2.6, *Step 0.* Consider the Cartesian diagram

$$\begin{array}{ccccc}
X & \xrightarrow{\widetilde{\Delta}_{H\backslash X}} & X \times H\backslash X & \xrightarrow{\widetilde{p}_1} & X \\
{\scriptstyle f}\downarrow & & \downarrow{\scriptstyle f\times\text{id}} & & \downarrow{\scriptstyle f} \\
H\backslash X & \xrightarrow{\Delta_{H\backslash X}} & H\backslash X \times H\backslash X & \xrightarrow{p_1} & H\backslash X \\
& & \downarrow{\scriptstyle p_2} & & \\
& & H\backslash X. & &
\end{array}$$

For $\mathcal{F} \in \text{D-mod}_\lambda(X)$, the object $\text{Ps-Id}_{H\backslash X} \circ \text{Av}_*^H(\mathcal{F})$ identifies with

$$(p_2)_*\left((\Delta_{H\backslash X})_!(k_{H\backslash X}) \overset{!}{\otimes} (p_1^! \circ f_*(\mathcal{F}))\right) \simeq$$

$$(p_2)_*\left((\Delta_{H\backslash X})_!(k_{H\backslash X}) \overset{!}{\otimes} ((f \times \text{id})_* \circ \widetilde{p}_1^!(\mathcal{F}))\right) \simeq$$

$$\simeq (p_2 \circ (f \times \text{id}))_*\left((f \times \text{id})^! \circ (\Delta_{H\backslash X})_!(k_{H\backslash X}) \overset{!}{\otimes} \widetilde{p}_1^!(\mathcal{F})\right) \overset{\text{smooth base change}}{\simeq}$$

$$\simeq (p_2 \circ (f \times \text{id}))_*\left((\widetilde{\Delta}_{H\backslash X})_! \circ f^!(k_{H\backslash X}) \overset{!}{\otimes} \widetilde{p}_1^!(\mathcal{F})\right) \simeq$$

$$\simeq (\widetilde{p}_2)_*\left((\widetilde{\Delta}_{H\backslash X})_!(k_X) \overset{!}{\otimes} \widetilde{p}_1^!(\mathcal{F})\right)[2\dim(H)],$$

where $\widetilde{p}_2 = p_2 \circ (f \times \text{id})$, and we have used the fact that

$$f^!(k_{H\backslash X}) \simeq k_X[2\dim(H)].$$

Note also that

$$\text{Av}_!^H(\mathcal{F}) \simeq f_!(\mathcal{F})[2\dim(H)] = (\widetilde{p}_2)_! \circ (\widetilde{\Delta}_{H\backslash X})_!(\mathcal{F})[2\dim(H)] =$$

$$= (\widetilde{p}_2)_! \circ (\widetilde{\Delta}_{H\backslash X})_! \circ (\widetilde{\Delta}_{H\backslash X})^* \circ \widetilde{p}_1^*(\mathcal{F})[2\dim(H)] \overset{\text{projection formula}}{\simeq}$$

$$\simeq (\widetilde{p}_2)_!\left((\widetilde{\Delta}_{H\backslash X})_!(k_X) \overset{*}{\otimes} \widetilde{p}_1^*(\mathcal{F})\right)[2\dim(H)].$$

3.2.8. *Proof of Theorem* 3.2.6, *Step 1.* Thus we are reduced to considering the diagram

$$\begin{array}{ccccc} X & \xrightarrow{\widetilde{\Delta}_{H\backslash X}} & X \times H\backslash X & \xrightarrow{\widetilde{p}_1} & X \\ & & \Big\downarrow{\scriptstyle \widetilde{p}_2} & & \\ & & H\backslash X. & & \end{array}$$

Since X was assumed proper, the map $\widetilde{p}_2$ is proper, hence $(\widetilde{p}_2)_! \simeq (\widetilde{p}_2)_*$. The map $\widetilde{p}_1$ is smooth of relative dimension $\dim(X) - \dim(H)$, so

$$\widetilde{p}_1^*(\mathcal{F}) \simeq \widetilde{p}_1^!(\mathcal{F})[-2(\dim(X) - \dim(H))].$$

Hence, it suffices to construct a natural transformation

$$(3.2) \quad (\widetilde{\Delta}_{H\backslash X})_!(k_X) \overset{*}{\otimes} \widetilde{p}_1^!(\mathcal{F}) \to (\widetilde{\Delta}_{H\backslash X})_!(k_X) \overset{!}{\otimes} \widetilde{p}_1^!(\mathcal{F})[2(2\dim(X) - \dim(H))]$$

and show that it is an isomorphism if $\mathcal{F}$ is ULA with respect to f.

3.2.9. *Proof of Theorem* 3.2.6, *Step 2.* Consider the morphism

$$(f \times \text{id}) : X \times H\backslash X \to H\backslash X \times H\backslash X.$$

For any $\mathcal{F}_1 \in \text{D-mod}_{-\lambda,\lambda}(H\backslash X \times H\backslash X)$ and $\mathcal{F}_2 \in \text{D-mod}_{-\lambda,\lambda}(X \times H\backslash X)$ we have the canonical map (see [**Ga1**, Sect. 2.3]):

$$(f \times \text{id})^*(\mathcal{F}_1) \overset{*}{\otimes} \mathcal{F}_2 \to (f \times \text{id})^*(\mathcal{F}_1) \overset{!}{\otimes} \mathcal{F}_2[2(2\dim(X) - \dim(H))].$$

This morphism is an isomorphism for $\mathcal{F}_2$ that is ULA with respect to $f \times \text{id}$. Since all the functors involved are continuous, this remains true if $\mathcal{F}_2$ is a colimit of ULA objects. Further, since the functors involved have a bounded cohomological amplitude, the same is true if the cohomologies of $\mathcal{F}_2$ have this property.

We take $\mathcal{F}_2 = \widetilde{p}_1^!(\mathcal{F})$. The assumption that $\mathcal{F}$ has cohomologies that are ULA with respect to f implies that $\mathcal{F}_2$ has the same property with respect to $f \times \mathrm{id}$.

We take $\mathcal{F}_1 = (\Delta_{H\backslash X})_!(k_{H\backslash X})$. Then

$$(f \times \mathrm{id})^*(\mathcal{F}_1) \simeq (\widetilde{\Delta}_{H\backslash X})_!(k_X).$$

This yields the desired (iso)morphism (3.2). □

REMARK 3.2.10. In Theorem 3.2.6, we could have taken X to be a smooth and proper scheme, and $f : X \to \mathcal{Y}$ a smooth map (not necessarily a quotient map). Of course, Av_*^H would mean f_*, while $\mathrm{Av}_!^H$ would mean the (partially defined) left adjoint of f^*, i.e. $f_![2(\dim(X) - \dim(\mathcal{Y}))]$.

3.3. A variant.

3.3.1. Let $H' \subset H$ be a subgroup. Consider the forgetful functor

$$\mathbf{oblv}_{H/H'} : \text{D-mod}_\lambda(H\backslash X) \to \text{D-mod}_\lambda(H'\backslash X).$$

Its right adjoint $\mathrm{Av}_*^{H/H'}$ is given by *-direct image along

$$H'\backslash X \to H\backslash X,$$

and the partially defined left adjoint $\mathrm{Av}_!^{H/H'}$ is given by !-direct image along the above morphism, shifted by $2(\dim(H) - \dim(H'))$.

3.3.2. Let H'_{red} denote the reductive quotient of H'. We have:

THEOREM 3.3.3.

(a) *There exists a canonically defined natural transformation*

$$\mathrm{Av}_!^{H/H'} \to \text{Ps-Id}_{H\backslash X} \circ \mathrm{Av}_*^{H/H'}[2\dim(X) - \dim(H'_{\mathrm{red}})], \tag{3.3}$$

(where the left-hand side is a partially defined functor).

(b) *The map (3.3) is an isomorphism when evaluated on objects from* $\mathrm{Ind}^\wedge(\text{D-mod}_\lambda(H'\backslash X)^{\mathrm{ULA}})$, *where the ULA condition is taken with respect to the projection* $f : H'\backslash X \to H\backslash X$.

3.3.4. The proof repeats verbatim that of Theorem 3.3.3 with the following modification:

LEMMA 3.3.5. *Let X be a proper scheme acted on by a group H'. Let p denote the projection $H'\backslash X \to \mathrm{pt}$. Then we have a canonical isomorphism of functors*

$$p_* \simeq p_![-2\dim(H') + \dim(H'_{\mathrm{red}})].$$

PROOF. Factor the map p as

$$H'\backslash X \to H'\backslash \mathrm{pt} \to \mathrm{pt}.$$

The first arrow is proper, and this reduces the assertion of the lemma to the case $X = \mathrm{pt}$. In the latter case, this is an easy verification. □

REMARK 3.3.6. In the above lemma, it is crucial that we understand p_* as the *renormalized* direct image functor of [**DrGa1**, Sect. 9.3] (i.e., the continuous extension of the restriction of the usual p_* to compact objects).

3.4. First applications. In this subsection we will take G to be a reductive group, X its flag variety, and $H = N$ the unipotent radical of a Borel.

3.4.1. We are going to show:

THEOREM 3.4.2. *The functor* $\text{Ps-Id}_{N\backslash X}$ *induces a self-equivalence of* $\text{D-mod}_\lambda(N\backslash X)$. *Moreover,* $\text{Ps-Id}_{N\backslash X}[2\dim(X)]$ *identifies with the composite*

$$\text{D-mod}_\lambda(N\backslash X) \xrightarrow{\Upsilon^{-1}} \text{D-mod}_\lambda(N^-\backslash X) \xrightarrow{(\Upsilon^-)^{-1}} \text{D-mod}_\lambda(N\backslash X).$$

The proof is based on the following assertion, proved below:

PROPOSITION 3.4.3. *Any object in the essential image of the forgetful functor*

$$\mathbf{oblv}_{N^-} : \text{D-mod}_\lambda(N^-\backslash X)^c \to \text{D-mod}_\lambda(X)$$

is ULA with respect to $X \to N\backslash X$.

PROOF OF THEOREM 3.4.2. Since, by Proposition 1.4.2, the functors Υ and Υ^- are equivalences with

$$(\Upsilon^-)^{-1} \simeq \text{Av}_!^N \circ \mathbf{oblv}_{N^-},$$

it suffices to establish an isomorphism

$$\text{Ps-Id}_{N\backslash X} \circ \text{Av}_*^N \circ \mathbf{oblv}_{N^-}[2\dim(X)] \simeq \text{Av}_!^N \circ \mathbf{oblv}_{N^-}.$$

However, the latter follows from Proposition 3.4.3 and Theorem 3.2.6. □

REMARK 3.4.4. The fact that $\text{Ps-Id}_{N\backslash X}$ is an equivalence is also a special case of Theorem 3.1.4.

3.4.5. *Proof of Theorem* 2.4.3. We start with the (iso)morphism of Theorem 3.2.6 and compose it with $\text{Av}_*^{N^-} \circ \mathbf{oblv}_N$. We obtain a map

$$\text{Av}_*^{N^-} \circ \mathbf{oblv}_N \circ \text{Av}_!^N \to \text{Av}_*^{N^-} \circ \mathbf{oblv}_N \circ \text{Ps-Id}_{N\backslash X} \circ \text{Av}_*^N[2\dim(X)],$$

which is an isomorphism on objects whose cohomologies are ULA with respect to $X \to N\backslash X$.

We claim that the RHS, i.e.,

$$\text{Av}_*^{N^-} \circ \mathbf{oblv}_N \circ \text{Ps-Id}_{N\backslash X} \circ \text{Av}_*^N[2\dim(X)],$$

is canonically isomorphic to

$$\text{Av}_!^{N^-} \circ \mathbf{oblv}_N \circ \text{Av}_*^N.$$

In fact, we claim that there is a canonical isomorphism

$$\text{Av}_*^{N^-} \circ \mathbf{oblv}_N \circ \text{Ps-Id}_{N\backslash X}[2\dim(X)] \simeq \text{Av}_!^{N^-} \circ \mathbf{oblv}_N \tag{3.4}$$

as functors $\text{D-mod}_\lambda(N\backslash X) \to \text{D-mod}_\lambda(N^-\backslash X)$.

Since the functors $\text{Av}_*^N \circ \mathbf{oblv}_{N^-} = \Upsilon$ and $\text{Av}_!^{N^-} \circ \mathbf{oblv}_N$ are mutually inverse, the latter isomorphism is equivalent to

$$\Upsilon^- \circ \text{Ps-Id}_{N\backslash X}[2\dim(X)] \simeq \Upsilon^{-1},$$

while the latter is the assertion of Theorem 3.4.2. □

3.4.6. As another application of Theorem 3.2.6, we will now prove:

THEOREM 3.4.7. *let $\mathcal{F} \in \text{D-mod}(X)$ be (N^-, ψ)-equivariant, where $\psi : N^- \to \mathbb{G}_a$ is a non-degenerate character. Then there exists a functorial isomorphism (depending on a certain choicc)*

$$\text{Av}_!^N(\mathcal{F}) \simeq \text{Av}_*^N(\mathcal{F})[2\dim(X)].$$

PROOF. By a variant of Proposition 3.4.3 (where we replace equivariance by twisted equivariance), we have a canonical isomorphism

$$\text{Ps-Id}_{N\backslash X} \circ \text{Av}_*^N(\mathcal{F})[2\dim(X)] \simeq \text{Av}_!^N(\mathcal{F}).$$

By Theorem 3.4.2, the left-hand side can be further rewritten as

$$\text{Av}_!^N \circ w_0 \cdot \text{Av}_!^N \circ w_0 \cdot \text{Av}_*^N(\mathcal{F}),$$

where $w_0 \cdot -$ is the functor of translation by (a representative of) the longest element of the Weyl group.

We claim that there is a functorial isomorphism

$$\text{Av}_!^N \circ w_0 \cdot \text{Av}_*^N(\mathcal{F}) \simeq \text{Av}_*^N(\mathcal{F})[\dim(X)],$$

depending on a certain choice.

Namely, it is known that objects of the form $\text{Av}_*^N(\mathcal{F})$ for $\mathcal{F} \in \text{D-mod}(X)^{N^-,\psi}$ are *canonically* of the form

$$\mathcal{M} \underset{\text{End}(\Xi)}{\otimes} \Xi, \quad \mathcal{M} \in \text{End}(\Xi)^{\text{op}}\text{-mod},$$

for a choice of the "big projective" $\Xi \in (\text{D-mod}(X)^N)^\heartsuit$. (Indeed, such objects are right-orthogonal to the other indecomposable projectives in $(\text{D-mod}(X)^N)^\heartsuit$).

Denote

$$\Xi' := \text{Av}_!^N \circ w_0(\Xi)[-\dim(X)].$$

We obtain

$$\text{Av}_!^N \circ w_0 \cdot \text{Av}_*^N(\mathcal{F}) \simeq \text{Av}_!^N \circ w_0(\mathcal{M} \underset{\text{End}(\Xi)}{\otimes} \Xi) \simeq \mathcal{M} \underset{\text{End}(\Xi)}{\otimes} \Xi'[\dim(X)].$$

Now, it is also known that

$$\Xi' := \text{Av}_!^N \circ w_0(\Xi)[-\dim(X)]$$

is *non-canonically* isomorphic again to Ξ, in a way compatible with the action of $\text{End}(\Xi)$.

A choice of such an isomorphism gives rise to an identification

$$\mathcal{M} \underset{\text{End}(\Xi)}{\otimes} \Xi'[\dim(X)] \simeq \mathcal{M} \underset{\text{End}(\Xi)}{\otimes} \Xi[\dim(X)] \simeq \text{Av}_*^N(\mathcal{F}),$$

as desired. □

3.5. Transversality and the proof of Proposition 3.4.3.

3.5.1. Let H_1 and H_2 be two groups acting on a smooth variety X. We shall say that these two actions are *transversal* if for every point $x \in X$, the orbits

$$H_1 \cdot x \subset X \supset H_2 \cdot x$$

are transversal at x.

LEMMA 3.5.2. *The actions of H_1 and H_2 on X are transversal if and only if the map*

$$H_1 \times H_2 \times X \to X \times X, \quad (h_1, h_2, x) \mapsto (h_1 \cdot x, h_2 \cdot x) \tag{3.5}$$

is smooth.

3.5.3. *Example.* It is easy to see that for X being the flag variety of G and $H_1 = N$, and $H_2 = N^-$, then the corresponding actions are transversal.

3.5.4. We have the following generalization of Proposition 3.4.3:

PROPOSITION 3.5.5. *Let the actions of H_1 and H_2 on X be transversal. Then for any (twisted) D-module $\mathcal{F}$ on $H_1\backslash X$, if the (twisted) D-module $\mathbf{oblv}_{H_1}(\mathcal{F})$ on X is coherent, it is ULA with respect to the projection $X \to H_2\backslash X$.*

PROOF. Consider the Cartesian diagram

$$\begin{array}{ccccc}
H_1 \times H_2 \times X & \longrightarrow & H_1 \times X & \xrightarrow{\mathrm{act}_1} & X \\
\downarrow & & \downarrow{\scriptstyle \mathrm{pr}_1} & & \downarrow{\scriptstyle f_1} \\
H_2 \times X & \xrightarrow{\mathrm{pr}_2} & X & \xrightarrow{f_1} & H_1\backslash X \\
\downarrow{\scriptstyle \mathrm{act}_2} & & \downarrow{\scriptstyle f_2} & & \\
X & \xrightarrow{f_2} & H_2\backslash X. & &
\end{array}$$

The property of being ULA is smooth-local with respect to the base, so it is enough to show that the pullback of $\mathbf{oblv}_{H_1}(\mathcal{F})$ to $H_2 \times X$ is ULA with respect to the map act_2.

The ULA property is also smooth-local with respect to the source. Hence, it suffices to show that the further pullback of $\mathcal{F}$ to $H_1 \times H_2 \times X$ is ULA with respect to the composite left vertical arrow.

However, the latter arrow factors as

$$H_1 \times H_2 \times X \to X \times X \xrightarrow{p_2} X \to H_1\backslash X.$$

Since the map $H_1 \times H_2 \times X \to X \times X$ is smooth, it suffices to show that the pullback of $\mathcal{F}' := \mathbf{oblv}_{H_1}(\mathcal{F})$ along $X \times X \xrightarrow{p_2} X$ is ULA with respect to p_1. However, this is true for any coherent object $\mathcal{F}'$. $\square$

4. The case of a symmetric pair

4.1. Adjusting the previous framework.

4.1.1. In this section we will take G equipped with an involution θ; set $K := G^\theta$. Let P be a minimal parabolic compatible with θ (i.e., minimal among parabolics for which $\theta(P)$ is opposite to P); denote $P^- := \theta(P)$.

We change the notations, and in this section denote by N (resp., N^-) the unipotent radical of P (resp., P^-) and

$$M_K := P \cap P^- \cap K.$$

We have the following basic assertion:

LEMMA 4.1.2.

(a) *The groups K, $M_K \cdot N$ and $M_K \cdot N^-$ act on X with finitely many orbits.*

(b) *The actions of $M_K \cdot N$ and K on X, as well as the actions of $M_K \cdot N$ and $M_K \cdot N^-$ on X, are transversal.*

4.1.3. The discussion in Sect. 2 needs to be modified as follows: instead of the functor

$$\mathbf{oblv}_N : \mathcal{C}^N \to \mathcal{C}$$

and its right and (partially defined) left adjoints Av_*^N and $\mathrm{Av}_!^N$, we consider the analogous functor

$$\mathbf{oblv}_{M_K\cdot N/M_K} : \mathcal{C}^{M_K\cdot N} \to \mathcal{C}^{M_K}$$

and its right and (partially defined) left adjoints $\mathrm{Av}_*^{M_K\cdot N/M_K}$ and $\mathrm{Av}_!^{M_K\cdot N/M_K}$. However, by abuse of notation, we will still write $\mathbf{oblv}_N$ instead of $\mathbf{oblv}_{M_K\cdot N/M_K}$, etc.

In Proposition 2.1.4 we take Υ to be the functor

$$\mathcal{C}^{M_K\cdot N^-} \to \mathcal{C}^{M_K\cdot N}$$

given by $\mathrm{Av}_*^N \circ \mathbf{oblv}_{N^-}$.

The key fact is that this functor is an equivalence for $\mathcal{C} = \text{D-mod}_\lambda(X)$ (with the same proof), and hence also for $\mathfrak{g}\text{-mod}_\chi$. It's inverse is again given by $\Upsilon^{-1} = \mathrm{Av}_!^{N^-} \circ \mathbf{oblv}_N$.

The assertions of Theorems 2.4.2 and 2.4.3 should be modified as follows. Let

$$(\mathfrak{g}\text{-mod}_\chi^{M_K})^{\mathfrak{n}\text{-f.g.}} \subset \mathfrak{g}\text{-mod}_\chi^{M_K}$$

be the full subcategory equal to the preimage of $\mathfrak{g}\text{-mod}_\chi^{\mathfrak{n}\text{-f.g.}} \subset \mathfrak{g}\text{-mod}_\chi$ under the forgetful functor $\mathbf{oblv}_{M_K} : \mathfrak{g}\text{-mod}_\chi^{M_K} \to \mathfrak{g}\text{-mod}_\chi$.

THEOREM 4.1.4.

(a) *There is a canonically defined natural transformation of functors*

$$\mathrm{Av}_*^{N^-} \circ \mathbf{oblv}_N \circ \mathrm{Av}_!^N \to \mathrm{Av}_!^{N^-} \circ \mathbf{oblv}_N \circ \mathrm{Av}_*^N \simeq J,$$

considered as functors $\mathfrak{g}\text{-mod}_\chi^{M_K} \to \mathfrak{g}\text{-mod}_\chi^{M_K\cdot N^-}$.

(b) *The above natural transformation is an isomorphism when evaluated on objects from $\mathrm{Ind}^\wedge((\mathfrak{g}\text{-mod}_\chi^{M_K})^{\mathfrak{n}\text{-f.g.}})$.*

THEOREM 4.1.5.

(a) *There is a canonically defined natural transformation of functors*

$$\mathrm{Av}_*^{N^-} \circ \mathbf{oblv}_N \circ \mathrm{Av}_!^N \to \mathrm{Av}_!^{N^-} \circ \mathbf{oblv}_N \circ \mathrm{Av}_*^N \simeq J,$$

considered as functors

$$\text{D-mod}_\lambda(M_K\backslash X) \to \text{D-mod}_\lambda(M_K \cdot N^-\backslash X).$$

(b) *The above natural transformation is an isomorphism on* $\text{Ind}^\wedge(\text{D-mod}_\lambda(M_K\backslash X)^{\text{ULA}})$, *where the ULA condition is taken with respect to the projection* $M_K\backslash X \to M_K N\backslash X$.

4.1.6. Theorem 4.1.5 is proved in the same manner as Theorem 2.4.3, using Proposition 3.3.3 (with $H = M_K \cdot N$ and $H' = M_K$) and the following analog of Theorem 3.4.2:

THEOREM 4.1.7. *The functor* $\text{Ps-Id}_{M_K\cdot N\backslash X}$ *is a self-equivalence of* $\text{D-mod}_\lambda(M_K\cdot N\backslash X)$. *Moreover,* $\text{Ps-Id}_{M_K\cdot N\backslash X}[2\dim(X)-\dim(M_K)]$ *identifies with the composite*

$$\text{D-mod}_\lambda(M_K\cdot N\backslash X) \overset{\Upsilon^{-1}}{\longrightarrow} \text{D-mod}_\lambda(M_K\cdot N^-\backslash X) \overset{(\Upsilon^-)^{-1}}{\longrightarrow} \text{D-mod}_\lambda(M_K\cdot N\backslash X).$$

Theorem 4.1.4(a) is proved as Theorem 2.4.2(a). Theorem 4.1.4(b) is proved using the following version of Proposition 2.4.5(a):

PROPOSITION 4.1.8.
(a) *The functor* Loc *sends*

$$\text{Ind}^\wedge((\mathfrak{g}\text{-mod}_\chi^{M_K})^{\mathfrak{n}\text{-f.g.}}) \to \text{Ind}^\wedge(\text{D-mod}_\lambda(M_K\backslash X)^{\text{ULA}}).$$

(b) *The functor* Γ *sends*

$$\text{D-mod}_\lambda(M_K\backslash X)^{\text{ULA}} \to (\mathfrak{g}\text{-mod}_\chi^{M_K})^{\mathfrak{n}\text{-f.g.}}$$

and

$$\text{Ind}^\wedge(\text{D-mod}_\lambda(M_K\backslash X)^{\text{ULA}}) \to \text{Ind}^\wedge((\mathfrak{g}\text{-mod}_\chi^{M_K})^{\mathfrak{n}\text{-f.g.}}).$$

PROOF. For point (a), it is enough to see that Loc sends an object $\mathcal{F}$ in $(\mathfrak{g}\text{-mod}_\chi^{M_K})^{\mathfrak{n}\text{-f.g.}}$ to an object whose cohomologies are ULA with respect to $M_K\backslash X \to M_K\cdot N\backslash X$.

Being ULA is smooth-local on the source, hence it is enough to see that the cohomologies of $\mathbf{oblv}_{M_K}\circ \text{Loc}(\mathcal{F})$ are ULA with respect to $X \to M_K\cdot N\backslash X$.

For this, it is enough to see that the cohomologies of $\mathbf{oblv}_{M_K}\circ\text{Loc}(\mathcal{F}) \simeq \text{Loc}\circ\mathbf{oblv}_{M_K}(\mathcal{F})$ are ULA with respect to $X \to N\backslash X$, which is the case by Proposition 2.4.5.

Point (b) is proved similarly. □

4.2. The Casselman-Jacquet functor for $(\mathfrak{g}, K)$-modules.
4.2.1. In this subsection we will prove:

THEOREM 4.2.2.
(a) *The functor*

$$\mathfrak{g}\text{-mod}_\chi^K \overset{\mathbf{oblv}_{K/M_K}}{\longrightarrow} \mathfrak{g}\text{-mod}_\chi^{M_K} \overset{J}{\to} \mathfrak{g}\text{-mod}_\chi^{M_K\cdot N^-}$$

identifies canonically with

$$\text{Av}_*^{N^-}\circ\mathbf{oblv}_N\circ\text{Av}_!^N\circ\mathbf{oblv}_{K/M_K}$$

and is t-exact.

(b) *The functor*

$$\text{D-mod}_\lambda(X)^K \overset{\mathbf{oblv}_{K/M_K}}{\longrightarrow} \text{D-mod}_\lambda(X)^{M_K} \overset{J}{\to} \text{D-mod}_\lambda(X)^{M_K \cdot N^-}$$

identifies canonically with

$$\text{Av}_*^{N^-} \circ \mathbf{oblv}_N \circ \text{Av}_!^N \circ \mathbf{oblv}_{K/M_K}$$

and is t-exact.

4.2.3. We will first prove:

PROPOSITION 4.2.4.
(a) *The functor* $\mathbf{oblv}_{K/M_K}$ *maps* $(\mathfrak{g}\text{-mod}_\chi^K)^c$ *to* $(\mathfrak{g}\text{-mod}_\chi^{M_K})^{\mathfrak{n}\text{-f.g.}}$.
(b) *The functor* $\mathbf{oblv}_{K/M_K}$ *maps* $\text{D-mod}_\lambda(K\backslash X)^c$ *to objects in* $\text{D-mod}_\lambda(M_K\backslash X)^{\text{ULA}}$*, where the ULA condition is taken with respect to* $M_K\backslash X \to M_K N\backslash X$.
(c) *The functor* $\mathbf{oblv}_K$ *maps* $\text{D-mod}_\lambda(K\backslash X)^c$ *to objects in* $\text{D-mod}_\lambda(X)$ *that are holonomic.*

PROOF. Point (a) is well-known: it is enough to show that objects of the form

$$(U(\mathfrak{g}) \underset{U(\mathfrak{k})}{\otimes} \rho) \underset{Z(\mathfrak{g})}{\otimes} k, \quad \rho \in \text{Rep}(K)^{\text{f.d.}}$$

where $Z(\mathfrak{g}) \to k$ is given by χ, belong to $\mathfrak{g}\text{-mod}_\chi^{\mathfrak{n}\text{-f.g.}}$. For that it suffices to show that $U(\mathfrak{g}) \underset{U(\mathfrak{k})}{\otimes} \rho$ is finitely generated over $U(\mathfrak{n}) \otimes Z(\mathfrak{g})$, and this follows from the corresponding assertion at the associated graded level.

To show point (b), since the property of being ULA is smooth-local on the source, it is enough to show that $\mathbf{oblv}_K$ maps $\text{D-mod}_\lambda(K\backslash X)^c$ to objects in $\text{D-mod}_\lambda(X)$ that are ULA with respect to $X \to M_K \cdot N\backslash X$. This follows from Proposition 3.5.5 and Lemma 4.1.2(b).

Alternative proof: change the twisting λ by an integral amount to make Γ an equivalence. Then point (b) follows from point (a), combined with Proposition 4.1.8.

Finally, point (c) follows from the fact that the group K has finitely many orbits on X. □

4.3. Proof of Theorem 4.2.2.

4.3.1. First, we note that the fact that $J \circ \mathbf{oblv}_{K/M_K}$ is isomorphic to

$$\text{Av}_*^{N^-} \circ \mathbf{oblv}_N \circ \text{Av}_!^N \circ \mathbf{oblv}_{K/M_K}$$

follows from Theorem 4.1.4(b) and Proposition 4.2.4(a) (resp., Theorem 4.1.5(b) and Proposition 4.2.4(b)).

To prove the t-exactness, we proceed as follows:

4.3.2. *Step 1.* First, we claim that the functor $J \circ \mathbf{oblv}_{K/M_K}$ is left t-exact for $\text{D-mod}_\lambda(K\backslash X)$.

This statement is insensitive to changing λ by an integral twisting. Hence, we can assume that λ is such that Γ is an equivalence. Since Γ is t-exact, the assertion now follows from Propositions 4.2.4(a) and 2.3.8.

4.3.3. *Step 2.* We now claim that the functor $J \circ \mathbf{oblv}_{K/M_K}$ is right t-exact, still for $\text{D-mod}_\lambda(K\backslash X)$.

Indeed, this follows by Verdier duality from the previous step, using Proposition 4.2.4(c).

4.3.4. *Step 3.* It remains to show that $J \circ \mathbf{oblv}_{K/M_K}$ is right t-exact on $\mathfrak{g}\text{-mod}_\chi^K$. Since λ was chosen so that Γ is t-exact, the functor Loc is right t-exact.

We have:

$$J \circ \mathbf{oblv}_{K/M_K} \simeq \Gamma \circ J \circ \mathbf{oblv}_{K/M_K} \circ \mathrm{Loc},$$

where the right-hand side is a composition of t-exact and right t-exact functors.

□[Theorem 2.4.2]

4.4. The "2nd adjointness" conjecture. In the previous subsection we studied the functors

$$J \circ \mathbf{oblv}_{K/M_K} : \mathfrak{g}\text{-mod}_\chi^K \to \mathfrak{g}\text{-mod}_\chi^{M_K \cdot N^-}$$

and

$$J \circ \mathbf{oblv}_{K/M_K} : \text{D-mod}_\lambda(K\backslash X) \to \text{D-mod}_\lambda(M_K \cdot N^-\backslash X).$$

In this subsection we will study functors in the opposite direction, namely,

$$\mathrm{Av}_!^{K/M_K} \text{ and } \mathrm{Av}_*^{K/M_K}$$

that go from $\mathfrak{g}\text{-mod}_\chi^{M_K \cdot N}$ (or $\mathfrak{g}\text{-mod}_\chi^{M_K \cdot N^-}$) to $\mathfrak{g}\text{-mod}_\chi^K$ and from $\text{D-mod}_\lambda(M_K \cdot N\backslash X)$ (or $\text{D-mod}_\lambda(M_K \cdot N^-\backslash X)$) to $\text{D-mod}_\lambda(K\backslash X)$, respectively.

4.4.1. First, we note the following consequence of Lemma 4.1.2, Proposition 3.5.5 and Theorem 3.3.3:

COROLLARY 4.4.2. *We have a canonical isomorphism*

$$\mathrm{Av}_!^{K/M_K} \simeq \text{Ps-Id}_{K\backslash X} \circ \mathrm{Av}_*^{K/M_K}[2\dim(X) - \dim(M_K)]$$

as functors $\text{D-mod}_\lambda(M_K \cdot N\backslash X) \to \text{D-mod}_\lambda(K\backslash X)$.

Combining with Proposition 1.2.6, we obtain:

COROLLARY 4.4.3. *The partially defined functor* $\mathrm{Av}_!^{K/M_K}$ *is defined on the essential image of*

$$\mathbf{oblv}_N : \mathfrak{g}\text{-mod}_\chi^{M_K \cdot N} \to \mathfrak{g}\text{-mod}_\chi^K.$$

4.4.4. For a group H, let us write

$$\mathfrak{l}_H = \Lambda^{\dim(H)}(\mathfrak{h})[\dim H].$$

For a pair of groups $H_1 \subset H_2$, set $\mathfrak{l}_{H_2/H_1} = \mathfrak{l}_{H_2} \otimes \mathfrak{l}_{H_1}^{-1}$.

The same symbols will also stand for the functors of tensoring by those lines. Set $\mathcal{C} = \mathfrak{g}\text{-mod}_\chi$ or $\mathcal{C} = \text{D-mod}_\lambda(X)$. We propose the following conjecture:

CONJECTURE 4.4.5. *There exists a canonical isomorphism*

$$\mathrm{Av}_*^{K/M_K} \simeq \mathfrak{l}_{K/M_K}^{-1} \circ \mathrm{Av}_!^{K/M_K} \circ \Upsilon$$

as functors

$$\mathcal{C}^{M_K \cdot N^-} \to \mathcal{C}^K.$$

In light of Corollary 4.4.2, in the case of $\mathcal{C} = \text{D-mod}_\lambda(X)$, we can reformulate Conjecture 4.4.5 as follows:

CONJECTURE 4.4.6. *The following diagram of functors commutes:*

$$\begin{array}{ccc} \mathcal{C}^K & \xleftarrow{\text{Av}_*^{K/M_K}} & \mathcal{C}^{M_K\cdot N} \\ {\scriptstyle \iota^{-1}_{K/M_K}\circ\text{Ps-Id}_{K\backslash X}[2\dim(X)-\dim(M_K)]}\downarrow & & \downarrow{\scriptstyle \Upsilon^{-1}} \\ \mathcal{C}^K & \xleftarrow{\text{Av}_*^{K/M_K}} & \mathcal{C}^{M_K\cdot N^-}. \end{array}$$

Note that Conjecture 4.4.6 can be thought of as a sort of functional equation for the functor Av^{K/M_K}, cf. [**Ga3**].

4.4.7. Note that we have two adjoint pairs of functors

$$\text{Av}_!^{N^-} : \mathcal{C}^K \rightleftarrows \mathcal{C}^{M_K\cdot N^-} : \text{Av}_*^{K/M_K}$$

and

$$\text{Av}_!^{K/M_K} : \mathcal{C}^{M_K\cdot N^-} \rightleftarrows \mathcal{C}^K : \text{Av}_*^{N^-}.$$

We obtain that Conjecture 4.4.5 is equivalent to the following one:

CONJECTURE 4.4.8. *The* right *adjoint functor to*

$$\text{Av}_*^{K/M_K} \circ \mathbf{oblv}_{N^-} : \mathcal{C}^{M_K\cdot N^-} \to \mathcal{C}^K$$

is given by

$$J \circ \mathbf{oblv}_{K/M_K} \circ \iota_{K/M_K}.$$

4.4.9. Combining with Theorem 4.1.4, we further obtain that Conjecture 4.4.8 is equivalent to:

CONJECTURE 4.4.10. *The* right *adjoint functor to*

$$\text{Av}_*^{K/M_K} \circ \mathbf{oblv}_{N^-} : \mathcal{C}^{M_K\cdot N^-} \to \mathcal{C}^K$$

is given by

$$\Upsilon^- \circ \text{Av}_!^N \circ \mathbf{oblv}_{K/M_K} \circ \iota_{K/M_K}.$$

4.4.11. We regard Conjecture 4.4.10 as an analog of Bernstein's *2nd adjointness theorem* for $\mathfrak{p}$-adic groups.

Recall that the latter says that in addition to the tautological adjunction (the 1st adjointness)

$$r : \mathbf{G}\text{-mod} \rightleftarrows \mathbf{M}\text{-mod} : i$$

(here we denote by G a $\mathfrak{p}$-adic group, by M its Levi subgroup, by i the normalized parabolic induction functor, and by r the Jacquet functor), we also have an adjunction

$$i : \mathbf{M}\text{-mod} \rightleftarrows \mathbf{G}\text{-mod} : \overline{r},$$

where $\overline{r}$ is the Jacquet functor with respect to the opposite parabolic.

Here is the table of analogies/points of difference between $\mathfrak{p}$-adic groups and symmetric pairs:

- The analog of $\mathbf{G}$-mod is the category $\mathfrak{g}\text{-mod}_\chi^K$;
- The analog of $\mathbf{M}$-mod is *not* $\mathfrak{m}\text{-mod}_\chi^{M_K}$, but rather $\mathfrak{g}\text{-mod}_\chi^{M_K\cdot N}$ (or $\mathfrak{g}\text{-mod}_\chi^{M_K\cdot N^-}$); note that this category explicitly depends on the choice of the parabolic or its opposite.

- The analog of the tautological identification $\mathbf{M}$-mod $=$ $\mathbf{M}$-mod is the intertwining functor Υ;
- The analog of the induction functor i is Av_*^{K/M_K};
- The analog of the Jacquet functor r (resp., $\overline{r}$) is $\mathrm{Av}_!^N$ (resp., $\mathrm{Av}_!^{N^-}$).

With these analogies, Conjecture 4.4.10 says that the right adjoint to the induction functor Av_*^{K/M_K} is isomorphic to the Jacquet functor $\mathrm{Av}_!^N$, up to replacing N by N^-, inserting the intertwining functor, and a twist.

References

[BB1] A. Beĭlinson and J. Bernstein, *Localisation de g-modules* (French, with English summary), C. R. Acad. Sci. Paris Sér. I Math. **292** (1981), no. 1, 15–18. MR610137

[BB2] A. Beĭlinson and J. Bernstein, *A generalization of Casselman's submodule theorem*, Representation theory of reductive groups (Park City, Utah, 1982), Progr. Math., vol. 40, Birkhäuser Boston, Boston, MA, 1983, pp. 35–52. MR733805

[BG] A. Braverman and D. Gaitsgory, *Geometric Eisenstein series*, Invent. Math. **150** (2002), no. 2, 287–384, DOI 10.1007/s00222-002-0237-8. MR1933587

[BK] R. Bezrukavnikov and D. Kazhdan, *Geometry of second adjointness for p-adic groups*, Represent. Theory **19** (2015), 299–332, DOI 10.1090/ert/471. With an appendix by Yakov Varshavsky, Bezrukavnikov and Kazhdan. MR3430373

[CY] T.-H. Chen and A. Yom Din, *A formula for the geometric Jacquet functor and its character sheaf analogue*, Geom. Funct. Anal. **27** (2017), no. 4, 772–797, DOI 10.1007/s00039-017-0413-z. MR3678501

[DrGa1] V. Drinfeld and D. Gaitsgory, *On some finiteness questions for algebraic stacks*, Geom. Funct. Anal. **23** (2013), no. 1, 149–294, DOI 10.1007/s00039-012-0204-5. MR3037900

[DrGa2] V. Drinfeld and D. Gaitsgory, *On a theorem of Braden*, Transform. Groups **19** (2014), no. 2, 313–358, DOI 10.1007/s00031-014-9267-8. MR3200429

[ENV] M. Emerton, D. Nadler, and K. Vilonen, *A geometric Jacquet functor*, Duke Math. J. **125** (2004), no. 2, 267–278, DOI 10.1215/S0012-7094-04-12523-0. MR2096674

[FG] E. Frenkel and D. Gaitsgory, *Local geometric Langlands correspondence and affine Kac-Moody algebras*, Algebraic geometry and number theory, Progr. Math., vol. 253, Birkhäuser Boston, Boston, MA, 2006, pp. 69–260, DOI 10.1007/978-0-8176-4532-8_3. MR2263193

[Ga1] D. Gaitsgory, *Functors given by kernels, adjunctions and duality*, J. Algebraic Geom. **25** (2016), no. 3, 461–548, DOI 10.1090/jag/654. MR3493590

[Ga2] D. Gaitsgory, *Sheaves of categories and the notion of 1-affineness*, Stacks and categories in geometry, topology, and algebra, Contemp. Math., vol. 643, Amer. Math. Soc., Providence, RI, 2015, pp. 127–225, DOI 10.1090/conm/643/12899. MR3381473

[Ga3] D. Gaitsgory, *A "strange" functional equation for Eisenstein series and miraculous duality on the moduli stack of bundles* (English, with English and French summaries), Ann. Sci. Éc. Norm. Supér. (4) **50** (2017), no. 5, 1123–1162, DOI 10.24033/asens.2641. MR3720026

[Ga4] D. Gaitsgory, *Ind-coherent sheaves* (English, with English and Russian summaries), Mosc. Math. J. **13** (2013), no. 3, 399–528, 553. MR3136100

[GR] D. Gaitsgory and N. Rozenblyum, *Crystals and D-modules*, Pure Appl. Math. Q. **10** (2014), no. 1, 57–154, DOI 10.4310/PAMQ.2014.v10.n1.a2. MR3264953

[Yo] A. Yom Din, *On properties of the Casselman-Jacquet functor*, arXiv: 1609.02523.

Department of Mathematics, University of Chicago, Chicago, Illinois 60637

Department of Mathematics, Harvard University, Cambridge, Massachusetts 02138

Department of Mathematics, California Institute of Technology, Pasadena, California 91125

Proceedings of Symposia in Pure Mathematics
Volume **101**, 2019
https://doi.org/10.1090/pspum/101/01792

Periods and theta correspondence

Wee Teck Gan

to Professor Joseph Bernstein,
with admiration

1. Introduction

One important application of the theory of theta correspondence is that it gives a way of relating certain periods of one member of a dual pair with certain periods on the other member of the dual pair. It thus serves as a complement to the method of the relative trace formula. While the relative trace formula may have a wider range of applicability, one advantage of the theta correspondence approach for such problems is that the analysis of the local and global period problems are often completely parallel, so that one tends to obtain quite complete results at both the local and global level with essentially the same amount and type of work.

More precisely, suppose that $G \times H$ is a reductive dual pair, with a Weil representation ω, over a local field F. Let H' be a (typically unipotent) subgroup of H equipped with a character μ. Then for an irreducible representation π of G, one has:

$$\mathrm{Hom}_{H'}(\Theta(\pi), \mu) = \mathrm{Hom}_{G \times H'}(\omega, \pi \boxtimes \mu) = \mathrm{Hom}_G(\omega_{H',\mu}, \pi).$$

Thus, if we could compute the twisted coinvariant $\omega_{H',\mu}$ of the Weil representation in some other way, we will obtain a description of $\mathrm{Hom}_{H'}(\Theta(\pi), \mu)$. In turns out that frequently, one has

$$\omega_{H',\mu} \cong \mathrm{ind}_{G'}^{G} \nu$$

as G-modules (or contains the RHS as a submodule whose associated quotient is "spectrally negligible"). Then an application of Frobenius reciprocity will give a natural isomorphism

$$\mathrm{Hom}_{H'}(\Theta(\pi), \mu) \cong \mathrm{Hom}_{G'}(\pi^\vee, \nu^{-1}).$$

Thus, we will obtain a statement of the form:

$$\Theta(\pi) \text{ has nonzero } (H', \mu)\text{-period} \Longleftrightarrow \pi^\vee \text{ has nonzero } \nu^{-1}\text{-period}$$

at the local level. In the above argument, the key step is thus the independent computation of $\omega_{H',\mu}$.

At the global level, if one has a cuspidal representation π of G, then we may consider its global theta lift which is spanned by the automorphic forms

$$\theta(\phi, f)(h) = \int_{G(k)\backslash G(\mathbb{A})} \theta(\phi)(gh) \cdot \overline{f(g)}\, dg$$

as ϕ varies over elements in the global Weil representation ω and $f \in \pi$. One is interested in computing the (H', μ)-period integral

$$\begin{aligned}\theta(\phi, f)_{H',\mu}(1) &= \int_{H'(k)\backslash H'(\mathbb{A})} \overline{\mu(h)} \cdot \theta(\phi, f)(h)\, dh \\ &= \int_{G(k)\backslash G(\mathbb{A})} \overline{f(g)} \cdot \left(\int_{H'(k)\backslash H'(\mathbb{A})} \overline{\mu(h)} \cdot \theta(\phi)(gh)\, dh \right) dg\end{aligned}$$

where we have exchanged the order of integration. Thus, the key is now to compute the inner integral, which is the (H', μ)-period of the theta function $\theta(\phi)$ and is the global analog of the computation of $\omega_{H',\mu}$ in the local setting. As in the local situation, the result of this computation frequently takes the form

$$\theta(\phi)_{H',\mu}(g) = \sum_{\gamma \in G'(k)\backslash G(k)} l(\gamma g \phi)$$

where $0 \neq l \in \mathrm{Hom}_{G'(\mathbb{A})}(\omega, \nu)$ for some subgroup $G' \subset G$ equipped with an automorphic character ν. Hence,

$$\begin{aligned}\theta(\phi, f)_{H',\chi}(1) &= \int_{G(k)\backslash G(\mathbb{A})} \overline{f(g)} \cdot \sum_{\gamma \in G'(k)\backslash G(k)} l(\gamma g \phi)\, dg \\ &= \int_{G'(k)\backslash G(\mathbb{A})} \overline{f(g)} \cdot l(g\phi)\, dg \\ &= \int_{G'(\mathbb{A})\backslash G(\mathbb{A})} l(g\phi) \cdot \left(\int_{G'(k)\backslash G'(\mathbb{A})} \nu(g') \cdot \overline{f(g'g)}\, dg' \right) dg\end{aligned}$$

where the inner integral is the complex conjugate of the (G', ν)-period of f. In this way, one deduces the statement

the (H', μ)-period integral is nonzero on $\Theta(\pi)$

$\Longleftrightarrow$ the (G', ν)-period integral is nonzero on π.

Thus one sees the parallel between the local and global computation.

One of the first instance of the above computation is the computation of the Jacquet modules of $\Theta(\pi)$, so that $H' = N$ is the nilpotent radical of a maximal parabolic subgroup $P = MN$ of H, and μ is the trivial character. In the local setting, this computation was carried out in the paper [**K1**] of Kudla, whereas in the global setting, it was carried out by Rallis [**R**]. Another instance is the computation of the Whittaker periods of $\Theta(\pi)$ where H' is the unipotent radical of a Borel subgroup and μ is a generic character.

As we mentioned, the above discussion typically applies to the case when H' is unipotent. When H' is reductive, however, one can sometimes study the (H', μ)-period of $\Theta(\pi)$ by using a so-called see-saw diagram:

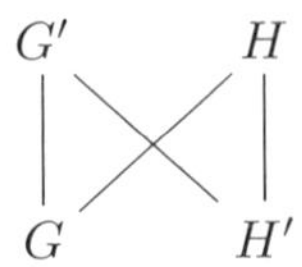

where $H' \times G'$ is another dual pair contained in the same ambient group, with $G \subset G'$ and $H' \subset H$. For $\pi \in \mathrm{Irr}(G)$, the see-saw identity then gives

$$\mathrm{Hom}_{H'}(\Theta(\pi), \mu) \cong \mathrm{Hom}_{G \times H'}(\Omega, \pi \boxtimes \mu) \cong \mathrm{Hom}_{G'}(\Theta(\mu), \pi).$$

If one can determine the theta lift $\Theta(\mu)$ of μ to G' precisely, the see-saw identity allows one to reduce the determination of the (H', μ)-periods of $\Theta(\pi)$ to the co-period $\mathrm{Hom}_{G'}(\Theta(\mu), \pi)$ of π. Typically, $\Theta(\mu)$ is a simpler representation, such as a degenerate principal series representation of G', and the above space of co-periods has a better chance of being understood. In the global setting, $\Theta(\mu)$ may be the special value of an Eisenstein series, and its pairing with π may be a global zeta integral which represents an automorphic L-function of π.

In this paper, we shall consider two instances of such transfer of periods between the members of a dual reductive pair, illustrating the two techniques above:

(a) the relation of Shalika periods and linear periods; this is the case when H' is the unipotent radical of a Siegel parabolic subgroup and μ is a nondegenerate character;

(b) the torus periods for GL_2; this revisits the Tunnell-Saito theorem, for which we give a different proof from those in the literature.

Another very nice illustration of the transfer of periods by theta correspondence are the recent papers [**GZ1**, **GZ2**] of R. Gomez and C.B. Zhu. There, they consider the so-called Kawanaka or generalized Gelfand-Graev models associated to unipotent orbits and showed that the theta correspondence transfer such models from one member of the dual pair to another, in a way which is controlled by the geometry of a natural moment map. For more details, the reader can consult the paper of Zhu [**Z**] in these proceedings.

Acknowledgments: The author is partially supported by a Singapore government MOE Tier 2 grant R-146-000-175-112. He would like to thank the editors of the Bernstein proceedings for agreeing to consider his very late submission, as well as Erez Lapid, Gordan Savin and Alberto Minguez for helpful conversations and feedback. Thanks are also due to the referee for his/her meticulous work and useful comments. It is a pleasure and privilege to dedicate this paper to Professor Joseph Bernstein whose mathematical outlook, writings and results have never ceased to be a source of inspiration. It should be clear that his paper [**BZ**] with Zelevinsky supplies most of the foundational results and techniques used in this paper.

2. Shalika and Linear Periods

In this section, we consider the simplest example of a relation of periods between the two members of a dual pair. When we specialise the discussion of this section to the case of the so-called Type II dual pair (i.e. for $\mathrm{GL}(m) \times \mathrm{GL}(n)$), this relation becomes one between the Shalika period and the linear period on $\mathrm{GL}(n)$. This relation is a folkloric result which is probably well-known to some experts; various special cases in the setting of symplectic-orthogonal dual pairs have been shown by M. Hanzer [**H**] and B.Y. Liu [**L**]. We first set up some notations.

Let F be a non-archimedean local field of characteristic not 2 and let E be a separable F algebra with $\dim_F E = 1$ or 2. Fix a nontrivial additive character ψ of F. Let W be a skew-Hermitian space and V a Hermitian space over E, with isometry groups $\mathrm{U}(W)$ and $\mathrm{U}(V)$ respectively. Then the group $\mathrm{U}(W) \times \mathrm{U}(V)$ has an associated Weil representation $\Omega = \omega_\psi$ which depends also on a pair of characters

(χ_V, χ_W) of $E^\times$ (see [**GI**, §4] for details). For an irreducible representation π of $\mathrm{U}(V)$, we have its big theta lift $\Theta(\pi)$ which is a smooth representation of $\mathrm{U}(W)$, so that

$$\mathrm{Hom}_{\mathrm{U}(V)}(\omega_\psi, \pi \boxtimes \Sigma) \cong \mathrm{Hom}_{\mathrm{U}(W)}(\Theta(\pi), \Sigma)$$

for any smooth representation Σ of $\mathrm{U}(W)$.

Assume in this section that W is split, in the sense that there is a maximal isotropic subspace $X \subset W$ of dimension $\frac{1}{2} \cdot \dim_E W$, so that

$$W = X \oplus X^*$$

for any isotropic space X^* which is in perfect duality with X. The stabilizer of X is a maximal (Siegel) parabolic subgroup

$$P(X) = \mathrm{GL}(X) \cdot N(X)$$

where the unipotent radical $N(X) \subset \mathrm{Hom}(X^*, X) = X \otimes X$ is abelian and may be identified with the set of Hermitian forms on X^*. For a Hermitian form A on X^*, we shall often write $n(A)$ for the corresponding element in $N(X)$. Using the fixed additive character ψ, the Pontrjagin dual of $N(X)$ can then be identified with $N(X^*) \subset \mathrm{Hom}(X, X^*)$: the element $B \in N(X^*)$ gives the character

$$\psi_B(n(A)) = \psi(Tr_X(A \circ B)) \quad \text{for } A \in N(X).$$

Note that B may be regarded as a Hermitian form on X and in the following, we shall consider the Hermitian space (X, B), and simply denote it by B.

The purpose of this section is to determine

$$\Theta(\pi)^*_{N(X),\psi_B} = \mathrm{Hom}_{N(X)}(\Theta(\pi), \psi_B) = \mathrm{Hom}_{\mathrm{U}(V)\times N(X)}(\Omega, \pi \boxtimes \psi_B)$$

for $\pi \in \mathrm{Irr}(\mathrm{U}(V))$ and $B \in N(X^*)$ a nondegenerate Hermitian form. We first note that

$$\Theta(\pi)^*_{N(X),\psi_B} = \mathrm{Hom}_{\mathrm{U}(V)}(\Omega_{N(X),\psi_B}, \pi).$$

The following proposition determines $\Omega_{N(X),\psi_B}$.

PROPOSITION 2.1. *As a representation of* $\mathrm{U}(V)$, $\Omega_{N(X),\psi_B} = 0$ *if there is no embedding of Hermitian spaces* $B \hookrightarrow V$. *If there is an embedding* $j : B \hookrightarrow V$ *of Hermitian spaces, then we may write* $V = j(B) \oplus j(B)^\perp$ *and*

$$\Omega_{N(X),\psi_B} \cong \mathrm{ind}_{U(j(B)^\perp)}^{\mathrm{U}(V)} \chi_W \circ \det_{j(B)^\perp}.$$

PROOF. We work with the Schrodinger model of the Weil representation Ω. This is realized on the space $\mathcal{S}(X^* \otimes V)$ of Schwarz functions on $X^* \otimes V = \mathrm{Hom}(X, V)$. The action of $P(X)$ is given as follows:

(2.2) $(h \cdot \phi)(T) = \chi_W(\det_V(h)) \cdot \phi(h^{-1}(T)) \quad$ for $h \in \mathrm{U}(V)$;

(2.3) $(m \cdot \phi)(T) = \chi_V(\det_X(m)) \cdot |\det_X(m)|^{\frac{1}{2}\dim V} \phi(m^{-1}T) \quad$ for $m \in \mathrm{GL}(X)$;

(2.4) $(n(A) \cdot \phi)(T) = \psi_{T^*(V)}(A) \cdot \phi(T) \quad$ for $n(A) \in N(X)$.

Here χ_V is some unitary character which will not concern us here, and $T^*(V)$ is the Hermitian form on X obtained by pulling back the Hermitian form on V using $T \in \mathrm{Hom}(X, V)$.

From this, one deduces that there is an isomorphism of $\mathrm{U}(V)$-modules:

$$\Omega_{N(X),\psi_B} \cong \mathcal{S}(\mathcal{O}_B)$$

where $\mathcal{O}_B \subset \mathrm{Hom}(X, V)$ is the Zariski closed subset

$$\mathcal{O}_B = \{T : T^*(V) = B\} = \{\text{embeddings of Hermitian space } B \hookrightarrow V\}.$$

Here, note that $\mathrm{U}(V)$ preserves $\mathcal{O}_B$ and its action on $\mathcal{S}(\mathcal{O}_B)$ is geometric. The natural projection $\Omega \to \Omega_{N(X),\psi_B}$ is simply given by the restriction of functions from $X^* \otimes V$ to the subset $\mathcal{O}_B$.

Hence, if $\mathcal{O}_B$ is empty, i.e. if there is no embedding $B \hookrightarrow V$, then $\Omega_{N(X),\psi_B} = 0$, as desired. On the other hand, if $\mathcal{O}_B$ is nonempty, then Witt's theorem says that $\mathrm{U}(V)$ acts transitively on $\mathcal{O}_B$. If we fix a base point $j \in \mathcal{O}_B$, and write $V = j(B) \oplus j(B)^\perp$, then the stabilizer of j in $\mathrm{U}(V)$ is equal to $\mathrm{U}(j(B)^\perp)$. As a consequence, we deduce that

$$\Omega_{N(X),\psi_B} \cong \mathrm{ind}_{\mathrm{U}(j(B)^\perp)}^{\mathrm{U}(V)} \chi_W \circ \det_{j(B)^\perp}$$

as desired. □

From the proposition, we see that

$$\begin{aligned}\mathrm{Hom}_{N(X)}(\Theta(\pi), \psi_B) &\cong \mathrm{Hom}_{\mathrm{U}(V)}(\mathrm{ind}_{U(j(B)^\perp)}^{\mathrm{U}(V)} \chi_W \circ \det_{j(B)^\perp}, \pi) \\ &\cong \mathrm{Hom}_{\mathrm{U}(j(B)^\perp)}(\pi^\vee, \chi_W^{-1} \circ \det_{j(B)^\perp})\end{aligned}$$

as vector spaces. In fact, there are some extra symmetries here. More precisely, the stabilizer in $\mathrm{GL}(X)$ of the character ψ_B is the subgroup $\mathrm{U}(B)$ and $\Omega_{N(X),\psi_B}$ is naturally a representation of $\mathrm{U}(B) \times \mathrm{U}(V)$. Keeping track of the $\mathrm{U}(B)$-action in the proof of the above proposition, one sees that as a $\mathrm{U}(B) \times \mathrm{U}(V)$-module,

$$\Omega_{N(X),\psi_B} \cong \mathrm{ind}_{\mathrm{U}(B)^\Delta \times \mathrm{U}(j(B)^\perp)}^{\mathrm{U}(B)\times\mathrm{U}(V)} (\chi_V \circ \det_B) \boxtimes (\chi_W \circ \det_{j(B)^\perp})$$

where $\mathrm{U}(B)^\Delta$ is the diagonally embedded subgroup

$$\mathrm{U}(B) \xrightarrow{\ \Delta\ } \mathrm{U}(B) \times (\mathrm{U}(B) \times \mathrm{U}(j(B)^\perp)) \longrightarrow \mathrm{U}(B) \times \mathrm{U}(V).$$

Hence we obtain:

THEOREM 2.5. *Fix a nondegenerate Hermitian form B on X. Given $\pi \in \mathrm{Irr}(\mathrm{U}(V))$ and $\sigma \in \mathrm{Irr}(\mathrm{U}(B))$, there are isomorphisms of vector spaces*

$$\begin{aligned}&\mathrm{Hom}_{\mathrm{U}(B)\ltimes N(X)}(\Theta(\pi), \sigma \boxtimes \psi_B) \\ &\qquad \cong \mathrm{Hom}_{\mathrm{U}(B)\times\mathrm{U}(j(B)^\perp)}(\pi^\vee, \sigma \cdot (\chi_V^{-1} \circ \det_B) \boxtimes (\chi_W^{-1} \circ \det_{j(B)^\perp})).\end{aligned}$$

Note that in the case of symplectic-orthogonal dual pairs, the characters $\chi_V \circ \det_B$ and $\chi_W \circ \det_{j(B)^\perp}$ are trivial, but they need not be trivial in the case of unitary dual pairs. Moreover, one can exchange the adjectives "Hermitian" and "skew-Hermitian" in the above discussion.

When we specialise the theorem to the case where $\sigma = \mathbb{C}$ is the trivial representation, then the space

$$\mathrm{Hom}_{\mathrm{U}(B)\ltimes N(X)}(\Theta(\pi), \mathbb{C} \boxtimes \psi_B)$$

is the space of Shalika periods (relative to B) for the representation $\Theta(\pi)$ of $\mathrm{U}(W)$, whereas we will call the space

$$\mathrm{Hom}_{\mathrm{U}(B)\times\mathrm{U}(j(B)^\perp)}(\pi^\vee, (\chi_V^{-1} \circ \det_B) \boxtimes (\chi_W^{-1} \circ \det_{j(B)^\perp}))$$

the space of linear periods (relative to (j, B)) of the representation $\pi^\vee$ of $\mathrm{U}(V)$. This special case of the theorem thus relates Shalika periods of $\Theta(\pi)$ with linear periods of $\pi^\vee$. We thus have the following corollary:

COROLLARY 2.6. *For* $\pi^\vee \in \mathrm{Irr}(\mathrm{U}(V))$, *one has:*

$$\pi^\vee \text{ has nonzero linear period} \Longleftrightarrow \Theta(\pi) \text{ has nonzero Shalika period.}$$

Thus,

$$\theta(\pi) \text{ has nonzero Shalika period} \Longrightarrow \pi^\vee \text{ has nonzero linear period.}$$

In particular, if $\Theta(\pi) = \theta(\pi)$, *then*

$$\theta(\pi) \text{ has nonzero Shalika period} \Longleftrightarrow \pi^\vee \text{ has nonzero linear period.}$$

This corollary highlights the importance of determining whether $\Theta(\pi) = \theta(\pi)$ in such applications. In the case of symplectic-orthogonal or unitary dual pairs, when $\dim W = \dim V$, it has been shown in [**GS**, **GI**] that $\Theta(\pi) = \theta(\pi)$ when π is tempered.

3. Type II Dual Pairs

In principle, one can specialize Corollary 2.6 to the case when $E = F \times F$ is the split quadratic algebra, so that $\mathrm{U}(V)$ and $\mathrm{U}(W)$ are general linear groups. However, this specialisation process may not be so transparent or easy to work out, so we will give an independent treatment of this special case in this section. We shall consider the case $\dim_E V = \dim_E W = 2n$, so that

$$\mathrm{U}(V) \cong \mathrm{U}(W) \cong \mathrm{GL}_{2n}(F)$$

and Corollary 2.6 relates the linear and Shalika periods on $\mathrm{GL}_{2n}(F)$.

3.1. Weil representation. Changing notation from the previous section, we let V and W be $2n$-dimensional vector spaces over F. The Weil representation Ω_{2n} of $\mathrm{GL}(V) \times \mathrm{GL}(W)$ is given by

$$\Omega_{2n} = |\det_V|^{-n} \otimes |\det_W|^n \otimes \Omega_{2n}^\natural$$

where $\Omega_{2n}^\natural$ is the natural action of $\mathrm{GL}(V) \times \mathrm{GL}(W)$ on $\mathcal{S}(W^* \otimes V) = \mathcal{S}(\mathrm{Hom}(W, V))$ given by

$$(g, h)\phi(X) = \phi(g^{-1} \circ X \circ h) \quad \text{for } (g, h) \in \mathrm{GL}(V) \times \mathrm{GL}(W).$$

One knows from [**Mi**] that for $\pi \in \mathrm{Irr}(\mathrm{GL}(V))$, $\theta(\pi) \cong \pi^\vee$.

Note in particular that this Weil representation Ω_{2n} is not exactly the specialisation of the one considered in §2 to the case of $E = F \times F$. That specialisation would produce a theta lifting which is the identity map here (since it preserves central character), rather than the contragredient. Hence, we have implicitly twisted the specialisation from §2 by an outer automorphism for one member of the dual pair. In any case, we shall start with the representation Ω_n defined here and give an independent treatment.

3.2. Schrodinger model. To produce a Schrodinger model for Ω_n, we write:

$$V = V_1 + V_2 \quad \text{with } \dim V_i = n.$$

Define a partial Fourier transform

$$\mathcal{F} : \mathcal{S}(W^* \otimes V) \longrightarrow \mathcal{S}(W \otimes V_1^*) \otimes \mathcal{S}(W^* \otimes V_2)$$

by

$$\mathcal{F}(\phi)(T, X) = \int_{V_1 \otimes W^*} \phi(Y, X) \cdot \psi(\mathrm{Tr}_W(T \circ Y))\, dY.$$

Then we may transport the action of $\mathrm{GL}(V) \times \mathrm{GL}(W)$ to the new model using $\mathcal{F}$. In this new model, we may write down the action of $P(V_1) \times \mathrm{GL}(W)$, $P(V_1) = (\mathrm{GL}(V_1) \times \mathrm{GL}(V_2)) \cdot N(V_1)$ is the parabolic subgroup of $\mathrm{GL}(V)$ stabilizing V_1, so that $N(V_1) \cong \mathrm{Hom}(V_2, V_1)$. With $f = \mathcal{F}(\phi)$, the action of $P(V_1) \times \mathrm{GL}(W)$ is then given by:

$$\begin{cases} (h \cdot f)(T, X) = f(h^{-1} \circ T, X \circ h) \qquad \text{for } h \in \mathrm{GL}(W); \\ ((g_1, g_2) \cdot f)(T, X) = (\det(g_1)/\det(g_2))^n \cdot f(T \circ g_1, g_2^{-1} \circ X) \\ \qquad\qquad\qquad\qquad\qquad\qquad\qquad \text{for } (g_1, g_2) \in \mathrm{GL}(V_1) \times \mathrm{GL}(V_2); \\ (n(A) \cdot f)(T, X) = \psi(Tr_{V_2}(XTA)) \cdot f(T, X) \qquad \text{for } A \in \mathrm{Hom}(V_2, V_1). \end{cases}$$

This new model is thus the Schrodinger model in this case.

3.3. Shalika period. Now given $B \in \mathrm{Isom}(V_1, V_2)$, we have the character $\psi_B : n(A) \mapsto \psi(\mathrm{Tr}_{V_2}(BA))$ of $N(V_1)$, whose stabilizer in $\mathrm{GL}(V_1) \times \mathrm{GL}(V_2)$ is the diagonally embedded subgroup

$$G_B = \{(g, BgB^{-1}) : g \in \mathrm{GL}(V_1)\} \subset \mathrm{GL}(V_1) \times \mathrm{GL}(V_2).$$

Then using the Schrodinger model, one sees that

$$\Omega_{N(V_1),\psi_B} \cong \mathcal{S}(\mathcal{O}_B)$$

where

$$\mathcal{O}_B = \{(T, X) \in \mathrm{Hom}(V_1, W) \times \mathrm{Hom}(W, V_2) : XT = B\}.$$

It is easy to see that $G_B \times \mathrm{GL}(W)$ acts transitively on $\mathcal{O}_B$. If we fix $(T_0, X_0) \in \mathcal{O}_B$, so that

$$W = \mathrm{Im}(T_0) \oplus \mathrm{Ker}(X_0) =: W_1 \oplus W_2,$$

then the stabilizer of (T_0, X_0) in $G_B \times \mathrm{GL}(W)$ is the subgroup

$$H_B = \{((g, BgB^{-1}), (T_0 g T_0^{-1}|_{W_1}, h)) : g \in \mathrm{GL}(V_1), h \in \mathrm{GL}(W_2)\}$$

It follows from the above discussion that

$$\Omega_{N(V_1),\psi_B} \cong \mathrm{ind}_{H_B}^{G_B \times \mathrm{GL}(W)} \mathbb{C}$$

Hence, we have the following restatement of Theorem 2.5 in the context of GL-dual pairs (of equal rank):

THEOREM 3.1. *For any $\pi \in \mathrm{Irr}(\mathrm{GL}(W))$ and $\sigma \in \mathrm{Irr}(G_B)$ (with $B \in \mathrm{Isom}(V_1, V_2)$), one has*

$$\mathrm{Hom}_{G_B}(\Theta(\pi), \sigma \boxtimes \psi_B) \cong \mathrm{Hom}_{\mathrm{GL}(W_1) \times \mathrm{GL}(W_2)}(\pi^\vee, \sigma \boxtimes \mathbb{C})$$

where we have regarded σ naturally as a representation of $\mathrm{GL}(W_1)$.

Taking σ to be the trivial representation, the above theorem relates the Shalika period of $\Theta(\pi)$ to the linear period of $\pi^\vee$. In the next section, we shall investigate whether $\Theta(\pi)$ is irreducible.

4. Big Theta Lift

In this section, we study the irreducibility of the big theta lift. Changing notation from the above, we consider the Weil representation Ω_n of $\mathrm{GL}(V)\times\mathrm{GL}(W)\cong\mathrm{GL}_n(F)\times\mathrm{GL}_n(F)$ on $\mathcal{S}(V\otimes W^*)$ as defined in the previous section (or as in [**Mi**]). The goal of the section is to show the following result.

THEOREM 4.1. *Let π be an irreducible generic representation of* $\mathrm{GL}(V)$. *Then*

$$\Theta(\pi)=\theta(\pi)\cong\pi^\vee\in\mathrm{Irr}(\mathrm{GL}(W)).$$

We note that Theorem 4.1 has also been shown in a recent paper of Fang-Sun-Xue [**FSX**, Theorem 1.7] (including archimedean local fields as well); we thank Alberto Minguez for bringing this preprint to our attention. We will later comment briefly on the difference in the techniques used in the proof of [**FSX**] and the one we give below.

As a consequence of Theorem 4.1 and Theorem 3.1, we have:

THEOREM 4.2. *For an irreducible generic representation π of* $\mathrm{GL}_{2n}(F)$

$$\pi \text{ has nonzero Shalika period} \Longleftrightarrow \pi \text{ has nonzero linear period.}$$

We note that Theorem 4.2 has also been shown by Matringe (see [**M1**, Thm. 5.1] and [**M3**, Cor. 1.1]) using different means. Further, Matringe showed that the existence of these periods for π is equivalent to the requirement that the L-parameter of π be a symplectic parameter [**M2**, Cor. 3.15].

The remainder of this section is devoted to the proof of Theorem 4.1. The proof is similar to that for the analogous results for Type I dual pairs of the same rank, as given in [GS] for the case of $\mathrm{Mp}_{2n}\times\mathrm{O}_{2n+1}$ and in [GI] for $\mathrm{U}_n\times\mathrm{U}_n$ and $\mathrm{O}_{2n}\times\mathrm{Sp}_{2n}$. It uses two main ingredients:

(a) a standard result of Kudla [**K1**] on the Jacquet modules of the Weil representation. For Type II dual pairs, a very nice reference is the paper [**Mi**] of Minguez; see [**Mi**, Prop. 3.2].
(b) the invariance of the Weil representation Ω_n under the simultaneous twisting by the outer automorphisms of $\mathrm{GL}(V)$ and $\mathrm{GL}(W)$.

For the sake of comparison, it is interesting to note that the proof of Theorem 4.1 given in [**FSX**] makes use of the theory of Godement-Jacquet zeta integrals and the analytic properties of Rankin-Selberg L-functions, as well as the ingredient (b) above. Our use of (a) means that it will be difficult to extend our proof to the case of archimedean local fields.

Let us elaborate on (b). It will be easiest to think of Ω_n as the action of $\mathrm{GL}_n(F)\times\mathrm{GL}_n(F)$ on $\mathcal{S}(M_n(F))$ and to consider the outer automorphism $c(g,h)=({}^tg^{-1},{}^th^{-1})$ of $\mathrm{GL}_n(F)$. Then the c-twisted representation Ω_n^c is $(g,h)\mapsto\Omega_n(c(g,h))$.

LEMMA 4.3. *The representations Ω_n and Ω_n^c are isomorphic.*

PROOF. An isomorphism of the two representations in question is given by the Fourier transform. More precisely, the choice of an isomorphism $a: V\longrightarrow V^*$ gives rise to an outer automorphism of $\mathrm{GL}(V)$ given by $g\mapsto a^{-1}(g^*)^{-1}a$, where $g^*\in\mathrm{GL}(V^*)$ is naturally induced by g. Likewise, an isomorphism $b: W\longrightarrow W^*$ gives rise to an outer automorphism on $\mathrm{GL}(W)$. So the choice of a and b gives an outer automorphism c of $\mathrm{GL}(V)\times\mathrm{GL}(W)$. The isomorphism

$$a\otimes b^{-1}: V\otimes W^*\longrightarrow V^*\otimes W$$

intertwines the c-twisted action of $\mathrm{GL}(V) \times \mathrm{GL}(W)$ on $V \otimes W^*$ with the standard action on $V^* \otimes W$, i.e. there is a commutative diagram

$$\begin{array}{ccc} V \otimes W^* & \xrightarrow{c(g,(h^*)^{-1})} & V \otimes W^* \\ {\scriptstyle a\otimes b^{-1}}\downarrow & & \downarrow{\scriptstyle a\otimes b^{-1}} \\ V^* \otimes W & \xrightarrow{((g^*)^{-1},h)} & V^* \otimes W \end{array}$$

for every $(g, h) \in \mathrm{GL}(V) \times \mathrm{GL}(W)$. This induces an isomorphism of Ω_n^c (realised on $\mathcal{S}(V \otimes W^*)$) with the Weil representation of $\mathrm{GL}(V) \times \mathrm{GL}(W)$ on $\mathcal{S}(V^* \otimes W)$. Then the Fourier transform

$$\Omega_n = \mathcal{S}(V \otimes W^*) \longrightarrow \mathcal{S}(W \otimes V^*) \cong \Omega_n^c$$

gives an equivariant isomorphism. □

As a consequence of this lemma, we have:

COROLLARY 4.4. *If* $\pi \in \mathrm{Irr}(\mathrm{GL}(W))$, *then*

$$\Theta(\pi^\vee) \cong \Theta(\pi)^c.$$

Hence, $\Theta(\pi)$ *is irreducible if and only if* $\Theta(\pi^\vee)$ *is irreducible.*

4.1. The case $n = 1$. We can now begin the proof of Theorem 4.1. It will be given by induction on n and so we first need to treat the base case: $n = 1$. Here, we are basically considering the natural action of $F^\times \times F^\times$ on $\mathcal{S}(F)$ (i.e. the representation $\Omega_1^\natural$). Since one has a short exact sequence

$$0 \longrightarrow \mathcal{S}(F^\times) \longrightarrow \mathcal{S}(F) \longrightarrow \mathbb{C} \longrightarrow 0,$$

it is clear that $(\Omega_1^\natural)_{F^\times,\chi} \cong \chi \boxtimes \chi^{-1}$ if $\chi \neq 1$. On the other hand, if $\chi = 1$ is the trivial character, then one a priori has $1 \leq \dim \mathcal{S}(F^\times)_1 \leq 2$. However, the analysis in Tate's thesis (see also the expository paper [**K2**] of Kudla where this precise issue was addressed) shows that the Haar measure of $F^\times$, regarded as an invariant linear functional of $\mathcal{S}(F^\times)$ does not extend to an invariant linear functional on $\mathcal{S}(F)$. This implies the desired result for the trivial character.

As pointed out to us by Gordan Savin, one can avoid the appeal to Tate's thesis and simply use the ingredient (b) above. Namely, one has the short exact sequences:

$$0 \longrightarrow \mathcal{S}(F^\times) \longrightarrow \Omega_1 \longrightarrow |-|^{-1/2} \longrightarrow 0$$

$$0 \longrightarrow \mathcal{S}(F^\times) \longrightarrow \Omega_1^c \longrightarrow |-|^{1/2} \longrightarrow 0$$

Since $\Omega_1 \cong \Omega_1^c$, it follows that for any character χ, $\Theta(\chi) = \chi^{-1}$.

We shall henceforth assume that $n \geq 2$.

4.2. Discrete series case. Suppose first that $\pi \in \mathrm{Irr}(\mathrm{GL}(W))$ is an essentially discrete series representation. Because of Corollary 4.4, we may assume that the central character ω_π satisfies

$$|\omega_\pi| = |\det|^\alpha \quad \text{with } \alpha < 1/2.$$

Actually, Corollary 4.4 allows us to assume that $\alpha \leq 0$ but the proof below works as long as $\alpha < 1/2$. Recall that $\Theta(\pi)$ also has central character ω_π. Consider a Jordan-Holder sequence

$$\Theta(\pi) = \Sigma_0 \supset \Sigma_1 \supset \ldots \supset \Sigma_m \supset \Sigma_{m+1} = 0,$$

and set $\Pi_i = \Sigma_i/\Sigma_{i+1}$. We shall first show that each Π_i is an essentially discrete series representation of $\mathrm{GL}(V)$. This will show that $\Theta(\pi)$ is essentially square-integrable (with a central character) and hence is semisimple. By the Howe duality conjecture (a theorem of Minguez [**Mi**] in this case), one then deduces that $\Theta(\pi)$ is irreducible.

To show that Π_i is an essentially discrete series representation for each i, suppose for the sake of contradiction that this is not the case, and let k be the minimal index such that Π_k is not discrete series, so that it is a Langlands subrepresentation of the contragredient of a standard module. In particular, for some maximal parabolic subgroup $P_{t,n-t}$ of $\mathrm{GL}(V) = \mathrm{GL}_n(F)$ stabilizing a t-dimensional subspace and with Levi factor $L_{t,n-t} = \mathrm{GL}_t(F) \times \mathrm{GL}_{n-t}(F)$, the normalised Jacquet module $R_{P_{t,n-t}}(\Pi_k)$ has a quotient

$$R_{P_{t,n-t}}(\Pi_k) \twoheadrightarrow \tau|\det|^a \boxtimes \sigma|\det|^b$$

with τ a discrete series representation with unitary central character, σ is an irreducible representation with unitary central character and $a \leq b$. Consideration of the central character shows that

$$at + b(n-t) = n\alpha < n/2.$$

Since $a \leq b$, we see that

$$a < 1/2.$$

Moreover, the 1-dimensional central torus of L_t consisting of the elements

$$\lambda(z) := \mathrm{diag}(z^{n-t} \cdot I_t, z^{-t} \cdot I_{n-t})$$

acts on $\tau|\det|^a \boxtimes \sigma|\det|^b$ by a character

$$\lambda(z) \mapsto |z|^{t(n-t)(a-b)},$$

where the exponent is non-positive.

By the exactness of the Jacquet functor, we have:

$$R_{P_{t,n-t}}(\Sigma_0) \supset \ldots. \supset R_{P_{t,n-t}}(\Sigma_m) \supset 0$$

with $R_{P_{t,n-t}}(\Pi_i) = R_{P_{t,n-t}}(\Sigma_i)/R_{P_{t,n-t}}(\Sigma_{i+1})$. This gives a short exact sequence

$$0 \longrightarrow \tau|\det|^a \boxtimes \sigma|\det|^b \longrightarrow A \longrightarrow B \longrightarrow 0$$

with A a quotient of $R_{P_{t,n-t}}(\Theta(\pi))$ and $B = R_{P_{t,n-t}}(\Theta(\pi))/R_{P_{t,n-t}}(\Sigma_k)$. We claim that this short exact sequence splits.

For this, we note that for $i < k$, Π_i is discrete series and B is an iterated extensions of these $R_{P_{t,n-t}}(\Pi_i)$ for $i < k$. By Casselman's square-integrability criterion, for any irreducible subquotient of $R_{P_{t,n-t}}(\Pi_i)$ (with $i < k$), the central torus Z acts (up to a unitary character) by

$$\lambda(z) \mapsto |z|^\delta \quad \text{with } \delta > 0.$$

This shows that the above short exact sequence splits and hence there is an equivariant map

$$R_{P_{t,n-t}}(\Theta(\pi)) \twoheadrightarrow \tau|\det|^a \boxtimes \sigma|\det|^b.$$

This gives a nonzero equivariant map

$$R_{P_{t,n-t}}(\Omega) \longrightarrow \pi \otimes (\tau|\det|^a \boxtimes \sigma|\det|^b)$$

so that

$$\pi^* \hookrightarrow \mathrm{Hom}_{L_t}(R_{P_{t,n-t}}(\Omega), \tau|\det|^a \boxtimes \sigma|\det|^b).$$

At this point, we shall appeal to the determination of the normalized Jacquet module of Ω given in [**Mi**, Prop. 3.2]. This gives an equivariant filtration (the so-called Kudla filtration) of $R_{P_{t,n-t}}(\Omega)$:

$$R_{P_{t,n-t}}(\Omega) = R_0 \supset R_1 \supset \ldots \supset R_t \supset R_{t+1} = 0,$$

whose successive quotients are described in [**Mi**, Prop. 3.2] (though note that [**Mi**, Prop. 3.2] works with $\Omega_n^\natural$ rather than Ω_n) . Arguing as in [**GS**, Lemma 7.4], we see that

$$\mathrm{Hom}_{L_{t,n-t}}(R_i/R_{i+1}, \tau|\det|^a \boxtimes \sigma|\det|^b) = 0 \qquad \text{for } i < t.$$

It is here that we use the fact that $a < 1/2$. Indeed, by [**Mi**, Prop. 3.2], we have:

$$R_i/R_{i+1} \cong \mathrm{ind}_{P_{t-i,i} \times \mathrm{GL}_{n-t} \times Q_{i,n-i}}^{\mathrm{GL}_t \times \mathrm{GL}_{n-t} \times \mathrm{GL}_n} \Xi_i \quad \text{(normalized induction)}$$

for some module Ξ_i on which GL_{t-i} acts by the character $|\det|^{(t-i)/2}$, and where we have used $Q_{i,n-i}$ to denote a maximal parabolic subgroup of $\mathrm{GL}(W) \cong \mathrm{GL}_n$ stabilizing an i-dimensional subspace. Hence,

$$\mathrm{Hom}_{L_{t,n-t}}(R_i/R_{i+1}, \tau|\det|^a \boxtimes \sigma|\det|^b)$$
$$\cong \mathrm{Hom}_{\mathrm{GL}_{t-i} \times \mathrm{GL}_i \times \mathrm{GL}_{n-t}}\left(\mathrm{ind}_{Q_{i,n-i}}^{\mathrm{GL}_n} \Xi_i, R_{\overline{P}_{t-i,i}}(|\det|^a \cdot \tau) \boxtimes |\det|^b \cdot \sigma\right)$$

Now

$$R_{\overline{P}_{t-i,i}}(|\det|^a \cdot \tau_1) = \delta_1|\det|^{t_1+a} \boxtimes \delta_2|\det|^{t_2+a} \quad \text{(if it is nonzero)}$$

with δ_1 and δ_2 unitary discrete series representations (with unitary central characters),

$$t_1 \le t_2 \quad \text{and} \quad t_1(t-i) + t_2 i = 0,$$

so that

$$t_1 \le 0 \quad \text{and} \quad t_1 + a < 1/2,$$

Since $(t-i)/2 \ge 1/2$ if $i < t$, we see that the above Hom space is 0, as desired.

From this, we see that

$$\pi^* \hookrightarrow \mathrm{Hom}_{L_{t,n-t}}(R_t, \tau|\det|^a \boxtimes \sigma|\det|^b).$$

Now,

$$R_t \cong \mathrm{Ind}_{L_{t,n-t} \times Q_{t,n-t}}^{L_{t,n-t} \times \mathrm{GL}_n} (\mathcal{S}(\mathrm{GL}_t) \otimes \Omega_{n-t}),$$

so that

$$\mathrm{Hom}_{L_{t,n-t}}(R_t, \tau|\det|^a \boxtimes \sigma|\det|^b) = \left(\tau^\vee|\det|^{-a} \times \Theta_{n-t}(\sigma)|\det|^{-b}\right)^*.$$

This implies that

$$\pi^\vee \hookrightarrow \left(\tau^\vee|\det|^{-a} \times \Theta_{n-t}(\sigma)|\det|^{-b}\right)^\vee = \tau|\det|^a \times \Theta_{n-t}(\sigma)^\vee|\det|^b$$

so that there is a nonzero map

$$R_{P_{t,n-t}}(\pi^\vee) \longrightarrow \tau|\det|^a \boxtimes \Theta_{n-t}(\sigma)^\vee|\det|^b.$$

Since $\pi^\vee$ is also a discrete series representation and $a - b \le 0$, this contradicts Casselman's square-integrability criterion. With this contradiction, we have shown the desired result when π is an essentially discrete series representation such that $|\omega_\pi| = |\det|^\alpha$ with $\alpha < 1/2$. As mentioned earlier, this implies the result for any essentially discrete series representation, in view of Corollary 4.4.

4.3. General case. We can now complete the proof of Theorem 4.1 in the general case by induction. Suppose that $n \geq 2$ and $\pi \in \mathrm{Irr}(\mathrm{GL}(W))$ is generic. Then by [**Ze**, Thm. 9.7]

$$\pi \cong \tau_1 |\det|^{s_1} \times ... \times \tau_r |\det|^{s_r} \quad \text{with } s_1 \leq s_2 \leq .... \leq s_r$$

with τ_i discrete series representations of GL_{n_i} with unitary central characters on GL_{n_i}. Since we have already resolved the case of essentially discrete series representations, we may assume that $r \geq 2$, i.e. π is not essentially discrete series. Moreover, Corollary 4.4 allows us to replace π by $\pi^\vee$. Hence, there is no loss of generality in assuming that

$$s_1 \leq 0.$$

Writing

$$\pi \cong \tau_1 |\det|^{s_1} \times \sigma$$

where $\sigma = \tau_2 |\det|^{s_2} \times ... \times \tau_r |\det|^{s_r}$, we see by induction hypothesis that $\Theta(\sigma) = \sigma^\vee$ is irreducible.

With $P = P_{n_1, n-n_1}$ the relevant maximal parabolic subgroup, we have

$$\Theta(\pi)^* \cong \mathrm{Hom}_{\mathrm{GL}_n}(\Omega_n, \tau_1 |\det|^{s_1} \times \sigma) \cong \mathrm{Hom}_L(R_P(\Omega_n), \tau_1 |\det|^{s_1} \boxtimes \sigma)$$

Again, we may appeal to the result of [**Mi**, Prop. 3.2] for the computation of the Jacquet module of Ω_n. Since $s_1 \leq 0$, we deduce as in our earlier application of Kudla's filtration that only the bottom piece of the filtration can contribute to the Hom space. This implies that

$$\Theta_{n-n_1}(\sigma) \times \tau_1^\vee |\det|^{-s_1} \twoheadrightarrow \Theta(\pi) \neq 0.$$

But the representation on the LHS is

$$\sigma^\vee \times \tau_1 |\det|^{-s_1} \cong \pi^\vee.$$

Hence we have shown that $\Theta(\pi) = \pi^\vee = \theta(\pi)$, as desired. This completes the proof of Theorem 4.1.

5. A Theorem of Tunnell and Saito

In this section, we shall consider our second example of the transfer of periods under the theta correspondence; this will allow us to give another proof of the Tunnell-Saito theorem. Hence we will be considering torus periods for representations of $\mathrm{GL}_2(F)$ and its inner form $D^\times$, where D is a quaternion division F-algebra. As the periods here are over reductive subgroups, the manner in which torus periods are transferred are of a different nature from that in the previous two sections: it is effected by a see-saw diagram. As usual, let us recall some of the relevant notions.

5.1. Torus periods. Let B be a quaternion F-algebra (possibly spit) with norm map N_B and trace map Tr_B. Let B_0 be the subspace of trace 0 elements in B. Consider the quadratic space $V_B = (B_0, -N_B)$, so that $\mathrm{SO}(V) \cong PB^\times$ with an element of $PB^\times$ acting on B_0 by conjugation. Taking $x_a \in B_0$ with $x^2 = a$, we obtain an algebra embedding

$$F_a = F[t]/(t^2 - a) \hookrightarrow B.$$

The stabilizer of x_a in $\mathrm{SO}(V_B)$ is equal to the 1-dimensional torus

$$T_a = \text{image of } F(a)^\times \text{ in } PB^\times.$$

5.2. Theorem of Tunnell-Saito. The following is a classic theorem of Tunnell and Saito:

THEOREM 5.1. *Let π be an infinite-dimensional representation of $\mathrm{GL}_2(F)$ and χ a character of the torus $T_a \cong F(a)^\times$ such that $\chi|_{F^\times} = \omega_\pi$. Let π_D be the Jacquet-Langlands transfer of π to $D^\times$ (where D is the quaternion division F-algebra). Then we have:*

$$\dim \mathrm{Hom}_{T_a}(\pi, \chi) + \dim \mathrm{Hom}_{T_a}(\pi_D, \chi) = 1.$$

Moreover, if $\pi(\chi)$ is the dihedral representation of $\mathrm{GL}_2(F)$ associated to χ, then

$$\mathrm{Hom}_{T_a}(\pi, \chi) \neq 0 \Longleftrightarrow \epsilon(1/2, \pi \times \pi(\chi)^\vee, \psi) \cdot \chi_a(-1) = 1,$$

and

$$\mathrm{Hom}_{T_a}(\pi_D, \chi) \neq 0 \Longleftrightarrow \epsilon(1/2, \pi \otimes \pi(\chi)^\vee, \psi) \cdot \chi_a(-1) = -1,$$

This theorem is the simplest case of the local Gross-Prasad conjecture for special orthogonal groups (which has been shown by Waldspurger). It was first proved by Tunnell [**T**] by a direct K-type computation, and then given a completely different proof by H. Saito [**S**] using the base change character identities. Yet another proof was later given by D. Prasad [**P**], using a global-to-local argument. There is also a global analog of this theorem which is a well-known result of Waldspurger [**W**] relating global torus periods to certain central critical L-values. This global result of Waldspurger was shown using theta correspondence and is an input in Prasad's global-to-local argument. In the following, we shall give a fourth proof based on the theta correspondence and the properties of the local Rankin-Selberg integrals for $\mathrm{GL}(2) \times \mathrm{GL}(2)$. The proof we give below is naturally parallel to Waldspurger's proof of the global analog of the theorem. But as the theorem has been proved thrice already, our rendition will be slightly sketchy. We remark that an analogous situation in a higher rank setting is the relation between trilinear forms on GL_2 and the triple product epsilon factor (i.e. D. Prasad's thesis); see [**G**] for a treatment based on theta correspondence.

5.3. Dual pairs. With B a quaternion F-algebra, we set $\epsilon(B) = +1$ or -1 depending on whether B splits or not. Consider the 4-dimensional quadratic space (B, N_B). Then the identity component of the associated similitude group

$$\mathrm{GSO}(B) \cong (B^\times \times B^\times)/\Delta\mathbb{G}_m$$

with (b_1, b_2) on the RHS acting on B by

$$(b_1, b_2) : b \mapsto b_1 \cdot b \cdot b_2^{-1}.$$

The similitude group $\mathrm{GO}(B)$ is generated by $\mathrm{GSO}(B)$ together with the element $b \mapsto \overline{b}$ which has determinant -1.

With W denoting the 2-dimensional symplectic space, one has a dual pair $\mathrm{Sp}(W) \times \mathrm{O}(B)$ in $\mathrm{Sp}(W \otimes B)$. For the purpose of proving the Tunnell-Saito theorem, it is necessary to work with a similitude version of dual pairs: $\mathrm{GSp}(W) \times \mathrm{GO}(B)$.

Given a quadratic F-algebra E, with associated quadratic character χ_E of $F^\times$, fix an embedding $E \hookrightarrow B$ if it exists, and write

$$B = E \oplus E \cdot b,$$

where the conjugation action of b on E is the Galois action. Then we note that

$$\epsilon(B) = \chi_E(-N_B(b)).$$

The two dimensional quadratic subspaces E and $E \cdot b$ have discriminant algebra equal to E. For such a quadratic space V_E, we set $\epsilon(V_E) = +1$ or -1 according to whether V_E is isomorphic to (E, N_E) or not.

In any case, $\mathrm{GSO}(E) \cong \mathrm{GSO}(E \cdot b) \cong E^\times$, with $E^\times$ acting by left multiplication on E and $E \cdot b$. There is a natural embedding

$$\iota : (\mathrm{GSO}(E) \times \mathrm{GSO}(E \cdot b))^0 \hookrightarrow \mathrm{GSO}(B) \cong (B^\times \times B^\times)/\Delta F^\times,$$

where the LHS consists of elements $(h_1, h_2) \in E^\times \times E^\times$ such that $N(h_1) = N(h_2)$. The image of this embedding is the subgroup

$$(E^\times \times E^\times)/\Delta F^\times \subset (B^\times \times B^\times)/\Delta F^\times.$$

More precisely, if

$$(h_1, h_2) \mapsto (b_1, b_2) \in (E^\times \times E^\times)/\Delta F^\times,$$

then

$$h_1 = b_1/b_2 \quad \text{and} \quad h_2 = b_1/b_2^\sigma$$

so that

$$h_1/h_2 = b_2^\sigma/b_2 \quad \text{and} \quad h_2^\sigma/h_1 = b_1^\sigma/b_1.$$

In particular, observe that for a given character χ of $E^\times$

$$(\chi \boxtimes \chi^{-1}) \cdot \iota = \chi \boxtimes 1.$$

5.4. A see-saw diagram. In view of the above discussion, one has the following see-saw diagram:

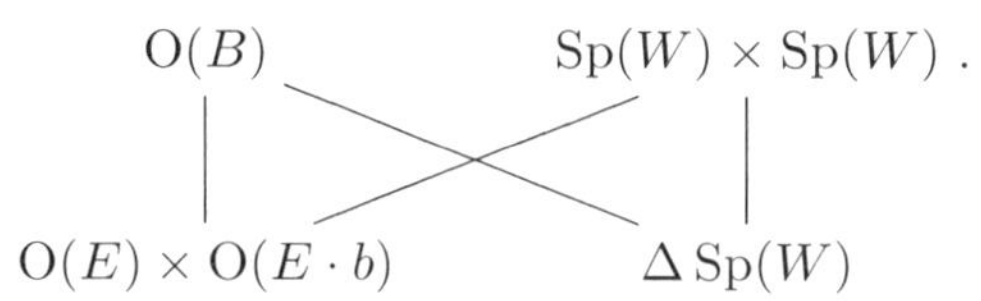

In fact, we shall consider its similitude version:

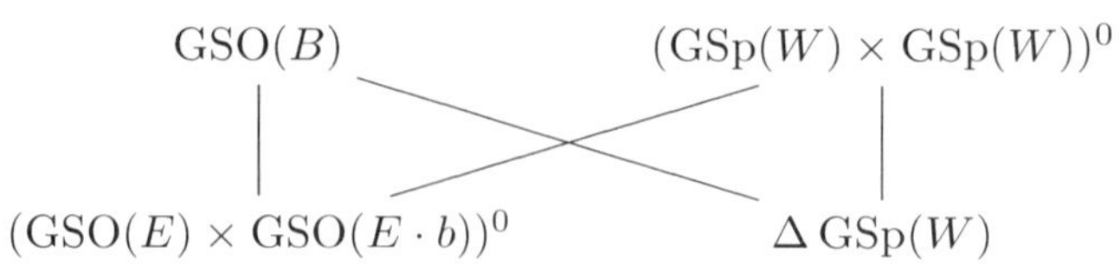

Since we will be using similitude theta correspondence, we shall review it briefly.

5.5. Similitude theta correspondence. For the dual pair $\mathrm{Sp}(W) \times \mathrm{O}(B)$, the Weil representation $\omega_{\psi,W,B}$ can be naturally extended to a slightly bigger group:

$$R_{W,B} = \{(g,h) \in \mathrm{GSp}(W) \times \mathrm{GO}(B) : \det(g) = \mathrm{sim}(h)\} \subset \mathrm{GSp}(W) \times \mathrm{GO}(B),$$

where sim denotes the similitude character on $\mathrm{GO}(B)$. Note here that the maps det and sim are both surjective onto $F^\times$. Then one may form the compactly-induced representation

$$\Omega_{W,B} = \mathrm{ind}_{R_{W,B}}^{\mathrm{GSp}(W)\times \mathrm{GO}(B)} \omega_{\psi,W,B}$$

of $\mathrm{GSp}(W) \times \mathrm{GO}(B)$, which is independent of ψ. We may thus consider the theta correspondence for this similitude dual pair.

PROPOSITION 5.2. *If π is an infinite-dimensional representation of* $\mathrm{GSp}(W)$, *then*

$$\Theta_{W,B}(\pi) = \pi_B \boxtimes \pi_B^\vee$$

as a representation of $(B^\times \times B^\times)/\Delta F^\times$.

Similarly, for the pair $\mathrm{GO}(E) \times \mathrm{GSp}(W)$ (or $\mathrm{GO}(E \cdot b) \times \mathrm{GSp}(W)$), the Weil representation extends to

$$R_{W,E} = \{(g,h) \in \mathrm{GSp}(W) \times \mathrm{GO}(E) : \det(g) = \mathrm{sim}(h)\}.$$

Note however that the norm map on $E^\times$ is not surjective onto $F^\times$ if E is a field, so that $R_{W,E}$ is actually a subgroup of $\mathrm{GSp}(W)^+ \times \mathrm{GO}(E)$ where

$$\mathrm{GSp}(W)^+ = \{g \in \mathrm{GSp}(W) : \det(g) \in N_{E/F}E^\times\}.$$

In this case, we consider the representation of $\mathrm{GSp}(W)^+ \times \mathrm{GO}(E)$ defined by:

$$\Omega^+_{W,E} = \mathrm{ind}_{R_{W,E}}^{\mathrm{GSp}(W)^+\times \mathrm{GO}(E)} \omega_{\psi,W,E},$$

and similarly for $\Omega^+_{W,E\cdot b}$. Using these, one may consider theta correspondences for these similitude dual pairs.

It is important to note that we do not induce the Weil representation all the way up to $\mathrm{GSp}(W) \times \mathrm{GO}(E)$ when E is a field, but to a subgroup of index 2. This may seem like a minor technicality, but this annoying minor detail is the *raison d'être* for the dichotomy of torus period in the local theorem, at least from the point of view of our proof. In any case, the representation $\Omega^+_{W,E}$ depends on the $N_{E/F}E^\times$-orbit of ψ. If ψ' is an additive character which lies in the other $N_{E/F}E^\times$-orbit, then we set $\Omega^-_{W,E}$ for the induction of $\omega_{\psi',W,E}$. Note that

$$\Omega^+_{W,Eb} \cong \Omega^{+\chi_E(N_B(b))}_{W,E} \quad \text{and} \quad \Omega^-_{W,Eb} \cong \Omega^{-\chi_E(N_B(b)}_{W,E}.$$

PROPOSITION 5.3. *Consider the theta correspondence for* $\mathrm{GSp}(W)^+ \times \mathrm{GSO}(E)$ *under $\Omega^\pm_{W,E}$. If χ is a character of $E^\times \cong$* $\mathrm{GSO}(E)$ *and $\pi(\chi)$ the associated dihedral representation of* $\mathrm{GL}_2(F)$, *then*

$$\pi(\chi)|_{\mathrm{GSp}(W)^+} \cong \Theta^+_{W,E}(\chi) \oplus \Theta^-_{W,E}(\chi).$$

Moreover, the constituent $\Theta^+_{W,E}(\chi)$ is characterised as the unique ψ-generic summand.

We shall set

$$\pi(\chi)^+ = \Theta^+_{W,E}(\chi) \qquad \text{and} \qquad \pi(\chi)^- = \Theta^-_{W,E}(\chi).$$

On the other hand, one has:

PROPOSITION 5.4. *Consider the principal series representation* $I(1, \chi_E) = 1 \times \chi_E$ *of* $\mathrm{GSp}(W) \cong \mathrm{GL}_2(F)$, *and write*

$$I(1, \chi_E) = I(1, \chi_E)^+ \oplus I(1, \chi_E)^-$$

with $I(1, \chi_E)^+$ *the unique* ψ*-generic summand. Then under the theta correspondence for* $\mathrm{GSp}(W)^+ \times \mathrm{GSO}(Eb)$, *one has*

$$\Theta^+_{W,Eb}(1) = I(1, \chi_E)^{+\chi_E(N_B(b))} \text{ and } \quad \Theta^-_{W,Eb}(1) = I(1, \chi_E)^{-\chi_E(N_B(b))}.$$

5.6. Local see-saw identity. We can now begin the proof of Theorem 5.1. Consider the space

$$\Sigma_B := \mathrm{Hom}_{\mathrm{GSp}(W) \times H}(\Omega_{W,B}, \pi \boxtimes (\chi \boxtimes 1)),$$

where

$$H = (\mathrm{GSO}(E) \times \mathrm{GSO}(Eb))^0.$$

On one hand, it is equal to

$$\mathrm{Hom}_H(\Theta_{W,B}(\pi), \chi \boxtimes 1) \cong \mathrm{Hom}_{E^\times}(\pi_B, \chi) \otimes \mathrm{Hom}_{E^\times}(\pi_B^\vee, \chi^\vee),$$

which is precisely the space we want to understand. On the other hand, by induction in stages,

$$\Omega_{W,B} = \mathrm{ind}_{\mathrm{GSp}(W)^+ \times H}^{\mathrm{GSp}(W) \times H} \mathrm{ind}_{(\mathrm{GSp}(W) \times H)^0}^{\mathrm{GSp}(W)^+ \times H} \omega_{\psi,W,B}.$$

Thus, by Frobenius reciprocity,

$$\Sigma_B = \mathrm{Hom}_{\mathrm{GSp}(W)^+ \times H}(\mathrm{ind}_{(\mathrm{GSp}(W) \times H)^0}^{\mathrm{GSp}(W)^+ \times H} \omega_{\psi,W,E} \otimes \omega_{\psi,W,Eb}, \pi \boxtimes \chi \boxtimes 1)$$

which is in turn equal to

$$\mathrm{Hom}_{\mathrm{GSp}(W)^+}(\Theta^+_{W,E}(\chi) \boxtimes \Theta^+_{W,Eb}(1), \pi).$$

By Frobenius reciprocity again, we obtain

$$\Sigma_B = \mathrm{Hom}_{\mathrm{GSp}(W)}(\mathrm{Ind}_{\mathrm{GSp}(W)^+}^{\mathrm{GSp}(W)} (\Theta^+_{W,E}(\chi) \boxtimes \Theta^+_{W,Eb}(1)), \pi).$$

The see-saw diagram thus allows us to transfer a period problem from one side of the see-saw to the other. Now we note:

LEMMA 5.5. *As a representation of* $\mathrm{GSp}(W)$,

$$\begin{aligned}\Pi_B &:= \mathrm{Ind}_{\mathrm{GSp}(W)^+}^{\mathrm{GSp}(W)} (\Theta^+_{W,E}(\chi) \boxtimes \Theta^+_{W,Eb}(1)) \\ &= (\Theta^+_{W,E}(\chi) \boxtimes \Theta^+_{W,Eb}(1)) \oplus (\Theta^-_{W,E}(\chi) \boxtimes \Theta^-_{W,Eb}(1)).\end{aligned}$$

Thus, in view of Proposition 5.3, Proposition 5.4 and Lemma 5.5, if we sum over the two possible B's, we obtain

$$\bigoplus_B \Pi_B = \pi(\chi) \otimes I(1, \chi_E).$$

Hence, we have:

$$\begin{aligned}&\bigoplus_B \mathrm{Hom}_{E^\times}(\pi_B, \chi) \otimes \mathrm{Hom}_{E^\times}(\pi_B^\vee, \chi^{-1}) \\ \cong &\bigoplus_B \mathrm{Hom}_{\mathrm{GSp}(W)}(\Pi_B, \pi) \\ = &\mathrm{Hom}_{\mathrm{GSp}(W)}(\pi(\chi) \otimes I(1, \chi_E) \otimes \pi^\vee, \mathbb{C}).\end{aligned}$$

This latter space of trilinear forms can be rather easily shown to be 1-dimensional (by Mackey theory). Since it is not hard to see that $Hom_{E^\times}(\pi_B, \chi) \cong \mathrm{Hom}_{E^\times}(\pi_B^\vee, \chi^{-1})$, we obtain

$$\sum_B (\dim \mathrm{Hom}_{E^\times}(\pi_B, \chi))^2 = 1,$$

and hence

$$\sum_B \dim \mathrm{Hom}_{E^\times}(\pi_B, \chi) = 1.$$

This proves the first part of Theorem 5.1. Moreover, we have seen that

$$\mathrm{Hom}_{E^\times}(\pi_B, \chi) \neq 0 \Longleftrightarrow \mathrm{Hom}_{\mathrm{GSp}(W)}(\Pi_B, \pi) \neq 0.$$

5.7. Local Rankin-Selberg integral. To prove the epsilon dichotomy part of Theorem 5.1, we need to introduce the local Rankin-Selberg integrals for the local factors of $\mathrm{GL}(2) \times \mathrm{GL}(2)$, which was studied by Jacquet [**J**].

Start with irreducible generic representations π_1 and π_2 of $\mathrm{GSp}(W) \cong \mathrm{GL}_2(F)$ and let

$$l_1 \in \mathrm{Hom}_N(\pi_1, \psi) \quad \text{and} \quad l_2 \in \mathrm{Hom}_N(\pi_2, \psi^{-1})$$

be nonzero Whittaker functionals. Consider also the family of principal series representations

$$I(s, \chi_1, \chi_2) := \pi(\chi_1 |-|^s, \chi_2 |-|^{-s})$$

and assume that

$$\chi_1 \cdot \chi_2 \cdot \omega_{\pi_1} \cdot \omega_{\pi_2} = 1.$$

For $v_i \in \pi_i$ and $\Phi(s) \in I(s, \chi, \chi_2)$, the associated local Rankin-Selberg integral is given by:

$$Z(s, f_1, f_2, \Phi) = \int_{N(F)\backslash \mathrm{PGL}_2(F)} l_1(gv_1) \cdot l_2(gv_2) \cdot \Phi(s, g)\, dg.$$

This converges for $Re(s)$ sufficiently large and has meromorphic continuation to all of $\mathbb{C}$. The local Rankin-Selberg L-factor $L(s, \pi_1 \times \pi_2 \times \chi_1)$ is defined to be the GCD of this family of zeta integrals, as $\Phi(s)$ ranges over "good" sections of $I(s, \chi_1, \chi_2)$. Thus the normalized zeta integral

$$Z^*(s, f_1, f_2, \Phi) = \frac{Z(s, f_1, f_2, \Phi)}{L(s, \pi_1 \times \pi_2 \times \chi_1)}$$

is thus an entire function, which is nonzero at any given $s_0 \in \mathbb{C}$ for some choice of data. In particular, for each $s \in \mathbb{C}$, one obtains a nonzero $\mathrm{GL}_2(F)$-invariant map

$$Z^*(s, -, -, -) : \pi_1 \otimes \pi_2 \otimes I(s, \chi_1, \chi_2) \longrightarrow \mathbb{C}.$$

The local zeta integral satisfies a local functional equation which we shall describe presently. Let

$$M(s, \chi_1, \chi_2) : I(s, \chi_1, \chi_2) \longrightarrow I(-s, \chi_2, \chi_1)$$

be the standard intertwining operator. Then for an appropriate normalized version $M^*_\psi(s, \chi_1, \chi_2)$, one has the identity

$$Z^*(-s, v_1, v_2, M^*_\psi(s, \chi_1, \chi_2)\Phi) = \epsilon(s + \frac{1}{2}, \pi_1 \times \pi_2 \times \chi_1, \psi) \cdot Z^*(s, v_1, v_2, \Phi).$$

The constant of proportionality $\epsilon(s, \pi_1 \times \pi_2 \times \chi_1, \psi)$ is defined to be the Rankin-Selberg epsilon factor attached to $\pi_1 \times \pi_2 \times \chi_1$.

5.8. The case at hand. We shall now specialize the above description to the case at hand, namely, set

$$\pi_1 = \pi(\chi), \quad \pi_2 = \pi^\vee, \quad I(s,\chi_1,\chi_2) = I(s,1,\chi_E).$$

The local zeta integral $Z^*(0)$ is a nonzero element of the 1-dimensional vector space

$$\mathrm{Hom}_{\mathrm{GSp}(W)}(\pi(\chi)\otimes I(1,\chi_E)\otimes\pi^\vee,\mathbb{C}).$$

Recalling that as a $\mathrm{GSp}(W)$-module,

$$\pi(\chi)\otimes I(1,\chi_E) = \bigoplus_B \Pi_B,$$

we deduce that

$$\mathrm{Hom}_{E^\times}(\pi_B,\chi)\neq 0 \Longleftrightarrow Z^*(0) \text{ is nonzero when restricted to } \Pi_B\otimes\pi^\vee.$$

To study the effect of $Z^*(0)$ on the submodule Π_B, we take note of the following two lemmas.

LEMMA 5.6. *(i) Via the restriction of functions, we have a* $\mathrm{GL}_2(F)^+$*-equivariant isomorphism*

$$I(s,1,\chi_E)\longrightarrow J(s):=\mathrm{Ind}_{B^+}^{\mathrm{GL}_2^+}|-|^s\times|-|^{-s}.$$

In particular, at $s=0$, one has a $\mathrm{GL}_2(F)^+$*-equivariant isomorphism*

$$I(1,\chi_E)\longrightarrow J(0)=\mathrm{Ind}_{B^+}^{\mathrm{GL}_2^+}1.$$

Similarly, the restriction of functions gives an identification

$$I(\chi_E,1)\longrightarrow J(0).$$

Let $J(0)^\pm$ be the image of $I(1,\chi_E)^\pm$ under the restriction map, so that $J(0)^+$ is ψ-generic.
(ii) The normalized intertwining operator $M_\psi^*(0,1,\chi)$ induces a $\mathrm{GL}_2(F)^+$*-equivariant non-scalar map*

$$M_\psi^*(0): J(0)\longrightarrow J(0).$$

satisfying $M_\psi^*(0)^2=1$. Further:

$$M_\psi^*(0)=\begin{cases}+1 \text{ on } J(0)^+;\\ -1 \text{ on } J(0)^-.\end{cases}$$

LEMMA 5.7. *Consider the modified zeta integrals*

$$Z_+(s,v_1,v_2,\Phi)=\int_{N(F)\backslash\mathrm{PGL}_2(F)^+} l_1(gv_1)\cdot l_2(gv_2)\cdot\Phi_s(g)\,dg.$$

Then the family

$$\{Z_+(s,v_1,v_2,\Phi): v_1\in\pi(\chi)^+, v_2\in\pi^\vee \text{ and } \Phi\in I(s) \text{ arbitrary}\}$$

is identical to the family of Rankin-Selberg zeta integrals.

PROOF. Let $c\in\mathrm{GL}_2(F)\smallsetminus\mathrm{GL}_2(F)^+$. Then we can write

$$Z(s,v_1,v_2,\Phi)=$$

$$\int_{N(F)\backslash\mathrm{PGL}_2(F)^+} l_1(gv_1)\cdot l_2(gv_2)\cdot\Phi_s(g)\,dg+\int_{N(F)\backslash\mathrm{PGL}_2(F)^+} l_1(gcv_1)\cdot l_2(gcv_2)\cdot\Phi_s(g)\,dg.$$

Now observe that for $v_1 \in \pi(\chi)^+$,

$$l_1(gcv_1) = 0 \quad \text{for } v_1 \in \pi(\chi)^+ \text{ and } g \in \mathrm{GL}_2(F)^+,$$

so that the second term in the above identity vanishes and

$$Z(s, v_1, v_2, \Phi) = Z_+(s, v_1, v_2, \Phi).$$

Similarly, if $v_1 \in \pi(\chi)^-$, then the first term vanishes, so that

$$Z(s, v_1, v_2, \Phi) = Z_+(s, c \cdot v_1, v_2, \Phi).$$

This proves the lemma. □

5.9. Proof of Theorem 5.1. We can now complete the proof of Theorem 5.1. We have:

$$\begin{aligned}
&\mathrm{Hom}_{E^\times}(\pi_B, \chi) \neq 0 \\
\Longleftrightarrow\, &\mathrm{Hom}_{\mathrm{GL}_2}(\pi^\vee \otimes \Pi_B, \mathbb{C}) \neq 0 \\
\Longleftrightarrow\, &Z^*(0) \text{ is nonzero on } \pi^\vee \otimes \Pi_B \\
\Longleftrightarrow\, &Z_+^*(0) \text{ is nonzero on } \pi^\vee \otimes \pi(\chi)^+ \otimes J(0)^{\epsilon(B)\cdot\chi_E(-1)}.
\end{aligned}$$

Using the local functional equation

$$Z_+^*(0) \circ (1_{\pi(\chi)^+} \otimes 1_{\pi^\vee} \otimes M_\psi^*(0)) = \epsilon(1/2, \pi(\chi) \times \pi^\vee, \psi) \cdot Z_+^*(0)$$

and taking note of Lemma 5.6 and Lemma 5.7, we see that

$$Z_+^*(0) \text{ is nonzero on } \pi^\vee \otimes \pi(\chi)^+ \otimes I(0)^{\epsilon(B)\cdot\chi_E(-1)}$$

if and only if

$$\epsilon(1/2, \pi(\chi) \times \pi^\vee, \psi) = \epsilon(B) \cdot \chi_E(-1).$$

This proves Theorem 5.1.

References

[BZ] J. Bernstein and A. Zelevinsky, *Representations of the group $GL(n, F)$, where F is a local non-Archimedean field* (Russian), Uspehi Mat. Nauk **31** (1976), no. 3(189), 5–70. MR0425030

[FSX] Y. Fang, B. Sun, and H. Xue, *Godement-Jacquet L-functions and full theta lifts*, Math. Z. **289** (2018), no. 1-2, 593–604, DOI 10.1007/s00209-017-1967-z. MR3803804

[G] W. T. Gan, *Trilinear forms and triple product epsilon factors*, Int. Math. Res. Not. IMRN **15** (2008), Art. ID rnn058, 15, DOI 10.1093/imrn/rnn058. MR2438071

[GI] W. T. Gan and A. Ichino, *Formal degrees and local theta correspondence*, Invent. Math. **195** (2014), no. 3, 509–672, DOI 10.1007/s00222-013-0460-5. MR3166215

[GS] W. T. Gan and G. Savin, *Representations of metaplectic groups I: epsilon dichotomy and local Langlands correspondence*, Compos. Math. **148** (2012), no. 6, 1655–1694, DOI 10.1112/S0010437X12000486. MR2999299

[GZ1] R. Gomez and C.-B. Zhu, *Local theta lifting of generalized Whittaker models associated to nilpotent orbits*, Geom. Funct. Anal. **24** (2014), no. 3, 796–853, DOI 10.1007/s00039-014-0276-5. MR3213830

[GZ2] R. Gomez and C. B. Zhu, *Local theta lifting of generalized Whittaker models associated to nilpotent orbits II*, in preparation.

[H] M. Hanzer, *Generalized Shalika model on $\mathrm{SO}_{4n}(F)$, symplectic linear model on $\mathrm{Sp}_{4n}(F)$ and theta correspondence* (English, with English and Croatian summaries), Rad Hrvat. Akad. Znan. Umjet. Mat. Znan. **19(523)** (2015), 55–68. MR3456592

[J] H. Jacquet, *Automorphic forms on* GL(2). *Part II*, Lecture Notes in Mathematics, Vol. 278, Springer-Verlag, Berlin-New York, 1972. MR0562503

[K1] S. S. Kudla, *On the local theta-correspondence*, Invent. Math. **83** (1986), no. 2, 229–255, DOI 10.1007/BF01388961. MR818351

[K2] S. S. Kudla, *Tate's thesis*, An introduction to the Langlands program (Jerusalem, 2001), Birkhäuser Boston, Boston, MA, 2003, pp. 109–131. MR1990377

[L] B. Liu, *Model transition under local theta correspondence*, J. Algebra **448** (2016), 431–445, DOI 10.1016/j.jalgebra.2015.09.041. MR3438317

[M1] N. Matringe, *Linear and Shalika local periods for the mirabolic group, and some consequences*, J. Number Theory **138** (2014), 1–19, DOI 10.1016/j.jnt.2013.11.012. MR3168918

[M2] N. Matringe, *On the local Bump-Friedberg L-function*, J. Reine Angew. Math. **709** (2015), 119–170, DOI 10.1515/crelle-2013-0083. MR3430877

[M3] N. Matringe, *Shalika periods and parabolic induction for $GL(n)$ over a non-archimedean local field*, Bull. Lond. Math. Soc. **49** (2017), no. 3, 417–427, DOI 10.1112/blms.12020. MR3723627

[Mi] A. Mínguez, *Correspondance de Howe explicite: paires duales de type II* (French, with English and French summaries), Ann. Sci. Éc. Norm. Supér. (4) **41** (2008), no. 5, 717–741, DOI 10.24033/asens.2080. MR2504432

[P] D. Prasad, *Relating invariant linear form and local epsilon factors via global methods*, Duke Math. J. **138** (2007), no. 2, 233–261, DOI 10.1215/S0012-7094-07-13823-7. With an appendix by Hiroshi Saito. MR2318284

[R] S. Rallis, *On the Howe duality conjecture*, Compositio Math. **51** (1984), no. 3, 333–399. MR743016

[S] H. Saito, *On Tunnell's formula for characters of* GL(2), Compositio Math. **85** (1993), no. 1, 99–108. MR1199206

[T] J. B. Tunnell, *Local ϵ-factors and characters of* GL(2), Amer. J. Math. **105** (1983), no. 6, 1277–1307, DOI 10.2307/2374441. MR721997

[W] J.-L. Waldspurger, *Sur les valeurs de certaines fonctions L automorphes en leur centre de symétrie* (French), Compositio Math. **54** (1985), no. 2, 173–242. MR783511

[Ze] A. V. Zelevinsky, *Induced representations of reductive $\mathfrak{p}$-adic groups. II. On irreducible representations of* GL(n), Ann. Sci. École Norm. Sup. (4) **13** (1980), no. 2, 165–210. MR584084

[Z] C.B. Zhu, *Vanishing and non-vanishing in local theta correspondence*, these proceedings.

Department of Mathematics, National University of Singapore, Block S17, 10 Lower Kent Ridge Road, Singapore 119076

Email address: matgwt@nus.edu.sg

Proceedings of Symposia in Pure Mathematics
Volume **101**, 2019
https://doi.org/10.1090/pspum/101/01793

Generalized and degenerate Whittaker quotients and Fourier coefficients

Dmitry Gourevitch and Siddhartha Sahi

ABSTRACT. The study of Whittaker models for representations of reductive groups over local and global fields has become a central tool in representation theory and the theory of automorphic forms. However, only generic representations have Whittaker models. In order to encompass other representations, one attaches a degenerate (or a generalized) Whittaker model $\mathcal{W}_{\mathcal{O}}$, or a Fourier coefficient in the global case, to any nilpotent orbit.

In this note we survey some classical and some recent work in this direction - for Archimedean, p-adic and global fields. The main results concern the existence of models. For a representation π, call the set of maximal elements of the set of orbits $\mathcal{O}$ with $\mathcal{W}_{\mathcal{O}}$ that includes π the *Whittaker support* of π. The two main questions discussed in this note are:

(1) What kind of orbits can appear in the Whittaker support of a representation?
(2) How does the Whittaker support of a given representation π relate to other invariants of π, such as its wave-front set?

CONTENTS

2010 *Mathematics Subject Classification.* Primary 20G05, 20G20, 20G25, 20G30, 20G35, 22E27, 22E46, 22E50, 22E55, 17B08.

Key words and phrases. Fourier coefficient, wave-front set, oscillator representation, Heisenberg group, metaplectic group, admissible orbit, distinguished orbit, cuspidal representation, automorphic form.

References

1. Introduction

The study of Whittaker and degenerate Whittaker models for representations of reductive groups over local fields evolved in connection with the theory of automorphic forms (via their Fourier coefficients), and has found important applications in both areas. See for example [**GK75**, **Sha74**, **NPS73**, **Kos78**, **CHM00**] for Whittaker models and [**Kaw85**, **Yam86**, **Wall88**, **MW87**, **Mat87**, **Gin06**, **Jia07**, **GRS11**, **Gin14**, **GZ14**, **JLS16**, **CFGK**] for degenerate and generalized Whittaker models and Fourier coefficients.

In this note we survey some classical results and some recent work on degenerate and generalized Whittaker models and Fourier coefficients, including some recent results of the authors obtained in collaboration with R. Gomez and others.

Let F be either $\mathbb{R}$ or a finite extension of $\mathbb{Q}_p$, and let G be a finite central extension of the group of F-points of a connected reductive algebraic group defined over F[1]. Let $\mathrm{Rep}^\infty(G)$ denote the category of smooth representations of G (see §1.1 below). Let $\mathfrak{g}$ denote the Lie algebra of G and $\mathfrak{g}^*$ denote its dual space. To every coadjoint nilpotent orbit $\mathcal{O} \subset \mathfrak{g}^*$ and every $\pi \in \mathrm{Rep}^\infty(G)$ we associate a certain generalized Whittaker quotient $\pi_{\mathcal{O}}$ (see §2.1 below). Let $\mathrm{WO}(\pi)$ denote the set of all nilpotent orbits $\mathcal{O}$ with $\pi_{\mathcal{O}} \neq 0$ and $\mathrm{WS}(\pi)$ denote the set of maximal orbits in $\mathrm{WO}(\pi)$ with respect to the closure ordering. We call $\mathrm{WS}(\pi)$ the *Whittaker support of* π.

Denote by $\mathcal{M}(G)$ the category of admissible (finitely-generated) representations (see §1.1 below). For $\pi \in \mathcal{M}(G)$, and a nilpotent orbit $\mathcal{O} \subset \mathfrak{g}^*$, [**How74**, **HC77**, **BV80**] define a coefficient $c_{\mathcal{O}}(\pi)$ using the asymptotics of the character of π at $1 \in G$ (see §4 below). Denote by $\mathrm{WF}(\pi)$ the set of maximal elements in the set of orbits with non-zero coefficients. For non-Archimedean F this set coincides with $\mathrm{WS}(\pi)$.

THEOREM A ([**MW87**, Proposition I.11, Theorem I.16 and Corollary I.17], and [**Var14**]). *Assume that F is non-Archimedean and G is algebraic*[2] *and let $\pi \in \mathcal{M}(G)$. Then*

(i) $\mathrm{WF}(\pi) = \mathrm{WS}(\pi)$.
(ii) *For any $\mathcal{O} \in \mathrm{WF}(\pi)$, $c_{\mathcal{O}}(\pi) = \dim \pi_{\mathcal{O}}$.*

We conjecture that the same holds for Archimedean F. In §4 we define the wave-front set $\mathrm{WF}(\pi)$, sketch the proof of Theorem A and survey some partial results in the Archimedean case, based on [**GW80**, **Mat87**, **Mat90a**, **GGS17**, **GGS**, **GSS18**].

The next question that arises is what orbits can appear in Whittaker supports of representations. Based on the Kirillov orbit method one may conjecture that they are all admissible. This notion has to do with splitting of a certain metaplectic double cover of the centralizer G_φ for any φ in the orbit (see §3.1 below). Based on the Langlands correspondence it makes sense to conjecture that if G is algebraic then these orbits are special in the sense of Lusztig ([**Lus79**], [**Spa82**, Ch. III]).

[1] We view complex reductive groups as a special case of real reductive groups.
[2] It seems that the assumption that G is algebraic is not used in the proof.

More precisely, an orbit is said to be special if the corresponding orbit over the algebraic closure is special. For non-Archimedean F these conjectures might hold, but for Archimedean F they do not. For example, the minimal orbit of $G_2(\mathbb{R})$ is not special and appears in $\mathrm{WS}(\pi)$ for some irreducible unitary π by [**Vog94**]. Also, the minimal orbit of $U(2,1)$ is non-admissible and appears in $\mathrm{WS}(\pi)$ for some π. On the positive side, the following holds for all F.

THEOREM B ([**GGS**, Theorem A], following [**Moe96**, **GRS99**, **Gin06**, **JLS16**]). *Let $\pi \in \mathrm{Rep}^\infty(G)$ and let $\mathcal{O} \in \mathrm{WS}(\pi)$. Then $\mathcal{O}$ is a quasi-admissible orbit (see* §3.1 *below).*

Also, for non-Archimedean F and any $\pi \in \mathrm{Rep}^\infty(G_2(F))$ all $\mathcal{O} \in \mathrm{WS}(\pi)$ are special by [**JLS16**, **LS08**].

Let us summarize the relations between the above notions for classical groups.

THEOREM C ([**GGS**, §6], based on [**Moe96**, **Nev99**, **Oht91a**]). *Let $\mathcal{O} \subset \mathfrak{g}^*$ be a nilpotent orbit.*

(i) *If G is $\mathrm{GL}_n(F)$ or $\mathrm{SL}_n(F)$ then all orbits are admissible and special.*
(ii) *If G is $U(V)$ or $SU(V)$ for a hermitian space V then all orbits are quasi-admissible and special. If $F \neq \mathbb{R}$ then they are all admissible.*
(iii) *If G if either $\mathrm{Sp}_{2n}(F)$, or $O(V)$ or $\mathrm{SO}(V)$ (for a quadratic space V over F), then the following are equivalent:*

   (a) $\mathcal{O}$ is admissible (b) $\mathcal{O}$ is quasi-admissible (c) $\mathcal{O}$ is special

(iv) *If $F = \mathbb{R}$ and G is a classical group not listed above then all orbits are admissible.*

It is possible that the notions of admissible and quasi-admissible are equivalent for all G in the case when F is non-archimedean. These notions differ for $U(p,q)$ and for $SU(p,q)$. They also differ for the split real forms of E_7 and E_8, though we do not know whether the non-admissible quasi-admissible orbits appear in Whittaker supports of representations. It is also possible that all special orbits are quasi-admissible for all groups. If G is algebraic, F is Archimedean and $\pi \in \mathcal{M}(G)$ has integral infinitesimal character then all $\mathcal{O} \in \mathrm{WF}(\pi)$ are special, cf. [**BV82**, Theorem D] and [**BV83a**, Theorem 1.1].

Under additional assumptions on π one can show that all the orbits in $\mathrm{WS}(\pi)$ are F-distinguished, i.e. do not intersect the Lie algebras of proper Levi F-subgroups of G.

THEOREM D ([**Moe96**, **Har12**, **GGS**]). *Let $\pi \in \mathrm{Rep}^\infty(G)$, and let $\mathcal{O} \subset \mathfrak{g}^*$ be a nilpotent orbit.*

(i) *If F is non-Archimedean, $\mathcal{O} \in \mathrm{WS}(\pi)$ and π is quasi-cuspidal then $\mathcal{O}$ is F-distinguished.*
(ii) *If π is admissible and tempered, $\mathcal{O} \in \mathrm{WF}(\pi)$ and either G is classical or F is Archimedean then $\mathcal{O}$ is F-distinguished.*

Note that if G is semi-simple then $\mathcal{O}$ is F-distinguished if and only if all reductive subgroups of the centralizer of any element of $\mathcal{O}$ are compact. Thus, such orbits are sometimes called compact. In the Archimedean case, compact orbits were classified in [**PT04**].

Over Archimedean fields, all the orbits in $\mathrm{WF}(\pi)$ lie in the same complex orbit by [**Ros95**, Theorem D]. This is conjectured to hold for p-adic F as well, but so far this conjecture is known only for the group $\mathrm{GL}_n(F)$ (see [**MW87**, §II.2]). Over finite fields, an analogous result is proven in [**Lus92**], as well as some analogues of Theorems A and B.

In [**GZ14, LoMa15, Prz**] it is shown that the Whittaker support and the wave-front set have the expected behaviour under θ-correspondence.

The behaviour of wave-front set under parabolic and cohomological induction is studied in [**MW87, BB99, BV83b**] (see §4.1 below).

In the non-Archimedean case, the functor $\pi \mapsto \pi_{\mathcal{O}}$ is exact. We conjecture that the same holds in the Archimedean case, at least on the subcategory of admissible representations π with $\mathrm{WO}(\pi)$ lying in the closure of $\mathcal{O}$. This is shown for principal nilpotent $\mathcal{O}$ in [**CHM00**]. For some partial results in this direction for other $\mathcal{O}$ see [**Lyn79, Wall88, Mat90a, Wall88, AGSah15b, GGS17**].

One can define more general models and quotients. Namely, let $S \in \mathfrak{g}$ such that the adjoint action of S is diagonalizable over $\mathbb{Q}$, and let $\varphi \in \mathfrak{g}^*$ such that $ad^*(S)(\varphi) = -2\varphi$. In §2.1 below we define a *degenerate Whittaker model* $\mathcal{W}_{S,\varphi}$, and define a degenerate Whittaker quotient of π to be the coinvariants $\pi_{S,\varphi} := (\mathcal{W}_{S,\varphi} \otimes \pi)_G$. In §2.3 we discuss an epimorphism $\mathcal{W}_{\mathcal{O}} \twoheadrightarrow \mathcal{W}_{S,\varphi}$ constructed in [**GGS17**], under the condition that the orbit $\mathcal{O}$ intersects the closure of the orbit of φ under the centralizer of S in G. The existence of this epimorphism sheds some light on the non-maximal orbits in $\mathrm{WO}(\pi)$. In particular, it is shown that for $\pi \in \mathrm{Rep}^{\infty}(\mathrm{GL}_n(F))$, we have $\mathcal{O} \in \mathrm{WO}(\pi)$ if and only if there exists an orbit $\mathcal{O}' \in \mathrm{WS}(\pi)$ with $\mathcal{O} \subset \overline{\mathcal{O}'}$. The analogous statement for other groups is not known in general. For $\pi \in \mathcal{M}(\mathrm{GL}_n(F))$ we also have $\mathrm{WS}(\pi) = \mathrm{WF}(\pi)$. Also, for $\mathcal{O} \in \mathrm{WS}(\pi)$, the induced map $\pi_{\mathcal{O}} \twoheadrightarrow \pi_{S,\varphi}$ is non-zero and is an isomorphism for non-archimedean F. This fact is used in the proof of Theorems B and D(i).

Over the adeles one defines the Whittaker support using period integrals rather than quotients (see §2.2 below for more details). These period integrals are called *Fourier coefficients*. We refer the reader to [**Gin06**] for some intriguing open questions on Fourier coefficients of automorphic forms. Most of the questions described above work analogously over the adeles. Let K be a global field, $\mathbf{G}$ be a reductive group defined over K, and π be an automorphic representation of the adelic points $G := \mathbf{G}(\mathbb{A})$.

THEOREM E ([**GRS03, Gin06, JLS16, Shen16, GGS**]). *Let $\mathcal{O} \in \mathrm{WS}(\pi)$ be a nilpotent $\mathbf{G}(K)$-orbit. Then*

(i) *$\mathcal{O}$ is quasi-admissible. Furthermore, if $\mathbf{G}$ is classical then $\mathcal{O}$ is special.*
(ii) *If π is cuspidal then φ does not belong to the Lie algebra of any Levi subgroup of $\mathbf{G}(K)$ defined over K.*

COROLLARY F. *Let $\mathcal{O} \in \mathrm{WS}(\pi)$ be a nilpotent $\mathbf{G}(K)$-orbit. Assume that π is cuspidal and fix $\varphi \in \mathcal{O}$. Then any Levi subgroup of the stabilizer of φ in $\mathbf{G}(K)$ is K-anisotropic. Moreover, assume that $\mathbf{G}$ is classical, and let λ be the partition corresponding to φ. Then*

(i) *If $\mathbf{G} = \mathrm{GL}_n$ or $\mathbf{G} = \mathrm{SL}_n$ then λ consists of one part, i.e. π is generic.*
(ii) *If $\mathbf{G} = \mathrm{Sp}_n$ then λ is totally even.*
(iii) *If $\mathbf{G} = \mathrm{SO}_n$ or $\mathbf{G} = \mathrm{O}_n$ then λ is totally odd, i.e. consists of odd parts only.*

The study of Fourier coefficients of automorphic representations has applications in string theory, cf. [**GMV15**] and [**FGKP18**, Ch. 2]. A special role is played by Fourier coefficients of minimal and next-to-minimal representations of the split simply laced groups $\mathrm{SL}(n), SO(5,5), E_6, E_7, E_8$. In [**GGKPS**] we express any minimal or next-to-minimal automorphic form for these groups through its Whittaker-Fourier coefficients, *i.e.* period integrals over the nilradical of a Borel subgroup of G against a character of this subgroup, following [**MS12, AGKLP**]. This is important since the latter integral is often Eulerian, which allows it to be computed explicitly. On the other hand, the analogous statement does not hold for $Sp(4)$. We do not expect it to hold for higher symplectic groups either.

The ability to express automorphic forms through Whittaker-Fourier coefficients also implies that minimal and next-to-minimal representations cannot appear in the cuspidal spectrum. For split classical groups of rank bigger than 2 the latter follows from Theorem E(ii) and Corollary F.

The following table compares our notation to the notation of several papers regarding invariants of local and global representations.

Present paper	[**MW87**]	[**BV80, Ros95**]	[**Gin06**]	[**JLS16**]	[**GGS17**]
WO	$\mathcal{N}_{Wh}$			WF	
WS			$\mathcal{O}_G$		
WF	maximal elements in $\mathcal{N}_{\mathrm{tr}}$				
$\overline{\mathrm{WF}}$		AS			WF

Acknowledgments. We thank Joseph Bernstein, Yuanqing Cai, David Ginzburg, Raul Gomez, Henrik Gustafsson, David Kazhdan, Axel Kleinschmidt, Erez Lapid, Baiying Liu, Daniel Persson, Gordan Savin, Eitan Sayag, and David Soudry for fruitful discussions.

D.G. partially supported by ERC StG grant 637912, and ISF grant 249/17. S.S. was partially supported by Simons Foundation grant 509766.

1.1. Notation. Let F be either $\mathbb{R}$ or a finite extension of $\mathbb{Q}_p$ and let $\mathfrak{g}$ be a reductive Lie algebra over F. We say that an element $S \in \mathfrak{g}$ is *rational semi-simple* if its adjoint action on $\mathfrak{g}$ is diagonalizable with eigenvalues in $\mathbb{Q}$. For a rational semi-simple element S and a rational number r we denote by $\mathfrak{g}^S_r$ the r-eigenspace of the adjoint action of S and by $\mathfrak{g}^S_{\geq r}$ the sum $\bigoplus_{r' \geq r} \mathfrak{g}^S_{r'}$. We will also use the notation $(\mathfrak{g}^*)^S_r$ and $(\mathfrak{g}^*)^S_{\geq r}$ for the corresponding grading and filtration of the dual Lie algebra $\mathfrak{g}^*$. For $X \in \mathfrak{g}$ or $X \in \mathfrak{g}^*$ we denote by $\mathfrak{g}_X$ the centralizer of X in $\mathfrak{g}$, and by G_X the centralizer of X in G.

If (f, h, e) is an $\mathfrak{sl}_2$-triple in $\mathfrak{g}$, we will say that e is a nil-positive element for h, f is a nil-negative element for h, and h is a neutral element for e. For a representation V of (f, h, e) we denote by V^e the space spanned by the highest-weight vectors and by V^f the space spanned by the lowest-weight vectors. We refer to [**Bou75**, §11] or [**Kos59**] for standard facts on $\mathfrak{sl}_2$-triples. We will say that $h \in \mathfrak{g}$ is a *neutral* element for $\varphi \in \mathfrak{g}^*$ if h can be completed to an $\mathfrak{sl}_2$-triple (f, h, e) such that φ is given by the Killing form pairing with f. By the Jacobson-Morozov theorem such h exists and is unique up to conjugation by the centralizer of φ.

From now on we fix a non-trivial unitary additive character $\chi : F \to \mathrm{S}^1$ such that

(1) if $F = \mathbb{R}$ we have $\chi(x) = \exp(2\pi i x)$ and
if F is non-Archimedean the kernel of χ is the ring of integers.

For non-archimedean F we will work with l-groups, *i.e.* Hausdorff topological groups having a basis for the topology at the identity consisting of open compact subgroups. This generality includes F-points of algebraic groups defined over F, and their finite covers (see [**BZ76**]).

For $F = \mathbb{R}$ we will work with affine Nash groups, *i.e.* Lie groups that are given in $\mathbb{R}^n$ by semi-algebraic equations, as well as the graphs of their multiplication maps. This generality includes $\mathbb{R}$-points of algebraic groups defined over $\mathbb{R}$, and their finite covers (see [**dCl91, AG08, Sun15, FS16**]).

NOTATION 1.1. If G is an l-group, we denote by $\mathrm{Rep}^\infty(G)$ the category of smooth representations of G in complex vector spaces. For $V, W \in \mathrm{Rep}^\infty(G)$, $V \otimes W$ will denote the usual tensor product over $\mathbb{C}$. We denote by $\mathcal{M}(G) \subset \mathrm{Rep}^\infty(G)$ the subcategory consisting of representations of finite length.

If G is an affine Nash group, we denote by $\mathrm{Rep}^\infty(G)$ the category of smooth nuclear Fréchet representations of G of moderate growth. This is essentially the same definition as in [**dCl91**, §1.4] with the additional assumption that the representation spaces are nuclear (see e.g. [**Tre67**, §50]). For $V, W \in \mathrm{Rep}^\infty(G)$, $V \otimes W$ will denote the completed projective tensor product. We denote by $\mathcal{M}(G) \subset \mathrm{Rep}^\infty(G)$ the subcategory consisting of admissible finitely generated representations, see [**Wall92**, Ch. 11]. This category is abelian, see [**Wall92**, Ch. 11] or [**Cas89**].

For $\pi \in \mathrm{Rep}^\infty(G)$, denote by π_G the space of coinvariants, i.e. quotient of π by the intersection of kernels of all G-invariant functionals. Explicitly,

$$\pi_G = \pi / \overline{\{\pi(g)v - v \mid v \in \pi,\, g \in G\}},$$

where the closure is needed only for Archimedean F. In the latter case, for connected G we have $\pi_G = \pi/\overline{\mathfrak{g}_{\mathbb{C}}\pi}$ which in turn is equal to the quotient of $\mathrm{H}_0(\mathfrak{g}, \pi)$ by the closure of zero.

DEFINITION 1.2. If G is an l-group, $H \subset G$ a closed subgroup and $\pi \in \mathrm{Rep}^\infty(H)$, we denote by $\mathrm{ind}_H^G(\pi)$ the smooth compactly-supported induction as in [**BZ76**, §2.22].

If G is an affine Nash group, $H \subset G$ a closed Nash subgroup and $\pi \in \mathrm{Rep}^\infty(H)$, we denote by $\mathrm{ind}_H^G(\pi)$ the Schwartz induction as in [**dCl91**, §2].

This induction has the expected properties of small induction, such as induction by stages and Frobenius reciprocity.

2. Degenerate Whittaker models and Fourier coefficients

2.1. Definitions. Let G be a finite central extension of the group G^{alg} of F-points of a reductive algebraic group defined over F. Let G^{ad} denote the corresponding adjoint algebraic group.

LEMMA 2.1 ([**MW95**, Appendix I]). *Let $U \subset G^{\mathrm{alg}}$ be a unipotent subgroup, and $\hat{U}$ be the preimage of U in G. Then there exists a unique open subgroup $U' \subset \hat{U}$ that projects isomorphically onto U.*

We will therefore identify the unipotent subgroups of G^{alg} with their liftings in G.

DEFINITION 2.2. Let W_n denote the 2n-dimensional F-vector space $(F^n)^* \oplus F^n$ and let ω be the standard symplectic form on W_n. The Heisenberg group H_n is the algebraic group with underlying algebraic variety $W_n \times F$ with the group law given by

$$(w_1, z_1)(w_2, z_2) = (w_1 + w_2, z_1 + z_2 + 1/2\omega(w_1, w_2)).$$

Note that $H_0 = F$.

DEFINITION 2.3. Let χ be the additive character of F, as in (1). Extend χ trivially to a character of the commutative subgroup $0 \oplus F^n \oplus F \subset H_n$. The oscillator representation ϖ_χ is the unitary induction of χ from $0 \oplus F^n \oplus F$ to H_n. Define the smooth oscillator representation σ_χ to be the space of smooth vectors in ϖ_χ. One can show that $\sigma_\chi = \mathrm{ind}_{0\oplus F^n \oplus F}^{H_n}(\chi)$.

DEFINITION 2.4. (i) A *Whittaker pair* is an ordered pair (S, φ) such that $S \in \mathfrak{g}$ is rational semi-simple, and $\varphi \in (\mathfrak{g}^*)^S_{-2}$. Given such a Whittaker pair, we define the *degenerate Whittaker model* $\mathcal{W}_{S,\varphi}$ in the following way: let $\mathfrak{u} := \mathfrak{g}^S_{\geq 1}$. Define an anti-symmetric form ω_φ on $\mathfrak{g}$ by $\omega_\varphi(X, Y) := \varphi([X, Y])$. Let $\mathfrak{n}$ be the radical of $\omega_\varphi|_\mathfrak{u}$. Note that $\mathfrak{u}, \mathfrak{n}$ are nilpotent subalgebras of $\mathfrak{g}$, and $[\mathfrak{u}, \mathfrak{u}] \subset \mathfrak{g}^S_{\geq 2} \subset \mathfrak{n}$. Let $U := \mathrm{Exp}(\mathfrak{u})$ and $N := \mathrm{Exp}(\mathfrak{n})$ be the corresponding nilpotent subgroups of G. Let $\mathfrak{n}' := \mathfrak{n} \cap \mathrm{Ker}(\varphi)$, $N' := \mathrm{Exp}(\mathfrak{n}')$. If $\varphi = 0$ we define

$$\mathcal{W}_{S,0} := \mathrm{ind}_U^G(\mathbb{C}). \tag{2}$$

Assume now that φ is non-zero. Then U/N' has a natural structure of a Heisenberg group, and its center is N/N'. Let χ_φ denote the unitary character of N/N' given by $\chi_\varphi(\exp(X)) := \chi(\varphi(X))$. Let σ_φ denote the oscillator representation of U/N' with central character χ_φ, and σ'_φ denote its trivial lifting to U. Define

$$\mathcal{W}_{S,\varphi} := \mathrm{ind}_U^G(\sigma'_\varphi). \tag{3}$$

(ii) For a nilpotent element $\varphi \in \mathfrak{g}^*$, define the *generalized Whittaker model* $\mathcal{W}_\varphi$ corresponding to φ to be $\mathcal{W}_{S,\varphi}$, where S is a neutral element for φ if $\varphi \neq 0$ and $S = 0$ if $\varphi = 0$. We will also call $\mathcal{W}_{S,\varphi}$ *neutral degenerate Whittaker model*. Since all neutral elements for φ are conjugate by the centralizer of φ, $\mathcal{W}_\varphi$ depends only on the coadjoint orbit of φ, and does not depend on the choice of S. Thus we will also use the notation $\mathcal{W}_\mathcal{O}$ for a nilpotent coadjoint orbit $\mathcal{O} \subset \mathfrak{g}^*$. See [**GGS17**, §5] for a formulation of this definition without choosing S.

(iii) To $\pi \in \mathrm{Rep}^\infty(G)$ associate the degenerate and generalized Whittaker quotients by

$$\pi_{S,\varphi} := (\mathcal{W}_{S,\varphi} \otimes \pi)_G \text{ and } \pi_\varphi := (\mathcal{W}_\varphi \otimes \pi)_G. \tag{4}$$

Slightly different degenerate Whittaker models are considered in [**GGS17**] and denoted $\mathcal{W}_{S,\varphi}(\pi)$. They relate to $\pi_{S,\varphi}$ by $\mathcal{W}_{S,\varphi}(\pi) = \pi^*_{S,\varphi}$.

2.2. Fourier coefficients. Let K be a number field and let $\mathbb{A} = \mathbb{A}_K$ be its ring of adeles. In this section we let χ be a non-trivial unitary character of $\mathbb{A}$, which is trivial on K. Then χ defines an isomorphism between $\mathbb{A}$ and $\hat{\mathbb{A}}$ via the map $a \mapsto \chi_a$, where $\chi_a(b) = \chi(ab)$ for all $b \in \mathbb{A}$. This isomorphism restricts to an isomorphism

$$\widehat{\mathbb{A}/K} \cong \{\psi \in \hat{\mathbb{A}} \,|\, \psi|_K \equiv 1\} = \{\chi_a \,|\, a \in K\} \cong K. \tag{5}$$

Given an algebraic group $\mathbf{G}$ defined over K we will denote its Lie algebra by $\mathfrak{g}$ and we will denote the group of its adelic (resp. K-rational) points by $\mathbf{G}(\mathbb{A})$ (resp. $\mathbf{G}(K)$). We will also define the Lie algebras $\mathfrak{g}(\mathbb{A})$ and $\mathfrak{g}(K)$ in a similar way.

Given a Whittaker pair (S,φ) on $\mathfrak{g}(K)$, we set $\mathfrak{u}=\mathfrak{g}^S_{\geq 1}$ and $\mathfrak{n}$ to be the radical of the form $\omega_\varphi|_\mathfrak{u}$, where $\omega_\varphi(X,Y)=\varphi([X,Y])$, as before. Let $\mathbf{U}=\exp\mathfrak{u}$, and $\mathbf{N}=\exp\mathfrak{n}$. Define a character χ_φ on $\mathbf{U}(\mathbb{A})$ by $\chi_\varphi(\exp X)=\chi(\varphi(X))$ and note that it is automorphic, *i.e.* trivial on $\mathbf{U}(K)$. Let G be a finite central extension of $\mathbf{G}(\mathbb{A})$, such that the cover $G\twoheadrightarrow\mathbf{G}(\mathbb{A})$ splits over $\mathbf{G}(K)$. Fix a discrete subgroup $\Gamma\subset G$ that projects isomorphically onto $\mathbf{G}(K)$. Note that $\mathbf{U}(\mathbb{A})$ has a canonical lifting into G, see e.g. [**MW95**, Appendix I].

DEFINITION 2.5. Let (S,φ) be a Whittaker pair for $\mathfrak{g}(K)$ and let $\mathbf{U},\mathbf{N}$, and χ_φ be as above. For an automorphic form f, we define its (S,φ)–*Fourier coefficient* to be

$$\mathcal{F}_{S,\varphi}(f):=\int_{N(\mathbb{A})/N(K)}\chi_\varphi(n)^{-1}f(n)dn. \tag{6}$$

Observe that $\mathcal{F}_{S,\varphi}$ defines a linear functional on the space of automorphic forms. For a subrepresentation π of the space of automorphic forms on G, we denote the restriction of $\mathcal{F}_{S,\varphi}$ to π by $\mathcal{F}_{S,\varphi}(\pi)$. We denote by $\mathrm{WO}(\pi)$ the set of $\mathbf{G}(K)$-orbits of all $\varphi\in\mathfrak{g}(K)$ with $\mathcal{F}_\varphi(\pi)\neq 0$, and by $\mathrm{WS}(\pi)$ the set of maximal orbits in $\mathrm{WO}(\pi)$. In [**GGS17**], $\mathcal{F}_{S,\varphi}$ is denoted $\mathcal{WF}_{S,\varphi}$.

REMARK 2.6. (i) The set $\mathrm{WS}(\pi)$ depends on the embedding of π into the space of automorphic forms. Indeed, [**GGJ02**] construct an example of an automorphic representation of the split $G_2(\mathbb{A})$ that has a regular orbit in $\mathrm{WS}(\pi)$ for one embedding and does not for other embeddings.

(ii) One can define abstract (degenerate) Whittaker quotients for automorphic representations, following Definition 2.4 and [**HS16**, §5.2]. The non-vanishing of such a quotient follows from the non-vanishing of the corresponding Fourier coefficient (for some realization), but in general does not imply it. Indeed, [**GS87**] construct an automorphic cuspidal representation of the metaplectic double cover of $\mathrm{SL}_2(\mathbb{A})$ that has an abstract Whittaker model corresponding to some non-zero nilpotent φ, but with vanishing $\mathcal{F}_\varphi$. One usually considers Fourier coefficients rather than abstract quotients since the former have number-theoretic and string theoretic applications.

(iii) To the best of our knowledge, the Whittaker support and the non-vanishing of abstract quotients are the only ways to attach nilpotent orbits to automorphic representations. Some authors call the $\mathrm{WO}(\pi)$ *the wave-front set* just by definition.

2.3. Comparison between different Whittaker pairs. We start with a corollary of the Stone-von-Neumann theorem.

COROLLARY 2.7 (See e.g. [**GGS17**, Corollary 2.4.5]). *Let $W:=(F^n)^*\oplus F^n$ and let H be the corresponding Heisenberg group and σ_χ its smooth oscillator representation. Let $L\subset W$ be a Lagrangian subspace. Extend χ trivially from F to the abelian subgroup $L\oplus F\subset H_n$. Then*

$$\mathrm{ind}_{L\oplus F}^{H_n}\chi\cong\sigma_\chi.$$

Using a version of Frobenius reciprocity, and induction by stages we get:

LEMMA 2.8 (See e.g. [**GGS**, Lemma 2.5.3]). *Let $\mathfrak{l} \subset \mathfrak{g}^S_{\geq 1}$ be a maximal isotropic subalgebra and $L := \operatorname{Exp}(\mathfrak{l})$. Let $\pi \in \operatorname{Rep}^\infty(G)$. Then*

$$\mathcal{W}_{S,\varphi} \cong \operatorname{ind}_L^G \chi_\varphi \text{ and } \pi_{S,\varphi} \cong (\pi \otimes \chi_\varphi)_L.$$

This lemma is the central tool for the comparison of different degenerate Whittaker models.

REMARK 2.9. (i) The previous lemma implies that for any two maximal isotropic subalgebras $\mathfrak{l}, \mathfrak{r} \subset \mathfrak{g}^S_{\geq 1}$ we have an isomorphism

$$\operatorname{ind}^G_{\operatorname{Exp}(\mathfrak{l})} \chi_\varphi \simeq \operatorname{ind}^G_{\operatorname{Exp}(\mathfrak{r})} \chi_\varphi. \tag{7}$$

(ii) A similar isomorphism was introduced in [**GRS99**, Lemma 2.2] and called *root exchange* (see also [**GRS11**, Lemma 7.11] or [**LaMao14**, Lemma A.1]).

(iii) Observe that the isomorphism (7) can be realized explicitly as an integral transform: given $f \in \operatorname{ind}^G_{\operatorname{Exp}(\mathfrak{l})} \chi_\varphi$ we can define $\check{f} \in \operatorname{ind}^G_{\operatorname{Exp}(\mathfrak{r})} \chi_\varphi$ simply by setting

(8)
$$\check{f}(g) = \int_{\operatorname{Exp}(\mathfrak{l}\cap\mathfrak{r}\cap\operatorname{Ker}\varphi)\backslash \operatorname{Exp}(\mathfrak{r}\cap\operatorname{Ker}\varphi)} f(ng)\,dn = \int_{\operatorname{Exp}(\mathfrak{l}\cap\mathfrak{r})\backslash \operatorname{Exp}(\mathfrak{r})} \chi_\varphi(n)^{-1} f(ng)\,dn.$$

The previous results imply that the map $f \mapsto \check{f}$ defines an isomorphism $\operatorname{ind}^G_{\operatorname{Exp}(\mathfrak{l})} \chi_\varphi \xrightarrow{\sim} \operatorname{ind}^G_{\operatorname{Exp}(\mathfrak{r})} \chi_\varphi$. The corresponding formula in [**GGS17**, Remark 3.2.2] has a typo.

EXAMPLE 2.10. Let us now give several examples for $\operatorname{GL}_n(F)$. We identify $\mathfrak{g}$ with $\mathfrak{g}^*$ using the trace form. For $1 \leq i, j \leq n$ let E_{ij} denote the corresponding elementary matrix. Let B denote the group of upper-triangular matrices and $B' = [B, B]$ denote its unipotent radical.

(i) First, let $n := 2$. Then there are two nilpotent orbits: the zero one and the regular one. For a regular nilpotent $\varphi \in \mathfrak{g}^*$, any corresponding degenerate Whittaker model is the classical Whittaker model, for any G. For the zero φ, there are always several choices. For example, we can always choose $S = 0$ and have $U = L = \{\operatorname{Id}\}$. We can also choose S to be any rational semi-simple element. For $\operatorname{GL}_2(F)$ we can take $S = \operatorname{diag}(1, -1)$ and get $U = L = B'$. The corresponding Whittaker quotients are different both in the real and in the p-adic case. Indeed, $\pi_0 = \pi$ always, while $\pi_{S,0}$ is finite-dimensional for all admissible π of finite length. For non Archimedean F and cuspidal π it even vanishes.

(ii) Let us now take $G = \operatorname{GL}_3(F)$, and let $\varphi := E_{21}$. Then we can take $h := \operatorname{diag}(1, -1, 0)$ and $S := \operatorname{diag}(2, 0, -2)$. Then (h, φ) is a neutral Whittaker pair, and (S, φ) is a Whittaker pair. For S we have $L = U = B'$ and for h we can choose

$$\mathfrak{l} = \begin{pmatrix} 0 & * & * \\ 0 & 0 & 0 \\ 0 & 0 & 0 \end{pmatrix} \subset \mathfrak{u} = \begin{pmatrix} 0 & * & * \\ 0 & 0 & 0 \\ 0 & * & 0 \end{pmatrix} \tag{9}$$

(iii) For $G = \operatorname{GL}_4(F)$, and let $\varphi := E_{21} + E_{43}$. Then we can take $h := \operatorname{diag}(1, -1, 1, -1)$ and $S := \operatorname{diag}(3, 1, -1, -3)$. Then (h, φ) is a neutral

Whittaker pair, and (S,φ) is a Whittaker pair. For S we have $L=U=B'$ and for h we have

$$\mathfrak{l}=\mathfrak{u}=\begin{pmatrix} 0 & * & 0 & * \\ 0 & 0 & 0 & 0 \\ 0 & * & 0 & * \\ 0 & 0 & 0 & 0 \end{pmatrix}.$$

Here we see a new phenomenon: this subalgebra does not lie in the Lie algebra of B', and cannot be conjugated into it by the centralizer of φ. However, it is still possible to present the corresponding degenerate Whittaker model as a quotient of $\mathcal{W}_\varphi$, as we explain below.

THEOREM 2.11 ([**GGS17**, Theorem A] and [**GGS**, Theorem D]). *Let (S,φ) be a Whittaker pair. Let $\overline{G_S\varphi}$ denote the closure of the orbit of φ under the centralizer of S. Then*

(i) *For any $\psi\in\overline{G_S\varphi}$ there exists a natural surjection $\nu:\mathcal{W}_\psi\twoheadrightarrow\mathcal{W}_{S,\varphi}$.*
(ii) *Let $\pi\in\operatorname{Rep}^\infty(G)$ such that $G\cdot\varphi\in\operatorname{WS}(\pi)$. Then $\pi_{S,\varphi}\neq 0$. Moreover, if F is non-Archimedean then the epimorphism $\pi_\varphi\twoheadrightarrow\pi_{S,\varphi}$ induced by ν is an isomorphism.*

REMARK 2.12. Let us comment on the Archimedean case of (ii). Already from Example 2.10(i) we see that π_φ is not isomorphic to $\pi_{S,\varphi}$ in this case. We conjecture though that they will be isomorphic for unitary π. Also, in a work in progress we will show that they will be isomorphic if we replace the usual quotients $\pi_{h,\varphi}$ and $\pi_{S,\varphi}$ by the inverse limits $\lim\limits_{\leftarrow}\pi/(\mathfrak{l}_\varphi)^i\pi$, where $\mathfrak{l}$ is a maximal isotropic subalgebra of $\mathfrak{g}^h_{\geq 1}$ (respectively $\mathfrak{g}^h_{\geq 1}$) and $\mathfrak{l}_\varphi$ is the ideal of the universal enveloping algebra generated by $\mathfrak{l}\otimes\varphi$.

Let us explain the construction of ν for the case $\psi=\varphi$. One can show that S can be presented as $h+Z$, where h is a neutral element for φ and Z commutes with h and centralizes φ. Consider a deformation $S_t=h+tZ$, and denote by $\mathfrak{u}_t$ the sum of eigenspaces of $ad(S_t)$ with eigenvalues at least 1. We call a rational number $0<t<1$ *regular* if $\mathfrak{u}_t=\mathfrak{u}_{t+\varepsilon}$ for any small enough rational ε, and *critical* otherwise. Note that there are finitely many critical numbers, and denote them by $t_1<\cdots<t_n$. Denote also $t_0:=0$ and $t_{n+1}:=1$. Choose a Lagrangian $\mathfrak{m}\subset\mathfrak{g}^Z_0\cap\mathfrak{g}^S_1$. For each t we define two subalgebras $\mathfrak{l}_t,\mathfrak{r}_t\subset\mathfrak{u}_t$ by

$$\mathfrak{l}_t:=\mathfrak{m}+(\mathfrak{u}_t\cap\mathfrak{g}^Z_{<0})+\operatorname{Ker}(\omega|_{\mathfrak{u}_t})\text{ and }\mathfrak{r}_t:=\mathfrak{m}+(\mathfrak{u}_t\cap\mathfrak{g}^Z_{>0})+\operatorname{Ker}(\omega|_{\mathfrak{u}_t}).\tag{10}$$

Both $\mathfrak{l}_t$ and $\mathfrak{r}_t$ are maximal isotropic subspaces with respect to the form ω_φ, and thus the restrictions of φ to $\mathfrak{l}_t$ and $\mathfrak{r}_t$ define characters of these subalgebras. Let $L_t:=\operatorname{Exp}(\mathfrak{l}_t)$ and $R_t:=\operatorname{Exp}(\mathfrak{r}_t)$ denote the corresponding subgroups and χ_φ denote their characters defined by φ. By Lemma 2.8 we have

$$\mathcal{W}_{S_t,\varphi}\simeq\operatorname{ind}^G_{L_t}(\chi_\varphi)\simeq\operatorname{ind}^G_{R_t}(\chi_\varphi).$$

We show that for any $0\leq i\leq n$, $\mathfrak{r}_{t_i}\subset\mathfrak{l}_{t_{i+1}}$. This gives a natural epimorphism

$$\mathcal{W}_{S_{t_i},\varphi}\simeq\operatorname{ind}^G_{L_{t_i}}(\chi_\varphi)\twoheadrightarrow\operatorname{ind}^G_{R_{t_i}}(\chi_\varphi)\simeq\mathcal{W}_{S_{t_{i+1}},\varphi}.$$

Altogether, we get

$$\mathcal{W}_{h,\varphi}=\mathcal{W}_{S_{t_0},\varphi}\twoheadrightarrow\mathcal{W}_{S_{t_1},\varphi}\twoheadrightarrow\cdots\twoheadrightarrow\mathcal{W}_{S_{t_{n+1}},\varphi}=\mathcal{W}_{S,\varphi}.\tag{11}$$

EXAMPLE 2.13. Let $G := \mathrm{GL}(4, F)$ and let S be the diagonal matrix $\mathrm{diag}(3, 1, -1, -3)$. Identify $\mathfrak{g}$ with $\mathfrak{g}^*$ using the trace form and let $f := \varphi := E_{21} + E_{43}$, where E_{ij} are elementary matrices. Then we have $S = h + Z$ with $h = \mathrm{diag}(1, -1, 1, -1)$ and $Z = \mathrm{diag}(2, 2, -2, -2)$. Thus $S_t = \mathrm{diag}(1 + 2t, -1 + 2t, 1 - 2t, -1 - 2t)$ and the weights of S_t are as follows:

$$\begin{pmatrix} 0 & 2 & 4t & 4t+2 \\ -2 & 0 & 4t-2 & 4t \\ -4t & -4t+2 & 0 & 2 \\ -4t-2 & -4t & -2 & 0 \end{pmatrix}.$$

The critical numbers are 1/4 and 3/4. For $t \geq 3/4$, the degenerate Whittaker model $\mathcal{W}_{S_t,\varphi}$ is the induction $\mathrm{ind}_{B'}^{G} \chi_\varphi$, where B' is the group of upper-unitriangular matrices. The sequence of inclusions $\mathfrak{r}_0 \subset \mathfrak{l}_{1/4} \sim \mathfrak{r}_{1/4} \subset \mathfrak{l}_{3/4} = \mathfrak{r}_{3/4}$ is:

(12)
$$\begin{pmatrix} 0 & - & 0 & - \\ 0 & 0 & 0 & 0 \\ 0 & - & 0 & - \\ 0 & 0 & 0 & 0 \end{pmatrix} \subset \begin{pmatrix} 0 & - & a & - \\ 0 & 0 & 0 & a \\ 0 & * & 0 & - \\ 0 & 0 & 0 & 0 \end{pmatrix} \sim \begin{pmatrix} 0 & - & * & - \\ 0 & 0 & 0 & * \\ 0 & 0 & 0 & - \\ 0 & 0 & 0 & 0 \end{pmatrix} \subset \begin{pmatrix} 0 & - & - & - \\ 0 & 0 & * & - \\ 0 & 0 & 0 & - \\ 0 & 0 & 0 & 0 \end{pmatrix}$$

Here, both $*$ and $-$ denote arbitrary elements. $-$ denotes the entries in $\mathfrak{v}_t$ and $*$ those in $\mathfrak{w}_t = \mathfrak{g}_1^{S_t}$. The letter a denotes an arbitrary element, but the two appearances of a denote the same numbers. The passage from $\mathfrak{l}_{1/4}$ to $\mathfrak{r}_{1/4}$ is denoted by $\sim$. At 3/4 we have $\mathfrak{l}_{3/4} = \mathfrak{r}_{3/4}$.

Let us now sketch the proof of part (ii) of Theorem 2.11. The sequence of epimorphisms (11) naturally defines a sequence of epimorphisms

$$\pi_{h,\varphi} = \pi_{S_{t_0},\varphi} \twoheadrightarrow \pi_{S_{t_1},\varphi} \twoheadrightarrow \cdots \twoheadrightarrow \pi_{S_{t_{n+1}},\varphi} = \pi_{S,\varphi}. \tag{13}$$

We see that for each i, $\pi_{S_{t_{i+1}},\varphi}$ is the quotient of $\pi_{S_{t_i},\varphi}$ by the group $A_i := L_{t_{i+1}}/R_t$, that we show to be commutative. Using Fourier transform on this group, one shows that in order to prove the theorem it is enough to show that $\pi_{S_{t_i},\varphi}$ is a non-generic representation of A_i. For that purpose we show that every unitary character of A_i is given by some $\varphi' \in \mathfrak{g}^*$ with $ad^*(S_{t_{i+1}})\varphi' = -\varphi'$ such that φ' does not lie in the tangent space to $\mathcal{O}$ at φ. We then define a quasi-Whittaker quotient $\pi_{S_{t_{i+1}},\varphi,\varphi'}$, and show that its dual is the space of $(A_i, \chi_{\varphi'})$-equivariant functionals on $\pi_{S_{t_i},\varphi}$. Then we generalize (13) to quasi-Whittaker quotients, construct some additional epimorphisms and deduce the vanishing of $\pi_{S_{t_{i+1}},\varphi,\varphi'}$ from the vanishing of $\pi_{\mathcal{O}'}$ for all $\mathcal{O}' \neq \mathcal{O}$ with $\mathcal{O} \subset \overline{\mathcal{O}'}$.

Let us now follow this argument in the setting of Example 2.13. Let $\pi \in \mathrm{Rep}^\infty(G)$ with $G\varphi \in \mathrm{WS}(\pi)$. The sequence of epimorphisms (13) is given by the sequence of inclusions (12). To see that these epimorphisms are non-zero (and are isomorphisms for $F \neq \mathbb{R}$) we need to analyze the dual spaces to $(\mathfrak{g}_1^{S_{1/4}})^f$ and $(\mathfrak{g}_1^{S_{3/4}})^f$. These spaces are spanned by $E_{13} + E_{24}$ and by E_{23} respectively. Thus, the dual spaces are spanned by $E_{31} + E_{42}$ and by E_{32} respectively. Note that the joint centralizer of h, Z and φ in G acts on these spaces by scalar multiplications, identifying all non-trivial elements. It is enough to show that $\pi_{S_{1/4},\varphi,E_{31}+E_{42}} = 0$ and $\pi_{S_{3/4},\varphi,E_{32}} = 0$.

First assume by way of contradiction that $\pi_{S_{3/4},\varphi,E_{32}} \neq 0$. Note that $E_{32} \in \mathfrak{g}^{S_1}_{-2}$ and that $\mathfrak{w}_1 = 0$. Thus $\mathfrak{u}_1 = \mathfrak{l}_1 = \mathfrak{r}_{3/4}$ and

$$\pi_{S_1,\varphi+E_{32}} \simeq \pi_{S_{3/4},\varphi,E_{32}} \neq 0.$$

Note that $\Phi := \varphi + E_{32} = E_{21} + E_{43} + E_{32}$ is a regular nilpotent element, and $S_1 = S = \mathrm{diag}(3, 1, -1, -3)$ is a neutral element for it. Thus $\pi_\Phi \neq 0$, contradicting the assumption that Φ is maximal in $\mathrm{WS}(\pi)$.

Now assume by way of contradiction that $\pi_{S_{1/4},\varphi,E_{31}+E_{42}} \neq 0$. Note that $E_{31} + E_{42} \in \mathfrak{g}^{S_{1/2}}_{-2}$ and that $\mathfrak{w}_{1/2} = 0$. Thus $\mathfrak{l}_{1/2} = \mathfrak{u}_{1/2} = \mathfrak{r}_{1/4}$ and

$$\pi_{S_{1/2},\varphi+E_{31}+E_{42}} \simeq \pi_{S_{1/4},\varphi,E_{31}+E_{42}} \neq 0.$$

Note that $\Psi := \varphi+E_{31}+E_{42} = E_{21}+E_{43}+E_{31}+E_{42}$ is a regular nilpotent element, and $S_{1/2} = \mathrm{diag}(2, 0, 0, -2)$ is a neutral element for it. Thus $\pi_\Psi \neq 0$, contradicting the assumption that $G \cdot \varphi$ is maximal in $\mathrm{WO}(\pi)$.

One can deduce from Theorem 2.11 that for non-Archimedean F and quasi-cuspidal $\pi \in \mathrm{Rep}^\infty(G)$, the orbits in $\mathrm{WS}(\pi)$ are F-distinguished.

SKETCH OF PROOF OF THEOREM D(i). Let π be quasi-cuspidal and let $\mathcal{O} \in \mathrm{WS}(\pi)$. Suppose by way of contradiction that $\mathcal{O}$ is not F-distinguished. Thus there exists a proper parabolic subgroup $P \subset G$, a Levi subgroup $L \subset P$ and a nilpotent $f \in \mathfrak{l}$ such that $\varphi \in \mathcal{O}$, where $\varphi \in \mathfrak{g}^*$ is given by the Killing form pairing with f. Let h be a neutral element for f in $\mathfrak{l}$. Choose a rational-semisimple element $Z \in \mathfrak{g}$ such that L is the centralizer of Z, $\mathfrak{p} := \mathfrak{g}^Z_{\geq 0}$ is the Lie algebra of P, and all the positive eigenvalues of Z are bigger than all the eigenvalues of h by at least 2. Note that $\mathfrak{n} := \mathfrak{g}^Z_{>0}$ is the nilradical of $\mathfrak{p}$. Let $S := h + Z$. By construction we have $\mathfrak{n} \subset \mathfrak{g}^S_{>2}$ and thus the degenerate Whittaker quotient $\pi_{S,\varphi}$ is a quotient of $r_P\pi$. By Theorem 2.11, the maximality of $\mathcal{O}$ implies $\pi_{S,\varphi} \simeq \pi_\varphi$. Thus $r_P\pi$ does not vanish, in contradiction with the condition that π is quasi-cuspidal. □

In the global case, we have the following analog of Theorem 2.11.

THEOREM 2.14 ([**GGS17**, Theorem C] and [**GGS**, Theorem 8.0.3]). *Let $(S, \varphi) \in \mathfrak{g}(K) \times \mathfrak{g}^*(K)$ be a Whittaker pair. Let π be an automorphic representation of G. Then*

(i) *The functional $\mathcal{F}_{S,\varphi}(\pi)$ can be obtained by a series of integral transforms from $\mathcal{F}_\varphi(\pi)$.*
(ii) *If $\mathbf{G}(K)\varphi \in \mathrm{WS}(\pi)$ then the functional $\mathcal{F}_\varphi(\pi)$ can be obtained by a series of integral transforms from $\mathcal{F}_{S,\varphi}(\pi)$.*

To prove this analog, we argue in the same way, but replace Lemma 2.8 by an explicit integral transform, in the spirit of Remark 2.9.

This theorem can be generalized. Namely, for (ii) we show in [**GGKPS**] that if $\varphi \notin \mathrm{WS}(\pi)$ then $\mathcal{F}_\varphi(\pi)$ can be obtained by a series of integral transforms from $\mathcal{F}_{S,\varphi}(\pi)$ and from $\{\mathcal{F}_\psi \mid \mathbf{G}(K)\psi \in \mathrm{WS}(\pi)\}$. For (i) let $\gamma = (e, h, f)$ be an $\mathfrak{sl}_2$-triple in $\mathfrak{g}$, let Z commute with γ and let $f' \in \mathfrak{g}^e \cap \mathfrak{g}^{h+Z}_{-2}$. Let $\varphi, \psi \in \mathfrak{g}^*$ be given by Killing form pairings with $f, f + f'$ respectively. We show in [**GGS**] that $\mathcal{F}_\varphi(\pi)$ can be obtained by a series of integral transforms from $\mathcal{F}_{K+h,\psi}(\pi)$. We deduce that for $\mathbf{G} = \mathrm{GL}_n$, the set $\mathrm{WO}(\pi)$ is closed under the natural ordering on orbits.

2.4. The Slodowy slice. Let us define an important geometric notion used in several proofs in [**GGS17**, **GGS**]. For an $\mathfrak{sl}_2$-triple (e, h, f) in $\mathfrak{g}$, define the Slodowy slice at f to the orbit $\mathcal{O}$ of f to be the affine space $f + \mathfrak{g}^e$. This space is strongly transversal to $\mathcal{O}$. Indeed, the tangent space at f to $\mathcal{O}$ is $[f, \mathfrak{g}]$ and $\mathfrak{g} = [f, \mathfrak{g}] + \mathfrak{g}^e$.

Quantizing the Slodowy slice one gets a finite W-algebra (see *e.g.* [**PrSk02**, **GG02**]). The representation theory of finite W-algebras seems to be closely related to the theory of generalized Whittaker models over Archimedean fields.

3. Admissible and quasi-admissible orbits

3.1. Definitions. Let $\gamma = (e, h, f)$ be an $\mathfrak{sl}_2$-triple in $\mathfrak{g}$ and let $\varphi \in \mathfrak{g}^*$ be given by the Killing form pairing with f. Let G_γ denote the joint centralizer of the three elements of γ. It is well known that G_γ is a Levi subgroup of G_φ. Recall that φ induces a non-degenerate symplectic form ω_φ on $\mathfrak{g}_1^h$ and note that G_γ acts on $\mathfrak{g}_1^h$ preserving the symplectic form. That is, there is a natural map $G_\gamma \to \mathrm{Sp}(\mathfrak{g}_1^h) = \mathrm{Sp}(\omega_\varphi)$. Let $\widetilde{\mathrm{Sp}(\omega_\varphi)} \to \mathrm{Sp}(\omega_\varphi)$ be the metaplectic double covering, and set

$$\widetilde{G_\gamma} = G_\gamma \times_{\mathrm{Sp}(\omega_\varphi)} \widetilde{\mathrm{Sp}(\omega_\varphi)}.$$

Observe that the natural map $\widetilde{G_\gamma} \to G_\gamma$ defines a double cover of G_γ. We denote by M_γ the subgroup of G_γ generated by the unipotent elements. Let $\widetilde{M_\gamma}$ denote the preimage of M_γ under the projection $\widetilde{G_\gamma} \to G_\gamma$. Note that different choices of γ with the same f lead to conjugate groups G_γ and M_γ. One can also define a covering $\widetilde{G_\varphi}$ of the group G_φ, using the symplectic form defined by φ on $\mathfrak{g}/\mathfrak{g}^\varphi$. It is easy to see that this cover splits over the unipotent radical of G_φ, and that the preimage of G_γ in $\widetilde{G_\varphi}$ is isomorphic to $\widetilde{G_\gamma}$, see *e.*g. [**Nev99**].

DEFINITION 3.1. Let $\mathbf{H}$ be a linear algebraic group defined over F, and fix an embedding $\mathbf{H} \hookrightarrow \mathrm{GL}_\mathbf{n}$. Denote by $\mathfrak{h}$ the Lie algebra of $\mathbf{H}(F)$ and by H_0 the open normal subgroup of $\mathbf{H}(F)$ generated by the image of the exponential map $\mathfrak{h} \to \mathbf{H}(F)$.

Note that H_0 does not depend on the embedding of $\mathbf{H}$ into GL_n. Note also that if $\mathbf{H}$ is semi-simple then $H_0 = \mathbf{H}(F)$ and if $F = \mathbb{R}$ then H_0 is the connected component of $\mathbf{H}(F)$. For H' a finite central extension of H, we define H'_0 to be the preimage of H_0 under the projection $H' \twoheadrightarrow H$.

DEFINITION 3.2 ([**Nev99**]). We say that a nilpotent orbit $\mathcal{O} \subset \mathfrak{g}^*$ is *admissible* if for some (equivalently, for any) choice of $\varphi \in \mathcal{O}$, the covering $\widetilde{G_\varphi} \to G_\varphi$ splits over $(G_\varphi)_0$.

As observed in [**Nev99**], this definition of admissibility is compatible with Duflo's original definition for the Archimedean case, given in [**Duf80**].

DEFINITION 3.3. We say that a nilpotent orbit $\mathcal{O} \subset \mathfrak{g}^*$ is *quasi-admissible* if for some (equivalently, for any) $\varphi \in \mathcal{O}$, the covering $\widetilde{G_\varphi} \to G_\varphi$ admits a finite dimensional *genuine* representation, that is, a finite dimensional representation on which the non-trivial element ε in the preimage of $1 \in G_\varphi$ acts by $-\mathrm{Id}$.

3.2. On the proof of Theorem B. Let $\gamma = (f, h, e)$ be an $\mathfrak{sl}_2$-triple. Let $\varphi \in \mathfrak{g}^*$ be given by the Killing form pairing with f. Let us define the action of $\widetilde{G_\gamma}$ on π_φ. We will use the notation of Definition 2.4. Since the adjoint action of G_γ preserves $\mathfrak{g}_1^h$ and the symplectic form on it, it preserves U/N'. Since σ_φ is the unique smooth irreducible representation of U/N' with central character χ_φ, we have a projective action of G_γ on σ_φ. By [**Wei64**] this action lifts to a genuine representation of $\widetilde{G_\gamma}$. This gives raise to an action of $\widetilde{G_\gamma}$ on $\mathcal{W}_\varphi$ by $(\tilde{g}f)(x) = \tilde{g}(f(g^{-1}x))$. This action commutes with the action of G and thus defines an action of $\widetilde{G_\gamma}$ on $\pi_\varphi = (\mathcal{W}_\varphi \otimes \pi)_G$.

The technique described in §2.3 implies the following theorem.

THEOREM 3.4 ([**GGS**, Theorem C]). *Let $\pi \in \operatorname{Rep}^\infty(G)$ and assume that $G\cdot\varphi \in \operatorname{WS}(\pi)$. Then the action of $\widetilde{M_\gamma}$ on the dual space π_φ^* is locally finite.*

Since the action of $\widetilde{M_\gamma}$ on π_φ is genuine, this theorem implies Theorem B after some geometric considerations.

4. Wave-front sets

4.1. Definition. Let $\pi \in \mathcal{M}(G)$. Let χ_π be the character of π. It is a generalized function on G and it defines a generalized function ξ_π on a neighborhood of zero in $\mathfrak{g}_n$, by restriction to a neighborhood of $1 \in G$ and applying logarithm. In the non-Archimedean case ξ_π is a combination of Fourier transforms of G-measures of nilpotent coadjoint orbits ([**How74**],[**HC77**, p. 180]). The measures extend to $\mathfrak{g}^*$ by [**RR72**]. In the Archimedean case, the leading term of the asymptotic expansion of ξ_π near 0 is equal to such a linear combination [**BV80**, Theorems 1.1 and 4.1]. For each nilpotent orbit $\mathcal{O}$ denote by $c_\mathcal{O}(\pi)$ the coefficient of the Fourier transform of the appropriately normalized G-invariant measure of $\mathcal{O}$ in the decomposition of ξ_π.

Define the wave front set $\operatorname{WF}(\pi)$ to be the set of orbits $\mathcal{O}$ such that $c_\mathcal{O}(\pi) \neq 0$ and $c_{\mathcal{O}'} = 0$ for every orbit $\mathcal{O}'$ that includes $\mathcal{O}$ in its closure. Denote by $\overline{WF}(\pi)$ the closure of the union of all the orbits in $\operatorname{WF}(\pi)$.

The behaviour of $\operatorname{WF}(\pi)$ under induction is studied in [**MW87, BB99**]. It corresponds to the induction of nilpotent orbits defined in [**Spa82**, §II.3]. Namely, let $P \subset G$ be a parabolic subgroup and L be the reductive quotient of P. Let $\mathfrak{l}$ and $\mathfrak{p}$ denote the Lie algebras of L and P. Let $\mathcal{O} \subset \mathfrak{l}^* \subset \mathfrak{p}^*$ be a nilpotent orbit and let $\hat{\mathcal{O}}$ be its preimage under the restriction map $\mathfrak{g}^* \twoheadrightarrow \mathfrak{p}^*$. A G-orbit $\mathcal{O}' \subset \mathfrak{g}$ is said to be induced from $\mathcal{O} \subset \mathfrak{l}$ if $\mathcal{O}'$ intersects $\hat{\mathcal{O}}$ by an open subset. All the induced orbits are conjugate over the algebraic closure of F.

The behaviour of $\operatorname{WF}(\pi)$ under cohomological induction is studied in [**BV83b**].

It follows from Theorem A and from [**SV00**] that for any $\mathcal{O} \in \operatorname{WF}(\pi)$, $c_\mathcal{O}$ is a natural number (at least for algebraic G). In the p-adic case it is shown in [**Var14**, Corollary 1] that $c_\mathcal{O} \in \mathbb{Q}$ for all $\mathcal{O}$. Moreover, for $\operatorname{GL}_n(F)$ $c_\mathcal{O} \in \mathbb{Z}$ for all $\mathcal{O}$ by [**How74**], [**DeB04**, 10.3]. If $\pi \in \mathcal{M}(\operatorname{GL}_n(F))$ is irreducible, then $c_\mathcal{O} = 1$ for all $\mathcal{O} \in \operatorname{WF}(\pi)$ by [**MW87**, §II.2].

4.2. On the proof of Theorem A. We give a sketch of the proof, which closely follows [**MW87, Var14**] but also highlights some novel features suggested by [**Say02**] (cf. [**AGSay15**, §5.2]). This is part of ongoing joint work with E. Sayag, R. Gomez and A. Kemarsky.

Let (H, φ) be a Whittaker pair such that there exists a morphism $\nu : F^\times \to G$ with $d_1\nu(1) = H$. Any neutral Whittaker pair has this property. Let $\mathcal{O} := G \cdot \varphi$. Let $\pi \in \mathcal{M}(G)$ and assume that

$$\text{for any } \mathcal{O}' \in \mathrm{WF}(\pi) \text{ with } \mathcal{O} \subseteq \overline{\mathcal{O}'}, \text{ we have } \mathcal{O}' = \mathcal{O}. \tag{14}$$

It is enough to show that under this assumption we have $\dim \mathcal{W}_{H,\varphi} = c_{\mathcal{O}}$.

Assume for simplicity that all the eigenvalues of $ad(H)$ are even, and that $\varphi \neq 0$.

The starting point of the proof is the technique of approximation of unipotent subgroups of G by open compact subgroups of G, as in Jacquet's proof of exactness of his parabolic reduction functors. More precisely, let $U \subset G$ and χ_φ be the nilpotent subgroup and its character constructed in Definition 2.4. Then [**MW87**, **Var14**] construct, following [**How77**, **Rod75**], a descending sequence of open compact subgroups K_n and their characters χ_n such that $\bigcap K_n$ is trivial, $U \subset \bigcup t^n K_n t^{-n}$, and for each n, and each $u \in t^n K_n t^{-n} \cap U$, $\chi_\varphi(u) = \chi_n(t^{-n} u t^n)$.

Let e_n denote the Hecke algebra elements given by integration on K_n versus the product of the Haar measure by χ_n. For n big enough, K_n lies in the image of the exponential map, and we consider the lifting $\exp^*(e_n)$ to $\mathfrak{g}$ and its Fourier transform $\mathcal{F}(\exp^*(e_n))$. Let $p : \mathfrak{g}^* \setminus \{0\} \to \mathbb{P}(\mathfrak{g}^*)$ denote the natural projection to the projective space. Let $\mu_{\mathcal{O}}$ denote the appropriately normalized G-invariant measure on $\mathcal{O}$. The e_n are constructed such that

(a) $\mathcal{F}(\exp^*(e_n))$ is the characteristic function of an open compact subset $B_n \subset \mathfrak{g}^*$.
(b) $\mu_{\mathcal{O}}(\mathcal{O} \cap B_n) = 1$
(c) $p(Ad(t)^n B_n)$ converge to $\{p(\varphi)\}$ as $n \to \infty$.

Let W_n be the image of $\pi(e_n)$ and $W'_n := \pi(t^n) W_n$. By these properties, assumption (14) and the definitions of the character χ_π and the wave-front set we have, for n big enough,

$$\begin{aligned}\dim W'_n &= \dim W_n = \mathrm{Tr}(\pi(e_n)) = \chi_\pi(e_n) = \sum_{\mathcal{O}'} c_{\mathcal{O}'} \mu_{\mathcal{O}'}(\mathcal{F}(\exp^*(e_n)) = \\ &= \sum_{\mathcal{O}'} c_{\mathcal{O}'} \mu_{\mathcal{O}'}(B_n \cap \mathcal{O}') = 0 + \sum_{\mathcal{O}' \subset \overline{\mathcal{O}}} c_{\mathcal{O}'} \mu_{\mathcal{O}'}(B_n \cap \mathcal{O}') = 0 + c_{\mathcal{O}} + 0 = c_{\mathcal{O}}.\end{aligned} \tag{15}$$

It is now left to prove that for n big enough, W'_n projects isomorphically onto $\pi_{H,\varphi}$. For this purpose it is shown that for n big enough, $\pi(t^{n+1} e_{n+1} t^{-n-1})$ defines an embedding $W'_n \hookrightarrow W'_{n+1}$, that is compatible with the projections to $\pi_{H,\varphi}$, and that these projections are also embeddings.

(16)

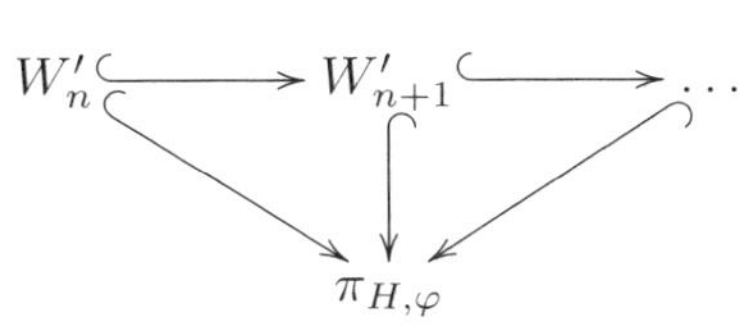

Since $\dim W'_n = c_{\mathcal{O}}$, the images of these embeddings coincide and have dimension $c_{\mathcal{O}}$. Since $t^n K_n t^{-n}$ "approximate" U, and χ_n are compatible with χ_φ, this image is the whole space $\pi_{H,\varphi}$.

4.3. Archimedean case. In this subsection we fix G to be a real reductive group, and consider the relationship between wave-front set, Whittaker support, and several additional invariants. One of these invariants is the associated variety of the annihilator of π, that we call for brevity *annihilator variety* and denote $\mathcal{V}(\pi)$. It is defined for all $\pi \in \mathrm{Rep}^\infty(G)$, as the zero set in $\mathfrak{g}_{\mathbb{C}}^*$ of the ideal in the symmetric algebra $S(\mathfrak{g}_{\mathbb{C}})$, which is generated by the symbols of the annihilator ideal of π in the universal enveloping algebra $U(\mathfrak{g}_{\mathbb{C}})$. This invariant is more coarse than $\mathrm{WF}(\pi)$, as the following theorem shows.

THEOREM 4.1 ([**Ros95**, Theorem D], using [**BB85, Jos85**]). *Let $\pi \in \mathcal{M}(D)$. Then $\mathcal{V}(\pi)$ is the Zariski closure of $\overline{\mathrm{WF}(\pi)}$. Moreover, if π is irreducible then $\mathcal{V}(\pi)$ is the closure of a single complex nilpotent orbit.*

In particular, if G is a complex reductive group or $G = \mathrm{GL}_n(\mathbb{R})$ we have $\overline{\mathrm{WF}}(\pi) = \mathcal{V}(\pi) \cap \mathfrak{g}^*$. Thus, in these cases $\mathrm{WF}(\pi)$ consists of a single orbit for all irreducible admissible π. However, for $G = \mathrm{SL}_2(\mathbb{R})$ this is not the case. Indeed, this group has two regular real nilpotent orbits. For irreducible principal series representations, WF is the union of these orbits, while for discrete series WF consists of one of these orbits.

Let us now discuss the relation of $\mathcal{V}(\pi)$ and $\mathrm{WF}(\pi)$ to $\mathrm{WO}(\pi)$.

THEOREM 4.2 ([**Mat87**, Corollary 4]). *Let $\pi \in \mathrm{Rep}^\infty(G)$, and $\mathcal{O} \in \mathrm{WO}(\pi)$. Then $\mathcal{O} \subset \mathcal{V}(\pi)$.*

Let us now list several results in the other direction.

THEOREM 4.3 ([**Mat90a**]). *Suppose that G is a complex reductive group and let $\pi \in \mathcal{M}(G)$ have regular infinitesimal character. Let (S, φ) be a Whittaker pair such that the orbit $G\varphi$ lies in $\mathrm{WF}(\pi)$, and intersects the nilradical of the parabolic subgroup defined by S in a dense subset. Then $0 < \dim \pi_{S,\varphi} < \infty$.*

The result in [**Mat90a**] includes also the vanishing of the corresponding higher homologies.

THEOREM 4.4 ([**GGS17**, §3.3], based on [**GS15**]). *Let G be quasi-split and algebraic, and let $\pi \in \mathcal{M}(G)$. Let (S, φ) be a Whittaker pair such that φ is a principal nilpotent element of a Levi subalgebra of $\mathfrak{g}$. Suppose $\varphi \in \overline{\mathrm{WF}}(\pi)$. Then there exists $g \in G_{\mathbb{C}}$ such that $ad(g)$ preserves $\mathfrak{g}$ and $\pi_{ad(g)(S),ad(g)(\varphi)} \neq 0$. Moreover, for complex G and for $G = \mathrm{GL}_n(\mathbb{R})$ we have $\pi_{S,\varphi} \neq 0$.*

The sets WF and WS coincide also for representations induced from finite-dimensional representations of parabolic subgroups, by [**BB99, GSS18**]. For further results on the equality of WF and WS see [**GW80, Yam01, Prz91, Mat90b, Mat92**].

For $\mathrm{GL}_n(F)$ one can express $\pi_{\mathcal{O}}$ through the Bernstein-Zelevinsky derivatives and their Archimedean analogs, see [**BZ77, AGSah15a**] for the definitions and [**GGS17**, Theorem E] for their relation to $\pi_{\mathcal{O}}$. This implies the following theorem, for all local fields F with $char F = 0$.

THEOREM 4.5 ([**MW87**, §II.2] and [**GGS17**, Corollary G]). *Let $\pi \in \mathcal{M}(\mathrm{GL}_n(F))$ be an irreducible unitarizable representation, and $\mathcal{O} \in \mathrm{WF}(\pi)$. Then $\dim \pi_{\mathcal{O}} = 1$.*

This type of unique models also plays an important role in a family of Rankin-Selberg integrals that represent tensor product L-functions for classical groups (cf. [**CFGK**, **CFGK2**]).

In [**How81**], $\overline{\mathrm{WF}(\pi)}$ is defined for any continuous representation π of any Lie group in a Hilbert space. By [**How81**, Theorem 1.8] and [**Ros95**, Theorem 3.4], this definition extends the one given above for reductive G and $\pi \in \mathcal{M}(G)$ (via realizing π as the space of smooth vectors in a Hilbert space representation). The equality of $\overline{\mathrm{WF}(\pi)}$ to several other analytic invariants is proven in [**Ros95**, Theorem B].

Let us add that for irreducible unitary representations of type I classical reductive groups, the Howe rank ([**How82**]) is determined by the maximal among the ranks of the matrices lying in the annihilator variety, by [**He08**].

One can also attach to any smooth admissible representation π of a real reductive group G and to any maximal compact subgroup $K \subset G$ an algebraic invariant: *the associated variety* of the module $\pi^{(K)}$ of K-finite vectors. This variety is a set of nilpotent K-orbits in $\mathfrak{k}^{\perp} \subset \mathfrak{g}^*$, with multiplicities, see [**Vog91**, **Vog17**]. The dimension of this variety equals the Gelfand-Kirillov dimension of π and is also equal to half of the dimension of the annihilator variety of π.

By [**SV00**], if G is algebraic then the maximal orbits in this set correspond to WF under the Kostant-Sekiguchi correspondence, and the multiplicities are equal to the coefficients $c_{\mathcal{O}}$. The Kostant-Sekiguchi correspondence maps K-orbits in $\mathfrak{k}^{\perp}$ to G-orbits in $\mathfrak{g}^*$ that lie in the same orbit of the complexification $G_{\mathbb{C}}$ on $\mathfrak{g}^*_{\mathbb{C}}$, preserving the closure order.

References

[AG08] A. Aizenbud and D. Gourevitch, *Schwartz functions on Nash manifolds*, Int. Math. Res. Not. IMRN **5** (2008), Art. ID rnm 155, 37, DOI 10.1093/imrn/rnm155. MR2418286

[AGKLP] O. Ahlén, H. P. A. Gustafsson, A. Kleinschmidt, B. Liu, and D. Persson, *Fourier coefficients attached to small automorphic representations of $SL_n(\mathbb{A})$*, J. Number Theory **192** (2018), 80–142, DOI 10.1016/j.jnt.2018.03.022. MR3841547

[AGSay15] A. Aizenbud, D. Gourevitch, and E. Sayag, *$\mathfrak{z}$-finite distributions on p-adic groups*, Adv. Math. **285** (2015), 1376–1414, DOI 10.1016/j.aim.2015.07.033. MR3406530

[AGSah15a] A. Aizenbud, D. Gourevitch, and S. Sahi, *Derivatives for smooth representations of $GL(n,\mathbb{R})$ and $GL(n,\mathbb{C})$*, Israel J. Math. **206** (2015), no. 1, 1–38, DOI 10.1007/s11856-015-1149-9. MR3319629

[AGSah15b] A. Aizenbud, D. Gourevitch, and S. Sahi, *Twisted homology for the mirabolic nilradical*, Israel J. Math. **206** (2015), no. 1, 39–88, DOI 10.1007/s11856-014-1150-3. MR3319630

[BB99] D. Barbasch and M. Božičević, *The associated variety of an induced representation*, Proc. Amer. Math. Soc. **127** (1999), no. 1, 279–288, DOI 10.1090/S0002-9939-99-04482-2. MR1458862

[BB85] W. Borho and J.-L. Brylinski, *Differential operators on homogeneous spaces. III. Characteristic varieties of Harish-Chandra modules and of primitive ideals*, Invent. Math. **80** (1985), no. 1, 1–68, DOI 10.1007/BF01388547. MR784528

[BV80] D. Barbasch and D. A. Vogan Jr., *The local structure of characters*, J. Funct. Anal. **37** (1980), no. 1, 27–55, DOI 10.1016/0022-1236(80)90026-9. MR576644

[BV82] D. Barbasch and D. Vogan, *Primitive ideals and orbital integrals in complex classical groups*, Math. Ann. **259** (1982), no. 2, 153–199, DOI 10.1007/BF01457308. MR656661

[BV83a] D. Barbasch and D. Vogan, *Primitive ideals and orbital integrals in complex exceptional groups*, J. Algebra **80** (1983), no. 2, 350–382, DOI 10.1016/0021-8693(83)90006-6. MR691809

[BV83b] D. Barbasch and D. Vogan, *Weyl group representations and nilpotent orbits*, Representation theory of reductive groups (Park City, Utah, 1982), Progr. Math., vol. 40, Birkhäuser Boston, Boston, MA, 1983, pp. 21–33. MR733804

[Bou75] N. Bourbaki: *Groupes et algebres de Lie*, Chap. 7 et 8. fasc. XXXVIII. Paris; Hermann (1975).

[BZ76] I. N. Bernšteĭn and A. V. Zelevinskiĭ, *Representations of the group $GL(n, F)$, where F is a local non-Archimedean field* (Russian), Uspehi Mat. Nauk **31** (1976), no. 3(189), 5–70. MR0425030

[BZ77] I. N. Bernstein and A. V. Zelevinsky, *Induced representations of reductive $\mathfrak{p}$-adic groups. I*, Ann. Sci. École Norm. Sup. (4) **10** (1977), no. 4, 441–472. MR0579172

[Car85] R. W. Carter, *Finite groups of Lie type*, Pure and Applied Mathematics (New York), John Wiley & Sons, Inc., New York, 1985. Conjugacy classes and complex characters; A Wiley-Interscience Publication. MR794307

[Cas89] W. Casselman, *Canonical extensions of Harish-Chandra modules to representations of G*, Canad. J. Math. **41** (1989), no. 3, 385–438, DOI 10.4153/CJM-1989-019-5. MR1013462

[CFGK] Y. Cai, S. Friedberg, D. Ginzburg, E. Kaplan *Doubling Constructions for Covering Groups and Tensor Product L-Functions,* arXiv:1601.08240.

[CFGK2] Y. Cai, S. Friedberg, D. Ginzburg, E. Kaplan *Doubling Constructions and Tensor Product L-Functions: the linear case,* arXiv:1710.00905 .

[CHM00] W. Casselman, H. Hecht, and D. Miličić, *Bruhat filtrations and Whittaker vectors for real groups*, The mathematical legacy of Harish-Chandra (Baltimore, MD, 1998), Proc. Sympos. Pure Math., vol. 68, Amer. Math. Soc., Providence, RI, 2000, pp. 151–190, DOI 10.1090/pspum/068/1767896. MR1767896

[CoMG93] D. H. Collingwood and W. M. McGovern, *Nilpotent orbits in semisimple Lie algebras*, Van Nostrand Reinhold Mathematics Series, Van Nostrand Reinhold Co., New York, 1993. MR1251060

[DeB04] S. DeBacker, *Lectures on harmonic analysis for reductive p-adic groups*, Representations of real and p-adic groups, Lect. Notes Ser. Inst. Math. Sci. Natl. Univ. Singap., vol. 2, Singapore Univ. Press, Singapore, 2004, pp. 47–94. MR2090869

[dCl91] F. du Cloux, *Sur les représentations différentiables des groupes de Lie algébriques* (French), Ann. Sci. École Norm. Sup. (4) **24** (1991), no. 3, 257–318. MR1100992

[Duf80] M. Duflo, *Construction de représentations unitaires d'un groupe de Lie* (French, with English summary), Harmonic analysis and group representations, Liguori, Naples, 1982, pp. 129–221. MR777341

[FS16] Y. Fang and B. Sun, *Chevalley's theorem for affine Nash groups*, J. Lie Theory **26** (2016), no. 2, 359–369. MR3417410

[FGKP18] P. Fleig, H. P. A. Gustafsson, A. Kleinschmidt, and D. Persson, *Eisenstein series and automorphic representations*, Cambridge Studies in Advanced Mathematics, vol. 176, Cambridge University Press, Cambridge, 2018. With applications in string theory. MR3793195

[GG02] W. L. Gan and V. Ginzburg, *Quantization of Slodowy slices*, Int. Math. Res. Not. **5** (2002), 243–255, DOI 10.1155/S107379280210609X. MR1876934

[GGJ02] W. T. Gan, N. Gurevich, and D. Jiang, *Cubic unipotent Arthur parameters and multiplicities of square integrable automorphic forms*, Invent. Math. **149** (2002), no. 2, 225–265, DOI 10.1007/s002220200210. MR1918673

[GGS17] R. Gomez, D. Gourevitch, S. Sahi: *Generalized and degenerate Whittaker models,* Compositio Mathematica **153**, n. 2, pp. 223-256 (2017). DOI:10.1112/S0010437X16007788.

[GGS] R.Gomez, D. Gourevitch, S. Sahi: *Whittaker supports for representations of reductive groups,* arXiv:1610.00284v4.

[GGKPS] D. Gourevitch, H. Gustafsson, A. Kleinschmidt, D. Persson, S. Sahi: *Expressing automorphic forms in next-to-minimal representations through their Whittaker-Fourier coefficients,* arXiv:1811.05966.

[Gin06] D. Ginzburg, *Certain conjectures relating unipotent orbits to automorphic representations*, Israel J. Math. **151** (2006), 323–355, DOI 10.1007/BF02777366. MR2214128

[Gin14] D. Ginzburg, *Towards a classification of global integral constructions and functorial liftings using the small representations method*, Adv. Math. **254** (2014), 157–186, DOI 10.1016/j.aim.2013.12.007. MR3161096

[GK75] I. M. Gel′fand and D. A. Kajdan, *Representations of the group* $\mathrm{GL}(n, K)$ *where* K *is a local field*, Lie groups and their representations (Proc. Summer School, Bolyai János Math. Soc., Budapest, 1971), Halsted, New York, 1975, pp. 95–118. MR0404534

[GMV15] M. B. Green, S. D. Miller, and P. Vanhove, *Small representations, string instantons, and Fourier modes of Eisenstein series*, J. Number Theory **146** (2015), 187–309, DOI 10.1016/j.jnt.2013.05.018. MR3267116

[GRS99] D. Ginzburg, S. Rallis, and D. Soudry, *On a correspondence between cuspidal representations of* GL_{2n} *and* $\widetilde{\mathrm{Sp}}_{2n}$, J. Amer. Math. Soc. **12** (1999), no. 3, 849–907, DOI 10.1090/S0894-0347-99-00300-8. MR1671452

[GRS03] D. Ginzburg, S. Rallis, and D. Soudry, *On Fourier coefficients of automorphic forms of symplectic groups*, Manuscripta Math. **111** (2003), no. 1, 1–16, DOI 10.1007/s00229-003-0355-7. MR1981592

[GRS11] D. Ginzburg, S. Rallis, and D. Soudry, *The descent map from automorphic representations of* $\mathrm{GL}(n)$ *to classical groups*, World Scientific Publishing Co. Pte. Ltd., Hackensack, NJ, 2011. MR2848523

[GS15] D. Gourevitch and S. Sahi, *Degenerate Whittaker functionals for real reductive groups*, Amer. J. Math. **137** (2015), no. 2, 439–472, DOI 10.1353/ajm.2015.0008. MR3337800

[GS87] S. Gelbart and D. Soudry, *On Whittaker models and the vanishing of Fourier coefficients of cusp forms*, Proc. Indian Acad. Sci. Math. Sci. **97** (1987), no. 1-3, 67–74 (1988), DOI 10.1007/BF02837815. MR983606

[GSS18] D. Gourevitch, S. Sahi, E. Sayag: *Analytic continuation of equivariant distributions,* arXiv:1608.03442, International Mathematics Research Notices, rnx326 (2018), https://doi.org/10.1093/imrn/rnx326.

[GW80] R. Goodman and N. R. Wallach, *Whittaker vectors and conical vectors*, J. Funct. Anal. **39** (1980), no. 2, 199–279, DOI 10.1016/0022-1236(80)90013-0. MR597811

[GZ14] R. Gomez and C.-B. Zhu, *Local theta lifting of generalized Whittaker models associated to nilpotent orbits*, Geom. Funct. Anal. **24** (2014), no. 3, 796–853, DOI 10.1007/s00039-014-0276-5. MR3213830

[Har12] B. Harris, *Tempered representations and nilpotent orbits*, Represent. Theory **16** (2012), 610–619, DOI 10.1090/S1088-4165-2012-00414-9. MR3001468

[HC77] Harish-Chandra, *The characters of reductive p-adic groups*, Contributions to algebra (collection of papers dedicated to Ellis Kolchin), Academic Press, New York, 1977, pp. 175–182. MR0492086

[He08] H. He, *Associated varieties and Howe's* N*-spcctrum*, Pacific J. Math. **237** (2008), no. 1, 97–119, DOI 10.2140/pjm.2008.237.97. MR2415210

[How74] R. Howe, *The Fourier transform and germs of characters (case of* Gl_n *over a* p*-adic field)*, Math. Ann. **208** (1974), 305–322, DOI 10.1007/BF01432155. MR0342645

[How77] R. E. Howe, *Kirillov theory for compact* p*-adic groups*, Pacific J. Math. **73** (1977), no. 2, 365–381. MR0579176

[How81] R. Howe, *Wave front sets of representations of Lie groups*, Automorphic forms, representation theory and arithmetic (Bombay, 1979), Tata Inst. Fund. Res. Studies in Math., vol. 10, Tata Inst. Fundamental Res., Bombay, 1981, pp. 117–140. MR0633659

[How82] R. Howe, *On a notion of rank for unitary representations of the classical groups*, Harmonic analysis and group representations, Liguori, Naples, 1982, pp. 223–331. MR777342

[HS16] J. Hundley and E. Sayag, *Descent construction for GSpin groups*, Mem. Amer. Math. Soc. **243** (2016), no. 1148, v+124, DOI 10.1090/memo/1148. MR3517154

[Jia07] D. Jiang, *Periods of automorphic forms*, Proceedings of the International Conference on Complex Geometry and Related Fields, AMS/IP Stud. Adv. Math., vol. 39, Amer. Math. Soc., Providence, RI, 2007, pp. 125–148. MR2338623

[JLS16] D. Jiang, B. Liu, and G. Savin, *Raising nilpotent orbits in wave-front sets*, Represent. Theory **20** (2016), 419–450, DOI 10.1090/ert/490. MR3564676

[Jos85] A. Joseph, *On the associated variety of a primitive ideal*, J. Algebra **93** (1985), no. 2, 509–523, DOI 10.1016/0021-8693(85)90172-3. MR786766

[Kaw85] N. Kawanaka, *Generalized Gel′fand-Graev representations and Ennola duality*, Algebraic groups and related topics (Kyoto/Nagoya, 1983), Adv. Stud. Pure Math., vol. 6, North-Holland, Amsterdam, 1985, pp. 175–206. MR803335

[Kos59] B. Kostant, *The principal three-dimensional subgroup and the Betti numbers of a complex simple Lie group*, Amer. J. Math. **81** (1959), 973–1032, DOI 10.2307/2372999. MR0114875

[Kos78] B. Kostant, *On Whittaker vectors and representation theory*, Invent. Math. **48** (1978), no. 2, 101–184, DOI 10.1007/BF01390249. MR507800

[LaMao14] E. Lapid and Z. Mao, *Model transition for representations of metaplectic type*, Int. Math. Res. Not. IMRN **19** (2015), 9486–9568, DOI 10.1093/imrn/rnu225. With an appendix by Marko Tadić. MR3431601

[LoMa15] H. Y. Loke and J. Ma, *Invariants and K-spectrums of local theta lifts*, Compos. Math. **151** (2015), no. 1, 179–206, DOI 10.1112/S0010437X14007520. MR3305312

[LS08] H. Y. Loke and G. Savin, *On minimal representations of Chevalley groups of type D_n, E_n and G_2*, Math. Ann. **340** (2008), no. 1, 195–208, DOI 10.1007/s00208-007-0144-9. MR2349773

[Lus79] G. Lusztig, *A class of irreducible representations of a Weyl group*, Nederl. Akad. Wetensch. Indag. Math. **41** (1979), no. 3, 323–335. MR546372

[Lus92] G. Lusztig, *A unipotent support for irreducible representations*, Adv. Math. **94** (1992), no. 2, 139–179, DOI 10.1016/0001-8708(92)90035-J. MR1174392

[Lyn79] T. E. Lynch, *Generalized Whittaker vectors and representation theory*, ProQuest LLC, Ann Arbor, MI, 1979. Thesis (Ph.D.)–Massachusetts Institute of Technology. MR2940885

[Mat87] H. Matumoto, *Whittaker vectors and associated varieties*, Invent. Math. **89** (1987), no. 1, 219–224, DOI 10.1007/BF01404678. MR892192

[Mat90a] H. Matumoto, *$C^{-\infty}$-Whittaker vectors for complex semisimple Lie groups, wave front sets, and Goldie rank polynomial representations*, Ann. Sci. École Norm. Sup. (4) **23** (1990), no. 2, 311–367. MR1046499

[Mat90b] H. Matumoto, *Whittaker modules associated with highest weight modules*, Duke Math. J. **60** (1990), no. 1, 59–113, DOI 10.1215/S0012-7094-90-06002-8. MR1047117

[Mat92] H. Matumoto, *$C^{-\infty}$-Whittaker vectors corresponding to a principal nilpotent orbit of a real reductive linear Lie group, and wave front sets*, Compositio Math. **82** (1992), no. 2, 189–244. MR1157939

[MS12] S. D. Miller and S. Sahi, *Fourier coefficients of automorphic forms, character variety orbits, and small representations*, J. Number Theory **132** (2012), no. 12, 3070–3108, DOI 10.1016/j.jnt.2012.05.032. MR2965210

[Moe96] C. Mœglin, *Front d'onde des représentations des groupes classiques p-adiques* (French, with French summary), Amer. J. Math. **118** (1996), no. 6, 1313–1346. MR1420926

[MW87] C. Mœglin and J.-L. Waldspurger, *Modèles de Whittaker dégénérés pour des groupes p-adiques* (French), Math. Z. **196** (1987), no. 3, 427–452, DOI 10.1007/BF01200363. MR913667

[MW95] C. Mœglin and J.-L. Waldspurger, *Spectral decomposition and Eisenstein series*, Cambridge Tracts in Mathematics, vol. 113, Cambridge University Press, Cambridge, 1995. Une paraphrase de l'Écriture [A paraphrase of Scripture]. MR1361168

[MT16] K. Maktouf and P. Torasso, *Restriction de la représentation de Weil à un sous-groupe compact maximal* (French, with English summary), J. Math. Soc. Japan **68** (2016), no. 1, 245–293, DOI 10.2969/jmsj/06810245. MR3454560

[Nev99] M. Nevins, *Admissible nilpotent coadjoint orbits of p-adic reductive Lie groups*, Represent. Theory **3** (1999), 105–126, DOI 10.1090/S1088-4165-99-00072-2. MR1698202

[Nev02] M. Nevins, *Admissible nilpotent orbits of real and p-adic split exceptional groups*, Represent. Theory **6** (2002), 160–189, DOI 10.1090/S1088-4165-02-00134-6. MR1915090

[NPS73] M. E. Novodvorskiĭ and I. I. Pjateckiĭ-Šapiro, *Generalized Bessel models for the symplectic group of rank* 2 (Russian), Mat. Sb. (N.S.) **90(132)** (1973), 246–256, 326. MR0338279

[Oht91a] T. Ohta, *Classification of admissible nilpotent orbits in the classical real Lie algebras*, J. Algebra **136** (1991), no. 2, 290–333, DOI 10.1016/0021-8693(91)90049-E. MR1089302

[PrSk02] A. Premet, *Special transverse slices and their enveloping algebras*, Adv. Math. **170** (2002), no. 1, 1–55, DOI 10.1006/aima.2001.2063. With an appendix by Serge Skryabin. MR1929302

[PT04] V. L. Popov and E. A. Tevelev, *Self-dual projective algebraic varieties associated with symmetric spaces*, Algebraic transformation groups and algebraic varieties, Encyclopaedia Math. Sci., vol. 132, Springer, Berlin, 2004, pp. 131–167, DOI 10.1007/978-3-662-05652-3_8. MR2090673

[Prz91] T. Przebinda, *Characters, dual pairs, and unipotent representations*, J. Funct. Anal. **98** (1991), no. 1, 59–96, DOI 10.1016/0022-1236(91)90091-I. MR1111194

[Prz] T. Przebinda, *The character and the wave front set correspondence in the stable range*, J. Funct. Anal. **274** (2018), no. 5, 1284–1305, DOI 10.1016/j.jfa.2018.01.002. MR3778675

[Ros95] W. Rossmann, *Picard-Lefschetz theory and characters of a semisimple Lie group*, Invent. Math. **121** (1995), no. 3, 579–611, DOI 10.1007/BF01884312. MR1353309

[RR72] R. Ranga Rao, *Orbital integrals in reductive groups*, Ann. of Math. (2) **96** (1972), 505–510, DOI 10.2307/1970822. MR0320232

[Rod75] F. Rodier, *Modèle de Whittaker et caractères de représentations* (French), Noncommutative harmonic analysis (Actes Colloq., Marseille-Luminy, 1974), Springer, Berlin, 1975, pp. 151–171. Lecture Notes in Math., Vol. 466. MR0393355

[Say02] E. Sayag: *A Generalization of Harish-Chandra Regularity Theorem*, Thesis submitted for the degree of "Doctor of Philosophy", Tel-Aviv University (2002).

[Sha74] J. A. Shalika, *The multiplicity one theorem for* GL_n, Ann. of Math. (2) **100** (1974), 171–193, DOI 10.2307/1971071. MR0348047

[SV00] W. Schmid and K. Vilonen, *Characteristic cycles and wave front cycles of representations of reductive Lie groups*, Ann. of Math. (2) **151** (2000), no. 3, 1071–1118, DOI 10.2307/121129. MR1779564

[Shen16] X. Shen, *Top Fourier coefficients for cuspidal representations of symplectic groups*, Int. Math. Res. Not. IMRN **10** (2017), 2909–2947, DOI 10.1093/imrn/rnw073. MR3658128

[Spa82] N. Spaltenstein, *Classes unipotentes et sous-groupes de Borel* (French), Lecture Notes in Mathematics, vol. 946, Springer-Verlag, Berlin-New York, 1982. MR672610

[Sun15] B. Sun, *Almost linear Nash groups*, Chin. Ann. Math. Ser. B **36** (2015), no. 3, 355–400, DOI 10.1007/s11401-015-0915-7. MR3341164

[Tre67] F. Trèves, *Topological vector spaces, distributions and kernels*, Academic Press, New York-London, 1967. MR0225131

[Var14] S. Varma, *On a result of Moeglin and Waldspurger in residual characteristic 2*, Math. Z. **277** (2014), no. 3-4, 1027–1048, DOI 10.1007/s00209-014-1292-8. MR3229979

[Vog91] D. A. Vogan Jr., *Associated varieties and unipotent representations*, Harmonic analysis on reductive groups (Brunswick, ME, 1989), Progr. Math., vol. 101, Birkhäuser Boston, Boston, MA, 1991, pp. 315–388. MR1168491

[Vog94] D. A. Vogan Jr., *The unitary dual of* G_2, Invent. Math. **116** (1994), no. 1-3, 677–791, DOI 10.1007/BF01231578. MR1253210

[Vog17] D. A. Vogan Jr., *The size of infinite-dimensional representations*, Jpn. J. Math. **12** (2017), no. 2, 175–210, DOI 10.1007/s11537-017-1648-z. MR3694931

[Wall88] N. R. Wallach, *Lie algebra cohomology and holomorphic continuation of generalized Jacquet integrals*, Representations of Lie groups, Kyoto, Hiroshima, 1986, Adv. Stud. Pure Math., vol. 14, Academic Press, Boston, MA, 1988, pp. 123–151. MR1039836

[Wall92] N. R. Wallach, *Real reductive groups. I*, Pure and Applied Mathematics, vol. 132, Academic Press, Inc., Boston, MA, 1988. MR929683

[Wei64] A. Weil, *Sur certains groupes d'opérateurs unitaires* (French), Acta Math. **111** (1964), 143–211, DOI 10.1007/BF02391012. MR0165033

[Yam86] H. Yamashita, *On Whittaker vectors for generalized Gel′fand-Graev representations of semisimple Lie groups*, J. Math. Kyoto Univ. **26** (1986), no. 2, 263–298, DOI 10.1215/kjm/1250520922. MR849220

[Yam01] H. Yamashita, *Cayley transform and generalized Whittaker models for irreducible highest weight modules* (English, with English and French summaries), Astérisque **273** (2001), 81–137. Nilpotent orbits, associated cycles and Whittaker models for highest weight representations. MR1845715

[Zel80] A. V. Zelevinsky, *Induced representations of reductive* $\mathfrak{p}$*-adic groups. II. On irreducible representations of* $\mathrm{GL}(n)$, Ann. Sci. École Norm. Sup. (4) **13** (1980), no. 2, 165–210. MR584084

Department of Mathematics, Weizmann Institute of Science, Rehovot 7610001 Israel

Email address: dmitry.gourevitch@weizmann.ac.il

Department of Mathematics, Rutgers University, Hill Center - Busch Campus, 110 Frelinghuysen Road, Piscataway, New Jersey 08854-8019

Email address: sahi@math.rugers.edu

Proceedings of Symposia in Pure Mathematics
Volume **101**, 2019
https://doi.org/10.1090/pspum/101/01794

Geometric approach to the fermionic Fock space, via flag varieties and representations of algebraic (super)groups

Caroline Gruson and Vera Serganova

To Joseph Bernstein

Introduction

The famous Borel-Weil-Bott theorem (see [**1**]) relates the cohomology groups of line bundles over flag varieties and irreducible representations of reductive algebraic groups in characteristic zero. More precisely, any line bundle over a flag variety, say G/P, has non-zero cohomology in at most one degree, and the corresponding cohomology group turns out to be an irreducible representation of G. In more general settings, for instance if the base field is of positive characteristic or if we consider supergroups, such cohomology groups give representations which are no longer necessarily irreducible, moreover, the cohomology groups in general are not concentrated in one degree. These representations are of essential importance, in the super case they lead to what is called Euler characteristic virtual modules, which have an easily computable character, and turn out to be a key ingredient in the character formulae for irreducible representations of the supergroup.

The present paper is inspired by Bott's philosophy: on one hand he relates the cohomology of line bundles over a flag manifold G/P and the cohomology of the nilradical of the Lie algebra of the parabolic subgroup $P \subset G$ (Theorem I in [**1**]) and on the other hand he explores the G-module structure on the cohomology of a line bundle over G/P (Theorem IV in [**1**]). This led us to consider two functors, called geometric restriction and geometric induction, in the $\mathbb{Z}/2\mathbb{Z}$-graded setting in characteristic zero, and to explore their relation. Those functors will be properly defined as soon as we have enough notation.

This paper has another source of inspiration.

In a book published in 1981 ([**8**]), Andrei Zelevinsky categorified an infinite-rank PSH-algebra in terms of representations of the collection of all $GL(n, \mathbb{F})$ where $\mathbb{F}$ is a finite field. He did this using a pair of adjoint functors, the parabolic induction and its adjoint.

We intend, in this paper, to apply the same set of ideas to the categorification of a infinite Clifford algebra acting on the Fock space of semi-infinite forms, in terms of representations of the collection of all classical supergroups $SOSP(2m+1, 2n)$, using the geometric induction functor and its adjoint called geometric restriction.

Let us start with the preliminary example of classical groups.

Let $(G_n)_{n\geq 1}$ be a family of complex classical Lie groups, G_n of rank n, together with inclusions

$$G_1 \subset G_2 \subset \ldots \subset G_n \subset \ldots$$

in such a way that $G_{n-1}\times \mathbb{C}^*$ is the reductive part of a maximal parabolic subgroup denoted by P_n of G_n, and we denote the maximal unipotent subgroup of P_n by U_n. For instance, consider $G_n = GL(n, \mathbb{C})$. We use gothic letters for the corresponding Lie algebras. We denote by $\mathcal{F}_n$ the category of finite-dimensional G_n-modules, it is a semi-simple category and we denote by $\mathcal{K}_n$ its Grothendieck group.

We use the functors Γ_i^a and H_b^j defined as follows:

$$\Gamma_i^a : \mathcal{F}_n \to \mathcal{F}_{n+1},$$

$$\Gamma_i^a(M) := H^i(G_{n+1}/P_{n+1}, \mathcal{L}(\mathbb{C}_{a+n} \boxtimes M)^*)^*$$

where $\mathbb{C}_a$ is the one-dimensional representation of $\mathbb{C}^*$ with character $a \in \mathbb{Z}$; we assume that U_n acts trivially and $\mathcal{L}(\mathbb{C}_a \boxtimes M)^*$ is the induced vector bundle $G_n \times_{P_{n+1}} (\mathbb{C}_a \boxtimes M)^*$.

$$H_b^j : \mathcal{F}_n \to \mathcal{F}_{n-1},$$

$$H_b^j(M) := \mathrm{Hom}_{\mathbb{C}^*}(\mathbb{C}_{b+n}, H^j(\mathfrak{u}_n, M)).$$

At the level of Grothendieck groups we obtain linear maps

$$\gamma^a : \mathcal{K}_n \to \mathcal{K}_{n+1}$$

$$[M] \mapsto \sum_i (-1)^i [H^i(G_{n+1}/P_{n+1}, \mathcal{L}(\mathbb{C}_{a+n} \boxtimes M)^*)^*],$$

$$\eta_b : \mathcal{K}_n \to \mathcal{K}_{n-1}$$

$$[M] \mapsto \sum_j (-1)^j [\mathrm{Hom}_{\mathbb{C}^*}(\mathbb{C}_{b+n}, H^j(\mathfrak{u}_n, M))].$$

We set $\mathcal{K} := \oplus_n \mathcal{K}_n$ and extend those maps to $\mathcal{K}$. Then, applying Borel-Weil-Bott theorem, we obtain the following relations, for all a and b in $\mathbb{Z}$:

$$\gamma^a\gamma^b + \gamma^b\gamma^a = 0, \tag{1}$$

$$\eta_a\eta_b + \eta_b\eta_a = 0, \tag{2}$$

$$\gamma^a\eta_b + \eta_b\gamma^a = \delta_{a,b} Id. \tag{3}$$

We recognise those relations as the ones of the infinite dimensional Clifford algebra **C**. Furthermore, we see $\mathcal{K}$ as an irreducible representation of **C** which is induced by the trivial representation of the subalgebra of **C** generated by $(\eta_b)_{b\in\mathbb{Z}}$.

This provides a categorification of the Clifford algebra **C** by the family of classical groups $(G_n)_{n\geq 1}$.

We follow the same scheme for the family of classical Lie supergroups $SOSP(2m+1, 2n)$ when m and n vary, in this case we categorify the representation of the infinite Clifford algebra in the Fock space of semi-infinite forms. In the last section, we explain how our previous categorification work on orthosymplectic Lie superagebras ([**4**]) can be understood in this context.

We would also like to mention the work of Michael Ehrig and Catharina Stroppel [**2**], who used quantized symmetric pairs in order to refine our previous results on the category of finite dimensional modules over orthosymplectic Lie superalgebras and obtain a diagrammatic description of the endomorphism algebras of projective generators.

It would be very interesting now to construct a canonical basis in the Fock space of semi-infinite forms.

Finally, we would like to emphasize that what we do here can easily be done for all series of classical Lie supergroups, with minor changes only.

We are grateful to A. Sergeev who suggested to look at Fock spaces, in relation to orthosymplectic groups. The second author acknowledges support of the NSF grant DMS-1701532.

1. Basic setting

We work over the field of complex numbers in the category of $\mathbb{Z}/2\mathbb{Z}$-graded spaces. The reader should keep it in mind when we consider symmetric and exterior powers.

We denote by $\mathfrak{g}_{m,n}$ the Lie superalgebra $\mathfrak{osp}(2m+1,2n)$ and

$$\mathfrak{g}_{\infty,\infty} = \varinjlim_{m,n\to\infty} \mathfrak{g}_{m,n}.$$

Furthermore, we fix an embedding $\mathfrak{g}_{m,n} \subset \mathfrak{g}_{\infty,\infty}$.

We also fix a Cartan subalgebra $\mathfrak{h} \subset \mathfrak{g}_{\infty,\infty}$ and the standard basis $\{\varepsilon_i, \delta_j\}_{i,j\in\mathbb{Z}_{>0}}$. The roots of $\mathfrak{g}_{\infty,\infty}$ in this basis are:

$$(\pm\varepsilon_i),\quad (\pm\delta_j),$$

$$(\pm\varepsilon_i \pm \delta_j),\quad (\pm 2\delta_j),$$

$$(\pm\varepsilon_i \pm \varepsilon_j),\quad (\pm\delta_i \pm \delta_j),$$

where i,j vary from 1 to ∞, and in the last line, $i \neq j$.

Then the roots of $\mathfrak{g}_{m,n}$ lie in the subspace generated by $(\varepsilon_i)_{1\leq i\leq m}$ and $(\delta_j)_{1\leq j\leq n}$.

We fix a Borel subalgebra $\mathfrak{b}_0$ of $(\mathfrak{g}_{\infty,\infty})_0$ with the set of positive roots

$$\{\varepsilon_i, 2\delta_j,\ (i,j>0),\quad \varepsilon_i \pm \varepsilon_j,\ \delta_i \pm \delta_j\ (i>j>0)\}.$$

Inside $\mathfrak{g}_{m,n}$, we denote by $\mathfrak{p}_{\underline{m},n}$ (resp $\mathfrak{p}_{m,\underline{n}}$) the unique parabolic subalgebra containing $\mathfrak{b}_0$ with semi-simple part $\mathfrak{g}_{m-1,n}$ (resp $\mathfrak{g}_{m,n-1}$).

We denote by $G_{m,n}$ the supergroup $SOSP(2m+1,2n)$ and by $T_{m,n}$ the maximal torus of $G_{m,n}$ with Lie algebra $\mathfrak{h}\cap\mathfrak{g}_{m,n}$.

For fixed m and n, we denote by $\mathcal{F}_{m,n}$ the category of finite dimensional $G_{m,n}$-modules and by $\mathcal{K}_{m,n}$ its Grothendieck group.

Let: $\mathcal{F} := \oplus_{m,n}\mathcal{F}_{m,n}$ and $\mathcal{K} := \oplus_{m,n}\mathcal{K}_{m,n}$.

Let B be a Borel subgroup of $G_{m,n}$ with Lie superalgebra $\mathfrak{b}$ containing $\mathfrak{b}_0\cap\mathfrak{g}_{m,n}$ and let Δ_1^+ (resp. Δ_0^+) be the set of odd (resp. even) positive roots of $\mathfrak{g}_{m,n}$. Set

$$\rho_B = \frac{1}{2}\sum_{\alpha\in\Delta_0^+}\alpha - \frac{1}{2}\sum_{\alpha\in\Delta_1^+}\alpha.$$

Let $\Lambda_{m,n}$ be the set of weights λ such that $\lambda - \rho_B$ is a character of $T_{m,n}$. Independently of the choice of B we have

$$\Lambda_{m,n} := \{\lambda = a_1\varepsilon_1 + \cdots + a_m\varepsilon_m + b_1\delta_1 + \cdots + b_n\delta_n \,|\, a_i, b_j \in \frac{1}{2} + \mathbb{Z}\}.$$

We set

$$\Lambda_{m,n}^+ := \{\lambda \in \Lambda_{m,n} \,|\, a_i, b_j \in \frac{1}{2} + \mathbb{N},\ a_1 < \cdots < a_m\,, b_1 < \cdots < b_n\}.$$

Let ν be a character of $T_{m,n}$. We denote by $\mathcal{L}_\nu$ the corresponding line bundle over $G_{m,n}/B$.

Recall the definition of the *Euler characteristic*. For every $\lambda \in \Lambda_{m,n}$ we set

$$\mathcal{E}(\lambda) := \sum_{i \geq 0} (-1)^i \left[H^i(G_{m,n}/B, \mathcal{L}^*_{\lambda - \rho_B})^* \right] \in \mathcal{K}_{m,n}.$$

Recall also that the character of this virtual module is easy to compute, namely

$$Ch(\mathcal{E}(\lambda)) = \frac{D_1}{D_0} \sum_{w \in W_{m,n}} \varepsilon(w) e^{w(\lambda)},$$

where $W_{m,n}$ is the Weyl group of $SO(2m+1) \times SP(2n)$, $D_0 = \Pi_{\alpha \in \Delta_0^+}(e^{\alpha/2} - e^{-\alpha/2})$, $D_1 = \Pi_{\alpha \in \Delta_1^+}(e^{\alpha/2} + e^{-\alpha/2})$.

(Remark: We avoided indexes m and n in this formula since one can easily recover them from the shape of λ).

Note that if we change our choice of B containing $B_0 \cap G_{m,n}$, the character of $\mathcal{E}(\lambda)$ doesn't change, thus the class in $\mathcal{K}_{m,n}$ remains the same, see [**4**].

For $w \in W_{m,n}$, notice that

$$\mathcal{E}(w(\lambda)) = \varepsilon(w)\mathcal{E}(\lambda). \tag{4}$$

PROPOSITION 1. *(see [**4**]) The set*

$$\{\mathcal{E}(\lambda), \lambda \in \Lambda^+_{m,n}\}$$

gives a linearly independent family in $\mathcal{K}_{m,n}$, and we denote by $\mathcal{K}(\mathcal{E})_{m,n}$ the subgroup generated by this family. We also set $\mathcal{K}(\mathcal{E}) := \oplus_{m,n}\mathcal{K}(\mathcal{E})_{m,n}$.

2. Fock space

Let V be a countable dimensional vector space together with a basis $(v_i)_{i \in \frac{1}{2}+\mathbb{Z}}$ and similarly W with a basis $(w_i)_{i \in \frac{1}{2}+\mathbb{Z}}$ with a non-degenerate pairing such that (v_i) and (w_i) are dual bases.

Let $Cl(V \oplus W)$ be the *Clifford algebra* of $V \oplus W$, namely if we denote by $T(V \oplus W)$ the tensor algebra of $V \oplus W$,

$$Cl(V \oplus W) =$$

$$= T(V \oplus W)/(v \otimes v' + v' \otimes v, w \otimes w' + w' \otimes w, v \otimes w + w \otimes v - (v, w), v, v' \in V, w, w' \in W).$$

The *Fock space* of semi-infinite forms, $\mathbf{F}$, is the vector space generated by

$$v_{i_1} \wedge \ldots \wedge v_{i_k} \wedge \ldots,$$

for $i_1 > \ldots\ldots > i_k > \ldots$ such that, for n large enough, $i_n = i_{n-1} - 1$.

There is a natural linear action of $Cl(V \oplus W)$ on $\mathbf{F}$, denoted by $\bullet$, given by:

$$\forall v \in V,\ v \bullet v_{i_1} \wedge \ldots \wedge v_{i_k} \wedge v_{i_{k+1}} \cdots = v \wedge v_{i_1} \wedge \ldots \wedge v_{i_k} \wedge v_{i_{k+1}} \cdots$$

$$\forall w \in W,\ w \bullet v_{i_1} \wedge \ldots \wedge v_{i_k} \wedge v_{i_{k+1}} \cdots = \sum_j (-1)^{j-1}(w, v_{i_j}) v_{i_1} \wedge \ldots \wedge \hat{v}_{i_j} \wedge \ldots$$

Define the *vacuum vector* in $\mathbf{F}$ as

$$| > := v_{-\frac{1}{2}} \wedge v_{-\frac{3}{2}} \wedge \ldots$$

then, for $i < 0$, v_i acts on $| >$ by 0 as w_j for $j > 0$.

We can also see $\mathbf{F}$ as an induced module the following way. Denote by $Cl^+(V \oplus W)$ the subalgebra generated by $\{v_i, i < 0,\ w_j, j > 0\}$, consider its trivial module and induce to the whole $Cl(V \oplus W)$: this gives another construction of $\mathbf{F}$.

Let $\lambda = \sum_{1 \leq i \leq m, 1 \leq j \leq n} a_i \varepsilon_i + b_j \delta_j \in \Lambda^+_{m,n}$. We define a $\mathbb{Z}$-linear map f: $\mathcal{K}(\mathcal{E}) \longrightarrow \mathbf{F}$ such that for any $\mathcal{E}(\lambda) \in \mathcal{K}(\mathcal{E})_{m,n}$:

$$\mathcal{E}(\lambda) \mapsto v_{a_m} \wedge \ldots \wedge v_{a_1} \wedge \ldots \wedge \hat{v}_{-b_1} \wedge \ldots \wedge \hat{v}_{-b_n} \wedge \ldots$$

3. Duality between geometric induction and restriction

In this section we will consider 3 different Grothendieck groups for $G_{m,n}$ namely $\mathcal{K}(P)_{m,n}$ generated by the indecomposable projective modules, $\mathcal{K}(\mathcal{E})_{m,n}$ which we already met and $\mathcal{K}(L)_{m,n} := \mathcal{K}_{m,n}$ generated by the simple modules. After tensoring by the rational numbers $\mathbb{Q}$, $\mathcal{K}(P)_{m,n} \otimes \mathbb{Q}$ and $\mathcal{K}(\mathcal{E})_{m,n} \otimes \mathbb{Q}$ coincide (see [**4**]). We consider the natural pairing between $\mathcal{K}(P)_{m,n}$ and $\mathcal{K}(L)_{m,n}$, $\langle [P], [L] \rangle := \dim \operatorname{Hom}(P, L)$. The restriction of this pairing to $\mathcal{K}(P)_{m,n} \times \mathcal{K}(P)_{m,n}$ is symmetric (and therefore it is a scalar product): indeed $\dim \operatorname{Hom}(P_1, P_2) = \dim \operatorname{Hom}(P_2^*, P_1^*)$ and in this case projective modules happen to be self-dual (see [**7**]).

PROPOSITION 2. *Let us extend the scalar product from $\mathcal{K}(P)_{m,n}$ to $\mathcal{K}(P)_{m,n} \otimes \mathbb{Q}$. Then the set of $\mathcal{E}(\lambda)$, when λ varies in $\Lambda^+_{m,n}$, form an orthonormal basis of $\mathcal{K}(P)_{m,n} \otimes \mathbb{Q}$.*

PROOF. Let $L(\lambda)$ denote the simple module with highest weight λ and $P(\lambda)$ denote its projective cover. Consider the decompositions

$$[P(\lambda)] = \sum_{\mu} b_{\lambda,\mu} \mathcal{E}(\mu), \quad \mathcal{E}(\mu) = \sum_{\nu} a_{\mu,\nu} [L(\nu)].$$

By the weak BGG reciprocity, [**4**], we have $b_{\lambda,\mu} = a_{\mu,\lambda}$. Now, we write

$$\mathcal{E}(\mu) = \sum_{\lambda} c_{\mu,\lambda} [P(\lambda)].$$

Then, clearly, we have the following relation

$$\sum_{\lambda} c_{\mu,\lambda} b_{\lambda,\nu} = \sum_{\lambda} c_{\mu,\lambda} a_{\nu,\lambda} = \delta_{\mu,\nu}.$$

On the other hand,

$$\langle [P(\lambda)], [L(\kappa)] \rangle = \delta_{\lambda,\kappa}.$$

Therefore

$$\langle \mathcal{E}(\mu), \mathcal{E}(\nu) \rangle = \sum_{\lambda,\kappa} c_{\mu,\lambda} a_{\nu,\kappa} \langle [P(\lambda)], [L(\kappa)] \rangle = \sum_{\lambda} c_{\mu,\lambda} a_{\nu,\lambda} = \delta_{\mu,\nu}.$$

□

Let G be a quasireductive algebraic supergroup, which is an algebraic supergroup with reductive even part (see [**7**] for information on their representation theory). It is important that the category of finite-dimensional representations of G has enough projective objects, and every projective object is also injective. Let $Q \subset G$ be a parabolic subgroup with quasireductive part K. Let $\mathfrak{g}, \mathfrak{q}, \mathfrak{k}$ denote the respective Lie superalgebras, and let $\mathfrak{r}$ denote the nil-radical of $\mathfrak{q}$. Consider the following derived functors $\Gamma_i : K - \text{mod} \longrightarrow G - \text{mod}$ and $H^i : G - \text{mod} \longrightarrow K - \text{mod}$ defined by

$$\Gamma_i(M) := H^i(G/Q, \mathcal{L}(M^*))^*, \quad H^i(N) := H^i(\mathfrak{r}, N).$$

Here we denote by $\mathcal{L}(M^*)$ the vector bundle on G/Q induced from M^*. The collection of functors Γ_i is referred to as *geometric induction* while that of H^i is referred to as *geometric restriction.*

The following observation is due to Penkov [**6**].

PROPOSITION 3. *For any K-module M we have*

$$\sum_i (-1)^i [\Gamma_i(M)] = \sum_i (-1)^i [H^i(G_0/Q_0, \mathcal{L}(S^\bullet(\mathfrak{r}_1) \otimes M^*))^*].$$

Note that $\Gamma_i(M) = 0$ for $i > \dim G_0/Q_0$ and that $S^\bullet(\mathfrak{r}_1)$ is finite dimensional.

PROPOSITION 4. *For every projective G-module P, every injective K-module M and $i \geq 0$ there is a canonical isomorphism*

$$\mathrm{Hom}_G(\Gamma_i(M), P) \simeq \mathrm{Hom}_K(M, H^i(P)).$$

PROOF. This result is a slight generalization of Proposition 1 in [**4**]. We consider an injective resolution $0 \to R^0 \to R^1 \to \dots$ of M in the category of Q-modules. Since $\mathrm{Hom}_G(P, \cdot)$ is an exact functor, $\mathrm{Hom}_G(P, H^i(G/Q, M))$ is given by the i-th cohomology group of the complex

$$0 \to \mathrm{Hom}_G(P, H^0(G/Q, R^0)) \to \mathrm{Hom}_G(P, H^0(G/Q, R^1)) \to \dots.$$

The Frobenius reciprocity implies

$$\mathrm{Hom}_G(P, H^0(G/Q, R^j)) \simeq \mathrm{Hom}_Q(P, R^j).$$

Thus, we obtain the isomorphism

$$\mathrm{Hom}_G(P, H^i(G/Q, M)) \simeq \mathrm{Ext}^i_Q(P, M).$$

To compute $\mathrm{Ext}^i_Q(P, M)$ consider the injective resolution of M in the category of Q-modules given by the Koszul complex

$$C^i = \Gamma_{\mathfrak{q}} \,\mathrm{Coind}^{\mathfrak{q}}_{\mathfrak{k}}(M \otimes \Lambda^i(\mathfrak{r}^*)),$$

where $\Gamma_{\mathfrak{q}}$ denote the functor of $\mathfrak{q}$-finite vectors. Then $\mathrm{Ext}^i_Q(P, M)$ is by definition the i-th homology of the complex

$$\mathrm{Hom}_Q(P, C^\bullet) = \mathrm{Hom}_K(P, M \otimes \Lambda^\bullet(\mathfrak{r}^*)) = \mathrm{Hom}_K(P \otimes \Lambda^\bullet(\mathfrak{r}), M).$$

By injectivity of M, the i-th homology group is $\mathrm{Hom}_K(H_i(\mathfrak{r}, P), M)$.

Now we use the double dualization and the fact that P^* is also projective:

$$\mathrm{Hom}_G(\Gamma_i(M), P) \simeq \mathrm{Hom}_G(P^*, H^i(G/Q, M^*)) \simeq \mathrm{Hom}_K(H_i(\mathfrak{r}, P^*), M^*) \simeq$$
$$\simeq \mathrm{Hom}_K(M, H^i(P)).$$

Hence the statement. □

Next we prove the following lemma.

LEMMA 1. *The restricted module* $\mathrm{Res}_K P$ *is projective in the category* $K-\mathrm{mod}$.

PROOF. Note that P is a direct summand of the induced module $\mathrm{Ind}^{\mathfrak{g}}_{\mathfrak{g}_0} S$ for some semisimple $\mathfrak{g}_0$-module S. Using the isomorphism

$$\mathrm{Res}_K \mathrm{Ind}^{\mathfrak{g}}_{\mathfrak{g}_0} S \simeq \mathrm{Ind}^{\mathfrak{k}}_{\mathfrak{k}_0} S \otimes S^\bullet(\mathfrak{g}_1/\mathfrak{k}_1),$$

we obtain that P is a direct summand of some module induced from a semisimple $\mathfrak{k}_0$-module. Therefore P is projective as a K-module. □

PROPOSITION 5. *Let P be a projective G-module and M an arbitrary K-module. For every $i \geq 0$ there is a canonical isomorphism*

$$\mathrm{Hom}_G(\Gamma_i(M), P) \simeq H^i(\mathfrak{q}, \mathfrak{k}, M^* \otimes P),$$

where $H^i(\mathfrak{q}, \mathfrak{r}, \cdot)$ denotes the relative cohomology group.

PROOF. According to the proof of Proposition 4 we have to show

$$\mathrm{Ext}^i_Q(M, P) \simeq H^i(\mathfrak{q}, \mathfrak{k}, M^* \otimes P).$$

We have

$$\mathrm{Ext}^i_Q(M, P) \simeq \mathrm{Ext}^i_Q(\mathbb{C}, M^* \otimes P).$$

Since P is projective and hence also injective in the category of K-modules, $M^* \otimes P$ is also injective. Thus, the Koszul complex $C^i = \Gamma_{\mathfrak{q}} \operatorname{Coind}^{\mathfrak{q}}_{\mathfrak{k}}(M^* \otimes P \otimes \Lambda^i(\mathfrak{r}^*))$ gives an injective resolution of $M^* \otimes P$ in the category of Q-modules. The complex $\mathrm{Hom}_Q(\mathbb{C}, C^\bullet)$ is the relative cochain complex which computes the relative cohomology $H^i(\mathfrak{q}, \mathfrak{r}, M^* \otimes P)$. □

4. Two functors on $\mathcal{F}$

We choose a parabolic subalgebra $\mathfrak{p}$ which can be either $\mathfrak{p}_{\underline{m},n}$ or $\mathfrak{p}_{m,\underline{n}}$ in $\mathfrak{g}_{m,n}$, where:

$\mathfrak{p}_{\underline{m},n} = \mathfrak{g}_{m-1,n} \oplus \mathbb{C}z \oplus \mathfrak{r}_{\underline{m},n}$, $\mathfrak{g}_{m,n} = \mathfrak{p}_{\underline{m},n} \oplus \mathfrak{r}^-_{\underline{m},n}$

$\mathfrak{p}_{m,\underline{n}} = \mathfrak{g}_{m,n-1} \oplus \mathbb{C}z \oplus \mathfrak{r}_{m,\underline{n}}$, and $\mathfrak{g}_{m,n} = \mathfrak{p}_{m,\underline{n}} \oplus \mathfrak{r}^-_{m,\underline{n}}$.

Denote by Z the center of the reductive part of the parabolic subgroup P corresponding to the parabolic subalgebra we chose above (Lie algebra of Z is $\mathbb{C}z$). For any $a \in \mathbb{Z}$ we denote by $\mathbb{C}_a$ the corresponding character of Z. Since Z is a one-parameter subgroup of the torus $T_{m,n}$, if $\varpi : Z \to \mathbb{C}^*$ is the generator of the character ring of Z , we denote by $\mathbb{C}_{a\varpi}$ the associated $T_{m,n}$-module (in our case, ϖ is either ε_n or δ_n). Now, if $M \in \mathcal{F}_{m-1,n}$ or $\mathcal{F}_{m,n-1}$, denote $\mathbb{C}_a \boxtimes M$ the P-module with trivial action of the corresponding nilradical $\mathfrak{r}$ and the given action of $Z \times G_{m-1,n}$, or $Z \times G_{m,n-1}$ depending on the way the parabolic is chosen.

DEFINITION 1. We define the following functors:

$$\Gamma^a_i : \mathcal{F} \to \mathcal{F}, a \in \frac{1}{2} + \mathbb{Z}$$

if $a > 0$, if $M \in \mathcal{F}_{m-1,n}$, $\Gamma^a_i(M) := H^i(G_{m,n}/P_{\underline{m},n}, \mathcal{L}(\mathbb{C}_{(a-(m-n-\frac{1}{2}))\varepsilon_m} \boxtimes M)^*)^*$,
if $a < 0$, if $M \in \mathcal{F}_{m,n-1}$, $\Gamma^a_i(M) := H^i(G_{m,n}/P_{m,\underline{n}}, \mathcal{L}(\mathbb{C}_{(-a-(n-m-\frac{1}{2}))\delta_n} \boxtimes M)^*)^*$.

$$H^j_b : \mathcal{F} \to \mathcal{F}, b \in \frac{1}{2} + \mathbb{Z}$$

if $b > 0$, if $M \in \mathcal{F}_{m,n}$, $H^j_b(M) := \mathrm{Hom}_Z(\mathbb{C}_{(b-(m-n-\frac{1}{2}))\varepsilon_m}, H^j(\mathfrak{r}_{\underline{m},n}, M)) \in \mathcal{F}_{m-1,n}$
if $b < 0$, if $M \in \mathcal{F}_{m,n}$, $H^j_b(M) := \mathrm{Hom}_Z(\mathbb{C}_{(-b-(n-m-\frac{1}{2}))\delta_n}, H^j(\mathfrak{r}_{m,\underline{n}}, M)) \in \mathcal{F}_{m,n-1}$.

Now, we consider the following operators in $\mathcal{K}$: if $M \in \mathcal{F}_{m,n}$, denoting the sign of a half-integer x by $sgn(x)$:

$$\gamma^a([M]) := sgn(a)^m \sum_{i \geq 0} (-1)^i [\Gamma^a_i(M)]$$

$$\eta_b([M]) := sgn(b)^m \sum_{j \geq 0} (-1)^j [H^j_b(M)].$$

We can identify the Grothendieck ring with the ring of characters of finite dimensional modules (cf [**4**]) and so we will check the relations we need at the level of characters.

We recall the following formula ([**3**], prop. 1): for $P \subset G$ a parabolic subgroup of a quasireductive supergroup with Levi part L,

$$\sum_i (-1)^i Ch(H^i(G/P, \mathcal{L}(M^*))^*) = D \sum_{w \in W} \varepsilon(w) w \left(\frac{e^\rho Ch(M)}{\Pi_{\alpha \in \Delta^+_{1,\mathfrak{l}}} (1 + e^{-\alpha})} \right),$$

where $D := \frac{D_0}{D_1}$, $D_0 = \Pi_{\alpha \in \Delta_0^+} (e^{\alpha/2} - e^{-\alpha/2})$, $D_1 = \Pi_{\alpha \in \Delta_1^+} (e^{\alpha/2} + e^{-\alpha/2})$, where Δ_0^+, Δ_1^+ denote the sets of positive even and odd roots of $\mathfrak{g}$ respectively and $\Delta^+_{1,\mathfrak{l}}$ denotes the set of positive odd roots of $\mathfrak{l}$.

In what follows we use the notation

$$(a_m, \ldots, a_1 | b_1, \ldots, b_n) := \sum_{i=1}^m a_i \varepsilon_i + \sum_{j=1}^n b_j \delta_j.$$

PROPOSITION 6. *Let* $\nu = (a_m, \ldots, a_1 | b_1, \ldots, b_n) \in \Lambda^+_{m,n}$. *Then one has:*

(1) $a > 0$, *if* $\exists i$ *s.t.* $a_{i+1} > a > a_i$,

$$\gamma^a(\mathcal{E}(\nu)) = (-1)^{m-i} \mathcal{E}(a_m, \ldots, a_{i+1}, a, a_i, \ldots a_1 | b_1, \ldots, b_n),$$

and $\gamma^a(\mathcal{E}(\nu)) = 0$ *if* $\exists i$, $a = a_i$.

(2) $a < 0$, *if* $\exists i$ *s.t.* $b_i < -a < b_{i+1}$,

$$\gamma^a(\mathcal{E}(\nu)) = (-1)^{n-i} \mathcal{E}(a_m, \ldots a_1 | b_1, \ldots, b_i, -a, b_{i+1}, \ldots, b_n),$$

and $\gamma^a(\mathcal{E}(\nu)) = 0$ *if* $\exists i$, $a = -b_i$.

(3) $b > 0$, *if* $\exists i$ *s.t.* $b = a_i$

$$\eta_b(\mathcal{E}(\nu)) = (-1)^{m-i} \mathcal{E}(a_m, \ldots, a_{i+1}, a_{i-1}, \ldots, a_1 | b_1, \ldots, b_n)$$

if $b \neq a_i \forall i$, $\eta_b(\mathcal{E}(\nu)) = 0$.

(4) $b < 0$, *if* $\exists i$ *s.t.* $b = -b_i$

$$\eta_b(\mathcal{E}(\nu)) = (-1)^{n-i} \mathcal{E}(a_m, \ldots, a_1 | b_1, \ldots, b_{i-1}, b_{i+1}, \ldots, b_n)$$

if $b \neq -b_i \forall i$, $\eta_b(\mathcal{E}(\nu)) = 0$.

PROOF. First we prove (1). Let us use [**3**], Theorem 1: one has, if M is a B-module,

$$\sum_{i,j} (-1)^{i+j} [H^i(G_{m,n}/P_{\underline{m},n}, \mathcal{L}(H^j(P_{\underline{m},n}/B, \mathcal{L}(M^*)))^*]$$
$$= \sum_k (-1)^k [H^k(G_{m,n}/B, \mathcal{L}(M^*))^*].$$

We take for M the 1-dimensional representation $\mathbb{C}_\lambda$ with $\lambda + \rho_B = (a, a_m, \ldots, a_1 | b_1, \ldots, b_n)$. Then, using the equation (4), and the definition of γ^a, we get

$$\gamma^a(\mathcal{E}(\nu)) = \mathcal{E}(\lambda) = (-1)^{m-i} (a_m, \ldots, a_{i+1}, a, a_i, \ldots, a_1 | b_1, \ldots b_n)$$

for the index i of the statement. Hence the statement.

Now we will prove (3). Recall the symmetric pairing $\mathcal{K}(P) \times \mathcal{K}(P) \to$ defined by

$$\langle [M], [P] \rangle := \dim \operatorname{Hom}_{G_{m,n}}(M, P),$$

introduced in Section 3. Proposition 4 implies that for every projective $G_{m-1,n}$-module M and every projective $G_{m,n}$-module P, we have an isomorphism

$$\mathrm{Hom}_{G_{m,n}}(\Gamma_i^a(M), P) \simeq \mathrm{Hom}_{G_{m-1,n}}(M, H_a^i(P)).$$

At the level of Grothendieck groups we get

$$\langle \gamma^a([M]), [P] \rangle := \langle [M], \eta_a([P]) \rangle.$$

Since $\mathcal{E}(\nu)$ is a $\mathbb{Q}$-linear combination of classes of projective modules, we can rewrite the above inequality in the from:

$$\langle \gamma^a(\mathcal{E}(\nu)), \mathcal{E}(\mu) \rangle = \langle \mathcal{E}(\nu), \eta_a(\mathcal{E}(\mu)) \rangle.$$

Now (3) follows from the fact that $\mathcal{E}(\lambda)$ form an orthonormal basis in $\mathcal{K}(P) \otimes \mathbb{Q}$ with respect to the form $\langle \cdot, \cdot \rangle$.

The proof of assertions (2) and (4) is similar and we leave it to the reader. □

COROLLARY 1. *The subspace $\mathcal{K}(P) \otimes \mathbb{Q} \subset \mathcal{K}(L) \otimes \mathbb{Q}$ is γ^a and η_b-stable. Moreover, the linear operators γ^a and η_a in $\mathcal{K}(P) \otimes \mathbb{Q}$ are mutually adjoint with respect to the pairing $\langle \cdot, \cdot \rangle$.*

5. Link with the Clifford algebra

Let us now interpret the map f of section 2 in terms of the functors described in the previous section. The proposition 6 has the following immediate corollary:

COROLLARY 2. *One has:*
$f \circ \gamma^a = v_a \circ f$ *for* $a > 0$,
$f \circ \gamma^a = w_a \circ f$ *for* $a < 0$,
$f \circ \eta_b = w_b \circ f$ *for* $b > 0$,
$f \circ \eta_b = v_b \circ f$ *for* $b < 0$,
where v_a, w_b stand for the action on the Fock space of the corresponding elements of the Clifford algebra.

This gives us an action of the Clifford algebra on the Grothendieck group $\mathcal{K}(\mathcal{E})$.

THEOREM 1. *The operators γ^a and η_b ($a, b \in \frac{1}{2} + \mathbb{Z}$) in the Grothendieck group $\mathcal{K}$ satisfy the Clifford relations:*

$$\eta_a \eta_b + \eta_b \eta_a = 0, \quad \gamma^a \gamma^b + \gamma^b \gamma^a = 0, \quad \gamma^a \eta_b + \eta_b \gamma^a = \delta_{a,b}.$$

PROOF. Let a and b be half-integers. We first show that

$$\eta_a \eta_b + \eta_b \eta_a = 0.$$

The arguments involved in the proof depend on the signs of a and b, we will take care of the cases $a, b > 0$ and $a > 0, b < 0$, leaving $a < 0, b < 0$ to the reader.

Assume first that $a > 0, b > 0$, let M be a $\mathfrak{g}_{m,n}$ module, we consider the following increasing chain of Lie superalgebras:

$$\mathfrak{g}_{m-2,n} \subset \mathfrak{p}_{\underline{m-1},n} \subset \mathfrak{g}_{m-1,n} \subset \mathfrak{p}_{\underline{m},n} \subset \mathfrak{g}_{m,n}.$$

Let $\mathfrak{q}$ be the parabolic subalgebra with reductive part equal to the direct sum of $\mathfrak{g}_{m-2,n}$ and the two-dimensional center $Z_{\mathfrak{q}}$, and the nilradical $\mathfrak{r} = \mathfrak{r}_{\underline{m},n} + \mathfrak{r}_{\underline{m-1},n}$. Then using the Hochschild–Serre spectral sequence for the pair $\mathfrak{r}_{\underline{m-1},n} \subset \mathfrak{r}$ we obtain

$$\eta_a \eta_b [M] = \sum_i (-1)^i [\mathrm{Hom}_{Z_{\mathfrak{q}}}(\mathbb{C}_{(b-(n-m-1/2))\varepsilon_m + (a-(n-m+1/2))\varepsilon_{m-1}}, H^i(\mathfrak{r}, M))],$$

and

$$\eta_b\eta_a[M] = \sum_i (-1)^i [\mathrm{Hom}_{Z_{\mathfrak{q}}}(\mathbb{C}_{(a-(n-m-1/2))\varepsilon_m+(b-(n-m+1/2))\varepsilon_{m-1}}, H^i(\mathfrak{r}, M))].$$

Let us note that in the above two formulae as well as in all formulae in this proof only finitely many terms in summation are not zero. This follows immediately from the fact that the centre of the reductive subalgebra of any parabolic subalgebra acts with positive weights on the nil-radical.

Now we consider the one-dimensional root subalgebra $\mathfrak{s} := \mathfrak{g}_\beta \subset \mathfrak{r}$ for the root $\beta = \varepsilon_m - \varepsilon_{m-1}$. Note that $\mathfrak{s}$ is the nilradical of a Borel subalgebra of the $\mathfrak{sl}(2)$ generated by $\mathfrak{g}_\beta$ and $\mathfrak{g}_{-\beta}$. Hence by the Kostant theorem we have for any $\mathfrak{sl}(2)$-module N

$$[\mathrm{Hom}_{Z_{\mathfrak{q}}}(\mathbb{C}_{(a-(n-m-1/2))\varepsilon_m+(b-(n-m+1/2))\varepsilon_{m-1}}, H^p(\mathfrak{s}, N))] =$$
$$[\mathrm{Hom}_{Z_{\mathfrak{q}}}(\mathbb{C}_{(b-(n-m-1/2))\varepsilon_m+(a-(n-m+1/2))\varepsilon_{m-1}}, H^q(\mathfrak{s}, N))]$$

for $(p, q) = (0, 1)$ or $(1, 0)$.

Once again we apply the Hochschild–Serre spectral sequence for the pair $\mathfrak{s} \subset \mathfrak{r}$ to get

$$\eta_a\eta_b[M] =$$
$$= \sum_i (-1)^{i+j} [\mathrm{Hom}_{Z_{\mathfrak{q}}}(\mathbb{C}_{(b-(n-m-1/2))\varepsilon_m+(a-(n-m+1/2))\varepsilon_{m-1}}, H^i(\mathfrak{s}, \Lambda^j(\mathfrak{r}/\mathfrak{s})^* \otimes M))],$$

$$\eta_b\eta_a[M] =$$
$$= \sum_i (-1)^{i+j} [\mathrm{Hom}_{Z_{\mathfrak{q}}}(\mathbb{C}_{(a-(n-m-1/2))\varepsilon_m+(b-(n-m+1/2))\varepsilon_{m-1}}, H^i(\mathfrak{s}, \Lambda^j(\mathfrak{r}/\mathfrak{s})^* \otimes M))].$$

This implies the relation.

Now let $a > 0, b < 0$. Let M be a $G_{m,n}$-module. Set

$$\mathfrak{r} := \mathfrak{r}_{\underline{m},n} + \mathfrak{r}_{m-1,\underline{n}}, \quad \mathfrak{r}' := \mathfrak{r}_{m,\underline{n}} + \mathfrak{r}_{\underline{m},n-1}.$$

Let $Z \subset T_{m,n}$ be the centralizer of $\mathfrak{g}_{m-1,n-1}$. Using Hochschild–Serre spectral sequence we obtain

$$\eta_b\eta_a[M] = \sum_i (-1)^{m-1+i} [\mathrm{Hom}_Z(\mathbb{C}_{(a-(m-n-1/2))\varepsilon_m-(b+n-m+1/2)\delta_n}, H^i(\mathfrak{r}, M))]$$

and

$$\eta_a\eta_b[M] = \sum_i (-1)^{m+i} [\mathrm{Hom}_Z(\mathbb{C}_{(a-(m-n+1/2))\varepsilon_m-(b+n-m-1/2)\delta_n}, H^i(\mathfrak{r}', M))].$$

Let $\alpha = \varepsilon_m - \delta_n$. Consider the root subalgebras $\mathfrak{g}_\alpha, \mathfrak{g}_{-\alpha} \subset \mathfrak{g}_{m,n}$. Note that $\mathfrak{s} := \mathfrak{r} \cap \mathfrak{r}'$ is an ideal of codimension 1 in both $\mathfrak{r}$ and $\mathfrak{r}'$ and that

$$\mathfrak{r} = \mathfrak{s} + \mathfrak{g}_\alpha, \quad \mathfrak{r}' = \mathfrak{s} + \mathfrak{g}_{-\alpha}.$$

Therefore by Hochschild–Serre spectral sequence we have

$$\eta_b\eta_a =$$
$$= \sum_{i,j} (-1)^{i+j+m-1} [\mathrm{Hom}_Z(\mathbb{C}_{(a-(m-n-1/2))\varepsilon_m-(b+n-m+1/2)\delta_n}, \Lambda^j(\mathfrak{g}_{-\alpha}) \otimes H^i(\mathfrak{s}, M))],$$

$$\eta_a\eta_b[M] =$$
$$= \sum_{i,j}(-1)^{i+j+m}[\mathrm{Hom}_Z(\mathbb{C}_{(a-(m-n+1/2))\varepsilon_m-(b+n-m-1/2)\delta_n}, \Lambda^j(\mathfrak{g}_\alpha)\otimes H^i(\mathfrak{s}, M))].$$

Taking into account that

$$\sum_j(-1)^j Ch(\Lambda^j(\mathfrak{g}_\alpha)) = \frac{1}{1+e^\alpha} = \frac{e^{-\alpha}}{1+e^{-\alpha}} = e^{-\alpha}\left(\sum_j(-1)^j Ch(\Lambda^j(\mathfrak{g}_{-\alpha}))\right)$$

we obtain

$$\sum_j(-1)^j Ch(\Lambda^j(\mathfrak{g}_\alpha)\otimes H^i(\mathfrak{s}, M)) = e^{-\alpha}\left(\sum_j(-1)^j Ch(\Lambda^j(\mathfrak{g}_{-\alpha})\otimes H^i(\mathfrak{s}, M))\right).$$

Therefore

$$Ch(\eta_a\eta_b[M]) = -Ch(\eta_b\eta_a[M]),$$

which proves the relation.

Note that the relation

$$\gamma^a\gamma^b + \gamma^b\gamma^a = 0.$$

follows from the relation for η_a, η_b by adjointness proven in Proposition 4.

Let us now show that if $a > 0$ and $b < 0$, then

$$\gamma^a\eta_b + \eta_b\gamma^a = 0.$$

The case $a < 0$ and $b > 0$ is similar and we leave it to the reader.

One should keep in mind the following diagram

$$\begin{array}{ccc} & \eta_b & \\ \mathcal{F}_{m,n} & \to & \mathcal{F}_{m,n-1} \\ \gamma^a\downarrow & & \downarrow\gamma^a \\ \mathcal{F}_{m+1,n} & \to & \mathcal{F}_{m+1,n-1} \\ & \eta_b & \end{array}$$

because we follow it to keep tracks of the weights.

Let us denote ChM_γ the character of $\mathrm{Hom}_Z(\mathbb{C}_\gamma, M)$. Then one has:

$$\gamma^a\eta_b Ch(M) = \sum_{i,j}(-1)^{i+j+m}$$
$$Ch(\Gamma_j(G_{m+1,n-1}/P_{\underline{m+1},n-1}, (\mathbb{C}_{(a-(m-n+1/2))\varepsilon_{m+1}}\boxtimes\Lambda^i(\mathfrak{r}^*_{m,\underline{n}})\otimes M)_{(-b-(n-m-1/2))\delta_n})),$$

$$\eta_b\gamma^a Ch(M) = \sum_{i,j}(-1)^{i+j+m+1}$$
$$Ch((\Lambda^i(\mathfrak{r}^*_{m+1,\underline{n}})\otimes\Gamma_j(G_{m+1,n}/P_{\underline{m+1},n}, \mathbb{C}_{(a-(m-n-1/2))\varepsilon_{m+1}}\boxtimes M))_{(-b-(n-m-3/2))\delta_n}).$$

We use Proposition 3. For any $G_{m,k}$-module N, the following holds:

$$\sum_j(-1)^j\Gamma_j(G_{m+1,k}/P_{\underline{m+1},k}, N) = \sum_j(-1)^j\Gamma_j(G_{m+1,0}/P_{\underline{m+1},0}, N\otimes S^\bullet((\mathfrak{r}^*_{\underline{m+1},k})_{\bar{1}}).$$

Then if we set:

$$X := \sum_i(-1)^i\Lambda^i(\mathfrak{r}^*_{m,\underline{n}})\otimes S^\bullet((\mathfrak{r}^*_{\underline{m+1},n-1})_{\bar{1}})$$

and

$$Y := \sum_i (-1)^i \Lambda^i(\mathfrak{r}^*_{m+1,\underline{n}}) \otimes S^\bullet((\mathfrak{r}^*_{\underline{m+1},n})_{\bar{1}}),$$

we get

$$\gamma^a \eta_b(Ch(M)) = \sum_j (-1)^{j+m}$$
$$Ch(\Gamma_j(G_{m+1,0}/P_{\underline{m+1},0}, \mathbb{C}_{(a-(m-n+1/2))\varepsilon_{m+1}} \boxtimes X \otimes M)_{(-b-(n-m-1/2))\delta_n})$$

and

$$\eta_b \gamma^a(Ch(M)) = \sum_j (-1)^{j+m+1}$$
$$Ch(\Gamma_j(G_{m+1,0}/P_{\underline{m+1},0}, \mathbb{C}_{(a-(m-n-1/2))\varepsilon_{m+1}} \boxtimes Y \otimes M)_{(-b-(n-m-3/2))\delta_n}).$$

Next we compute the quotient $Ch(X)/Ch(Y)$. One has

$$Ch(X) = \frac{(1-e^{-2\delta_n})\prod_{i=1}^{n-1}(1-e^{-\delta_n \pm \delta_i})\prod_{j=1}^{n-1}(1+e^{\pm\delta_j - \varepsilon_{m+1}})}{(1+e^{-\delta_n})\prod_{j=1}^{m}(1+e^{-\delta_n \pm \varepsilon_j})},$$

$$Ch(Y) = \frac{(1-e^{-2\delta_n})\prod_{i=1}^{n-1}(1-e^{-\delta_n \pm \delta_i})\prod_{j=1}^{n}(1+e^{\pm\delta_j - \varepsilon_{m+1}})}{(1+e^{-\delta_n})\prod_{j=1}^{m+1}(1+e^{-\delta_n \pm \varepsilon_j})},$$

and the quotient turns out to be

$$Ch(X)/Ch(Y) = e^{\varepsilon_{m+1} - \delta_n}.$$

The result follows.

Let us show finally that for $a, b > 0$ one has

$$\gamma^a \eta_b + \eta_b \gamma^a = \delta_{a,b}$$

where $\delta_{a,b}$ stands for the Kronecker symbol. The proof we provide lacks functoriality at the moment, but we intend to improve it.

Let $R : \mathcal{F}_{m,n} \to \mathcal{F}_{m,0}$ be the restriction functor and denote by the same letter the corresponding map of the Grothendieck groups. Then it follows from Proposition 3 that for any $M \in \mathcal{F}_{m,n}$,

$$R(\gamma^a[M]) = \gamma^{a+n}([S^\bullet((\mathfrak{r}^*_{\underline{m+1},n})_{\bar{1}})]R[M]).$$

On the other hand, for any Lie superalgebra $\mathfrak{r}$ and $\mathfrak{r}$-module M we have

$$\sum_i (-1)^i [H^i(\mathfrak{r}, M)] = \sum_{k,l} (-1)^{k+l} [\Lambda^l(\mathfrak{r}^*_{\bar{1}})][H^k(\mathfrak{r}_{\bar{0}}, M)].$$

Therefore

$$R(\eta_a[M]) = \sum_k (-1)^k \eta_{a+n}([\Lambda^k((\mathfrak{r}^*_{\underline{m},n})_{\bar{1}})]R[M]).$$

Therefore for $M \in \mathcal{F}_{m,n}$ we have

$$R(\gamma^a \eta_b[M]) = \sum_k (-1)^k \gamma^{n+a}\left([S^\bullet((\mathfrak{r}^*_{\underline{m},n})_{\bar{1}})]\eta_{b+n}([\Lambda^k((\mathfrak{r}^*_{\underline{m},n})_{\bar{1}})]R[M])\right),$$

$$R(\eta_b \gamma^a[M]) = \sum_k (-1)^k \eta_{b+n}\left([\Lambda^k((\mathfrak{r}^*_{\underline{m+1},n})_{\bar{1}})]\gamma^{n+a}([S^\bullet((\mathfrak{r}^*_{\underline{m+1},n})_{\bar{1}})]R[M])\right).$$

Let us denote by U the standard representation of $\mathfrak{sp}(2n) \subset \mathfrak{osp}(2m+1, 2n)$ and consider it as purely odd superspace. Then

$$Ch((\mathfrak{r}^*_{\underline{m},n})_{\bar{1}}) = e^{-\varepsilon_m} Ch(U).$$

Therefore the above expressions can be rewritten in the form

$$R(\gamma^a \eta_b[M]) = \sum_{k,l} (-1)^k \gamma^{n+a-l} \left([S^l(U)]\eta_{b+n+k}([\Lambda^k(U)]R[M])\right),$$

$$R(\eta_b \gamma^a[M]) = \sum_{k} (-1)^k \eta_{b+n+k} \left([\Lambda^k(U)]\gamma^{n+a-l}([S^l(U)]R[M])\right).$$

Now we note that the action of $G_{m,0}$ on U is trivial, hence multiplication with its exterior and symmetric powers commute with γ^a and η_b. Thus, we have

$$R(\gamma^a \eta_b[M]) = \sum_{k,l} (-1)^k [S^l(U)][\Lambda^k(U)]\gamma^{n+a-l}\eta_{b+n+k}(R[M]),$$

$$R(\eta_b \gamma^a[M]) = \sum_{k} (-1)^k ([\Lambda^k(U)][S^l(U)]\eta_{b+n+k}\gamma^{n+a-l}(R[M]).$$

Since $\mathcal{F}_{m,0}$ is the category of representations of a purely even reductive group, we have $\mathcal{K}(\mathcal{E})_{m,0} = \mathcal{K}(L)_{m,0}$. Therefore Proposition 6 implies that for any $N \in \mathcal{F}_{m,0}$

$$\gamma^a \eta_b[N] + \eta_b \gamma^a[N] = \delta_{a,b}[N].$$

Hence,

$$(\gamma^a \eta_b + \eta_b \gamma^a)(R[M]) = \sum_{l,k} (-1)^k [S^l(U)][\Lambda^k(U)]\delta_{a+n-k,b+n+l} R[M],$$

and hence

$$(\gamma^a \eta_b + \eta_b \gamma^a)(R[M]) = \sum_{k+l=a-b} (-1)^k [S^l(U)][\Lambda^k(U)]R[M].$$

Since the Koszul complex is acyclic except in the zero degree we have the identity

$$\sum_{k+l=p} (-1)^k [\Lambda^k(U)][S^l(U)] = \begin{cases} 1 \; if \; p=0 \\ 0 \; otherwise \end{cases}.$$

Hence, the sum we compute has only one non-zero term, namely we get:

$$(\gamma^a \eta_b + \eta_b \gamma^a)(R[M]) = \delta_{a,b} R[M].$$

Since the map R is injective this proves the result for $\gamma^a \eta_b + \eta_b \gamma^a$, $a, b > 0$. The case $a, b < 0$ is similar and we leave it to the reader. □

6. Translation functors

We would like to link this approach with the results on translation functors in [**4**].

Recall the Lie algebra $\mathfrak{gl}(\infty)$ which is embedded in $Cl(V \oplus W)$ as the span of $v_a w_b$, $a, b \in \frac{1}{2} + \mathbb{Z}$. The subalgebra $\mathfrak{gl}(\frac{\infty}{2})$ is generated by $v_a w_b + v_{-a} w_{-b}$, $a, b \in \frac{1}{2} + \mathbb{N}$.

Inside the Fock space $\mathbf{F}$, we consider the subspace $\mathbf{F}_{m,n}$ which is the image of $\mathcal{K}(\mathcal{E})_{m,n}$ under the map f, defined at the end of section 2.

REMARK 1. The space $\mathbf{F}_{m,n}$ is stable under the action of $\mathfrak{gl}(\frac{\infty}{2})$. Furthermore, it is not difficult to see that $\mathbf{F}_{m,n}$ is isomorphic to $\Lambda^m(V_+) \otimes \Lambda^n(W_+)$ as an $\mathfrak{sl}(\frac{\infty}{2})$-module, where V_+ and W_+ are respectively the standard and costandard module of $\mathfrak{gl}(\frac{\infty}{2})$.

Consider the Cartan subalgebra $\mathfrak{t}$ of $\mathfrak{gl}(\frac{\infty}{2})$ with basis $t_a := v_a w_a + v_{-a} w_{-a}$ for all $a \in \frac{1}{2} + \mathbb{N}$, then $\mathbf{F}$ is a semi-simple $\mathfrak{t}$-module. We denote by ω the $\mathfrak{t}$-weight of the vacuum vector: $\omega(t_a) = 1$ for all $a \in \frac{1}{2} + \mathbb{N}$. Let $\beta_a \in \mathfrak{t}^*$ be such that $\beta_a(t_b) = \delta_{a,b}$. If $\lambda = (a_m, \dots, a_1 | b_1, \dots, b_n)$, then the $\mathfrak{t}$-weight of $f(\mathcal{E}(\lambda))$ equals

$$\beta(\lambda) := \omega + \beta_{a_1} + \dots + \beta_{a_m} - \beta_{b_1} - \dots - \beta_{b_n}.$$

LEMMA 2. *Let $\mathcal{E}(\lambda), \mathcal{E}(\mu) \in \mathcal{K}(\mathcal{E})_{m,n}$. Then $\mathcal{E}(\lambda)$ and $\mathcal{E}(\mu)$ are in the same block of $\mathcal{F}_{m,n}$ if and only if the $\mathfrak{t}$-weights of $f(\mathcal{E}(\lambda))$ and $f(\mathcal{E}(\mu))$ coincide.*

PROOF. The statement follows from the remark 1 after comparing with the weights denoted by $\gamma(\lambda)$ in [**4**] (we do not keep this notation here because we have introduced a γ^a which is not related). The relation between those $\mathfrak{t}$-weights is $\beta(\lambda) = \omega + \gamma(\lambda)$. □

Consider now the Chevalley generators of $\mathfrak{gl}(\frac{\infty}{2})$, $E_{a,a+1}$ and $E_{a+1,a}$ for all $a \in \frac{1}{2} + \mathbb{N}$. As it was shown in [**4**], the categorification of the action of these generators in $\Lambda^m(V_+) \otimes \Lambda^n(W_+)$ is given by the translation functors:

$$T_{a+1,a}(M) := (M \otimes \mathbf{E})_{\beta+\beta_{a+1}-\beta_a}, \quad T_{a,a+1}(M) := (M \otimes \mathbf{E})_{\beta+\beta_a-\beta_{a+1}},$$

where $\mathbf{E}$ is the standard $\mathfrak{g}_{m,n}$-module, we assume that the $\mathfrak{g}_{m,n}$-module-M belongs to the block corresponding to the $\mathfrak{t}$-weight β, and by $(N)_{\beta'}$ we denote the projection of the $\mathfrak{g}_{m,n}$-module N onto the block corresponding to the $\mathfrak{t}$-weight β'. By abuse of notations we denote also by $T_{a+1,a}$ and $T_{a,a+1}$ the corresponding linear operators in $\mathcal{K}(\mathcal{E})_{m,n}$.

The following statement is an immediate consequence of the remark 1 and Lemma 4 in [**4**].

PROPOSITION 7. *For all $a \in \frac{1}{2} + \mathbb{N}$ we have*

$$f \circ T_{a+1,a} = E_{a+1,a} \circ f, \quad f \circ T_{a,a+1} = E_{a,a+1} \circ f.$$

References

[1] R. Bott, *Homogeneous vector bundles*, Ann. of Math. (2) **66** (1957), 203–248, DOI 10.2307/1969996. MR0089473

[2] M. Ehrig and C. Stroppel, *On the category of finite-dimensional representations of* $\mathrm{OSp}(r|2n)$*: Part I*, Representation theory—current trends and perspectives, EMS Ser. Congr. Rep., Eur. Math. Soc., Zürich, 2017, pp. 109–170. MR3644792

[3] C. Gruson and V. Serganova, *Cohomology of generalized supergrassmannians and character formulae for basic classical Lie superalgebras*, Proc. Lond. Math. Soc. (3) **101** (2010), no. 3, 852–892, DOI 10.1112/plms/pdq014. MR2734963

[4] C. Gruson and V. Serganova, *Bernstein-Gelfand-Gelfand reciprocity and indecomposable projective modules for classical algebraic supergroups* (English, with English and Russian summaries), Mosc. Math. J. **13** (2013), no. 2, 281–313, 364. MR3134908

[5] V. G. Kac, *Lie superalgebras*, Advances in Math. **26** (1977), no. 1, 8–96, DOI 10.1016/0001-8708(77)90017-2. MR0486011

[6] I. B. Penkov, *Borel-Weil-Bott theory for classical Lie supergroups* (Russian), Current problems in mathematics. Newest results, Vol. 32, Itogi Nauki i Tekhniki, Akad. Nauk SSSR, Vsesoyuz. Inst. Nauchn. i Tekhn. Inform., Moscow, 1988, pp. 71–124. Translated in J. Soviet Math. **51** (1990), no. 1, 2108–2140. MR957752

[7] V. Serganova, *Quasireductive supergroups*, New developments in Lie theory and its applications, Contemp. Math., vol. 544, Amer. Math. Soc., Providence, RI, 2011, pp. 141–159, DOI 10.1090/conm/544/10753. MR2849718
[8] A. V. Zelevinsky, *Representations of finite classical groups: A Hopf algebra approach*, Lecture Notes in Mathematics, vol. 869, Springer-Verlag, Berlin-New York, 1981. MR643482

Université de Lorraine, CNRS, IECL, F-54000 Nancy, France
Email address: caroline.gruson@univ-lorraine.fr

Department of Mathematics, University of California, Berkeley, California 94720-3840
Email address: serganov@math.berkeley.edu

Proceedings of Symposia in Pure Mathematics
Volume **101**, 2019
https://doi.org/10.1090/pspum/101/01795

Representations of a p-adic group in characteristic p

G. Henniart and M.-F. Vignéras

For Joseph Bernstein

ABSTRACT. Let F be a locally compact non-archimedean field of residue characteristic p, $\mathbf{G}$ a connected reductive group over F, and R a field of characteristic p. When R is algebraically closed, the irreducible admissible R-representations of $G = \mathbf{G}(F)$ are classified in [J. Amer. Math. Soc. **30** (2017), no. 2, 495–559] in term of supersingular R-representations of the Levi subgroups of G and parabolic induction; there is a similar classification for the simple modules of the pro-p Iwahori Hecke R-algebra $H(G)_R$ in [N. Abe, DOI:10.1515/crelle-2016-0043]. In this paper, we show that both classifications hold true when R is not algebraically closed.

CONTENTS

I. Introduction

I.1 In this paper, p is a prime number, F is a locally compact non-archimedean field of residual characteristic p, $\mathbf{G}$ is a connected reductive group over F and finally R is a field.

Recent applications of automorphic forms to number theory have imposed the study of smooth representations of $G = \mathbf{G}(F)$ on R-vector spaces; indeed one expects a strong relation, à la Langlands, with R-representations of the Galois group of F. The most intriguing case is when the characteristic of R is p - the only established case, however, is that of $GL(2, \mathbb{Q}_p)$.

The first focus is on irreducible representations. When R is algebraically closed of characteristic p, the irreducible admissible R-representations of G have been classified in terms of parabolic induction of supersingular R-representations of Levi subgroups of G [**AHHV**]. But the restriction to algebraically closed R is undesirable: for example, in the work of Breuil and Colmez on $GL(2, \mathbb{Q}_p)$, R is often finite. Here we extend to any R the classification of [**AHHV**] and its consequences.

Let B be a minimal parabolic subgroup of G and I a compatible pro-p Iwahori subgroup of G. If W is a smooth R-representation of G, the space W^I of I-fixed elements is a right module over the Hecke ring $H(G)$ of I in G; it is non-zero if W is, and finite dimensional if W is admissible. Even though W^I might not be simple over $H(G)$ when W is irreducible, it is important to study simple $R \otimes H(G)$-modules. When R is algebraically closed of characteristic p, they have been classified ([**Abe**], see also [**AHenV2**, Cor:4.30]) in terms of supersingular $R \otimes H(M)$-modules, where M is a Levi subgroup of G and $H(M)$ the Hecke ring of $I \cap M$ in M. The classification uses a parabolic induction process from $H(M)$-modules to $H(G)$-modules. Again we extend that classification to any R of characteristic p.

I.2 Before we state our main results more precisely, let us describe our principal tools for reducing them to the known case where R is algebraically closed - those tools are developed in section II.

The idea is to introduce an algebraic closure R^{alg} of R, and study scalar extension from R-representations of G to R^{alg}-representations of G, or from $R \otimes H(G)$-modules to $R^{alg} \otimes H(G)$-modules. The important remark is that when W is an irreducible admissible R-representation of G, or a simple $R \otimes H(G)$-module, its commutant has finite dimension over R. The following result examines what happens for more general extensions R' of R.

THEOREM I.1. [Decomposition theorem] *Let R be a field, A an R-algebra*[1]*, and V a simple A-module with commutant $D = \mathrm{End}_A V$ of finite dimension over R. Let E denote the center of the skew field D, δ the reduced degree of D over E, E_{sep}/R the maximal separable subextension of E/R.*

(1) Let E' be a finite separable extension of E splitting D, L/R the normal closure of E'/R and R' an extension of L. Then the scalar extension $V_{R'}$ of V to R' has length $\delta[E : R]$ and is a direct sum

$$V_{R'} \simeq \oplus^{\delta} \oplus_{j \in \mathrm{Hom}_R(E_{sep}, R')} W'_j$$

of δ copies of a direct sum of $[E_{sep} : R]$ modules W'_j with commutant the local artinian ring $R' \otimes_{j, E_{sep}} E$ which has residue field R'. For each j, the $A_{R'}$-module

[1] all our algebras are associative with unit

W'_j is indecomposable of length $[E : E_{sep}]$, its simple subquotients are all isomorphic to the $A_{R'}$-module $V'_j = R' \otimes_{(R' \otimes_{E_{sep}} E)} W'_j$ which has commutant R', and descend to the finite extension L/R.

If R'/R is normal, the isomorphism classes of the $A_{R'}$-modules W'_j, resp. V'_j, for $j \in \mathrm{Hom}_R(E_{sep}, R')$ form an $\mathrm{Aut}_R(R')$-orbit of cardinality $[E_{sep} : R]$.

(2) Let R^{alg}/R be an algebraic closure. The map which to V as above associates the $\mathrm{Aut}_R(R^{alg})$-orbit of a simple subquotient V' of $V_{R^{alg}}$ induces a bijection

- from the set of isomorphism classes $[V]$ of simple A-modules V with commutant of finite dimension over R,

- onto the set of $\mathrm{Aut}_R(R^{alg})$-orbits of isomorphism classes $[V']$ of absolutely simple $A_{R^{alg}}$-modules V' descending to a finite extension of R.

We note that the $\mathrm{Aut}_R(R^{alg})$-orbit of $[V']$ is finite when V' descends to a finite extension of R. Part (1) of the theorem implies:

COROLLARY I.2. *For any extension R'/R, the length of the $A_{R'}$-module $V_{R'}$ is $\leq \delta[E : R]$, and the dimension over R' of the commutant of any subquotient of $V_{R'}$ is finite.*

When the field R is perfect (example: R finite or of characteristic 0), every algebraic extension of R is separable [**Lang**, VII §7 Cor. 7.8]. In that simple case, the $A_{R'}$-modules W'_j, are absolutely simple in Thm. I.1; in fact, for any extension R'/R, $V_{R'}$ is semi-simple [**BkiA8**, §12 n° 1 Prop.1].

The second theorem is a criterion, inspired by [**AHenV1**, Lemma 3.11], for a functor to preserve the lattice of submodules of a module W. If W is an object in an abelian category, we write $\mathcal{L}_W$ for the lattice of its subobjects; if W has finite length, that length is written $\lg(W)$.

THEOREM I.3. [Lattice isomorphism] *Let $F : \mathcal{C} \to \mathcal{D}$ be a functor between abelian categories having a right adjoint G; write $\epsilon : \mathrm{id} \to GF$ for the unit of the adjunction, and $\eta : FG \to \mathrm{id}$ for the counit.*

(a) Let W be a finite length object in $\mathcal{C}$ such that

(i) $F(Y)$ and $GF(Y)$ are simple for any simple subquotient Y of W;

(ii) $F(W)$ and $GF(W)$ have finite length $\lg(F(W)) = \lg(GF(W)) = \lg(W)$. Then for any subquotient Y of W, $F(Y)$ and $GF(Y)$ have finite length $\lg(F(Y)) = \lg(GF(Y)) = \lg(Y)$, and $\epsilon_Y : Y \to GF(Y)$ is an isomorphism; in addition the maps $Y \mapsto F(Y) : \mathcal{L}_W \to \mathcal{L}_{F(W)}$ and $X \mapsto \epsilon_W^{-1}(G(X)) : \mathcal{L}_{F(W)} \to \mathcal{L}_W$ are lattice isomorphisms, inverse to each other.

(b) Let V be a finite length object in $\mathcal{D}$ such that

(i) $G(X)$ and $FG(X)$ are simple for any simple subquotient X of V;

(ii) $G(V)$ and $FG(V)$ have finite length $\lg(G(V)) = \lg(FG(V)) = \lg(V)$. Then for any subquotient X of V, $G(X)$ and $FG(X)$ have finite length $\lg(G(X)) = \lg(FG(X)) = \lg(X)$, and $\eta_X : FG(X) \to X$ is an isomorphism. In addition, the maps $X \mapsto G(X) : \mathcal{L}_V \to \mathcal{L}_{G(V)}$ and $Y \mapsto \eta_V(F(Y)) : \mathcal{L}_{G(V)} \to \mathcal{L}_V$ are lattice isomorphisms, inverse to each other.

The present formulation and its proof in §II.4 owe much to the referee. We get (b) from (a) by reversing the arrows.

COROLLARY I.4. *Assume that F is fully faithful. Let W be a finite length object in $\mathcal{C}$ such that*

(i) $F(Y)$ is simple for any simple subquotient Y of W;

(ii) $F(W)$ has finite length $\lg(F(W)) = \lg(W)$.
Then $Y \mapsto F(Y) : \mathcal{L}_W \to \mathcal{L}_{F(W)}$ is a lattice isomorphism.

We end §II with another lattice isomorphism inspired by [**Abe**, Lemma 5.3]. Let R be a field, A an R-algebra, and V a simple A-module with commutant R. We have the tensor product $- \otimes_R V : \mathcal{C} \to \mathcal{D}$ from the abelian category $\mathcal{C}$ of R-vector spaces to the abelian category $\mathcal{D}$ of A-modules; it has a right adjoint $\mathrm{Hom}_A(V, -)$.

THEOREM I.5. [Lattice isomorphism and tensor product]
(i) For any R-vector space W, $W \otimes_R V$ is an isotypic A-module of type V and the map $Y \mapsto Y \otimes_R V : \mathcal{L}_W \to \mathcal{L}_{W \otimes_R V}$ is a lattice isomorphism. Moreover, the natural map
$$W \xrightarrow{\epsilon_W} \mathrm{Hom}_A(V, W \otimes_R V) \quad \epsilon_W(w) : v \mapsto w \otimes v$$
is an isomorphism of R-vector spaces.
(ii) For $b_W \in \mathrm{End}_R(W)$, $b_V \in \mathrm{End}_R(V)$ and an R-subspace Y of W, we have:
$b_W(Y) \subset Y$ implies $b_W(Y) \otimes_R b_V(V) \subset Y \otimes_R V$ and the converse is true provided that $b_V \neq 0$.

In our applications, the action of A on V extends to an R-algebra A' containing A, and there is an R-basis B of A contained in an R-basis B' of A' such that no element of $B' \setminus B$ acts by 0 on V, the action of B on W by the identity extends to an action of A' and the diagonal action of B' on $W \otimes_R V$ yields an A'-module structure. On $X = V, W$ or $W \otimes_R V$, $b \in B'$ acts via an endomorphism written b_X.

COROLLARY I.6. *In the above situation, in Theorem* I.5*:*
The map $Y \mapsto Y \otimes_R V$ yields a lattice isomorphism $\mathcal{L}'_W \to \mathcal{L}'_{W \otimes_R V}$ between the lattices of A'-submodules of W and of $W \otimes_R V$. The A'-module structure on $\mathrm{Hom}_A(V, W \otimes_R V)$ such that the isomorphism $W \xrightarrow{\epsilon_W} \mathrm{Hom}_A(V, W \otimes_R V)$ is A'-equivariant, satisfies for all $f \in \mathrm{Hom}_A(V, W \otimes_R V)$,
if $b \in B$ then $b(f) = f$, and if $b \in B' \setminus B$ acts invertibly on V then $b(f) = b_{W \otimes_R V} \circ f \circ b_V^{-1}$.

In that situation the natural map $\mathrm{Hom}_A(V, W \otimes_R V) \otimes_R V \to W \otimes_R V$ is also an isomorphism of A'-modules for $b \in B'$ acting diagonally.

I.3 In §III, for a field R of characteristic p, we prove the classification of the irreducible admissible R-representations of G in terms of supersingular irreducible admissible R-representations of Levi subgroups of G.

We always take our parabolic subgroups to contain a minimal one $B = ZU$ in good position with respect to I. An R-triple (P, σ, Q) of G consists of a parabolic subgroup $P = MN$ of G, a smooth R-representation σ of M, and a parabolic subgroup Q of G satisfying $P \subset Q \subset P(\sigma)$, where $P(\sigma) = M(\sigma)N(\sigma)$ is the maximal parabolic subgroup of G to which σ extends trivially on N; the restriction to M_Q of that extension is denoted by $e_Q(\sigma)$. By definition

$$(0.1) \qquad I_G(P, \sigma, Q) = \mathrm{Ind}_{P(\sigma)}^G(\mathrm{St}_Q^{M(\sigma)}(\sigma)) \quad \text{where}$$

$$(0.2) \qquad \mathrm{St}_Q^{M(\sigma)}(\sigma) = \mathrm{Ind}_Q^{M(\sigma)}(e_Q(\sigma)) / \sum_{Q \subsetneq Q' \subset P(\sigma)} \mathrm{Ind}_{Q'}^{M(\sigma)}(e_{Q'}(\sigma)),$$

is the generalized Steinberg R-representation of $M(\sigma)$ and $\mathrm{Ind}_Q^{M(\sigma)}$ stands for the parabolic smooth induction functor $\mathrm{Ind}_{Q \cap M(\sigma)}^{M(\sigma)}$. In §III.3 we show that $I_G(P, -, Q)$

and scalar extension are compatible: for any R-triple (P,σ,Q) of G, we have $R'\otimes_R I_G(P,\sigma,Q)\simeq I_G(P,R'\otimes_R\sigma,Q)$ for any field extension R'/R and $I_G(P,\sigma,Q)$ descends to a subfield of R if and only if σ does (Prop.III.13).

What does supersingular mean for an irreducible admissible R-representation π of G ? We know what it means to be a supersingular $H(G)_R=R\otimes_{\mathbb{Z}}H(G)$-module: for all $P\neq G$ containing B, a certain central element T_P of the pro-p Iwahori Hecke ring $H(G)$ should act locally nilpotently [**Vig17**]. We say that π is **supersingular** if the I-invariant module π^I is supersingular as a right $H(G)_R$-module (Definition III.14 in §III.4). In §III.4, we show that supersingularity is compatible with scalar extension (Lemma III.16) and that π^I is supersingular if and only if π^I contains a non-zero supersingular element (Theorem III.17). When R is algebraically closed, this definition is equivalent to the one in [**AHHV**], by [**OV**].

THEOREM I.7. [Classification theorem for G]

For any R-triple (P,σ,Q) of G with σ irreducible admissible supersingular, $I_G(P,\sigma,Q)$ is an irreducible admissible R-representation of G.

If $I_G(P,\sigma,Q)\simeq I_G(P_1,\sigma_1,Q_1)$ for two R-triples (P,σ,Q) and (P_1,σ_1,Q_1) of G with σ,σ_1 irreducible admissible supersingular and P,P_1 containing B, then $P=P_1,Q=Q_1$ and $\sigma\simeq\sigma_1$.

For any irreducible admissible R-representation π of G, there exists an R-triple (P,σ,Q) of G with σ irreducible admissible supersingular and P containing B, such that $\pi\simeq I_G(P,\sigma,Q)$.

When R is algebraically closed, this is the classification theorem of [**AHHV**]. In §III.5 we descend the classification theorem from R^{alg} to R by a formal proof using the decomposition theorem (Thm.I.1) and a lattice isomorphism $\mathcal{L}_{\sigma_{R^{alg}}}\simeq\mathcal{L}_{I_G(P,\sigma_{R^{alg}},Q)}$ when σ is irreducible admissible supersingular and $\sigma_{R^{alg}}$ its scalar extension to R^{alg} (Prop.III.10 in §III.3, Remark III.18 in §III.4).

I.4 In §IV, for a field R of characteristic p we prove a similar classification for the simple right $H(G)_R$-modules. As in [**AHenV2**] when R is algebraically closed, this classification uses for a parabolic subgroup $P=MN$ of G containing B, the parabolic induction functor

$$\mathrm{Ind}_P^{H(G)}:\mathrm{Mod}_R(H(M))\to\mathrm{Mod}_R(H(G))$$

from right $H(M)_R$-modules to right $H(G)_R$-modules, analogue of the parabolic smooth induction: indeed $(\mathrm{Ind}_P^G\sigma)^I$ is naturally isomorphic to $\mathrm{Ind}_P^{H(G)}(\sigma^{I\cap M})$ for a smooth R-representation σ of G [**OV**]. An R-triple $(P,\mathcal{V},Q)$ of $H(G)$ consists of parabolic subgroups $P=MN\subset Q$ of G containing B and of a right $H(M)_R$-module $\mathcal{V}$ with $Q\subset P(\mathcal{V})$ (Definition IV.8), for an R-triple $(P,\mathcal{V},Q)$ of $H(G)$ we define a right $H(G)_R$-module $I_{H(G)}(P,\mathcal{V},Q)$ as for the group.

In Proposition IV.12, we prove that $I_{H(G)}(P,-,Q)$ and scalar extension are compatible, as in the group case (Prop. III.13).

THEOREM I.8. [Classification theorem for $H(G)$]

For any R-triple $(P,\mathcal{V},Q)$ of $H(G)$ with $\mathcal{V}$ simple supersingular, $I_{H(G)}(P,\mathcal{V},Q)$ is a simple $H(G)_R$-module.

If $I_{H(G)}(P,\mathcal{V},Q)\simeq I_{H(G)}(P_1,\mathcal{V}_1,Q_1)$ for R-triples $(P,\mathcal{V},Q)$ and $(P_1,\mathcal{V}_1,Q_1)$ of $H(G)$ with $\mathcal{V}$ and $\mathcal{V}_1$ simple supersingular, then $P=P_1,Q=Q_1$ and $\mathcal{V}\simeq\mathcal{V}_1$.

Any simple right $H(G)_R$-module $\mathcal{X}$ is isomorphic to $I_{H(G)}(P,\mathcal{V},Q)$ for some R-triple $(P,\mathcal{V},Q)$ of $H(G)$ with $\mathcal{V}$ simple supersingular.

The proof follows the same pattern as for the group G, by a descent to R of the classification over R^{alg} [**AHenV2**].

Assuming that R contains a root of unity of order the exponent of the quotient Z_k of the parahoric subgroup of Z by its pro-p Sylow subgroup, the simple supersingular $H(G)_R$-modules are classified [**Oss**], [**Vig17**, Thm.1.6]; in particular when G is semisimple and simply connected, they have dimension 1. With Thm. I.8, we have a complete classification of the $H(G)_R$-modules.

The ring $H(M)$ does not embed in the ring $H(G)$ and different inductions from $\mathrm{Mod}_R(H(M))$ to $\mathrm{Mod}_R(H(G))$ are possible. When R is algebraically closed, Abe proved the classification theorem (Thm.I.8) using one of them, the parabolic coinduction[2] [**Abe**]. In the appendix we define and compare 8 natural inductions following [**Abeparind**]; the classification theorem can be expressed with any these 8 inductions instead of the parabolic induction (for the parabolic coinduction [**AHenV2**, Cor. 4.24]).

I.5 In §V, still with R of characteristic p we give applications (Thm. I.9, I.10, I.12, I.13) of the classification for G (Thm. I.7) and for $H(G)$ (Thm. I.8); they were already known when R is algebraically closed, except for the parts (ii),(iii) of Theorem I.10 below.

THEOREM I.9. [Vanishing of the smooth dual] *The smooth dual of an infinite dimensional irreducible admissible R-representation of G is* 0.

This was proved by different methods when the characteristic of F is 0 in [**Kohl**] and when R is algebraically closed in [**AHenV2**, Thm.6.4]. In §V.1 we deduce easily the theorem from the theorem over R^{alg} using scalar extension to R^{alg} (Thm. I.1).

[Description of $\mathrm{Ind}_P^G\sigma$ for an irreducible admissible R-representation σ of M, and of $\mathrm{Ind}_P^{H(G)}\mathcal{V}$ for a simple $H(M)_R$-module $\mathcal{V}$] Here $P = MN$ is a fixed parabolic subgroup of G.

We write $\mathcal{L}_\pi$ for the lattice of subrepresentations of an R-representation π of G, and $\mathcal{L}_\mathcal{X}$ for the lattice of submodules of an $H(G)_R$-module $\mathcal{X}$.

Recall that for a set X, an upper set in $\mathcal{P}(X)$ is a set $\mathcal{Q}$ of subsets of X, such that if $X_1 \subset X_2 \subset X$ and $X_1 \in \mathcal{Q}$ then $X_2 \in \mathcal{Q}$. Write $\mathcal{L}_{\mathcal{P}(X),\geq}$ for the lattice of upper sets in $\mathcal{P}(X)$. For two subsets X_1, X_2 of X write $X_1 \setminus X_2$ for the complementary set of $X_1 \cap X_2$ in X_1.

By the classification theorems, $\sigma \simeq I_M(P_1 \cap M, \sigma_1, Q \cap M)$ with (P_1, σ_1, Q) an R-triple of G, $Q \subset P$ and σ_1 irreducible admissible supersingular and $\mathcal{V} \simeq I_{H(M)}(P_1 \cap M, \mathcal{V}_1, Q \cap M)$ with $(P_1, \mathcal{V}_1, Q)$ an R-triple of $H(G)$, $Q \subset P$, and $\mathcal{V}_1$ simple supersingular.

With these notations we have:

THEOREM I.10. [Lattices $\mathcal{L}_{\mathrm{Ind}_P^G\sigma}$ and $\mathcal{L}_{\mathrm{Ind}_P^{H(G)}\mathcal{V}}$]

(i) The R-representation $\mathrm{Ind}_P^G\sigma$ of G is multiplicity free of irreducible subquotients $I_G(P_1, \sigma_1, Q')$ for R-triples (P_1, σ_1, Q') of G with $Q' \cap P = Q$.

Sending $I_G(P_1, \sigma_1, Q')$ to $\Delta_{Q'} \cap (\Delta_{P(\sigma_1)} \setminus \Delta_P)$ gives a lattice isomorphism[3]

$$\mathcal{L}_{\mathrm{Ind}_P^G\sigma} \to \mathcal{L}_{\mathcal{P}(\Delta_{P(\sigma_1)}\setminus\Delta_P),\geq}.$$

[2]The parabolic coinduction is the induction used by Abe

[3]see the discussion in [**He**]§11 on the lattice of submodules of a multiplicity free module

(ii) The $H(G)_R$-module $\mathrm{Ind}_P^{H(G)} \mathcal{V}$ satisfies the analogue of (i).

(iii) If $\sigma^{I\cap M}$ is simple and the natural surjective map $\sigma^{I\cap M} \otimes_{H(M)} \mathbb{Z}[(I \cap M)\backslash M] \to \sigma$ is bijective, then the I-invariant functor $(-)^I$ and its left adjoint $- \otimes_{H(G)} \mathbb{Z}[I\backslash G]$ give lattice isomorphisms between $\mathcal{L}_{\mathrm{Ind}_P^G(\sigma)}$ and $\mathcal{L}_{\mathrm{Ind}_P^{H(G)}(\sigma^{I\cap M})}$ inverse of each other.

When R is algebraically closed (i) is proved in [**AHenV1**, §3.2]. In §V.2 we prove (i) and (ii); (iii) follows from (i), (ii), Corollary I.4 and the commutativity of the parabolic inductions with $(-)^I$ and $- \otimes_{H(G)} \mathbb{Z}[I\backslash G]$ [**OV**].

Corollary I.11. *1. The socle and the cosocle of $\mathrm{Ind}_P^G \sigma$ are irreducible; $\mathrm{Ind}_P^G \sigma$ is irreducible if and only if P contains $P(\sigma_1)$. The same is true for $\mathrm{Ind}_P^{H(G)}\mathcal{V}$.*

2. Let π be an irreducible admissible R-representation of G; we write $\pi \simeq I_G(P,\sigma,Q)$ with σ irreducible admissible supersingular.

If $\sigma^{I\cap M}$ is simple and the natural map $\sigma^{I\cap M} \otimes_{H(M)} \mathbb{Z}[(I \cap M)\backslash M] \to \sigma$ is bijective, then π^I is simple and $\pi \simeq \pi^I \otimes_{H(G)} \mathbb{Z}[I\backslash G]$.

The first assertion for σ and R is algebraically closed is proved in [**AHenV1**, Cor. 3.3 and 4.4].

[Computation of the left adjoint and the right adjoint of the parabolic induction, of π^I for an irreducible admissible R-representation π of G, and of $\mathcal{X} \otimes_{H(G)} \mathbb{Z}[I\backslash G]$ for a simple $H(G)_R$-module $\mathcal{X}$]

For a parabolic subgroup P_1 of G, write $L_{P_1}^G$ for the left adjoint of $\mathrm{Ind}_{P_1}^G$, $R_{P_1}^G$ for its right adjoint [**Vigadjoint**], and $L_{P_1}^{H(G)}$ for the left adjoint of $\mathrm{Ind}_{P_1}^{H(G)}$, $R_{P_1}^{H(G)}$ for its right adjoint [**VigpIwst**].

Theorem I.12. [Adjoint functors of the parabolic induction and of the I-invariant]

(i) $L_{P_1}^G(\pi)$ and $R_{P_1}^G(\pi)$ are 0 or irreducible admissible.

$L_{P_1}^G(\pi) \neq 0 \Leftrightarrow P_1 \supset P, \langle P_1, Q\rangle \supset P(\sigma) \Leftrightarrow L_{P_1}^G(\pi) \simeq I_{M_1}(P \cap M_1, \sigma, Q \cap M_1)$.

$R_{P_1}^G(\pi) \neq 0 \Leftrightarrow P_1 \supset Q \Leftrightarrow R_{P_1}^G(\pi) \simeq I_{M_1}(P \cap M_1, \sigma, Q \cap M_1)$.

(ii) $L_{P_1}^{H(G)}(\mathcal{X})$ and $R_{P_1}^{H(G)}(\mathcal{X})$ satisfy (i) with the obvious modifications.

(iii) We have natural isomorphisms $\pi^I \simeq I_{H(G)}(P, \sigma^{I\cap M}, Q)$ and $\mathcal{X} \otimes_{H(G)_R} R[I\backslash G]$

$\simeq I_G(P, \mathcal{V} \otimes_{H(M)_R} R[(I \cap M)\backslash M], Q)$.

Example: $L_{P(\sigma)}^G(I_G(P,\sigma,Q)) \simeq R_{P(\sigma)}^G(I_G(P,\sigma,Q)) \simeq \mathrm{St}_Q^{M(\sigma)}(\sigma)$ and the analogous for $I_{H(G)}(P,\mathcal{V},Q)$.

Proving Theorem I.12 from the classification theorem needs no new techniques (§V.3).

[Equivalence between supersingularity, supercuspidality and cuspidality]

An irreducible admissible R-representation π of G is said to be

- **supercuspidal** if it is not a subquotient of $\mathrm{Ind}_P^G \tau$ with $\tau \in \mathrm{Mod}_R^\infty(M)$ irreducible admissible for any parabolic subgroups $P = MN \subsetneq G$.
- **cuspidal** if $L_P^G(\pi) = R_P^G(\pi) = 0$ for all parabolic subgroups $P \subsetneq G$.

Theorem I.13. *Let π be an irreducible admissible R-representation of G. Then π is supersingular if and only if its is supercuspidal if and only if it is cuspidal.*

The equivalence of supersingular with supercuspidal, resp. cuspidal, follows from Thm. I.10, resp. Thm. I.12. When R is algebraically closed, the first equivalence was proved in [**AHHV**, Thm. VI.2] and the second one in [**AHenV1**, Cor.6.9].

An irreducible admissible R-representation π admits a supercuspidal support: the parabolic subgroup $P = MN$ containing B and the isomorphism class of the irreducible admissible supercuspidal R-representation of σ of M such that π is a subquotient of $\mathrm{Ind}_P^G(\sigma)$ are unique; this follows from Thm. I.7 and Thm. I.13.

Acknowledgments We thank the CNRS, the IMJ Paris-Diderot University, the Paris-Sud University, and the Weizmann Institute where part of our work was carried out.

II. Some general algebra

II.1. Review on scalar extension. We consider a field R and an R-algebra A (always associative with unit).

For an extension R' of R (which we see as a field R' containing R), the scalar extension functor $R' \otimes_R - : \mathrm{Mod}_R \to \mathrm{Mod}_{R'}$ from R to R', also denoted $(-)_{R'}$, is faithful exact and left adjoint to the restriction functor from R' to R.

The scalar extension $A_{R'}$ of A is an R'-algebra and if W a (left or right) A-module, $W_{R'}$ is an $A_{R'}$-module. An $A_{R'}$-module W' isomorphic to such a $W_{R'}$ is said to descend to R or to be defined over R, and W is called an R-structure for W' (more precisely the isomorphism $W' \simeq W_{R'}$ is an R-structure for W').

REMARK II.1. Let R^{alg} be an algebraic closure of R. If A is a finitely generated R-algebra, an $A_{R^{alg}}$-module W of finite dimension over R^{alg} descends to a finite extension of R. Indeed, if (w_i) is an R^{alg}-basis of W, (a_j) a finite set of generators of A, and $a_j w_i = \sum_k r_{j,i,k} w_k$, the extension R'/R generated by the coefficients $r_{j,i,k} \in R^{alg}$ is finite and the $A_{R'}$-module $\oplus_i R' w_i$ gives an R'-structure for W.

REMARK II.2. If V, W are A-modules, the natural map

$$(\mathrm{Hom}_A(V,W))_{R'} \to \mathrm{Hom}_{A_{R'}}(V_{R'}, W_{R'}) \tag{1.1}$$

is injective [**BkiA2**, II §7 n°7 Prop.16] and bijective if R'/R is finite [**BkiA8**, §12, n°2 Lemme 1], or if V is a finitely generated A-module (proof as in [**Pask**, Lemma 5.1])[4].

Let V be a simple A-module; we write D for the commutant $\mathrm{End}_A(V)$, so that D is a division algebra, and E for the center of D. Since V is finitely generated, the commutant of $V_{R'}$ is $D_{R'}$ and its center is $E_{R'}$, by Remark II.2. That V is simple has further consequences:

(P1) As an A-module, $V_{R'}$ is a direct sum of A-modules isomorphic to V, i.e. V-isotypic of type V [**BkiA8**, §4, n°4, Prop.1].

(P2) The map $\mathfrak{A} \mapsto \mathfrak{A} V_{R'}$ is a lattice isomorphism of the lattice of right ideals $\mathfrak{A}$ of $D_{R'}$ onto the lattice of $A_{R'}$-submodules W of $V_{R'}$, with inverse $W \mapsto \{d \in D_{R'}, dV_{R'} \subset W\}$ [**BkiA8**, §12, n°2, Thm.2b)].

(P3) For any right ideal $\mathfrak{A}$ of $D_{R'}$, via the isomorphism $V_{R'} \simeq D_{R'} \otimes_D V$, $\mathfrak{A} V_{R'}$ corresponds to $\mathfrak{A} \otimes_D V$. As the functor $X \mapsto X \otimes_D V$ from right D-vector spaces to A-modules is exact, if $\mathfrak{B} \subset \mathfrak{A}$ are right ideals of $D_{R'}$, then $\mathfrak{A} V_{R'} / \mathfrak{B} V_{R'}$ is isomorphic to $(\mathfrak{A}/\mathfrak{B}) \otimes_D V_{R'}$.

[4] We are grateful to the referee for that reference

(P4) If the extension R'/R is finite, $V_{R'}$ has finite length as an A-module, so also as an $A_{R'}$-module; then $D_{R'}$ is left and right artinian and $E_{R'}$ is artinian [**BkiA8**, §12, n°5, Prop.5 a)]. If moreover R'/R is separable, $V_{R'}$ is semisimple [**BkiA8**, §12, n°5, Cor.].

(P5) If $\dim_R D$ is finite, then $\dim_{R'} D_{R'} = \dim_R D$ and the length of the $A_{R'}$-module $V_{R'}$ is $\leq [D : R]$ by (P2); the best bound is given in Thm. I.1.

REMARK II.3. A non-zero A-module W is called absolutely simple if $W_{R'}$ is simple for any extension R'/R.

A simple A-module V is absolutely simple if and only if $\operatorname{End}_A V = R$. For $\Rightarrow$ [**BkiA8**, §3,n°2,Cor.2, p.44]. For $\Leftarrow$ follows from (P5). If R is algebraically closed of cardinal $> \dim_R V$, then $D = R$ [**BkiA8**, §3,n°2,Thm.1, p.43].

II.2. A bit of ring and module theory. We examine the tensor product $L \otimes_R E$ of two field extensions L/R and E/R. Seeing the commutative ring $L \otimes_R E$ as a module over itself, its simple subquotients are isomorphic to simple $L \otimes_R E$-modules, that is to simple quotients.

LEMMA II.4. *Let E/R be a finite extension and L/E an extension.*

(1) If E/R is purely inseparable, then $L \otimes_R E$ is a commutative artinian local ring with residue field L.

(2) If E/R is separable and L contains a Galois closure of E/R, then

$$L \otimes_R E \simeq \prod_{j \in \operatorname{Hom}_R(E,L)} L \otimes_{j,E} E \simeq L^{[E:R]},$$

and if F/R is a subextension of E/R, the restriction $\operatorname{Hom}_R(E, L) \to \operatorname{Hom}_R(F, L)$ *is surjective.*

(3) If L/R is normal, then $\operatorname{Aut}_R(L)$ *acts transitively on* $\operatorname{Hom}_R(E, L)$.

(4) If E/R is normal, the ring homomorphism

$$x \otimes y \mapsto (xj(y))_j \ : \ E \otimes_R E \to \prod_{j \in \operatorname{Aut}_R(E)} E$$

is surjective of kernel the Jacobson radical of $E \otimes_R E$.

PROOF. As E/R is finite, the commutative ring $L \otimes_R E$ has finite dimension over L, hence is Artinian. Let R' be a field quotient of $L \otimes_R E$. The quotient map $\varphi : L \otimes_R E \to R', \varphi(x \otimes y) = \varphi_1(x)\varphi_2(y)$, is given by non zero R-homomorphisms $\varphi_1 : L \to R', \varphi_1(x) = \varphi(x \otimes 1)$, and $\varphi_2 : E \to R', \varphi_2(y) = \varphi(1 \otimes y)$.

If E/R is purely inseparable, φ_2 is the restriction of φ_1 to E thus we have (1).

Let $J = \operatorname{Hom}_R(E, L)$ and for $j \in J$, let f_j the surjective map $L \otimes_R E \to L \otimes_{j,E} E \xrightarrow{\simeq} L$. If $j \neq j'$ are distinct in J, and $x \in E$ with $j(x) \neq j'(x)$, we have

$$f_j(j(x) \otimes 1 - 1 \otimes x) = 0, \quad f_{j'}(j(x) \otimes 1 - 1 \otimes x) = j(x) - j'(x) \neq 0.$$

Hence $\operatorname{Ker} f_j \neq \operatorname{Ker} f_{j'}$. By the Chinese Remainder Theorem,

$$\prod f_j : L \otimes_R E \to \prod_{j \in J} L \otimes_{j,E} E \xrightarrow{\simeq} L^J \tag{2.2}$$

is surjective. It is injective if and only if $[E : R] = |J|$.

If E/R is separable and L contains a Galois closure of E/R, then $[E : R] = |J|$ (and conversely), and for any subextension F/R of E/R, F/R and E/F are separable and L contains a Galois closure of F/R and of E/F, thus the restriction

$\mathrm{Hom}_R(E, L) \to \mathrm{Hom}_R(F, L)$ of kernel $\mathrm{Hom}_F(E, L)$ is surjective by a counting argument since $[E : R] = [E : F][F : R]$. This gives (2).

Let E_{sep}/R be the maximal separable subextension of E/R. The extension E/E_{sep} is purely inseparable and the restriction $\mathrm{Hom}_R(E, L) \to \mathrm{Hom}_R(E_{sep}, L)$ is injective.

If L/R is normal, (3) is true as $\mathrm{Hom}_R(E, L) \to \mathrm{Hom}_R(E_{sep}, L)$ is injective and (3) is true when E/R is separable by Galois theory. If $L = E$, for $j \in \mathrm{Hom}_R(E, E) = \mathrm{Aut}_R(E)$ and $x, y \in E$, we have $f_j(x \otimes y) = xj(y)$. If R' is a field quotient of $E \otimes_R E$, the quotient map satisfies $\varphi(x \otimes y) = \varphi_1(x)\varphi_2(y)$ for φ_1, φ_2 in $\mathrm{Hom}_R(E, R')$. If moreover E/R is normal, then $R' = E$ and $\varphi = \varphi_1 \circ f_j$ where $\varphi_2 = \varphi_1 \circ j$ in $\mathrm{Aut}_R(E)$. This gives (4). □

LEMMA II.5. *Let R'/R be a normal field extension, A and R-algebra and V' a simple $A_{R'}$-module descending to a finite extension of R. Then V' is isomorphic to a submodule of the scalar extension $V_{R'}$ from R to R' of a simple A-module V. For any such V, $\dim_R V$ is finite if $\dim_{R'} V'$ is, and $\dim_R \mathrm{End}_A V$ is finite if $\dim_{R'} \mathrm{End}_{A_{R'}} V'$ is.*

PROOF. a) Assume first that the normal extension R'/R is finite. Then $A_{R'}$ is a (free) finitely generated A-module, so V' as an A-module is finitely generated, and in particular has a simple quotient V: $\mathrm{Hom}_A(V', V) \neq 0$. By Remark II.2, $\mathrm{Hom}_{A_{R'}}(V'_{R'}, V_{R'}) \neq 0$.

The $A_{R'}$-module $V'_{R'}$ admits a finite filtration of quotients V'_j for $j \in \mathrm{Aut}_R(R')$, where V'_j is isomorphic to V' with the j-twisted action $(y \otimes a)v' = j(y)av'$ for $y \in R', a \in A, v \in V'$. Indeed, $V'_{R'} = R' \otimes_R V' \simeq (R' \otimes_R R') \otimes_{R'} V'$, the artinian commutative ring $R' \otimes_R R'$ admits a finite filtration with quotients isomorphic to simple $R' \otimes_R R'$-modules, and the simple $R' \otimes_R R'$-modules are R'_j for $j \in \mathrm{Aut}_R(R')$, where R'_j is isomorphic to R' with $x \otimes y \in R' \otimes_R R'$ acting by multiplication by $xj(y)$ by Lemma II.4 (4).

We deduce that $\mathrm{Hom}_{A_{R'}}(V'_j, V_{R'}) \neq 0$ for some $j \in \mathrm{Aut}_R(R')$. But $V_{R'}$ is isomorphic to its j-twists $(V_{R'})_j$ for all $j \in \mathrm{Aut}_R(R')$, so we have $\mathrm{Hom}_{A_{R'}}(V', V_{R'}) \neq 0$.

Let V be any simple A-module with $\mathrm{Hom}_{A_{R'}}(V', V_{R'}) \neq 0$. Then $\mathrm{Hom}_A(V', V) \neq 0$ as $V_{R'}$ as an A-module is V-isotypic, so $\dim_R V$ is finite if $\dim_{R'} V'$ is. Put $D = \mathrm{End}_A(V)$ and $D' = \mathrm{End}_{A_{R'}}(V')$ and let W be the maximal V'-isotypic submodule of $V_{R'}$. Then W is $D_{R'}$-stable and we get a homomorphism $D_{R'} \to \mathrm{End}_{A_{R'}} W$ which is necessarily injective on D, since D is a division algebra. By (P4), $V_{R'}$ has finite length, so W also has finite length and $\mathrm{End}_{A_{R'}} W$ is a matrix algebra over D'; it follows that if $\dim_{R'} D'$ is finite, so is $\dim_{R'}(\mathrm{End}_{A_{R'}} W)$ hence also $\dim_R(\mathrm{End}_{A_{R'}} W)$, $\dim_{R'}(D_{R'})$ and $\dim_R D$.

b) Let us treat the general case. By assumption there is a finite normal subextension L of R in R' and an A_L-module U such that $V' = R' \otimes_L U$ - then U is necessarily simple. By a) $\mathrm{Hom}_{A_L}(U, V_L) \neq 0$ for some simple A-module V and by Remark II.2, $\mathrm{Hom}_{A_{R'}}(V', V_{R'}) \neq 0$.

Conversely, if V is some simple A-module with $\mathrm{Hom}_{A_{R'}}(V', V_{R'}) \neq 0$ then by Remark II.2 again $\mathrm{Hom}_{A_L}(U, V_L) \neq 0$, so the other assertions follow from a). □

We pursue with an easy application of Morita theory in the special case of a matrix ring.

LEMMA II.6. *Let A, B be two rings and n a positive integer.*

1) Let W be an A-module. A ring isomorphism $\mathrm{End}_A W \simeq M(n, B)$ induces an A-module isomorphism $W \simeq \oplus^n V$ for some A-module V with commutant B.

2) If B is a commutative artinian local ring of residue field R, then $M(n, B)$ is left Artinian, and as a left module over itself, all its simple subquotients are isomorphic to R^n.

PROOF. 1) If V is a B-module, then V^n is naturally an $M(n, B)$-module, and the functor $V \mapsto V^n$ is an equivalence from the category of B-modules to the category of $M(n, B)$-modules; that is the elementary case of Morita theory. By that equivalence, if V is a left (A, B) bimodule, then V^n is left $(A, M(n, B))$ bimodule, and any left $(A, M(n, B))$ bimodule structure of V^n comes in that way from a left (A, B) bimodule structure on V. As $\mathrm{End}_A(V^n)$ identifies with $M(n, \mathrm{End}_A(V))$, the condition $\mathrm{End}_A(V^n) = M(n, B)$ is the same as $\mathrm{End}_A(V) = B$, and 1) follows.

2) As a left module over itself, $M(n, B)$ is isomorphic to the direct sum of n copies of B^n (let $M(n, B)$ act on the column vectors). By the equivalence recalled in the proof of 1), the $M(n, B)$-module B^n has the same length as B over itself and its simple subquotients are isomorphic to R^n, hence 2). □

II.3. Proof of the decomposition theorem (Thm.I.1 and Cor.I.2). Let V be a simple A-module with commutant $D = \mathrm{End}_A V$ of finite dimension $\dim_R D$ over R. Let E denote the center of the skew field D, δ the reduced degree of D over E, E_{sep}/R the maximal separable subextension of E/R.

Two well-known properties will be used in the proof:

(P6) A finite extension E'/E splits D, i.e. $E' \otimes_E D \simeq M(\delta, E')$, if and only if E' is isomorphic to a maximal subfield of a matrix algebra over D [**BkiA8**, §15, n°3, Prop.5]. The field D contains a maximal subfield, which a separable extension E'/E of degree δ [**CR**, 7.24 Prop] or [**BkiA8**, loc.cit. and §14, n°7].

(P7) For a finite separable extension E'/E and E'_{sep}/R the maximal separable subextension of E'/R, the natural map $x \otimes y \mapsto xy : E'_{sep} \otimes_{E_{sep}} E \to E'$ is an isomorphism (because always surjective and the dimension over E'_{sep} of both sides is the same $[E : E_{sep}]$ by [**Lang**, VII §7, Cor. 7.5] applied to the finite extensions E'_{sep}/E_{sep} separable and E/E_{sep} purely inseparable).

Proof of Thm.I.1 (1).

Let R'/R be an extension containing a normal closure of a finite separable extension E'/E splitting D. For example, R' can be an algebraic closure R^{alg}/R. Let $J = \mathrm{Hom}_R(E_{sep}, R')$. By Lemma II.4 (1), we have a ring isomorphism

$$R' \otimes_R E_{sep} \simeq \prod_{j \in J} R' \otimes_{j, E_{sep}} E_{sep} \simeq R'^{[E_{sep}:R]}.$$

We denote by e_j the idempotents of $(E_{sep})_{R'}$ associated to this decomposition. Tensoring on the right by E, D, or V over E_{sep} and we get product decompositions

$$E_{R'} = \prod_{j \in J} e_j E_{R'}, \ D_{R'} = \prod_{j \in J} e_j D_{R'}, \ V_{R'} = \oplus_{j \in J} e_j V_{R'}$$

where $e_j E_{R'} \simeq R' \otimes_{j, E_{sep}} E$, $e_j D_{R'} \simeq R' \otimes_{j, E_{sep}} D$, $e_j V_{R'} \simeq R' \otimes_{j, E_{sep}} V$. By Lemma II.4 and (P7), for $j \in J$ there exists $j' \in \mathrm{Hom}_R(E'_{sep}, R')$ of restriction $j'|_{E_{sep}} = j$, and

$$R' \otimes_{j, E_{sep}} E \simeq R' \otimes_{j', E'_{sep}} E'_{sep} \otimes_{E_{sep}} E \simeq R' \otimes_{j', E'_{sep}} E'$$

is a commutative artinian local ring of residue field R'. We obtain ring isomorphisms

$$R' \otimes_{j,E_{sep}} D \simeq R' \otimes_{j,E_{sep}} E \otimes_E D$$
$$\simeq R' \otimes_{j',E'_{sep}} E' \otimes_E D \simeq R' \otimes_{j',E'_{sep}} M(\delta, E') \simeq M(\delta, R' \otimes_{j,E_{sep}} E).$$

By Lemma II.6, there exists an $A_{R'}$-module W'_j such that

$$R' \otimes_{j,E_{sep}} V \simeq \oplus^\delta W'_j, \quad \mathrm{End}_{A_{R'}} W'_j \simeq R' \otimes_{j,E_{sep}} E.$$

By Remark II.2, for $j \in J$, the commutant of the $A_{R'}$-module $e_j V_{R'} = R' \otimes_{j,E_{sep}} V$ is $e_j D = R' \otimes_{j,E_{sep}} D$. Applying (P2) and (P3), the map $\mathfrak{A} \mapsto \mathfrak{A} e_j V_{R'}$ is a lattice isomorphism of the lattice of right ideals $\mathfrak{A}$ of $e_j D_{R'}$ onto the lattice of $A_{R'}$-submodules of $e_j V_{R'}$, and for two right ideals $\mathfrak{A} \subset \mathfrak{B}$ of $e_j D_{R'}$, the $A_{R'}$-module $\mathfrak{B} e_j V_{R'} / \mathfrak{A} e_j V_{R'}$ is isomorphic to $(\mathfrak{B}/\mathfrak{A}) \otimes_{e_j D} e_j V_{R'}$. As $e_j D_{R'} \simeq M(\delta, e_j E)$ and $e_j E$ is a commutative artinian local ring of residue field R, by Lemma II.6, the $A_{R'}$-module W'_j is indecomposable of length $[E : E_{sep}]$ and its simple subquotients are all isomorphic to the $A_{R'}$-module

$$V'_j = R' \otimes_{(R' \otimes_{j,E_{sep}} E)} W'_j$$

with commutant R', hence absolutely simple by Remark II.3.

The group $\mathrm{Aut}_R(R')$ of R-automorphisms of R' acts on the $A_{R'}$-modules, fixing the isomorphism class of the scalar extension from R to R' of an A-module. If R'/R is normal, it acts transitively on the set J by Lemma II.4 (3), and for $g \in \mathrm{Aut}_R(R')$ we have $g(e_j) = e_{g \circ j}$. By Krull-Remak-Schmidt's theorem, $g(W'_j) \simeq W'_{g \circ j}$. The same is true for the simple subquotients: $g(V'_j) \simeq V'_{g \circ j}$.

The dimension over R' of the commutant of any subquotient of the $A_{R'}$-module

$$R' \otimes_R V = \oplus^\delta \oplus_{j \in J} W'_j$$

is finite (because the length of $R' \otimes_R V$ is finite and R' is the commutant of any of its simple subquotients).

Let L be the normal closure of E'/R in R'/R. These results applied to R'/R and to L/R, show that scalar extension from L to R' induces a lattice isomorphism $\mathcal{L}_{V_L} \to \mathcal{L}_{V_{R'}}$. This ends the proof of Thm.I.1 (1).

Proof of Thm.I.1 (2).

Thm.I.1 1) applies to $R' = R^{alg}$ an algebraic closure of R. It shows that for any simple A-module V with $\dim_R V$ finite, the simple subquotients of $V_{R^{alg}}$ are absolutely simple, descend to a finite subextension of R^{alg}/R and their isomorphism classes form a finite $\mathrm{Aut}_R(R^{alg})$-orbit.

Conversely, let V' be an absolutely simple $A_{R^{alg}}$-module descending to a finite extension L of R. We prove that the $\mathrm{Aut}_R(R^{alg})$-orbit $\mathrm{Aut}_R(R^{alg})[V']$ of the isomorphism class $[V']$ of V' is finite. Let W' denote an A_L-module with scalar extension $W_{R^{alg}} = V'$ to R^{alg}. Necessarily, W is absolutely simple. By Lemma II.5, W' is contained in the scalar extension V_L from R to L of a simple A-module V with $\dim_R V$ finite. We proved that $V_{R^{alg}}$ has finite length and that the isomorphism classes of its simple subquotients form an $\mathrm{Aut}_R(R^{alg})$-orbit. Hence $\mathrm{Aut}_R(R^{alg})[V']$ is finite, and the map $[V] \to \mathrm{Aut}_R(R^{alg})[V']$ in Thm.I.1 (2) is surjective. It is also injective because $V_{R^{alg}}$ is V-isotypic as an A-module (by P1), so the same is true for V'. This ends the proof of Thm.I.1 (2).

Proof of Corollary I.2.

Let L/R be any extension and L^{alg} an algebraic closure of L. The scalar extension from R to L^{alg} is the scalar extension of R to L followed by the scalar extension from L to L^{alg}.

(i) The length of the $A_{L^{alg}}$-module $V_{L^{alg}}$ is $\delta[E:R]$ by part 1) of Thm.I.1, hence the length of the A_L-module V_L is $\leq \delta[E:R]$.

(ii) Let W be a subquotient of V_L. We show that the commutant of W has finite dimension over L. As $W_{L^{alg}}$ is a subquotient of $V_{L^{alg}}$, by part 2) of Thm.I.1, the dimension over L^{alg} of the commutant of $W_{L^{alg}}$ is finite. By (i) the A_L-module W has finite length hence is finitely generated and by Remark II.2, $\dim_{L^{alg}}(\mathrm{End}_{A_{L^{alg}}} W_{L^{alg}}) = \dim_L(\mathrm{End}_{A_L} W)$. This ends the proof of Corollary I.2.

II.4. Proof of the lattice theorems (Thm. I.3, I.5 and Cor. I.4, I.6). Our overall reference for abelian categories is [**KS**, Chapter 8].

Let $\mathcal{C}$ be an abelian category and W an object in $\mathcal{C}$. A subobject of W is an isomorphism class of monomorphisms $f: Y \to W$ [**KS**, Def. 1.2.18]. The ordered set $\mathcal{L}_W$ of subobjects of W is a bounded lattice: the meet of two subobjects $f: Y \to W$ and $f': Y' \to W$ is the kernel of $(f, -f'): Y \oplus Y' \to W$ and their join is its image. As in module categories[5], we write $Y \cap Y'$ for the meet, $Y + Y'$ for the join [**KS**, 8.3.10]; we note the exact sequence

$$0 \to (Y \cap Y') \to (Y \oplus Y') \to Y + Y' \to 0.$$

We define the lattice $\overline{\mathcal{L}}_W$ of quotients of W: it is the lattice of subobjects of W in the opposite category of $\mathcal{C}$. The map which to a subobject Y of W associates its cokernel (written W/Y) yields a lattice isomorphism $\mathcal{L}_W \to \overline{\mathcal{L}}_W$.

If $\mathcal{D}$ is an abelian category and $F: \mathcal{C} \to \mathcal{D}$ a left exact functor, then $Y \mapsto F(Y): \mathcal{L}_W \to \mathcal{L}_{F(W)}$ is an ordered preserving map; if F is not left exact, $F(Y)$ might not be a subobject of $F(W)$ if Y is a subobject of W.

LEMMA II.7. *Let $F: \mathcal{C} \to \mathcal{D}$ be a functor between abelian categories which is left or right exact, and let W be a finite length object of $\mathcal{C}$* [**KS**, Ex. 8.20, p. 205].

(i) Assume that $F(Y)$ is 0 or simple (that is, $\lg(F(Y)) \leq 1$) for any simple subquotient Y of W. Then, $F(Y)$ has finite length $\lg(F(Y)) \leq \lg(Y)$ for any subquotient Y of W.

(ii) If moreover $\lg(F(W)) = \lg(W)$, then for any subquotient Y of W, $\lg(F(Y)) = \lg(Y)$ and an exact sequence $0 \to Y' \to Y \to Y'' \to 0$ in $\mathcal{C}$ yields via F an exact sequence $0 \to F(Y') \to F(Y) \to F(Y'') \to 0$ in $\mathcal{D}$; in addition $Y \mapsto F(Y)$ gives an injective morphism of bounded lattices $\mathcal{L}_W \to \mathcal{L}_{F(W)}$.

PROOF. (i) Our proof proceeds by induction on the length of $\lg(Y)$ of a subquotient Y of W. By assumption $\lg(F(Y)) \leq \lg(Y)$ if $\lg(Y) \leq 1$. If $\lg(Y) \geq 2$, we choose an exact sequence $0 \to Y' \to Y \to Y'' \to 0$ in $\mathcal{C}$ with non-zero Y', Y''. If F is left exact, $0 \to F(Y') \to F(Y) \to F(Y'')$ is exact, if F is right exact, $F(Y') \to F(Y) \to F(Y'') \to 0$ is exact; in either case we get $\lg(F(Y)) \leq \lg(F(Y')) + \lg(F(Y''))$ which by induction is $\leq \lg(Y') + \lg(Y'') = \lg(Y)$.

(ii) For any subobject Y of W, the exact sequence $0 \to Y \to W \to W/Y \to 0$ gives $\lg(F(W)) \leq \lg(F(Y)) + \lg(F(W/Y))$ as above; applying (i), $\lg(F(Y)) \leq \lg(Y)$, $\lg(F(W/Y)) \leq \lg(W/Y)$. By assumption $\lg(F(W)) = \lg(W) = \lg(Y) + \lg(W/Y)$ so we get equalities throughout: $\lg(F(Y)) = \lg(Y)$ and $\lg(F(W/Y)) = \lg(W/Y)$.

[5] In any case, all our applications are to module categories

For any subquotient Y of W we repeat the argument to get $\lg(F(Y)) = \lg(Y)$. An exact sequence $0 \to Y' \to Y \to Y'' \to 0$ in $\mathcal{C}$ yields a sequence in $\mathcal{D}$

$$0 \to F(Y') \to F(Y) \to F(Y'') \to 0$$

which is exact on one side by the exactness property of F, and on the other side by length count.

It remains to prove the last assertion; if Y is a subobject of W we already know that $F(Y)$ is a subobject of $F(W)$ and that the map $Y \mapsto F(Y) : \mathcal{L}_W \to \mathcal{L}_{F(W)}$ is order preserving. It certainly sends the largest element W of $\mathcal{L}_W$ to the largest element $F(W)$ of $\mathcal{L}_{F(W)}$ and similarly for the smallest elements (the 0 elements). Let us verify that it preserves meets and joins. So let Y, Y' be two objects in $\mathcal{C}$. The two natural monomorphisms $Y \to Y \oplus Y', Y' \to Y \oplus Y'$, upon applying F, give a map $F(Y) \oplus F(Y') \to F(Y \oplus Y')$. If F is right exact, it is an isomorphism [**KS**, line after Prop.3.3.3]. If F is left exact, the map $F(Y \times Y') \to F(Y) \times F(Y')$ coming from the two maps $Y \times Y' \to Y$ and $Y \times Y' \to Y'$, is also an isomorphism [**KS**, Prop.3.3.3]; using the natural isomorphisms $Y \oplus Y' \to Y \times Y'$ and $F(Y) \oplus F(Y') \to F(Y) \times F(Y')$ in the abelian categories $\mathcal{C}$ and $\mathcal{D}$, we see that $F(Y) \oplus F(Y') \to F(Y \oplus Y')$ is an isomorphism too. Applying this to W and W, we see that $\lg(F(W \oplus W)) = 2\lg(F(W)) = 2\lg(W) = \lg(W \oplus W)$. Now let $f : Y \to W, f' : Y' \to W$ be subobjects of W; then applying the results obtained so far to the subobject $(f, -f') : Y \oplus Y' \to W \oplus W$ of $W \oplus W$, we see that the sequence in $\mathcal{D}$

$$0 \to F(Y \cap Y') \to F(Y \oplus Y') \to F(Y + Y') \to 0$$

is exact. But the composite $F(Y) \oplus F(Y') \to F(Y \oplus Y') \to F(Y + Y') \to F(W)$ is obtained from f, f' via F, and we see that $F(Y \cap Y') = F(Y) \cap F(Y')$ and $F(Y + Y') = F(Y) + F(Y')$. If Y, Y' satisfy $F(Y) = F(Y')$ then $F(Y + Y') = F(Y) = F(Y')$ so $\lg(Y + Y') = \lg(Y) = \lg(Y')$, which implies $Y = Y'$, hence the injectivity. □

REMARK II.8. [**KS**, Prop. 1.5.6]:
For any adjunction (F, G, η, ϵ) between two categories,
- F is fully faithful if and only if the unit ϵ is an isomorphism,
- G is fully faithful if and only if the counit η is an isomorphism,
- the following equivalent properties imply that F, G are quasi-inverses of each other:
    - F and G are fully faithful,
    - F is an equivalence,
    - G is an equivalence.

We are now ready to prove Theorem I.3 and Corollary I.4.

We prove Thm. I.3 (a). We can apply Lemma II.7 to F and W by the assumptions. As above any simple subquotient X of $F(W)$ is isomorphic to $F(Y)$ for some simple subquotient Y of W; thus we can apply Lemma II.7 to G and $F(W)$. Let Y be a subquotient of W; by induction on $\lg(Y)$ we prove now that ϵ_Y is an isomorphism. Through adjunction ϵ_Y corresponds to the identity map $F(Y) \to F(Y)$, in particular ϵ_Y is not 0 if $F(Y)$ is not 0. If Y is simple then $GF(Y)$ is simple and the non-zero map $\epsilon_Y : Y \mapsto GF(Y)$ is an isomorphism. If $\lg(Y) \geq 2$, we choose an exact sequence $0 \to Y' \to Y \to Y'' \to 0$ in $\mathcal{C}$ with non-zero Y', Y''. Applying F

then G gives a commutative diagram

$$\begin{array}{ccccccccc}
0 & \longrightarrow & Y' & \longrightarrow & Y & \longrightarrow & Y'' & \longrightarrow & 0 \\
 & & \downarrow{\scriptstyle \epsilon_{Y'}} & & \downarrow{\scriptstyle \epsilon_{Y}} & & \downarrow{\scriptstyle \epsilon_{Y''}} & & \\
0 & \longrightarrow & (G \circ F)(Y') & \longrightarrow & (G \circ F)(Y) & \longrightarrow & (G \circ F)(Y'') & \longrightarrow & 0
\end{array}$$

where the lines are exact. By induction $\epsilon_{Y'}, \epsilon_{Y''}$ are isomorphisms, and so is ϵ_Y. From Lemma II.7 we obtain injective lattice morphisms $\mathcal{L}_W \to \mathcal{L}_{F(W)}$ and $\mathcal{L}_{F(W)} \to \mathcal{L}_{GF(W)}$ whose composite coincides with $Y \mapsto \epsilon_W(Y)$, so they are both bijective and consequently lattice isomorphisms. Hence Thm. I.3 (a).

To prove Theorem I.3 (b) we "reverse the arrows" i.e. consider F and G as functors between the opposite categories to $\mathcal{C}$ and $\mathcal{D}$. Applying (a) we get a lattice isomorphism $U \mapsto G(U) : \overline{\mathcal{L}}_V \to \overline{\mathcal{L}}_{G(V)}$; then $X \mapsto G(X) : \mathcal{L}_V \to \mathcal{L}_{G(V)}$ is an isomorphism because $G(V/X)$ is isomorphic to $G(V)/G(X)$ for a subobject X of V.

By Remark II.8, if F is fully faithful then $\epsilon_Y : Y \to GF(Y)$ is an isomorphism for any object Y of $\mathcal{C}$. Thus Corollary I.4 is an immediate consequence of Theorem I.3 (a).

REMARK II.9. The referee noted that if we assume, for W of finite length in $\mathcal{C}$

(i) $F(Y)$ is simple for any simple subquotient $F(Y)$ of W,

(ii) $\lg(F(W)) = \lg(W)$ and ϵ_W is an isomorphism, then ϵ_Y is an isomorphism for any **subobject** Y of W, and $X \mapsto \epsilon_W^{-1}(G(X))$ provides a left inverse to $Y \mapsto F(Y) : \mathcal{L}_W \to \mathcal{L}_{F(W)}$.

REMARK II.10. Let $F : \mathcal{C} \to \mathcal{D}$ be a functor between abelian categories and W a finite length object of $\mathcal{C}$ satisfying:

$$X \mapsto F(X) : \mathcal{L}_W \to \mathcal{L}_{F(W)} \quad \text{is a lattice isomorphism.}$$

Then any subquotient of W satisfies the same property. Indeed, this is clear for a subobject W' of W. For any exact sequence $0 \to W_1 \to W_2 \to W_3 \to 0$ in $\mathcal{C}$ with W_2 a subobject of W, the sequence $0 \to F(W_1) \to F(W_2) \to F(W_3) \to 0$ in $\mathcal{D}$ is exact by length count. Let $\mathcal{L}_{W_2}(W_1)$ denote the lattice of subobjects Y of W_2 containing W_1. The map $Y \mapsto F(Y) : \mathcal{L}_{W_2}(W_1) \to \mathcal{L}_{F(W_2)}(F(W_1))$ is a lattice isomorphism. Taking the cokernels, it corresponds to a lattice isomorphism $Z \mapsto F(Z) : \mathcal{L}_{W_3} \to \mathcal{L}_{F(W_3)}$.

We now prove the second lattice theorem I.5.

(i) This is classical. See [**BkiA8**, §4 n°4 Prop. 3 b) and n°5 Def. 3 and Thm. 2 a)].

(ii) The first statement is obvious. Assume $b(Y) \otimes_R b(V) \subset Y \otimes_R V$ and let $y \in Y$ and $v \in V$. Any R-linear form λ on V defines a linear map $Y \otimes_R V \to Y$ sending $b(y) \otimes b(v)$ to $\lambda(b(v))b(y)$. If $b_V \neq 0$ we can choose $v \in V$ and λ such that $\lambda(b(v)) \neq 0$ and then $b(y) \in Y$.

We finally prove Corollary I.6. Clearly the lattice isomorphism $Y \mapsto Y \otimes_R V$ in Thm. I.5 (i) sends an A'-submodule of W to an A'-submodule of $W \otimes_R V$. If an A-submodule $Y \otimes_R V$ of $W \otimes_R V$ is A'-stable, then Thm. I.5 (ii) implies that Y is an A'-submodule of W because no element in $B' \setminus B$ acts by 0 on V, as every element of $B' \setminus B$ acts invertibly on V.

The structure of A'-module on W induces a structure of A'-module on $\mathrm{Hom}_A(V, W \otimes_R V)$ such that the isomorphism ϵ_W of Thm. I.5 (i) is A'-equivariant. For $f \in \mathrm{Hom}_A(V, W \otimes_R V)$, we have $b(f) = f$ if $b \in B$ as B acts by the identity on W. If $b \in B \setminus B'$, for all $w \in W$ we have $b(\epsilon_W(w)) = \epsilon_W(b(w))$, meaning that for all $v \in V$, $b(\epsilon_W(w))(v) = b(w) \otimes v = (b_{W\otimes_R V} \circ \epsilon_W(w) \circ b_V^{-1})(v)$ as b_V is invertible. Therefore $b(f) = b_{W\otimes_R V} \circ f \circ b_V^{-1}$ for all $f \in \mathrm{Hom}_A(V, W \otimes_R V)$, if $b \in B \setminus B'$.

III. Classification theorem for G

III.1. Admissibility, K-invariants, and scalar extension. In this section III, R is any field and G is a locally profinite group. An $R[G]$-module π is smooth if $\pi = \cup_K \pi^K$ with K running through the open compact subgroups of G, and is admissible if it is smooth and $\dim_R \pi^K$ is finite for all K. If π^K generates π then $\mathrm{End}_{R[G]} \pi \subset \mathrm{End}_R \pi^K$. Fix such a K for the rest of §III.1.

The category $\mathrm{Mod}_R(G)$ of $R[G]$-modules and the subcategory $\mathrm{Mod}_R^\infty(G)$ of smooth $R[G]$-modules are abelian, but not the additive subcategory $\mathrm{Mod}_R(G)^a$ of admissible $R[G]$-modules in general (when F has characteristic p). The subcategory $\mathrm{Mod}_R^K(G)$ of $R[G]$-modules π generated by π^K is additive with a generator $R[K\backslash G]$ but is not abelian in general [6]. The commutant of $R[K\backslash G]$ is the Hecke R-algebra

$$\mathrm{End}_{R[G]} R[K\backslash G] \simeq R \otimes_{\mathbb{Z}} H(G,K) = H(G,K)_R$$

(the Hecke ring $H(G,K)$ is $\mathrm{End}_{\mathbb{Z}[G]} \mathbb{Z}[K\backslash G]$). We have the abelian category $\mathrm{Mod}_R(H(G,K))$ of right $H(G,K)_R$-modules (which we also call $H(G,K)$-modules over R). The functor

$$\mathcal{T} := - \otimes_{H(G,K)} \mathbb{Z}[K\backslash G] : \mathrm{Mod}_R(H(G,K)) \to \mathrm{Mod}_R(G)$$

with image $\mathrm{Mod}_R^K(G)$, is left adjoint to the K-invariant functor $(-)^K : \mathrm{Mod}_R(G) \to \mathrm{Mod}_R(H(G,K))$.

The unit $\epsilon : \mathrm{id}_{\mathrm{Mod}_R(H(G,K))} \to (-)^K \circ \mathcal{T}$ and the counit $\eta : \mathcal{T} \circ (-)^K \to \mathrm{id}_{\mathrm{Mod}_R^K(G)}$ of the adjunction correspond to the natural maps $\mathcal{X} \xrightarrow{\epsilon_{\mathcal{X}}} \mathcal{T}(\mathcal{X})^K, \epsilon_{\mathcal{X}}(x) = x \otimes 1$ for $x \in \mathcal{X} \in \mathrm{Mod}_R(H(G,K))$ and $\mathcal{T}(\pi^K) \xrightarrow{\eta_\pi} \pi, \eta_\pi(v \otimes Kg) = gv$ for $g \in G, v \in \pi^K, \pi \in \mathrm{Mod}_R(G)$.

LEMMA III.1. *(i) If π is generated by π^K and $\dim_R \pi^K < \infty$ (in particular if π is irreducible admissible and $\pi^K \neq 0$), then $\dim_R \mathrm{End}_{R[G]} \pi$ is finite.*

(ii) Let R'/R be an extension. The adjoint functors $\mathcal{T}$, $(-)^K$, the unit ϵ and the counit η commute with scalar extension: there are natural isomorphisms

$$\mathcal{T}(\mathcal{X})_{R'} \simeq \mathcal{T}(\mathcal{X}_{R'}),\ (\pi^K)_{R'} \simeq (\pi_{R'})^K,\ (\epsilon_{\mathcal{X}})_{R'} \simeq \epsilon_{\mathcal{X}_{R'}},\ (\eta_\pi)_{R'} \simeq \eta_{\pi_{R'}}.$$

In particular, π is admissible if and only if $\pi_{R'}$ is admissible.

(iii) Let R'/R be an extension and π a smooth irreducible R-representation of G generated by π^K. Then, any subquotient π' of $\pi_{R'}$ is generated by π'^K.

PROOF. (i) and (ii) are clear. We prove (iii).

Assume that π is generated by π^K. It is clear that $\pi_{R'}$ is generated by $(\pi^K)_{R'}$, hence by $(\pi_{R'})^K = (\pi^K)_{R'}$ (Lemma III.1). Let π' be a subquotient of $\pi_{R'}$ and $\mathfrak{A} \subset \mathfrak{B}$ be right ideals of $D_{R'} = \mathrm{End}_{R'[G]}(\pi_{R'})$ such that

$$\pi' \simeq (\mathfrak{B}/\mathfrak{A}) \otimes_D \pi_{R'}$$

[6] If $\mathrm{Mod}_R^K(G)$ is abelian and G second countable, $\mathrm{Mod}_R^K(G)$ is a Grothendieck category (same proof than for $\mathrm{Mod}_R(G)$ [**Vigadjoint**, lemma 3.2])

(apply (P2) and (P3) in §II.1 to π seen as a simple $R[G]$-module). If $v \in \pi'$, then v is a finite sum $v = \sum_{x \in \mathfrak{B}/\mathfrak{A}, w \in \pi_{R'}} x \otimes w$ and each w is a finite sum $w = \sum_{g \in G, u \in \pi_{R'}^K} gu$; as $x \otimes gu = g(x \otimes u)$ and $x \otimes u \in \pi'^K$, the representation π' is generated by π'^K. $\square$

We deduce that if ϵ or η is an isomorphism of functors, then it is also true if we replace R by a subfield. Recalling Remark II.8:

LEMMA III.2. *If the K-invariant functor $(-)^K : \mathrm{Mod}_R^K(G) \to \mathrm{Mod}_R(H(G,K))$ over R is an equivalence, then it is an equivalence over any subfield R' of R. If $\pi \in \mathrm{Mod}_R^K(G)$ and π^K is defined over R', then π is defined over R'.*

REMARK III.3. Assume that R is a field of characteristic p and K is a pro-p-Iwahori subgroup. The functor $(-)^K$ of Lemma III.2 is an equivalence if $G = GL(2, \mathbb{Q}_p)$ and $p \neq 2$, or if $G = SL(2, \mathbb{Q}_p)$.

Indeed, for $GL(2, \mathbb{Q}_p)$ this is proved under the extra-hypothesis that R contains a $(p-1)$-th root of 1 ([**O**] plus [**K**]), that we can remove with Lemma III.2. For $G = SL(2, \mathbb{Q}_p)$, see [**OS**, Prop. 3.25].

III.2. Decomposition Theorem for G. Let G be a locally profinite group, R'/R a field extension and R^{alg}/R an algebraic closure. We apply Lemma II.5, Theorem I.1 and Corollary I.2 to the group ring $A = R[G]$ and to a smooth R-representation π of G, seen as an A-module V.

We keep the same notations as in §II.1. If π is a smooth irreducible R-representation of G, the scalar extension of π to R' is a smooth R'-representation $\pi_{R'}$ of G. When the commutant $D = \mathrm{End}_{R[G]} \pi$ of π has finite dimension over R, we denote E the center of D, δ the reduced degree of D over E, E'/E a finite separable field extension splitting D, L/R a normal closure of E'/R.

THEOREM III.4. *1) If $\dim_R \mathrm{End}_{R[G]} \pi$ is finite and R'/R is normal and contains L, then*

$$\pi_{R'} \simeq \oplus^{\delta} \oplus_{i \in \mathrm{Hom}_R(E_{sep}, R')} W_i'$$

has length $\delta[E:R]$, W_i' is an indecomposable smooth R'-representation of G. All irreducible subquotients of W_i' have commutant R' and have the same isomorphism class $[V_i']$; the $[V_i']$ form a single orbit under $\mathrm{Aut}_R(R')$.

The map $[\pi] \to \mathrm{Aut}_R(R^{alg})[\pi']$ where π' is an irreducible subquotient of $\pi_{R^{alg}}$, is a bijection from the set of isomorphism classes $[\pi]$ of smooth irreducible R-representations π of G with $\dim_R \mathrm{End}_{R[G]} \pi < \infty$ onto the set of $\mathrm{Aut}_R(R^{alg})$-orbit of isomorphism classes $[\pi']$ of smooth absolutely irreducible R^{alg}-representations π' of G descending to some finite extension of R.

2) If $\dim_R \mathrm{End}_{R[G]} \pi$ is finite, $\pi_{R'}$ has length $\leq \delta[E:R]$. For any non-zero subquotient π' of $\pi_{R'}$ we have $\dim_{R'} \mathrm{End}_{R'[G]} \pi' < \infty$ and π' admissible is equivalent to π admissible.

3) If R'/R is normal, a smooth irreducible R'-representation π' of G descending to a finite extension of R is isomorphic to a subrepresentation of $\pi_{R'}$ for some smooth irreducible R-representation π of G. For any such π, $\dim_R \pi$, resp. $\dim_R \mathrm{End}_{R[G]} \pi$, is finite if $\dim_{R'} \pi'$, resp. $\dim_{R'} \mathrm{End}_{R'[G]} \pi'$, is.

PROOF. 1), 3) and the first assertion of 2) follow from Lemma II.5, Theorem I.1 and Corollary I.2. Let us prove the claims about admissibility in 2). Take an algebraic closure R'^{alg} of R' containing R^{alg}. Then $\pi_{R'^{alg}} \simeq (\pi_{R'})_{R'^{alg}} \simeq (\pi_{R^{alg}})_{R'^{alg}}$

and one of the representations $\pi, \pi_{R'}, \pi_{R^{alg}}, \pi_{R'^{alg}}$ is admissible if and only if the other ones are (Lemma III.1 (ii)).

Applying 1), $\pi_{R^{alg}}$ has finite length, its irreducible subquotients are $\mathrm{Aut}_R(R^{alg})$-conjugate, isomorphic to subrepresentations and scalar extension induces a bijection from the isomorphism classes of irreducible subquotients of $\pi_{R^{alg}}$ onto those of $\pi_{R'^{alg}}$. So some irreducible subquotient of $\pi_{R^{alg}}$ is admissible if and only if π is admissible if and only if all irreducible subquotients of $\pi_{R^{alg}}$ are admissible, if and only if all irreducible subquotients of $\pi_{R'^{alg}}$ are admissible.

In a finite length representation, if all irreducible subquotients are admissible, then all subquotients are admissible. So π is admissible if and only if some non-zero subquotient of $\pi_{R'^{alg}}$ is admissible if and only if all subquotients of $\pi_{R'^{alg}}$ are admissible.

Let π' be a non-zero subquotient of $\pi_{R'}$. Then $\pi'_{R'^{alg}}$ is a non-zero subquotient of $\pi_{R'^{alg}}$. As π' is admissible if and only if $\pi'_{R'^{alg}}$ is, we deduce that π' is admissible if and only if π is admissible. □

Let K be an open compact subgroup of G, $R \subset R'$ a field extension, R^{alg} an algebraic closure of R and π an irreducible admissible R'-representation of G with $\pi^K \neq 0$. The rationality field $R[\pi]$ of π is the subfield of R' fixed by the $\mathrm{Aut}(R')$-stabilizer $H[\pi] = \{\sigma \in \mathrm{Aut}(R') \mid R' \otimes_\sigma \pi \simeq \pi\}$ of the isomorphism class $[\pi]$ of π.

PROPOSITION III.5. *(i) Any finite dimensional R^{alg}-representation of $H(G,K)$ descends to a finite extension of R, when the Hecke ring $H(G,K)$ is finitely generated (see Lemma* III.7 *below).*

(ii) If the $H(G,K)_{R'}$-module π^K descends to R, then

a) π descends to R if the pro-order of K is invertible in R.

b) π descends to the subfield of R' fixed by $\mathrm{Aut}_{R[\pi]}(R')$ if the commutant of π^K is R'.

c) π descends to a finite extension of $R[\pi]$ if R is finite and $R' = R^{alg}$.

PROOF. (i) follows from [**Viglivre**, II.4.7]: Let (e_i) be a basis of $\mathcal{M}$ and (T_j) a finite set of generators of the ring $H(G,K)$. There are finitely many elements $c_{i,j,k} \in R^{alg}$ such that $e_i T_j = \sum_k c_{i,j,k} e_k$. Let L/R be the finite extension generated by all the $c_{i,j,k}$ and $\mathcal{M}_L$ the L-vector subspace of basis (e_i). Then $\mathcal{M}_L$ is $H(G,K)$-stable and the natural map $R^{alg} \otimes_L \mathcal{M}_L \to \mathcal{M}$ is an $R^{alg}[H(G,K)]$-isomorphism.

(ii) We suppose that $\pi^K \neq 0$ descends to R; we choose an $H(G,K)_R$-stable submodule $(\pi^K)_R \subset \pi^K$ generated over R by an R'-basis of π^K; put π_R for the R-subrepresentation of π generated by $(\pi^K)_R$.

a) By assumption the pro-order of K is invertible in R, By [**Viglivre**] one can put on the space $H(G)_R$ of locally constant compactly supported functions from G to R a structure of convolution algebra such that the characteristic function $e = e_K$ of K is an idempotent; then $H(G,K)_R$ appears as $eH(G)_R e$. A smooth R-representation π of G is naturally a $H(G)_R$-module and $H(G,K)_R$ acts on $\pi^K = e\pi$ via the inclusion $eH(G)_R e \subset H(G)_R$. Since π is an irreducible admissible R'-representation of G with $\pi^K \neq 0$, π^K is a simple $H(G,K)_{R'}$-module [**Viglivre**] and π can be recovered from π^K. Indeed, following [**BK**, 4.2.3 Prop.], if $\mathcal{X}$ is a simple $H(G,K)_{R'}$-module, then $\mathcal{X} \otimes_{\mathbb{Z}} \mathbb{Z}[K\backslash G]$ has a maximal subrepresentation killed by

e, the corresponding quotient X is irreducible and the quotient map induces an $H(G,K)_{R'}$-isomorphism $\mathcal{X} \simeq Xe$. If $\mathcal{X} = \pi^K$ then $X = \pi$.

Since π^K is a simple $H(G,K)_{R'}$-module, $(\pi^K)_R$ is a simple $H(G,K)_R$-module. Applying the above procedure over R, we consider the quotient ρ of $(\pi^K)_R \otimes_{\mathbb{Z}} \mathbb{Z}[K\backslash G]$ by its maximal subrepresentation W killed by e; it is an irreducible and admissible R-representation of G. We have the exact sequence

$$0 \to R' \otimes_R W \to R' \otimes_R (\pi^K)_R \otimes_{\mathbb{Z}} \mathbb{Z}[K\backslash G] \to R' \otimes_R \rho \to 0.$$

Clearly $(R' \otimes_R W)e = 0$ and $R' \otimes_R \rho$ isomorphic to a direct sum of copies of ρ as an $R[G]$-module has no non-zero subrepresentation killed by e. It follows that $R' \otimes_R W$ is the maximal subrepresentation of $R' \otimes_R (\pi^K)_R \otimes_{\mathbb{Z}} \mathbb{Z}[K\backslash G]$ killed by e, hence $\pi \simeq R' \otimes_R \rho$ descends to R.

b) and c) The set $\{gv \mid g \in G\}$ certainly generates π as an R'-vector space, so we can extract a basis $\{g_i v \mid i \in I\}$. For $g \in G$ we express $gv = \sum_{i\in I} \lambda_i g_i v$ with unique $\lambda_i \in R'$, almost all 0. We will show:

(*) $\sigma(\lambda_i) = \lambda_i$ for all $i \in I$ and

- for all $\sigma \in \mathrm{Aut}_{R[\pi]}(R')$ if $\mathrm{End}_{R'[G]}\, \pi^K = R'$,

- for all $\sigma \in \mathrm{Aut}_L(R')$ for some finite extension $L/R[\pi]$ if R is finite and $R' = R^{alg}$.

This will imply that for all $i \in I$, λ_i lies in the subfield L of R' fixed by $\mathrm{Aut}_{R[\pi]}(R')$ if $\mathrm{End}_{R'[G]}\, \pi^K = R'$, and in a finite extension $L/R[\pi]$ if R is finite and $R' = R^{alg}$. Thus, the L-vector subspace V of π of basis $(g_i v)_{i\in I}$ is stable by G, it is an L-subrepresentation π_L of π such that the natural isomorphism $R' \otimes_L \pi_L \to \pi$ is an $R'[G]$-isomorphism.

To prove (*) it suffices to find for all σ in (*) an intertwining operator $A_\sigma : \pi \to R' \otimes_\sigma \pi$ such that $A_\sigma(v) = 1 \otimes v$. Indeed, for such an operator $A = A_\sigma$,

$$\begin{aligned} 1 \otimes gv = A(gv) &= A\Big(\sum_{i\in I} \lambda_i g_i v\Big) = \sum_{i\in I} \lambda_i A(g_i v) \\ &= \sum_{i\in I} \lambda_i (1 \otimes g_i v) = \sum_{i\in I} 1 \otimes \sigma(\lambda_i) g_i v = 1 \otimes \sum_{i\in I} \sigma(\lambda_i) g_i v; \end{aligned}$$

so $\sum_{i\in I} \lambda_i g_i v = \sum_{i\in I} \sigma(\lambda_i) g_i v$, that is, $\sigma(\lambda_i) = \lambda_i$ for all $i \in I$.

To find A_σ, we note that for $\sigma \in \mathrm{Aut}_{R[\pi]}(R')$, the natural map $f : (\pi^K)_R \to R' \otimes_\sigma \pi^K$ sending x to $1 \otimes x$ extends to an intertwining operator $\pi^K \to R' \otimes_\sigma \pi^K$.

- If $\mathrm{End}_{R'[G]}\, \pi^K = R'$, then any intertwining operator $\pi \to R' \otimes_\sigma \pi$ restricts on $(\pi^K)_R$ to a multiple of f, hence we can find A_σ. This ends the proof of (iii) in the case b).

- If R is finite and $R' = R^{alg}$, we choose a (topological) generator τ of the (pro)cyclic group $\mathrm{Aut}_{R[\pi]}(R^{alg})$ and an R'-basis of π^K contained in $(\pi^K)_R$; the restriction $A_\tau^K : \pi^K \to R^{alg} \otimes_\tau \pi^K$ of A_τ to π^K has a matrix $\mathrm{Mat}(A_\tau^K)$ on this basis. The coefficients $\mathrm{Mat}(A_\tau^K)$ are fixed by τ^m for some positive integer m. For any positive integer k, we have the intertwining operator $A_{\tau^{mk}} = (\tau^{m-1}(A_\tau) \dots \tau(A_\tau) A_\tau)^k : \pi \to R^{alg} \otimes_{\tau^{mk}} \pi$ with restriction $A_{\tau^{mk}}^K = (\tau^{m-1}(A_\tau^K) \dots \tau(A_\tau^K) A_\tau^K)^k$ to π^K of matrix $\mathrm{Mat}(A_{\tau^m}^K)^k$. As the order of $\mathrm{Mat}(A_{\tau^m}^K)$ is finite, we can choose k_o such that $\mathrm{Mat}(A_{\tau^{mk_o}}^K)$ is the identity. Then $A_{\tau^{mk_o}}(v) = 1 \otimes v$. Therefore the subfield of R^{alg} fixed by τ^{mk_o} is a finite extension $R'/R[\pi]$ such that $A_\sigma(v) = 1 \otimes v$ for all $\sigma \in \mathrm{Aut}_{R'}(R^{alg})$. This ends the proof of (iii) in the case c). ☐

REMARK III.6. If the K-invariant functor $(-)^K : \mathrm{Mod}^K_{R'}(G) \to \mathrm{Mod}_{R'}(H(G,K))$ over R' is an equivalence (Lemma III.2), then π^K descends to R if and only if π does.

III.3. The representations $I_G(P, \sigma, Q)$. Until the end of the article G is a p-adic reductive group (in the following sense).

The base field F is locally compact non-archimedean of residue characteristic p. A linear algebraic group over F is written with a boldface letter like $\mathbf{H}$, and its group of F-points by the corresponding ordinary letter $H = \mathbf{H}(F)$. We fix an arbitrary connected reductive F-group $\mathbf{G}$, a maximal F-split torus $\mathbf{T}$ in $\mathbf{G}$ and a minimal F-parabolic subgroup $\mathbf{B}$ of $\mathbf{G}$ containing $\mathbf{T}$; we write $\mathbf{Z}$ for the centralizer of $\mathbf{T}$ in $\mathbf{G}$ and $\mathbf{U}$ for the unipotent radical of $\mathbf{B}$. We denote by $\mathbf{G}^{is}$ the product of the isotropic simple components of the simply connected cover of the derived group of $\mathbf{G}$.

Let Φ^+ denote the set of roots of $\mathbf{T}$ in $\mathbf{U}$, $\Delta \subset \Phi^+$ the set of simple roots. We say that P is a **parabolic subgroup** of G and write $P = MN$ to mean that $\mathbf{P}$ is an F-parabolic subgroup of $\mathbf{G}$ containing $\mathbf{B}$, $\mathbf{M}$ the Levi subgroup containing $\mathbf{Z}$ and $\mathbf{N}$ the unipotent radical; the parabolic subgroups P of G are in bijection $P \mapsto \Delta_P = \Delta_M$ with the subsets Δ. For $J \subset \Delta$ we write $P_J = M_J N_J$ for the corresponding parabolic subgroup; for a singleton $J = \{\alpha\}$ we rather write $P_\alpha = M_\alpha N_\alpha$. We have $G = M\langle {}^G N\rangle$ for the normal subgroup $\langle {}^G N\rangle$ of G generated by N.

The image of G^{is} in G is the normal subgroup G' of G generated by U, and $G = ZG'$. Set P^{is} for the parabolic subgroup of G^{is} of image $P \cap G'$ in G.

LEMMA III.7. *Let K be an open compact subgroup of G. The Hecke ring $H(G,K) = \mathrm{End}_{\mathbb{Z}[G]} \mathbb{Z}[K\backslash G]$ is finitely generated, if K is a normal subgroup of a special parahoric subgroup of G and admits an Iwahori decomposition*[7].

PROOF. It is only proved that $\mathbb{Z}[1/p] \otimes_{\mathbb{Z}} H(G,K)$ is finitely generated in [**Viglivre**, II.2.13 Prop.].

When G is compact, the lemma is obvious as the set $K\backslash G/K$ is finite.

When G is compact modulo its centre Z_G, this is also clear as the set $K\backslash G/KZ_G$ is finite and the group $Z_G/(Z_G \cap K)$ is commutative and finitely generated. One can choose a finite set of representatives g_i such that all the double classes of G modulo K are of the form $Kg_i zK$ for $z \in Z_G$ and representatives z_j of a finite set of generators of $Z_G/(Z_G \cap K)$. The product of Kg_iK and of $Kz_jK = Kz_j = z_jK$ is Kg_iz_jK, and the ring $H(G,K)$ is generated by the Kg_iz_jK.

For G general, the same arguments imply that the ring $H(Z^+, K\cap Z)$ is finitely generated (Z^+ is the positive monoid cf.§IV.1). When K has an Iwahori decomposition and is a normal subgroup of a special parahoric subgroup K_0 of G, the map $(K \cap Z)z(K \cap Z) \mapsto KzK : H(Z^+, K \cap Z) \to H(Z^+, K)$ is a ring embedding of image the subring of $H(G,K)$ generated by the elements KzK for $z \in Z^+$ [**VigSelecta**, II.4], and moreover the Cartan decomposition [**HV1**, 6.3 Prop.] implies $H(G,K) = H(K_0,K)H(Z^+,K)H(K_0,K)$ [**Viglivre**, II.2.13 Prop.]. Thus, the ring $H(G,K)$ is finitely generated. □

[7] K is called " bien placé par rapport à (B, Z, U)" in [**Viglivre**, II.1.3 (vi)]

REMARK III.8. If K is an Iwahori or a pro-p Iwahori subgroup of G, then $H(G, K)$ is a finite module over its centre and the centre is finitely generated [**VigpIwc**].

Until the end of the article R is a field of characteristic p. We are interested in irreducible admissible R-representations of G.

For a parabolic group $P = MN$ of G, the smooth parabolic induction functor $\mathrm{Ind}_P^G : \mathrm{Mod}_R^\infty(M) \to \mathrm{Mod}_R^\infty(G)$ is fully faithful, and admits a left adjoint L_P^G and a right adjoint R_P^G [**Vigadjoint**]. The right adjoint R_P^G respects admissibility[**AHenV1**, Cor. 4.13] hence is equal on admissible representation to the Emerton's $\overline{P}$-ordinary part functor $\mathrm{Ord}_{\overline{P}}^G$ where $\overline{\mathbf{P}}$ is the opposite of $\mathbf{P}$ with respect to $\mathbf{B}$ [**Eme**, 3.1.9 Definition].

For a pair of parabolic subgroups $Q \subset P$ of G, write Ind_Q^M for $\mathrm{Ind}_{Q\cap M}^M$ and consider the Steinberg R-representation $\mathrm{St}_Q^M(R)$ of M, quotient of $\mathrm{Ind}_Q^M(R)$ (R stands for the trivial R-representation $\mathrm{Triv}_{Q\cap M}$ of $Q \cap M$) by the sum $\sum_{Q'} \mathrm{Ind}_{Q'}^M(R)$, Q' running through the parabolic subgroups of G with $Q \subsetneq Q' \subset P$. The R-representation $\mathrm{St}_Q^M(R)$ of M is absolutely irreducible and admissible [**Ly**], and $\mathrm{St}_Q^M(R) \simeq R \otimes_{\mathbb{Z}} \mathrm{St}_Q^M$ where $\mathrm{St}_Q^M = \mathrm{St}_Q^M(\mathbb{Z})$.

Writing $P_2 = M_2N_2$ for the parabolic subgroup corresponding to $\Delta_P \setminus \Delta_Q$, the inflation to M_2^{is} of the restriction of St_Q^M to M_2' is $\mathrm{St}_{(Q\cap M_2)^{is}}^{M_2^{is}}(R)$ ([**AHHV**, II.8 Proof of Proposition and Remark] when R is algebraically closed, but the proofs do not use this hypothesis). Therefore the action of M_2' on $\mathrm{St}_Q^M(R)$ is absolutely irreducible.

To an R-representation σ of M are associated the following parabolic subgroups of G:

a) $P_\sigma = M_\sigma N_\sigma$ corresponding to the set Δ_σ of $\alpha \in \Delta \setminus \Delta_M$ such that $Z \cap M'_\alpha$ acts trivially on σ.

b) $P(\sigma) = M(\sigma)N(\sigma)$ corresponding to $\Delta(\sigma) = \Delta_P \cup \Delta_\sigma$. By [**AHHV**, II.7 Proposition and Remark 2] which remain valid when R is not algebraically closed, there exists an extension $e(\sigma)$ to $P(\sigma)$ of σ trivial on N; we write also $e(\sigma)$ for its restriction to $M(\sigma)$. For $P \subset Q \subset P(\sigma)$, the generalized Steinberg representation $\mathrm{St}_Q^{M(\sigma)}(\sigma)$ of $M(\sigma)$ defined in §I (0.2), is admissible and isomorphic to $e(\sigma) \otimes_{\mathbb{Z}} \mathrm{St}_Q^{M(\sigma)}$.

c) $P_{min} = M_{min}N_{min} \subset P$ the smallest parabolic subgroup of G such that σ is extended from an R-representation σ_{min} of M_{min} trivially on $N_{min} \cap M$ [**AHenV1**, Lemma 2.9]. Then $\Delta(\sigma_{min}) = \Delta(\sigma)$, $e_Q(\sigma) = e_Q(\sigma_{min})$, and $\Delta_{\sigma_{min}}$, $\Delta_{\sigma_{min}} \setminus \Delta_{P_{min}}$ are orthogonal [**AHenV1**, Lemma 2.10]. This implies that $M(\sigma) = M_{min}M'_{\sigma_{min}}$, M_{min} normalizes $M'_{\sigma_{min}}$, and that $e(\sigma)$ is trivial on $M'_{\sigma_{min}}$.

DEFINITION III.9. An R-triple (P, σ, Q) of G consists of a parabolic subgroup $P = MN$ of G, a smooth R-representation σ of M, and a parabolic subgroup Q of G with $P \subset Q \subset P(\sigma)$. The smooth R-representation of G defined by an R-triple (P, σ, Q) of G is

$$I_G(P, \sigma, Q) = \mathrm{Ind}_{P(\sigma)}^G(\mathrm{St}_Q^{M(\sigma)}(\sigma)).$$

The representation $I_G(P, \sigma, Q)$ is equal to $I_G(P_{min}, \sigma_{min}, Q)$ [**AHenV1**, Lemma 2.11]; it is admissible when σ is admissible [**AHenV1**, Thm.4.21].

PROPOSITION III.10. *Let (P, σ, Q) be an R-triple of G such that, σ is admissible of finite length, $P(\sigma) = P(\tau)$ and $I_G(P, \tau, Q)$ is irreducible for each irreducible subquotient τ of σ. Then $P(\sigma) = P(\sigma')$ for any non-zero subrepresentation σ' of σ, and the map $\sigma' \mapsto I_G(P, \sigma', Q) : \mathcal{L}_\sigma \to \mathcal{L}_{I_G(P,\sigma,Q)}$ is a lattice isomorphism.*

PROOF. Clearly $P(\sigma) \subset P(\sigma')$. As σ' has finite length, it contains an irreducible subrepresentation τ. From $P(\sigma) \subset P(\sigma') \subset P(\tau)$ and $P(\sigma) = P(\tau)$, we get $P(\sigma) = P(\sigma')$.

We are in the situation of Corollary I.6 for $A = R[M'_\sigma] \subset A' = R[M(\sigma)]$ and the $R[M(\sigma)]$-modules $W = e(\sigma)$ and $V = \mathrm{St}_Q^{M(\sigma)}(R)$, with the basis $B = M'_\sigma$ of A acting by the identity on W and the basis $B' = M(\sigma)$ of A' acting invertibly on V. Applying Cor.I.6, the natural maps

$$e(\sigma) \to \mathrm{Hom}_{R[M'_\sigma]}(\mathrm{St}_Q^{M(\sigma)}(R), \mathrm{St}_Q^{M(\sigma)}(\sigma)),$$
$$\mathrm{Hom}_{R[M'_\sigma]}(\mathrm{St}_Q^{M(\sigma)}(R), \mathrm{St}_Q^{M(\sigma)}(\sigma)) \otimes_R \mathrm{St}_Q^{M(\sigma)}(R) \to \mathrm{St}_Q^{M(\sigma)}(\sigma)$$

are $R[M(\sigma)]$-isomorphisms and $\sigma' \mapsto \mathrm{St}_Q^{M(\sigma)}(\sigma') : \mathcal{L}_\sigma \to \mathcal{L}_{\mathrm{St}_Q^{M(\sigma)}(\sigma)}$ is a lattice isomorphism. In particular, $\mathrm{St}_Q^{M(\sigma)}(\sigma)$ has finite length, $\lg(\mathrm{St}_Q^{M(\sigma)}(\sigma)) = \lg(\sigma)$, and the irreducible subquotients $\mathrm{St}_Q^{M(\sigma)}(\sigma)$ are $\mathrm{St}_Q^{M(\sigma)}(\tau)$ for the irreducible subquotients τ of σ. As $I_G(P, \tau, Q)$ is irreducible and equal to $\mathrm{Ind}_{P(\sigma)}^G \mathrm{St}_Q^{M(\sigma)}(\tau))$ for each τ, we are in the situation of Corollary I.4 for the fully faithful exact functor $F = \mathrm{Ind}_{P(\sigma)}^G : \mathrm{Mod}_R(M(\sigma)) \to \mathrm{Mod}_R(G)$ having a right adjoint $G = R_P^G$, and $W = \mathrm{St}_Q^{M(\sigma)}(\sigma)$. We deduce that the map $\sigma' \mapsto I_G(P, \sigma', Q) : \mathcal{L}_\sigma \to \mathcal{L}_{I_G(P,\sigma,Q)}$ is a lattice isomorphism. □

REMARK III.11. $I_G(P, \sigma, Q)$ determines the isomorphism class of $e(\sigma)$ because

$$e(\sigma) \simeq \mathrm{Hom}_{R[M'_\sigma]}(\mathrm{St}_Q^{P(\sigma)}(R), R_{P(\sigma)}^G(I_G(P, \sigma, Q)))$$

(proof of Prop. III.10 and $R_{P(\sigma)}^G(I_G(P, \sigma, Q)) \simeq \mathrm{St}_Q^{P(\sigma)}(\sigma)$).

Let R' be a field containing R. Scalar extension from R to R' commutes with the different steps in the construction of $I_G(P, \sigma, Q)$:

PROPOSITION III.12. *(i) The parabolic induction functor Ind_P^G commutes with the scalar restriction from R' to R and with the scalar extension from R to R'. The left adjoint L_P^G (resp. right adjoint R_P^G) of the parabolic induction commutes with scalar extension (resp. restriction).*

(ii) If $\pi \in \mathrm{Mod}_R^\infty(G)$ is such that $\pi_{R'} \simeq \mathrm{Ind}_P^G(\sigma')$ with $\sigma' \in \mathrm{Mod}_{R'}^\infty(M)$, then σ' is isomorphic to $(L_P^G \pi)_{R'}$.

PROOF. (i) Choosing a continuous section $P\backslash G \to G$, $\mathrm{Ind}_P^G \sigma$ identifies with $\sigma \otimes_{\mathbb{Z}} C_c^\infty(P\backslash G, \mathbb{Z})$ as an R-module [**AHenV1**]; this implies the first assertions, and the next sentence follows by adjunction. Part (ii) follows because Ind_P^G is fully faithful. □

PROPOSITION III.13. [Strong compatibility of $I_G(P,-,Q)$ with scalar extension]

(i) Let (P,σ,Q) be an R-triple of G. Then

$P(\sigma)=P(\sigma_{R'})$, $(P,\sigma_{R'},Q)$ is an R'-triple of G, and if σ is irreducible and σ' a non-zero subquotient of $\sigma_{R'}$, then $P(\sigma)=P(\sigma')$. Moreover,
$(e(\sigma))_{R'}=e(\sigma_{R'})$, $(\mathrm{St}_Q^{P(\sigma)}(\sigma))_{R'}\simeq\mathrm{St}_Q^{P(\sigma_{R'})}(\sigma_{R'})$ and $I_G(P,\sigma,Q)_{R'}\simeq I_G(P,\sigma_{R'},Q)$.

(ii) Let (P,σ,Q) be an R'-triple of G. If $e(\sigma)$ or $\mathrm{St}_Q^{P(\sigma)}(\sigma)$ or $I_G(P,\sigma,Q)$ descends to R, then σ descends to R.

Precisely, if $e(\sigma)=\tau_{R'}$ or $\mathrm{St}_Q^{P(\sigma)}(\sigma)=\rho_{R'}$ or $I_G(P,\sigma,Q)=\pi_{R'}$ for R-representations τ of $M(\sigma)$ or ρ of $M(\sigma)$ or π of G,

then σ is the scalar extension of the natural R-representation of M on τ, or $\mathrm{Hom}_{R[M'_\sigma]}(\mathrm{St}_Q^{P(\sigma)}(R),\rho)$, or $\mathrm{Hom}_{R[M'_\sigma]}(\mathrm{St}_Q^{P(\sigma)}(R),L^G_{P(\sigma)}\pi)$.

PROOF. (i) $\sigma_{R'}$ is a direct sum of $R[M]$-modules isomorphic to σ. If σ is irreducible, any subquotient σ' of $\sigma_{R'}$ is σ-isotypic. For $\alpha\in\Delta-\Delta_P$, $Z\cap M'_\alpha$ acts trivially on an $R'[M]$-module τ if and only if it acts trivially on τ seen as an $R[M]$-module. So $P(\sigma)=P(\sigma_{R'})$ (hence $(P,\sigma_{R'},Q)$ is an R'-triple of G), and if σ is irreducible $P(\sigma)=P(\sigma')$. It is clear from the definition that the extension commutes with scalar extension $R'\otimes_R e(\sigma)=e(R'\otimes_R\sigma)$. The scalar extension of $\mathrm{St}_Q^{P(\sigma)}(\sigma)=e(\sigma)\otimes_{\mathbb{Z}}\mathrm{St}_Q^{P(\sigma)}$ from R to R' is $R'\otimes_R\mathrm{St}_Q^{P(\sigma)}(\sigma)=R'\otimes_R e(\sigma)\otimes_{\mathbb{Z}}\mathrm{St}_Q^{P(\sigma)}\simeq e(R'\otimes_R\sigma)\otimes_{\mathbb{Z}}\mathrm{St}_Q^{P(\sigma)}\simeq\mathrm{St}_Q^{P(\sigma)}(\sigma_{R'})=\mathrm{St}_Q^{P(\sigma_{R'})}(\sigma_{R'})$. The scalar extension of $I_G(P,\sigma,Q)=\mathrm{Ind}^G_{P(\sigma)}(\mathrm{St}_Q^{P(\sigma)}(\sigma))$ from R to R' is $R'\otimes_R I_G(P,\sigma,Q)\simeq\mathrm{Ind}^G_{P(\sigma)}(\mathrm{St}_Q^{P(\sigma)}(\sigma_{R'}))=\mathrm{Ind}^G_{P(\sigma_{R'})}(\mathrm{St}_Q^{P(\sigma_{R'})}(\sigma_{R'}))=I_G(P,\sigma_{R'},Q)$.

(ii) If $I_G(P,\sigma,Q)=\pi_{R'}$, we have $\mathrm{St}_Q^{P(\sigma)}(\sigma)\simeq(L^G_{P(\sigma)}\pi)_{R'}$ (Proposition III.12 (ii)).

If $\mathrm{St}_Q^{P(\sigma)}(\sigma)\simeq\rho_{R'}$, then $e(\sigma)\simeq\mathrm{Hom}_{R'[M'_\sigma]}(\mathrm{St}_Q^{P(\sigma)}(R'),\rho_{R'})$ (Remark III.11); as $\mathrm{St}_Q^{P(\sigma)}(R')=\mathrm{St}_Q^{P(\sigma)}(R)_{R'}$, is irreducible, $\mathrm{Hom}_{R'[M'_\sigma]}(\mathrm{St}_Q^{P(\sigma)}(R'),\rho_{R'})\simeq\mathrm{Hom}_{R[M'_\sigma]}(\mathrm{St}_Q^{P(\sigma)}(R),\rho)_{R'}$ (Remark II.2).

If $e(\sigma)\simeq\tau_{R'}$ then $\sigma\simeq(\tau|_M)_{R'}$ because the restriction to M commutes with scalar extension. □

III.4. Supersingular representations.

We keep the notations of §III.3. When R is algebraically closed and π is an irreducible admissible R-representation of G, in [**AHHV**] the definition of supersingularity uses the Hecke algebras defined by the irreducible smooth R-representations of the special parahoric subgroups of G. Two equivalent simpler criterions using the pro-p Iwahori Hecke R-algebra of G are given in [**OV**, Thm. 5.3]. We will use these equivalent criterions to extend the definition of supersingularity to the situation where R is not algebraically closed, and π is a non-zero smooth representation generated by its pro-p Iwahori invariants.

Let I be a pro-p Iwahori subgroup of G **compatible with** B, so that $I\cap M$ is a pro-p Iwahori subgroup of M for any parabolic subgroup $P=MN$ (we recall that P contains $B=ZU$ and M contains Z). Let Z_0 be the unique parahoric subgroup of Z and Z_1 the pro-p Sylow subgroup of Z_0. We defined in §III.1 the pro-p Iwahori Hecke ring $H(G,I)=H(G)$, the pro-p Iwahori Hecke R-algebra $H(G)_R$ and the categories $\mathrm{Mod}_R(H(G))$ and $\mathrm{Mod}_R^\infty(G)$. The elements in $H(G)$ with support in G'

form a subring $H(G')$ normalized by a subring of $H(G)$ isomorphic to $\mathbb{Z}[\Omega]$ for a commutative finitely generated subgroup Ω, $H(G)$ is the product of $H(G')$ by $\mathbb{Z}[\Omega]$ and

$$H(G') \cap \mathbb{Z}[\Omega] \simeq \mathbb{Z}[Z'_k], \; Z'_k = (Z_0 \cap G')/(Z_1 \cap G').$$

To M is associated a certain element T_M in $H(G')$ which is central in $H(G)$ [**Vig17**].

DEFINITION III.14. 1. An non-zero element v in a right $H(G)_R$-module is called supersingular if $vT_M^n = 0$ for all $M \neq G$ and some positive integer n. A non-zero $H(G)_R$-module is called supersingular if its non-zero elements are supersingular.

2. A non-zero smooth R-representation π of G generated by π^I is called supersingular if the right $H(G)_R$-module π^I is supersingular.

Any non-zero R-representation of G has a non-zero I-invariant vector, as the characteristic of R is p, hence any irreducible smooth R-representation π of G is generated by π^I. As explained above, when π is irreducible admissible and R algebraically closed, our definition of supersingularity is equivalent to the definition given in [**AHHV**] by [**OV**, Thm. 5.3].

REMARK III.15. 1. Let $0 \to \mathcal{V}' \to \mathcal{V} \to \mathcal{V}'' \to 0$ be an exact sequence of $H(G)_R$-modules. Then $\mathcal{V}$ is supersingular if and only if $\mathcal{V}'$ and $\mathcal{V}''$ are supersingular.

2. When R contains a root of unity of order the exponent of $Z_k = Z_0/(Z_0 \cap I)$, the simple supersingular $H(G)_R$-modules are classified [**Vig17**, Thm. 6.18]; as $H(G')_R$-modules, they are sums of supersingular characters.

3. The group $\mathrm{Aut}(R)$ of automorphisms of R acts on $\mathrm{Mod}_R(G)$ and on $\mathrm{Mod}_R(H(G))$. Clearly, the action of $\mathrm{Aut}(R)$ commutes with the I-invariant functor, and respects supersingularity, irreducibility, and admissibility.

Supersingularity commutes with scalar extension:

LEMMA III.16. *Let R'/R an extension.*

1) A (right) $H(G)_R$-module $\mathcal{X}$ is supersingular if and only if $\mathcal{X}_{R'}$ is; a smooth R-representation π of G generated by π^I is supersingular if and only if $\pi_{R'}$ is.

2) Let π be a smooth irreducible R-representation π of G with $\dim_R \mathrm{End}_{R[G]} \pi < \infty$ and π' be a non-zero subquotient of $\pi_{R'}$. Then π is supersingular if and only if π' is supersingular.

PROOF. 1) In $\mathcal{X}_{R'} = R' \otimes_R \mathcal{X}$, we have $(r' \otimes x)T_M = r' \otimes x T_M$ for $r' \in R', x \in \mathcal{X}$; clearly the non-zero elements of $\mathcal{X}_{R'}$ are supersingular if and only if the non-zero elements of $\mathcal{X}$ are supersingular.

If π is generated by π^I, then $\pi_{R'}$ is generated by $\pi^I_{R'} = (\pi^I)_{R'}$ (Lemma III.1 (iii)). By the previous case, π is supersingular if and only if $\pi_{R'}$ is.

2) Any non-zero subquotient π' of $\pi_{R'}$ is generated by π'^I because π is (Lemma III.1 (iii)). The proof that π is supersingular if and only if π' is supersingular is the same as for admissible. Applying Thm. III.4 2) and Remark III.15, we can replace "admissible" by "supersingular" in the proof of Thm. III.4 3). □

As an application, supersingularity for an irreducible admissible R-representation of G can be detected on a weaker property, as in the case where R is algebraically closed:

THEOREM III.17. *Let π be an irreducible admissible R-representation of G. Then π is supersingular if and only if π^I contains a non-zero supersingular element.*

PROOF. Suppose that π^I contains a non-zero supersingular element. By Lemma III.1, $(\pi^I)_{R^{alg}} = (\pi_{R^{alg}})^I$. By Lemma III.16 and [**OV**, Thm. 5.3], $(\pi^I)_{R^{alg}}$ is supersingular. By Thm III.4, $\pi_{R^{alg}}$ has finite length. The irreducible subrepresentations of $\pi_{R^{alg}}$ are supersingular. By Lemma III.16, π is supersingular. The converse is obvious. □

REMARK III.18. *The scalar extension to R^{alg} of a R-triple (P,σ,Q) of G where σ is irreducible admissible supersingular, is an R^{alg}-triple $(P,\sigma_{R^{alg}},Q)$ of G satisfying the hypotheses of Proposition* III.10: the irreducible subquotients τ of $\sigma_{R^{alg}}$ are supersingular (Lemma III.16), $P(\tau) = P(\sigma) = P(\sigma_{R^{alg}})$ (Prop.III.13 (i)), and $I_G(P,\tau,Q)$ is irreducible (Classification theorem for G over R^{alg} [**AHHV**]).

III.5. Classification of irreducible admissible R-representations of G. We prove in this section the classification theorem for G (Thm. I.7). The arguments are formal and rely on:

1 The decomposition theorem for G (Thm.III.4).

2 The classification theorem for G (Thm.I.7) over an algebraic closure R^{alg} of R [**AHHV**].

3 The compatibility of scalar extension from R to R^{alg} with supersingularity (Lemma III.16) and the strong compatibility with $I_G(P,-,Q)$ (Prop.III.13).

4 The lattice isomorphism $\mathcal{L}_{\sigma_{R^{alg}}} \to \mathcal{L}_{I_G(P,\sigma_{R^{alg}},Q)}$ for the scalar extension $\sigma_{R^{alg}}$ to R^{alg} of an irreducible admissible supersingular R-representation σ (Prop.III.10 and Rem.III.18).

We start the proof with an R-triple $(P = MN,\sigma,Q)$ be of G with σ irreducible admissible supersingular. We show that $I_G(P,\sigma,Q)$ is irreducible. By the decomposition theorem for M, $\sigma_{R^{alg}}$ has finite length, $I_G(P,\sigma_{R^{alg}},Q)$ also by the lattice isomorphism $\mathcal{L}_{\sigma_{R^{alg}}} \to \mathcal{L}_{I_G(P,\sigma_{R^{alg}},Q)}$, and $I_G(P,\sigma,Q)_{R^{alg}} \simeq I_G(P,\sigma_{R^{alg}},Q)$ by compatibility of the scalar extension with $I_G(P,-,Q)$; as the scalar extension is faithful and exact, $I_G(P,\sigma,Q)$ has also finite length. Let π be an irreducible R-subrepresentation of $I_G(P,\sigma,Q)$. As $I_G(P,\sigma,Q)$ is admissible, π is admissible. The scalar extension $\pi_{R^{alg}}$ is isomorphic to a subrepresentation of $I_G(P,\sigma,Q)_{R^{alg}} \simeq I_G(P,\sigma_{R^{alg}},Q)$. By the lattice isomorphism $\mathcal{L}_{\sigma_{R^{alg}}} \to \mathcal{L}_{I_G(P,\sigma_{R^{alg}},Q)}$, $\pi_{R^{alg}} \simeq I_G(P,\rho,Q)$ for a subrepresentation ρ of $\sigma_{R^{alg}}$. The representation ρ descends to R because $I_G(P,\rho,Q)$ does, by the strong compatibility of $I_G(P,-,Q)$ with scalar extension. But $\sigma_{R^{alg}}$ has no proper subrepresentation descending to R by the decomposition theorem for G, so $\rho = \sigma_{R^{alg}}$ and $\pi_{R^{alg}} = I_G(P,\sigma_{R^{alg}},Q) \simeq I_G(P,\sigma,Q)_{R^{alg}}$, or equivalently, $\pi \simeq I_G(P,\rho,Q)$.

Next, let (P,σ,Q) and (P_1,σ_1,Q_1) be two R-triples of G with σ,σ_1 irreducible admissible supersingular and $I_G(P,\sigma,Q) \simeq I_G(P_1,\sigma_1,Q_1)$. By scalar extension $I_G(P,\sigma_{R^{alg}},Q) \simeq I_G(P_1,(\sigma_1)_{R^{alg}},Q_1)$. The classification theorem over R^{alg} implies $P = P_1, Q = Q_1$ and some irreducible subquotient σ^{alg} of $\sigma_{R^{alg}}$ is isomorphic to some irreducible subquotient σ_1^{alg} of $(\sigma_1)_{R^{alg}}$. As R-representations of G, σ^{alg} is σ-isotypic and σ_1^{alg} is σ_1-isotypic, hence σ,σ_1 are isomorphic.

Finally, let π be an arbitrary irreducible admissible R-representation of G. By the decomposition theorem for G, its scalar extension $\pi_{R^{alg}}$ has finite length; we choose an irreducible subrepresentation π^{alg} of $\pi_{R^{alg}}$. By the decomposition theorem for G, π^{alg} is admissible, descends to a finite extension of R. By the classification theorem over R^{alg},

$$\pi^{alg} \simeq I_G(P, \sigma^{alg}, Q)$$

for an R^{alg}-triple $(P = MN, \sigma^{alg}, Q)$ of G with σ^{alg} irreducible admissible supersingular. By the strong compatibility of $I_G(P, -, Q)$ with scalar extension, σ^{alg} descends to a finite extension of R. By the decomposition theorem for M, σ^{alg} is contained in the scalar extension $\sigma_{R^{alg}}$ of an irreducible admissible R-representation σ. By compatibility of scalar extension with supersingularity and $I_G(P, -, Q)$, (P, σ, Q) is an R-triple of G, σ is supersingular and $I_G(P, \sigma_{R^{alg}}, Q) \simeq I_G(P, \sigma, Q)_{R^{alg}}$. By the lattice isomorphism $\mathcal{L}_{\sigma_{R^{alg}}} \to \mathcal{L}_{I_G(P, \sigma_{R^{alg}}, Q)}$, $I_G(P, \sigma^{alg}, Q)$ is contained in $I_G(P, \sigma_{R^{alg}}, Q)$. The irreducible representation π^{alg} is isomorphic to an irreducible subrepresentation of $I_G(P, \sigma, Q)_{R^{alg}}$. The decomposition theorem for G implies that

$$\pi \simeq I_G(P, \sigma, Q).$$

This ends the proof of the classification theorem for G (Theorem I.7).

IV. Classification theorem for $H(G)$

Let R be a field of characteristic p and G a p-adic reductive group, as in §III.3. Let I be a pro-p Iwahori subgroup of G compatible with B, $H(G)$ the pro-p Iwahori Hecke ring, $H(G)_R = R \otimes_{\mathbb{Z}} H(G)$, Z_1 the pro-p Sylow of the unique parahoric subgroup Z_0 of Z and $Z_k = Z_0/Z_1$, as in §III.4.

In this section we prove results analogous to those of Section §III but for right $H(G)_R$-modules. Although the I-invariant functor and its left adjoint relate R-representations of $H(G)$ and G, the relation in characteristic p is weaker than in the complex case and does not permit to deduce the case of the pro-p Iwahori Hecke algebra from the case of the group: similar results for $H(G)$ and G have to be proved separately.

IV.1. Pro-p Iwahori Hecke ring. The center $Z(H(G))$ of the pro-p Iwahori Hecke ring $H(G)$ is a finitely generated subring and $H(G)$ is a finitely generated module over its center; the same is true for the center of $H(G)_R$ [**VigpIwc**]. This implies that the dimension over R of a simple $H(G)_R$-module is finite [**Hn**, 2.8 Prop.].

Let $P = MN$ be a parabolic subgroup of G. The pro-p Iwahori Hecke ring $H(M)$ of M for the pro-p Iwahori subgroup $I \cap M$ does not embed in the ring $H(G)$. However we are in the good situation where $H(M)$ is a localization of a subring $H(M^+)$ (of elements supported in the positive monoid $M^+ := \{m \in M \mid m(I \cap N)m^{-1} \subset I \cap N\}$) which embeds in $H(G)$. We explain this in more detail after introducing more notations than in §III.3 and §III.4; our main reference is [**VigpIw**].

An upper or lower index M indicates an object defined for M; for G we suppress the index. We write $\mathcal{N}_M$ for the F-points of the normalizer of $\mathbf{T}$ in $\mathbf{M}$, $\mathbb{W}_M = \mathcal{N}_M/Z$, $W_M = \mathcal{N}_M/Z_1$, $W_{M'}$ for the image of $M' \cap \mathcal{N}_M$ in W_M, $\Lambda = Z/Z_1$, $\lg_M$ for the length of W_M, Ω_M for the image in W_M of the $\mathcal{N}_M$-normalizer of $(I \cap M)$; Ω_M is also the set of $u \in W_M$ of length $\lg_M(u) = 0$ (the group $\Omega = \Omega_G$ was introduced in §III.4).

The natural map $W_M \to (I \cap M)\backslash M/(I \cap M)$ is bijective, $W_{M'}$ is a normal subgroup W_M and a quotient of $W_{M^{is}}$ (via the quotient map $M^{is} \to M'$), and we have $W_M = W_{M'}\Omega_M, W_{M'} \cap \Omega_M = W_{M'} \cap Z_k$.

For $m \in M$ and $w = w(m) \in W_M$ image of $m_1 \in \mathcal{N}_M$ such that $(I \cap M)m(I \cap M) = (I \cap M)m_1(I \cap M)$ (denoted also $(I \cap M)w(I \cap M)$), the characteristic function of $(I \cap M)m(I \cap M)$ seen as an element of $H(M)$ is written $T^M(m)$ or $T^M(w)$; we have also $T^{M,*}(m) = T^{M,*}(w)$ in $H(M)$ defined by $T^{M,*}(w)T^M(w^{-1}) = [(I \cap M)w(I \cap M) : (I \cap M)]$ [**VigpIw**, Prop.4.13]. For $u \in \Omega_M$, $T^{M,*}(u) = T^M(u)$ is invertible of inverse $T^M(u^{-1})$. The $\mathbb{Z}$-module $H(M)$ is free with a natural basis $(T^M(w))_{w \in W_M}$, and another basis $(T^{M,*}(w))_{w \in W_M}$, called the $*$-basis. The $\mathbb{Z}$-submodule of basis $(T^M(u) = T^{M,*}(u))_{u \in Z_k}$ is the subring $H(Z_0 \cap M)$ of elements supported on Z_0. The relations satisfied by the natural basis and the $*$-basis are the braid relations for $w_1, w_2 \in W_M$ such that $\lg_M(w_1w_2) = \lg_M(w_1) + \lg_M(w_2)$:

$$T^M(w_1)T^M(w_2) = T^M(w_1w_2), \quad T^{M,*}(w_1)T^{M,*}(w_2) = T^{M,*}(w_1w_2),$$

and the quadratic relations with a change of sign for $s \in W_{M'}, \lg_M(s) = 1$:

$$T^M(s)^2 = q_s + c_sT^M(s), \quad T^{M,*}(s)^2 = q_s - c_sT^{M,*}(s)$$

where $q_s = [(I \cap M)s(I \cap M) : (I \cap M)]$ and $c_s \in H(Z_0 \cap M')$ the subring of elements supported on $Z_0 \cap M'$, satisfy the congruences $q_s \equiv 0$ modulo p and $c_s \equiv -1$ modulo the ideal of $H(Z_0 \cap M')$ generated by p and $T(u) - 1$ for $u \in Z_k \cap W_{M'}$ [**VigpIw**]. Both q_s and c_s do not depend on M but $\lg_M$ depends on M. The quotient map $W_{M^{is}} \to W_{M'}$ respects the length and the coefficients of the quadratic relations, the surjective natural linear map from $H(M^{is})$ to the subring $H(M')$ of elements supported on M', is a ring homomorphism sending $T^{M^{is}}(w)$ to $T^M(w')$ and $T^{M^{is},*}(w)$ to $T^{M,*}(w')$ if $w' \in W_{M'}$ is the image of $w \in W_{M^{is}}$.

The injective linear maps associated to the bases

$$T^M(m) \mapsto T(m) : H(M) \xrightarrow{\theta_M^G} H(G), \quad T^{M,*}(m) \mapsto T^*(m) : H(M) \xrightarrow{\theta_M^{G,*}} H(G),$$

generally do not respect the product but their restrictions to the subrings $H(M^+)$ and $H(M^-)$ (of elements supported on the inverse monoid M^- of M^+) do.

REMARK IV.1. 1. For $P = MN \subset Q = M_QN_Q$, we have inclusions for $\epsilon \in \{+, -\}$:

$$M^\epsilon \subset M_Q^\epsilon, \quad \theta_M^G(H(M^\epsilon)) \subset \theta_{M_Q}^G(H(M_Q^\epsilon)), \quad \theta_M^{G,*}(H(M^\epsilon)) \subset \theta_{M_Q}^{G,*}(H(M_Q^\epsilon)).$$

2. When Δ_M and $\Delta \setminus \Delta_M$ are orthogonal, the situation is simpler. For $P_2 = M_2N_2$ the parabolic subgroup of G corresponding to $\Delta \setminus \Delta_M$:

G' is the direct product of M' and of M_2', $G = MM_2'$, $W' = W_{M_2'}W_{M'}$, $W_{M_2'} \cap W_{M'}\Omega = W_{M_2'} \cap Z_k$, $W = W_{M'}W_{M_2'}\Omega$ and for $w \in W_{M'}, w_2 \in W_{M_2'}, u \in \Omega$, $\lg(ww_2u) = \lg_M(w) + \lg_{M_2}(w_2)$. The braid and quadratic relations satisfied by $T(w) = T^G(w)$ for $w \in W_M$ are the same as for $T^M(w)$, the same is true for $T(w)$ and for M_2. Moreover, $\theta_M^G = \theta_M^{G,*}$, $M' \subset M^+ \cap M^-$ and $H(M') \times H(M_2') \xrightarrow{\theta_M^G \times \theta_{M_2}^G} H(G')$ is a ring isomorphism.

IV.2. Parabolic induction $\mathrm{Ind}_P^{H(G)}$. For a parabolic subgroup $P = MN$ of G, the parabolic inductions for the pro-p Iwahori Hecke rings and for the groups

$$\mathrm{Ind}_P^{H(G)} := - \otimes_{H(M^+),\theta_M^G} H(G) : \mathrm{Mod}_R(H(M)) \to \mathrm{Mod}_R(H(G)),$$
$$\mathrm{Ind}_P^G : \mathrm{Mod}_R^\infty(M) \to \mathrm{Mod}_R^\infty(G)$$

are compatible with the pro-p Iwahori invariant functor and its left adjoint: [**OV**, Prop.4.4, Prop.4.6] gives natural isomorphisms:

$$(2.1)\qquad (-)^I \circ \mathrm{Ind}_P^G \simeq \mathrm{Ind}_P^{H(G)} \circ (-)^{I\cap M},$$

$$(- \otimes_{H(G)} \mathbb{Z}[I\backslash G]) \circ \mathrm{Ind}_P^{H(G)} \simeq \mathrm{Ind}_P^G \circ (- \otimes_{H(M)} \mathbb{Z}[(I\cap M)\backslash M]).$$

The parabolic induction $\mathrm{Ind}_P^{H(G)}$ for the pro-p Iwahori Hecke rings has a right adjoint $R_P^{H(G)}$ and a left adjoint $L_P^{H(G)}$ as for the groups, [**VigpIwst**]. As $-\otimes_{H(M^+),\theta_M^G} H(G) \simeq \mathrm{Hom}_{H(M^+),\theta_M^{G,*}}(H(G),-)$ (Proposition VI.1 in the appendix below):

$$(2.2)\qquad L_P^{H(G)} \simeq - \otimes_{H(M^+),\theta_M^{G,*}} H(M), \quad R_P^{H(G)} = \mathrm{Hom}_{H(M^+),\theta_M^G}(H(M),-).$$

The right adjoint functors R_P^G and $R_P^{H(G)}$ are compatible with the pro-p Iwahori invariant functor but the left adjoint functors are not [**OV**, Cor.4.13].

REMARK IV.2. For the pro-p Iwahori Hecke algebra, the left adjoint $L_P^{H(G)}$ being a localization is exact but for the group, the left adjoint L_P^G is not exact.

PROPOSITION IV.3. *Let $P = MN, P_1 = M_1N_1$ be two parabolic subgroups of G. We have:*
(i) $R_{P_1}^{H(G)} \circ \mathrm{Ind}_P^{H(G)} \simeq \mathrm{Ind}_{P\cap P_1}^{H(M_1)} \circ R_{P\cap P_1}^{H(M)}$.
(ii) $L_{P_1}^{H(G)} \circ \mathrm{Ind}_P^{H(G)} \simeq \mathrm{Ind}_{P\cap P_1}^{H(M_1)} \circ L_{P\cap P_1}^{H(M)}$.
(iii) The parabolic induction functor $\mathrm{Ind}_P^{H(G)}$ *is fully faithful.*

PROOF. (i) is proved for the parabolic coinduction and its right adjoint in [**Abeparind**, Prop. 5.1][8]. Using the relation between the parabolic induction and coinduction given in the appendix we get (i).

(ii) follows from (i) by left adjunction and exchanging P, P_1.

(iii) The isomorphism (i) is described in the proof [**Abeparind**, Lemma 5.2]. For $P_1 = P$, one checks that it is given by the unit $\mathrm{id} \to R_P^{H(G)} \circ \mathrm{Ind}_P^{H(G)}$ of the adjunction. Applying Remark II.8, the functor $\mathrm{Ind}_P^{H(G)}$ is fully faithful. $\square$

IV.3. The $H(G)_R$-module $\mathrm{St}_Q^{H(G)}(\mathcal{V})$. The "trivial" representation of $H(G)$ is $\mathrm{Triv}_{H(G)} = (\mathrm{Triv}_G)^I$ where Triv_G is the trivial $\mathbb{Z}$-representation of G. Let $P = MN$ be a parabolic subgroup of G and $\mathrm{St}_P^{H(G)} := (\mathrm{St}_P^G)^I$. Put $\mathrm{Triv}_{H(G)_R} = R\otimes_{\mathbb{Z}} \mathrm{Triv}_{H(G)}$ and $\mathrm{St}_P^{H(G)}(R) := R\otimes_{\mathbb{Z}} \mathrm{St}_P^{H(G)}$; they are $H(G)_R$-modules. The $H(G)_R$-module $\mathrm{Ind}_P^{H(G)}(\mathrm{Triv}_{H(M)_R}) = \mathrm{Ind}_Q^{H(G)}(R)$ is isomorphic to $(\mathrm{Ind}_Q^G(R))^I$ (§IV.2). By [**Ly**], $\mathrm{St}_P^{H(G)}(R)$ is absolutely simple and isomorphic to the cokernel of the natural map

$$(3.3)\qquad \oplus_{P\subsetneq Q\subset G}(\mathrm{Ind}_Q^G(R))^I \to (\mathrm{Ind}_P^G(R))^I.$$

One knows that $T^*(z)$ acts trivially on $\mathrm{Ind}_P^{H(G)}(\mathbb{Z})$ and on $\mathrm{St}_P^{H(G)}$ for $z\in Z\cap M'$ [**AHenV2**, Ex.3.14].

Let $\mathcal{V}$ be a non-zero right $H(M)_R$-module, and $P_{\mathcal{V}} = M_{\mathcal{V}}N_{\mathcal{V}}$, $P(\mathcal{V}) = M(\mathcal{V})N(\mathcal{V})$ the parabolic subgroups of G corresponding to:

$$\Delta_{\mathcal{V}} = \{\alpha\in\Delta \ \text{ orthogonal to } \Delta_M, v = vT^{M,*}(z) \text{ for all } \ v\in\mathcal{V}, z\in Z\cap M'_\alpha\},$$

[8] What we call parabolic coinduction is denoted by I_P in [**Abeparind**, §4] and called parabolic induction

$\Delta(\mathcal{V}) = \Delta_M \cup \Delta_{\mathcal{V}}$ [**Abe**] [**AHenV2**, Def.4.12]. Different consequences for $M(\mathcal{V})$ of the orthogonality of Δ_M and $\Delta_{\mathcal{V}}$ are described in Remark IV.1 2.

DEFINITION IV.4. There is a unique right $H(M(\mathcal{V}))_R$-module $e(\mathcal{V})$ equal to $\mathcal{V}$ as an R-vector space, where $T^{M(\mathcal{V}),*}(m)$ acts by $T^{M,*}(m)$ for $m \in M$ and by the identity for $m \in M'_{\mathcal{V}}$ [**AHenV2**, Def.3.8 and remark before Cor. 3.9]; we say that $e(\mathcal{V})$ is the extension of $\mathcal{V}$ to $H(M(\mathcal{V}))$ or that $\mathcal{V}$ is the restriction of $e(\mathcal{V})$ to $H(M)$.

REMARK IV.5. Extension to $H(M(\mathcal{V}))$ gives a lattice isomorphism $\mathcal{L}_{\mathcal{V}} \to \mathcal{L}_{e(\mathcal{V})}$.

For $P = MN \subset Q = M_Q N_Q \subset P(\mathcal{V})$, we define similarly the extension $e_Q(\mathcal{V})$ of $\mathcal{V}$ to $H(M_Q)$. When $P \subset Q = M_Q N_Q$, we write $\mathrm{St}_P^{H(Q)} := \mathrm{St}_{P\cap M_Q}^{H(M_Q)}$.

LEMMA IV.6. *Assume that Δ_M is orthogonal to $\Delta\setminus\Delta_M$ and that we have right $H(G)_R$-modules $\mathcal{X}$ extending an $H(M)_R$-module and $\mathcal{Y}$ extending an $H(M_2)_R$-module, where $P_2 = M_2N_2$ is the parabolic subgroup of G corresponding to $\Delta\setminus\Delta_M$.*

Then, there is a structure of right $H(G)_R$-module on $\mathcal{X}\otimes_R\mathcal{Y}$ where $T^*(w)$ and $T(w)$ for $w \in W$ act diagonally, and on $\mathrm{Hom}_{\theta_{M_2}^{G,*}(H(M_2'))}(\mathcal{Y}, \mathcal{X}\otimes_R\mathcal{Y})$, where $T^*(w)$ acts by the identity for $w \in W_{M_2'}$ and by

$$(T^*(w)_{\mathcal{X}} \otimes T^*(w)_{\mathcal{Y}}) \circ - \circ (T^*(w)_{\mathcal{Y}})^{-1} \quad \textit{for } w \in W_{M'}\Omega,$$

where $T^*(w)_{\mathcal{X}}$ and $T^*(w)_{\mathcal{Y}}$ are the actions of $T^*(w)$ on $\mathcal{X}$ and $\mathcal{Y}$.

PROOF. For $\mathcal{X}\otimes_R\mathcal{Y}$ see [**AHenV2**, Prop.3.15, Cor.3.17].

Put $\mathcal{Z} = \mathrm{Hom}_{\theta_{M_2}^{G,*}(H(M_2'))}(\mathcal{Y}, \mathcal{X}\otimes_R\mathcal{Y})$; we check that the action $T^*(w)_{\mathcal{Z}}$ of $T^*(w)$ on $\mathcal{Z}$ for $w \in W$ defined in the lemma, respects the braid and quadratic relations (§IV.1). The braid relations follow from $W = W_{M_2'}W_{M'}\Omega$ and $T^*(ww_2u) = T^*(w)T^*(w_2)T^*(u)$ if $w \in W_{M'}, w_2 \in W_{M_2'}, u \in \Omega$ (Remark IV.1 2). For the quadratic relations, let $s_2 \in W_{M_2'}$ and $s \in W_{M'}$ of length 1. Then $T^*(s_2)_{\mathcal{X}}, T^*(s_2)_{\mathcal{Z}}$ and $T^*(s)_{\mathcal{Y}}$ are the identity. As $-c(s_2)_{\mathcal{Z}}$ is the identity and the characteristic of R is p, $T^*(s_2)_{\mathcal{Z}}$ verifies the quadratic relation; $T^*(s)_{\mathcal{Z}}(-) = (T^*(s)_{\mathcal{X}} \otimes \mathrm{id}_{\mathcal{Y}}) \circ -$ satisfies the quadratic relation because $T^*(s)_{\mathcal{X}}$ does (§IV.1). ☐

Assume $P \subset Q \subset P(\mathcal{V}) = G$, in particular Δ_M and $\Delta\setminus\Delta_M$ are orthogonal. We have $(\mathrm{St}_Q^G)^I = (\mathrm{St}_Q^G)^{I\cap M_2'}$ [**AHenV2**, §4.2, proof of theorem 4.7], the right $H(G)_R$-modules:

$$e(\mathcal{V}) \otimes_R \mathrm{Ind}_Q^{H(G)}(R), \quad \mathrm{St}_Q^{H(G)}(\mathcal{V}) = e(\mathcal{V}) \otimes_R \mathrm{St}_Q^{H(G)}(R),$$
$$\mathrm{Hom}_{H(M_2')_R}(e(\mathcal{V}), \mathrm{St}_Q^{H(G)}(\mathcal{V}))$$

where $T^*(w)$ acts diagonally for $w \in W$ on the first and second ones, and for the third one, the map $\theta_{M_2}^G = \theta_{M_2}^{G,*}$ embeds $H(M_2')$ in $H(G)$ (Remark IV.1 2), $T^*(w)$ acts by the identity for $w \in W_{M_2'}$ and by $T^*(w) \circ - \circ T^*(w)^{-1}$ for $w \in W_{M'}\Omega$ (Lemma IV.6).

From the $H(G)_R$-isomorphism

$$\mathrm{Ind}_Q^{H(G)}(e_Q(\mathcal{V})) \simeq e(\mathcal{V}) \otimes (\mathrm{Ind}_Q^G(R))^I$$

explicated in ([**AHenV2**, Prop.4.5], and the inclusion $(\mathrm{Ind}_{Q_1}^G(R))^I \subset (\mathrm{Ind}_Q^G(R))^I$ for $P \subset Q \subset Q_1$, we obtain an injective $H(G)_R$-isomorphism

$\mathrm{Ind}_{Q_1}^{H(G)}(e_{Q_1}(\mathcal{V})) \xrightarrow{\iota^G(Q,Q_1)} \mathrm{Ind}_Q^{H(G)}(e_Q(\mathcal{V}))$ and an $H(G)_R$-map

$$\oplus_{Q\subsetneq Q_1\subset G}\mathrm{Ind}_{Q_1}^{H(G)}(e_{Q_1}(\mathcal{V})) \xrightarrow{\oplus_{Q\subsetneq Q_1\subset G}\iota^G(Q,Q_1)} \mathrm{Ind}_Q^{H(G)}(e_Q(\mathcal{V})) \tag{3.4}$$

of cokernel isomorphic to $\mathrm{St}_Q^{H(G)}(\mathcal{V})$ [**AHenV2**, Cor.4.6].

PROPOSITION IV.7. *Assume $P\subset Q\subset P(\mathcal{V})=G$.*

(i) The natural maps $e(\mathcal{V})\to \mathrm{Hom}_{H(M'_2)_R}(\mathrm{St}_Q^{H(G)}(R), e(\mathcal{V})\otimes_R \mathrm{St}_Q^{H(G)}(R))$ and $\mathrm{Hom}_{H(M'_2)_R}(\mathrm{St}_Q^{H(G)}(R), \mathrm{St}_Q^{H(G)}(\mathcal{V}))\otimes_R \mathrm{St}_Q^{H(G)}(R)\to \mathrm{St}_Q^{H(G)}(\mathcal{V})$ are $H(G)_R$-isomorphisms.

(ii) The map $Y\mapsto Y\otimes_R \mathrm{St}_Q^{H(G)}(R):\mathcal{L}_{e(\mathcal{V})}\to\mathcal{L}_{\mathrm{St}_Q^{H(G)}(\mathcal{V})}$ is a lattice isomorphism of inverse $X\to\{y\in e(\mathcal{V}),\ y\otimes_{\mathbb{Z}}\mathrm{St}_Q^{H(G)}\subset X\}$.

PROOF. We are in the setting of Cor. I.6 for $A=H(M'_2)_R\subset A'=H(G)_R$ (the inclusion is via $\theta^G_{M_2}=\theta^{G,*}_{M_2}$), the bases $B=(T^*(w))_{w\in W_{M'_2}}$ and $B'=(T^*(w))_{w\in W}$, the right A-module $\mathcal{V}$, and the right A'-module $V=\mathrm{St}_Q^{H(G)}(R)=e(\mathrm{St}_Q^{H(M_2)}(R))$, absolutely simple as an A-module where T^*_w for $w\in W\setminus W_{M'_2}$ (contained in $W_{M'}\Omega$) acts invertibly. □

IV.4. The module $I_{H(G)}(P,\mathcal{V},Q)$.

DEFINITION IV.8. An R-triple $(P,\mathcal{V},Q)$ of $H(G)$ consists of a parabolic subgroup $P=MN$ of G, a right $H(M)_R$-module $\mathcal{V}$, a parabolic subgroup Q of G with $P\subset Q\subset P(\mathcal{V})$. To an R-triple $(P,\mathcal{V},Q)$ of $H(G)$ is attached a right $H(G)_R$-module

$$I_{H(G)}(P,\mathcal{V},Q)=\mathrm{Ind}_{P(\mathcal{V})}^{H(G)}(\mathrm{St}_Q^{H(M(\mathcal{V}))}(\mathcal{V}))$$

isomorphic to the cokernel of the $H(G)_R$-homomorphism

$$\oplus_{Q\subsetneq Q_1\subset P(\mathcal{V})}\mathrm{Ind}_{Q_1}^{H(G)}(e_{Q_1}(\mathcal{V})) \xrightarrow{\oplus_{Q\subsetneq Q_1\subset P(\mathcal{V})}\iota^G(Q_1,Q)} \mathrm{Ind}_Q^{H(G)}(e_Q(\mathcal{V}))$$

where $\iota^G(Q_1,Q)=\mathrm{Ind}_{P(\mathcal{V})}^{H(G)}(\iota^{M(\mathcal{V})}(Q\cap M(\mathcal{V}),Q_1\cap M(\mathcal{V})))$.

We can recover $\mathrm{St}_Q^{H(M(\mathcal{V}))}(\mathcal{V})$ and $e(\mathcal{V})$ from $I_{H(G)}(P,\mathcal{V},Q)$ and $P(\mathcal{V})$:

$$\mathrm{St}_Q^{H(M(\mathcal{V}))}(\mathcal{V})\simeq L^{H(G)}_{H(M(\mathcal{V}))}(I_{H(G)}(P,\mathcal{V},Q))) \tag{4.5}$$

by Proposition IV.3(ii) and

$$e(\mathcal{V})\simeq \mathrm{Hom}_{H(M'_{\mathcal{V}})}(\mathrm{St}_Q^{H(M(\mathcal{V}))}(R), L^{H(G)}_{H(M(\mathcal{V}))}(I_{H(G)}(P,\mathcal{V},Q))) \tag{4.6}$$

by Proposition IV.7(i).

PROPOSITION IV.9. *Let $(P,\mathcal{V},Q)$ be an R-triple of $H(G)$ with $\mathcal{V}$ of finite length and such that for each irreducible subquotient $\mathcal{X}$ of $\mathcal{V}$, $P(\mathcal{V})=P(\mathcal{X})$ and $I_{H(G)}(P,\mathcal{X},Q)$ is simple. Then $P(\mathcal{V})=P(\mathcal{V}')$ for any non-zero $H(M)_R$-submodule $\mathcal{V}'$ of $\mathcal{V}$; moreover the map $\mathcal{V}'\mapsto I_{H(G)}(P,\mathcal{V}',Q):\mathcal{L}_{\mathcal{V}}\to\mathcal{L}_{I_{H(G)}(P,\mathcal{V},Q)}$ is a lattice isomorphism.*

PROOF. $P(\mathcal{V})=P(\mathcal{V}')$ is proved as in Proposition III.10. We are in the situation of Corollary I.6 (proof of Prop.IV.7 for $M(\mathcal{V})$ instead of G). So $\mathrm{St}_Q^{H(M(\mathcal{V}))}(\mathcal{V})$

has finite length, and its irreducible subquotients are $\mathrm{St}_Q^{H(M(\mathcal{V}))}(\mathcal{X})$ for the irreducible subquotients $\mathcal{X}$ of $\mathcal{V}$. If $I_G(P,\mathcal{X},Q) = \mathrm{Ind}_{P(\mathcal{V})}^G(\mathrm{St}_Q^{M(\mathcal{V})}(\mathcal{X}))$ is irreducible for all $\mathcal{X}$, we are in the situation of Corollary I.4 for $F = \mathrm{Ind}_{P(\mathcal{V})}^{H(G)}$ and $W = \mathrm{St}_Q^{H(M(\mathcal{V}))}(\mathcal{V})$ because $\mathrm{Ind}_{P(\mathcal{V})}^{H(G)}$ has a right adjoint and is exact fully faithful (Proposition IV.3 (iii)) so the map $\mathcal{V}' \mapsto I_G(P,\mathcal{V}',Q) : \mathcal{L}_{\mathcal{V}} \to \mathcal{L}_{I_G(P,\mathcal{V},Q)}$ is a lattice isomorphism. □

REMARK IV.10. *The scalar extension to R^{alg} of a R-triple $(P,\mathcal{V},Q)$ of $H(G)$ where $\mathcal{V}$ is simple supersingular, is an R^{alg}-triple $(P,\sigma_{R^{alg}},Q)$ of $H(G)$ satisfying the hypotheses of Proposition* IV.9, as for the group (Remark III.18). By the decomposition theorem and Lemma III.16, $\mathcal{V}_{R^{alg}}$ has finite length and its irreducible subquotients $\mathcal{X}$ are supersingular, $P(\mathcal{X}) = P(\mathcal{V}) = P(\mathcal{V}_{R^{alg}})$ (Prop.IV.12 (ii)), and $I_{H(G)}(P,\mathcal{X},Q)$ is irreducible by the classification theorem for $H(G)$ over R^{alg} (Thm.I.8 [**AHenV2**]).

We now check the compatibility of $I_{H(G)}(P,\mathcal{V},Q)$ with scalar extension, as for the group (Propositions III.12 and III.13). Let R'/R be a field extension.

PROPOSITION IV.11. *(i) The parabolic induction commutes with the scalar restriction from R' to R and with the scalar extension from R to R'. Hence the left (resp. right) adjoint of the parabolic induction commutes with scalar extension (resp. restriction).*

(ii) An $H(M)_{R'}$-module $\mathcal{V}'$ and an $H(G)_R$-module $\mathcal{X}$ such that $\mathrm{Ind}_P^{H(G)}\mathcal{V}' \simeq \mathcal{X}_{R'}$, we have $\mathcal{V}' \simeq (L_P^{H(G)}\mathcal{X})_{R'}$.

PROOF. As for the group (Proposition III.12). Note that (i) is valid for commutative rings $R \subset R'$. □

PROPOSITION IV.12. *(i) Let $(P,\mathcal{V},Q)$ be an R-triple of $H(G)$. Then $P(\mathcal{V}) = P(\mathcal{V}_{R'})$; if $\mathcal{V}$ is simple and $\mathcal{V}'$ is a subquotient of $\mathcal{V}_{R'}$, then $P(\mathcal{V}) = P(\mathcal{V}')$ and*

$$(e(\mathcal{V}))_{R'} = e(\mathcal{V}_{R'}),\ \mathrm{St}_Q^{H(M(\mathcal{V}))}(\mathcal{V})_{R'} \simeq \mathrm{St}_Q^{H(M(\mathcal{V}))}(\mathcal{V}_{R'}),$$
$$I_{H(G)}(P,\mathcal{V},Q)_{R'} \simeq I_{H(G)}(P,\mathcal{V}_{R'},Q).$$

(ii) Let $(P,\mathcal{V}',Q)$ be an R'-triple of $H(G)$ such that $e(\mathcal{V}')$, resp. $\mathrm{St}_Q^{H(M(\mathcal{V}'))}(\mathcal{V}')$, resp. $I_{H(G)}(P,\mathcal{V},'Q)$, descend to R. Then $\mathcal{V}'$ descends to R.

Precisely, if $e(\mathcal{V}')$, resp. $\mathrm{St}_Q^{H(M(\mathcal{V}'))}(\mathcal{V}')$, resp. $I_{H(G)}(P,\mathcal{V}',Q)$, is the scalar extension from R to R' of $\mathcal{X}$, resp. $\mathcal{Y}$, resp. $\mathcal{Z}$, then $\mathcal{V}'$ is the scalar extension from R to R' of the natural action of $H(M)_R$ on $\mathcal{X}$, resp. $\mathrm{Hom}_{H(M'_{\mathcal{V}'})_R}(\mathrm{St}_Q^{H(M(\mathcal{V}'))}(R),\mathcal{Y})$, resp. $\mathrm{Hom}_{H(M'_{\mathcal{V}'})_R}(\mathrm{St}_Q^{H(M(\mathcal{V}'))}(R), L_{P(\mathcal{V}')}^{H(G)}\mathcal{Z})$.

PROOF. (i) As for the group (Proposition III.13).

(ii) If $I_{H(G)}(P,\mathcal{V}',Q) = \mathcal{Z}_{R'}$ then $\mathrm{St}_Q^{H(M(\mathcal{V}'))}(\mathcal{V}') = \mathcal{Y}_{R'}$ where $\mathcal{Y} \simeq L_{H(M(\mathcal{V}'))}^{H(G)}(\mathcal{Z})$ by (i) and (4.5).

If $\mathrm{St}_Q^{H(M(\mathcal{V}'))}(\mathcal{V}') = \mathcal{Y}_{R'}$, then $e(\mathcal{V}') = \mathcal{X}_{R'}$ where

$$\mathcal{X} \simeq \mathrm{Hom}_{H(M'_{\mathcal{V}'})_R}(\mathrm{St}_Q^{H(M(\mathcal{V}'))}(R),\mathcal{Y})$$

as $e(\mathcal{V}') \simeq \mathrm{Hom}_{H(M'_{\mathcal{V}'})_{R'}}(\mathrm{St}_Q^{H(M(\mathcal{V}'))}(R'), \mathcal{Y}_{R'})$ (Prop. IV.7) and $\mathrm{St}_Q^{H(M(\mathcal{V}'))}(R') \simeq (\mathrm{St}_Q^{H(M(\mathcal{V}'))}(R))_{R'}$.

If $e(\mathcal{V}') = \mathcal{X}_{R'}$ then $T^{M(\mathcal{V}'),*}(m)$ acts trivially on $\mathcal{X}_{R'}$ for $m \in M'_{\mathcal{V}'}$ hence also on $\mathcal{X}$ and $\mathcal{V}'$ is the scalar extension to R' of $\mathcal{X}$ seen as a $H(M)_R$-module. □

IV.5. Classification of simple modules over the pro-p Iwahori Hecke algebra. As in §III.5 for the group, the classification theorem for $H(G)$ over R^{alg} (Thm.I.8) descends to R by a formal proof relying on:

1 The decomposition theorem for $H(G)$ (Thm.I.1).

2 The classification theorem for $H(G)$ over R^{alg} (Thm.I.8 [**AHenV2**]).

3 The strong compatibility of scalar extension with $I_{H(G)}(P, -, Q)$ (Prop. IV.12) and supersingularity (Lemma III.16).

4 The lattice isomorphism $\mathcal{L}_{\mathcal{V}_{R^{alg}}} \to \mathcal{L}_{I_{H(G)}(P, \mathcal{V}_{R^{alg}}, Q)}$ for the scalar extension $\mathcal{V}_{R^{alg}}$ to R^{alg} of a simple supersingular $H(M)_R$-module $\mathcal{V}$ (Prop.IV.7 and Remark IV.10).

We start the proof with an R-triple $(P, \mathcal{V}, Q)$ of $H(G)$ with $\mathcal{V}$ simple supersingular and we prove that $I_{H(G)}(P, \mathcal{V}, Q)$ is simple. By the decomposition theorem, the $H(G)_{R^{alg}}$-module $\mathcal{V}_{R^{alg}}$ has finite length, and $I_{H(G)}(P, \mathcal{V}_{R^{alg}}, Q)$ also by the lattice isomorphism $\mathcal{L}_{\mathcal{V}_{R^{alg}}} \to \mathcal{L}_{I_{H(G)}(P, \mathcal{V}_{R^{alg}}, Q)}$. Scalar extension is faithful and exact and $I_{H(G)}(P, \mathcal{V}, Q)_{R^{alg}} \simeq I_{H(G)}(P, \mathcal{V}_{R^{alg}}, Q)$ so $I_{H(G)}(P, \mathcal{V}, Q)$ has also finite length. We choose a simple $H(G)_R$-submodule $\mathcal{X}$ of $I_{H(G)}(P, \mathcal{V}, Q)$. The $H(G)_{R^{alg}}$-module $\mathcal{X}_{R^{alg}}$ is contained in $I_{H(G)}(P, \mathcal{V}, Q)_{R^{alg}}$ hence $\mathcal{X}_{R^{alg}} \simeq I_{H(G)} P, \mathcal{V}', Q)$ for an $H(M)_{R^{alg}}$-submodule $\mathcal{V}'$ of $\mathcal{V}_{R^{alg}}$ by (5.8) and the lattice isomorphism $\mathcal{L}_{\mathcal{V}_{R^{alg}}} \to \mathcal{L}_{I_{H(G)}(P, \mathcal{V}_{R^{alg}}, Q)}$. As $I_{H(G)}(P, \mathcal{V}', Q)$ descends to R, $\mathcal{V}'$ is also by the strong compatibility of $I_{H(G)}(P, -, Q)$ with scalar extension. But no proper $H(M)_{R^{alg}}$-submodule of $\mathcal{V}_{R^{alg}}$ descends to R by the decomposition theorem for $H(G)$, so $\mathcal{V}' = \mathcal{V}_{R^{alg}}$, $\mathcal{X}_{R^{alg}} = I_{H(G)}(P, \mathcal{V}_{R^{alg}}, Q)$ and $\mathcal{X}_{R^{alg}} \simeq I_{H(G)}(P, \mathcal{V}, Q)_{R^{alg}}$ by compatibility of scalar extension with $I_{H(G)}(P, -, Q)$. So $\mathcal{X} \simeq I_{H(G)}(P, \mathcal{V}, Q)$ and $I_{H(G)}(P, \mathcal{V}, Q)$ is simple.

Next, let $(P, \mathcal{V}, Q)$ and $(P_1, \mathcal{V}_1, Q_1)$ be two R-triples of $H(G)$ with $\mathcal{V}, \mathcal{V}_1$ simple supersingular and $I_{H(G)}(P, \mathcal{V}, Q) \simeq I_{H(G)}(P_1, \mathcal{V}_1, Q_1)$. The scalar extensions to R^{alg} are isomorphic $(I_{H(G)}(P, \mathcal{V}, Q))_{R^{alg}} \simeq (I_{H(G)}(P_1, (\mathcal{V}_1), Q_1))_{R^{alg}}$. The classification theorem for $H(G)$ over R^{alg} and (5.8) imply $P = P_1, Q = Q_1$ and some simple $H(M)_{R^{alg}}$-subquotient $\mathcal{V}^{alg}$ of $\mathcal{V}_{R^{alg}}$ is isomorphic to some simple $H(M)_{R^{alg}}$-subquotient $\mathcal{V}_1^{alg}$ of $(\mathcal{V}_1)_{R^{alg}}$. As $\mathcal{V}^{alg}$ is $\mathcal{V}$-isotypic and $\mathcal{V}_1^{alg}$ is $\mathcal{V}_1$-isotypic as $H(M)_R$-module, $\mathcal{V}$ and $\mathcal{V}_1$ are isomorphic.

Finally, let $\mathcal{X}$ be an arbitrary simple $H(G)_R$-module. By the decomposition theorem, the $H(G)_{R^{alg}}$-module $\mathcal{X}_{R^{alg}}$ has finite length; we choose a simple submodule $\mathcal{X}^{alg}$ of $\mathcal{X}_{R^{alg}}$. By the classification theorem over R^{alg},

$$\mathcal{X}^{alg} \simeq I_{H(G)}(P, \mathcal{V}^{alg}, Q) \tag{5.7}$$

for an R^{alg}-triple $(P = MN, \mathcal{V}^{alg}, Q)$ of $H(G)$ where $\mathcal{V}^{alg}$ is a simple supersingular $H(M)_{R^{alg}}$-module. By the decomposition theorem, $\mathcal{X}^{alg}$ descends to a finite extension of R, and also $\mathcal{V}^{alg}$ by strong compatibility of scalar extension with $I_{H(G)}(P, -, Q)$. By the decomposition theorem, $\mathcal{V}^{alg}$ is contained in the scalar extension $\mathcal{V}_{R^{alg}}$ to R^{alg} of a simple $H(M)_R$-module $\mathcal{V}$. By compatibility of scalar extension with $I_{H(G)}(P,-,Q)$ and supersingularity, $\mathcal{V}$ is supersingular, $(P,\mathcal{V},Q)$ is an R-triple of G and

$$I_{H(G)}(P,\mathcal{V}_{R^{alg}},Q)\simeq I_{H(G)}(P,\mathcal{V},Q)_{R^{alg}}. \tag{5.8}$$

We have $I_{H(G)}(P,\mathcal{V}^{alg},Q)\subset I_{H(G)}(P,\mathcal{V}_{R^{alg}},Q)$ by the lattice isomorphism $\mathcal{L}_{\mathcal{V}_{R^{alg}}}\to \mathcal{L}_{I_{H(G)}(P,\mathcal{V}_{R^{alg}},Q)}$. The decomposition theorem and $\mathcal{X}^{alg}\subset I_{H(G)}(P,\mathcal{V},Q)_{R^{alg}}$ imply

$$\mathcal{X}\simeq I_{H(G)}(P,\mathcal{V},Q).$$

This ends the proof of the classification theorem for $H(G)$ (Thm.I.8). □

V. Applications

Let R be a field of characteristic p and G a reductive p-adic group as in §III.3.

V.1. Vanishing of the smooth dual. The dual of $\pi\in \mathrm{Mod}_R(G)$ is $\mathrm{Hom}_R(\pi,R)$ with the contragredient action of G, that is, $(gf)(gx)=f(x)$ for $g\in G, f\in \mathrm{Hom}_R(\pi,R), x\in\pi$. The smooth dual of π is $\pi^\vee:=\cup_K \mathrm{Hom}_R(\pi,R)^K$ where K runs through the open compact subgroups of G.

A finite dimensional smooth R-representation of G is fixed by an open compact subgroup, and its smooth dual is equal to its dual.

We prove Theorem I.9. Let R^{alg}/R be an algebraic closure and let π be a non-zero irreducible admissible R-representation π of G. By Remark II.2, $(\pi^\vee)_{R^{alg}}\subset(\pi_{R^{alg}})^\vee$. Assume that $\pi^\vee\neq 0$. Then, $(\pi^\vee)_{R^{alg}}\neq 0$, hence $(\pi_{R^{alg}})^\vee\neq 0$. We know that $\pi_{R^{alg}}$ has finite length (Thm. III.4), so $\rho^\vee\neq 0$ for some irreducible subquotient ρ of $\pi_{R^{alg}}$. By the theorem over R^{alg} [**AHenV2**, Thm.6.4], the R^{alg}-dimension of ρ is finite. The R^{alg}-dimension is constant on the $\mathrm{Aut}_R(R^{alg})$-orbit of ρ. By the decomposition theorem (Thm. III.4), the R^{alg}-dimension of $\pi_{R^{alg}}$ is finite. It is equal to the R-dimension of π. So we proved that $\pi^\vee\neq 0$ implies that the R-dimension of π is finite. □

V.2. Lattice of submodules (Proof of Theorem I.10).

V.2.1. We recall some properties of the I-invariant functor and of its left adjoint. Let σ be a smooth R-representation of M.

1. The parabolic induction commutes with $(-)^I$ and its left adjoint $-\otimes_{H(G)}\mathbb{Z}[I\backslash G]$ (§IV.2 (2.1)).
2. If the natural surjective $R[G]$-map (counit of the adjunction) $\sigma^{I\cap M}\otimes_{H(M,I\cap M)}\mathbb{Z}[(I\cap M)\backslash M]\to\sigma$ is an $R[M]$-isomorphism, it follows from 1 and the full faithfulness of Ind_P^G that $(\mathrm{Ind}_P^G(\sigma))^I\otimes_{H(G,I)}\mathbb{Z}[I\backslash G]$ is isomorphic to $\mathrm{Ind}_P^G(\sigma)$[9].
3. The natural $R[G]$-map $(\mathrm{Triv}_{H(G)}\otimes_{H(G)}\mathbb{Z}[I\backslash G])^I\to\mathrm{Triv}_G$ where Triv_G is the trivial R-representation of G and $\mathrm{Triv}_{H(G)}=(\mathrm{Triv}_G)^I$ [**OV**, end of the proof of Lemma 2.25].
4. $I_G(P,\sigma,Q)^I\simeq I_{H(G)}(P,\sigma^{I\cap M},Q)$ if $\sigma=\sigma_{min}$ (§III.3) and $P(\sigma)=P(\sigma^{I\cap M})$ [**AHenV2**, Thm. 4.17].

LEMMA V.1. *Let σ be an irreducible admissible supersingular R-representation of M. Then $\sigma=\sigma_{min}$, $P(\sigma)=P(\sigma^{I\cap M})$, so $I_G(P,\sigma,Q)^I\simeq I_{H(G)}(P,\sigma^{I\cap M},Q)$.*

[9]One can check that the natural surjective map (counit of the adjunction) $(\mathrm{Ind}_P^G(\sigma))^I\otimes_{H(G,I)}\mathbb{Z}[I\backslash G]\to\mathrm{Ind}_P^G(\sigma)$ is an $R[G]$-isomorphism

PROOF. The equality $\sigma = \sigma_{min}$ follows from the classification (Thm.I.7) because σ is supersingular (§III.4). When $\sigma = \sigma_{min}$, then Δ_σ is orthogonal to Δ_M (§III.3). As σ being irreducible is generated by $\sigma^{I\cap M}$, $P(\sigma) = P(\sigma^{I\cap M})$ [**AHenV2**, Thm.3.13]. □

5. $I_{H(G)}(P,\mathcal{V},Q)\otimes_{H(G)}\mathbb{Z}[I\backslash G]\simeq I_G(P,\mathcal{V}\otimes_{H(M)}\mathbb{Z}[(I\cap M)\backslash M],Q)$ if $\mathcal{V}$ is a simple supersingular $H(M)_R$-module (more generally, if $P(\mathcal{V})=P(\mathcal{V}\otimes_{H(M)}\mathbb{Z}[(I\cap M)\backslash M])$ when $\mathcal{V}\otimes_{H(M)}\mathbb{Z}[(I\cap M)\backslash M]\neq 0$) [**AHenV2**, Cor. 5.12, 5.13].

PROPOSITION V.2. *Let σ be an irreducible admissible supersingular R-representation of M such that $\sigma^{I\cap M}$ simple and the map $\sigma^{I\cap M}\otimes_{H(M)}\mathbb{Z}[(I\cap M)\backslash M]\to\sigma$ is bijective. Then,*

$\mathrm{Ind}_P^G(\sigma)$ *has multiplicity* 1 *and irreducible subquotients $I_G(P,\sigma,Q)$ for $P\subset Q\subset P(\sigma)$.*

$(\mathrm{Ind}_P^G\sigma)^I \simeq \mathrm{Ind}_P^{H(G)}(\sigma^{I\cap M})$ *has multiplicity* 1 *and simple subquotients $I_G(P,\sigma,Q)^I\simeq I_{H(G)}(P,\sigma^{I\cap M},Q)$ for $P\subset Q\subset P(\sigma)$.*

$\mathrm{Ind}_P^{H(G)}(\sigma^{I\cap M})\otimes_{H(G)}\mathbb{Z}[I\backslash G]\simeq\mathrm{Ind}_P^G(\sigma,Q)$ *and* $I_{H(G)}(P,\sigma^{I\cap M},Q)\otimes_{H(G)}\mathbb{Z}[I\backslash G]\simeq I_G(P,\sigma,Q)$ *for $P\subset Q\subset P(\sigma)$.*

PROOF. This follows from the above properties 1 to 5, Lemma V.1, the classification theorems I.7, I.8 and from [**AHHV**, III.24 Prop., the proof is valid for R not algebraically closed]. □

V.2.2. $\mathrm{Ind}_P^G(R)$ *and* $\mathrm{Ind}_P^{H(G)}(R)$. By [**Ly**, §9], the R-representation $\mathrm{Ind}_P^G(\mathrm{Triv}_M) = \mathrm{Ind}_P^G(R)$ of G is multiplicity free of irreducible subquotients $\mathrm{St}_Q^G(R)$ for $P\subset Q\subset G$. The $H(G)_R$-module $\mathrm{Ind}_P^{H(G)}(R)=(\mathrm{Ind}_P^G R)^I$ has a filtration with subquotients $\mathrm{St}_Q^G(R)^I=\mathrm{St}_Q^{H(G)}(R)$ for $P\subset Q\subset G$. By the classification theorem, the $\mathrm{St}_Q^{H(G)}(R)$ are simple not isomorphic. So $\mathrm{Ind}_P^{H(G)}(R)$ is multiplicity free of simple subquotients $\mathrm{St}_Q^{H(G)}(R)$ for $P\subset Q\subset G$.

Applying 1, 2 and 3 in §V.2.1, we see that $\mathrm{Ind}_P^{H(G)}(R)\otimes_{H(G)}\mathbb{Z}[I\backslash G]$ and $\mathrm{Ind}_P^G(R)$ are isomorphic; this implies that $\mathrm{St}_P^{H(G)}(R)\otimes_{H(G)}\mathbb{Z}[I\backslash G]$ and $\mathrm{St}_P^G(R)$ are also isomorphic.

We can apply Thm. I.3 (b) to the functor $F=-\otimes_{H(G)}\mathbb{Z}[I\backslash G]:\mathrm{Mod}_R(H(G))\to\mathrm{Mod}_R(G)$ of right adjoint $G=(-)^I$, and the $H(G)_R$-module $V=\mathrm{Ind}_P^{H(G)}(R)$. So $(-\otimes_{H(G)}\mathbb{Z}[I\backslash G],(-)^I)$ give lattice isomorphisms between $\mathcal{L}_{\mathrm{Ind}_P^{H(G)}(R)}$ and $\mathcal{L}_{\mathrm{Ind}_P^G(R)}$.

For $P\subset Q\subset G$, the subrepresentation of $\mathrm{Ind}_P^G(R)$ with cosocle $\mathrm{St}_Q^G(R)$ is $\mathrm{Ind}_Q^G(R)$, and sending $\mathrm{St}_Q^G(R)$ for $P\subset Q$ to $\Delta_Q\setminus\Delta_P$ induces a lattice isomorphism from $\mathcal{L}_{\mathrm{Ind}_P^G(R)}$ onto the set of upper sets in $\mathcal{P}(\Delta\setminus\Delta_P)$; to an upper set in $\mathcal{P}(\Delta\setminus\Delta_P)$ is associated the subrepresentation $\sum_J \mathrm{Ind}_{P_{J\cup\Delta_P}}^G(R)$ for J in the upper set [**AHenV1**, Prop.3.6].

V.2.3. $\mathrm{Ind}_P^G(\mathrm{St}_Q^M(R))$ *and* $\mathrm{Ind}_P^{H(G)}(\mathrm{St}_Q^{H(M)}(R))$ *for $Q\subset P$.* This case is a direct consequence of §V.2.2 because $\mathrm{Ind}_P^G(\mathrm{St}_Q^M(R))$ is a quotient of $\mathrm{Ind}_Q^G(R)$:

$$\mathrm{Ind}_P^G(\mathrm{St}_Q^M(R))=\mathrm{Ind}_Q^G(R)/\sum_{Q\subsetneq Q_1\subset P}\mathrm{Ind}_{Q_1}^G(R).$$

We deduce from §V.2.2 that $\mathrm{Ind}_P^G(\mathrm{St}_Q^M(R))$ is multiplicity free of irreducible subquotients $\mathrm{Ind}_P^G(\mathrm{St}_{Q'}^G(R))$ for $Q\subset Q'$ but Q' does not contain any Q_1 such that

$Q \subsetneq Q_1 \subset P$, that is, $Q = Q' \cap P$. The subrepresentation $\mathrm{Ind}_P^G(\mathrm{St}_{Q'}^M(R))$ of $\mathrm{Ind}_P^G(\mathrm{St}_Q^M(R))$ has cosocle $\mathrm{St}_{Q'}^G$. Sending $\mathrm{St}_{Q'}^G(R)$ to $\Delta_{Q'} \cap (\Delta \setminus \Delta_P)$ gives a lattice isomorphism from $\mathcal{L}_{\mathrm{Ind}_P^G(\mathrm{St}_Q^M(R))}$ onto the lattice of upper sets in $\mathcal{P}(\Delta \setminus \Delta_P)$ (which does not depend on Q). We deduce also from §V.2.2 and Remark II.10 that $-\otimes_{H(G)} \mathbb{Z}[I\backslash G]$ and $(-)^I$ give lattice isomorphisms between $\mathcal{L}_{\mathrm{Ind}_P^{H(G)}(\mathrm{St}_Q^{H(M)}(R))}$ and $\mathcal{L}_{\mathrm{Ind}_P^G(\mathrm{St}_Q^M(R))}$.

V.2.4. $\mathrm{Ind}_P^G \sigma$ *for* σ *irreducible admissible supersingular and* $\mathrm{Ind}_P^{H(G)} \mathcal{V}$ *for* $\mathcal{V}$ *simple supersingular.* $\mathrm{Ind}_P^G \sigma$ admits a filtration with quotients $I_G(P,\sigma,Q) = \mathrm{Ind}_{P(\sigma)}^G(\mathrm{St}_Q^{M(\sigma)}(\sigma))$ for $P \subset Q \subset P(\sigma)$, and by the classification theorem the $I_G(P,\sigma,Q)$ are irreducible and not isomorphic; so $\mathrm{Ind}_P^G(\sigma)$ is multiplicity free of irreducible subquotients $I_G(P,\sigma,Q)$ for $P \subset Q \subset P(\sigma)$. The maps

$$X \mapsto e(\sigma)\otimes_R X \mapsto \mathrm{Ind}_{P(\sigma)}^G(e(\sigma)\otimes_R X) : \mathcal{L}_{\mathrm{Ind}_P^{M(\sigma)}(R)} \to \mathcal{L}_{e(\sigma)\otimes_R \mathrm{Ind}_P^{M(\sigma)}(R)} \to \mathcal{L}_{\mathrm{Ind}_P^G(\sigma)}$$

are lattice isomorphisms: this follows from the lattice theorems and the classification theorem (Thm.I.3, Thm.I.5, Thm.I.7), as in Proposition III.10 (for R algebraically closed [**AHenV1**, Prop.3.8]).

The same arguments show that $\mathrm{Ind}_P^{H(G)}(\mathcal{V})$ is multiplicity free of simple subquotients $I_{H(G)}(P,\mathcal{V},Q)$ for $P \subset Q \subset P(\mathcal{V})$ and that the maps

$$Y \mapsto e(\mathcal{V}) \otimes_R Y \mapsto \mathrm{Ind}_{P(\mathcal{V})}^{H(G)}(e(\mathcal{V}) \otimes_R Y) : \mathcal{L}_{\mathrm{Ind}_P^{H(M(\mathcal{V}))}(R)} \to \mathcal{L}_{e(\mathcal{V})\otimes_R \mathrm{Ind}_P^{H(M(\mathcal{V}))}(R)} \to \mathcal{L}_{\mathrm{Ind}_P^{H(G)}(\mathcal{V})}$$

are lattice isomorphisms, by applying Thm.I.3, Thm.I.5, Thm.I.8, as in Proposition IV.9.

V.2.5. $\mathrm{Ind}_P^G(\mathrm{St}_Q^M(\sigma_1))$ *and* $\mathrm{Ind}_P^{H(G)}(\mathrm{St}_Q^{H(M)}(\mathcal{V}_1))$ *for an* R*-triple* (P_1,σ_1,P) *of* G*,* $P_1 \subset Q \subset P$*,* σ_1 *irreducible admissible supersingular and similarly for* $\mathcal{V}_1$. This is a direct consequence of §V.2.4 because

$$\mathrm{Ind}_P^G(\mathrm{St}_Q^M(\sigma_1)) = (\mathrm{Ind}_Q^G e_Q(\sigma_1))/(\sum_{Q\subsetneq Q_1 \subset P} \mathrm{Ind}_{Q_1}^G e_{Q_1}(\sigma_1))$$

is a subquotient of $\mathrm{Ind}_{P_1}^G(\sigma_1)$ as $e_Q(\sigma_1) \subset \mathrm{Ind}_{P_1}^{M_Q}(\sigma_1)$ and similarly for $\mathcal{V}_1$. We have $\mathrm{Ind}_{Q_1}^G e_{Q_1}(\sigma_1) \simeq \mathrm{Ind}_{P(\sigma_1)}^G(e(\sigma_1) \otimes_R \mathrm{ind}_{Q_1}^{M(\sigma_1)}(R))$, and a lattice isomorphism (§V.2.4):

$$X \mapsto \mathrm{Ind}_{P(\sigma_1)}^G(e(\sigma_1) \otimes_R X) : \mathcal{L}_{\mathrm{Ind}_{P_1}^{M(\sigma_1)}(R)} \to \mathcal{L}_{\mathrm{Ind}_{P_1}^G(\sigma_1)}$$

inducing a lattice isomorphism (Remark II.10):

$$\mathcal{L}_{\mathrm{Ind}_P^{M(\sigma_1)}(\mathrm{St}_Q^M(R))} \to \mathcal{L}_{\mathrm{Ind}_P^G(\mathrm{St}_Q^M(\sigma_1))}.$$

The R-representation $\mathrm{Ind}_P^G(\mathrm{St}_Q^M(\sigma_1))$ is multiplicity free of irreducible subquotients $I_G(P_1,\sigma_1,Q')$ for the R-triples (P_1,σ_1,Q') of G with $Q' \cap P = Q$ (§V.2.3). And similarly for $\mathcal{V}_1$ with the same arguments and references.

V.2.6. $\mathrm{Ind}_P^G \sigma$ *for* σ *irreducible admissible and* $\mathrm{Ind}_P^{H(G)} \mathcal{V}$ *for* $\mathcal{V}$ *simple.* By the classification theorem, there exists an R-triple (P_1,σ_1,Q) of G with $Q \subset P$, σ_1 irreducible admissible supersingular such that

$$\sigma \simeq I_M(P_1 \cap M, \sigma_1, Q\cap M) = \mathrm{Ind}_{P(\sigma_1)\cap M}^M(\mathrm{St}_{Q\cap M}^{M(\sigma_1)\cap M}(\sigma_1)).$$

The transitivity of the induction implies $\mathrm{Ind}_P^G\sigma \simeq \mathrm{Ind}_{P(\sigma_1)\cap P}^G(\mathrm{St}_Q^{M(\sigma_1)\cap M}(\sigma_1))$. This is the case §V.2.5 with $P(\sigma_1)\cap P$. The R-representation $\mathrm{Ind}_P^G\sigma$ of G is multiplicity free of irreducible subquotients $I_G(P_1,\sigma_1,Q')$ for the R-triples (P_1,σ_1,Q') of G with $Q'\cap P=Q$ (note that $Q'\subset P(\sigma_1), Q\subset P$). The map

$$X\mapsto \mathrm{Ind}_{P(\sigma_1)}^G(e(\sigma_1)\otimes_R X):\mathcal{L}_{\mathrm{Ind}_{P(\sigma_1)\cap P}^{M(\sigma_1)}(\mathrm{St}_Q^{M(\sigma_1)\cap M}(R))}\to\mathcal{L}_{\mathrm{Ind}_P^G(\sigma)}$$

is a lattice isomorphism. And similarly for $\mathcal{V}$ with the same arguments and references.

V.2.7. *Invariants by the pro-p Iwahori subgroup.* We start with an irreducible admissible R-representation σ of M and we keep the notations of §V.2.6. The classification theorem shows that

$$\sigma^{I\cap M} \text{ is simple } \Leftrightarrow \sigma_1^{I\cap M_1} \text{ is simple}$$

because $\sigma^{I\cap M}\simeq I_{H(M)}(P_1\cap M,\sigma_1^{I\cap M_1},Q\cap M)$ (§V.2.1) and $\sigma_1^{I\cap M_1}$ is supersingular of finite length.

Put $\mathcal{V}_1=\sigma_1^{I\cap M_1}$, and assume first that $P(\sigma_1)=P(\mathcal{V}_1)$. In §V.2.3 we saw that the maps

$$(2.1)\qquad X\mapsto X^{I\cap M(\sigma_1)},\quad Y\mapsto Y\otimes_{H(M(\sigma_1))}\mathbb{Z}[I\cap M(\sigma_1)\backslash M(\sigma_1)]$$

between $\mathcal{L}_{\mathrm{Ind}_{P(\sigma_1)\cap P}^{M(\sigma_1)}(\mathrm{St}_Q^{M(\sigma_1)\cap M}(R))}$ and $\mathcal{L}_{\mathrm{Ind}_{P(\sigma_1)\cap P}^{H(M(\sigma_1))}(\mathrm{St}_Q^{H(M(\sigma_1)\cap M)}(R))}$, are lattice isomorphisms. They induce lattice isomorphisms between $\mathcal{L}_{\mathrm{Ind}_P^G(\sigma)}$ and $\mathcal{L}_{\mathrm{Ind}_P^{H(G)}(\mathcal{V})}$:

(2.2)
$$\mathrm{Ind}_{P(\sigma_1)}^G(e(\sigma_1)\otimes_R X)\mapsto \mathrm{Ind}_{P(\mathcal{V}_1)}^{H(G)}(e(\mathcal{V}_1)\otimes_R X^{I\cap M(\sigma_1)}),$$
(2.3)
$$\mathrm{Ind}_{P(\mathcal{V}_1)}^{H(G)}(e(\mathcal{V}_1)\otimes_R Y)\mapsto \mathrm{Ind}_{P(\sigma_1)}^G(e(\sigma_1)\otimes_R (Y\otimes_{H(M(\sigma_1))}\mathbb{Z}[(I\cap M(\sigma_1))\backslash M(\sigma_1)])).$$

by the lattice isomorphisms of §V.2.6 with $\mathcal{L}_{\mathrm{Ind}_P^G(\sigma)}$ and $\mathcal{L}_{\mathrm{Ind}_P^{H(G)}(\mathcal{V})}$.

We assume now that $\sigma^{I\cap M}$ is simple and the natural map $\sigma^{I\cap M}\otimes_{H(M)}\mathbb{Z}[(I\cap M)\backslash M]\to\sigma$ bijective, and we prove that the map $Y\mapsto Y\otimes_{H(G)}\mathbb{Z}[I\backslash G]:\mathcal{L}_{\mathrm{Ind}_P^{H(G)}(\sigma^{I\cap M})}\to\mathcal{L}_{\mathrm{Ind}_P^G(\sigma)}$ is a lattice isomorphism. By Lemma V.1, $P(\sigma_1)=P(\mathcal{V}_1)$. By Remark II.10, it is enough to prove it when $\sigma=\sigma_1$, that is, σ is supersingular. For that, we use Thm. I.3 (b) with $F=-\otimes_{H(G)}\mathbb{Z}[I\backslash G]:\mathrm{Mod}_R(H(G))\to\mathrm{Mod}_R(G)$ of right adjoint $(-)^I$ and $V=\mathrm{Ind}_P^G\sigma$ which satisfy the hypotheses by Prop.V.2. This ends the proof of Thm. I.10.

V.3. Proof of Theorem I.12. Proving Theorem I.12 from the classification theorem needs no new techniques. It suffices to quote for $R_{P_1}^G(\pi)$ [**AHenV1**, Corollary 6.5], for $L_{P_1}^G(\pi)$ [**AHenV1**, Cor. 6.2, 6.8], for $L_{P_1}^{H(G)}(\mathcal{X})$ and $R_{P_1}^{H(G)}(\mathcal{X})$ ([**Abeparind**, Thm. 5.20] when R is algebraically closed, but this hypothesis is not used), for π^I and $\mathcal{X}\otimes_{H(G)}\mathbb{Z}[I\backslash G]$ [**AHenV2**, Thm.4.17, Thm.5.11].

VI. Appendix: Eight inductions $\mathrm{Mod}_R(H(M))\to\mathrm{Mod}_R(H(G))$

For a commutative ring R and a parabolic subgroup $P=MN$ of G, eight different inductions $\mathrm{Mod}_R(H(M))\to\mathrm{Mod}_R(H(G))$

$$-\otimes_{H(M^\epsilon),\theta^\eta}H(G)\quad \textit{and}\quad \mathrm{Hom}_{H(M^\epsilon),\theta^\eta}(H(G),-)\quad \textit{for } \epsilon\in\{+,-\},\ \eta\in\{\ ,*\}$$

are associated to the elements of $\{\otimes, \mathrm{Hom}\} \times \{+,-\} \times \{\theta, \theta^*\}$ where $\theta := \theta^G_M$ and $\{\theta^\eta, \theta^{*\eta}\} = \{\theta, \theta^*\}$ as sets (see IV.1). The triple $(\otimes, +, \theta)$ corresponds to the parabolic induction $\mathrm{Ind}_P^{H(G)}(-) = - \otimes_{H(M^+),\theta} H(G)$ and the triple $(\mathrm{Hom}, -, \theta^*)$ corresponds to $\mathrm{Hom}_{H(M^-),\theta^*}(H(G), -)$ that we call parabolic coinduction. Before comparing these eight inductions, we define the " twist by $n_{w_G w_M}$" and the involution $\iota^M_{\lg - \lg_M}$.

Twist by $n_{w_G w_M}$. We choose an injective homomorphism $w \mapsto n_w : \mathbb{W} \to W$ from the Weyl group $\mathbb{W}$ of Δ to W satisfying the braid relations (there is no canonical choice).

Put $w_M = w_P$ for the longest element of the finite Weyl group $\mathbb{W}_M$ of M (see §IV.1), and $P^{op} = M^{op} N^{op}$ for the parabolic subgroup of G corresponding to $\Delta_{M^{op}} = \Delta_{P^{op}} = w_G w_P(\Delta_P) = w_G(-\Delta_P)$ (it is contained in Δ and is the image of Δ_P by the opposition involution $\alpha \mapsto w_G(-\alpha)$ [**T**, 1.5.1]). The conjugation $w \mapsto n_{w_G w_M} w n^{-1}_{w_G w_M} : W_M \to W_{M^{op}}$ by $n_{w_G w_M}$ is a group isomorphism inducing the ring isomorphism "twist by $n_{w_G w_M}$":

$$H(M) \to H(M^{op}), \quad T^M_w \to T^{M^{op}}_{n_{w_G w_M} w n^{-1}_{w_G w_M}} \qquad (w \in W_M)$$

sending also $T^{M,*}_w)$ to $T^{M^{op},*}_{n_{w_G w_M} w n^{-1}_{w_G w_M}}$ [**Abe**, §4.3]. It restricts to an isomorphism $H(M^\epsilon) \to H(M^{op,-\epsilon})$ [**VigpIwst**, Prop.2.20], and its inverse is the twist by $n_{w_G w_{M^{op}}}$, because $n_{w_G w_{P^{op}}} = n_{w_P w_G} = n^{-1}_{w_G w_P}$.

We have the functor "twist by $n_{w_G w_M}$":

$$\mathrm{Mod}_R(H(M)) \xrightarrow{n_{w_G w_M}(-)} \mathrm{Mod}_R(H(M^{op})),$$

where the spaces of $\mathcal{V} \in \mathrm{Mod}_R(H(M))$ and $n_{w_G w_M}(\mathcal{V}) \in \mathrm{Mod}_R(H(M^{op}))$ are the same and $vT^M_w = vT^{M^{op}}_{n_{w_G w_M} w n^{-1}_{w_G w_M}}$ for $v \in \mathcal{V}, w \in W_M$.

Involution $\iota^M_{\lg - \lg_M}$ [**Abeparind**, §4.1]. The two commuting involutions ι^M and $\iota_{\lg - \lg_M}$ of the ring $H(M)$:

$(T^M_w, T^{M,*}_w) \xrightarrow{\iota^M} (-1)^{\lg_M(w)}(T^{M,*}_w, T^M_w)$ [**VigpIw**, Prop. 4.23],

$(T^M_w, T^{M,*}_w) \xrightarrow{\iota_{\lg - \lg_M}} (-1)^{\lg(w) - \lg_M(w)}(T^M_w, T^{M,*}_w)$ [**Abeparind**, Lemmas 4.2, 4.3, 4.4, 4.5].

give by composition an involution $\iota^M_{\lg - \lg_M}$ of $H(M)$

$(T^M_w, T^{M,*}_w) \xrightarrow{\iota^M_{\lg - \lg_M}} (-1)^{\lg(w)}(T^{M,*}_w, T^M_w)$.

The twist by $n_{w_G w_M}$ and the involution $\iota^M_{\lg - \lg_M}$ commute, and the image of T^M_w for $w \in W_M$ by

$$n_{w_G w_M}(-) \circ \iota^M_{\lg - \lg_M} = \iota^{M^{op}}_{\lg - \lg_{M^{op}}} \circ n_{w_G w_M}(-) : H(M) \to H(M^{op})$$

is

$$(-1)^{\lg(n_{w_G w_M} w n^{-1}_{w_G w_M})} T^{M^{op},*}_{n_{w_G w_M} w n^{-1}_{w_G w_M}} = (-1)^{\lg(w)} T^{M^{op},*}_{n_{w_G w_M} w n^{-1}_{w_G w_M}}$$

(the length $\lg_M$ of W_M is invariant by conjugation by w_M, and $\lg(n_{w_G w_M} w n^{-1}_{w_G w_M})$ $= \lg(n_{w_G} n_{w_M} w n^{-1}_{w_M} n^{-1}_{w_G}) = \lg(n_{w_M} w n^{-1}_{w_M}) = \lg(w)$). By functoriality, we get a functor

$$\mathrm{Mod}_R(H(M)) \xrightarrow{(-)^{\iota^M_{\lg - \lg_M}}} \mathrm{Mod}_R(H(M)).$$

When $M = G$, we write simply ι^G.

We are now ready for the comparison of the eight inductions, which follows from different propositions in [**Abeparind**] and [**Abeinv**]. Let $\mathcal{V}$ be any right $H(M^\epsilon)_R$-module. .

PROPOSITION VI.1. *Exchanging* $+,-$ *corresponds to the twist by* $n_{w_G w_M}$,

$$\mathcal{V} \otimes_{H(M^\epsilon),\theta^\eta} H(G) \simeq n_{w_G w_M}(\mathcal{V}) \otimes_{H(M^{op,-\epsilon}),\theta^\eta} H(G), \tag{0.1}$$

$$\operatorname{Hom}_{H(M^\epsilon),\theta^\eta}(H(G),\mathcal{V}) \simeq \operatorname{Hom}_{H(M^{op,-\epsilon}),\theta^\eta}(H(G), n_{w_G w_M}(\mathcal{V})). \tag{0.2}$$

Exchanging θ, θ^* *corresponds to the involutions* $\iota^M_{\lg - \lg_M}$ *and* ι^G.

$$(\mathcal{V} \otimes_{H(M^\epsilon),\theta^\eta} H(G))^{\iota^G} \simeq \mathcal{V}^{\iota^M_{\lg - \lg_M}} \otimes_{H(M^\epsilon),\theta^{*\eta}} H(G), \tag{0.3}$$

$$\operatorname{Hom}_{H(M^\epsilon),\theta^\eta}(H(G),\mathcal{V})^{\iota^G} \simeq \operatorname{Hom}_{H(M^\epsilon),\theta^{*\eta}}(H(G), \mathcal{V}^{\iota^M_{\lg - \lg_M}}). \tag{0.4}$$

Exchanging $\otimes, \operatorname{Hom}$ *corresponds to the involutions* $\iota^M_{\lg - \lg_M}$ *and* ι^G,

$$(\mathcal{V} \otimes_{H(M^\epsilon),\theta^\eta} H(G))^{\iota^G} \simeq \operatorname{Hom}_{H(M^\epsilon),\theta^\eta}(H(G), \mathcal{V}^{\iota^M_{\lg - \lg_M}}). \tag{0.5}$$

REMARK VI.2. By (0.3) and (0.5), exchanging $(\otimes, \theta^\eta)$ and $(\operatorname{Hom}, \theta^{*\eta})$ respects the isomorphism class:

$$\mathcal{V} \otimes_{H(M^\epsilon),\theta^\eta} H(G) \simeq \operatorname{Hom}_{H(M^\epsilon),\theta^{*\eta}}(H(G),\mathcal{V}). \tag{0.6}$$

In Propositions $\simeq$ mean that there are natural isomorphisms described in [**Abeparind**] and [**Abeinv**]

Duality Put ζ for the anti-involution of $H(G)$ defined by $\zeta(T_w) = T_{w^{-1}}$ for $w \in W$; we have also $\zeta(T^*_w) = T^*_{w^{-1}}$ [**VigpIwst**, Remark 2.12]. The dual of a right $H(G)_R$-module $\mathcal{X}$ is $\mathcal{X}^* = \operatorname{Hom}_R(\mathcal{X}, R)$ where $h \in H(G)_R$ acts on $f \in \mathcal{X}^*$ by $(fh)(x) = f(x\zeta(h))$ [**Abeinv**, Introduction].

PROPOSITION VI.3. *The dual exchanges* $(\otimes, +)$ *and* $(\operatorname{Hom}, -)$*:*

$$(\mathcal{V} \otimes_{H(M^\epsilon),\theta^\eta} H(G))^* \simeq \operatorname{Hom}_{H(M^{-\epsilon}),\theta^\eta}(H(G), \mathcal{V}^*), \tag{0.7}$$

$$\mathcal{V}^* \otimes_{H(M^\epsilon),\theta^\eta} H(G) \simeq (\operatorname{Hom}_{H(M^{-\epsilon}),\theta^\eta}(H(G),\mathcal{V}))^*. \tag{0.8}$$

PROOF. Applying (0.6), an upper isomorphism (0.7) for any $(\epsilon, \theta^\eta, \mathcal{V})$ is equivalent to a lower isomorphism (0.8) for any $(\epsilon, \theta^\eta, \mathcal{V})$. It suffices to prove (0.7).

An isomorphism (0.7) for $(+,\theta)$ and any $\mathcal{V}$ is implicit in [**Abeinv**, §4.1]. Using (0.1) (0.2), we get an isomorphism (0.7) for $(-,\theta)$ and any $\mathcal{V}$; so we proved (0.7) for θ and any $\epsilon, \mathcal{V}$. The image by ι^G of an isomorphism (0.7) for $(\theta, \epsilon, \mathcal{V}^{\iota^M_{\lg - \lg_M}})$ is an isomorphism (0.7) for $(\epsilon, \theta^*, \mathcal{V})$, because the anti-involution ζ_M of $H(M)$ commutes with the involution $\iota^M_{\lg - \lg_M}$, and their composite sends $(T^M_w, T^{M,*}_w)$ to $(-1)^{\lg(w)}(T^{M,*}_{w^{-1}}, T^M_{w^{-1}})$ for $w \in W_M$, as $\lg(w) = \lg(w^{-1})$. This ends the proof of (0.7). □

References

[Abe] N. Abe. *Modulo p parabolic induction of pro-p-Iwahori Hecke algebra,* J. Reine Angew. Math., DOI:10.1515/crelle-2016-0043.

[Abeparind] N. Abe. *Parabolic induction for pro-p Iwahori Hecke algebras,* arXiv:1612.01312.

[Abeinv] N. Abe. *Involutions on pro-p-Iwahori Hecke algebras,* arXiv:1704.00408v1.

[AHHV] N. Abe, G. Henniart, F. Herzig, and M.-F. Vignéras, *A classification of irreducible admissible* mod p *representations of p-adic reductive groups*, J. Amer. Math. Soc. **30** (2017), no. 2, 495–559, DOI 10.1090/jams/862. MR3600042

[AHenV1] N. Abe, G. Henniart, M.-F. Vigneras *Modulo p representations of reductive p-adic groups: Functorial properties*, Transactions of the AMS (2018) (to appear).

[AHenV2] N. Abe, G. Henniart, M.-F. Vigneras *On pro-p-Iwahori invariants of R-representations of reductive p-adic groups*, arXiv:1703.10384 (2017), submitted.

[BkiA2] N. Bourbaki. *Algèbre Chapitres 1 à 3*, Hermann 1970.

[BkiA8] N. Bourbaki. *Algèbre Chapitre 8* , Springer, 2012.

[BK] C. J. Bushnell and P. C. Kutzko, *The admissible dual of* $\mathrm{GL}(N)$ *via compact open subgroups*, Annals of Mathematics Studies, vol. 129, Princeton University Press, Princeton, NJ, 1993. MR1204652

[CR] C. W. Curtis and I. Reiner, *Methods of representation theory. Vol. I*, John Wiley & Sons, Inc., New York, 1981. With applications to finite groups and orders; Pure and Applied Mathematics; A Wiley-Interscience Publication. MR632548

[Eme] M. Emerton, *Ordinary parts of admissible representations of p-adic reductive groups I. Definition and first properties* (English, with English and French summaries), Astérisque **331** (2010), 355–402. MR2667882

[Hn] G. Henniart, *Sur les représentations modulo p de groupes réductifs p-adiques* (French, with English and French summaries), Automorphic forms and L-functions II. Local aspects, Contemp. Math., vol. 489, Amer. Math. Soc., Providence, RI, 2009, pp. 41–55, DOI 10.1090/conm/489/09546. MR2533002

[HV1] G. Henniart and M.-F. Vignéras, *A Satake isomorphism for representations modulo p of reductive groups over local fields*, J. Reine Angew. Math. **701** (2015), 33–75, DOI 10.1515/crelle-2013-0021. MR3331726

[He] F. Herzig, *The classification of irreducible admissible mod p representations of a p-adic* GL_n, Invent. Math. **186** (2011), no. 2, 373–434, DOI 10.1007/s00222-011-0321-z. MR2845621

[KS] M. Kashiwara and P. Schapira, *Categories and sheaves*, Grundlehren der Mathematischen Wissenschaften [Fundamental Principles of Mathematical Sciences], vol. 332, Springer-Verlag, Berlin, 2006. MR2182076

[K] K. Kozioł, *Pro-p-Iwahori invariants for* SL_2 *and L-packets of Hecke modules*, Int. Math. Res. Not. IMRN **4** (2016), 1090–1125, DOI 10.1093/imrn/rnv162. MR3493443

[Kohl] J. Kohlhaase, *Smooth duality in natural characteristic*, Adv. Math. **317** (2017), 1–49, DOI 10.1016/j.aim.2017.06.038. MR3682662

[Lang] S. Lang, *Algebra*, 2nd ed., Addison-Wesley Publishing Company, Advanced Book Program, Reading, MA, 1984. MR783636

[Lam] T. Y. Lam, *Lectures on modules and rings*, Graduate Texts in Mathematics, vol. 189, Springer-Verlag, New York, 1999. MR1653294

[Ly] T. Ly, *Représentations de Steinberg modulo p pour un groupe réductif sur un corps local* (French, with English and French summaries), Pacific J. Math. **277** (2015), no. 2, 425–462, DOI 10.2140/pjm.2015.277.425. MR3402357

[Oss] R. Ollivier, *Compatibility between Satake and Bernstein isomorphisms in characteristic p*, Algebra Number Theory **8** (2014), no. 5, 1071–1111, DOI 10.2140/ant.2014.8.1071. MR3263136

[O] R. Ollivier, *Le foncteur des invariants sous l'action du pro-p-Iwahori de* $\mathrm{GL}_2(F)$ (French, with English summary), J. Reine Angew. Math. **635** (2009), 149–185, DOI 10.1515/CRELLE.2009.078. MR2572257

[OS] R. Ollivier and P. Schneider, *A canonical torsion theory for pro-p Iwahori-Hecke modules*, Adv. Math. **327** (2018), 52–127, DOI 10.1016/j.aim.2017.06.013. MR3761991

[OV] R. Ollivier, M.-F. Vignéras. *Parabolic induction in characteristic p,* ArXiv 1703.04921v1. Selecta 2018 (to appear).

[Pask] V. Paškūnas, *The image of Colmez's Montreal functor*, Publ. Math. Inst. Hautes Études Sci. **118** (2013), 1–191, DOI 10.1007/s10240-013-0049-y. MR3150248

[Viglivre] M.-F. Vignéras, *Représentations l-modulaires d'un groupe réductif p-adique avec* $l \neq p$ (French, with English summary), Progress in Mathematics, vol. 137, Birkhäuser Boston, Inc., Boston, MA, 1996. MR1395151

[T] J. Tits, *Classification of algebraic semisimple groups*, Algebraic Groups and Discontinuous Subgroups (Proc. Sympos. Pure Math., Boulder, Colo., 1965), Amer. Math. Soc., Providence, R.I., 1966, 1966, pp. 33–62. MR0224710

[Vigadjoint] M.-F. Vignéras, *The right adjoint of the parabolic induction*, Arbeitstagung Bonn 2013, Progr. Math., vol. 319, Birkhäuser/Springer, Cham, 2016, pp. 405–425. MR3618059

[VigpIw] M.-F. Vigneras, *The pro-p-Iwahori Hecke algebra of a reductive p-adic group I*, Compos. Math. **152** (2016), no. 4, 693–753, DOI 10.1112/S0010437X15007666. MR3484112

[VigpIwc] M.-F. Vignéras, *The pro-p-Iwahori-Hecke algebra of a reductive p-adic group, II*, Münster J. Math. **7** (2014), no. 1, 363–379. MR3271250

[Vig17] M.-F. Vigneras, *The pro-p-Iwahori Hecke algebra of a reductive p-adic group III (spherical Hecke algebras and supersingular modules)*, J. Inst. Math. Jussieu **16** (2017), no. 3, 571–608, DOI 10.1017/S1474748015000146. MR3646282

[VigpIwst] M.-F. Vignéras, *The pro-p Iwahori Hecke algebra of a reductive p-adic group, V (parabolic induction)*, Pacific J. Math. **279** (2015), no. 1-2, 499–529, DOI 10.2140/pjm.2015.279.499. MR3437789

[VigSelecta] M.-F. Vignéras, *Induced R-representations of p-adic reductive groups*, Selecta Math. (N.S.) **4** (1998), no. 4, 549–623, DOI 10.1007/s000290050040. MR1668044

Laboratoire de Mathématiques d'Orsay, Université de Paris-Sud, CNRS, Université Paris-Saclay, Orsay cedex F-91405 France

Email address: Guy.Henniart@math.u-psud.fr

Institut de Mathématiques de Jussieu-Paris Rive Gauche, 4 Place Jussieu, Paris 75005 France

Email address: marie-france.vigneras@imj-prg.fr

Proceedings of Symposia in Pure Mathematics
Volume **101**, 2019
https://doi.org/10.1090/pspum/101/01796

On the support of matrix coefficients of supercuspidal representations of the general linear group over a local non-archimedean field

Erez Lapid

To Joseph Bernstein

ABSTRACT. We derive an upper bound on the support of matrix coefficients of suprecuspidal representations of the general linear group over a non-archimedean local field. The results are in par with those which can be obtained from the Bushnell–Kutzko classification of supercuspidal representations, but they are proved independently.

1. Introduction

Throughout, let F be a local non-archimedean field, $\mathcal{O}$ its ring of integers, and let $G = \mathrm{GL}_r(F)$ be the general linear group of rank r with center Z. Let $\mathbf{K} = \mathrm{GL}_r(\mathcal{O})$ be the standard maximal compact subgroup of G and for any $n \geq 1$ let $K(n) = K_r(n)$ be the principal congruence subgroup, i.e., the kernel of the canonical map $\mathbf{K} \to \mathrm{GL}_r(\mathcal{O}/\varpi^n\mathcal{O})$ where ϖ is a uniformizer of F. Denote by $\mathcal{B}(n)$ the ball

$$\mathcal{B}(n) = \{g \in G : \|g\|, \|g^{-1}\| \leq q^n\}$$

where q is the size of the residue field of F and $\|g\| = \max |g_{i,j}|_F$ where $g_{i,j}$ are the entries of g. Thus, $\{\mathcal{B}(n) : n \geq 1\}$ is an open cover of G by compact sets.

The purpose of this short paper is to give a new proof of the following result.[1]

THEOREM 1.1. *Let (π, V) be a supercuspidal representation of G and let $(\pi^\vee, V^\vee)$ be its contragredient. Let $v \in V$, $v^\vee \in V^\vee$ and assume that v and $v^\vee$ are fixed under $K(n)$ for some $n \geq 1$. Then the support of the matrix coefficient $g \mapsto (\pi(g)v, v^\vee)$ is contained in $Z\mathcal{B}(c(r)n)$ where $c(r)$ is an explicit constant depending on r only.*

As explained in [**FLM12**], the theorem is a direct consequence of the classification of irreducible supercuspidal representations by Bushnell–Kutzko [**BK93**], or more precisely, of the fact that every such representation is induced from a representation of an open subgroup of G which is compact modulo Z. This is in fact known for many other cases of reductive groups over local non-archimedean fields and in these cases it implies the analogue of Theorem 1.1. We refer the reader to [**FLM12**] for more details.

[1] Throughout, by a representation of G we always mean a complex, smooth representation.

In contrast, the proof given here is independent of the classification. It is based on two ingredients. The first, which is special to the general linear group, is basic properties of local Rankin–Selberg integrals for $G \times G$ which were defined and studied by Jacquet–Piatetski-Shapiro–Shalika. In particular, we use an argument of Bushnell–Henniart, originally used to give an upper bound on the conductor of Rankin–Selberg local factors [**BH97**]. The second ingredient is Howe's result on the integrality of the formal degree with respect to a suitable Haar measure [**How74**], a result which was subsequently extended to any reductive group [**HC99**, **Vig90**, **SS97**]. The two ingredients are linked by the fact, which also follows from properties of Rankin–Selberg integrals, that the formal degree is essentially the conductor of $\pi \times \pi^\vee$, a feature that admits a conjectural generalization for any reductive group [**HII08**]. (For another relation between formal degrees and support of matrix coefficients see [**Key92**].)

As explained in [**FL18**, **FL19**], Theorem 1.1 is of interest for the problem of limit multiplicity.

I would like to thank Joseph Bernstein, Stephen DeBacker, Tobias Finis, Atsushi Ichino, Hervé Jacquet, Julee Kim and Simon Marshall for useful discussions and suggestions. I am especially grateful to Guy Henniart for his input leading to Remark 2.4.

2. A variant for Whittaker functions

It is advantageous to formulate a variant of Theorem 1.1 for the Whittaker model. Throughout, fix a character ψ of F which is trivial on $\mathcal{O}_F$ but non-trivial on $\varpi^{-1}\mathcal{O}_F$. Let N be the subgroup of upper unitriangular matrices in G. If π is a generic irreducible representation of G, we write $\mathcal{W}^\psi(\pi)$ for its Whittaker model with respect to the character ψ_N of N given by $u \mapsto \psi(u_{1,2} + \cdots + u_{r-1,r})$. Recall that every irreducible supercuspidal representation of G is generic [**GK75**].

Let A be the diagonal torus of G. For all $n \geq 1$ let $A(n)$ be the open subset

$$A(n) = \{\operatorname{diag}(t_1, \dots, t_r) \in A : q^{-n} \leq |t_i/t_{i+1}| \leq q^n, \quad i = 1, \dots, r-1\}.$$

Clearly, $ZA(n) = A(n)$ and $A(n)$ is compact modulo Z.

THEOREM 2.1. *There exists a constant $c = c(r)$ with the following property. Let π be an irreducible supercuspidal representation of G with Whittaker model $\mathcal{W}^\psi(\pi)$ and $n \geq 1$. Then the support of any $W \in \mathcal{W}^\psi(\pi)^{K(n)}$ is contained in $NA(cn)\mathbf{K}$.*

In order to prove Theorem 2.1 we set some more notation. For any function W on G let

$$M_W = \sup\{\operatorname{val}(\det g) : W(g) \neq 0 \text{ and } \|g_r\| = 1\},$$
$$m_W = \inf\{\operatorname{val}(\det g) : W(g) \neq 0 \text{ and } \|g_r\| = 1\},$$

(including possibly $\pm\infty$) where g_r is the last row of g and $\|(x_1, \dots, x_r)\| = \max |x_i|$.

Recall that by a standard argument (cf. [**CS80**, Proposition 6.1]) for any right $K(n)$-invariant and left (N, ψ_N)-equivariant function W on G, if $W(tk) \neq 0$ for some $t = \operatorname{diag}(t_1, \dots, t_n) \in A$ and $k \in \mathbf{K}$ then $|t_i/t_{i+1}| \leq q^n$, $i = 1, \dots, r-1$. Hence, $m_W \geq -\binom{r}{2}n$.

For any $W \in \mathcal{W}^\psi(\pi)$ let $\widetilde{W} \in \mathcal{W}^{\psi^{-1}}(\pi^\vee)$ be given by $\widetilde{W}(g) = W(w_r\, {}^t g^{-1})$ where $w_r = \begin{pmatrix} & & 1 \\ & \iddots & \\ 1 & & \end{pmatrix}$.

Let $t = t(\pi)$ be the order of the group of unramified characters χ of F^* such that $\pi \otimes \chi \simeq \pi$. Clearly, t divides r.

Let $f = f(\pi \times \pi^\vee) \in \mathbb{Z}$ be the conductor of the pair $\pi \times \pi^\vee$ (see below).

Theorem 2.1 is an immediate consequence of the following two results.

PROPOSITION 2.2. *For any* $0 \neq W \in \mathcal{W}^\psi(\pi)$ *we have*

$$M_W + m_{\widetilde{W}} = r - t - f.$$

In particular, if $W \in \mathcal{W}^\psi(\pi)^{K(n)}$ *then*

$$M_W \leq \binom{r}{2} n + r - t - f.$$

PROPOSITION 2.3. *We have*

$$f \geq (r+1)r - 2(t + v_q(t))$$

where $v_q(t)$ *is the maximal power of* q *dividing* t. *Moreover,* f *is even if* q *is not a square.*

PROOF OF PROPOSITION 2.2. The argument is inspired by [**BH97**].

We may assume without loss of generality that π is unitarizable. For any $\Phi \in \mathcal{S}(F^r)$ consider the local Rankin–Selberg integral

$$A^\psi(s, W, \Phi) = \int_{N\backslash G} |W(g)|^2 \, \Phi(g_r) \, |\det g|^s \; dg,$$

a Laurent series in $x = q^{-s}$ which represents a rational function in x [**JPSS83**]. Note that if $\Phi(0) = 0$ then $A^\psi(s, W, \Phi)$ is a Laurent polynomial in x since W is compactly supported modulo ZN. Also note that for any $\lambda \in F^*$ we have

$$A^\psi(s, W, \Phi(\lambda\cdot)) = |\lambda|^{-rs} A^\psi(s, W, \Phi).$$

Recall the functional equation ([**JPSS83**, Theorem 2.7 and Proposition 8.1] together with [**BH99**])

$$q^{f(\frac{1}{2}-s)}(1 - q^{-ts})A^\psi(s, W, \Phi) = (1 - q^{t(s-1)})A^{\psi^{-1}}(1 - s, \widetilde{W}, \tilde{\Phi})$$

where

$$\tilde{\Phi}(x) = \int_{F^n} \Phi(y)\psi(x\,{}^t y) \; dy$$

is the Fourier transform of Φ and $f \in \mathbb{Z}$ is the conductor.

Now let Φ_0 be the characteristic function of the standard lattice $\{\xi \in F^r : \|\xi\| \leq 1\}$ and set $A_0^\psi(s, W) = A^\psi(s, W, \Phi_0)$. Then $\tilde{\Phi}_0 = \Phi_0$ and we obtain

$$q^{f(\frac{1}{2}-s)}(1 - q^{-ts})A_0^\psi(s, W) = (1 - q^{t(s-1)})A_0^{\psi^{-1}}(1 - s, \widetilde{W})$$

Let $\Phi_1 = \Phi_0 - \Phi_0(\varpi^{-1}\cdot)$ be the characteristic function of the $\mathbf{K}$-invariant set $\{\xi \in F^r : \|\xi\| = 1\}$ of primitive vectors. Then $A_1^\psi(s, W) := A^\psi(s, W, \Phi_1) = (1 - q^{-rs})A_0^\psi(s, W)$ and thus,

$$q^{f(\frac{1}{2}-s)} \frac{1 - q^{-ts}}{1 - q^{-rs}} A_1^\psi(s, W) = \frac{1 - q^{t(s-1)}}{1 - q^{r(s-1)}} A_1^{\psi^{-1}}(1 - s, \widetilde{W}).$$

Recall that $A_1^\psi(s, W)$ is a Laurent polynomial $P_w^\psi(x)$ in $x = q^{-s}$. We get an equality of Laurent polynomials

$$q^{\frac{1}{2}f} x^f \frac{1 - x^t}{1 - x^r} P_W^\psi(x) = \frac{1 - y^t}{1 - y^r} P_{\widetilde{W}}^{\psi^{-1}}(y) \tag{1}$$

where $y = q^{-1}x^{-1}$. Note that the fact that $P_W^{\psi}(x)$ is divisible by $\frac{1-x^r}{1-x^t}$ amounts to saying that the integral

$$\int_{g\in N\backslash G:\|g_r\|=1,\mathrm{val}(\det g)\equiv a \pmod{r/t}} |W(g)|^2 \, dg$$

is independent of a, which in turn follows from (and in fact, equivalent to) the fact that W is orthogonal to $W\,|\det|^{\frac{2\pi i j}{r\log q}}$ unless j is divisible by r/t. Also note that P_W^{ψ} has non-negative coefficients and the degree of P_W^{ψ} is M_W. Likewise, the degree of $P_W^{\psi}(y)$ as a Laurent polynomial in x (i.e., the order of pole of P_W^{ψ} at 0) is $-m_W$. Comparing degrees in (1) we obtain Proposition 2.2. □

Proof of Proposition 2.3. By an argument based on properties of Rankin-Selberg integrals, the formal degree, with respect to a suitable choice of Haar measure, is related to f by the formula

$$d_\pi = \frac{t}{r} q^{\frac{1}{2}f} \frac{1-q^{-1}}{1-q^{-t}}$$

([**ILM17**, Theorem 2.1]). Comparing it to the formal degree of the Steinberg representation St with respect to the same measure we get

$$\frac{d_\pi}{d_{\mathrm{St}}} = t\cdot q^{t-\binom{r+1}{2}+\frac{1}{2}f}\cdot\frac{q^r-1}{q^t-1}.$$

On the other hand, $\frac{d_\pi}{d_{\mathrm{St}}}$ (which is independent of the choice of Haar measure) is a (positive) integer (cf. [**How74**], [**Rog81**], [**Hen84**, Appendice 3]). The lemma follows. □

Remark 2.4. As explained to me by Guy Henniart, the lower bound in Proposition 2.3 is not sharp. In fact, a precise formula for $f(\pi\times\pi^\vee)$, and more generally for $f(\pi_1\times\pi_2)$ for an arbitrary pair of irreducible supercuspidal representations π_i of $\mathrm{GL}_{n_i}(F)$, $i=1,2$ is given in [**BHK98**]. The expression is in terms of the Bushnell–Kutzko description of supercuspidal representations. Using this, one can sharpen Proposition 2.3 as follows.

First note that $f(\pi\times\pi^\vee)$ is insensitive to twisting π by a character. Suppose that π is minimal under twists, i.e., $f(\pi\times\chi)\ge f(\pi)$ for any character χ of F^* where $f(\pi)$ is the conductor of π. Then by [**BH17**, Lemma 3.5] and its proof we have $f(\pi\times\pi^\vee)\ge r^2-t+\frac{1}{2}r(f(\pi)-r)$ with equality if $f(\pi)=r$, i.e., if π has level 0 (that is, if π has a non-zero vector invariant under $K(1)$, in which case $t=r$). Thus, if $f(\pi)>r+1$ then we get $f(\pi\times\pi^\vee)\ge r(r+1)-t$. On the other hand, we always have $f(\pi)\ge r$ [**Bus87**, (5.1)] and if $f(\pi)=r+1$, i.e., if π is epipelagic then $t=1$ and $f(\pi\times\pi^\vee)=(r-1)(r+2)$. More generally, if $f(\pi)$ is coprime to r, i.e., if π is a Carayol representation, then $t=1$ and $f(\pi\times\pi^\vee)=(r-1)(f(\pi)+1)$. For instance, this follows from [**BHK98**, (6.1.1),(6.1.2)] and [ibid., Theorem 6.5(i)] where in its notation we have $n=e=d=r$ and $\mathfrak{c}(\beta_1)=m(r-1)$ – cf. second paragraph ("minimal case") of [ibid., p. 727] with $k=m$, $e(\gamma)=d=r$.

To conclude, for any supercuspidal π we have

$$f(\pi\times\pi^\vee)\ge r(r+1)-2t \tag{2}$$

with equality if and only if π is a twist of either a representation of level 0 or an epipelagic representation.

Alternatively, one could infer (2) and the conditions for equality from the local Langlands correspondence for G (cf. [**GR10**]). Details will appear in the upcoming thesis of Kilic.

The results of [**BHK98**] also give that $f(\pi \times \pi^\vee)$ is even. (Details will be given elsewhere.) In the Galois side this follows from a result of Serre [**Ser71**].

I am grateful to Guy Henniart for providing me this explanation and allowing me to include it here.

For our purpose, the precise lower bound on $f(\pi \times \pi^\vee)$ is immaterial – it is sufficient to have the inequality $f \geq 0$ (or even, $f \geq c'n$ for some fixed c' depending only on r). The point is that we do not use either the local Langlands correspondence or the classification of supercuspidal representations (but of course, we do use the non-trivial analysis of [**How74**] which depends on [**How77**]).

Nonetheless, it would be interesting to prove (2) (and perhaps the evenness of $f(\pi \times \pi^\vee)$) without reference to the classification or to the local Langlands correspondence.

REMARK 2.5. As was pointed out to me (independently) by Joseph Bernstein, Hervé Jacquet and Simon Marshall, Theorem 2.1 immediately gives the bound

$$\sum_\pi \dim \pi^{K(n)} \leq cn^{r-1}q^{n\binom{r+1}{2}} \tag{3}$$

for a suitable constant c (depending only on r) and all $n \geq 1$, where π ranges over the irreducible supercuspidal representations of G up to a twist by an unramified character. Moreover, since the restriction to the mirabolic subgroup is injective on $\mathcal{W}^\psi(\pi)$, we also get $\dim \pi^{K(n)} \leq cn^{r-1}q^{n\binom{r}{2}}$ for any (supercuspidal) π and n. We recall that by [**Rod75**] we have

$$\lim_{n\to\infty} q^{-\binom{r}{2}n} \dim \pi^{K(n)} = \prod_{i=1}^{r} \frac{1-q^{-i}}{1-q^{-1}}$$

for any irreducible generic π. On the other hand by [**BH07**], the number of non-zero summands in (3) is of the order of magnitude of q^{nr}.

3. Proof of main result

In order to deduce Theorem 1.1 from Theorem 2.1 we make the argument of [**LM15**, Proposition 2.11] effective in the case of $G = \mathrm{GL}_r$.

For any $t = \mathrm{diag}(t_1, \dots, t_r) \in A$ consider the compact open subgroup

$$N(t) = N \cap t\mathbf{K}t^{-1} = \{u \in N : \mathrm{val}(u_{i,j}) \geq \mathrm{val}(t_i) - \mathrm{val}(t_j) \text{ for all } i < j\}$$

of N. Set

$$t_0 = \mathrm{diag}(\varpi, \varpi^2, \dots, \varpi^{2^{r-1}}) \in A.$$

PROPOSITION 3.1. *Let f be a compactly supported continuous function on G. Assume that f is bi-invariant under $K_{r-1}(n)$ for some $n \geq 1$. Then*

$$\int_N f(u)\psi_N(u)\, du = \int_{N(t_0^n)} f(u)\psi_N(u)\, du.$$

In particular, $\int_N f(u)\psi_N(u)\, du = 0$ if f vanishes on $\mathcal{B}((2^{r-1}-1)n)$.

We first need some more notation. Write $N = U \ltimes V$ where $U = N_{r-1}$ is the group of unitriangular matrices in GL_{r-1} embedded in GL_r in the upper left corner and V is the unipotent radical of the parabolic subgroup of type $(r-1,1)$, i.e., the $r-1$-dimensional abelian group

$$V = \{u \in N : u_{i,j} = 0 \text{ for all } i < j < r\}.$$

For any $\underline{\alpha} = (\alpha_1, \dots, \alpha_{r-1}) \in F^{r-1}$ let $x(\underline{\alpha})$ be the $r \times r$-matrix that is the identity except for the $(r-1)$-th row which is $(\alpha_1, \dots, \alpha_{r-1}+1, 0)$. Of course x is not a group homomorphism because of the diagonal entry.

For $t \in A$ write $U(t) = U \cap N(t)$ and $V(t) = V \cap N(t)$ so that $N(t) = U(t) \ltimes V(t)$. The following is elementary.

Lemma 3.2. *For any $n \geq 1$ let $L(n)$ be the lattice of F^{r-1} given by*

$$L(n) = \{(\alpha_1, \dots, \alpha_{r-1}) : \mathrm{val}(\alpha_j) \geq n(2^{r-1} - 2^{j-1}),\ j = 1, \dots, r-1\}.$$

Then

(1) *$x(\underline{\alpha}), x(\underline{\alpha})^u \in K_{r-1}(n)$ for any $\underline{\alpha} \in L(n)$ and $u \in U(t_0^n)$.*
(2) *For any $v \in V$ we have*

$$\mathrm{vol}(L(n))^{-1} \int_{L(n)} \psi_N(v^{x(\underline{\alpha})})\, d\underline{\alpha} = \begin{cases} \psi_N(v) & v \in V(t_0^n), \\ 0 & \textit{otherwise.} \end{cases}$$

Proof of Proposition 3.1. We prove the statement by induction on r. The case $r = 1$ is trivial. For the induction step, note that if f is bi-invariant under $K_{r-1}(n)$ then the function $h = \int_V f(\cdot v)\psi_N(v)\, dv$ on GL_{r-1} is bi-$K_{r-2}(n)$-invariant (since GL_{r-2} normalizes the character $\psi_N|_V$). Therefore, by induction hypothesis we have

$$\begin{aligned}\int_N f(u)\psi_N(u)\, du = \int_U h(u)\psi_N(u)\, du &= \int_{U(t_0^n)} h(u)\psi_N(u)\, du \\ &= \int_V \int_{U(t_0^n)} f(uv)\psi_N(uv)\, du\, dv.\end{aligned}$$

Now we use Lemma 3.2. By part 1, Since f is bi-$K_{r-1}(n)$-invariant, for any $\underline{\alpha} \in L(n)$ we can write the above as

$$\int_V \int_{U(t_0^n)} f(ux(\underline{\alpha})vx(\underline{\alpha})^{-1})\psi_N(uv)\, du\, dv = \int_V \int_{U(t_0^n)} f(uv)\psi_N(uv^{x(\underline{\alpha})})\, du\, dv.$$

Averaging over $\underline{\alpha} \in L(n)$ and using part 2, we may replace the integration over V by integration over $V(t_0^n)$. This yields the induction step. □

Let $\Pi_\psi = \mathrm{ind}_N^G \psi$. For $\varphi \in \Pi_\psi$ and $\varphi^\vee \in \Pi_{\psi^{-1}}$ let

$$(\varphi, \varphi^\vee)_{N\backslash G} = \int_{N\backslash G} \varphi(g)\varphi^\vee(g)\, dg.$$

Also set,

$$A^\circ(n) = A \cap \mathcal{B}(n) = \{\mathrm{diag}(t_1, \dots, t_r) \in A : q^{-n} \leq |t_i| \leq q^n,\ \ i = 1, \dots, r\}.$$

Proposition 3.1, together with the argument of [**LM15**, Proposition 2.12], which was communicated to us by Jacquet, yield the following.

COROLLARY 3.3. *There exists a constant c, depending only on r with the following property. Assume that $\varphi \in \Pi_{\psi}^{K(n)}$ and $\varphi^{\vee} \in \Pi_{\psi^{-1}}^{K(n)}$ are both supported in $NA^{\circ}(n)\mathbf{K}$ for some $n \geq 1$. Then $(\Pi_{\psi}(\cdot)\varphi, \varphi^{\vee})_{N\backslash G}$ is supported in the ball $\mathcal{B}(cn)$.*

Indeed, the function $M(g) = (\Pi_{\psi}(\cdot)\varphi, \varphi^{\vee})_{N\backslash G}$ is clearly bi-$K(n)$-invariant. Let f be a compactly supported bi-$K(n)$-invariant function on G. By Fubini's theorem

$$\begin{aligned}\int_G M(g)f(g)\,dg &= \int_G \int_{N\backslash G} \varphi(xg)\varphi^{\vee}(x)f(g)\,dx\,dg \\ &= \int_{N\backslash G}\int_G \varphi(xg)\varphi^{\vee}(x)f(g)\,dg\,dx = \int_{N\backslash G}\int_G \varphi(g)\varphi^{\vee}(x)f(x^{-1}g)\,dg\,dx \\ &= \int_{N\backslash G}\int_{N\backslash G} \varphi^{\vee}(x)\varphi(y)K_f(x,y)\,dy\,dx\end{aligned}$$

where

$$K_f(x,y) = \int_N f(x^{-1}ny)\psi_N(n)\,dn.$$

From Proposition 3.1 we infer that there exists c, depending only on r, such that if f vanishes on $\mathcal{B}(cn)$ then $K_f(x,y) = 0$ for all $x, y \in \mathcal{B}(n)$. The corollary follows.

Finally, we prove Theorem 1.1.

PROOF OF THEOREM 1.1. Let π be a supercuspidal irreducible representation of G. Suppose that $W \in \mathcal{W}^{\psi}(\pi)^{K(n)}$ and $W^{\vee} \in \mathcal{W}^{\psi^{-1}}(\pi^{\vee})^{K(n)}$. By Theorem 2.1, both W and $W^{\vee}$ are supported in $NA(cn)\mathbf{K}$ for suitable c. Upon modifying c, we may write $W(g) = \int_Z W_0(zg)\omega_{\pi}^{-1}(z)\,dz$ (with $\mathrm{vol}(Z \cap \mathbf{K}) = 1$) where $W_0 \in \Pi_{\psi}^{K(n)}$ is supported in $NA^{\circ}(cn)\mathbf{K}$. For instance we can take $W_0 = W1_X$ where X is the set

$$X = \{g \in G : 0 \leq \mathrm{val}(\det g) < r\}.$$

Similarly, write $W^{\vee}(g) = \int_Z W_0^{\vee}(zg)\omega_{\pi}(z)\,dz$ where $W_0^{\vee} \in \Pi_{\psi^{-1}}^{K(n)}$ is supported in $NA^{\circ}(cn)\mathbf{K}$. Then up to a scalar

$$(\pi(g)W, W^{\vee}) = \int_{ZN\backslash G} W(xg)W^{\vee}(x)\,dx = \int_Z (\Pi_{\psi}(zg)W_0, W_0^{\vee})_{N\backslash G}\,dz.$$

The result therefore follows from Corollary 3.3. □

References

[BH97] C. J. Bushnell and G. Henniart, *An upper bound on conductors for pairs*, J. Number Theory **65** (1997), no. 2, 183–196, DOI 10.1006/jnth.1997.2142. MR1462836

[BH99] Colin J. Bushnell and Guy Henniart, *Calculs de facteurs epsilon de paires pour* GL_n *sur un corps local. I* (French, with English and French summaries), Bull. London Math. Soc. **31** (1999), no. 5, 534–542, DOI 10.1112/S0024609399005974. MR1703873

[BH07] Colin J. Bushnell and Guy Henniart, *Counting the discrete series for* $\mathrm{GL}(n)$, Bull. Lond. Math. Soc. **39** (2007), no. 1, 133–137, DOI 10.1112/blms/bdl024. MR2303528

[BH17] C. J. Bushnell and G. Henniart, *Strong exponent bounds for the local Rankin-Selberg convolution*, Bull. Iranian Math. Soc. **43** (2017), no. 4, 143–167. MR3711826

[BHK98] Colin J. Bushnell, Guy M. Henniart, and Philip C. Kutzko, *Local Rankin-Selberg convolutions for* GL_n*: explicit conductor formula*, J. Amer. Math. Soc. **11** (1998), no. 3, 703–730, DOI 10.1090/S0894-0347-98-00270-7. MR1606410

[BK93] Colin J. Bushnell and Philip C. Kutzko, *The admissible dual of* $\mathrm{GL}(N)$ *via compact open subgroups*, Annals of Mathematics Studies, vol. 129, Princeton University Press, Princeton, NJ, 1993. MR1204652

[Bus87] Colin J. Bushnell, *Hereditary orders, Gauss sums and supercuspidal representations of* GL_N, J. Reine Angew. Math. **375/376** (1987), 184–210, DOI 10.1515/crll.1987.375-376.184. MR882297

[CS80] W. Casselman and J. Shalika, *The unramified principal series of p-adic groups. II. The Whittaker function*, Compositio Math. **41** (1980), no. 2, 207–231. MR581582

[FL18] Tobias Finis and Erez Lapid, *An approximation principle for congruence subgroups II: application to the limit multiplicity problem*, Math. Z. **289** (2018), no. 3-4, 1357–1380, DOI 10.1007/s00209-017-2002-0. MR3830253

[FL19] Tobias Finis and Erez Lapid, *On the analytic properties of intertwining operators II: local degree bounds and limit multiplicities*, Israel J. Math. **to appear**, arXiv:1705.08191.

[FLM12] Tobias Finis, Erez Lapid, and Werner Müller, *On the degrees of matrix coefficients of intertwining operators*, Pacific J. Math. **260** (2012), no. 2, 433–456, DOI 10.2140/pjm.2012.260.433. MR3001800

[GK75] I. M. Gel′fand and D. A. Kajdan, *Representations of the group* $\mathrm{GL}(n, K)$ *where* K *is a local field*, Lie groups and their representations (Proc. Summer School, Bolyai János Math. Soc., Budapest, 1971), Halsted, New York, 1975, pp. 95–118. MR0404534

[GR10] Benedict H. Gross and Mark Reeder, *Arithmetic invariants of discrete Langlands parameters*, Duke Math. J. **154** (2010), no. 3, 431–508, DOI 10.1215/00127094-2010-043. MR2730575

[HC99] Harish-Chandra, *Admissible invariant distributions on reductive p-adic groups*, University Lecture Series, vol. 16, American Mathematical Society, Providence, RI, 1999. With a preface and notes by Stephen DeBacker and Paul J. Sally, Jr. MR1702257

[Hen84] Guy Henniart, *La conjecture de Langlands locale pour* GL(3) (French, with English summary), Mém. Soc. Math. France (N.S.) **11-12** (1984), 186. MR743063

[HII08] Kaoru Hiraga, Atsushi Ichino, and Tamotsu Ikeda, *Formal degrees and adjoint* γ*-factors*, J. Amer. Math. Soc. **21** (2008), no. 1, 283–304, DOI 10.1090/S0894-0347-07-00567-X. MR2350057

[How74] Roger Howe, *The Fourier transform and germs of characters (case of* Gl_n *over a p-adic field)*, Math. Ann. **208** (1974), 305–322, DOI 10.1007/BF01432155. MR0342645

[How77] Roger E. Howe, *Some qualitative results on the representation theory of* Gl_n *over a p-adic field*, Pacific J. Math. **73** (1977), no. 2, 479–538. MR0492088

[ILM17] Atsushi Ichino, Erez Lapid, and Zhengyu Mao, *On the formal degrees of square-integrable representations of odd special orthogonal and metaplectic groups*, Duke Math. J. **166** (2017), no. 7, 1301–1348, DOI 10.1215/00127094-0000001X. MR3649356

[JPSS83] H. Jacquet, I. I. Piatetskii-Shapiro, and J. A. Shalika, *Rankin-Selberg convolutions*, Amer. J. Math. **105** (1983), no. 2, 367–464, DOI 10.2307/2374264. MR701565

[Key92] C. David Keys, *A bound for the formal degree of a supercuspidal representation*, Amer. J. Math. **114** (1992), no. 6, 1257–1268, DOI 10.2307/2374762. MR1198303

[LM15] Erez Lapid and Zhengyu Mao, *A conjecture on Whittaker-Fourier coefficients of cusp forms*, J. Number Theory **146** (2015), 448–505, DOI 10.1016/j.jnt.2013.10.003. MR3267120

[Rod75] F. Rodier, *Modèle de Whittaker et caractères de représentations* (French), Noncommutative harmonic analysis (Actes Colloq., Marseille-Luminy, 1974), Springer, Berlin, 1975, pp. 151–171. Lecture Notes in Math., Vol. 466. MR0393355

[Rog81] Jonathan D. Rogawski, *An application of the building to orbital integrals*, Compositio Math. **42** (1980/81), no. 3, 417–423. MR607380

[Ser71] Jean-Pierre Serre, *Conducteurs d'Artin des caractères réels* (French), Invent. Math. **14** (1971), 173–183, DOI 10.1007/BF01418887. MR0321908

[SS97] Peter Schneider and Ulrich Stuhler, *Representation theory and sheaves on the Bruhat-Tits building*, Inst. Hautes Études Sci. Publ. Math. **85** (1997), 97–191. MR1471867

[Vig90] Marie-France Vignéras, *On formal dimensions for reductive p-adic groups*, Festschrift in honor of I. I. Piatetski-Shapiro on the occasion of his sixtieth birthday, Part I (Ramat Aviv, 1989), Israel Math. Conf. Proc., vol. 2, Weizmann, Jerusalem, 1990, pp. 225–266. MR1159104

Department of Mathematics, Weizmann Institute of Science, Rehovot 7610001, Israel

Email address: erez.m.lapid@gmail.com

Proceedings of Symposia in Pure Mathematics
Volume **101**, 2019
https://doi.org/10.1090/pspum/101/01797

On the generalized Springer correspondence

G. Lusztig

Dedicated to Joseph Bernstein for his 72nd birthday

Introduction

0.1. Let $\mathbf{k}$ be an algebraically closed field of characteristic $p \geq 0$. Let G be a connected reductive group over $\mathbf{k}$. Let $\mathfrak{A}$ be the set of all pairs $(\mathbf{c}, \mathcal{E})$ where $\mathbf{c}$ is a unipotent class in G and $\mathcal{E}$ is an irreducible G-equivariant local system on $\mathbf{c}$ defined up to isomorphism; let $\overline{\mathfrak{A}}$ be the set of triples $(P, C, \mathcal{S}_0)$ (up to G-conjugacy) where P is a parabolic subgroup of G, C is a unipotent class of the reductive quotient $\bar{P}$ of P and $\mathcal{S}_0$ is an irreducible $\bar{P}$-equivariant cuspidal local system on C. According to [**L1**, 6.5], there is a canonical surjective map $\mathfrak{A} \to \overline{\mathfrak{A}}$ (whose fibres are called blocks) such that the block corresponding to $(P, C, \mathcal{S}_0) \in \overline{\mathfrak{A}}$ is in natural bijection ("generalized Springer correspondence") with the set $\mathrm{Irr}\mathcal{W}$ of irreducible representations (up to isomorphism) of the finite group $\mathcal{W} := NL/L$ where NL is the normalizer in G of a Levi subgroup L of P. (One can show that $\mathcal{W}$ is naturally a Weyl group depending on the block.) The case considered originally by Springer, see [**Sp**], with some restrictions on p, involves the block corresponding to $(B, \{1\}, \bar{\mathbf{Q}}_l)$ where B is a Borel subgroup of G. The problem of determining explicitly the generalized Springer correspondence can be reduced to the case where G is almost simple, simply connected. For such G the explicit bijection was determined in [**L1**], [**S2**] and the references there, for any block except for

(a) two blocks for G of type E_6 with $p \neq 3$, with $\mathcal{W}$ of type G_2 and

(b) two blocks for G of type E_8 with $p = 3$, with $\mathcal{W}$ of type G_2;

for these blocks the method of [**S2**] (based mostly on the restriction theorem [**L1**, 8.3]) had the following gap: it gave the explicit bijection only up to composition with a permutation of $\mathrm{Irr}\mathcal{W}$ which interchanges the two 2-dimensional irreducible representations of $\mathcal{W}$ and keeps fixed all the other irreducible representations.

0.2. Let $(\mathbf{c}, \mathcal{E}) \in \mathfrak{A}$ and let $(\mathbf{c}', \mathcal{E}') \in \mathfrak{A}$ be such that $\mathbf{c}'$ is contained in the closure $\bar{\mathbf{c}}$ of $\mathbf{c}$. For an integer k let $m^k_{(\mathbf{c}', \mathcal{E}'), (\mathbf{c}, \mathcal{E})}$ be the multiplicity of $\mathcal{E}'$ in the local system on $\mathbf{c}'$ obtained by restricting to $\mathbf{c}'$ the k-th cohomology sheaf of the intersection cohomology complex $IC(\bar{\mathbf{c}}, \mathcal{E})$. Let

$$m_{(\mathbf{c}', \mathcal{E}'), (\mathbf{c}, \mathcal{E})} = \sum_{k \geq 0} m^k_{(\mathbf{c}', \mathcal{E}'), (\mathbf{c}, \mathcal{E})} q^{k/2} \in \mathbf{N}[q^{1/2}]$$

Supported in part by the National Science Foundation.

where $q^{1/2}$ is an indeterminate. In [**L2**, 24.8] it is shown (assuming that p is not a bad prime for G) that $m_{(\mathbf{c}',\mathcal{E}'),(\mathbf{c},\mathcal{E})} \in \mathbf{N}[q]$, that $m_{(\mathbf{c}',\mathcal{E}'),(\mathbf{c},\mathcal{E})}$ is 0 if $(\mathbf{c}',\mathcal{E}'),(\mathbf{c},\mathcal{E})$ are not in the same block and that $m_{(\mathbf{c}',\mathcal{E}'),(\mathbf{c},\mathcal{E})}$ is explicitly computable in terms of the generalized Springer correspondence if $(\mathbf{c}',\mathcal{E}'),(\mathbf{c},\mathcal{E})$ are in the same block.

0.3. In this subsection we consider a block of G as in 0.1(a),(b) where p not a bad prime for G; thus we must be in case 0.1(a) and $p \notin \{2,3\}$. An attempt to close the gap in this case was made in [**L2**, 24.10], based on computing the polynomials $m_{(\mathbf{c}',\mathcal{E}'),(\mathbf{c},\mathcal{E})}$ (see 0.2) for $(\mathbf{c}',\mathcal{E}'),(\mathbf{c},\mathcal{E})$ in this block. Unfortunately, the attempt in [**L2**] contained a calculation error. (I thank Frank Lübeck for pointing this out to me.) As a result, the gap remained. In this paper we refine the analysis in [**L2**, 24.10] and close the gap for the blocks in 0.1(a) with $p \notin \{2,3\}$ (see Theorem 5.5). We now describe our strategy. Using the algorithm of [**L2**, 24.10] (based on the two possible scenarios corresponding to the two possible inputs for the generalized Springer correspondence) we compute the polynomials $m_{(\mathbf{c}',\mathcal{E}'),(\mathbf{c},\mathcal{E})}$ for $(\mathbf{c}',\mathcal{E}'),(\mathbf{c},\mathcal{E})$ in the block. We get two sets of polynomials one in each of the two scenarios. In fact the algorithm gives only rational functions in q which, by a miracle, turn out to be in $\mathbf{N}[q]$ in both scenarios. So by this computation one cannot rule out one of the scenarios and a further argument is needed. Let $u \in G$ be a unipotent element of G such that the Springer fibre at u is 3-dimensional (such u is unique up to conjugacy) and let $\mathbf{X}_u$ be the "generalized Springer fibre" at u (attached to the block). It is known from [**L1**] that $\mathcal{W}$ acts naturally on $H^j_c(\mathbf{X}_u,?)$ where ? is a suitable local system on $\mathbf{X}_u$ attached to the block. By a known result, from the knowledge of the polynomials $m_{(\mathbf{c}',\mathcal{E}'),(\mathbf{c},\mathcal{E})}$ one can extract the trace of a simple reflection of $\mathcal{W}$ on $\sum_j(-1)^j H^j_c(\mathbf{X}_u,?)$. It turns out that this trace is 1 in one scenario and -1 in the other scenario. We show that this trace is equal to the Euler characteristic of a certain explicit open subvariety of the Springer fibre at u. We then try to compute from first principles this Euler characteristic; the computation occupies most of the paper. The computation does not give the exact value of the Euler characteristic; it only shows that it is one of the numbers $0,1,2$. Since we already know that it is 1 or -1 we deduce that it is 1; moreover, this determines which scenario is real and which one is not.

0.4. Note that Theorem 5.5 completes the explicit determination of the generalized Springer correspondence for any G and any block, assuming that p is not a bad prime for G.

In the case where p is a bad prime for G, a gap remains, and in 6.2 we state a conjecture about it.

0.5. *Notation.* All algebraic varieties are assumed to be over $\mathbf{k}$ and all algebraic groups are assumed to be affine.

For any connected algebraic group H let U_H be the unipotent radical of H and let $\bar{H} = H/U_H$, a connected reductive group; let $\pi_H : H \to \bar{H}$ be the obvious homomorphism. Let $\mathcal{Z}^0_{\bar{H}}$ be the connected centre of $\bar{H}$. Let $\mathcal{B}^H$ (resp. $\mathcal{B}^{\bar{H}}$) be the variety of Borel subgroups of H (resp. $\bar{H}$). Note that $\beta \mapsto \pi_H^{-1}(\beta)$ is an isomorphism $\mathcal{B}^{\bar{H}} \xrightarrow{\sim} \mathcal{B}^H$. Now H (resp. $\bar{H}$) acts by simultaneous conjugation on $\mathcal{B}^H \times \mathcal{B}^H$ (resp. $\mathcal{B}^{\bar{H}} \times \mathcal{B}^{\bar{H}}$); the orbits of this action are naturally parametrized by the Weyl group W^H (resp. $W^{\bar{H}}$) of H (resp. $\bar{H}$). The identification $\mathcal{B}^{\bar{H}} \times \mathcal{B}^{\bar{H}} \leftrightarrow \mathcal{B}^H \times \mathcal{B}^H$ via $\pi_H \times \pi_H$, induces an identification $W^H = W^{\bar{H}}$.

If H is a connected reductive group we denote by H_{ad} the adjoint group of H. In this paper G is a fixed connected reductive group. We write $\mathcal{B} = \mathcal{B}^G$, $W = W^G$. Let $\mathcal{O}_w$ be the G-orbit on $\mathcal{B} \times \mathcal{B}$ indexed by $w \in W$. The simple reflections in W are denoted as $\{s_i; i \in I\}$ where I is an indexing set. For any $J \subset I$ let W_J be the subgroup of W generated by $\{s_i; i \in J\}$. Let $\epsilon_J : W_J \to \{\pm 1\}$ be the homomorphism given by $\epsilon_J(s_i) = -1$ for all $i \in J$. The set $\mathcal{P}$ of parabolic subgroups of G is naturally partitioned as $\mathcal{P} = \sqcup_{J \subset I} \mathcal{P}_J$ where for $J \subset I$, $\mathcal{P}_J$ consists of parabolic subgroups P with the following property: for $w \in W$, we have $(B, B') \in \mathcal{O}_w$ for some B, B' in $\mathcal{B}$, $B \subset P$, $B' \subset P$ if and only if $w \in W_J$. Note that $\mathcal{B} = \mathcal{P}_\emptyset$.

Now let $J \subset I$ and let $P \in \mathcal{P}_J$. Then $\mathcal{B}^P$ is equal to the closed subvariety $\{B \in \mathcal{B}; B \subset P\}$ of $\mathcal{B}$ and the obvious map $W^P \to W$ identifies W^P with the subgroup W_J of W. The set of simple reflections of $W^P = W^{\bar{P}}$ becomes the set $\{s_i; i \in J\}$.

For $i \in I$ we write $\mathcal{P}_i$ instead of $\mathcal{P}_{\{i\}}$ and $\mathcal{P}^i$ instead of $\mathcal{P}_{I - \{i\}}$, a class of maximal parabolic subgroups. For any $B \in \mathcal{B}$ there is a unique $P \in \mathcal{P}^i$ such that $B \subset P$; we set $P = B(i)$. A subvariety of $\mathcal{B}$ is said to be an i-line if it is of the form $\mathcal{B}^P$ for some $P \in \mathcal{P}_i$ (which is necessarily unique). If P, P' are parabolic subgroups of G we write $P \spadesuit P'$ whenever P, P' contain a common Borel subgroup.

Let l be a fixed prime number such that $l \neq p$. All local systems are assumed to be $\bar{\mathbf{Q}}_l$-local systems.

0.6. Most of this paper was written during a month long visit to the Institute of Mathematics of the Academia Sinica, Taipei, in the summer of 2016. I thank Shun-Jen Cheng for his hospitality.

Contents

1. Preliminaries

1.1. We fix a unipotent element $u \in G$; let $\mathcal{B}_u = \{B \in \mathcal{B}; u \in B\}$. This is a nonempty subvariety of $\mathcal{B}$. Let $\underline{\mathcal{B}}_u$ be the set of irreducible components of $\mathcal{B}_u$. According to Spaltenstein [**S1**], for $X \in \underline{\mathcal{B}}_u$, $d_u := \dim X$ depends only on u, not on X. For any $X \in \underline{\mathcal{B}}_u$ let J_X be the set of all $i \in I$ such that X is a union of i-lines. We show:

(a) *Let $J = J_X$, let $B \in X$ and let P be the unique subgroup in $\mathcal{P}_J$ such that $B \subset P$. Then $\mathcal{B}^P \subset X$.*

Let $B' \in X$. We can find a sequence $B = B_0, B_1, \ldots, B_t = B'$ in $\mathcal{B}$ such that for $k = 0, 1, \ldots, t-1$ we have $(B_k, B_{k+1}) \in \mathcal{O}_{s_{i_k}}$ where $i_k \in J$. We show by induction on k that $B_k \in X$. For $k = 0$ this holds by assumption. Assume now that $k \geq 1$. By the induction hypothesis we have $B_{k-1} \in X$. Since $i_k \in J_X$, the i_k-line containing B_{k-1} is contained in X. Since B_k is contained in this i_k-line we have $B_k \in X$. This completes the induction. We see that $B' \in X$. This proves (a).

We show:

(b) *In the setup of (a), assume in addition that d_u is equal to ν_J, the number of reflections in W_J. Then $\mathcal{B}^P = X$.*

Note that $\mathcal{B}^P$ is a closed irreducible subvariety of dimension ν_J of X and X is irreducible of dimension d_u. The result follows.

1.2. In this subsection we assume that $G_{ad} = \prod_{i\in I]}^m G_i$ where $G_i \cong PGL_2(\mathbf{k})$ for all $i \in I$. Assume that $B \in \mathcal{B}_u$ is such that for any $i \in I$, the i-line through B is contained in $\mathcal{B}_u$. We show that $u = 1$. We write $u = (u_i)$ where $u_i \in G_i$ for all i. Let $K = \{i \in I; u_i = 1\}$. Then $\mathcal{B}_u$ is a product of copies of $\mathbf{P}^1$ indexed by K. Moreover, $\mathcal{B}_u$ contains an i-line if and only if $i \in K$. Our assumption implies that $K = I$ hence $u = 1$, as asserted.

1.3. We shall need the following variant of 1.1(a),(b).

(a) *Let J be a subset of I such that $s_j s_{j'} = s_{j'} s_j$ for all j, j' in J. Let $B \in \mathcal{B}_u$ be such that for any $j \in J$ the j-line containing B is contained in $\mathcal{B}_u$. Let P be the unique subgroup in $\mathcal{P}_J$ such that $B \subset P$. Assume that $\sharp(J) = d_u$. Then $X = \mathcal{B}^P$ is an irreducible component of $\mathcal{B}_u$ and $J_X = J$.*

We apply 1.2 with G, u replaced by $\bar{P}, \pi_P(u)$. Note that $\bar{P}_{ad}$ is a product of copies of $PGL_2(\mathbf{k})$. We see that $\pi_P(u) = 1$ that is, $u \in U_P$. Then any Borel subgroup of P contains u. Now (a) follows.

1.4. Let $[\underline{\mathcal{B}}_u]$ be the $\bar{\mathbf{Q}}_l$-vector space with basis $\{X; X \in \underline{\mathcal{B}}_u\}$. Let ρ_u be the Springer representation of W on the vector space $[\underline{\mathcal{B}}_u]$. The following property of ρ_u appeared in a letter of the author to Springer (March 1978), see also [**Ho**]:

(a) *Let $i \in I$ and let $X \in \underline{\mathcal{B}}_u$. Then*

$s_i X = -X$ if $i \in J_X$;

$s_i X - X \in \sum_{X' \in \underline{\mathcal{B}}_u; i \in J_{X'}} \mathbf{Z}X'$ if $i \notin J_X$.

For $J \subset I$ let $(\underline{\mathcal{B}}_u)'_J = \{X \in \underline{\mathcal{B}}_u; J \subset J_X\}$ and let $[(\underline{\mathcal{B}}_u)'_J]$ be the subspace of $[\underline{\mathcal{B}}_u]$ spanned by $(\underline{\mathcal{B}}_u)'_J$. We have $(\underline{\mathcal{B}}_u)'_J = \cap_{i\in J}(\underline{\mathcal{B}}_u)'_{\{i\}}$ and $[(\underline{\mathcal{B}}_u)'_J] = \cap_{i\in J}[(\underline{\mathcal{B}}_u)'_{\{i\}}]$. From (a) we see that for any $i \in J$, we have $[(\underline{\mathcal{B}}_u)'_{\{i\}}] = \{v \in [\underline{\mathcal{B}}_u]; s_i v = -v\}$. Hence $[(\underline{\mathcal{B}}_u)'_J] = \cap_{i\in J}\{v \in [\underline{\mathcal{B}}_u]; s_i v = -v\} = \{v \in [\underline{\mathcal{B}}_u]; wv = \epsilon_J(w)v \forall w \in W_J v\}$. Thus, $\sharp(\underline{\mathcal{B}}_u)'_J$ is equal to the multiplicity $(\epsilon_J : \rho_u|_{W_J})$ of ϵ_J in the W_J-module $\rho_u|_{W_J}$. Let $(\underline{\mathcal{B}}_u)_J = \{X \in \underline{\mathcal{B}}_u; J = J_X\}$ We have $\sharp\underline{\mathcal{B}}_u)'_J = \sum_{J'; J \subset J' \subset I} \sharp(\underline{\mathcal{B}}_u)_{J'}$. Hence $\sharp\underline{\mathcal{B}}_u)_J = \sum_{J'; J\subset J' \subset I}(-1)^{|J'|-|J|}\sharp\underline{\mathcal{B}}_u)'_{J'}$ so that

(b) $\sharp(\underline{\mathcal{B}}_u)_J = \sum_{J'; J\subset J'\subset I}(-1)^{|J'|-|J|}(\epsilon_{J'} : \rho_u|_{W_{J'}})$.

1.5. In this subsection we assume that we are given $J \subset I$, $P \in \mathcal{P}_J$ and $\bar{u} \in \bar{P}$ such that $u \in P$, $\bar{u} = \pi_P(u)$ and such that, if C is the conjugacy class of u in G then

(a) $C \cap \pi_P^{-1}(\bar{u})$ is open dense in $\pi_P^{-1}(\bar{u})$.

Then C is induced (in the sense of [**LS**]) by the $\bar{P}$-conjugacy class of $\bar{u}$.

Let $\mathcal{B}^P, \mathcal{B}^{\bar{P}}$ be as in 0.4 and let $\mathcal{B}^{\bar{P}}_{\bar{u}} = \{\beta \in \mathcal{B}^{\bar{P}}; \bar{u} \in \beta\}$. Let $d_{\bar{u}} = \dim \mathcal{B}^{\bar{P}}_{\bar{u}}$. From [**LS**] it is known that

(b) $d_u = d_{\bar{u}}$.

For any irreducible component ξ of $\mathcal{B}^{\bar{P}}_{\bar{u}}$, the subset J_ξ of J is defined as in 1.1 (replacing G, I, u by $\bar{P}, J, \bar{u}$). Let $\tilde{\xi}$ be the image of ξ under $\beta \mapsto \pi_P^{-1}(\beta)$; note that $\tilde{\xi}$ is a closed irreducible subvariety of $\mathcal{B}_u$ of dimension $d_{\bar{u}}$ hence by (b) is an irreducible component of $\mathcal{B}_u$. From the definitions we see that

(c) $J_\xi = J_{\tilde{\xi}}$ where $J_{\tilde{\xi}}$ is defined as in 1.1.

1.6. Let $i \in I$. Let $\mathcal{B}_{u,i}$ be the set of all $B \in X$ such that the i-line though B is contained in $\mathcal{B}_u$ or equivalently such that if $P_i \in \mathcal{P}_i$ is defined by $B \subset P_i$ then $u \in U_{P_i}$.

For $X \in \underline{\mathcal{B}}_u$ let $X_i = X \cap \mathcal{B}_{u,i}$.

2. A trace computation

2.1. We fix $J \subset I$ such that $J \neq I$ and $P_J \in \mathcal{P}_J$. We write π_J instead of π_{P_J}. We fix a unipotent conjugacy class C of $\bar{P}_J$ and let $\mathcal{S}_0$ be an irreducible cuspidal $\bar{P}_J$-equivariant local system on C. Let $\mathfrak{S} = C\mathcal{Z}^0_{\bar{P}_J}$, a locally closed subvariety of $\bar{P}_J$; let $cl(\mathfrak{S})$ be the closure of $\mathfrak{S}$ in $\bar{P}_J$. Let $\mathcal{S}$ be the inverse image of $\mathcal{S}_0$ under $\mathfrak{S} \to C$ (taking unipotent part). Let $\mathcal{W} = N_W W_J / W_J$ where $N_W W_J$ is the normalizer of W_J in W. For any $i \in I - J$ let $N_{W_{J\cup i}} W_J$ be the normalizer of W_J in $W_{J\cup i}$. From [**L1**, 9.2] we see that $N_{W_{J\cup i}} W_J / W_J$ has order 2 and that $\mathcal{W}$ is a Coxeter group with simple reflections $\{\sigma_i; i \in I - J\}$ where σ_i is the unique nonidentity element of $N_{W_{J\cup i}} W_J / W_J$, viewed as an element of $\mathcal{W}$. Let

$$\mathbf{X} = \{(xP_J, g) \in G/P_J \times G; x^{-1}gx \in P_J, \pi_J(x^{-1}gx) \in \mathfrak{S}\}.$$

Let $\overline{\mathbf{Y}}$ be the set of all $g \in G$ such that for some $xP_J \in G/P_J$ we have $x^{-1}gx \in P_J$, $\pi_J(\xi^{-1}gx) \in cl(\mathfrak{S})$; this is a closed subset of G. Define $\tilde{\pi} : \mathbf{X} \to \overline{\mathbf{Y}}$ by $\tilde{\pi}(xP_J, g) = g$. We define a local system $\hat{\mathcal{S}}$ on $\mathbf{X}$ by requiring that $\hat{\mathcal{S}}_{(xP_J, g)} = \mathcal{S}_{\pi_J(x^{-1}gx)}$. Note that $\hat{\mathcal{S}}$ is well defined by the $\bar{P}_J$-equivariance of $\mathcal{S}_0$. Let $K = \tilde{\pi}_! \hat{\mathcal{S}} \in \mathcal{D}(\overline{\mathbf{Y}})$. From [**L1**, 4.5, 9.2] we see that K is an intersection cohomology complex on $\overline{\mathbf{Y}}$ and that we have canonically $\mathrm{End}(K) = \bar{\mathbf{Q}}_l[\mathcal{W}]$. Let $g \in \overline{\mathbf{Y}}$ and let

$$\mathbf{X}_g = \{xP_J \in G/P_J; x^{-1}gx \in P_J, \pi_J(x^{-1}gx) \in \mathfrak{S}\}.$$

Now $\mathbf{X}_g$ is a subvariety of $\mathbf{X}$ via $xP_J \mapsto (xP_J, g)$. We denote the restriction of $\hat{\mathcal{S}}$ from $\mathbf{X}$ to $\mathbf{X}_g$ again by $\hat{\mathcal{S}}$. Since $H^j_c(\mathbf{X}_g, \hat{\mathcal{S}})$ (for $j \in \mathbf{Z}$) is a stalk of a cohomology sheaf of K at g, we see that $H^j_c(\mathbf{X}_g, \hat{\mathcal{S}})$ is naturally a $\mathcal{W}$-module.

We now fix a subset J' of I such that $J \subset J'$. We define $P_{J'} \in \mathcal{P}_{J'}$ by $P_J \subset P_{J'}$. We write $U_{J'}, \pi_{J'}$ instead of $U_{P_{J'}}, \pi_{P_{J'}}$. We shall give an alternative description of the restriction of the $\mathcal{W}$-module $H^j_c(\mathbf{X}_g, \hat{\mathcal{S}})$ to the subgroup $\mathcal{W}'$ of $\mathcal{W}$ generated by $\{\sigma_i; i \in J' - J\}$.

We have a commutative diagram of algebraic varieties with cartesian squares

$$\begin{array}{ccccccc}
\tilde{M}_{J',g} & \xrightarrow{\tilde{j}} & M_{J'} & \xleftarrow{\tilde{\omega}} & E \times X & \xrightarrow{\tilde{q}} & X \\
{\scriptstyle \pi''}\downarrow & & {\scriptstyle \pi'}\downarrow & & {\scriptstyle 1\times\pi}\downarrow & & {\scriptstyle \pi}\downarrow \\
M_{J',g} & \xrightarrow{j} & M_{J'} & \xleftarrow{\omega} & E \times \bar{Y} & \xrightarrow{q} & \bar{Y}
\end{array}$$

where

$$X = \{(xP_J, zU_{J'}) \in (P_{J'}/P_J) \times (P_{J'}/U_{J'}); x^{-1}zx \in P_{\mathbf{J}}, \pi_J(x^{-1}zx) \in \mathfrak{S}\};$$

$\bar{Y}$ is the set of all $zU_{J'} \in P_{J'}/U_{J'}$ such that for some $xP_J \in P_{J'}/P_J$, we have $x^{-1}zx \in P_J$, $\pi_J(x^{-1}zx) \in cl(\mathfrak{S})$;

$\tilde{M}_{J'}$ is the set of all pairs $(xP_J, y(xU_{J'}x^{-1}))$ where $xP_J \in G/P_J$, $y(xU_{J'}x^{-1}) \in (xP_Jx^{-1})/(xU_{J'}x^{-1})$ are such that $\pi_J(x^{-1}yx) \in \mathfrak{S}$;

$M_{J'}$ is the set of all pairs $(xP_{J'}, y(xU_{J'}x^{-1}))$ where $xP_{J'} \in G/P_{J'}$, $y(xU_{J'}x^{-1}) \in (xP_{J'}x^{-1})/(xU_{J'}x^{-1})$ are such that for some $v \in P_{J'}$, we have $y \in xvP_Jv^{-1}x^{-1}$ and $\pi_J(v^{-1}x^{-1}yxv) \in cl(\mathfrak{S})$;

$$\tilde{M}_{J',g} = \{(xP_J, y(xU_{J'}x^{-1})) \in \tilde{M}_{J'}; g \in xP_Jx^{-1}, g^{-1}y \in xU_{J'}x^{-1}\};$$

$$M_{J',g} = \{(xP_{J'}, y(xU_{J'}x^{-1})) \in M_{J'}; g \in xP_{J'}x^{-1}, g^{-1}y \in xU_{J'}x^{-1}\};$$

$$E = G/U_{J'};$$

$j, \tilde{j}$ are the obvious imbeddings; $q, \tilde{q}$ are the obvious projections;

$$\omega(hU_{J'}, zU_{J'}) = (hP_{J'}, (hzh^{-1})(hU_{J'}h^{-1})),$$

$$\tilde{\omega}(hU_{J'}, xP_J, zU_{J'}) = (hxP_J, (hzh^{-1})(hxU_{J'}x^{-1}h^{-1}));$$

$$\pi(xP_J, zU_{J'}) = zU_{J'}, \pi'(xP_J, y(xU_{J'}x^{-1})) = (xP_{J'}, y(xU_{J'}x^{-1})),$$

$$\pi''(xP_J, y(xU_{J'}x^{-1})) = (xP_{J'}, y(xU_{J'}x^{-1})).$$

All maps in the diagram are compatible with the natural actions of $\bar{P}_{J'}$ where the action of $\bar{P}_{J'}$ on the four spaces on the left is trivial. Moreover, $\tilde{\omega}$ and ω are principal $\bar{P}_{J'}$-bundles. We define a local system $\bar{\mathcal{S}}$ on X by requiring that $\bar{\mathcal{S}}_{(xP_J, zU_{J'})} = \mathcal{S}_{\pi_J(x^{-1}zx)}$. We define a local system $\dot{\mathcal{S}}$ on $\tilde{M}_{J'}$ by requiring that $\dot{\mathcal{S}}_{(xP_J, y(xU_{J'}x^{-1}))} = \mathcal{S}_{\pi_J(x^{-1}yx)}$. Note that $\bar{\mathcal{S}}, \dot{\mathcal{S}}$ are well defined by the $\bar{P}_J$-equivariance of $\mathcal{S}_0$. Note also that $K' := \pi_! \bar{\mathcal{S}}$ is like K above (with G replaced by $\bar{P}_{J'}$) hence from [**L1**, 4.5, 9.2] we see that K' is an intersection cohomology sheaf on $\bar{Y}$ and we have canonically $\mathrm{End}_{\mathcal{D}(\bar{Y})}(K') = \mathrm{End}_{\mathcal{D}_{\bar{P}_{J'}}(\bar{Y})}(K') = \bar{\mathbf{Q}}_l[\mathcal{W}']$.

We have an isomorphism $\mathbf{X}_g \xrightarrow{\sim} \tilde{M}_{J',g}$ given by $xP_J \mapsto (xP_J, g(xU_{J'}x^{-1}))$, under which these two varieties are identified; then the local system $\tilde{j}^*\dot{\mathcal{S}}$ on $\tilde{M}_{J',g}$ becomes $\hat{\mathcal{S}}$. We have $\tilde{q}^*\bar{\mathcal{S}} = \tilde{\omega}^*\dot{\mathcal{S}}$ hence $q^*\pi_!\bar{\mathcal{S}} = \omega^*(\pi'_!\dot{\mathcal{S}})$. The functors

$$\mathcal{D}_{\bar{P}_{J'}}(\bar{Y}) \xrightarrow{q^*} \mathcal{D}_{\bar{P}_{J'}}(E \times \bar{Y}) \xleftarrow{\omega^*} \mathcal{D}(M_{J'}) \xrightarrow{j^*} \mathcal{D}(M_{J',g})$$

induce algebra homomorphisms

$$\mathrm{End}_{\mathcal{D}_{\bar{P}_{J'}}(\bar{Y})}(\pi_!\bar{\mathcal{S}}) \to \mathrm{End}_{\mathcal{D}_{\bar{P}_{J'}}(E\times\bar{Y})}(q^*\pi_!\bar{\mathcal{S}}) \leftarrow \mathrm{End}_{\mathcal{D}(M_{J'})}(\pi'_!\dot{\mathcal{S}})$$
$$\to \mathrm{End}_{\mathcal{D}(M_{J',g})}(\pi''_!\tilde{j}^*\dot{\mathcal{S}})$$

of which the second one is an isomorphism since ω is a principal $\bar{P}_{J'}$-bundle. Taking the composition of the first homomorphism with the inverse of the second one and with the third one and identifying

$$\mathrm{End}_{\mathcal{D}_{\bar{P}_{J'}}(\bar{Y})}(\pi_!\bar{\mathcal{S}}) = \mathrm{End}_{\mathcal{D}(\bar{Y})}(\pi_!\bar{\mathcal{S}}) = \bar{\mathbf{Q}}_l[\mathcal{W}']$$

we obtain an algebra homomorphism

$$\bar{\mathbf{Q}}_l[\mathcal{W}'] \to \mathrm{End}_{\mathcal{D}(M_{J',g})}(\pi''_!\tilde{j}^*\dot{\mathcal{S}}).$$

It follows that

$$H^j_c(\mathbf{X}_g, \hat{\mathcal{S}}) = H^j_c(\tilde{M}_{J',g}, \tilde{j}^*\dot{\mathcal{S}}) = H^j_c(M_{J',g}, \pi''_!\tilde{j}^*\dot{\mathcal{S}})$$

is naturally a module over the algebra $\mathrm{End}_{\mathcal{D}(M_{J',g})}(\pi''_!\tilde{j}^*\dot{\mathcal{S}})$ hence a module over $\bar{\mathbf{Q}}_l[\mathcal{W}']$. From the definitions we see that this $\mathcal{W}'$-module structure on $H^j_c(\mathbf{X}_g, \hat{\mathcal{S}})$ coincides with the restriction to $\mathcal{W}'$ of the $\mathcal{W}$-module structure on $H^j_c(\mathbf{X}_g, \hat{\mathcal{S}})$ considered above.

2.2. We now assume that $i \in I - J$ and that $J' = J \cup \{i\}$. Let

$$t_i(g) = \sum_j (-1)^j \mathrm{tr}(\sigma_i, H_c^j(\mathbf{X}_g, \hat{\mathcal{S}}))$$

where σ_i acts by the $\mathcal{W}$-action. By 2.1, we have

$$t_i(g) = \sum_j (-1)^j \mathrm{tr}(\sigma_i, H_c^j(M_{J',g}, \pi''_! \tilde{j}^* \dot{\mathcal{S}}))$$

where σ_i acts by the $\mathcal{W}'$-action. The σ_i action on $\pi_! \bar{\mathcal{S}}$ induces an σ_i-action on $\mathcal{H}^{j'}(\pi_! \bar{\mathcal{S}})$ and this induces an σ_i-action on $\mathcal{H}^{j'}(\pi_! \bar{\mathcal{S}})$ and an σ_i-action on

$$H_c^j(M_{J',g}, \mathcal{H}^{j'}(\pi''_! \tilde{j}^* \dot{\mathcal{S}})).$$

(Here $\mathcal{H}^{j'}()$ denotes the j'-th cohomology sheaf). We have a spectral sequence

$$H_c^j(M_{J',g}, \mathcal{H}^{j'}(\pi''_! \tilde{j}^* \dot{\mathcal{S}})) \implies H_c^{j+j'}(M_{J',g}, \pi''_! \tilde{j}^* \dot{\mathcal{S}})$$

which is compatible with the σ_i-actions. It follows that

(a) $$t_i(g) = \sum_{j,j'} (-1)^{j+j'} \mathrm{tr}(\sigma_i, H_c^j(M_{J',g}, \mathcal{H}^{j'}(\pi''_! \tilde{j}^* \dot{\mathcal{S}}))).$$

We now make the further assumption that g is unipotent and that s_i commutes with W_J. In this case we have

$$\mathbf{X}_g = \{xP_J \in G/P_J; x^{-1}gx \in P_J, \pi_J(x^{-1}gx) \in C\};$$

moreover, the isomorphism $W_{\{i\}} \times W_J \xrightarrow{\sim} W_{J'}$ induced by multiplication corresponds to a direct product decomposition $(\bar{P}_{J'})_{ad} = H' \times H''$ where $H' \cong PGL_2(\mathbf{k})$, $H'' \cong (\bar{P}_J)_{ad}$.

Let $\bar{C}$ be the image of C under $\bar{P}_J \to (\bar{P}_J)_{ad} = H''$. Let C^1 be the unipotent class in $\bar{P}_{J'}$ whose image in $(\bar{P}_{J'})_{ad} = H' \times H''$ is $\{1\} \times \bar{C}$. Let C^r be the unipotent class in $\bar{P}_{J'}$ whose image in $(\bar{P}_{J'})_{ad} = H' \times H''$ is (regular unipotent class in H')$\times \bar{C}$. Then $C^1 \cup C^r$ is exactly the set of unipotent elements in the image of $\pi : X \to \bar{Y}$. Now $\pi : X \to \bar{Y}$ restricts to $\pi^{-1}(C^1) \to C^1$ which is a $\mathbf{P}^1$-bundle and to $\pi^{-1}(C^r) \to C^r$ which is an isomorphism. There is a well defined local system $\mathcal{S}_1$ on C^1 whose inverse image under $\pi^{-1}(C^1) \to C^1$ is the restriction of $\bar{\mathcal{S}}$ to $\pi^{-1}(C^1)$. There is a well defined local system $\mathcal{S}_r$ on C^r whose inverse image under $\pi^{-1}(C^r) \to C^r$ is the restriction of $\bar{\mathcal{S}}$ to $\pi^{-1}(C^r)$. Let

$$M_{J',g,1} = \{(xP_{J'}, y(xU_{J'}x^{-1})) \in M_{J',g}; \pi_{J'}(x^{-1}gx) \in C^1\},$$

$$M_{J',g,r} = \{(xP_{J'}, y(xU_{J'}x^{-1})) \in M_{J',g}; \pi_{J'}(x^{-1}gx) \in C^r\}.$$

Using the cartesian squares in 2.1 we see that $\pi'' : \tilde{M}_{J',g} \to M_{J',g}$ restricts to $\pi''_1 : \pi''^{-1}(M_{J',g,1}) \to M_{J',g,1}$ which is a $\mathbf{P}^1$-bundle and to $\pi''_r : \pi''^{-1}(M_{J',g,r}) \to M_{J',g,r}$ which is an isomorphism. Moreover there is a well defined local system $\mathcal{S}_1$ on $M_{J',g,1}$ such that $\pi''^*_1 \mathcal{S}_1$ is the restriction $\dot{\mathcal{S}}_1$ of $\tilde{j}^* \dot{\mathcal{S}}$ to $\pi''^{-1}(M_{J',g,1})$; there is a well defined local system $\mathcal{S}_r$ on $M_{J',g,r}$ such that $\pi''^*_r \mathcal{S}_r$ is the restriction $\dot{\mathcal{S}}_r$ of $\tilde{j}^* \dot{\mathcal{S}}$ to $\pi''^{-1}(M_{J',g,r})$. Since $M_{J',g} = M_{J',g,1} \cup M_{J',g,r}$ is a partition and $M_{J',g,1}$

(resp. $M_{J',g,r}$) is closed (resp. open) in $M_{J',g}$ we see that (a) implies

$$t_i(g) = \sum_{j,j'}(-1)^{j+j'}\operatorname{tr}(\sigma_i, H_c^j(M_{J',g,1}, \mathcal{H}^{j'}(\pi''_{1!}\dot{\mathcal{S}}_1))) + \sum_{j,j'}(-1)^{j+j'}\operatorname{tr}(\sigma_i, H_c^j(M_{J',g,r}, \mathcal{H}^{j'}(\pi''_{r!}\dot{\mathcal{S}}_r))).$$

In the first sum over j, j' we can assume that $j' \in \{0,2\}$; the σ_i-action is multiplication by -1 if $j' = 2$ and by 1 if $j' = 0$; in the second sum over j, j' we can assume that $j' = 0$; moreover the σ_i-action is multiplication by 1. (Here we use the definition of the σ_i action on K'.) Thus, we have

$$\begin{aligned} t_i(g) &= -\sum_j(-1)^j \dim H_c^j(M_{J',g,1}, \mathcal{H}^2(\pi''_{1!}\dot{\mathcal{S}}_1))) \\ &+ \sum_j(-1)^j \dim H_c^j(M_{J',g,1}, \mathcal{H}^0(\pi''_{1!}\dot{\mathcal{S}}_1))) + \sum_j(-1)^j \dim H_c^j(M_{J',g,r}, \mathcal{H}^0(\pi''_{r!}\dot{\mathcal{S}}_r))) \\ &= -\sum_j(-1)^j \dim H_c^j(M_{J',g,1}, \mathcal{S}_1(-2)) + \sum_j(-1)^j \dim H_c^j(M_{J',g,1}, \mathcal{S}_1) \\ &+ \sum_j(-1)^j \dim H_c^j(M_{J',g,r}, \mathcal{S}_r). \end{aligned}$$

Note that the Tate twist does not affect the dimension; hence after cancellation we obtain

$$t_i(g) = \sum_j(-1)^j \dim H_c^j(M_{J',g,r}, \mathcal{S}_r) = \sum_j(-1)^j \dim H_c^j(\pi''^{-1}(M_{J',g,r}), \dot{\mathcal{S}}_r)$$

that is,

(b) $$t_i(g) = \chi(\pi''^{-1}(M_{J',g,r}), \dot{\mathcal{S}}_r),$$

where, for an algebraic variety X and a local system $\mathcal{E}$ on X we set $\chi(X,\mathcal{E}) = \sum_j(-1)^j \dim H_c^j(X,\mathcal{E})$. (We also set $\chi(X) = \chi(X, \bar{\mathbf{Q}}_l)$.)

3. Computations in certain groups of semisimple rank ≤ 5

3.1. Until the end of 3.14 we assume that G_{ad} is of type D_4 and that $u \in G$ (see 1.1) is such that $d_u = 3$. Note that u is unique up to conjugation. We can write $I = \{\alpha, \beta, \gamma, \omega\}$ where the numbering is chosen so that each of $s_\alpha s_\omega, s_\beta s_\omega, s_\gamma s_\omega$ has order 3.

The Springer representation ρ_u is a sum of two irreducible representations of W: one eight dimensional and one six dimensional. Using 1.4(b) we see that there is a unique $S \in \underline{\mathcal{B}}_u$ such that $J_S = \{\omega\}$, a unique $\hat{S} \in \underline{\mathcal{B}}_u$ such that $J_{\hat{S}} = \{\alpha,\beta,\gamma\}$, exactly two irreducible components $S_{\beta\gamma'}, S_{\beta'\gamma}$ of $\mathcal{B}_u$ such that $J_{S_{\beta\gamma'}} = J_{S_{\beta'\gamma}} = \{\alpha,\omega\}$, exactly two irreducible components $S_{\alpha\gamma}, S_{\alpha'\gamma'}$ of $\mathcal{B}_u$ such that $J_{S_{\alpha'\gamma'}} = J_{S_{\alpha'\gamma'}} = \{\beta,\omega\}$ and exactly two irreducible components $S_{\alpha\beta}, S_{\alpha'\beta'}$ of $\mathcal{B}_u$ such that $J_{S_{\alpha\beta}} = J_{S_{\alpha'\beta'}} = \{\gamma,\omega\}$.

For any $X \in \underline{\mathcal{B}}_u$ such that $\omega \in J_X$ and any $i \in \{\alpha,\beta,\gamma\}$ we set

$$X_i^* = \{B \in X;\ \text{any } B' \text{ on the same } \omega\text{-line as } B \text{ is in } X_i\}.$$

We will show:

(a) *there are well defined parabolic subgroups* $P^\alpha \neq \tilde{P}^\alpha$ *in* $\mathcal{P}^\alpha$, $P^\beta \neq \tilde{P}^\beta$ *in* $\mathcal{P}^\beta$, $P^\gamma \neq \tilde{P}^\gamma$ *in* $\mathcal{P}^\gamma$ *and* P^ω *in* $\mathcal{P}^\omega$ *such that*

$$P^\alpha\spadesuit P^\omega, \tilde{P}^\alpha\spadesuit P^\omega, P^\beta\spadesuit P^\omega, \tilde{P}^\beta\spadesuit P^\omega, P^\gamma\spadesuit P^\omega, \tilde{P}^\gamma\spadesuit P^\omega,$$

$$\hat{S} = \{B \in \mathcal{B}; B(\omega) = P^\omega\},$$
$$S_{\beta\gamma'} = \{B \in \mathcal{B}; B(\beta) = P^\beta, B(\gamma) = \tilde{P}^\gamma\},$$
$$S_{\beta'\gamma} = \{B \in \mathcal{B}; B(\beta) = \tilde{P}^\beta, B(\gamma) = P^\gamma\},$$
$$S_{\alpha\gamma} = \{B \in \mathcal{B}; B(\alpha) = P^\alpha, B(\gamma) = P^\gamma\},$$
$$S_{\alpha'\gamma'} = \{B \in \mathcal{B}; B(\alpha) = \tilde{P}^\alpha, B(\gamma) = \tilde{P}^\gamma\},$$
$$S_{\alpha\beta} = \{B \in \mathcal{B}; B(\alpha) = P^\alpha, B(\beta) = P^\beta\},$$
$$S_{\alpha'\beta'} = \{B \in \mathcal{B}; B(\alpha) = \tilde{P}^\alpha, B(\beta) = \tilde{P}^\beta\},$$

$$S^*_\alpha = \{B \in \mathcal{B}; B(\beta) = P^\beta, B(\gamma) = \tilde{P}^\gamma, B(\alpha)\spadesuit P^\omega\}$$
$$\sqcup \{B \in \mathcal{B}; B(\beta) = \tilde{P}^\beta, B(\gamma) = P^\gamma, B(\alpha)\spadesuit P^\omega\},$$

$$S^*_\beta = \{B \in \mathcal{B}; B(\alpha) = P^\alpha, B(\gamma) = P^\gamma, B(\beta)\spadesuit P^\omega\}$$
$$\sqcup \{B \in \mathcal{B}; B(\alpha) = \tilde{P}^\alpha, B(\gamma)\tilde{P}^\gamma, B(\beta)\spadesuit P^\omega\},$$

$$S^*_\gamma = \{B \in \mathcal{B}; B(\alpha) = P^\alpha, B(\beta) = P^\beta, B(\gamma)\spadesuit P^\omega\}$$
$$\sqcup \{B \in \mathcal{B}; B(\alpha) = \tilde{P}^\alpha, B(\beta) = \tilde{P}^\beta, B(\gamma)\spadesuit P^\omega\}.$$

(b) *Let* $Y = \{B \in S; B(\omega) = P^\omega\}$. *Then* $Y \subset S_\alpha \cap S_\beta \cap S_\gamma$. *Moreover,* Y *meets any* ω*-line in* S *in exactly one point.*

From (b) we deduce:

(c) *For* $i \in \{a, \beta, \gamma\}$ *we have* $S_i = S^*_i \cup Y$.

The inclusion $S^*_i \cup Y \subset S_i$ is clear. Conversely, let $B \in S_i$ be such that $B \notin Y$. Let L_B be the i-line containing B; we have $L_B \subset \mathcal{B}_u$. Let L be the ω-line containing B. We have $L \subset S$ hence $L \subset \mathcal{B}_u$. By (b) there exists $B' \in L$ such that $B' \in Y$; in particular we have $B' \in S_i$. Let $L_{B'}$ be the i-line containing B'; we have $L_{B'} \subset \mathcal{B}_u$. We define $\pi \in \mathcal{P}^{i,\omega}$ by the condition that $B \subset \pi$. Now $\mathcal{B}^\pi$ is naturally imbedded in $\mathcal{B}$; it is a flag manifold of type A_2 since $(s_\omega s_i)^3 = 1$. Note that $\bar{\pi}$ is a connected reductive group of type A_2 and $\mathcal{B}^\pi$ can be viewed as the flag manifold of $\bar{\pi}$. Moreover we have $L \subset \mathcal{B}^\pi, L_B \subset \mathcal{B}^\pi, L_{B'} \subset \mathcal{B}^\pi$. Since $u \in B$ we have $u \in \pi$ hence $Ad(u) : \mathcal{B}^\pi \to \mathcal{B}^\pi$ is well defined and its fixed point set contains $L \cup L_B \cup L_{B'}$. But a unipotent element in $PGL_3(\mathbf{k})$ whose fixed point set of the flag manifold contains three distinct lines must be the identity element. In particular, for any $B'' \in L$, Ad(u) acts as identity on the i-line through B''; thus the i-line through B'' is contained in $\mathcal{B}_u$. In other words we have $B \in S^*_i$. Thus we have $S_i \subset S^*_i \cup Y$. This proves (c) (assuming (b)).

We now deduce from (a),(b) the following statement.

(d) $$Y = S_\alpha \cap S_\beta \cap S_\gamma.$$

Assume that $B \in S_\alpha \cap S_\beta \cap S_\gamma$ and $B \notin Y$. Using (c) we deduce $B \in (S^*_\alpha \cup Y) \cap (S^*_\beta \cup Y) \cap (S^*_\gamma \cup Y)$. Since $B \notin Y$ we deduce that $B \in S^*_\alpha \cap S^*_\beta \cap S^*_\gamma$. But from (a) we see that

(e) $$S^*_\alpha \cap S^*_\beta \cap S^*_\gamma = \emptyset.$$

This contradiction shows that $S_\alpha \cap S_\beta \cap S_\gamma \subset Y$. The opposite inclusion is known from (b). This proves (d).

We now show, assuming (a):

(e) $\hat{S} \cap S_{\beta\gamma'}$ *is a single* α*-line;* $\hat{S} \cap S_{\beta'\gamma}$ *is a single* α*-line.*

From (a) we have

$$\hat{S} \cap S_{\beta\gamma'} = \{B \in \mathcal{B}; B(\omega) = P^\omega, B(\beta) = P^\beta, B(\gamma) = \tilde{P}^\gamma\}$$
$$= \{B \in \mathcal{B}; B \subset P^\omega, B \subset P^\beta, B \subset \tilde{P}^\gamma\} = \{B \in \mathcal{B}; B \subset P^\omega \cap P^\beta \cap \tilde{P}^\gamma\}.$$

Since $P^\beta \spadesuit P^\omega$, $\tilde{P}^\gamma \spadesuit P^\omega$ and $s_\beta s_\gamma = s_\gamma s_b$, the intersection $P^\omega \cap P^\beta \cap \tilde{P}^\gamma$ is a parabolic subgroup in $\mathcal{P}^\alpha$. This proves the first assertion of (e); the second assertion of (e) is proved in the same way.

3.2. To prove 3.1(a),(b) for G is the same as proving them for G_{ad}. Hence we can assume that G is the special orthogonal group associated to an 8-dimensional $\mathbf{k}$-vector space V with a given nondegenerate quadratic form $Q : V \to \mathbf{k}$ and associate symmetric bilinear form $(,) : V \times V \to \mathbf{k}$. Until the end of 3.14 we shall adhere to this assumption.

Now, $A := u - 1 : V \to V$ is nilpotent with Jordan blocks of sizes $3, 3, 1, 1$. More precisely, we can find a basis $\{e_i, e'_i; i \in \{0, 1, 2, 3\}\}$ of V such that

$Q(e_i) = Q(e'_i) = 0$ for $i \in \{0, 1, 2, 3\}$;
$(e_i, e_j) = (e'_i, e'_j) = 0$ for all i, j;
$(e_i, e'_j) = 1$ if $i \neq j$ are both odd or if $i = j$ are both even;
$(e_i, e'_j) = 0$ otherwise

and such that

$Ae_0 = 0, Ae'_0 = 0,$
$Ae_1 = e_2 + xe_3, Ae_2 = e_3, Ae_3 = 0,\ Ae'_1 = -e'_2 + x'e'_3, Ae'_2 = -e'_3, Ae'_3 = 0,$

where $x, x' \in \mathbf{k}$ satisfy $x + x' = 1$. A subspace U of V is said to be isotropic if $Q|_U = 0$. Let ${}^!L_4 = \text{span}(e_3, e'_3, e_0, e_2)$, ${}^!L'_4 = \text{span}(e_3, e'_3, e'_0, e'_2)$. $L^!_4 = \text{span}(e_3, e'_3, e'_0, e_2)$, $L'^!_4 = \text{span}(e_3, e'_3, e_0, e'_2)$. These subspaces are isotropic; in 3.4 we will show that they are intrinsic to u; they do not depend on the specific basis used to define u. We identify $\mathcal{P}^\alpha$ with the variety of all isotropic lines in V; $\mathcal{P}^\omega$ with the variety of all isotropic planes in V; $\mathcal{P}^\beta$ with the variety of all isotropic 4-spaces in V in the G-orbit of ${}^!L_4$ and ${}^!L'_4$; $\mathcal{P}^\gamma$ with the variety of all isotropic 4-spaces in V in the G-orbit of $L^!_4$ and $L'^!_4$. (In each case the identification attaches to an isotropic subspace its stabilizer in G.) We identify $\mathcal{B}$ with the variety of all sequences $(V_1, V_2, V_4, \tilde{V}_4) \in \mathcal{P}^\alpha \times \mathcal{P}^\omega \times \mathcal{P}^\beta \times \mathcal{P}^\gamma$ such that $V_1 \subset V_2 \subset V_4, V_2 \subset \tilde{V}_4$. (Such a sequence is identified with its stabilizer $\{g \in G; gV_1k = V_1, gV_2 = V_2, gV_4 = V_4, g\tilde{V}_4 = \tilde{V}_4\}$.)

We consider the following subspaces of V:

$I_2 = A^2V = \text{span}(e_3, e'_3)$, $I_4 = AV = \text{span}(e_2, e_3, e'_2, e'_3)$,
$K_4 = \ker(A) = \text{span}(e_0, e_3, e'_0, e'_3)$, $K_6 = \ker(A^2) = span(e_0, e_2, e_3, e'_0, e'_2, e'_3)$.

We have $I_2 \subset I_4 \subset K_6$, $I_2 \subset K_4 \subset K_6$, $K_4 \cap I_4 = I_2$.

3.3. We show:

(a) *There are exactly two subspaces* $V_4 \in \mathcal{P}^\beta$ *such that* $I_2 \subset V_4$ *and* $\dim(AV_4) \leq 1$. *They are* ${}^!L_4, {}^!L'_4$.

(b) *There are exactly two subspaces* $V_4 \in \mathcal{P}^\gamma$ *such that* $I_2 \subset V_4$ *and* $\dim(AV_4) \leq 1$. *They are* $L^!_4, L'^!_4$.

We prove (a). The subspaces $V_4 \in \mathcal{P}^\beta$ such that $I_2 \subset V_4$ are determined by their intersection with $L^!_4, L'^!_4$; this intersection is of the form $\text{span}(e_3, e'_3, ae_0 + be'_2)$ where

$(a,b)\in\mathbf{k}^2-\{0,0\}$ hence the required 4-subspaces are of the form $\mathrm{span}(e_3,e'_3,ae_0+be'_2,xe_0+x'e'_0+ze_2+z'e'_2)$ where $ae_0+be'_2,xe_0+x'e'_0+ze_2+z'e'_2$ are linearly independent and we have $ax'+bx=0,xx'+zz'=0$. Moreover the condition that $\dim(AV_4)\le 1$ is that $\dim\mathrm{span}(-be'_3,ze_3-z'e'_3)\le 1$, that is $bz=0$. Assume first that $a\ne,b\ne 0$; then $z=0$, $xx'=0$. From $ax'+bx=0,xx'=0$ we deduce that $x=x'=0$. Since $ae_0+be'_2,z'e'_2$ are linearly independent we see that $z'\ne 0$ and our V_4 is $\mathrm{span}(e_3,e'_3,ae_0+be'_2,e'_2)=\mathrm{span}(e_3,e'_3,e_0,e'_2)\in\mathcal{P}^\gamma$, a contradiction. Assume now that $a=0,b\ne 0$; then $z=0,x=0$. Since $be'_2,x'e'_0+z'e'_2$ are linearly independent we see that $x'\ne 0$ and our V_4 is $\mathrm{span}(e_3,e'_3,e'_2,x'e'_0+z'e'_2)=\mathrm{span}(e_3,e'_3,e'_2,e'_0)={}^!L'_4$. Assume now that $a\ne 0,b=0$; then $x'=0$ and $zz'=0$. Since $ae_0,xe_0+ze_2+z'e'_2$ are linearly independent we have either $z=0,z'\ne 0$ or $z\ne 0,z'=0$. If $z=0,z'\ne 0$ then our V_4 is $\mathrm{span}(e_3,e'_3,e_0,xe_0+z'e'_2)=\mathrm{span}(e_3,e'_3,e_0,e'_2)\in\mathcal{P}^\gamma$, a contradiction. If $z\ne 0,z'=0$ then our V_4 is $\mathrm{span}(e_3,e'_3,e_0,xe_0+ze_2)=\mathrm{span}(e_3,e'_3,e_0,e_2)={}^!L_4$. Thus $V_4\in\{{}^!L_4,{}^!L'_4\}$. Conversely it is clear that if $V_4\in\{{}^!L_4,{}^!L'_4\}$ then V_4 satisfies the requirements of (a). This proves (a).

The proof of (b) is entirely similar to that of (a).

3.4. We set $L_1=\mathrm{span}(e_3)\in\mathcal{P}^\alpha$, $L'_1=\mathrm{span}(e'_3)\in\mathcal{P}^\alpha$. We have $A({}^!L_4)=A(L_4^!)=L_1$, $A({}^!L'_4)=A(L_4'^!)=L'_1$. In particular, if V_4 is as in 3.3(a),(b), then $AV_4\in\{L_1,L'_1\}$ is a one dimensional (isotropic) subspace of I_2.

From 3.3(a),(b) we see that ${}^!L_4,L_4^!,{}^!L'_4,L_4'^!$ are intrinsic to u and do not depend on the specific basis used to define u. It follows also that L_1,L'_1 are intrinsic to u. Note that L_1,L'_1 are the two lines in I_2 which are isotropic for the quadratic form $I_2\to\mathbf{k}$ given by $v\mapsto Q(\tilde v)$ where $\tilde v$ is any vector in I_4 such that $A\tilde v=v$.

3.5. Note that Q induces a nondegenerate quadratic form on K_4/I_2 which has exactly two isotropic lines. Hence there are exactly two isotropic 3-spaces contained in K_4 and containing I_2. They are $\mathrm{span}(e_3,e'_3,e_0)={}^!L_4\cap L_4'^!$ and $\mathrm{span}(e_3,e'_3,e'_0)={}^!L'_4\cap L_4^!$.

3.6. Let

$$\tilde Y=\{(V_1,V_2,V_4,\tilde V_4)\in\mathcal{B};V_2=I_2\}.$$

Let $(V_1,V_2,V_4,\tilde V_4)\in\tilde Y$. We have clearly $AV_1=0$. Moreover, since V_4 is an isotropic subspace containing I_2 we must have $V_4\subset K_6=\ker(A^2)$ hence $AV_4\subset\ker(A)\cap AV=I_4\cap K_4=I_2\subset V_4$. Thus, $AV_4\subset V_4$. Similarly we have $A\tilde V_4\subset\tilde V_4$. We see that $(V_1,V_2,V_4,\tilde V_4)\in\mathcal{B}_u$. Thus, $\tilde Y\subset\mathcal{B}_u$. Note that $\tilde Y$ is isomorphic to $\mathbf{P}^1\times\mathbf{P}^1\times\mathbf{P}^1$ hence it is a closed irreducible subvariety of $\mathcal{B}_u$ of dimension 3. Thus $\tilde Y$ is an irreducible component of $\mathcal{B}_u$. It satisfies $J_{\tilde Y}=\{\alpha,\beta,\gamma\}$. Hence $\tilde Y=\hat S$.

3.7. Let

$$N_1=\{(V_1,V_2,V_4,\tilde V_4)\in\mathcal{B};V_4={}^!L_4,\tilde V_4=L_4'^!\},$$
$$N_2=\{(V_1,V_2,V_4,\tilde V_4)\in\mathcal{B};V_4={}^!L'_4,\tilde V_4=L_4^!\}.$$

We have $A({}^!L_4\cap L_4'^!)=0$, $A({}^!L'_4\cap L_4^!)=0$ hence if $(V_1,V_2,V_4,\tilde V_4)$ is in N_1 or N_2 then $AV_2=0$. Moreover, each of ${}^!L_4,L_4'^!,{}^!L'_4,L_4^!$ is A-stable. We see that $(V_1,V_2,V_4,\tilde V_4)\in\mathcal{B}_u$. Thus, $N_1\subset\mathcal{B}_u$, $N_2\subset\mathcal{B}_u$. Note that N_1 are N_2 are isomorphic to the flag manifold of $GL_3(\mathbf{k})$ hence they are closed irreducible subvarieties of $\mathcal{B}_u$ of dimension 3. Thus they are (distinct) irreducible components of $\mathcal{B}_u$. They satisfy $J_{N_1}=J_{N_2}=\{\alpha,\omega\}$. Hence N_1,N_2 are the same as $S_{\beta\gamma'},S_{\beta'\gamma}$ (up to order).

3.8. Let

$$N_3 = \{(V_1, V_2, V_4, \tilde V_4) \in \mathcal{B}; V_1 = L_1, \tilde V_4 = L_4^!\},$$

$$N_4 = \{(V_1, V_2, V_4, \tilde V_4) \in \mathcal{B}; V_1 = L'_1, \tilde V_4 = L_4'^!\}.$$

Let $(V_1, V_2, V_4, \tilde V_4) \in N_3$. Let $V_3 = V_4 \cap \tilde V_4$. Since $A(\tilde V_4) = L_1$ we have $A(V_3) \subset L_1 = V_1 \subset V_3$ that is $uV_3 = V_3$; moreover, $A(V_2) \subset L_1 = V_1$ hence $AV_2 \subset V_2$. Now V_4 is the only subspace in its G-orbit that contains V_3; since $uV_4 \supset uV_3 = V_3$ it follows that $uV_4 = V_4$. The inclusions $AV_1 \subset V_1$, $A(\tilde V_4) \subset \tilde V_4$ are obvious. We see that $(V_1, V_2, V_4, \tilde V_4) \in \mathcal{B}_u$. Thus, $N_3 \subset \mathcal{B}_u$. Note that N_3 is isomorphic to the space of pairs (V_2, V_3) where $V_2 \subset V_3$ are subspaces of $\tilde V_4$ with $\dim V_2 = 2, \dim V_3 = 3$; hence N_3 is isomorphic to the flag manifold of $GL_3(\mathbf{k})$. Thus N_3 is a closed irreducible subvariety of $\mathcal{B}_u$ of dimension 3. Thus $N_3 \in \underline{\mathcal{B}}_u$. We have $J_{N_3} = \{\beta, \omega\}$. Similarly, $N_4 \in \underline{\mathcal{B}}_u$ and $J_{N_4} = \{\beta, \omega\}$. Hence N_3, N_4 are the same as $S_{\alpha\gamma}, S_{\alpha'\gamma'}$ (up to order).

3.9. Let

$$N_5 = \{(V_1, V_2, V_4, \tilde V_4) \in \mathcal{B}; V_1 = L_1, V_4 = {}^!L_4\},$$

$$N_6 = \{(V_1, V_2, V_4, \tilde V_4) \in \mathcal{B}; V_1 = L'_1, V_4 = {}^!L'_4\}.$$

A proof completely similar to that in 3.8 shows that $N_5 \in \underline{\mathcal{B}}_u$, $N_6 \in \underline{\mathcal{B}}_u$. We have $J_{N_5} = J_{N_6} = \{\gamma, \omega\}$. Hence N_5, N_6 are the same as $S_{\alpha\beta}, S_{\alpha'\beta'}$ (up to order).

3.10. Let $X = \{(V_1, V_2, V_4, \tilde V_4) \in \mathcal{B}; V_1 \subset I_2 \subset V_4 \cap \tilde V_4, A(V_4 \cap \tilde V_4) \subset V_1\}$. We show:

(a) *X is a smooth irreducible projective variety of dimension* 3.

The fact that X is a projective variety is obvious. Now let Z be the variety of all pairs V_1, V_3 of isotropic subspaces of V of dimension 1 and 3 respectively such that $V_1 \subset I_2 \subset V_3$ and $AV_3 \subset V_1$. Clearly $\zeta : (V_1, V_2, V_4, \tilde V_4) \mapsto (V_1, V_4 \cap \tilde V_4)$ makes X into a $\mathbf{P}^1$-bundle over Z. Hence it is enough to show that Z is a smooth irreducible surface. Now the space of lines in I_2 is a projective line and the space of V_3 containing I_2 and contained in K_6 (without the isotropy condition) is a projective 3-space. Hence we can identify Z with

$$\{((z_0, z'_0, z_2, z'_2), (y, y')) \in \mathbf{P}^3 \times \mathbf{P}^1; z_0 z'_0 + z_2 z'_2 = 0, z_2 y' + z'_2 y = 0\}.$$

The subset of Z where $z'_2 \neq 0, y' \neq 0$ is

$$\{(z_0, z'_0, z_2; y) \in \mathbf{k}^3 \times \mathbf{k}; z_0 z'_0 + z_2 = 0, z_2 + y = 0\}$$
$$= \{(z_0, z'_0, z_2) \in \mathbf{k}^3; z_0 z'_0 + z_2 = 0\} \cong \mathbf{k}^2.$$

Similarly the subset of Z where $z_2 \neq 0, y \neq 0$ is $\cong \mathbf{k}^2$. The subset where $z_0 \neq 0, y \neq 0$ is

$$\{(z'_0, z_2, z'_2, y') \in \mathbf{k}^4; z'_0 + z_2 z'_2 = 0, z_2 y' + z'_2 = 0\} \cong \mathbf{k}^2.$$

The subset where $z_0 \neq 0, y' \neq 0$ is $\cong \mathbf{k}^2$; the subset where $z'_0 \neq 0, y \neq 0$ is $\cong \mathbf{k}^2$; the subset where $z'_0 \neq 0, y' \neq 0$ is $\cong \mathbf{k}^2$. Thus, Z is a union of 8 smooth irreducible surfaces (open in Z) hence Z is a smooth surface. Since the 8 open subsets above have a nonempty intersection, we see that Z is irreducible. This proves (a).

Now if $(V_1, V_2, V_4, \tilde V_4) \in X$ then setting $V_3 = V_4 \cap \tilde V_4$ we have $AV_1 = 0$ (since $V_1 \subset I_2$), $A(V_3) \subset V_1 \subset V_3$, $AV_2 \subset V_1 \subset V_2$ hence $uV_1 \subset V_1$, $uV_3 \subset V_3$, $uV_2 \subset V_2$; since $V_4 \in \mathcal{P}^\beta$, $\tilde V_4 \in \mathcal{P}^\gamma$ are determined uniquely by the conditions $V_3 \subset V_4$, $V_3 \subset \tilde V_4$, we have necessarily $uV_4 = V_4$, $u\tilde V_4 = \tilde V_4$. Thus, $X \subset \mathcal{B}_u$. Using this and (a) together with $\dim \mathcal{B}_u = 3$ we see that

(b) *X is an irreducible component of $\mathcal{B}_u$.*

It follows that the subsets $X_\alpha, X_\beta, X_\gamma$ of X are well defined (see 1.6).

We show:

(c) $\chi(X) = 12$.

Here $\chi()$ is as in 2.2. Since X is a $\mathbf{P}^1$-bundle over Z, an equivalent statement is:

(d) $\chi(Z) = 6$.

The space $\mathfrak{U}$ of isotropic 3-spaces in V that contain I_2 hence are contained in K_6 can be identified with the space of isotropic lines in K_6/I_2 with its obvious quadratic form hence it is a product $\mathbf{P}^1 \times \mathbf{P}^1$. The map $Z \to \mathfrak{U}$, $(V_1, V_3) \mapsto V_3$ is an isomorphism over the complement of two points in $\mathfrak{U}$, namely ${}^!L_4 \cap L_4'{}^!$ and ${}^!L_4' \cap L_4^!$ and has fibres $\mathbf{P}^1$ at each of those two points. It follows that $\chi(Z) = \chi(\mathbf{P}^1 \times \mathbf{P}^1) - 2 + 2\chi(\mathbf{P}^1) = 4 - 2 + 4 = 6$, as desired.

3.11. Clearly, X is a union of ω-lines. Hence for $i \in \{\alpha, \beta, \gamma\}$, X_i^* is defined as in 3.1. We have

$$X_\alpha = \{(V_1, V_2, V_4, \tilde{V}_4) \in X;\ \text{for any line } V_1' \text{ in } V_2 \text{ we have } AV_1' = 0\}$$

hence

$$X_\alpha = \{(V_1, V_2, V_4, \tilde{V}_4) \in X; V_2 \subset K_4\}.$$

We deduce that

$$X_\alpha^* = \{(V_1, V_2, V_4, \tilde{V}_4) \in X;\ \text{for any } V_2' \text{ with } V_1 \subset V_2' \subset V_4 \cap \tilde{V}_4 \text{ we have } V_2' \subset K_4\}.$$

Since the various V_2' such that $V_1 \subset V_2' \subset V_4 \cap \tilde{V}_4$ generate $V_4 \cap \tilde{V}_4$ we see that

$$X_\alpha^* = \{(V_1, V_2, V_4, \tilde{V}_4) \in X; I_2 \subset V_4 \cap \tilde{V}_4 \subset K_4\}$$

If $(V_1, V_2, V_4, \tilde{V}_4) \in X_\alpha^*$ then $A(V_4 \cap \tilde{V}_4) = 0$ hence $\dim A(V_4) \le 1$, $\dim A(\tilde{V}_4) \le 1$ (since $V_4 \cap \tilde{V}_4$ has codimension 1 in V_4 and in $\tilde{V}_4$). Using 3.3(a),(b) we deduce that $V_4 \in \{{}^!L_4, {}^!L_4'\}$, $\tilde{V}_4 \in \{L_4^!, L_4'{}^!\}$. Thus we have either $V_4 = {}^!L_4, \tilde{V}_4 = L_4'{}^!$ or $V_4 = {}^!L_4', \tilde{V}_4 = L_4^!$. We see that

$$X_\alpha^* = \{(V_1, V_2, V_4, \tilde{V}_4) \in X; V_4 = {}^!L_4, \tilde{V}_4 = L_4'{}^!\}$$
$$\sqcup \{(V_1, V_2, V_4, \tilde{V}_4) \in X; V_4 = {}^!L_4', \tilde{V}_4 = L_4^!\}.$$

(The right hand side is clearly contained in the left hand side.) Since $A({}^!L_4 \cap L_4'{}^!) = 0$, $A({}^!L_4' \cap L_4^!) = 0$, we have

$$X_\alpha^* = \{(V_1, V_2, V_4, \tilde{V}_4) \in \mathcal{B}; V_1 \subset I_2, V_4 = {}^!L_4, \tilde{V}_4 = L_4'{}^!\}$$
$$\text{(a)} \qquad \sqcup \{(V_1, V_2, V_4, \tilde{V}_4) \in \mathcal{B}; V_1 \subset I_2, V_4 = {}^!L_4', \tilde{V}_4 = L_4^!\}.$$

3.12. We have

$$X_\beta = \{(V_1, V_2, V_4, \tilde{V}_4) \in X;\ \text{for any } V_4' \in \mathcal{P}^\beta \text{ with } V_2 \subset V_4' \text{ we have } u(V_4') = V_4'\}.$$

Let $(V_1, V_2, V_4, \tilde{V}_4) \in X_\beta$. Then for any $V_4' \in \mathcal{P}^\beta$ with $V_2 \subset V_4'$ we have that $\tilde{V}_4 \cap V_4'$ is an isotropic 3-space containing V_2; it is u-stable (since $\tilde{V}_4$ and V_4' are u-stable). Moreover all isotropic 3-spaces containing V_2 and contained in $\tilde{V}_4$ are obtained in this way hence are all u-stable; it follows that u acts as 1 on $\tilde{V}_4/V_2$ that is, $A(tV_4) \subset V_2$. Thus, $X_\beta \subset \{(V_1, V_2, V_4, \tilde{V}_4) \in X; A(\tilde{V}_4) \subset V_2\}$. A similar argument shows the reverse inclusion. Thus

$$X_\beta = \{(V_1, V_2, V_4, \tilde{V}_4) \in X; A(\tilde{V}_4) \subset V_2\}.$$

We deduce that

$$X_\beta^* = \{(V_1, V_2, V_4, \tilde{V}_4) \in X;\ \text{for any } V_2' \text{ with } V_1 \subset V_2' \subset V_4 \cap \tilde{V}_4 \text{ we have } A(\tilde{V}_4) \subset V_2'\}.$$

Since the intersection of all V_2' such that $V_1 \subset V_2' \subset V_4 \cap \tilde{V}_4$ is V_1 we see that

$$X_\beta^* = \{(V_1, V_2, V_4, \tilde{V}_4) \in X; A(\tilde{V}_4) \subset V_1\}$$

that is,

$$X_\beta^* = \{(V_1, V_2, V_4, \tilde{V}_4) \in \mathcal{B}; V_1 \subset I_2 \subset V_4 \cap \tilde{V}_4, A(\tilde{V}_4) \subset V_1\}.$$

Using 3.3(b), we see that for $\tilde{V}_4$ in the right hand side we have $\tilde{V}_4 \in \{L_4^!, L_4'^!\}$ so that V_1 is necessarily L_1 (if $\tilde{V}_4 = L_4^!$) or L_1' (if $\tilde{V}_4 = L_4'^!$). Thus,

(a)
$$\begin{aligned} X_\beta^* &= \{(V_1, V_2, V_4, \tilde{V}_4) \in \mathcal{B}; I_2 \subset V_4, V_1 = L_1, \tilde{V}_4 = L_4^!\} \\ &\sqcup \{(V_1, V_2, V_4, \tilde{V}_4) \in \mathcal{B}; I_2 \subset V_4, V_1 = L_1', \tilde{V}_4 = L_4'^!\}. \end{aligned}$$

An entirely similar argument yields:

$$X_\gamma = \{(V_1, V_2, V_4, \tilde{V}_4) \in S; A(V_4) \subset V_2\},$$

(b)
$$\begin{aligned} X_\gamma^* &= \{(V_1, V_2, V_4, \tilde{V}_4) \in \mathcal{B}; I_2 \subset \tilde{V}_4, V_1 = L_1, V_4 = {}^!L_4\} \\ &\sqcup \{(V_1, V_2, V_4, \tilde{V}_4) \in \mathcal{B}; I_2 \subset \tilde{V}_4, V_1 = L_1', V_4 = {}^!L_4'\}. \end{aligned}$$

From 3.11(a) we see that X_α^* has two irreducible components, each one being a $\mathbf{P}^1$-bundle over $\mathbf{P}^1$ hence is two-dimensional; we see that $X_\alpha^* \neq X$ hence X is not a union of α-lines. Similarly from (a),(b) we see that $X_\beta^* \neq X$, $X_\gamma^* \neq X$ hence X is not a union of β-lines and X is not a union of γ-lines. Thus we have $J_X = \{\omega\}$. It follows that

(c)
$$X = S.$$

3.13. We have

$$\begin{aligned} Y &= \{(V_1, V_2, V_4, \tilde{V}_4) \in X; V_2 = I_2\} \\ &= \{(V_1, V_2, V_4, \tilde{V}_4) \in \mathcal{B}; V_2 = I_2, A(V_4 \cap \tilde{V}_4) \subset V_1\}. \end{aligned}$$

Assume that $(V_1, V_2, V_4, \tilde{V}_4) \in Y$. If V_1' is an isotropic line contained in I_2 then $AV_1' = 0$ (since $AI_2 = 0$). This shows that $Y \subset X_\alpha = S_\alpha$. If U is an isotropic 4-space containing I_2 then $U \subset K_6 = \ker(A^2)$ hence $AU \subset \ker(A) \cap AV = K_4 \cap I_4 = I_2 \subset U$; thus, $AU \subset U$. This shows that $Y \subset X_\beta = S_\beta$ and $Y \subset X_\gamma = S_\gamma$. This proves the first assertion of 3.1(b). Now let $(V_1, V_2, V_4, \tilde{V}_4) \in X$ and let L be the ω-line in X containing $(V_1, V_2, V_4, \tilde{V}_4)$. By the definition of X if V_2 is replaced by I_2 the resulting quadruple $(V_1, I_2, V_4, \tilde{V}_4)$ belongs to X (and even to Y). This proves the second assertion of 3.1(b).

3.14. From 3.11(a) and 3.12(a),(b),(c), we see that 3.1(a) holds. Here P^α (resp. $\tilde{P}^\alpha$) is the stabilizer of L_1 (resp. L_1'); P^β (resp. $\tilde{P}^\beta$) is the stabilizer of ${}^!L_4$ (resp. ${}^!L_4'$); P^γ (resp. $\tilde{P}^\gamma$) is the stabilizer of $L_4^!$ (resp. $L_4'^!$).

3.15. Until the end of 3.18 we assume that G_{ad} is of type D_5. In this case we can write $I = \{\alpha, \beta, \gamma, \delta, \omega\}$ where the numbering is chosen so that each of $s_\alpha s_\omega, s_\beta s_\omega, s_\gamma s_\omega, s_\alpha s_\delta$ has order 3. We shall assume that $u \in G$ (see 1.1) is such that $d_u = 3$; this determines u uniquely up to conjugacy. We can find $P \in \mathcal{P}^\delta$ such that $u \in P$ and $\bar{u} := \pi_P(u) \in \bar{P}$ satisfies $d_{\bar{u}} = 3$ and, if C is the conjugacy class of u in G, then $C \cap \pi_P^{-1}(\bar{u})$ is open dense in $\pi_P^{-1}(\bar{u})$. Since the adjoint group of $\bar{P}$ is of type D_4, the irreducible component S of $\mathcal{B}_{\bar{u}}^{\bar{P}}$ is defined as in 3.1 with G, u replaced by $\bar{P}, \bar{u}$. Note that the Weyl group of $\bar{P}$ is canonically identified with the subgroup of W generated by $\{s_\alpha, s_\beta, s_\gamma, s_\omega\}$. The imbedding $\mathcal{B}^{\bar{P}} \to \mathcal{B}$, $B_1 \mapsto \pi_P^{-1}(B_1)$ restricts to an imbedding $\mathcal{B}_{\bar{u}}^{\bar{P}} \to \mathcal{B}_u$ and to an isomorphism of S onto an irreducible component $\tilde{S}$ of $\mathcal{B}_u$. (See 1.5.) This imbedding carries any i-line in $\mathcal{B}^{\bar{P}}$ (where $i \in \{\alpha, \beta, \gamma, \omega\}$) to an i-line in $\mathcal{B}$.

Using 1.4(b), we see that there is a unique $E \in \underline{\mathcal{B}}_u$ such that $J_E = \{\beta, \gamma, \delta\}$.

The inverse images under π_P of the parabolic subgroups

$P^\alpha, \tilde{P}^\alpha, P^\beta, \tilde{P}^\beta, P^\gamma, \tilde{P}^\gamma, P^\omega$

as in 3.1(a) (with G, u replaced by $\bar{P}, \bar{u}$) are denoted again by the same letters; thus we have

$P^\alpha \neq \tilde{P}^\alpha$ in $\mathcal{P}^\alpha$, $P^\beta \neq \tilde{P}^\beta$ in $\mathcal{P}^\beta$, $P^\gamma \neq \tilde{P}^\gamma$ in $\mathcal{P}^\gamma$, $P^\omega \in \mathcal{P}^\omega$,

where $\mathcal{P}^i$ refers to G.

3.16. We will show that there is a well defined parabolic subgroup $\hat{P}^\alpha \in \mathcal{P}^\alpha$ such that

(a) $\hat{P}^\alpha \notin \{P^\alpha, \tilde{P}^\alpha\}$, $\hat{P}^\alpha \spadesuit P^\omega$ and

(b) $\tilde{S}_\delta = \{B \in \tilde{S}; B(\alpha) = \hat{P}^\alpha\}$,

(c) $E = \{B \in \mathcal{B}; B(\alpha) = \hat{P}^\alpha, B(\omega) = P^\omega\}$.

To prove this is the same as proving it for G_{ad}. Hence we can assume that G is the special orthogonal group associated to a 10-dimensional $\mathbf{k}$-vector space V with a given nondegenerate quadratic form $Q : V \to \mathbf{k}$ and associate symmetric bilinear form $(,) : V \times V \to \mathbf{k}$. Until the end of 3.18 we shall adhere to this assumption. Now $A := u - 1 : V \to V$ is nilpotent. If $p \neq 2$, A has Jordan blocks of sizes $5, 3, 1, 1$. If $p = 2$, A has Jordan blocks of sizes $4, 4, 1, 1$ and moreover we have $Q(A^2 x) \neq 0$ for some $x \in V$. These conditions describe completely the conjugacy class of u.

We identify $\mathcal{P}^\delta$ with the variety of all isotropic lines in V; $\mathcal{P}^\alpha$ with the variety of all isotropic planes in V; $\mathcal{P}^\omega$ with the variety of all isotropic 3-spaces in V. (In each case the identification attaches to an isotropic subspace its stabilizer in G.) We will describe later $\mathcal{P}^\beta$ and $\mathcal{P}^\gamma$.

We can find a basis $\{e_i, e'_i; i \in \{0, 1, 2, 3\}\} \sqcup \{f, f'\}$ of V such that

P is the stabilizer in G of the line $\mathbf{k}f$;

$Q(e_i) = Q(e'_i) = 0$ for $i \in \{0, 1, 2, 3\}$, $Q(f) = Q(f') = 0$;

$(e_i, e_j) = (e'_i, e'_j) = 0$ for all i, j and $(f, e_i) = (f', e_i) = 0$ for all i;

$(e_i, e'_j) = 1$ if $i \neq j$ are both odd or if $i = j$ are both even and $(f, f') = 1$;

$(e_i, e'_j) = 0$ otherwise,

and such that

$Af' = a_0 e_0 + a_1 e_1 + a_2 e_2 + a_3 e_3 + a'_0 e'_0 + a'_1 e'_1 + a'_2 e'_2 + a'_3 e'_3 + df$

$Ae_0 = b_0 f, Ae'_0 = b'_0 f,$

$Ae_1 = e_2 + x e_3 + b_1 f, Ae_2 = e_3 + b_2 f, Ae_3 = b_3 f,\ Ae'_1 = -e'_2 + x' e'_3 + b'_1 f,$

$Ae'_2 = -e'_3 + b'_2 f, Ae'_3 = b'_3 f, Af = 0,$

where $x, x' \in \mathbf{k}$ satisfy $x + x' = 1$, compare with 3.2. Here

$a_0, a_1, a_2, a_3, a'_0, a'_1, a'_2, a'_3, b_0, b_1, b_2, b_3, b'_0, b'_1, b'_2, b'_3, d$

are elements of $\mathbf{k}$ such that

$d + a_0a'_0 + a_1a'_3 + a_2a'_2 + a_3a'_1 = 0,\ a'_0 + b_0 = 0, a_0 + b'_0 = 0,$

$a'_3 + a'_2 + xa'_1 + b_1 = 0, a_3 - a_2 + x'a_1 + b'_1 = 0,\ a'_2 + a'_1 + b_2 = 0, a_2 - a_1 + b'_2 = 0,$

$a'_1 + b_3 = 0, a_1 + b'_3 = 0.$

(These equations express the fact that u is an isometry for Q.)

We identify $\mathcal{P}^\beta$ with the variety of all isotropic 5-spaces in V in the G-orbit of $\text{span}(f, e_3, e'_3, e_0, e_2)$; $\mathcal{P}^\gamma$ with the variety of all isotropic 5-spaces in V in the G-orbit of $\text{span}(f, e_3, e'_3, e'_0, e_2)$. (In each case the identification attaches to an isotropic subspace its stabilizer in G.) We identify $\mathcal{B}$ with the variety of all sequences $(V_1, V_2, V_3, V_5, \tilde{V}_5) \in \mathcal{P}^\delta \times \mathcal{P}^\alpha \times \mathcal{P}^\omega \times \mathcal{P}^\beta \times \mathcal{P}^\gamma$ such that $V_1 \subset V_2 \subset V_3 \subset V_5, V_3 \subset \tilde{V}_5$. (Such a sequence is identified with its stabilizer $\{g \in G; gV_1 = V_1, gV_2 = V_2, gV_3 = V_3, gV_5 = V_5, g\tilde{V}_5 = \tilde{V}_5\}$.)

We now compute the powers A^k for $k = 2, 3, 4$ (here $*$ denotes an element of $\mathbf{k}$).

$A^2 f' = a_1e_2 + (a_2 + xa_1)e_3 - a'_1e'_2 + (-a'_2 + x'a'_1)e'_3 + *f,$

$A^2 e_0 = 0,\ A^2 e_1 = e_3 + *f,\ A^2 e_2 = -a'_1 f,\ A^2 e_3 = 0,$

$A^2 e'_0 = 0,\ A^2 e'_1 = e'_3 + *f,\ A^2 e'_2 = -a_1 f,\ A^2 e'_3 = 0,\ A^2 f = 0;$

$A^3 f' = a_1e_3 + a'_1e'_3 + *f,\ A^3 e_0 = 0,\ A^3 e_1 = -a'_1 f,\ A^3 e_2 = 0,\ A^3 e_3 = 0,$ $A^3 e'_0 = 0,$

$A^3 e'_1 = -a_1 f,\ A^3 e'_2 = 0,\ A^3 e'_3 = 0,\ A^3 f = 0;$

$A^4 f' = -2a_1a'_1 f,\ A^4 e_i = A^4 e'_i = 0$ for $i = 0, 1, 2, 3,\ A^4 f = 0.$

We see that $A^4 = 0$ if $p = 2$ and $A^5 = 0$ without restriction on p. Since A has a Jordan block of size 5 (if $p \neq 2$) and one of size 4 (if $p = 2$) we see that $A^4 \neq 0$ (if $p \neq 2$) and $A^3 \neq 0$ (if $p = 2$). It follows that $a_1a'_1 \neq 0$ if $p \neq 2$ and $(a_1, a'_1) \neq (0, 0)$ in any case.

Let $\mathcal{I}$ be the image of the map $V \to \mathbf{k}$, $v \mapsto Q(A^2 v)$. Since

$$A^2 V = \text{span}(f, e_3, e'_3, a_1e_2 - a'_1e'_2),$$

$\mathcal{I}$ is the same as the image of the map $\mathbf{k}^4 \to \mathbf{k}$,

$$(x, y, x, t) \mapsto Q(xf + ye_3 + ze'_3 + t(a_1e_2 - a'_1e'_2) = Q(t(a_1e_2 - a'_1e'_2)) = -t^2 a_1a'_1.$$

Thus, if $p \neq 2$ we have $\mathcal{I} = \mathbf{k}$ since $a_1a'_1 \neq 0$; if $p = 2$ then we already know that $\mathcal{I} \neq 0$ hence we must have $a_1a'_1 \neq 0$. Thus, without assumption on p we have $a_1a'_1 \neq 0$ and $\mathcal{I} = \mathbf{k}$.

3.17. Let $I_3 = \text{span}(f, e_3, e'_3) \in \mathcal{P}^\omega$, $I'_2 = \text{span}(f, a_1e_3 - a'_1e'_3)$, $K_7 = \{v \in V; (v, V_3) = 0\} = \text{span}(f, e_3, e'_3, e_2, e'_2, e_0, e'_0)$.

Note that $I'_2 = I_3 \cap \ker(A)$; moreover I'_2, I_3 are isotropic subspaces. From the definitions we have

$$\tilde{S} = \{(V_1, V_2, V_3, V_5, \tilde{V}_5) \in \mathcal{B}; V_1 = \mathbf{k}f \subset V_2 \subset I_3 \subset V_5 \cap \tilde{V}_5, A(V_5 \cap \tilde{V}_5) \subset V_2\}.$$

From the definitions, $\tilde{S}_\delta$ is the set of all $(V_1, V_2, V_3, V_5, \tilde{V}_5) \in \tilde{S}$ such that for any isotropic line V'_1 contained in V_2 we have $AV'_1 \subset V'_1$ that is, $AV'_1 = 0$. Since V_2 is generated by such V'_1, we see that

$$\tilde{S}_\delta = \{(V_1, V_2, V_3, V_5, \tilde{V}_5) \in \tilde{S}; AV_2 = 0\}$$
$$= \{(V_1, V_2, V_3, V_5, \tilde{V}_5) \in \tilde{S}; V_2 \subset I_3 \cap \ker(A)\} = \{(V_1, V_2, V_3, V_5, \tilde{V}_5) \in \tilde{S}; V_2 = I'_2\}.$$

This shows that 3.16(b) holds where $\hat{P}^\alpha$ is the parabolic subgroup in $\mathcal{P}^\alpha$ which is the stabilizer of I'_2. From the definitions we see that P^α (resp. $\tilde{P}^\alpha$) is the stabilizer of $\text{span}(e_3, f)$ (resp. of $\text{span}(e'_3, f)$). Since $\text{span}(e_3, f), \text{span}(e'_3, f), \text{span}(a_1e_3 - a'_1e_3, f)$ are distinct (recall that $a_1a'_1 \neq 0$) we see that $\hat{P}^\alpha \notin \{P^\alpha, \tilde{P}^\alpha\}$. Since P^ω is the stabilizer of I_3 and $I'_2 \subset I_3$ we see that $\hat{P}^\alpha \spadesuit P^\omega$. Thus, 3.16(a) holds.

3.18. Let

$$E' = \{(V_1, V_2, V_3, V_5, \tilde{V}_5) \in \mathcal{B}; V_2 = I'_2, V_3 = I_3\}.$$

If $(V_1, V_2, V_3, V_5, \tilde{V}_5) \in E'$ then $AV_1 = AV_2 = AV_3 = 0$; moreover we have $V_5 \subset K_7$ hence $AV_5 \subset AK_7 \subset I_3 \subset V_5$. Similarly, $A\tilde{V}_5 \subset I_3 \subset V_5$. Thus we have $(V_1, V_2, V_3, V_5, \tilde{V}_5) \in \mathcal{B}_u$. We see that $E' \subset \mathcal{B}_u$. Now E' is isomorphic to $\mathbf{P}^1 \times \mathbf{P}^1 \times \mathbf{P}^1$ hence it is a closed irreducible 3-dimensional subvariety of $\mathcal{B}_u$. Thus E' is an irreducible component of $\mathcal{B}_u$. We have $J_{E'} = \{\delta, \beta, \gamma\}$. We see that $E' = E$. From the description of $E = E'$ given above we see that 3.16(c) holds.

3.19. Until the end of 3.23 we assume that G_{ad} is of type A_5 and that $u \in G$ (see 1.1) is such that $d_u = 3$ and the Springer representation ρ_u is the irreducible representation of W of dimension five appearing in the third symmetric power of the reflection representation of W. Note that u is unique up to conjugation. We can write $I = \{1, 2, 3, 4, 5\}$ where the numbering is chosen so that each of $s_1s_2, s_2s_3, s_3s_4, s_4s_5$ has order 3.

Using 1.4(b) we see that there is a unique $T \in \underline{\mathcal{B}}_u$ such that $J_T = \{3\}$ and a unique $M \in \underline{\mathcal{B}}_u$ such that $J_M = \{1, 3, 5\}$. We shall prove

(a) *there are well defined parabolic subgroups $P^2 \in \mathcal{P}^2$, $P^4 \in \mathcal{P}^4$ such that $M = \{B \in \mathcal{B}; B(2) = P^2, B(4) = P^4\}$;*

we have $T_1 = T_5 = \{B \in T; B(2) = P^2, B(4) = P^4\} = M \cap T$ (we denote it by T_{15}); this is a $\mathbf{P}^1$-bundle over $\mathbf{P}^1$ and is a union of 3-lines;

we have $T_2 = T_4$ (we denote it by T_{24}); it intersects any 3-line in T in exactly one point;

the intersection $T_{15} \cap T_{24}$ is isomorphic to $\mathbf{P}^1$;

there is a unique morphism $\vartheta : T_{24} \to T_{15} \cap T_{24}$ such that for any $B \in T_{24}$ we have $(B, \vartheta(B)) \in \mathcal{O}_w$ for some $w \in W_{24}$; moreover, ϑ is a $\mathbf{P}^1$-bundle.

From (a) we see that

(b) $\chi(T_{24}) = 4$, $\chi(T) = 8$.

where $\chi()$ is as in 2.2.

3.20. To prove 3.19(a) for G is the same as proving it for G_{ad}. Hence we can assume that $G = GL(V)$ where V is a 6-dimensional $\mathbf{k}$-vector space. Until the end of 3.23 we shall adhere to this assumption. We can write $u = A + 1$ where $A : V \to V$ is nilpotent with Jordan blocks of sizes 3, 3. Let $K_2 = \ker A = A^2(V)$, $K_4 = \ker A^2 = A(V)$. We have $K_2 \subset K_4$.

For $i \in I$, we identify $\mathcal{P}^i$ with the variety of all i-dimensional subspaces of V. (The identification attaches to a subspace its stabilizer in G.) We identify $\mathcal{B}$ with the variety of all sequences $(V_1, V_2, V_3, V_4, V_5) \in \mathcal{P}^1 \times \mathcal{P}^2 \times \mathcal{P}^3 \times \mathcal{P}^4 \times \mathcal{P}^5$ such that $V_1 \subset V_2 \subset V_3 \subset V_4 \subset V_5$. (Such a sequence is identified with its stabilizer $\{g \in G; gV_1 = V_1, gV_2 = V_2, gV_4 = V_4, g\tilde{V}_4 = \tilde{V}_4\}$.)

3.21. Let
$$M' = \{(V_1, V_2, V_3, V_4, V_5) \in \mathcal{B}; V_2 = K_2, V_4 = K_4\}.$$
If $(V_1, V_2, V_3, V_4, V_5) \in M$ then
$AV_5 \subset AV = K_4 \subset V_5$, $AV_3 \subset AK_4 = A\ker(A^2) \subset \ker(A) = K_2 \subset V_3$,
$AV_1 \subset AK_2 = 0$, $AV_4 = AK_4 = K_2 \subset V_4$, $AK_2 = 0$.
We see that $(V_1, V_2, V_3, V_4, V_5) \in \mathcal{B}_u$. Thus, $M' \subset \mathcal{B}_u$. Note that M' is isomorphic to $\mathbf{P}^1 \times \mathbf{P}^1 \times \mathbf{P}^1$ hence it is a closed irreducible subvariety of $\mathcal{B}_u$ of dimension 3. Thus it is an irreducible component of $\mathcal{B}_u$. It satisfies $J_{M'} = \{1, 3, 5\}$. Hence $M' = M$.

3.22. Let
$$T' = \{(V_1, V_2, V_3, V_4, V_5) \in \mathcal{B}; V_1 \subset K_2, K_4 \subset V_5, A^2V_5 = V_1, AV_4 = V_2, V_2 \subset AV_5 \subset V_4\},$$

$$Z = \{(V_1, V_2, V_4, V_5) \in \mathcal{P}^1 \times \mathcal{P}^2 \times \mathcal{P}^4 \times \mathcal{P}^5; V_1 \subset V_2, V_4 \subset V_5, V_1 \subset K_2, K_4 \subset V_5, A^2V_5 = V_1, AV_4 = V_2, V_2 \subset AV_5 \subset V_4\}.$$
The map $\zeta : T' \to Z$, $(V_1, V_2, V_3, V_4, V_5) \mapsto (V_1, V_2, V_4, V_5)$ is a $\mathbf{P}^1$-bundle. Let
$$Z' = \{(V_1, V_5) \in \mathcal{P}^1 \times \mathcal{P}^5; V_1 \subset K_2, K_4 \subset V_5, A^2V_5 = V_1\}.$$
Consider the map $\zeta' : Z \to Z'$, $(V_1, V_2, V_4, V_5) \mapsto (V_1, V_5)$. Its fibre at (V_1, V_5) can be identified with the projective line $\{V_4; AV_5 \subset V_4 \subset V_5\}$ of the 2-dimensional vector space V_5/AV_5 (note that V_2 is uniquely determined by V_4 via $V_2 = AV_5$ and it automatically satisfies $V_1 \subset V_2 \subset AV_5$ since $A^2V_5 \subset AV_4 \subset AV_5$); thus ζ' is a $\mathbf{P}^1$-bundle. Note also that Z' can be identified with the projective line $\{V_5; K_4 \subset V_5 \subset V\}$ of the 2-dimensional vector space V/K_4. We see that T' is a $\mathbf{P}^1$-bundle over a $\mathbf{P}^1$-bundle over a $\mathbf{P}^1$-bundle. In particular, T' is a closed smooth irreducible subvariety of dimension 3 of $\mathcal{B}$. We show that $T' \subset \mathcal{B}_u$. It is enough to show that if $(V_1, V_2, V_3, V_4, V_5) \in T'$, then $AV_i \subset V_i$ for $i \in I$. For $i = 5$ this follows from $AV_5 \subset V_4 \subset V_5$. For $i = 4$ this follows from $AV_4 = V_2 \subset V_4$. For $i = 3$ this follows from $AV_3 \subset AV_4 = V_2 \subset V_3$. For $i = 2$ this follows from $AV_2 \subset AV_4 = V_2$. For $i = 1$ this follows from $AV_1 \subset AK_2 = 0$. Since $d_u = 3$, we see that $T' \in \underline{\mathcal{B}}_u$. Hence T'_i is defined for $i \in I$.

3.23. Note that $T'_3 = T'$. From the definitions we have
$$T'_1 = \{(V_1, V_2, V_3, V_4, V_5) \in T'; V_2 = K_2\},$$
$$T'_2 = \{(V_1, V_2, V_3, V_4, V_5) \in T'; AV_3 \subset V_1\},$$
$$T'_4 = \{(V_1, V_2, V_3, V_4, V_5) \in T'; AV_5 \subset V_3\},$$
$$T'_5 = \{(V_1, V_2, V_3, V_4, V_5) \in T'; V_4 = K_4\}.$$
We show:

(a) $$T'_2 = T'_4,$$

(b) $$T'_1 = T'_5.$$

Let $(V_1, V_2, V_3, V_4, V_5) \in T'$.

If $AV_5 \subset V_3$ then, since $\dim AV_5 = 3$, we have $AV_5 = V_3$. Hence $AV_3 = A^2V_5 = V_1$. Thus $T'_4 \subset T'_2$.

If $AV_3 \subset V_1$ then $A^2V_3 \subset AV_1 = 0$ hence $V_3 \subset K_4$. Now $A : K_4 \to K_2$ is surjective and $V_1 \subset K_2$ hence $\{x \in K_4; Ax \in V_1\}$ is a 3-dimensional subspace of K_4 containing V_3 hence $\{x \in K_4; Ax \in V_1\} = V_3$. Now $AV_5 \subset \{x \in K_4; Ax \in V_1\}$ since $AV_5 \subset K_4$ and $A^2V_5 = V_1$. Hence $AV_5 \subset V_3$ (and even $AV_5 = V_3$). Thus $T'_2 \subset T'_4$.

Assume that $V_4 = K_4$. Now $AV_4 = V_2$, $AK_4 = K_2$ hence $V_2 = K_2$. Thus $T'_5 \subset T'_1$.

Assume that $V_2 = K_2$. We have $K_4 = A^{-1}(K_2)$, $V_4 \subset A^{-1}(V_2) = A^{-1}(K_2)$ so that $V_4 \subset K_4$. We see that $T'_1 \subset T'_5$.

This proves (a),(b). We set $T'_{24} = T'_2 = T'_4$, $T'_{15} = T'_1 = T'_5$. We have

$$T'_{24} = \{(V_1, V_2, V_3, V_4, V_5) \in \mathcal{B}; V_1 \subset K_2, K_4 \subset V_5, AV_5 = V_3, AV_3 = V_1, AV_4 = V_2\},$$

$$T'_{15} = \{(V_1, V_2, V_3, V_4, V_5) \in \mathcal{B}; V_2 = K_2, V_4 = K_4, A^2V_5 = V_1\}.$$

We set

$$\mathfrak{Z} := T'_{24} \cap T'_{15} = \{(V_1, V_2, V_3, V_4, V_5) \in \mathcal{B}; AV_5 = V_3, AV_3 = V_1, V_2 = K_2, V_4 = K_4\}.$$

Now $\vartheta : T'_{24} \to \mathfrak{Z}$, $(V_1, V_2, V_3, V_4, V_5) \mapsto (V_1, K_2, V_3, K_4, V_5)$ is a $\mathbf{P}^1$-bundle; its fibre at $(V_1, K_2, V_3, K_4, V_5)$ is $\{(V_2, V_4) \in \mathcal{P}^2 \times \mathcal{P}^4; V_1 \subset V_2 \subset V_3, V_3 \subset V_4 \subset V_5, AV_4 = V_2\}$ which is isomorphic to the projective line $\{V_4; V_3 \subset V_4 \subset V_5\}$ of the 2-dimensional vector space V_5/V_3.

Note that $\mathfrak{Z}$ is isomorphic to the projective line $\{V_5; K_4 \subset V_5 \subset V\}$ of the 2-dimensional vector space V/K_4. Thus T'_{24} is a $\mathbf{P}^1$-bundle over $\mathbf{P}^1$ and thus, having dimension 2, is $\neq T'$. Now

$$T'_{15} \to \{(V_1, V_2, V_4, V_5) \in \mathcal{P}^1 \times \mathcal{P}^2 \times \mathcal{P}^4 \times \mathcal{P}^5; V_2 = K_2, V_4 = K_4, A^2V_5 = V_1\}$$

is a $\mathbf{P}^1$-bundle whose base is isomorphic to the projective line $\{V_5; K_4 \subset V_5 \subset V\}$ of the 2-dimensional vector space V/K_4. Thus T'_{15} is a $\mathbf{P}^1$-bundle over $\mathbf{P}^1$ and thus, having dimension 2, is $\neq T'$. Also T'_{15} is a union of 3-lines (the fibres of the map above.)

We now see that $T'_i \neq T'$ for $i \in I - \{3\}$. It follows that $T' = T$, see 3.19.

Now any 3-line in T is the fibre of $\zeta : T' \to Z$ at some $(V_1, V_2, V_4, V_5) \in Z$. This fibre contains exactly one point in T'_{24} namely the one defined by $V_3 = AV_5$. This completes the proof of 3.19(a).

3.24. In this subsection we assume that G_{ad} is of type $A_2A_2A_1$. We can write $I = \{0, 1, 2, 4, 5\}$ where the notation is chosen so that s_1s_2, s_4s_5 have order 3. We shall assume that $u \in G$ (see 1.1) is such that the image of u in G_{ad} has a projection to each of the three factors a subregular element in that factor. We have $d_u = 3$. Now $\mathcal{B}_u$ has four irreducible components: C_1, C_2, C_3, C_4 where $J_{C_1} = \{1, 0, 4\}$, $J_{C_2} = \{2, 0, 4\}$, $J_{C_3} = \{2, 0, 5\}$, $J_{C_4} = \{1, 0, 5\}$. It is clear that there are well defined $Q^i \in \mathcal{P}^i$, $i = 1, 2, 4, 5$, such that $\cap_{i \in \{1,2,4,5\}} Q^i$ contains a Borel subgroup and

$$C_1 = \{B \in \mathcal{B}_u; B(2) = Q^2, B(5) = Q^5\},$$

$$C_2 = \{B \in \mathcal{B}_u; B(1) = Q^1, B(5) = Q^5\},$$

$$C_3 = \{B \in \mathcal{B}_u; B(1) = Q^1, B(4) = Q^4\}.$$

$$C_4 = \{B \in \mathcal{B}_u; B(2) = Q^2, B(4) = Q^4\},$$

3.25. In this subsection we assume that G_{ad} is of type $A_2A_1A_1$. We can write $I = \{0, 1, 3, 5\}$ where the notation is chosen so that s_0s_3 has order 3. We shall assume that $u \in G$ (see 1.1) is such that the image of u in G_{ad} has a projection to each of the three factors a subregular element in that factor. We have $d_u = 3$. Now $\mathcal{B}_u$ has two irreducible components: C'_1, C'_2 where $J_{C'_1} = \{1, 3, 5\}$, $J_{C'_2} = \{1, 0, 5\}$. It is clear that there are well defined parabolic subgroups $P^0 \in \mathcal{P}^0$, $P^3 \in \mathcal{P}^3$ such that $P^0 \cap P^3$ contains a Borel subgroup and

$$C'_1 = \{B \in \mathcal{B}_u; B(0) = P^0\}, \quad C'_2 = \{B \in \mathcal{B}_u; B(3) = P^3\}.$$

4. Euler characteristic computations

4.1. In this section we assume that G_{ad} is of type E_6. We can write $I = \{0, 1, 2, 3, 4, 5\}$ where the numbering is chosen so that $s_1s_2, s_2s_3, s_3s_4, s_4s_5, s_0s_3$ have order 3. We shall assume that $u \in G$ (see 1.1) is such that $d_u = 3$; this determines u uniquely up to conjugacy. The Springer representation ρ_u of W is a direct sum of two irreducible representations: one of dimension 30 (appearing in the third symmetric power of the reflection representation) and one of dimension 15 (appearing in the fifth symmetric power of the reflection representation). Using 1.4(b) and the knowledge of ρ_u we see that there are exactly

-two irreducible components X of $\mathcal{B}_u$ such that $J_X = \{3\}$ (we call them $\mathbf{S}, \mathbf{T}$);

-two irreducible components X of $\mathcal{B}_u$ such that $J_X = \{0, 3\}$ (we call them $X(03), X'(03)$);

-one irreducible component X of $\mathcal{B}_u$ such that $J_X = \{0, 1, 4\}$ (we call it $X(014)$);

-one irreducible component X of $\mathcal{B}_u$ such that $J_X = \{0, 2, 4\}$ (we call it $X(024)$);

-one irreducible component X of $\mathcal{B}_u$ such that $J_X = \{0, 2, 5\}$ (we call it $X(025)$);

-one irreducible component X of $\mathcal{B}_u$ such that $J_X = \{0, 1, 5\}$ (we call it $X(015)$);

-one irreducible component X of $\mathcal{B}_u$ such that $J_X = \{1, 3, 5\}$ (we call it $X(135)$).

From 1.1(b) we see that if K is one of $\{0, 1, 4\}$, $\{0, 2, 4\}$, $\{0, 2, 5\}$, $\{0, 1, 5\}$, $\{1, 3, 5\}$ then the exists a unique $P_K \in \mathcal{P}_K$ such that $X(K) = \{B \in \mathcal{B}; B \subset P_K\}$.

Now u is induced from a unipotent element $u_1 \in \bar{P}$ where $P \in \mathcal{P}_{01245}$ and $(u_1, \bar{P})$ is like (u, G) in 3.24; in particular we have $u \in P, u_1 = \pi_P(u)$. The four irreducible components of $\mathcal{B}^{\bar{P}}_{u_1}$ (see 3.24) give rise as in 1.5 to four irreducible components of $\mathcal{B}_u$ with the J_X being preserved; these irreducible components of $\mathcal{B}_u$ must be the same as $X(014)$, $X(024)$, $X(025)$, $X(015)$. Using 3.24, we see that there exist $Q^i \in \mathcal{P}^i$ (relative to G) where $i \in \{1, 2, 3, 4, 5\}$ such that

$$X(014) = \{B \in \mathcal{B}; B(2) = Q^2, B(3) = Q^3, B(5) = Q^5\},$$

$$X(024) = \{B \in \mathcal{B}; B(1) = Q^1, B(3) = Q^3, B(5) = Q^5\},$$

$$X(025) = \{B \in \mathcal{B}; B(1) = Q^1, B(3) = Q^3, B(4) = Q^4\},$$

$$X(015) = \{B \in \mathcal{B}; B(2) = Q^2, B(3) = Q^3, B(4) = Q^4\}.$$

(Note that these conditions determine Q^i uniquely and that $Q^3 = P$.) Moreover, $\cap_{i \in \{1,2,3,4,5\}} Q^i$ contains a Borel subgroup. Next we note that u is induced from a unipotent element $u_2 \in \bar{P}'$ where $P' \in \mathcal{P}_{0135}$ and $(u_2, \bar{P}')$ is like (u, G) in 3.25; in particular we have $u \in P', u_2 = \pi_{P'}(u)$. The two irreducible components of $\mathcal{B}^{\bar{P}'}_{u_2}$ (see 3.25) give rise as in 1.5 to two irreducible components of $\mathcal{B}_u$ with the J_X being preserved; these irreducible components of $\mathcal{B}_u$ must be the same as $X(135)$, $X(015)$.

Using 3.25 we see that there exist $\tilde{Q}^i \in \mathcal{P}^i$ (relative to G) where $i \in \{0,2,3,4\}$ such that

$$X(135) = \{B \in \mathcal{B}; B(0) = \tilde{Q}^0, B(2) = \tilde{Q}^2, B(4) = \tilde{Q}^4\},$$

$$X(015) = \{B \in \mathcal{B}; B(2) = \tilde{Q}^2, B(3) = \tilde{Q}^3, B(4) = \tilde{Q}^4\}.$$

(Note that these conditions determine $\tilde{Q}^i$ uniquely and that $\tilde{Q}^2$, $\tilde{Q}^4$ contain P'.) Moreover, $\cap_{i\in\{0,2,3,4\}}\tilde{Q}^i$ contains a Borel subgroup. Since we have

$$\begin{aligned} X(015) &= \{B \in \mathcal{B}; B(2) = Q^2, B(3) = Q^3, B(4) = Q^4\} \\ &= \{B \in \mathcal{B}_u; B(2) = \tilde{Q}^2, B_3 = \tilde{Q}^3, B(4) = \tilde{Q}^4\} \end{aligned}$$

we see that we have $Q^2 = \tilde{Q}^2$, $Q^3 = \tilde{Q}^3$, $Q^4 = \tilde{Q}^4$. We set $Q^0 = \tilde{Q}^0$. Then we have

$$X(135) = \{B \in \mathcal{B}; B(0) = Q^0, B(2) = Q^2, B(4) = Q^4\}.$$

From the previous argument we see also that $Q_0 := Q^1 \cap Q^2 \cap Q^3 \cap Q^4 \cap Q^5$ is a parabolic subgroup in $\mathcal{P}_0$ and $Q_{15} := Q^0 \cap Q^2 \cap Q^3 \cap Q^4$ is a parabolic subgroup in $\mathcal{P}_{15}$. Let $Q_{015} = Q^2 \cap Q^3 \cap Q^4$. We have $Q_{015} \in \mathcal{P}_{015}$ and $Q_0 \subset Q_{015}$, $Q_{15} \subset Q_{015}$. Since the adjoint group of $\bar{Q}_{015}$ is of type $A_1A_1A_1$, the intersection $Q_0 \cap Q_{15}$ is a Borel subgroup of Q_{015}. We see that $Q^0 \cap Q^1 \cap Q^2 \cap Q^3 \cap Q^4 \cap Q^5$ is a Borel subgroup of G. We show:

(a) *The intersection $X(135) \cap X(024)$ consists of exactly one Borel subgroup B_0.*

Indeed the condition that $B \in \mathcal{B}$ belongs to $X(135)\cap X(024)$ is the same as $B(0) = Q^0, B(2) = Q^2, B(4) = Q^4, B(1) = Q^1, B(3) = Q^3, B(5) = Q^5$ that is, $B \subset Q^0 \cap Q^1 \cap Q^2 \cap Q^3 \cap Q^4 \cap Q^5$. It remains to use that the last intersection is a Borel subgroup of G.

4.2. Now u is induced from a unipotent element $u' \in \bar{Q}'$ where $Q' \in \mathcal{P}_{12345} = \mathcal{P}^0$ and (u', Q') is like (u, G) in 3.19; in particular, we have $u \in Q', u' = \pi_{Q'}(u')$. The irreducible components M (resp. T) of $\mathcal{B}_{u'}^{\bar{Q}'}$ as in 3.19 (with G, u replaced by $\bar{Q}', u')$) give rise as in 1.5 to irreducible components of $\mathcal{B}_u$ with the same J_X which therefore must be equal to $X(135)$ (resp. to $X' \in \{\mathbf{S}, \mathbf{T}\}$; we arrange notation so that $X' = \mathbf{T}$). From 3.19(a) we see that the following holds:

(a) *there are well defined parabolic subgroups $P'^2 \in \mathcal{P}^2$, $P'^4 \in \mathcal{P}^4$ such that $X(135) = \{B \in \mathcal{B}; B(0) = Q, B(2) = P'^2, B(4) = P'^4\}$;*

we have $\mathbf{T} \subset \{B \in \mathcal{B}; B(0) = Q', B(1)\spadesuit P'^2, B(5)\spadesuit P'^4\}$;

we have $\mathbf{T}_1 = \mathbf{T}_5 = \{B \in \mathbf{T}; B(0) = Q', B(2) = P'^2, B(4) = P'^4\} = X(135) \cap \mathbf{T}$ (we denote it by $\mathbf{T}_{15}$); this is a $\mathbf{P}^1$-bundle over $\mathbf{P}^1$ and is a union of 3-lines;

we have $\mathbf{T}_2 = \mathbf{T}_4$ (we denote it by $\mathbf{T}_{24}$); it intersects any 3-line in $\mathbf{T}$ in exactly one point;

the intersection $\mathbf{T}_{15} \cap \mathbf{T}_{24}$ is isomorphic to $\mathbf{P}^1$;

there is a unique morphism $\vartheta : \mathbf{T}_{24} \to \mathbf{T}_{15} \cap \mathbf{T}_{24}$ such that for any $B \in \mathbf{T}_{24}$ we have $(B, \vartheta(B)) \in \mathcal{O}_w$ for some $w \in W_{24}$; moreover, ϑ is a $\mathbf{P}^1$-bundle.

Since we have also $X(135) = \{B \in \mathcal{B}; B(0) = Q^0, B(2) = Q^2, B(4) = Q^4\}$ we see that we must have $Q' = Q^0, P'^2 = Q^2, P'^4 = Q^4$. Hence we have $\mathbf{T} \subset \{B \in \mathcal{B}; B(0) = Q^0, B(1)\spadesuit Q^2, B(5)\spadesuit Q^4\}$.

4.3. Now u is induced from a unipotent element $u'' \in \bar{Q}''$ where $Q'' \in \mathcal{P}_{0234}$ and $(u'', \bar{Q}'')$ is like (u, G) in 3.1 with $\alpha = 0, \beta = 2, \gamma = 4, \omega = 3$; in particular we have $u \in Q'', u'' = \pi_{Q''}(u)$. The irreducible components $\hat{S}$, $S_{\beta\gamma'}, S_{\beta',\gamma}$ (resp. S) of $\mathcal{B}_{u''}^{\bar{Q}''}$ defined as in 3.1 (with G, u replaced by $\bar{Q}'', u''$)) give rise as in 1.5 to irreducible components of $\mathcal{B}_u$ with the same J_X which therefore must be equal to $X(024)$, $X(03)$, $X'(03)$ (resp. $X'' \in \{\mathbf{S}, \mathbf{T}\}$); since $\chi(\mathbf{S}) = \chi(S) = 12$ and $\chi(\mathbf{T}) = 8$ (see 3.10(c), 3.19(b)), we must have $X'' = \mathbf{S}$. We set $Q'' = P^1 \cap P^5$ where $P^1 \in \mathcal{P}^1, P^5 \in \mathcal{P}^5$. Let $P^0 \neq \tilde{P}^0$ in $\mathcal{P}^0$, $P^2 \neq \tilde{P}^2$ in $\mathcal{P}^2$, $P^4 \neq \tilde{P}^4$ in $\mathcal{P}^4$, $P^3 \in \mathcal{P}^3$ (with $\mathcal{P}^i$ referring to G) be the inverse images under $\pi_{Q''}$ of the parabolic subgroups with the same names in $\bar{Q}''$.

From 3.1(a) we see that the following holds:

(a) *We have*

$$X(024) = \{B \in \mathcal{B}; B(1) = P^1, B(3) = P^3, B(5) = P^5\}$$

$$X(03) = \{B \in \mathcal{B}; B(1) = P^1, B(2) = P^2, B(4) = \tilde{P}^4, B(5) = P^5\},$$

$$X'(03) = \{B \in \mathcal{B}; B(1) = P^1, B(2) = \tilde{P}^2, B(4) = P^4, B(5) = P^5\},$$

$$\mathbf{S} \subset \{B \in \mathcal{B}; B(1) = P^1, B(5) = P^5\}.$$

Since

$$X(024) = \{B \in \mathcal{B}; B(1) = Q^1, B(3) = Q^3, B(5) = Q^5\},$$

we see that $P^1 = Q^1$, $P^3 = Q^3$, $P^5 = Q^5$. Thus,

$$X(03) = \{B \in \mathcal{B}; B(1) = Q^1, B(2) = P^2, B(4) = \tilde{P}^4, B(5) = Q^5\},$$

$$X'(03) = \{B \in \mathcal{B}; B(1) = Q^1, B(2) = \tilde{P}^2, B(4) = P^4, B(5) = Q^5\}.$$

From 3.1(e) we deduce

(b) $X(024) \cap X(03)$ *is a* 0-*line;* $X(024) \cap X'(03)$ *is a* 0-*line.*

4.4. Now u is induced from a unipotent element $'u \in {}'\bar{Q}$ where $'Q \in \mathcal{P}_{01234} = \mathcal{P}^5$ and $('u, '\bar{Q})$ is like (u, G) in 3.15 with $\delta = 1, \alpha = 2, \beta = 0, \gamma = 4, \omega = 3$; in particular we have $u \in {}'Q, 'u = \pi_{'Q}(u)$. The irreducible components $\tilde{S}$, E of $\mathcal{B}_{'u}^{'\bar{Q}}$ defined in 3.15 (with G, u replaced by $'\bar{Q}, 'u$)) give rise as in 1.5 to the irreducible components $\mathbf{S}$ and $X(014)$ of $\mathcal{B}_u$. From 3.16(a),(b),(c) we deduce that there is a well defined $\hat{P}^2 \in \mathcal{P}^2$ such that

$$X(014) = \{B \in \mathcal{B}; B(2) = \hat{P}^2, B(3) = Q^3, B(5) = {}'Q\},$$

$$\hat{P}^2 \notin \{P^2, \tilde{P}^2\},$$

$$\mathbf{S}_1 = \{B \in \mathbf{S}; B(2) = \hat{P}^2.\}$$

Since we have also

$$X(014) = \{B \in \mathcal{B}; B(2) = Q^2, B(3) = Q^3, B(5) = Q^5\},$$

we see that $\hat{P}^2 = Q^2$, $'Q = Q^5$. Thus we have

$$Q^2 \notin \{P^2, \tilde{P}^2\},$$

$$\mathbf{S}_1 = \{B \in \mathbf{S}; B(2) = Q^2.\}$$

4.5. Now u is induced from a unipotent element $''u \in ''\bar{Q}$ where $''Q \in \mathcal{P}_{02345} = \mathcal{P}^1$ and $(''u, ''\bar{Q})$ is like (u, G) in 3.15 with $\delta = 5, \alpha = 4, \omega = 3, \beta = 2, \gamma = 0$; in particular we have $u \in ''Q, ''u = \pi_{''Q}(u)$. The irreducible components $\tilde{S}$, E of $\mathcal{B}^{''\bar{Q}}_{''u}$ defined in 3.15 (with G, u replaced by $''\bar{Q}, ''u$)) give rise as in 1.5 to the irreducible components $\mathbf{S}$ and $X(025)$ of $\mathcal{B}_u$. From 3.16(a),(b),(c) we deduce as in 4.4 that:

$$Q^4 \notin \{P^4, \tilde{P}^4\},$$

$$\mathbf{S}_5 = \{B \in \mathbf{S}; B(4) = Q^4.\}$$

4.6. We show:

(a) $X(135) \cap X(03) = \emptyset, X(135) \cap X'(03) = \emptyset.$

Recall that

$X(135) = \{B \in \mathcal{B}; B(0) = Q^0, B(2) = Q^2, B(4) = Q^4\},$

$X(03) = \{B \in \mathcal{B}; B(1) = Q^1, B(2) = P^2, B(4) = \tilde{P}^4, B(5) = Q^5\},$

$X'(03) = \{B \in \mathcal{B}; B(1) = Q^1, B(2) = \tilde{P}^2, B(4) = P^4, B(5) = Q^5\}.$

If $B \in X(135) \cap X(03)$ then $B(2) = Q^2 = P^2$. This contradicts $Q^2 \neq P^2$. Similarly if $B \in X(135) \cap X'(03)$ then $B(2) = Q^2 = \tilde{P}^2$. This contradicts $Q^2 \neq \tilde{P}^2$. This proves (a).

Using $\mathbf{T}_{15} = \mathbf{T} \cap X(135)$ we see that (a) implies:

(b) $\mathbf{T}_{15} \cap X(03) = \emptyset, \mathbf{T}_{15} \cap X'(03) = \emptyset.$

4.7. We show:

(a) *The intersection of a* 0*-line in* $\mathcal{B}$ *with* $\mathbf{T}$ *is either a point or is empty.*

Assume that this intersection contains two distinct points B', B''. We have

$(B', B'') \in \mathcal{O}_{s_0}$.

Since $\mathbf{T} \subset Q^0$, we have $B' \in Q^0, B'' \in Q^0$. Since

$Q^0 \in \mathcal{P}_{12345}$,

we have $(B', B'') \in \mathcal{O}_w$ where $w \in W_{12345}$. This contradicts $w = s_0$; (a) is proved.

We show:

(b) *The intersection* $\mathbf{T} \cap X(024) \cap X(03)$ *is either a point or is empty. The intersection* $\mathbf{T} \cap X(024) \cap X'(03)$ *is either a point or is empty.*

This follows from (a) since $X(024) \cap X(03)$ is a 0-line and $X(024) \cap X'(03)$ is a 0-line, see 4.3(b).

4.8. We write $X(135) \cap X(024) = \{B_0\}$ as in 4.1(a). We show:

(a) $\mathbf{T}_{15} \cap \mathbf{T}_{24} \cap \mathbf{T}_0 \subset \{B_0\}$.

Let $B \in \mathbf{T}_{24} \cap \mathbf{T}_0$. Let $P \in \mathcal{P}^{024}$ be such that $B \subset P$. Since $B \in \mathbf{T}_0 \cap \mathbf{T}_2 \cap \mathbf{T}_4$, we see from 1.3(a) that $X = \mathcal{B}^P$ is an irreducible component of $\mathcal{B}_u$ and that $J_X = \{0, 2, 4\}$. Thus we have $X = X(024)$. We see that $B \in X(024)$. Thus $\mathbf{T}_{24} \cap \mathbf{T}_0 \subset X(024)$. Recall that $\mathbf{T}_{15} = X(135) \cap \mathbf{T}$. Thus $\mathbf{T}_{15} \subset X(135)$ and $\mathbf{T}_{15} \cap \mathbf{T}_{24} \cap \mathbf{T}_0 \subset X(024) \cap X(135) = \{B_0\}$; now (a) follows.

4.9. We show:

(a) *The morphism* $\vartheta : \mathbf{T}_{24} \to \mathbf{T}_{24} \cap \mathbf{T}_{15}$*, see 3.19, restricts to a morphism* $\vartheta_0 : \mathbf{T}_{24} \cap \mathbf{T}_0 \to \mathbf{T}_{24} \cap \mathbf{T}_{15} \cap \mathbf{T}_0$*. Moreover,* ϑ_0 *is a* $\mathbf{P}^1$*-bundle.*

Let $B \in \mathbf{T}_{24} \cap \mathbf{T}_0$ and let $B' = \vartheta(B) \in \mathbf{T}_{24} \cap \mathbf{T}_{15}$. As in the proof of 4.8(a) we have $B \in X(024)$. Recall that $(B, B') \in \mathcal{O}_w$ for some $w \in W_{24}$. Since $B \in X(024)$ it follows that $B' \in X(024)$. Thus $B' \in \mathbf{T}_{24} \cap \mathbf{T}_{15} \cap \mathbf{T}_0$. Thus, ϑ_0 is well defined. Conversely, let $B' \in \mathbf{T}_{24} \cap \mathbf{T}_{15} \cap \mathbf{T}_0$ and let $B \in \vartheta^{-1}(B')$. Again we have $B' \in X(024)$

and $(B, B') \in \mathcal{O}_w$ for some $w \in W_{24}$. It follows that $B \in X(024)$ so that $B \in \mathbf{T}_{24} \cap \mathbf{T}_0$. We see that ϑ_0 is a $\mathbf{P}^1$-bundle. This proves (a).

We show:

(b) *If $B_0 \in \mathbf{T}$ (hence $B_0 \in \mathbf{T}_{24} \cap \mathbf{T}_{15} \cap \mathbf{T}_0$) then $\mathbf{T}_{24} \cap \mathbf{T}_0$ is a projective line whose intersection with any 3-line in $\mathbf{T}$ has exactly one point or is empty. If $B_0 \notin \mathbf{T}$, then $\mathbf{T}_{24} \cap \mathbf{T}_0 = \emptyset$.*

Using (a), we see that it is enough to show the following statement.

If B, B' are Borel subgroups in $\mathbf{T}_{24} \cap \mathbf{T}_0$ such that B, B' are on the same 3-line in $\mathcal{B}$ then $B = B'$.

Since $\vartheta(B) = \vartheta(B') = B_0$, we have $(B, B_0) \in \mathcal{O}_w$, $(B', B_0) \in \mathcal{O}_{w'}$, for some w, w' in W_{24}. It follows that $(B, B') \in \mathcal{O}_y$ for some $y \in W_{24}$. Since B, B' are on the same 3-line we have also $(B, B') \in \mathcal{O}_z$ for some $z \in W_3$. This forces $y = z = 1$. This proves (b).

4.10. We identify $Y \subset S$ with a subvariety of $\mathbf{S}$. We show:

(a) *If $B_0 \notin \mathbf{T}$ then $\mathbf{S} \cap \mathbf{T} = \emptyset$. If $B_0 \in \mathbf{T}$ then $\mathbf{S} \cap \mathbf{T}$ is the union of all 3-lines which intersect $\{B \in \mathbf{T}_{24} \cap T_0; \vartheta(B) = B_0\} \cap Y$.*

For $K \subset I$ we denote by $w(K)$ an element of W_K. Let $B \in \mathbf{S} \cap \mathbf{T}$. The 3-line through B must intersect Y in a unique point B_1 and it must intersect $\mathbf{T}_{24}$ in a unique point B_2. Let $B_3 = \vartheta(B_2) \in \mathbf{T}_{24} \cap \mathbf{T}_{15} \subset X(135)$. We have $(B_3, B_0) \in \mathcal{O}_{w(135)}$. We have $Y \subset X(024)$, hence $(B_1, B_0) \in \mathcal{O}_{w(024)}$. We have $(B, B_1) \in \mathcal{O}_{w(3)}$, hence $(B, B_0) \in \mathcal{O}_{w(3)w(024)}$.

We have $(B, B_2) \in \mathcal{O}_{w'(3)}$, $(B_2, B_3) \in \mathcal{O}_{w(24)}$, $(B_3, B_0) \in \mathcal{O}_{w(135)}$. If $w(24) \neq 1$ it follows that $(B, B_0) \in \mathcal{O}_{w'(3)w(24)w(135)}$ hence $w'(3)w(24)w(135) = w(3)w(024)$ so that $w(135) = 1, w'(3) = w(3), w(024) = w(24)$ and $B_3 = B_0$, $B_1 = B_2 \in \mathbf{T}_{24} \cap Y$, $(B_1, B_0) \in \mathcal{O}_{w(24)}$. Since $B_3 \in \mathbf{T}$ and $B_3 = B_0$ we have also $B_0 \in \mathbf{T}$ (hence $B_0 \in \mathbf{T}_{24} \cap \mathbf{T}_{15} \cap \mathbf{T}_0$) and $B_1 \in \{B' \in \mathbf{T}_{24} \cap T_0; \vartheta(B) = B_0\} \cap Y$.

If $w_{24} = 1$ we must have $B_2 = B_3$, $(B_3, B_0) \in \mathcal{O}_{w_1(3)}$. Since $B_3 \in \mathbf{T}$ and $\mathbf{T}$ is a union of 3-lines it follows that $B_0 \in \mathbf{T}$. From $(B, B_2) \in \mathcal{O}_{w'(3)}, (B_2, B_0) \in \mathcal{O}_{w_1(3)}$ we obtain $(B, B_0) \in \mathcal{O}_{w_2(3)}$.

Now (a) follows (we have used 4.9(b)).

4.11. We show:

(a) *Assume that $B_0 \in \mathbf{T}$. Let $U = \{B \in \mathbf{T}_{24} \cap T_0; \vartheta(B) = B_0\}$ (a projective line). Then either $U \cap Y \cong \mathbf{P}^1$ or $U \cap Y$ consists of two points or $U \cap Y$ consists of one point.*

Recall that Y can be viewed as the variety

$$\{(z_0, z_0', z_2, z_2'), (y, y')) \in \mathbf{P}^3 \times \mathbf{P}^1; z_0 z_0' + z_2 z_2' = 0, z_2 y' + z_2' y = 0\}$$

or setting $z_0 = ac, z_0' = bd, z_2 = -ad, z_2' = bc$, as the variety

$$\{((a, b), (c, d), (y, y')) \in \mathbf{P}^1 \times \mathbf{P}^1 \times \mathbf{P}^1; -ady' + bcy = 0\}.$$

Now U can be viewed as a subvariety of $\mathbf{P}^1 \times \mathbf{P}^1 \times \mathbf{P}^1$ of the form

$$\{((a, b), (c, d), (y, y')) \in \mathbf{P}^1 \times \mathbf{P}^1 \times \mathbf{P}^1; c = m_1 a + m_2 b, d = m_1' a + m_2' b, y = y_0, y' = y_0\}$$

where $(y_0, y_0') \in \mathbf{P}^1$ is fixed and $(m_1, m_2, m_1', m_2') \in \mathbf{k}^4$ is fixed such that $m_1 m_2' - m_2 m_1' \neq 0$. Then $U \cap Y$ becomes

$$\begin{aligned}
&\{((a,b),(c,d),(y,y')) \in \mathbf{P}^1 \times \mathbf{P}^1 \times \mathbf{P}^1;\\
&- a(m_1'a + m_2'b)y_0' + b(m_1 a + m_2 b)y_0 = 0, y = y_0, y' = y_0\}\\
&\cong \{(a,b) \in \mathbf{P}^1; -a(m_1'a + m_2'b)y_0' + b(m_1 a + m_2 b)y_0 = 0\}\\
&= \{(a,b) \in \mathbf{P}^1; -m_1'y_0'a^2 + (-m_2'y_0' + m_1 y_0)ab + m_2 y_0 b^2 = 0\}.
\end{aligned}$$

It follows that, if each of $-m_1'y_0', -m_2'y_0' + m_1 y_0, m_2 y_0$ is zero then $U \cap Y \cong \mathbf{P}^1$; otherwise, $U \cap Y$ consists of one or two points. This proves (a).

4.12. We set $\mathbf{S}^\diamond = \mathbf{S} - (\mathbf{S}_0 \cup \mathbf{S}_1 \cup \mathbf{S}_2 \cup \mathbf{S}_4 \cup \mathbf{S}_5)$. This is an open dense subset of $\mathbf{S}$. We have the following result.

PROPOSITION 4.13. *We have* $\chi(\mathbf{S}^\diamond) = 1$.

Under the isomorphism $S \xrightarrow{\sim} \mathbf{S}$, $\mathbf{S}_5$ (resp. $\mathbf{S}_1$) corresponds to a closed subset R_5 (resp. R_1) of S and $\mathbf{S}^\diamond$ corresponds to the open dense subset $S^\diamond := S - (S_0 \cup R_1 \cup S_2 \cup S_4 \cup R_5)$ of S. It is enough to prove that

(a) $$\chi(S^\diamond) = 1.$$

The proof is given in 4.14-4.16.

4.14. To prove that $\chi(S^\diamond) = 1$ we can identify our S with $X \subset \mathcal{B}$ as in 3.12(c) in such a way that $S_0 = X_\alpha, S_2 = X_\beta, S_4 = X_\gamma$ and, if
$A : V \to V,\ L_1, L_1', {}^!L_4, {}^!L_4', L_4^!, L_4'^!, I_2, K_6$
are as in 3.2, 3.4, then R_5 is identified with

$$\{(V_1, V_2, V_4, \tilde{V}_4) \in X; V_4 = \hat{V}_4^!\},$$

R_1 is identified with

$$\{(V_1, V_2, V_4, \tilde{V}_4) \in X; \tilde{V}_4 = {}^!\hat{V}_4\},$$

where $\hat{V}_4^!$ is an isotropic 4-space in V in the same family as $L_4^!, L_4'^!$ but distinct from each of them and ${}^!\hat{V}_4^!$ is an isotropic 4-space in V in the same family as ${}^!L_4, {}^!L_4'$ but distinct from each of them; moreover we have $I_2 \subset {}^!\hat{V}_4$, $I_2 \subset \hat{V}_4^!$. Recall that $X_\alpha = X_\alpha^* \cup Y$, $X_\beta = X_\beta^* \cup Y$, $X_\gamma = X_\gamma^* \cup Y$, where

$$\begin{aligned}
X_\alpha^* = &\{(V_1, V_2, V_4, \tilde{V}_4) \in \mathcal{B}; V_1 \subset I_2, V_4 = {}^!L_4, \tilde{V}_4 = L_4'^!\}\\
&\sqcup \{(V_1, V_2, V_4, \tilde{V}_4) \in \mathcal{B}; V_1 \subset I_2, V_4 = {}^!L_4', \tilde{V}_4 = L_4^!\},
\end{aligned}$$

$$\begin{aligned}
X_\beta^* = &\{(V_1, V_2, V_4, \tilde{V}_4) \in \mathcal{B}; I_2 \subset V_4, V_1 = L_1, \tilde{V}_4 = L_4^!\}\\
&\sqcup \{(V_1, V_2, V_4, \tilde{V}_4) \in \mathcal{B}; I_2 \subset V_4, V_1 = L_1', \tilde{V}_4 = L_4'^!\},
\end{aligned}$$

$$\begin{aligned}
X_\gamma^* = &\{(V_1, V_2, V_4, \tilde{V}_4) \in \mathcal{B}; I_2 \subset \tilde{V}_4, V_1 = L_1, V_4 = {}^!L_4\}\\
&\sqcup \{(V_1, V_2, V_4, \tilde{V}_4) \in \mathcal{B}; I_2 \subset \tilde{V}_4, V_1 = L_1', V_4 = {}^!L_4'\},
\end{aligned}$$

$$Y = \{(V_1, V_2, V_4, \tilde{V}_4) \in X; V_2 = I_2\}.$$

4.15. As in 3.10 let Z be the variety of all pairs V_1, V_3 of isotropic subspaces of V of dimension 1 and 3 respectively such that $V_1 \subset I_2 \subset V_3$ and $AV_3 \subset V_1$ and let $\mathfrak{U}$ be the space of isotropic 3-spaces in V that contain I_2 hence are contained in K_6; we have $\mathfrak{U} \cong \mathbf{P}^1 \times \mathbf{P}^1$. From 3.10(d) we have $\chi(Z) = 6$.

Let $Z_1 = \{(V_1, V_3) \in Z; V_3 \subset {}^!\hat{V}_4\}$, $Z_5 = \{(V_1, V_3) \in Z; V_3 \subset \hat{V}_4^!\}$. Let $\mathfrak{U}_1 = \{V_3 \in \mathfrak{U}; V_3 \subset {}^!\hat{V}_4\}$, $\mathfrak{U}_5 = \{V_3 \in \mathfrak{U}; V_3 \subset \hat{V}_4^!\}$. Note that $\mathfrak{U}_1 \cong \mathbf{P}^1$, $\mathfrak{U}_5 \cong \mathbf{P}^1$. By the argument in the proof of 3.10(d), the map $Z \to \mathfrak{U}$, $(V_1, V_3) \mapsto V_3$, restricts to an isomorphism $Z_1 \xrightarrow{\sim} \mathfrak{U}_1$ and to an isomorphism $Z_2 \xrightarrow{\sim} \mathfrak{U}_2$ (note that $\mathfrak{U}_1, \mathfrak{U}_2$ do not contain the exceptional points ${}^!L_4 \cap L_4'^!$ and ${}^!L_4' \cap L_4^!$). We see that $Z_1 \cong \mathbf{P}^1$, $Z_5 \cong \mathbf{P}^1$. It follows that

$$\chi(Z_1) = \chi(Z_5) = 2.$$

We have $Z_1 \cap Z_5 = \{(V_1, V_3) \in Z; V_3 \subset {}^!\hat{V}_4 \cap \hat{V}_4^!\}$. Since $I_2 \subset {}^!\hat{V}_4 \cap \hat{V}_4^!$, the intersection ${}^!\hat{V}_4 \cap \hat{V}_4^!$ is a 3-dimensional isotropic subspace; we see that $Z_1 \cap Z_5$ consists of a single element (V_1, V_3) where $V_3 = {}^!\hat{V}_4 \cap \hat{V}_4^!$ and V_1 is uniquely determined by V_3. In particular,

$$\chi(Z_1 \cap Z_5) = 1.$$

We have

$$\chi(Z_1 \cup Z_5) = \chi(Z_1) + \chi(Z_5) - \chi(Z_1 \cap Z_5) = 2 + 2 - 1 = 3.$$

Let

$$Z_\alpha^* = \{(V_1, V_3) \in Z; V_1 \subset I_2, V_3 = {}^!L_4 \cap L_4'^!\} \sqcup \{(V_1, V_3) \in Z; V_1 \subset I_2, V_3 = {}^!L_4' \cap L_4^!\},$$
$$Z_\beta^* = \{(V_1, V_3) \in Z; V_1 = L_1, V_3 \subset L_4^!\} \sqcup \{(V_1, V_3) \in Z; V_1 = L_1', V_3 \subset L_4'^!\},$$
$$Z_\gamma^* = \{(V_1, V_3) \in Z; V_1 = L_1, V_3 \subset {}^!L_4\} \sqcup \{(V_1, V_3) \in Z; V_1 = L_1', V_3 \subset {}^!L_4'\},$$

We have

$$Z_\alpha^* \cap Z_\beta^* = \{(L_1, {}^!L_4' \cap L_4^!)\} \sqcup \{(L_1', {}^!L_4 \cap L_4'^!)\},$$
$$Z_\alpha^* \cap Z_\gamma^* = \{(L_1, {}^!L_4 \cap L_4'^!)\} \sqcup \{(L_1', {}^!L_4' \cap L_4^!)\},$$
$$Z_\beta^* \cap Z_\gamma^* = \{(L_1, {}^!L_4 \cap L_4^!)\} \sqcup \{(L_1', {}^!L_4' \cap L_4'^!)\},$$
$$Z_\alpha^* \cap Z_\beta^* \cap Z_\gamma^* = \emptyset.$$

Thus, each of $Z_\alpha^*, Z_\beta^*, Z_\gamma^*$ is a disjoint union of two copies of $\mathbf{P}^1$; each of $Z_\alpha^* \cap Z_\beta^*$, $Z_\alpha^* \cap Z_\gamma^*$, $Z_\beta^* \cap Z_\gamma^*$ consists of two points. We see that

$$\chi(Z_\alpha^*) = \chi(Z_\beta^*) = \chi(Z_\gamma^*) = 4,$$
$$\chi(Z_\alpha^* \cap Z_\beta^*) = \chi(Z_\alpha^* \cap Z_\gamma^*) = \chi(Z_\beta^* \cap Z_\gamma^*) = 2$$

.

$$\chi(Z_\alpha^* \cap Z_\beta^* \cap Z_\gamma^*) = 0.$$

We have

$$\chi(Z_\alpha^* \cup Z_\beta^* \cup Z_\gamma^*) = \chi(Z_\alpha^*) + \chi(, Z_\beta^*) + \chi(Z_\gamma^*)$$
$$- \chi(Z_\alpha^* \cap Z_\beta^*) - \chi(Z_\alpha^* \cap Z_\gamma^*) - \chi(Z_\beta^* \cap Z_\gamma^*) + \chi(Z_\alpha^* \cap Z_\beta^* \cap Z_\gamma^*)$$
$$= 4 + 4 + 4 - 2 - 2 - 2 + 0 = 6.$$

We have

$$Z_\alpha^* \cap Z_1 = \emptyset, Z_\gamma^* \cap Z_1 = \emptyset,$$
$$Z_\beta^* \cap Z_1 = \{(L_1', {}^!\hat{V}_4 \cap L_4'^!)\} \sqcup \{(L_1, {}^!\hat{V}_4 \cap L_4^!)\},$$
$$Z_\alpha^* \cap Z_5 = \emptyset, Z_\beta^* \cap Z_5 = \emptyset,$$

$$Z_\gamma^* = \{(L_1, {}^!L_4 \cap \hat{V}_4^!)\} \sqcup \{(L_1', {}^!L_4' \cap \hat{V}_4^!)\}.$$

Thus

$$(Z_\alpha^* \cup Z_\beta^* \cup Z_\gamma^*) \cap Z_1 = \{(L_1', {}^!\hat{V}_4 \cap L_4'{}^!)\} \sqcup \{(L_1, {}^!\hat{V}_4 \cap L_4^!)\},$$
$$(Z_\alpha^* \cup Z_\beta^* \cup Z_\gamma^*) \cap Z_5 = \{(L_1, {}^!L_4 \cap \hat{V}_4^!)\} \sqcup \{(L_1', {}^!L_4' \cap \hat{V}_4^!)\}.$$

Hence

$$(Z_\alpha^* \cup Z_\beta^* \cup Z_\gamma^*) \cap (Z_1 \cap Z_5) = \emptyset.$$

We have

$$\chi((Z_\alpha^* \cup Z_\beta^* \cup Z_\gamma^*) \cap Z_1) = \chi((Z_\alpha^* \cup Z_\beta^* \cup Z_\gamma^*) \cap Z_5) = 2,$$
$$\chi((Z_\alpha^* \cup Z_\beta^* \cup Z_\gamma^*) \cap (Z_1 \cap Z_5)) = 0$$

hence

$$\begin{aligned}&\chi((Z_\alpha^* \cup Z_\beta^* \cup Z_\gamma^*) \cap (Z_1 \cup Z_5))\\ &= \chi((Z_\alpha^* \cup Z_\beta^* \cup Z_\gamma^*) \cap Z_1) + \chi((Z_\alpha^* \cup Z_\beta^* \cup Z_\gamma^*) \cap Z_5)\\ &- \chi((Z_\alpha^* \cup Z_\beta^* \cup Z_\gamma^*) \cap (Z_1 \cap Z_5)) = 2 + 2 - 0 = 4.\end{aligned}$$

We have

$$\begin{aligned}&\chi((Z_\alpha^* \cup Z_\beta^* \cup Z_\gamma^*) \cup (Z_1 \cup Z_5))\\ &= \chi(Z_\alpha^* \cup Z_\beta^* \cup Z_\gamma^*) + \chi(Z_1 \cup Z_5)\\ &- \chi((Z_\alpha^* \cup Z_\beta^* \cup Z_\gamma^*) \cap (Z_1 \cup Z_5)) = 6 + 3 - 4 = 5.\end{aligned}$$

4.16. We have a partition $Z = Z' \cup Z''$ where $Z' = Z_\alpha^* \cup Z_\beta^* \cup Z_\gamma^* \cup Z_1 \cup Z_5$ and $Z'' = Z - Z'$. Recall that $\zeta : (V_1, V_2, V_4, \tilde{V}_4) \mapsto (V_1, V_4 \cap \tilde{V}_4)$ makes X into a $\mathbf{P}^1$-bundle over Z and that

$$Y = \{(V_1, V_2, V_4, \tilde{V}_4) \in X; V_2 = I_2\}.$$

Recall that ζ restricts to an isomorphism $Y \xrightarrow{\sim} Z$. Hence setting $Y' = \zeta^{-1}(Z') \cap Y$, $Y'' = \zeta^{-1}(Z'') \cap Y$, we have a partition $Y = Y' \cup Y''$ and ζ restricts to isomorphisms $Y' \xrightarrow{\sim} Z'$, $Y'' \xrightarrow{\sim} Z''$. We have

$$\chi(Y'') = \chi(Z'') = \chi(Z) - \chi(Z') = 6 - 5 = 1.$$

We have

$$X_\alpha^* = \zeta^{-1}(Z_\alpha^*), \quad X_\beta^* = \zeta^{-1}(Z_\beta^*), \quad X_\gamma^* = \zeta^{-1}(Z_\gamma^*),$$
$$R_1 = \zeta^{-1}(Z_1), \quad R_5 = \zeta^{-1}(Z_5).$$

Hence

$$X_\alpha^* \cup X_\beta^* \cup X_\gamma^* \cup R_1 \cup R_5 = \zeta^{-1}(Z'),$$

so that

$$\chi(X_\alpha^* \cup X_\beta^* \cup X_\gamma^* \cup R_1 \cup R_5) = \chi(\zeta^{-1}(Z')) = 2\chi(Z') = 10.$$

Recall from 3.1(c) that

$$X_\alpha \cup X_\beta \cup X_\gamma \cup R_1 \cup R_5 = X_\alpha^* \cup X_\beta^* \cup X_\gamma^* \cup R_1 \cup R_5 \cup Y$$

hence we have a partition

$$X_\alpha \cup X_\beta \cup X_\gamma \cup R_1 \cup R_5 = (X_\alpha^* \cup X_\beta^* \cup X_\gamma^* \cup R_1 \cup R_5) \cup Y'',$$

so that

$$\chi(X_\alpha \cup X_\beta \cup X_\gamma \cup R_1 \cup R_5) = \chi(X_\alpha^* \cup X_\beta^* \cup X_\gamma^* \cup R_1 \cup R_5) + \chi(Y'') = 10 + 1 = 11.$$

We have $\chi(X)=2\chi(Z)=12$ (see 3.10(c),(d)) hence

$\chi(X-(X_\alpha\cup X_\beta\cup X_\gamma\cup R_1\cup R_5))=\chi(X)-\chi(X_\alpha\cup X_\beta\cup X_\gamma\cup R_1\cup R_5)=12-11=1.$

This proves 4.13(a). Proposition 4.13 is proved.

4.17. We show:

(a) *Let $\mathcal{L}$ be a 3-line in $\mathbf{T}$. Then either $\mathcal{L}\subset\mathbf{T}_0$ or $\mathcal{L}\cap\mathbf{T}_0$ is exactly one point.*

Let $B\in\mathcal{L}$. Let $P\in\mathcal{P}_{03}$ be such that $B\subset P$. We have $u\in P$. Let $u'=\pi_P(u)$. Let $\bar{\mathcal{L}}=\{\pi_P(B');B'\in\mathcal{L}\}$. Note that $\bar{\mathcal{L}}$ is a line contained in $\mathcal{B}^{\bar{P}}_{u'}$. It follows that u' is not regular unipotent in $\bar{P}$. Since $\bar{P}_{ad}$ is of type A_2, u' must be subregular or 1. If $u'=1$ then $\mathcal{B}^{\bar{P}}_{u'}=\mathcal{B}^{\bar{P}}$ so that for any $B'\in\mathcal{L}$ the 0-line through B' is contained in $\mathcal{B}^P$ hence is contained in $\mathcal{B}_u$; thus $\mathcal{L}\subset\mathbf{T}_0$. If u' is subregular then $\mathcal{B}^{\bar{P}}_{u'}$ is the union of two projective lines which intersect in a single point. We see that there is a unique $B'\in\mathcal{L}$ such that the 0-line through B' is contained in $\mathcal{B}_u$; thus $B'\in\mathbf{T}_0$. This proves (a).

4.18. We set $\mathbf{T}^\diamond=\mathbf{T}-(\mathbf{T}_0\cup\mathbf{T}_1\cup\mathbf{T}_2\cup\mathbf{T}_4\cup\mathbf{T}_5)$. This is an open dense subset of $\mathbf{T}$. We have the following result.

PROPOSITION 4.19. *If $B_0\notin\mathbf{T}$ then $\chi(\mathbf{T}^\diamond)=0$. If $B_0\in\mathbf{T}$ then $\chi(\mathbf{T}^\diamond)=1$.*

Let $\mathfrak{V}$ be the variety of all 3-lines in $\mathbf{T}$. We have a partition $\mathfrak{V}=\mathfrak{V}_{15}\sqcup\hat{\mathfrak{V}}\sqcup\mathfrak{V}'\sqcup\mathfrak{V}''$ where

$\mathfrak{V}_{15}$ is the variety of all 3-lines $\mathcal{L}$ such that $\mathcal{L}\subset\mathbf{T}_{15}$;

$\hat{\mathfrak{V}}$ is the variety of all 3-lines $\mathcal{L}$ such that $\mathcal{L}\subset\mathbf{T}-\mathbf{T}_{15}$, $\mathcal{L}\subset\mathbf{T}_0$;

$\mathfrak{V}'$ is the variety of all 3-lines $\mathcal{L}$ such that $\mathcal{L}\subset\mathbf{T}-\mathbf{T}_{15}$, $\mathcal{L}\cap\mathbf{T}_0$ is exactly one point and that point is in $\mathbf{T}_{24}$;

$\mathfrak{V}''$ is the variety of all 3-lines $\mathcal{L}$ such that $\mathcal{L}\subset\mathbf{T}-\mathbf{T}_{15}$, $\mathcal{L}\cap\mathbf{T}_0$ is exactly one point and that point is not in $\mathbf{T}_{24}$.

The fact that this partition is well defined follows from 3.19(a) and the fact that $\mathbf{T}_{15}$ is a union of 3-lines. Let $\xi:\mathbf{T}\to\mathfrak{V}$ be the map which associates to $B\in\mathbf{T}$ the 3-line containing B.

The inverse images of $\mathfrak{V}_{15},\hat{\mathfrak{V}},\mathfrak{V}',\mathfrak{V}'$ under ξ are denoted by $\mathbf{T}_{15},\hat{\mathbf{T}},\mathbf{T}',\mathbf{T}''$. (This agrees with our earlier definition of $\mathbf{T}_{15}$.) We have $\mathbf{T}=\mathbf{T}_{15}\sqcup\hat{\mathbf{T}}\sqcup\mathbf{T}'\sqcup\mathbf{T}''$. We set $\mathbf{T}'^\diamond=\mathbf{T}'\cap\mathbf{T}^\diamond$, $\mathbf{T}''^\diamond=\mathbf{T}''\cap\mathbf{T}^\diamond$. Clearly, we have $\mathbf{T}^\diamond=\mathbf{T}'^\diamond\sqcup\mathbf{T}''^\diamond$. Let $\mathbf{T}''^\diamond\to\mathfrak{V}''$ be the restriction of ξ. This is a fibration with each fibre being a projective line from which two points have been removed: one in $\mathbf{T}_{24}$ and one in $\mathbf{T}_0$. It follows that $\chi(\mathbf{T}''^\diamond)=0$. Let $\mathbf{T}'^\diamond\to\mathfrak{V}'$ be the restriction of ξ. This is a fibration with each fibre being a projective line from which one point has been removed: the one in $\mathbf{T}_{24}$ (which is also in $\mathbf{T}_0$). It follows that $\chi(\mathbf{T}'^\diamond)=\chi(\mathfrak{V}')$. We have

$$\chi(\mathbf{T}^\diamond)=\chi(\mathbf{T}''^\diamond)+\chi(\mathbf{T}'^\diamond)=0+\chi(\mathfrak{V}')=\chi(\mathfrak{V}').$$

If $B_0\notin\mathbf{T}$ then $\mathbf{T}_{24}\cap\mathbf{T}_0=\emptyset$ (see 4.9(b)) hence $\mathfrak{V}'=\emptyset$ and $\chi(\mathbf{T}^\diamond)=\chi(\mathfrak{V}')=0$. Now assume that $B_0\in\mathbf{T}$. Then by 4.9(b), $\mathfrak{V}'$ is isomorphic to $\mathbf{T}_{24}\cap\mathbf{T}_0-\mathbf{T}_{24}\cap\mathbf{T}_{15}\cap\mathbf{T}_0$ and this is a projective line minus a point. Thus, in this case, $\chi(\mathbf{T}^\diamond)=\chi(\mathfrak{V}')=1$.

4.20. We show:

(a) *If $B_0\notin\mathbf{T}$ then $\mathbf{S}\cap\mathbf{T}=\emptyset$; hence $\mathbf{S}^\diamond\cap\mathbf{T}^\diamond=\emptyset$ and $\chi(\mathbf{S}^\diamond\cap\mathbf{T}^\diamond)=0$.*

(b) *If $B_0\in\mathbf{T}\cap\mathbf{S}$ then $\chi(\mathbf{S}^\diamond\cap\mathbf{T}^\diamond)$ is 1 or 0.*

(c) *If $B_0\in\mathbf{T},B_0\notin\mathbf{S}$ then $\chi(\mathbf{S}^\diamond\cap\mathbf{T}^\diamond)$ is 2 or 1.*

Now (a) follows from 4.10(a). Next we assume that $B\in\mathbf{T}$. As in 4.11(a), let

$U=\{B\in\mathbf{T}_{24}\cap T_0;\vartheta(B)=B_0\}$. We have a fibration $\mathbf{S}^\diamond\cap\mathbf{T}^\diamond\to(U\cap Y)-\{B_0\}$. This associates to B the intersection of the 3-line through B with $\mathbf{T}_{24}$, see 4.2(a); each fibre is a projective line with a point removed. It follows that $\chi(\mathbf{S}^\diamond\cap\mathbf{T}^\diamond)=\chi((U\cap Y)-\{B_0\})$. This equals $\chi(U\cap Y)$ if $B_0\notin Y$ (that is if $B_0\notin\mathbf{S}$) and it equals $\chi(U\cap Y)-1$ if $B_0\in Y$ (that is if $B_0\in\mathbf{S}$). It remains to use 4.11(a).

4.21. We show:

(a) *If* $B_0\notin\mathbf{T}$ *then* $\chi(\mathbf{S}^\diamond\cup\mathbf{T}^\diamond)=1$.

(b) *If* $B_0\in\mathbf{T}$ *then* $\chi(\mathbf{S}^\diamond\cup\mathbf{T}^\diamond)\in\{0,1,2\}$.

We have $\chi(\mathbf{S}^\diamond\cup\mathbf{T}^\diamond)=\chi(\mathbf{S}^\diamond)+\chi(\mathbf{T}^\diamond)-\chi(\mathbf{S}^\diamond\cap\mathbf{T}^\diamond)$. It remains to use 4.13, 4.19 and 4.20(a),(b),(c).

4.22. Let $\mathcal{B}_u^\diamond=\mathcal{B}_u-(\mathcal{B}_{u,0}\cup\mathcal{B}_{u,1}\cup\mathcal{B}_{u,2}\cup\mathcal{B}_{u,4}\cup\mathcal{B}_{u,5})$. We show:

(a) $\mathcal{B}_u^\diamond=\mathbf{S}^\diamond\cup\mathbf{T}^\diamond$.

The inclusion $\mathbf{S}^\diamond\cup\mathbf{T}^\diamond\subset\mathcal{B}_u^\diamond$ is obvious. Conversely, let $B\in\mathcal{B}_u^\diamond$. It is enough to show that $B\in\mathbf{S}\cup\mathbf{T}$. We have $B\in X$ for some $X\in\underline{\mathcal{B}}$. If X is $\mathbf{S}$ or $\mathbf{T}$, we are done. Thus we can assume that $X\notin\{\mathbf{S},\mathbf{T}\}$. We have $J_X\neq\{3\}$ and since $J_X\neq\emptyset$ we have $i\in J_X$ for some $i\in I-\{3\}$ and, in particular, $X\subset\mathcal{B}_{u,0}\cup\mathcal{B}_{u,1}\cup\mathcal{B}_{u,2}\cup\mathcal{B}_{u,4}\cup\mathcal{B}_{u,5}$. Since $B\in X$, we have $B\in\mathcal{B}_{u,0}\cup\mathcal{B}_{u,1}\cup\mathcal{B}_{u,2}\cup\mathcal{B}_{u,4}\cup\mathcal{B}_{u,5}$. This contradicts $B\in\mathcal{B}_u^\diamond$. This proves (a).

From (a) and 4.21(a),(b) we deduce

(b) *We have* $\chi(\mathcal{B}_u^\diamond)\in\{0,1,2\}$.

4.23. Let $\mathcal{P}_u^{reg}$ be the set of all $P\in\mathcal{P}_{1245}$ such that $u\in P$ and $\pi_P(u)$ is a regular unipotent element of $\bar{P}$. We define a map $\phi:\mathcal{P}_u^{reg}\to\mathcal{B}_u$ by $P\mapsto B$ where B is the unique Borel subgroup such that $u\in B\subset P$. Let $\mathcal{B}_u^{reg}=\mathcal{B}_u-(\mathcal{B}_{u,1}\cup\mathcal{B}_{u,2}\cup\mathcal{B}_{u,4}\cup\mathcal{B}_{u,5})$. We show:

(a) ϕ *defines an isomorphism* $\phi_0:\mathcal{P}_u^{reg}\xrightarrow{\sim}\mathcal{B}_u^{reg}$.

If $P,P'\in\mathcal{P}_u^{reg}$ and $\phi(P)=\phi(P')=B$ then P,P' are parabolic subgroups in $\mathcal{P}_{1245}$ containing B, hence $P=P'$. Assume now that $P\in\mathcal{P}_u^{reg}$ and let $B=\phi(P)$. For $i\in\{1,2,4,5\}$ let P_i be the parabolic subgroup in $\mathcal{P}_i$ such that $B\subset P_i$. We have $P_i\subset P$ and $u\in U_B$. Since $\pi_P(u)$ is regular unipotent in $\bar{P}$ we have $\pi_P(u)\notin\pi_P(U_{P_i})$ that is, $u\notin U_{P_i}$; but this is the same as saying that $B\neq\mathcal{B}_{u,i}$. Thus, $\phi(P)\in\mathcal{B}_u^{reg}$. We see that ϕ restricts to an injective map $\phi_0:\mathcal{P}_u^{reg}\to\mathcal{B}_u^{reg}$. Now let $B\in\mathcal{B}_u^{reg}$. Let P be the unique parabolic subgroup in $\mathcal{P}_{1245}$ such that $B\subset P$. We have $u\in P$. Again, for $i\in\{1,2,4,5\}$, let P_i be the parabolic subgroup in $\mathcal{P}_i$ such that $B\subset P_i$. We have $P_i\subset P$ and $u\in U_B$. Since $B\notin\mathcal{B}_{u,i}$, we have $u\notin U_{P_i}$ hence $\pi_P(u)\notin\pi_P(U_{P_i})$. It follows that $\pi_P(u)$ is regular unipotent in $\bar{P}$. Thus ϕ_0 is surjective hence a bijection. We omit the proof of the fact that ϕ_0 is an isomorphism.

4.24. We show:

(a) *Let* $\mathcal{L}$ *be a* 0*-line in* $\mathcal{B}$ *such that* $\mathcal{L}\cap\mathcal{B}_u^{reg}\neq\emptyset$. *If* $\mathcal{L}\subset\mathcal{B}_u$, *then* $\mathcal{L}\subset\mathcal{B}_u^{reg}$. *If* $\mathcal{L}\not\subset\mathcal{B}_u$, *then* $\mathcal{L}\cap\mathcal{B}_u=\mathcal{L}\cap\mathcal{B}_u^{reg}$ *is a single point.*

Assume first that $\mathcal{L}\subset\mathcal{B}_u$. Let $B\in\mathcal{L}\cap\mathcal{B}_u^{reg}$. If $\mathcal{L}$ contains some $B'\neq B$ such that $B'\in\mathcal{B}_{u,i}$ for some $i\in\{1,2,4,5\}$ then, applying 1.1(a) with B replaced by B' and $J=\{0,i\}$ we see that $\mathcal{B}^P\subset\mathcal{B}_u$ where $P\in\mathcal{P}_{0,i}$ contains B'. We have $B\in\mathcal{B}^P$ and the line of type i through B is contained in $\mathcal{B}^P$ hence in $\mathcal{B}_u$, so that $B\in\mathcal{B}_{u,i}$; this contradicts $B\in\mathcal{B}_u^{reg}$. We see that $\mathcal{L}\subset\mathcal{B}_u^{reg}$.

Assume next that $\mathcal{L} \not\subset \mathcal{B}_u$. Clearly, the intersection of any 0-line with $\mathcal{B}_u$ is either empty, or $\mathcal{L}$, or a point. In our case $\mathcal{L} \cap \mathcal{B}_u$ is not $\mathcal{L}$ and is not empty hence it is a point. Hence $\mathcal{L} \subset \mathcal{B}_u^{reg}$ is either empty or a point; by assumption it is nonempty hence it is a point. This proves (a).

4.25. From 4.24(a) we see that we have a partition $\mathcal{B}_u^{reg} = {}'\mathcal{B}_u^{reg} \cup {}''\mathcal{B}_u^{reg}$ where ${}''\mathcal{B}_u^{reg}$ is the union of the 0-lines contained in $\mathcal{B}_u^{reg}$ and ${}'\mathcal{B}_u^{reg}$ is the set of all $B \in \mathcal{B}_u^{reg}$ such that the 0-line through B intersects $\mathcal{B}_u$ in exactly one point, B. Note that ${}'\mathcal{B}_u^{reg} = \mathcal{B}_u^{\diamond}$. Hence, using 4.22(b), we have

(a) $\chi({}'\mathcal{B}_u^{reg}) \in \{0, 1, 2\}$.

For any $P \in \mathcal{P}_{1245}$ we denote by Q_P the unique parabolic subgroup in $\mathcal{P}_{01245}$ such that $P \subset Q_P$; note that $(\overline{Q_P})_{ad} = \bar{P}_{ad} \times H_P$ (canonically) where $H_P \cong PGL_2(\mathbf{k})$ (we use that s_0 commutes with W_{1245}). Let ${}''\mathcal{P}_u^{reg}$ (resp. ${}'\mathcal{P}_u^{reg}$) be the set of all $P \in \mathcal{B}_u^{reg}$ such that the image of $u \in Q_P$ under the obvious composition $Q_P \to (\overline{Q_P})_{ad} \to H_P$ is $1 \in H_P$ (resp. a regular unipotent element of H_P).

From the definitions we see that under the isomorphism $\phi_0 : \mathcal{P}_u^{reg} \xrightarrow{\sim} \mathcal{B}_u^{reg}$ in 4.23(a), the subset ${}''\mathcal{P}_u^{reg}$ of $\mathcal{P}_u^{reg}$ corresponds to the subset ${}''\mathcal{B}_u^{reg}$ of $\mathcal{B}_u^{reg}$; hence ${}'\mathcal{P}_u^{reg} = \mathcal{P}_u^{reg} - {}''\mathcal{P}_u^{reg}$ corresponds under ϕ_0 to ${}'\mathcal{B}_u^{reg} = \mathcal{B}_u^{reg} - {}''\mathcal{B}_u^{reg}$. Using this and (a) we deduce

(b) $\chi({}'\mathcal{P}_u^{reg}) \in \{0, 1, 2\}$.

4.26. We now assume that G is simply connected and $p \neq 3$. Let $J = \{1, 2, 4, 5\} \subset I$. We fix $P_J \in \mathcal{P}_J$. Let C be the regular unipotent conjugacy class of $\bar{P}_J$. Let $\mathcal{S}_0$ be an irreducible cuspidal $\bar{P}_J$-equivariant local system on C. Up to isomorphism there are two such local systems, one for each nontrivial character of the group (cyclic of order 3) of connected components of the centralizer in $\bar{P}_J$ of an element in C. We shall use the notation and results of 2.1, 2.2 for this $G, J, P_J, \mathcal{S}_0$.

In our case $\mathcal{W}$ in 2.1 is a Weyl group of type G_2 with simple reflections σ_0, σ_3. Now $\mathbf{X}_u$ and the local system $\hat{\mathcal{S}}$ on it are defined as in 2.1 (with $g = u$). Recall from 2.1 that $H_c^j(\mathbf{X}_u, \hat{\mathcal{S}})$ is naturally a $\mathcal{W}$-module for $j \in \mathbf{Z}$. We have the following result.

(a) $$\sum_j (-1)^j \mathrm{tr}(\sigma_0, H_c^j(\mathbf{X}_u, \hat{\mathcal{S}})) \in \{0, 1, 2\}.$$

Let $J' = J \cup \{0\}$ and let $M_{J',g,r}, \pi'', \dot{\mathcal{S}}_r$ be as in 2.1, 2.2 with $g = u$, $i = 0$. We set $\pi''^{-1}(M_{J',u,r}) = \mathcal{M}_r$. From 2.2(b) we see that the left hand side of (a) is equal to $\chi(\mathcal{M}_r, \dot{\mathcal{S}}_r)$ From the definitions we see that there exists an unramified principal covering $\psi : \hat{\mathcal{M}}_r \to \mathcal{M}_r$ with group $\mathbf{Z}/3$ (with generator κ) and $\theta \in \bar{\mathbf{Q}}_l - \{1\}$ with $\theta^3 = 1$ such that the stalk of $\dot{\mathcal{S}}_r$ at any $x \in \mathcal{M}_r$ is equal to the vector space of functions $f : \psi^{-1}(x) \to \bar{\mathbf{Q}}_l$ such that $f(\kappa\tilde{x}) = \theta f(\tilde{x})$ for any $\tilde{x} \in \psi^{-1}(x)$. It follows that

$$\chi(\mathcal{M}_r, \dot{\mathcal{S}}_r) = \sum_{h=0}^{2} \sum_j (-1)^j \mathrm{tr}((\kappa^h)^*, H_c^j(\hat{\mathcal{M}}_r, \bar{\mathbf{Q}}_l))\theta^h/3,$$

$$\chi(\mathcal{M}_r) = \sum_{h=0}^{2} \sum_j (-1)^j \mathrm{tr}((\kappa^h)^*, H_c^j(\hat{\mathcal{M}}_r, \bar{\mathbf{Q}}_l))/3,$$

where $(\kappa^h)^*$ is induced by the action of κ^h on $\hat{\mathcal{M}}_r$. If $h \in \{1, 2\}$ then $\kappa^h : \hat{\mathcal{M}}_r \to \hat{\mathcal{M}}_r$ has no fixed points and it has order 3. Since $p \neq 3$ we can apply the fixed point formula in [**DL**, 3.2] to see that $\sum_j(-1)^j\mathrm{tr}((\kappa^h)^*, H_c^j(\hat{\mathcal{M}}_r, \bar{\mathbf{Q}}_l)) = 0$ for $h \in \{1,2\}$. It follows that

$$\chi(\mathcal{M}_r, \dot{\mathcal{S}}_r) = \sum_j (-1)^j \dim H_c^j(\hat{\mathcal{M}}_r, \bar{\mathbf{Q}}_l))/3,$$

$$\chi(\mathcal{M}_r) = \sum_j (-1)^j \dim H_c^j(\hat{\mathcal{M}}_r, \bar{\mathbf{Q}}_l))/3,$$

so that

$$\chi(\mathcal{M}_r, \dot{\mathcal{S}}_r) = \chi(\mathcal{M}_r).$$

Thus, to prove (a), it is enough to show that

$$\chi(\mathcal{M}_r) \in \{0,1,2\}.$$

From the definitions we see that the map $\mathcal{M}_r \to {}'\mathcal{P}_u^{reg}$, $(xP_J, y(xU_{J'}x^{-1})) \mapsto xP_Jx^{-1}$, is a isomorphism. Hence it is enough to show that $\chi({}'\mathcal{P}_u^{reg}) \in \{0,1,2\}$. But this is known from 4.25(b). This completes the proof of (a).

5. The main result

5.1. In this section we assume that G is simply connected of type E_6 and that $p \neq 3$. We write $I = \{0,1,2,3,4,5\}$ as in 4.1. Let $J = \{1,2,4,5\} \subset I$ and let $P_J \in \mathcal{P}_J$. Let C be the regular unipotent class in $\bar{P}_J$. There are exactly two irreducible cuspidal $\bar{P}_J$-equivariant local systems on C; we fix one of them, say $\mathcal{S}_0$. Let $\mathcal{W} = N_W W_J/W_J$ where $N_W W_J$ is the normalizer of W_J in W. This is a finite Coxeter group of type G_2 with simple reflections σ_0, σ_3 defined as in 4.1. Now the block defined by $(P_J, C, \mathcal{S}_0)$ consists of six pairs $(C_h, \mathcal{E}_h)$, $h \in H := \{0,1,3,4,9,12\}$, where C_h is a unipotent class of G such that, for $u_h \in C_h$ we have $d_{u_h} = h$ and $\mathcal{E}_h$ is an irreducible G-equivariant local system on C_h, uniquely determined up to isomorphism by h. In [**S2**], $C_0, C_1, C_3, C_4, C_9, C_{12}$ are denoted by $E_6, E_6(a_1), A_5A_1, A_5, 2A_2A_1, 2A_2$ respectively. We shall assume that $u \in G$ (see 1.1) belongs to C_3.

Let $\mathrm{Irr}\mathcal{W}$ be the set of irreducible representations (up to isomorphism) of $\mathcal{W}$. We have $\mathrm{Irr}\mathcal{W} = \{1, s, \epsilon, \epsilon', \rho, \rho'\}$ where $1, s, \epsilon, \epsilon'$ are one-dimensional, ρ, ρ' are two-dimensional, 1 is the unit representation, and we have $\rho \otimes \rho = \rho' \otimes \rho' = 1 + \rho' + s$, $\rho \otimes \rho' = \epsilon + \epsilon' + \rho$; these properties distinguish ρ from ρ' and s from $1, \epsilon, \epsilon'$ but not ϵ from ϵ'. We can distinguish ϵ from ϵ' by the following requirements: σ_0 acts as -1 on ϵ and as 1 on ϵ'; σ_3 acts as -1 on ϵ' and as 1 on ϵ. We have also $\epsilon \otimes \epsilon = \epsilon' \otimes \epsilon' = s \otimes s = 1$, $\epsilon \otimes \epsilon' = s$, $\epsilon \otimes s = \epsilon'$, $\epsilon' \otimes s = \epsilon$, $\rho \otimes s = \rho$, $\rho' \otimes s = \rho'$, $\rho \otimes \epsilon = \rho \otimes \epsilon' = \rho'$, $\rho' \otimes \epsilon = \rho' \otimes \epsilon' = \rho$.

The generalized Springer correspondence for our block provides a bijection $\mathrm{Irr}\mathcal{W} \leftrightarrow \{(C_h, \mathcal{E}_h); h \in H\}$. For $h \in H$ let $E_h \in \mathrm{Irr}\mathcal{W}$ be corresponding to $(C_h, \mathcal{E}_h)$ under this bijection. According to [**S2**], either (a) or (b) below holds:

(a) $E_0 = 1$, $E_1 = \epsilon$, $E_3 = \rho$, $E_4 = \rho'$, $E_9 = \epsilon'$, $E_{12} = s$;

(b) $E_0 = 1$, $E_1 = \epsilon$, $E_3 = \rho'$, $E_4 = \rho$, $E_9 = \epsilon'$, $E_{12} = s$.

5.2. Let q be an indeterminate. For any representation E of $\mathcal{W}$ we set

$$\Gamma_E = (1/12) \sum_{w \in \mathcal{W}} (q^2-1)(q^6-1)\mathrm{tr}(w, E)/\det(q-w) \in \mathbf{N}[q].$$

We have

$$\Gamma_1 = q^6, \Gamma_\epsilon = \Gamma_{\epsilon'} = q^3, \Gamma_\rho = q + q^5, \Gamma_{\rho'} = q^2 + q^4, \Gamma_s = 1.$$

It follows that the matrix $(\Gamma_{E_h\otimes E_{h'}})_{(h,h')\in H\times H}$ is equal to

$$\begin{matrix} q^6 & q^3 & q+q^5 & q^2+q^4 & q^3 & 1\\ q^3 & q^6 & q^2+q^4 & q+q^5 & 1 & q^3\\ q+q^5 & q^2+q^4 & 1+q^2+q^4+q^6 & q+2q^3+q^5 & q^2+q^4 & q+q^5\\ q^2+q^4 & q+q^5 & q+2q^3+q^5 & 1+q^2+q^4+q^6 & q+q^5 & q^2+q^4\\ q^3 & 1 & q^2+q^4 & q+q^5 & q^6 & q^3\\ 1 & q^3 & q+q^5 & q^2+q^4 & q^3 & q^6 \end{matrix}$$

if 5.1(a) holds and to

$$\begin{matrix} q^6 & q^3 & q^2+q^4 & q+q^5 & q^3 & 1\\ q^3 & q^6 & q+q^5 & q^2+q^4 & 1 & q^3\\ q^2+q^4 & q+q^5 & 1+q^2+q^4+q^6 & q+2q^3+q^5 & q+q^5 & q^2+q^4\\ q+q^5 & q^2+q^4 & q+2q^3+q^5 & 1+q^2+q^4+q^6 & q^2+q^4 & q+q^5\\ q^3 & 1 & q+q^5 & q^2+q^4 & q^6 & q^3\\ 1 & q^3 & q^2+q^4 & q+q^5 & q^3 & q^6 \end{matrix}$$

if 5.1(b) holds. It follows that the matrix $M := (q^{-h-h'}\Gamma_{E_h\otimes E_{h'}})_{(h,h')\in H\times H}$ is equal to

$$\begin{matrix} q^6 & q^2 & q^{-2}+q^2 & q^{-2}+1 & q^{-6} & q^{-12}\\ q^2 & q^4 & q^{-2}+1 & q^{-4}+1 & q^{-10} & q^{-10}\\ q^{-2}+q^2 & q^{-2}+1 & q^{-6}+q^{-4}+q^{-2}+1 & q^{-6}+2q^{-4}+q^{-2} & q^{-10}+q^{-8} & q^{-14}+q^{-10}\\ q^{-2}+1 & q^{-4}+1 & q^{-6}+2q^{-4}+q^{-2} & q^{-8}+q^{-6}+q^{-4}+q^{-2} & q^{-12}+q^{-8} & q^{-14}+q^{-12}\\ q^{-6} & q^{-10} & q^{-10}+q^{-8} & q^{-12}+q^{-8} & q^{-12} & q^{-18}\\ q^{-12} & q^{-10} & q^{-14}+q^{-10} & q^{-14}+q^{-12} & q^{-18} & q^{-18} \end{matrix}$$

if 5.1(a) holds and to

$$\begin{matrix} q^6 & q^2 & q^{-1}+q & q^{-3}+q & q^{-6} & q^{-12}\\ q^2 & q^4 & q^{-3}+q & q^{-3}+q^{-1} & q^{-10} & q^{-10}\\ q^{-1}+q & q^{-3}+q & q^{-6}+q^{-4}+q^{-2}+1 & q^{-6}+2q^{-4}+q^{-2} & q^{-11}+q^{-7} & q^{-13}+q^{-11}\\ q^{-3}+q & q^{-3}+q^{-1} & q^{-6}+2q^{-4}+q^{-2} & q^{-8}+q^{-6}+q^{-4}+q^{-2} & q^{-11}+q^{-9} & q^{-15}+q^{-11}\\ q^{-6} & q^{-10} & q^{-11}+q^{-7} & q^{-11}+q^{-9} & q^{-12} & q^{-18}\\ q^{-12} & q^{-10} & q^{-13}+q^{-11} & q^{-15}+q^{-11} & q^{-18} & q^{-18} \end{matrix}$$

if 5.1(b) holds.

5.3. We can write uniquely $M = {}^t\Pi A\Pi$ where Π, A are matrices indexed by $H\times H$ such that A is diagonal and ${}^t\Pi$ is upper triangular with 1 on diagonal. More precisely A is equal to

$$\begin{matrix} q^{-2}(q^2-1)(q^6-1) & 0 & 0 & 0 & 0 & 0\\ 0 & q^{-4}(q^2-1)(q^6-1) & 0 & 0 & 0 & 0\\ 0 & 0 & q^{-8}(q^2-1)(q^6-1) & 0 & 0 & 0\\ 0 & 0 & 0 & q^{-8}(q^6-1) & 0 & 0\\ 0 & 0 & 0 & 0 & q^{-18}(q^6-1) & 0\\ 0 & 0 & 0 & 0 & 0 & q^{-18} \end{matrix}$$

both in case 5.1(a) and in case 5.1(b); moreover ${}^t\Pi$ is equal to

$$\begin{matrix} 1 & 0 & q^2 & q^2 & q^6 & q^6\\ 0 & 1 & 0 & q^2 & 0 & q^8\\ 0 & 0 & 1 & 1 & q^4 & q^4+q^8\\ 0 & 0 & 0 & 1 & q^4 & q^4+q^6\\ 0 & 0 & 0 & 0 & 1 & 1\\ 0 & 0 & 0 & 0 & 0 & 1 \end{matrix}$$

if 5.1(a) holds and to

$$\begin{array}{cccccc} 1 & 0 & 0 & q^3 & q^6 & q^6 \\ 0 & 1 & q & q & 0 & q^8 \\ 0 & 0 & 1 & 1 & q^5 & q^5+q^7 \\ 0 & 0 & 0 & 1 & q^3 & q^3+q^7 \\ 0 & 0 & 0 & 0 & 1 & 1 \\ 0 & 0 & 0 & 0 & 0 & 1 \end{array}$$

if 5.1(b) holds.

Let $h \le h'$ in H; let $\Pi_{h',h}$ be the (h',h)-entry of Π or the (h,h')-entry of ${}^t\Pi$. We write $\Pi_{h',h} = \sum_{k\in\mathbf{N}} \Pi^k_{h',h} q^k \in \mathbf{N}[q]$ where $\Pi^k_{h',h} \in \mathbf{N}$.

Note that $C_{h'}$ is contained in the closure $\bar{C}_h$ of C_h. For $k' \in \mathbf{Z}$ let

$$\mathcal{H}^{k'} IC(\bar{C}_h, \mathcal{E}_h)|_{C_{h'}}$$

be the restriction to $C_{h'}$ of the k'-th cohomology sheaf of the intersection cohomology complex $IC(\bar{C}_{h'}, \mathcal{E}_{h'})$; this is a G-equivariant local system on C_h.

In the remainder of this subsection we assume that $p \notin \{2,3\}$. From [**L2**, 24.8] we have that

$$\mathcal{H}^{k'} IC(\bar{C}_h, \mathcal{E}_h)|_{C_{h'}} = 0$$

for k' odd and that for $k \in \mathbf{Z}$,

$\mathcal{H}^{2k} IC(\bar{C}_h, \mathcal{E}_h)|_{C_{h'}}$ is isomorphic to a direct sum of $\Pi^k_{h',h}$ copies of $\mathcal{E}_{h'}$.

5.4. In this subsection we assume that $p \notin \{2,3\}$.

Recall from 2.1, 2.2 that

$$\mathbf{X}_u = \{xP_J \in G/P_J; x^{-1}gx \in P_J, \pi_J(x^{-1}ux) \in C\}$$

carries a local system $\hat{\mathcal{S}}$ induced from $\mathcal{S}_0$ and that for any $j \in \mathbf{Z}$, $\mathcal{W}$ acts naturally on $H^j_c(\mathbf{X}_u, \hat{\mathcal{S}})$. (Note that $H^j_c(\mathbf{X}_u, \hat{\mathcal{S}}) = 0$ for odd j.) Let $m_{j,h}$ be the number of times the irreducible $\mathcal{W}$-module E_h appears in the $\mathcal{W}$-module $H^{2j}_c(\mathbf{X}_u, \hat{\mathcal{S}})$. Using [**L1**, 6.5], we see that for $h \le 3$ we have $\sum_j m_{j,h} q^j = \Pi_{3,h} q^h \in \mathbf{Z}[q]$. Using the tables in 5.3 we deduce that, as a $\mathcal{W}$-module, $H^6_c(\mathbf{X}_u, \hat{\mathcal{S}})$ is ρ if 5.1(a) holds and is ρ' if 5.1(b) holds; $H^4_c(\mathbf{X}_u, \hat{\mathcal{S}})$ is 1 if 5.1(a) holds and is ϵ if 5.1(b) holds; $H^j_c(\mathbf{X}_u, \hat{\mathcal{S}})$ is 0 if $j \ne \{4,6\}$. It follows that

(a) $\sum_j (-1)^j \mathrm{tr}(\sigma_0, H^j_c(\mathbf{X}_u, \hat{\mathcal{S}}))$ is equal to $\mathrm{tr}(\sigma_0, \rho) + 1 = 1$ if 5.1(a) holds and to $\mathrm{tr}(\sigma_0, \rho') - 1 = -1$ if 5.1(b) holds.

From 4.26 we see that $\sum_j (-1)^j \mathrm{tr}(\sigma_0, H^j_c(\mathbf{X}_u, \hat{\mathcal{S}})) = \chi('\mathcal{P}^{reg}_u)$ where $'\mathcal{P}^{reg}_u$ is as in 4.25. Thus, $\chi('\mathcal{P}^{reg}_u)$ is equal to 1 if 5.1(a) holds and to -1 if 5.1(b) holds. By 4.25(b) we have $\chi('\mathcal{P}^{reg}_u) \in \{0,1,2\}$. We deduce the following result in which we assume that $p \notin \{2,3\}$.

THEOREM 5.5. *(a) Statement 5.1(a) holds.*
(b) We have $\chi('\mathcal{P}^{reg}_u) = 1$.

5.6. One can show that the statement of 5.5 remains true when $p = 2$. In this case the references to 4.26 and 4.25(b) can still be used. Although the statements at the end of 5.3 are not known in this case, the statement 5.4(a) remains valid in this case.

6. Final comments

6.1. In this and the next subsection we assume that G is simply connected/adjoint of type E_8 and that $p = 3$. We write $I = \{0,1,2,3,4,5,6,7\}$ where the numbering is chosen so that $s_1s_2, s_2s_3, s_3s_4, s_4s_5, s_5s_6, s_6s_7, s_0s_3$ have order 3. Let $J = \{0,1,2,3,4,5\} \subset I$ and let $P_J \in \mathcal{P}_J$. Let C be the regular unipotent class in $\bar{P}_J$. There are exactly two irreducible cuspidal $\bar{P}_J$-equivariant local systems on C; we fix one of them, say $\mathcal{S}_0$. Let $\mathcal{W} = N_W W_J/W_J$ where $N_W W_J$ is the normalizer of W_J in W. This is a finite Coxeter group of type G_2 with simple reflections σ_6, σ_7 defined as in 4.1. Now the block defined by $(P_J, C, \mathcal{S}_0)$ consists of six pairs $(C_h, \mathcal{E}_h)$, $h \in H := \{0,1,3,4,9,12\}$, where C_h is a unipotent class of G such that, for $u_h \in C_h$ we have $d_{u_h} = h$ and $\mathcal{E}_h$ is an irreducible G-equivariant local system on C_h, uniquely determined up to isomorphism by h. In [**S2**], $C_0, C_1, C_3, C_4, C_9, C_{12}$ are denoted by $E_8, E_8(a_1), E_7A_1, E_7, E_6A_1, E_6$ respectively. We shall assume that $u \in G$ (see 1.1) belongs to C_3. We identify the present $\mathcal{W}$ with the group denoted by $\mathcal{W}$ in 5.1 by requiring that σ_6, σ_7 correspond respectively to σ_3, σ_0 in 5.1. Note that s_7 commutes with W_J.

The generalized Springer correspondence for our block provides a bijection $\mathrm{Irr}\mathcal{W} \leftrightarrow \{(C_h, \mathcal{E}_h); h \in H\}$. For $h \in H$ let $E_h \in \mathrm{Irr}\mathcal{W}$ be corresponding to $(C_h, \mathcal{E}_h)$ under this bijection. According to [**S2**], either (a) or (b) below holds:

(a) $E_0 = 1$, $E_1 = \epsilon$, $E_3 = \rho$, $E_4 = \rho'$, $E_9 = \epsilon'$, $E_{12} = s$;

(b) $E_0 = 1$, $E_1 = \epsilon$, $E_3 = \rho'$, $E_4 = \rho$, $E_9 = \epsilon'$, $E_{12} = s$.

(Compare with 5.1(a),(b).)

CONJECTURE 6.2. *Statement 6.1(a) holds.*

Recall from 2.1, 2.2 that

$$\mathbf{X}_u = \{xP_J \in G/P_J; x^{-1}gx \in P_J, \pi_J(x^{-1}ux) \in C\}$$

carries a local system $\hat{\mathcal{S}}$ induced from $\mathcal{S}_0$ and that for any $j \in \mathbf{Z}$, $\mathcal{W}$ acts naturally on $H_c^j(\mathbf{X}_u, \hat{\mathcal{S}})$. From [**L2**] one can deduce, using computations like those in 5.2, that

(a) $\sum_j (-1)^j \mathrm{tr}(\sigma_7, H_c^j(\mathbf{X}_u, \hat{\mathcal{S}}))$ is equal to $\mathrm{tr}(\sigma_7, \rho) + 1 = 1$ if 6.1(a) holds and to $\mathrm{tr}(\sigma_7, \rho') - 1 = -1$ if 6.1(b) holds.

Hence to prove the conjecture it is enough to show that

(b) $\sum_j (-1)^j \mathrm{tr}(\sigma_7, H_c^j(\mathbf{X}_u, \hat{\mathcal{S}})) \in \mathbf{N}$.

By arguments similar to those in 4.26 we see that the sum in (b) is equal to the Euler characteristic of

$$\mathcal{B}_u - (\mathcal{B}_{u,0} \cup \mathcal{B}_{u,1} \cup \mathcal{B}_{u,2} \cup \mathcal{B}_{u,3} \cup \mathcal{B}_{u,4} \cup \mathcal{B}_{u,5} \cup \mathcal{B}_{u,7})$$

with coefficient in a local system of rank 1 defined by $\mathcal{S}_0$. Hence it is enough to show that this Euler characteristic is in $\mathbf{N}$. It is not clear how to do this. One difficulty is that the argument in 4.26 (based on [**DL**]) is not applicable hence in the Euler characteristic above the local system cannot be replaced by $\bar{\mathbf{Q}}_l$.

References

[DL] P. Deligne and G. Lusztig, *Representations of reductive groups over finite fields*, Ann. of Math. (2) **103** (1976), no. 1, 103–161, DOI 10.2307/1971021. MR0393266

[Ho] R. Hotta, *On Springer's representations*, J. Fac. Sci. Univ. Tokyo Sect. IA Math. **28** (1981), no. 3, 863–876 (1982). MR656061

[L1] G. Lusztig, *Intersection cohomology complexes on a reductive group*, Invent. Math. **75** (1984), no. 2, 205–272, DOI 10.1007/BF01388564. MR732546

[L2] G. Lusztig, *Character sheaves. V*, Adv. in Math. **61** (1986), no. 2, 103–155, DOI 10.1016/0001-8708(86)90071-X. MR849848

[LS] G. Lusztig and N. Spaltenstein, *Induced unipotent classes*, J. London Math. Soc. (2) **19** (1979), no. 1, 41–52, DOI 10.1112/jlms/s2-19.1.41. MR527733

[S1] N. Spaltenstein, *On the fixed point set of a unipotent element on the variety of Borel subgroups*, Topology **16** (1977), no. 2, 203–204, DOI 10.1016/0040-9383(77)90022-2. MR0447423

[S2] N. Spaltenstein, *On the generalized Springer correspondence for exceptional groups*, Algebraic groups and related topics (Kyoto/Nagoya, 1983), Adv. Stud. Pure Math., vol. 6, North-Holland, Amsterdam, 1985, pp. 317–338. MR803340

[Sp] T. A. Springer, *Trigonometric sums, Green functions of finite groups and representations of Weyl groups*, Invent. Math. **36** (1976), 173–207, DOI 10.1007/BF01390009. MR0442103

Department of Mathematics, Massachusetts Institute of Technology, Cambridge, Massachusetts 02139

Proceedings of Symposia in Pure Mathematics
Volume **101**, 2019
https://doi.org/10.1090/pspum/101/01798

The modular pro-p Iwahori-Hecke Ext-algebra

Rachel Ollivier and Peter Schneider

Dedicated to J. Bernstein on the occasion of his 72nd birthday

Abstract. Let $\mathfrak{F}$ be a locally compact nonarchimedean field of positive residue characteristic p and k a field of characteristic p. Let G be the group of $\mathfrak{F}$-rational points of a connected reductive group over $\mathfrak{F}$ which we suppose $\mathfrak{F}$-split. Given a pro-p-Iwahori subgroup I of G, we consider the space $\mathbf{X}$ of k-valued functions with compact support on G/I. It is naturally an object in the category $\mathrm{Mod}(G)$ of all smooth k-representations of G.

We study the graded Ext-algebra $E^* = \mathrm{Ext}^*_{\mathrm{Mod}(G)}(\mathbf{X}, \mathbf{X})$. Its degree zero piece $\mathrm{Ext}^0_{\mathrm{Mod}(G)}(\mathbf{X}, \mathbf{X})$ is the usual pro-p Iwahori Hecke algebra H. We describe the product in E^* and provide an involutive anti-automorphism of E^*. When I is a Poincaré group of dimension d, the Ext-algebra E^* is supported in degrees $i \in \{0 \ldots d\}$ and we establish a duality theorem between E^i and E^{d-i}. Under the same hypothesis (and assuming that $\mathbf{G}$ is almost simple and simply connected), we compute $\mathrm{Ext}^d_{\mathrm{Mod}(G)}(\mathbf{X}, \mathbf{X})$ as an H-module on the left and on the right. We prove that it is a direct sum of the trivial character, and of supersingular modules.

Contents

2010 *Mathematics Subject Classification.* 20C08, 22E50, 16E30, 22D35, 20J06.

1. Introduction

Let $\mathfrak{F}$ be a locally compact nonarchimedean field with residue characteristic p, and let G be the group of $\mathfrak{F}$-rational points of a connected reductive group $\mathbf{G}$ over $\mathfrak{F}$. We suppose that $\mathbf{G}$ is $\mathfrak{F}$-split in this article.

Let k be a field and let $\mathrm{Mod}(G)$ denote the category of all smooth representations of G in k-vector spaces. When $k = \mathbb{C}$, by a theorem of Bernstein [**1**] Cor. 3.9.ii, one of the blocks of $\mathrm{Mod}(G)$ is equivalent to the category of modules over the Iwahori-Hecke algebra of G. This block is the subcategory of all representations which are generated by their Iwahori-fixed vectors. It does not contain any supercuspidal representation of G.

When k has characteristic p, it is natural to consider the Hecke algebra H of the pro-p Iwahori subgroup $I \subset G$. In this case the natural left exact functor

$$\begin{aligned} \mathfrak{h} : \mathrm{Mod}(G) &\longrightarrow \mathrm{Mod}(H) \\ V &\longmapsto V^I = \mathrm{Hom}_{k[G]}(\mathbf{X}, V) \end{aligned}$$

sends a nonzero representation onto a nonzero module. Its left adjoint is

$$\begin{aligned} \mathfrak{t} : \mathrm{Mod}(H) &\longrightarrow \mathrm{Mod}^I(G) \subseteq \mathrm{Mod}(G) \\ M &\longmapsto \mathbf{X} \otimes_H M \ . \end{aligned}$$

Here $\mathbf{X}$ denotes the space of k-valued functions with compact support on G/I with the natural left action of G. The functor $\mathfrak{t}$ has values in the category $\mathrm{Mod}^I(G)$ of all smooth k-representations of G generated by their I-fixed vectors. This category, which a priori has no reason to be an abelian subcategory of $\mathrm{Mod}(G)$, contains all irreducible representations including the supercuspidal ones. But in general $\mathfrak{h}$ and $\mathfrak{t}$ are not quasi-inverse equivalences of categories and little is known about $\mathrm{Mod}^I(G)$ and $\mathrm{Mod}(G)$ unless $G = \mathrm{GL}_2(\mathbb{Q}_p)$ or $G = \mathrm{SL}_2(\mathbb{Q}_p)$ ([**12**], [**18**], [**21**], [**22**]).

From now on we assume k has characteristic p. The functor $\mathfrak{h}$, although left exact, is not right exact since p divides the pro-order of I. It is therefore natural to consider the derived functor. In [**24**] the following result is shown: When $\mathfrak{F}$ is a finite extension of $\mathbb{Q}_p$ and I is a torsionfree pro-p-group, there exists a derived version of the functor $\mathfrak{h}$ and $\mathfrak{t}$ providing an equivalence between the derived category of smooth representations of G in k-vector spaces and the derived category of differential graded modules over a certain differential graded pro-p Iwahori-Hecke algebra $H^\bullet$.

The current article is largely motivated by this theorem. The derived categories involved are not understood, and in fact the Hecke differential graded algebra $H^\bullet$ itself has no concrete description yet. We provide here our first results on the structure of its cohomology algebra $\mathrm{Ext}^*_{\mathrm{Mod}(G)}(\mathbf{X},\mathbf{X})$:

- We describe explicitly the product in $\mathrm{Ext}^*_{\mathrm{Mod}(G)}(\mathbf{X},\mathbf{X})$ (Proposition 5.3).
- We deduce the existence of an involutive anti-automorphism of $\mathrm{Ext}^*_{\mathrm{Mod}(G)}(\mathbf{X},\mathbf{X})$ as a graded Ext-algebra (Proposition 6.1).
- When I is a Poincaré group of dimension d, the Ext algebra is supported in degrees 0 to d and we establish a duality theorem between its ith and $d-i^{\mathrm{th}}$ pieces (Proposition 7.18).
- Under the same hypothesis (and assuming that $\mathbf{G}$ is almost simple and simply connected), we compute $\mathrm{Ext}^d_{\mathrm{Mod}(G)}(\mathbf{X},\mathbf{X})$ as an H-module on the left and on the right (Corollary 8.7). We prove that it is a direct sum of the trivial character, and of supersingular modules.

We hope that these results illustrate that the pro-p Iwahori-Hecke Ext-algebra $\mathrm{Ext}^*_{\mathrm{Mod}(G)}(\mathbf{X},\mathbf{X})$ is a natural object whose structure is rich and interesting in itself, even beyond its link to the representation theory of p-adic reductive groups. As a derived version of the Hecke algebra of the I-equivariant functions on an (almost) affine flag variety, we suspect that it will contribute to relating the mod p Langlands program to methods that appear in the study of geometric Langlands.

Both authors thank the PIMS at UBC Vancouver for support and for providing a very stimulating atmosphere during the *Focus Period on Representations in Arithmetic*. The first author is partially funded by NSERC Discovery Grant.

2. Notations and preliminaries

Throughout the paper we fix a locally compact nonarchimedean field $\mathfrak{F}$ (for now of any characteristic) with ring of integers $\mathfrak{O}$, its maximal ideal $\mathfrak{M}$, and a prime element π. The residue field $\mathfrak{O}/\pi\mathfrak{O}$ of $\mathfrak{F}$ is $\mathbb{F}_q$ for some power $q=p^f$ of the residue characteristic p. We choose the valuation $\mathrm{val}_{\mathfrak{F}}$ on $\mathfrak{F}$ normalized by $\mathrm{val}_{\mathfrak{F}}(\pi)=1$ We let $G:=\mathbf{G}(\mathfrak{F})$ be the group of $\mathfrak{F}$-rational points of a connected reductive group $\mathbf{G}$ over $\mathfrak{F}$ which we always assume to be $\mathfrak{F}$-split.

We fix an $\mathfrak{F}$-split maximal torus $\mathbf{T}$ in $\mathbf{G}$, put $T:=\mathbf{T}(\mathfrak{F})$, and let T^0 denote the maximal compact subgroup of T and T^1 the pro-p Sylow subgroup of T^0. We also fix a chamber C in the apartment of the semisimple Bruhat-Tits building $\mathscr{X}$ of G which corresponds to $\mathbf{T}$. The stabilizer $\mathcal{P}^{\dagger}_C$ of C contains an Iwahori subgroup J. Its pro-p Sylow subgroup I is called the pro-p Iwahori subgroup and is the main player in this paper. We have $T\cap J=T^0$ and $T\cap I=T^1$. If $N(T)$ is the normalizer of T in G, then we define the group $\widetilde{W}:=N(T)/T^1$. In particular, it contains T^0/T^1. The quotient $W:=N(T)/T^0\cong\widetilde{W}/(T^0/T^1)$ is the extended affine Weyl group. The finite Weyl group is $W_0:=N(T)/T$. For any compact open subset $A\subseteq G$ we let char_A denote the characteristic function of A.

The coefficient field for all representations in this paper is an arbitrary field k of characteristic $p>0$. For any open subgroup $U\subseteq G$ we let $\mathrm{Mod}(U)$ denote the abelian category of smooth representations of U in k-vector spaces. As usual, $K(U)$ denotes the homotopy category of unbounded (cohomological) complexes in $\mathrm{Mod}(U)$ and $D(U)$ the corresponding derived category.

2.1. Elements of Bruhat-Tits theory. We consider the root data associated to the choice of the maximal $\mathfrak{F}$-split torus $\mathbf{T}$ and record in this section the notations and properties we will need. We follow the exposition of [**20**] §4.1–§4.4 which refers mainly to [**25**] I.1. Further references are given in [**20**].

2.1.1. The root datum $(\Phi, X^*(T), \check{\Phi}, X_*(T))$ is reduced because the group $\mathbf{G}$ is $\mathfrak{F}$-split. Recall that $X^*(T)$ and $X_*(T)$ denote respectively the group of algebraic characters and cocharacters of T. Similarly, let $X^*(Z)$ and $X_*(Z)$ denote respectively the group of algebraic characters and cocharacters of the connected center Z of G. The standard apartment $\mathscr{A}$ attached to T in the semisimple building $\mathscr{X}$ of G is denoted by $\mathscr{A}$. It can be seen as the vector space

$$\mathbb{R} \otimes_{\mathbb{Z}} (X_*(T)/X_*(\mathrm{Z}))$$

considered as an affine space on itself. We fix a hyperspecial vertex of the chamber C and, for simplicity, choose it to be the zero point in $\mathscr{A}$. Denote by $\langle\, .\,,.\,\rangle : X_*(T) \times X^*(T) \to \mathbb{Z}$ the natural perfect pairing, as well as its $\mathbb{R}$-linear extension. Each root $\alpha \in \Phi$ defines a function $x \mapsto \alpha(x)$ on $\mathscr{A}$. For any subset Y of $\mathscr{A}$, we write $\alpha(Y) \geq 0$ if α takes nonnegative values on Y. To α is also associated a coroot $\check{\alpha} \in \check{\Phi}$ such that $\langle \check{\alpha}, \alpha\rangle = 2$ and a reflection on $\mathscr{A}$ defined by

$$s_\alpha : x \mapsto x - \alpha(x)\check{\alpha} \mod X_*(\mathrm{Z}) \otimes_{\mathbb{Z}} \mathbb{R}\ .$$

The subgroup of the transformations of $\mathscr{A}$ generated by these reflections identifies with the finite Weyl group W_0. The finite Weyl group W_0 acts by conjugation on T and this induces a faithful linear action on $\mathscr{A}$. Thus W_0 identifies with a subgroup of the transformations of $\mathscr{A}$ and this subgroup is the one generated by the reflections s_α for all $\alpha \in \Phi$. To an element $g \in T$ corresponds a vector $\nu(g) \in \mathbb{R} \otimes_{\mathbb{Z}} X_*(T)$ defined by

$$\langle \nu(g), \chi\rangle = -\operatorname{val}_{\mathfrak{F}}(\chi(g)) \qquad \text{for any } \chi \in X^*(T).$$

The quotient of T by $\ker(\nu) = T^0$ is a free abelian group Λ with rank equal to $\dim(T)$, and ν induces an isomorphism $\Lambda \cong X_*(T)$. The group Λ acts by translation on $\mathscr{A}$ via ν. The actions of W_0 and Λ combine into an action of W on $\mathscr{A}$. The extended affine Weyl group W is the semi-direct product $W_0 \ltimes \Lambda$ if we identify W_0 with the subgroup of W that fixes any lift of x_0 in the extended building of G.

2.1.2. *Affine roots and root subgroups.* We now recall the definition of the affine roots and the properties of the associated root subgroups. To a root α is attached a unipotent subgroup $\mathfrak{U}_\alpha$ of G such that for any $u \in \mathfrak{U}_\alpha \setminus \{1\}$, the intersection $\mathfrak{U}_{-\alpha} u \mathfrak{U}_{-\alpha} \cap N(T)$ consists in only one element called $m_\alpha(u)$. The image in W of this element $m_\alpha(u)$ is the reflection at the affine hyperplane $\{x \in \mathscr{A},\ \alpha(x) = -\mathfrak{h}_\alpha(u)\}$ for a certain $\mathfrak{h}_\alpha(u) \in \mathbb{R}$. Denote by Γ_α the discrete unbounded subset of $\mathbb{R}$ given by $\{\mathfrak{h}_\alpha(u),\ u \in \mathfrak{U}_\alpha \setminus \{1\}\}$. Since our group $\mathbf{G}$ is $\mathfrak{F}$-split we have $\{\mathfrak{h}_\alpha(u),\ u \in \mathfrak{U}_\alpha \setminus \{1\}\} = \mathbb{Z}$. The affine functions

$$(\alpha, \mathfrak{h}) := \alpha(\,.\,) + \mathfrak{h} \qquad \text{for } \alpha \in \Phi \text{ and } \mathfrak{h} \in \mathbb{Z}$$

are called the affine roots.

We identify an element α of Φ with the affine root $(\alpha, 0)$ so that the set of affine roots Φ_{aff} contains Φ. The action of W_0 on Φ extends to an action of W on Φ_{aff}. Explicitly, if $w = w_0 t_\lambda \in W$ is the composition of the translation by $\lambda \in \Lambda$ with $w_0 \in W_0$, then the action of w on the affine root $(\alpha, \mathfrak{h})$

$$(w_0(\alpha), \mathfrak{h} + (\operatorname{val}_{\mathfrak{F}} \circ \alpha)(\lambda)) = (w_0(\alpha), \mathfrak{h} - \langle \nu(\lambda), \alpha\rangle)\ .$$

Define a filtration of $\mathfrak{U}_\alpha$, $\alpha \in \Phi$ by

$$\mathfrak{U}_{\alpha,r} := \{u \in \mathfrak{U}_\alpha \setminus \{1\},\ \mathfrak{h}_\alpha(u) \geq r\} \cup \{1\} \text{ for } r \in \mathbb{R}\ .$$

For $(\alpha, \mathfrak{h}) \in \Phi_{aff}$, we put $\mathfrak{U}_{(\alpha,\mathfrak{h})} := \mathfrak{U}_{\alpha,\mathfrak{h}}$. Obviously, for $r \in \mathbb{R}$ a real number, $\mathfrak{h} \geq r$ is equivalent to $\mathfrak{U}_{(\alpha,\mathfrak{h})} \subseteq \mathfrak{U}_{\alpha,r}$.

By abuse of notation we write throughout the paper wMw^{-1}, for some $w \in W$ and some subgroup $M \subseteq G$, whenever the result of this conjugation is independent of the choice of a representative of w in $N(T)$. For example, for $(\alpha, \mathfrak{h}) \in \Phi_{aff}$ and $w \in W$, we have

$$w\mathfrak{U}_{(\alpha,\mathfrak{h})}w^{-1} = \mathfrak{U}_{w(\alpha,\mathfrak{h})} \ . \tag{2.1}$$

For any non empty subset $\mathbf{Y} \subset \mathscr{A}$, define

$$\begin{aligned} f_{\mathbf{Y}} : \Phi &\longrightarrow \mathbb{R} \cup \{\infty\} \\ \alpha &\longmapsto -\inf_{x \in \mathbf{Y}} \alpha(x) \ . \end{aligned}$$

and the subgroup of G

$$\mathfrak{U}_{\mathbf{Y}} = < \mathfrak{U}_{\alpha, f_{\mathbf{Y}}(\alpha)}, \ \alpha \in \Phi > \tag{2.2}$$

generated by all $\mathfrak{U}_{\alpha, f_{\mathbf{Y}}(\alpha)}$ for $\alpha \in \Phi$. We have ([**25**] Page 103, point 3.)

$$\mathfrak{U}_{\mathbf{Y}} \cap \mathfrak{U}_{\alpha} = \mathfrak{U}_{\alpha, f_{\mathbf{Y}}(\alpha)} \ \text{ for any } \alpha \in \Phi \ . \tag{2.3}$$

2.1.3. *Positive roots and length* . The choice of the chamber C determines the subset Φ^+ of the positive roots, namely the set of $\alpha \in \Phi$ taking nonnegative values on C. Denote by Π a basis for Φ^+. Likewise, the set of positive affine roots Φ^+_{aff} is defined to be the set of affine roots taking nonnegative values on C. The set of negative affine roots is $\Phi^-_{aff} := -\Phi^+_{aff}$.

LEMMA 2.1. i. $\mathfrak{U}_{\alpha, f_C(\alpha)} = \mathfrak{U}_{\alpha,0}$ *for* $\alpha \in \Phi^+$, *and* $\mathfrak{U}_{\alpha, f_C(\alpha)} = \mathfrak{U}_{\alpha,1}$ *for* $\alpha \in \Phi^-$.

ii. *The pro-p Iwahori subgroup I is generated by T^1 and $\mathfrak{U}_C$, namely by T^1 and all root subgroups $\mathfrak{U}_A$ for $A \in \Phi^+_{aff}$.*

iii. *For $\alpha \in \Phi$, we have $I \cap \mathfrak{U}_\alpha = \mathfrak{U}_{\alpha, f_C(\alpha)}$.*

PROOF. i., ii. This is given by [**25**] Prop. I.2.2 (recalled in [**20**] Proof of Lemma 4.8) and [**20**] Proof of Lemma 4.2. iii. This also follows from [**25**] Prop. I.2.2. □

The finite Weyl group W_0 is a Coxeter system generated by the set $S := \{s_\alpha : \alpha \in \Pi\}$ of reflections associated to the simple roots Π. It is endowed with a length function denoted by ℓ. This length extends to W in such a way that the length of an element $w \in W$ is the cardinality of $\{A \in \Phi^+_{aff}, \ w(A) \in \Phi^-_{aff}\}$. For any affine root $(\alpha, \mathfrak{h})$, we have in W the reflection $s_{(\alpha,\mathfrak{h})}$ at the affine hyperplane $\alpha(\,.\,) = -\mathfrak{h}$. The affine Weyl group is defined as the subgroup W_{aff} of W generated by all s_A for all $A \in \Phi_{aff}$.

There is a partial order on Φ given by $\alpha \leq \beta$ if and only if $\beta - \alpha$ is a linear combination with (integral) nonnegative coefficients of elements in Π. Let

$$\begin{aligned} \Phi^{min} &:= \{\alpha \in \Phi : \alpha \text{ is minimal for } \leq\} \quad \text{and} \\ \Pi_{aff} &:= \Pi \cup \{(\alpha, 1) : \alpha \in \Phi^{min}\}. \end{aligned}$$

Note that $\Pi_{aff} \subseteq \Phi^+_{aff}$. Let $S_{aff} := \{s_A : A \in \Pi_{aff}\}$, then the pair (W_{aff}, S_{aff}) is a Coxeter system and the length function ℓ restricted to W_{aff} coincides with the length function of this Coxeter system. For any $s \in S_{aff}$ there is a unique positive affine root $A_s \in \Phi^+_{aff}$ such that $sA_s \in \Phi^-_{aff}$. In fact A_s lies in Π_{aff}.

We have the following formula, for every $A \in \Pi_{aff}$ and $w \in W$ ([**16**] §1):

$$\ell(ws_A) = \begin{cases} \ell(w)+1 & \text{if } w(A) \in \Phi^+_{aff}, \\ \ell(w)-1 & \text{if } w(A) \in \Phi^-_{aff}. \end{cases} \tag{2.4}$$

The Bruhat-Tits decomposition of G says that G is the disjoint union of the double cosets JwJ for $w \in W$. As in [**20**] §4.3 (see the references there) we will denote by Ω the abelian subgroup of W of all elements with length zero and recall that Ω normalizes S_{aff}. Furthermore, W is the semi-direct product $W = \Omega \ltimes W_{aff}$. The length function is constant on the double cosets $\Omega w \Omega$ for $w \in W$. The stabilizer $\mathcal{P}^\dagger_C$ of C in G is the disjoint union of the double cosets $J\omega J$ for $\omega \in \Omega$ ([**20**] Lemma 4.9). We have

$$I \subseteq J \subseteq \mathcal{P}^\dagger_C$$

and I, the pro-unipotent radical of the parahoric subgroup J, is normal in $\mathcal{P}^\dagger_C$ (see [**20**] §4.5). In fact, J is also normal in $\mathcal{P}^\dagger_C$ since J is generated by T^0 and I, and since the action by conjugation of $\omega \in \Omega$ on T normalizes T^0.

2.1.4. Recall that we denote by $\widetilde{W}$ the quotient of $N(T)$ by T^1 and obtain the exact sequence

$$0 \to T^0/T^1 \to \widetilde{W} \to W \to 0 \ .$$

The length function ℓ on W pulls back to a length function ℓ on $\widetilde{W}$ ([**33**] Prop. 1). The Bruhat-Tits decomposition of G says that G is the disjoint union of the double cosets IwI for $w \in \widetilde{W}$. We will denote by $\widetilde{\Omega}$ the preimage of Ω in $\widetilde{W}$. It contains T^0/T^1.

Obviously $\widetilde{W}$ acts by conjugation on its normal subgroup T^0/T^1, and we denote this action simply by $(w,t) \mapsto w(t)$. But, since T/T^1 is abelian the action in fact factorizes through $W_0 = \widetilde{W}/T$. On the other hand W_0, by definition, acts on T and this action stabilizes the maximal compact subgroup T^0 and its pro-p Sylow T^1. Therefore we again have an action of W_0 on T^0/T^1. These two actions, of course, coincide.

2.1.5. *On certain open compact subgroups of the pro-p Iwahori subgroup.* Let $g \in G$. We let

$$I_g := I \cap gIg^{-1} \ . \tag{2.5}$$

Since this definition depends only on gJ, we may consider $I_w := I \cap wIw^{-1}$ for any $w \in W$ or $w \in \widetilde{W}$. Since I is normal in $\mathcal{P}^\dagger_C$, we have $I_{w\omega} = I_w$ for any $\omega \in \Omega$ and any $w \in W$ (but in general not $I_{\omega w} = I_w$).

Note that if $\tilde{w} \in \widetilde{W}$ lifts $w \in W$, then $I_{\tilde{w}} = I_w$.

LEMMA 2.2. *Let $v, w \in W$ such that $\ell(vw) = \ell(v) + \ell(w)$. We have*

$$I_{vw} \subseteq I_v \tag{2.6}$$

and

$$I \subseteq I_{v^{-1}} wIw^{-1} \ . \tag{2.7}$$

PROOF. (2.6): The claim is clear when w has length 0 since we then have $I_{vw} = I_v$. By induction, it suffices to treat the case when $w = s \in S_{aff}$. Note first that adjoining vsC to a given minimal gallery of $\mathscr{X}$ between C and vC gives a minimal gallery between C and vsC.

Let $y \in I_{vs}$. The apartment $y\mathscr{A}$ contains vsC and C and therefore it contains any minimal gallery from C to vsC by [**5**] 2.3.6. In particular, it contains vC. Let F be the facet with codimension 1 which is contained in both the closures of the chambers vC and vsC. It is fixed by y and therefore the image C' of the chamber vC under the action of y also contains F in its closure. By [**5**] 1.3.6, only two chambers of $y\mathscr{A}$ contain F in their closure, so $C' = vC$. Therefore, $y \in v\mathcal{P}_C^\dagger v^{-1} \cap I = \mathcal{P}_{vC}^\dagger \cap I$ where $\mathcal{P}_{vC}^\dagger$ denotes the stabilizer in G of the chamber vC. By [**20**] Lemma 4.10, the intersection $\mathcal{P}_{vC}^\dagger \cap I$ is contained in the parahoric subgroup vJv^{-1} of $\mathcal{P}_{vC}^\dagger$, and since it is a pro-p group, it is contained in its pro-p Sylow subgroup vIv^{-1}. We have proved that y lies in I_v.

(2.7): Let $w \in W$. We prove the following statement by induction on $\ell(v)$: let $v \in W$ such that $\ell(vw) = \ell(v) + \ell(w)$; then any $A \in \Phi_{aff}^+$ such that $vA \in \Phi_{aff}^-$ satisfies $w^{-1}A \in \Phi_{aff}^+$. Using (2.1) and Lemma 2.1.ii, this means that for $v \in W$ such that $\ell(vw) = \ell(v) + \ell(w)$ and $A \in \Phi_{aff}^+$ we have $vA \in \Phi_{aff}^+$ and $\mathcal{U}_A \subseteq v^{-1}\mathcal{U}_C v^{-1} \cap I \subseteq I_{v^{-1}}$, or $vA \in \Phi_{aff}^-$ and $w^{-1}A \in \Phi_{aff}^+$ so $\mathcal{U}_A \subseteq w\mathcal{U}_C w^{-1} \subseteq wIw^{-1}$. This implies that for $v \in W$ such that $\ell(vw) = \ell(v)+\ell(w)$ we have $\mathcal{U}_C \subseteq I_{v^{-1}}wIw^{-1}$ and again using Lemma 2.1.ii, that $I \subseteq I_{v^{-1}}wIw^{-1}$. We now proceed to the proof of the claim by induction.

When v has length zero the claim is clear. Now let v such that $\ell(vw) = \ell(v) + \ell(w)$ and $s \in S_{aff}$ such that $\ell(sv) = \ell(v) - 1$. This implies that $\ell(svw) \leq \ell(sv) + \ell(w) = \ell(vw) - 1$ and therefore $\ell(svw) = \ell(vw) - 1 = \ell(sv) + \ell(w)$. In particular we have $(vw)^{-1}A_s \in \Phi_{aff}^-$ with the notation introduced in §2.1.3 (see [**16**] Section 1, recalled in [**20**] (4.2)). By induction hypothesis, any $A \in \Phi_{aff}^+$ such that $svA \in \Phi_{aff}^-$ satisfies $w^{-1}A \in \Phi_{aff}^+$. Now let $A \in \Phi_{aff}^+$ such that $vA \in \Phi_{aff}^-$. We need to show that $w^{-1}A \in \Phi_{aff}^+$. If $svA \in \Phi_{aff}^-$ then it follows from the induction hypothesis. Otherwise, it means that $vA = -A_s$ and therefore $w^{-1}A = -(vw)^{-1}A_s \in \Phi_{aff}^+$. □

LEMMA 2.3. *Let $w \in W$. The product map induces a bijection*

$$\prod_{\alpha\in\Phi^-} \mathcal{U}_{\alpha, f_{C\cup wC}(\alpha)} \times T^1 \times \prod_{\alpha\in\Phi^+} \mathcal{U}_{\alpha, f_{C\cup wC}(\alpha)} \xrightarrow{\sim} I_w \tag{2.8}$$

where the products on the left hand side are ordered in some arbitrarily chosen way.

PROOF. The multiplication in G induces an injective map

$$\prod_{\alpha\in\Phi^-} \mathcal{U}_\alpha \times T \times \prod_{\alpha\in\Phi^+} \mathcal{U}_\alpha \hookrightarrow G\ .$$

In the notation of [**25**] §I.2 we have $I = R_C$ and $wIw^{-1} = R_{wC}$. Therefore [**25**] Prop. I.2.2 says that the above map restricts to bijections

$$\prod_{\alpha\in\Phi^-} \mathcal{U}_{\alpha, f_C(\alpha)} \times T^1 \times \prod_{\alpha\in\Phi^+} \mathcal{U}_{\alpha, f_C(\alpha)} \xrightarrow{\sim} I \tag{2.9}$$

and

$$\prod_{\alpha\in\Phi^-} \mathcal{U}_{\alpha, f_{wC}(\alpha)} \times T^1 \times \prod_{\alpha\in\Phi^+} \mathcal{U}_{\alpha, f_{wC}(\alpha)} \xrightarrow{\sim} wIw^{-1}\ ,$$

and hence to the bijection

$$\prod_{\alpha\in\Phi^-} \mathcal{U}_{\alpha, f_C(\alpha)} \cap \mathcal{U}_{\alpha, f_{wC}(\alpha)} \times T^1 \times \prod_{\alpha\in\Phi^+} \mathcal{U}_{\alpha, f_C(\alpha)} \cap \mathcal{U}_{\alpha, f_{wC}(\alpha)} \xrightarrow{\sim} I_w\ .$$

Since, obviously, $f_{C\cup wC}(\alpha) = \max(f_C(\alpha), f_{wC}(\alpha))$ we have

$$\mathfrak{U}_{\alpha,f_C(\alpha)} \cap \mathfrak{U}_{\alpha,f_{wC}(\alpha)} = \mathfrak{U}_{\alpha,f_{C\cup wC}(\alpha)} .$$

□

REMARK 2.4. Let $\alpha \in \Phi$ and $w \in W$. Define $g_w(\alpha) := \min\{m \in \mathbb{Z}, (\alpha, m) \in \Phi^+_{aff} \cap w\Phi^+_{aff}\}$. We have $\mathfrak{U}_{\alpha,g_w(\alpha)} = \mathfrak{U}_{\alpha,f_{C\cup wC}(\alpha)}$. This is Lemma 2.1.i when $w = 1$.

PROOF. First note that $\Phi^+_{aff} \cap w\Phi^+_{aff}$ is the set of affine roots which are positive on $C \cup wC$. Let $\alpha \in \Phi$. Since $\alpha + g_w(\alpha) \geq 0$ on $C \cup wC$ we have $f_{C\cup wC}(\alpha) \leq g_w(\alpha)$ and $\mathfrak{U}_{\alpha,g_w(\alpha)} \subseteq \mathfrak{U}_{\alpha,f_{C\cup wC}(\alpha)}$. Now let u be an element in $\mathfrak{U}_{\alpha,f_{C\cup wC}(\alpha)} \setminus \{1\}$. It implies $\mathfrak{h}_\alpha(u) \geq f_{C\cup wC}(\alpha)$ so $\alpha + \mathfrak{h}_\alpha(u) \geq 0$ on $C \cup wC$, therefore $\mathfrak{h}_\alpha(u) \geq g_w(\alpha)$ so $u \in \mathfrak{U}_{\alpha,g_w(\alpha)}$. □

COROLLARY 2.5. *Let $v, w \in W$ and $s \in S_{aff}$ with respective lifts $\tilde{v}, \tilde{w}$ and $\tilde{s}$ in $\widetilde{W}$. We have:*

i. $|I/I_w| = q^{\ell(w)}$;
ii. *if $\ell(vw) = \ell(v) + \ell(w)$ then $I\tilde{v}I \cdot I\tilde{w}I = I\tilde{v}\tilde{w}I$;*
iii. *if $\ell(ws) = \ell(w) + 1$ then I_{ws} is a normal subgroup of I_w of index q.*

PROOF. Points i. and ii. are well known. Compare i. with [**10**] Prop. 3.2 and §I.5 and ii. with [**10**] Prop. 2.8(i). For the convenience of the reader we add the arguments.

i. We obtain the result by induction on $\ell(w)$.
Suppose that $\ell(ws) = \ell(w) + 1$. Again by [**16**] Section 1 (recalled in [**20**] (4.2)) we have $wA_s \in \Phi^+_{aff}$ and $\Phi^+_{aff} \cap ws\Phi^+_{aff} = (\Phi^+_{aff} \cap w\Phi^+_{aff}) \setminus \{wA_s\}$. So if we let $(\beta, m) := wA_s$, then using Remark 2.4 we have

$$\mathfrak{U}_{\alpha,f_{C\cup wC}(\alpha)} = \mathfrak{U}_{\alpha,f_{C\cup wsC}(\alpha)} \quad \text{for any } \alpha \in \Phi,\ \alpha \neq \beta \tag{2.10}$$

and

$$\mathfrak{U}_{\beta,f_{C\cup wC}(\beta)} = \mathfrak{U}_{wA_s} = \mathfrak{U}_{(\beta,m)} \quad \text{and} \quad \mathfrak{U}_{\beta,f_{C\cup wsC}(\beta)} = \mathfrak{U}_{(\beta,1+m)}. \tag{2.11}$$

Hence using Lemma 2.3 we deduce that $I_{ws} \subseteq I_w$ and

$$I_w/I_{ws} \simeq \mathfrak{U}_{\beta,f_{C\cup wC}(\beta)}/\mathfrak{U}_{\beta,f_{C\cup wsC}(\beta)} = \mathfrak{U}_{\beta,m}/\mathfrak{U}_{\beta,m+1} , \tag{2.12}$$

which has cardinality q by [**31**] 1.1.

ii. It suffices to treat the case $v = s \in S_{aff}$. The claim then follows by induction on $\ell(v)$. Using Lemma 2.1 we have $I\tilde{s}I = I\tilde{s}\mathfrak{U}_{A_s}$ since $sA \in \Phi^+_{aff}$ for any $A \in \Phi^+_{aff} \setminus \{A_s\}$. Now $\ell(sw) = \ell(w) + 1$ means that $w^{-1}A_s \in \Phi^+_{aff}$ therefore (again by Lemma 2.1.ii) $I\tilde{s}I\tilde{w}I = I\tilde{s}\mathfrak{U}_{A_s}\tilde{w}I = I\tilde{s}\tilde{w}\mathfrak{U}_{w^{-1}A_s}I = I\tilde{s}\tilde{w}I$.

iii. We first treat the case that $w = 1$. By i. we only need to show that I_s is normal in I. Let F be the 1-codimensional facet common to C and sC. The pro-unipotent radical I_F of the parahoric subgroup $\mathbf{G}^\circ_F(\mathfrak{O})$ attached to F is a normal subgroup of $\mathbf{G}^\circ_F(\mathfrak{O})$ ([**25**] I.2). It follows from [**25**] Prop. I.2.11 and its proof that I_F is a normal subgroup of $I = I_C$. Obviously $I_s \subseteq I_F$. Hence it suffices to show equality. By [**25**] Prop. I.2.2 (see also [**20**] Proof of Lemma 4.8), the product map induces a bijection

$$\prod_{\alpha\in\Phi^-} \mathfrak{U}_{\alpha,f^*_F(\alpha)} \times T^1 \times \prod_{\alpha\in\Phi^+} \mathfrak{U}_{\alpha,f^*_F(\alpha)} \xrightarrow{\sim} I_F$$

where $f_F^*(\alpha) = f_F(\alpha)$ if $\alpha \neq \alpha_0$ and $f_F^*(\alpha_0) = f_F(\alpha_0) + 1 = \epsilon + 1$ with $(\alpha_0, \epsilon) = A_s$. In view of Lemma 2.3 it remains to show that $\mathfrak{U}_{\alpha, f_F^*(\alpha)} = \mathfrak{U}_{\alpha, f_{C \cup sC}(\alpha)}$ for any $\alpha \in \Phi$. But this is immediate from (2.10) and (2.11) applied with $w = 1$.

Coming back to the general case we see that $wIw^{-1} \cap wsI(ws)^{-1}$ is normal in wIw^{-1}. Hence $I_w \cap I_{ws} = I \cap wIw^{-1} \cap wsI(ws)^{-1}$ is normal in I_w. But in the proof of i. we have seen that $I_{ws} \subseteq I_w$. The assertion about the index also follows from i. $\square$

2.1.6. *Chevalley basis and double cosets decompositions.* Let $\mathbf{G}_{x_0}$ denote the Bruhat-Tits group scheme over $\mathfrak{O}$ corresponding to the hyperspecial vertex x_0 (cf. [**31**]). As part of a Chevalley basis we have (cf. [**6**] 3.2), for any root $\alpha \in \Phi$, a homomorphism $\varphi_\alpha : \mathrm{SL}_2 \longrightarrow \mathbf{G}_{x_0}$ of $\mathfrak{O}$-group schemes which restricts to isomorphisms

$$\{\begin{pmatrix} 1 & * \\ 0 & 1 \end{pmatrix}\} \xrightarrow{\cong} \mathfrak{U}_\alpha \qquad \text{and} \qquad \{\begin{pmatrix} 1 & 0 \\ * & 1 \end{pmatrix}\} \xrightarrow{\cong} \mathfrak{U}_{-\alpha} \ .$$

Moreover, one has $\check{\alpha}(x) = \varphi_\alpha(\left(\begin{smallmatrix} x & 0 \\ 0 & x^{-1} \end{smallmatrix}\right))$. We let the subtorus $\mathbf{T}_{s_\alpha} \subseteq \mathbf{T}$ denote the image (in the sense of algebraic groups) of the cocharacter $\check{\alpha}$. We always view these as being defined over $\mathfrak{O}$ as subtori of $\mathbf{G}_{x_0}$. The group of $\mathbb{F}_q$-rational points $\mathbf{T}_{s_\alpha}(\mathbb{F}_q)$ can be viewed as a subgroup of $T^0/T^1 \xrightarrow{\cong} \mathbf{T}(\mathbb{F}_q)$ (and is abstractly isomorphic to $\mathbb{F}_q^\times$). Given $z \in \mathbb{F}_q^\times$, we consider $[z] \in \mathfrak{O}^\times$ the Teichmüller representative ([**29**] II.4 Prop. 8) and denote by $\check{\alpha}([\_])$ the composite morphism of groups

$$\check{\alpha}([\_]) \ : \ \mathbb{F}_q^\times \xrightarrow{[\_]} \mathfrak{O}^\times \xrightarrow{\check{\alpha}} \mathbf{T}(\mathfrak{O}) = T^0 \ . \tag{2.13}$$

We will denote its kernel by $\mu_{\check{\alpha}}$. Looking at the commutative diagram

$$\begin{array}{ccccccc}
\mathfrak{O}^\times & \xrightarrow{\check{\alpha}} & \mathbf{T}_{s_\alpha}(\mathfrak{O}) & \xrightarrow{\subseteq} & \mathbf{T}(\mathfrak{O}) & \xrightarrow{\mathrm{pr}} & T^0/T^1 \\
\downarrow{\scriptstyle red} & & \downarrow{\scriptstyle red} & & \downarrow{\scriptstyle red} & \swarrow{\scriptstyle red}\ {\scriptstyle \cong} & \\
\mathbb{F}_q^\times & \xrightarrow{\check{\alpha}} & \mathbf{T}_{s_\alpha}(\mathbb{F}_q) & \xrightarrow{\subseteq} & \mathbf{T}(\mathbb{F}_q) & &
\end{array}$$

of reduction maps and using that the Teichmüller map is a section of the reduction map $\mathfrak{O}^\times \twoheadrightarrow \mathbb{F}_q^\times$ we deduce that the reduction map induces isomorphisms

$$\check{\alpha}([\mathbb{F}_q^\times]) \xrightarrow{\cong} \check{\alpha}(\mathbb{F}_q^\times) \qquad \text{and} \qquad \mu_{\check{\alpha}} \xrightarrow{\cong} \ker(\check{\alpha}_{|\mathbb{F}_q^\times}) \ .$$

REMARK 2.6. In Propositions 5.6 and 8.2 we will need to differentiate between an element $t \in \check{\alpha}([\mathbb{F}_q^\times]) \subset T^0$ and its image in $T^0/T^1 \subset \widetilde{W}$ which we will denote by $\bar{t}$.

REMARK 2.7. If φ_α is not injective then its kernel is $\{\left(\begin{smallmatrix} a & 0 \\ 0 & a \end{smallmatrix}\right) : a = \pm 1\}$ (cf. [**11**] II.1.3(7)). It follows that $\mu_{\check{\alpha}}$ has cardinality 1 or 2.

For the following two lemmas below, recall that the notation for the action of W on T^0/T^1 was introduced in §2.1.4.

LEMMA 2.8. *Suppose that* $\mathbf{G}$ *is semisimple simply connected, then:*

i. $\check{\alpha}(\mathbb{F}_q^\times) = \mathbf{T}_{s_\alpha}(\mathbb{F}_q)$ *for any* $\alpha \in \Phi$*;*

ii. $T^0/T^1 = \widetilde{\Omega}$ *is generated by the union of its subgroups* $\mathbf{T}_{s_\alpha}(\mathbb{F}_q)$ *for* $\alpha \in \Pi$*;*

PROOF. By our assumption that $\mathbf{G}$ is semisimple simply connected the set $\{\check{\alpha} : \alpha \in \Pi\}$ is a basis of the cocharacter group $X_*(T)$. This means that

$$\prod_{\alpha \in \Pi} \mathbb{G}_m \xrightarrow{\prod_\alpha \check{\alpha}} \mathbf{T}$$

is an isomorphism of algebraic tori. It follows that multiplication induces an isomorphism

$$\prod_{\alpha \in \Pi} \mathbf{T}_{s_\alpha}(\mathbb{F}_q) \xrightarrow{\cong} \mathbf{T}(\mathbb{F}_q) \ .$$

This implies ii. But, since any root is part of some basis of the root system, we also obtain that $\Phi \cap 2X_*(T) = \emptyset$. Hence φ_α and $\check{\alpha}$ are injective for any $\alpha \in \Phi$ (cf. [**11**] II.1.3(7)). For ii. it therefore remains to notice that $\check{\alpha} : \mathbb{F}_q^\times \to \mathbf{T}_{s_\alpha}(\mathbb{F}_q)$ is a map between finite sets of the same cardinality. $\square$

LEMMA 2.9. *For any $t \in T^0/T^1$ and any $A = (\alpha, \mathfrak{h}) \in \Phi_{aff}$, we have $s_A(t)t^{-1} \in \check{\alpha}([\mathbb{F}_q^\times])$.*

PROOF. Recall from §2.1.4 that the action of s_A on t is inflated from the action of its image $s_\alpha \in W_0$. The action of W_0 on $X_*(T)$ is induced by its action on T, i.e., we have

$$(w(\xi))(x) = w(\xi(x)) \qquad \text{for any } w \in W_0,\ \xi \in X_*(T), \text{ and } x \in T.$$

On the other hand the action of s_α on $\xi \in X_*(T)$ is given by

$$s_\alpha(\xi) = \xi - \langle \xi, \alpha \rangle \check{\alpha} \ .$$

So for any $\xi \in X_*(T)$ and any $y \in \mathbb{F}_q^\times$ we have

$$s_\alpha(\xi([y])) = (s_\alpha(\xi))([y]) = \xi([y])\check{\alpha}([y])^{-\langle \xi, \alpha \rangle} \in \xi([y])\check{\alpha}([\mathbb{F}_q^\times]) \ .$$

It remains to notice that, if $\xi_1, \ldots, \xi_m$ is a basis of $X_*(T)$, then $\prod_i \xi([\mathbb{F}_q^\times]) = T^0/T^1$. $\square$

For any $\alpha \in \Phi$ we have the additive isomorphism $x_\alpha : \mathfrak{F} \xrightarrow{\cong} \mathcal{U}_\alpha$ defined by

$$x_\alpha(u) := \varphi_\alpha(\left(\begin{smallmatrix} 1 & u \\ 0 & 1 \end{smallmatrix}\right)) \ . \tag{2.14}$$

Let $(\alpha, \mathfrak{h}) \in \Pi_{aff}$ and $s = s_{(\alpha, \mathfrak{h})}$. We put

$$n_s := \varphi_\alpha(\begin{pmatrix} 0 & \pi^{\mathfrak{h}} \\ -\pi^{-\mathfrak{h}} & 0 \end{pmatrix}) \in N(T) \ .$$

We have $n_s^2 = \check{\alpha}(-1) \in T^0$ and $n_s T^0 = s \in W$. We set:

$$\tilde{s} = n_s T^1 \ \in \widetilde{W} \ . \tag{2.15}$$

From (2.12), Lemma 2.3 and [**31**] 1.1 we deduce that $\{x_\alpha(\pi^{\mathfrak{h}} u)\}$ is a system of representatives of I/I_s when u ranges over a system of representatives of $\mathfrak{O}/\pi\mathfrak{O}$. We have the decomposition

$$I n_s I = n_s I \,\dot{\cup} \dot{\bigcup}_{z \in \mathbb{F}_q^\times} x_\alpha(\pi^{\mathfrak{h}}[z]) n_s I \ . \tag{2.16}$$

Note that since $[-1] = -1 \in \mathfrak{O}$, we have $x_\alpha(\pi^{\mathfrak{h}}[z])^{-1} = x_\alpha(-\pi^{\mathfrak{h}}[z]) = x_\alpha(\pi^{\mathfrak{h}}[-z])$ and for $z \in \mathbb{F}_q^\times$, we compute, using φ_α, that

$$x_\alpha(\pi^{\mathfrak{h}}[z])\check{\alpha}([z]) n_s = n_s x_\alpha(\pi^{\mathfrak{h}}[-z^{-1}]) n_s x_\alpha(\pi^{\mathfrak{h}}[-z]) \in n_s I n_s I \tag{2.17}$$

because the Teichmüller map $[\_] : \mathbb{F}_q^\times \to \mathfrak{O}^\times$ is a morphism of groups. Since $In_sI = n_sI \dot\cup \dot\bigcup_{z\in\mathbb{F}_q^\times} x_\alpha(\pi^{\mathfrak{h}}[-z^{-1}])n_sI$, it follows that

$$n_sIn_sI = In_s^2 \dot\cup \dot\bigcup_{z\in\mathbb{F}_q^\times} x_\alpha(\pi^{\mathfrak{h}}[z])\check\alpha([z])n_sI \subset In_s^2 \dot\cup \bigcup_{z\in\mathbb{F}_q^\times} I\check\alpha([z])n_sI \tag{2.18}$$

and hence

$$In_sI \cdot In_sI = In_s^2 \dot\cup \dot\bigcup_{t\in\check\alpha([\mathbb{F}_q^\times])} Itn_sI\ . \tag{2.19}$$

REMARK 2.10. Let $s \in S_{aff}$ with lift $\tilde s \in \widetilde W$. Let $w \in \widetilde W$ such that $\ell(\tilde s w) = \ell(w) - 1$. From (2.19) and Cor. 2.5.ii, we deduce that

$$I\tilde sI \cdot IwI = I\tilde sI \cdot I\tilde sI \cdot I(\tilde s^{-1}w)I = I\tilde swI \dot\cup \dot\bigcup_{t\in\check\alpha([\mathbb{F}_q^\times])} ItwI\ . \tag{2.20}$$

LEMMA 2.11. *For $u, v, w \in \widetilde W$ satisfying $IuI \subseteq IvI \cdot IwI$, we have:*

$$|\ell(w) - \ell(v)| \le \ell(u) \le \ell(w) + \ell(v)\ .$$

PROOF. It is enough the prove that for $u, v, w \in W$ satisfying $JuJ \subseteq JvJ{\cdot}JwJ$, we have:

$$|\ell(w) - \ell(v)| \le \ell(u) \le \ell(w) + \ell(v)\ .$$

Let $w \in W$. We prove by induction with respect to $\ell(v)$ that for $u \in W$ such that $JuJ \subseteq JvJ \cdot JwJ$ we have

$$\ell(w) - \ell(v) \le \ell(u) \le \ell(w) + \ell(v)\ . \tag{2.21}$$

This will prove the lemma because $Ju^{-1}J \subseteq Jw^{-1}J \cdot Jv^{-1}J$ so we have $\ell(u) = \ell(u^{-1}) \ge \ell(v^{-1}) - \ell(w^{-1}) = \ell(v) - \ell(w)$. If v has length 0 then $JvJwJ = JvwJ$ since v normalizes J. Therefore $u = vw$ and $\ell(u) = \ell(w)$ so the claim is true. Now suppose v has length 1 meaning $v = \omega s$ for some $s \in S_{aff}$ and $\omega \in \Omega$. Recall that $s^2 = 1$. From (2.19) we deduce

$$JsJ \cdot JsJ = J \dot\cup JsJ\ , \tag{2.22}$$

since $J = T^0I$ contains n_s^2 and $\check\alpha([\mathbb{F}_q^\times])$, where $(\alpha, \mathfrak{h}) \in \Pi_{aff}$ is such that $s = s_{(\alpha,\mathfrak{h})}$.

If $\ell(sw) = \ell(w) + 1$, then $\ell(vw) = \ell(v) + \ell(w)$ and $JvJ \cdot JwJ = JvwJ$ using Cor. 2.5.ii. Therefore (2.21) is obviously satisfied when $u = vw$.

Otherwise, $\ell(sw) = \ell(w) - 1$ and $\ell(vw) = \ell(w) - 1$. By (2.22) we get

$$JsJ \cdot JwJ = JsJ \cdot JsJ \cdot JswJ = JswJ \dot\cup JwJ. \tag{2.23}$$

Therefore $JvJ \cdot JwJ = J\omega swJ \dot\cup J\omega wJ$ and we see that (2.21) is satisfied for all $u \in \{vw, \omega w\}$.

Now let $v \in W$ with length > 1 and $s \in S_{aff}$ such that $\ell(sv) = \ell(v) - 1$. By induction hypothesis, $JsvJ \cdot JwJ$ is the disjoint union of double cosets of the form $Ju'J$ with

$$\ell(w) - \ell(v) + 1 \le \ell(u') \le \ell(w) + \ell(v) - 1 \tag{2.24}$$

and, using the previous case, $JvJ \cdot JwJ = JsJ \cdot JsvJ \cdot JwJ$ is a union of double cosets of the form JuJ with

$$\ell(u') - 1 \le \ell(u) \le \ell(u') + 1\ . \tag{2.25}$$

Combining (2.24) and (2.25), we see that u satisfies $\ell(w) - \ell(v) \le \ell(u) \le \ell(w) + \ell(v)$. □

2.2. The pro-p Iwahori-Hecke algebra. We start from the compact induction $\mathbf{X} := \mathrm{ind}_I^G(1)$ of the trivial I-representation. It can be seen as the space of compactly supported functions $G \to k$ which are constant on the left cosets mod I. It lies in $\mathrm{Mod}(G)$. For Y a compact subset of G which is right invariant under I, we denote by char_Y the characteristic function of Y. It is an element of $\mathbf{X}$.

The pro-p Iwahori-Hecke algebra is defined to be the k-algebra

$$H := \mathrm{End}_{k[G]}(\mathbf{X})^{\mathrm{op}} .$$

We often will identify H, as a right H-module, via the map

$$\begin{aligned} H &\xrightarrow{\cong} \mathbf{X}^I \\ h &\longmapsto (\mathrm{char}_I)h \end{aligned}$$

with the submodule $\mathbf{X}^I$ of I-fixed vectors in $\mathbf{X}$. The Bruhat-Tits decomposition of G says that G is the disjoint union of the double cosets IwI for $w \in \widetilde{W}$. Hence we have the I-equivariant decomposition

$$\mathbf{X} = \oplus_{w\in\widetilde{W}} \mathbf{X}(w) \quad \text{with} \quad \mathbf{X}(w) := \mathrm{ind}_I^{IwI}(1) ,$$

where the latter denotes the subspace of those functions in $\mathbf{X}$ which are supported on the double coset IwI. In particular, we have $\mathbf{X}(w)^I = k\tau_w$ where $\tau_w := \mathrm{char}_{IwI}$ and hence

$$H = \oplus_{w\in\widetilde{W}} k\tau_w$$

as a k-vector space. If $g \in IwI$ we sometimes also write $\tau_g := \tau_w$. The defining relations of H are the braid relations (see [**33**] Thm. 1)

$$\tau_w \tau_{w'} = \tau_{ww'} \qquad \text{for } w, w' \in \tilde{W} \text{ such that } \ell(ww') = \ell(w) + \ell(w') \tag{2.26}$$

together with the quadratic relations which we describe now (compare with [**20**] §4.8, and see more references therein). We refer to the notation introduced in §2.1.6, see (2.13) in particular. To any $s = s_{(\alpha,\mathfrak{h})} \in S_{aff}$, we attach the following idempotent element:

$$\theta_s := -|\mu_{\check{\alpha}}| \sum_{t\in\check{\alpha}([\mathbb{F}_q^\times])} \tau_t \in H . \tag{2.27}$$

The quadratic relations in H are:

$$\tau_{n_s}^2 = -\tau_{n_s}\theta_s = -\theta_s\tau_{n_s} \qquad \text{for any } s \in S_{aff}, \tag{2.28}$$

Since in the existing literature the definition of θ_s is not correct we will give a proof further below. Recall that we defined $\tilde{s} = n_s T^1 \in \widetilde{W}$ in (2.15). The quadratic relation says that $\tau_{\tilde{s}}^2 = -\theta_s\tau_{\tilde{s}} = -\tau_{\tilde{s}}\theta_s$. A general element $w \in \widetilde{W}$ can be decomposed into $w = \omega\tilde{s_1}\dots\tilde{s_\ell}$ with $\omega \in \widetilde{\Omega}$, $s_i \in S_{aff}$, and $\ell = \ell(w)$. The braid relations imply $\tau_w = \tau_\omega\tau_{\tilde{s_1}}\dots\tau_{\tilde{s_\ell}}$.

The subgroup of G generated by all parahoric subgroups is denoted by G_{aff} as in [**20**] §4.5. It is a normal subgroup of G and we have $G/G_{aff} = \Omega$. By Bruhat decomposition, G_{aff} is the disjoint union of all IwI for w ranging over the preimage $\widetilde{W}_{aff}$ of W_{aff} in $\widetilde{W}$. The subalgebra of H of the functions with support in G_{aff} is denoted by H_{aff} and has basis the set of all τ_w, $w \in \widetilde{W}_{aff}$. When $\mathbf{G}$ is simply connected semisimple, we have $G = G_{aff}$ and $H = H_{aff}$.

Recall that there is a unique involutive automorphism of H satisfying

$$\iota(\tau_{n_s}) = -\tau_{n_s} - \theta_s \quad \text{and } \iota(\tau_\omega) = \tau_\omega \text{ for all } s \in S_{aff} \text{ and } \omega \in \widetilde{\Omega} \tag{2.29}$$

(see for example [**20**] §4.8). It restricts to an involutive automorphism of H_{aff}.

Proof of (2.28). First we notice that θ_s is indeed an idempotent because

$$\begin{aligned}\theta_s^2 = |\mu_{\check\alpha}|^2 \sum_{u,t\in\check\alpha([\mathbb{F}_q^\times])} \tau_{ut} &= |\mu_{\check\alpha}|^2 |\alpha([\mathbb{F}_q^\times])| \sum_{t\in\check\alpha([\mathbb{F}_q^\times])} \tau_t \\ &= (q-1)|\mu_{\check\alpha}| \sum_{t\in\check\alpha([\mathbb{F}_q^\times])} \tau_t = -|\mu_{\check\alpha}| \sum_{t\in\check\alpha([\mathbb{F}_q^\times])} \tau_t = \theta_s \ .\end{aligned}$$

The support of $\tau_{n_s}^2$ is contained in In_sIn_sI. Its value at $h \in G$ is equal to $|(In_sI \cap hIn_s^{-1}I)/I| \cdot 1_k$, and by (2.19) we need to consider the cases of $h = n_s^2$ and of $h = tn_s$ for $t \in \check\alpha([\mathbb{F}_q^\times])$. For $h = n_s^2$ this value is equal to $|In_sI/I| \cdot 1_k = |I/I_s| \cdot 1_k = q \cdot 1_k = 0$. From (2.18), we deduce that for $t = \check\alpha([\zeta])$ where $\zeta \in \mathbb{F}_q^\times$, we have

$$\begin{aligned} tn_sIn_s^{-1}I &= It \,\dot\cup \dot\bigcup_{z\in\mathbb{F}_q^\times} tx_\alpha(\pi^{\mathfrak{h}}[z])\check\alpha([z])n_s^{-1}I \\ &= I\check\alpha(\zeta) \,\dot\cup \dot\bigcup_{z\in\mathbb{F}_q^\times} \check\alpha(\zeta)x_\alpha(\pi^{\mathfrak{h}}[z])\check\alpha([z])n_s^{-1}I \ . \end{aligned}$$

For $z \in \mathbb{F}_q^\times$ we compute

$$\check\alpha([\zeta])x_\alpha(\pi^{\mathfrak{h}}[z])\check\alpha([z])n_s^{-1} = x_\alpha(\pi^{\mathfrak{h}}[\zeta^2 z])\check\alpha([-\zeta z])n_s \in I\check\alpha([-\zeta z])n_sI \ .$$

So

$$tn_sIn_s^{-1}I \cap In_sI = \dot\bigcup_{z\in\mathbb{F}_q^\times,\, \check\alpha([-\zeta z])=1} \check\alpha([\zeta])x_\alpha(\pi^{\mathfrak{h}}[z])\check\alpha([z])n_s^{-1}I$$

and $|tn_sIn_s^{-1}I \cap In_sI/I| = |\mu_{\check\alpha}|$. We have proved that

$$\tau_{n_s}^2 = |\mu_{\check\alpha}| \sum_{t\in\check\alpha([\mathbb{F}_q^\times])} \tau_{tn_s} = -\theta_s n_s \ .$$

Noticing that $\check\alpha([\mathbb{F}_q^\times])n_s = n_s\check\alpha([\mathbb{F}_q^\times])$, we then obtain $\tau_{n_s}^2 = -\tau_{n_s}\theta_s$. □

2.2.1. *Idempotents in H.* We now introduce idempotents in H (compare with [**21**] §3.2.4). To any k-character $\lambda : T^0/T^1 \to k^\times$ of T^0/T^1, we associate the following idempotent in H:

$$e_\lambda := -\sum_{t\in T^0/T^1} \lambda(t^{-1})\tau_t \ . \tag{2.30}$$

It satisfies

$$e_\lambda\,\tau_t = \tau_t\, e_\lambda = \lambda(t)\, e_\lambda \tag{2.31}$$

for any $t \in T^0/T^1$. In particular, when λ is the trivial character we obtain the idempotent element denoted by e_1. Let $s = s_{(\alpha,\mathfrak{h})} \in S_{aff}$ with the corresponding idempotent θ_s as in (2.27). Using (2.31), we easily see that $e_\lambda\theta_s = -\sum_{z\in\mathbb{F}_q^\times} \lambda(\check\alpha([z]))e_\lambda$ therefore

$$e_\lambda\theta_s = \begin{cases} e_\lambda & \text{if the restriction of } \lambda \text{ to } \check\alpha([\mathbb{F}_q^\times]) \text{ is trivial,} \\ 0 & \text{otherwise.}\end{cases} \tag{2.32}$$

The action of the finite Weyl group W_0 on T^0/T^1 gives an action on the characters $T^0/T^1 \to k$. This actions inflates to an action of $\widetilde{W}$ denoted by $(w, \lambda) \mapsto {}^w\lambda$. From the braid relations (2.26), one sees that for $w \in \widetilde{W}$, we have

$$\tau_w \, e_\lambda = e_{{}^w\lambda} \, \tau_w \ . \tag{2.33}$$

Suppose for a moment that $\mathbb{F}_q \subseteq k$. We then denote by $\widehat{T^0/T^1}$ the set of k-characters of T^0/T^1. The family $\{e_\lambda\}_\lambda \in \widehat{T^0/T^1}$ is a family of orthogonal idempotents with sum equal to 1. It gives the following ring decomposition

$$k[T^0/T^1] = \prod_{\lambda \in \widehat{T^0/T^1}} k e_\lambda \ . \tag{2.34}$$

Let Γ denote the set of W_0-orbits in $\widehat{T^0/T^1}$. To $\gamma \in \Gamma$ we attach the element $e_\gamma := \sum_{\lambda \in \gamma} e_\lambda$. It is a central idempotent in H (see (2.33)), and we have the obvious ring decomposition

$$H = \prod_{\gamma \in \Gamma} H e_\gamma \ . \tag{2.35}$$

2.2.2. *Characters of H and H_{aff}.* Recall that for $s \in S_{aff}$ we have $\tau_{n_s}(\tau_{n_s} + \theta_s) = 0$ where θ_s is an idempotent element. Therefore, we see that a character $H \to k$ (resp. $H_{aff} \to k$) takes value 0 or -1 at τ_{n_s}. In fact, the following morphisms of k-algebras $H \to k$ are well defined (compare with [**20**] Definition after Remark 6.13):

$$\chi_{triv} : \tau_{\tilde{s}} \longmapsto 0, \ \tau_\omega \longmapsto 1, \text{ for any } s \in S_{aff} \text{ and } \omega \in \widetilde{\Omega} \ , \tag{2.36}$$

$$\chi_{sign} : \tau_{\tilde{s}} \longmapsto -1, \ \tau_\omega \longmapsto 1, \text{ for any } s \in S_{aff} \text{ and } \omega \in \widetilde{\Omega} \ . \tag{2.37}$$

They satisfy in particular $\chi_{triv}(e_1) = \chi_{sign}(e_1) = 1$. They are called the trivial and the sign character of H, respectively. Notice that $\chi_{sign} = \chi_{triv} \circ \iota$ (see (2.29)). The restriction to H_{aff} of χ_{triv} (resp. χ_{sign}) is called the trivial (resp. sign) character of H_{aff}.

We call a twisted sign character of H_{aff} a character $\chi : H_{aff} \to k$ such that $\chi(\tau_{n_s}) = -1$ for all $s \in S_{aff}$. The precomposition by ι of a twisted sign character of H_{aff} is called a twisted trivial character. This definition given in [**34**] coincides with the definition given in [**19**] §5.4.2 but it is simpler and more concise.

REMARK 2.12.
i. A twisted sign character χ of H_{aff} satisfies $\chi(\theta_s) = 1$ for all $s = s_{(\alpha,\mathfrak{h})} \in S_{aff}$ which is equivalent to $\chi(\tau_t) = 1$ for all $t \in \check{\alpha}([\mathbb{F}_q^\times])$. This is also true for a twisted trivial character since the involution ι fixes τ_t for $t \in T^0/T^1$. Therefore, the twisted sign (resp. trivial) characters of H_{aff} are in bijection with the characters $\lambda : T^0/T^1 \to k$ which are equal to 1 on the subgroup $(T^0/T^1)'$ of T^0/T^1 generated by all $\check{\alpha}([\mathbb{F}_q^\times])$ for $\alpha \in \Phi$, equivalently by all $\check{\alpha}([\mathbb{F}_q^\times])$ for $\alpha \in \Pi$.
ii. The twisted trivial characters of H_{aff} are characterized by the fact that they send τ_w to 0 for all $w \in \widetilde{W}_{aff}$ with length > 0 and τ_t to 1 for all $t \in (T^0/T^1)'$.

iii. By Lemma 2.9, the action of $\widetilde{W}$ on T^0/T^1, which is inflated from the action of its quotient W_0, induces the trivial action on the quotient $(T^0/T^1)/(T^0/T^1)'$. This has the following consequences:
   - $(T^0/T^1)'$ is a normal subgroup of $\widetilde{W}$.
   - given $\lambda : T^0/T^1 \to k^\times$ a character which is equal to 1 on $(T^0/T^1)'$, the corresponding idempotent e_λ is central in H (use (2.33))
   - the natural action $(\omega, \chi) \mapsto \chi(\tau_\omega^{-1} \_ \tau_\omega)$ of $\widetilde{\Omega}$ on the characters of H_{aff} fixes the twisted trivial characters. Since ι fixes the elements τ_ω for $\omega \in \widetilde{\Omega}$, this action also fixes the twisted sign characters.

iv. When **G** is semisimple simply connected, then $H = H_{aff}$ and $(T^0/T^1)' = T^0/T^1$ (Lemma 2.8.i) so the trivial (resp. sign) character of $H = H_{aff}$ is the only twisted trivial (resp. sign) character.

As in [**34**] §1.4, we notice that the Coxeter system (W_{aff}, S_{aff}) is the direct product of the irreducible affine Coxeter systems $(W^i_{aff}, S^i_{aff})_{1\leq i\leq r}$ corresponding to the irreducible components $(\Phi^i, \Pi^i)_{1\leq i\leq r}$ of the based root system (Φ, Π). For $i \in \{1, \ldots, r\}$, the k-module of basis $(\tau_w)_{w\in\widetilde{W}^i_{aff}}$ is a subalgebra of H_{aff}. Remark that it contains $\{\tau_t, t \in T^0/T^1\}$. Remark also that the anti-involution ι restricts to an anti-involution of H^i_{aff}.

We call a twisted sign character of H^i_{aff} a character $\chi : H^i_{aff} \to k$ such that $\chi(\tau_{n_s}) = -1$ for all $s \in S^i_{aff}$. The precomposition by ι of a twisted sign character of H^i_{aff} is called a twisted trivial character. The algebras $(H^i_{aff})_{1\leq i\leq r}$ will be called the irreducible components of H_{aff}.

A character of $H_{aff} \to k$ will be called **supersingular** if, for every $i \in \{1, \ldots, r\}$, it does not restrict to a twisted trivial or sign character of H^i_{aff}. This terminology is justified in the following subsection.

2.3. Supersingularity. We refer here to definitions and results contained in [**19**] §2 and §5. Note that there the field k was algebraically closed, but is easy to see that the claims that we are going to use are valid when k is not necessarily algebraically closed. In fact, in [**34**], these definitions and results are generalized to the case where k is an arbitrary field of characteristic p and **G** is a general connected reductive $\mathfrak{F}$-group.

In [**19**] §2.3.1. a central subalgebra $\mathcal{Z}^0(H)$ of H is defined (it is denoted by $\mathcal{Z}_T$ in [**34**]). This algebra is isomorphic to the affine semigroup algebra $k[X_*^{dom}(T)]$, where $X_*^{dom}(T)$ denotes the semigroup of all dominant cocharacters of T ([**19**] Prop. 2.10). The cocharacters $\lambda \in X_*^{dom}(T) \setminus (-X_*^{dom}(T))$ generate a proper ideal of $k[X_*^{dom}(T)]$, the image of which in $\mathcal{Z}^0(H)$ is denoted by $\mathfrak{J}$ as in [**19**] §5.2 (it coincides with the ideal $\mathcal{Z}_{T,\ell>0}$ of [**34**]).

Generalizing [**19**] Prop.-Def. 5.10 and [**34**] Def. 6.10 we call an H-module M supersingular if any element in M is annihilated by a power of $\mathfrak{J}$. Recall that a finite length H-module is always finite dimensional (see for example [**20**] Lemma 6.9). Hence, if M has finite length, then it is supersingular if and only if it is annihilated by a power of $\mathfrak{J}$. Also note that supersingularity can be tested after an arbitrary extension of the coefficient field k.

The supersingular characters of H_{aff} were defined at the end of §2.2.2. The two notions of supersingularity are related by the following fact.

LEMMA 2.13. - *Let $\chi : H_{aff} \to k$ be a supersingular character of H_{aff}. The left (resp. right) H-module $H \otimes_{H_{aff}} \chi$ (resp. $\chi \otimes_{H_{aff}} H$) is annihilated by $\mathfrak{J}$. In particular, it is a supersingular H-module.*

- *Let $\chi : H_{aff} \to k$ be a twisted trivial or sign character of H_{aff}. The left (resp. right) H-module $H \otimes_{H_{aff}} \chi$ (resp. $\chi \otimes_{H_{aff}} H$) does not have any nonzero supersingular subquotient.*

PROOF. Since supersingularity can be tested after an arbitrary extension of the coefficient field k, we may assume in this proof that k is algebraically closed. We only need to show that the generator $1 \otimes 1$ of $H \otimes_{H_{aff}} \chi$ is annihilated by $\mathfrak{J}$. In that case, Theorem 5.14 in [**19**] (when the root system is irreducible) and Corollary 6.13 in [**34**] state that, if a simple (left) H-module M contains a supersingular character χ of H_{aff}, then it is a supersingular H-module (in fact they state that this is an equivalence). The proof of the statement consists in picking an element m in M supporting the character χ and proving that $\mathfrak{J}$ acts by zero on it, the simplicity of M being used only to ensure that $\mathfrak{J}$ acts by zero on the whole M. Therefore the arguments apply to the left H-module $H \otimes_{H_{aff}} \chi$ when χ is a supersingular character of H_{aff} (although this module may not even be of finite length).

Now let $\chi : H_{aff} \to k$ be a twisted trivial (resp. sign) character of H_{aff}. As an H_{aff}-module, $H \otimes_{H_{aff}} \chi$ is isomorphic to a direct sum of copies of χ (Remark 2.12.iii). A nonzero H-module which is a subquotient of $H \otimes_{H_{aff}} \chi$ is therefore also a direct sum of copies of χ as an H_{aff}-module. Suppose that $H \otimes_{H_{aff}} \chi$ has a nonzero supersingular subquotient. Then it has a nonzero supersingular finitely generated subquotient. Since the latter admits a nonzero simple quotient, the H-module $H \otimes_{H_{aff}} \chi$ has a nonzero simple supersingular subquotient M. This is not compatible with M being a direct sum of copies of χ as an H_{aff}-module, as proved in [**19**] Lemma 5.12 when the root system is irreducible or in [**34**] Corollary 6.13.

The proof is the same for right H-modules. □

Define the decreasing filtration

$$F^n H := \oplus_{\ell(w) \geq n} k\tau_w \qquad \text{for } n \geq 0 \tag{2.38}$$

of H as a bimodule over itself.

LEMMA 2.14. *Under the hypothesis that $\mathbf{G}$ is semisimple simply connected with irreducible root system, we have:*

i. *As an H-module on the left and on the right, $F^m H / F^{m+1} H$, for any $m \geq 1$, is annihilated by $\mathfrak{J}$; in particular:*

$$\mathfrak{J}^{m-1} \cdot F^1 H = F^1 H \cdot \mathfrak{J}^{m-1} \subset F^m H \ .$$

ii. *As an H-module on the left and on the right, $(1 - e_1) \cdot (F^0 H / F^1 H)$ is annihilated by $\mathfrak{J}$; in particular:*

$$\mathfrak{J}^m \cdot [(1 - e_1) \cdot F^0 H + F^1 H] \subset F^m H \ .$$

PROOF. In this proof we consider left H-modules. The arguments are the same for the structures of right modules. Recall that under the hypothesis of the lemma, we have $W = W_{aff}$ and $H = H_{aff}$. Furthermore, we may assume that $\mathbb{F}_q \subseteq k$ and that $\mathbf{G} \neq 1$.

i. We will show that $F^m H / F^{m+1} H$, in fact, is a direct sum of supersingular characters. Since $\mathbb{F}_q \subseteq k$, a basis for $F^m H / F^{m+1} H$ is given by all $e_\lambda \tau_{\tilde{w}}$ for $w \in W$

with length m and lift $\tilde{w} \in \widetilde{W}$ and all $\lambda \in \widehat{T^0/T^1}$ (notation in §2.2.1). For $s = s_{(\alpha,\mathfrak{h})} \in S_{aff}$ we pick the lift $\tilde{s}$ as in (2.15). Using (2.33), (2.26) and (2.28):

$$\tau_{\tilde{s}} \cdot e_\lambda \tau_{\tilde{w}} = \begin{cases} e_{^s\lambda}\, \tau_{\tilde{s}}^2\, \tau_{\tilde{s}^{-1}\tilde{w}} = -\theta_s\, e_{^s\lambda}\, \tau_{\tilde{w}} & \text{if } \ell(sw) = \ell(w) - 1, \\ e_{^s\lambda} \tau_{\tilde{s}\tilde{w}} & \text{if } \ell(sw) = \ell(w) + 1. \end{cases}$$

So in $F^m H/F^{m+1}H$, we have, using (2.32):

$$\tau_{\tilde{s}} \cdot e_\lambda \tau_{\tilde{w}} = \begin{cases} 0 & \text{if } \ell(sw) = \ell(w) - 1 \text{ and } \lambda|_{\check{\alpha}([\mathbb{F}_q^\times])} \neq 1, \\ -e_{^s\lambda}\tau_{\tilde{w}} & \text{if } \ell(sw) = \ell(w) - 1 \text{ and } \lambda|_{\check{\alpha}([\mathbb{F}_q^\times])} = 1, \\ 0 & \text{if } \ell(sw) = \ell(w) + 1. \end{cases}$$

Using Lemma 2.9, notice that if λ is trivial on $\check{\alpha}([\mathbb{F}_q^\times])$, then ${}^s\lambda = \lambda$. This proves that $e_\lambda \tau_{\tilde{w}}$ supports the character $\chi : H_{aff} \to k$ defined by $\chi(\tau_t) = \lambda(t)$ for all $t \in T^0/T^1$ and

$$\chi(\tau_{\tilde{s}}) = \begin{cases} 0 & \text{if } \ell(sw) = \ell(w) - 1 \text{ and } \lambda|_{\check{\alpha}([\mathbb{F}_q^\times])} \neq 1 \text{ or if } \ell(sw) = \ell(w) + 1, \\ -1 & \text{if } \ell(sw) = \ell(w) - 1 \text{ and } \lambda|_{\check{\alpha}([\mathbb{F}_q^\times])} = 1 \end{cases}$$

for $s \in S_{aff}$. If there is $s_{(\alpha,\mathfrak{h})} \in S_{aff}$ such that $\lambda|_{\check{\alpha}([\mathbb{F}_q^\times])} \neq 1$ then χ is supersingular (Remark 2.12). Otherwise λ is trivial on T^0/T^1 and we want to check that χ is not a twisted sign or a twisted trivial character. This is because $m \geq 1$ and the root system of $\mathbf{G}$ is irreducible of rank > 0. Therefore there are $s, s' \in S_{aff}$ such that $\ell(sw) = \ell(w) - 1$ and $\ell(s'w) = \ell(w) + 1$. Since H_{aff} has only one irreducible component we see that χ is a supersingular character. By Lemma 2.13, the left H-module $F^m H/F^{m+1}H$ then is annihilated by $\mathfrak{J}$, which concludes the proof of i.

ii. Again we will show that $(1 - e_1) \cdot (F^0H/F^1H)$, in fact, is a direct sum of supersingular characters. A basis for $(1 - e_1) \cdot (F^0H/F^1H)$ is given by all e_λ for all $\lambda \in \widehat{T^0/T^1} \setminus \{1\}$. But $\tau_{\tilde{s}} e_\lambda \in F^1H$ by (2.26). This proves that e_λ supports the character $\chi : H_{aff} \to k$ defined by $\chi(\tau_t) = \lambda(t)$ (see (2.31)) and $\chi(\tau_{\tilde{s}}) = 0$ for $s \in S_{aff}$. It is not a twisted sign or a twisted trivial character since λ is nontrivial on $(T^0/T^1)' = T^0/T^1$. As the root system is irreducible, it is a supersingular character. Conclude using point i. and Lemma 2.13. □

3. The Ext-algebra

3.1. The definition. In order to introduce the algebra in the title we again start from the compact induction $\mathbf{X} = \mathrm{ind}_I^G(1)$ of the trivial I-representation, which lies in $\mathrm{Mod}(G)$. We form the graded Ext-algebra

$$E^* := \mathrm{Ext}^*_{\mathrm{Mod}(G)}(\mathbf{X}, \mathbf{X})^{\mathrm{op}}$$

over k with the multiplication being the (opposite of the) Yoneda product. Obviously

$$H := E^0 = \mathrm{End}^*_{\mathrm{Mod}(G)}(\mathbf{X}, \mathbf{X})^{\mathrm{op}}$$

is the usual pro-p Iwahori-Hecke algebra over k. By using Frobenius reciprocity for compact induction and the fact that the restriction functor from $\mathrm{Mod}(G)$ to $\mathrm{Mod}(I)$ preserves injective objects we obtain the identification

$$E^* = \mathrm{Ext}^*_{\mathrm{Mod}(G)}(\mathbf{X}, \mathbf{X})^{\mathrm{op}} = H^*(I, \mathbf{X}) \ . \tag{3.1}$$

The only part of the multiplicative structure on E^* which is still directly visible on the cohomology $H^*(I, \mathbf{X})$ is the right multiplication by elements in $E^0 = H$, which is functorially induced by the right action of H on $\mathbf{X}$. It is one of the main technical issues of this paper to make the full multiplicative structure visible on $H^*(I,\mathbf{X})$. We recall that for $* = 0$ the above identification is given by

$$\begin{aligned} H &\xrightarrow{\cong} \mathbf{X}^I \\ \tau &\longmapsto (\mathrm{char}_I)\tau \ . \end{aligned}$$

3.2. The technique. The technical tool for studying the algebra E^* is the I-equivariant decomposition

$$\mathbf{X} = \oplus_{w\in\widetilde{W}} \mathbf{X}(w)$$

introduced in section 2.2. Noting that the cohomology of profinite groups commutes with arbitrary sums, we obtain

$$H^*(I,\mathbf{X}) = \oplus_{w\in\widetilde{W}} H^*(I,\mathbf{X}(w)) \ .$$

Similarly as we write IwI, since this double coset only depends on the coset $w \in N(T)/T^1$, we will silently abuse notation in the following whenever something only depends on the coset w. We have the isomorphism of I-representations

$$\begin{aligned} \mathrm{ind}_I^{IwI}(1) &\xrightarrow{\cong} \mathrm{ind}_{I_w}^I(1) \\ f &\longmapsto \phi_f(a) := f(aw) \ . \end{aligned}$$

This gives rise to the left hand cohomological isomorphism

$$H^*(I,\mathbf{X}(w)) \xrightarrow{\cong} H^*(I,\mathrm{ind}_{I_w}^I(1)) \xrightarrow{\cong} H^*(I_w,k)$$

which we may combine with the right hand Shapiro isomorphism. For simplicity we will call in the following the above composite isomorphism the Shapiro isomorphism and denote it by Sh_w. Equivalently, it can be described as the composite map

$$\mathrm{Sh}_w : H^*(I,\mathbf{X}(w)) \xrightarrow{\mathrm{res}} H^*(I_w,\mathbf{X}(w)) \xrightarrow{H^*(I_w,\mathrm{ev}_w)} H^*(I_w,k) \tag{3.2}$$

where

$$\begin{aligned} \mathrm{ev}_w : \mathbf{X}(w) &\longrightarrow k \\ f &\longrightarrow f(w) \ . \end{aligned}$$

We leave it as an exercise to the reader to check that the map

$$\mathrm{Sh}_w^{-1} : H^*(I_w,k) \xrightarrow{\mathrm{i}_w} H^*(I_w,\mathbf{X}(w)) \xrightarrow{\mathrm{cores}} H^*(I,\mathbf{X}(w)) \ , \tag{3.3}$$

where

$$\begin{aligned} \mathrm{i}_w : k &\longrightarrow \mathbf{X}(w) \\ a &\longrightarrow a\,\mathrm{char}_{wI} \ , \end{aligned}$$

is the inverse of the Shapiro isomorphism Sh_w.

3.3. The cup product. There is a naive product structure on the cohomology $H^*(I, \mathbf{X})$. By multiplying maps we obtain the G-equivariant map

$$\begin{aligned} \mathbf{X} \otimes_k \mathbf{X} &\longrightarrow \mathbf{X} \\ f \otimes f' &\longmapsto ff' \ . \end{aligned}$$

It gives rise to the cup product

$$H^i(I, \mathbf{X}) \otimes_k H^j(I, \mathbf{X}) \xrightarrow{\cup} H^{i+j}(I, \mathbf{X}) \tag{3.4}$$

which, quite obviously, has the property that

$$H^i(I, \mathbf{X}(v)) \cup H^j(I, \mathbf{X}(w)) = 0 \quad \text{whenever } v \neq w. \tag{3.5}$$

On the other hand, since $\mathrm{ev}_w(ff') = \mathrm{ev}_w(f)\,\mathrm{ev}_w(f')$ and since the cup product is functorial and commutes with cohomological restriction maps, we have the commutative diagrams

$$\begin{array}{ccc} H^i(I, \mathbf{X}(w)) \otimes_k H^j(I, \mathbf{X}(w)) & \xrightarrow{\cup} & H^{i+j}(I, \mathbf{X}(w)) \\ \Big\downarrow {\scriptstyle \mathrm{Sh}_w \otimes \mathrm{Sh}_w} & & \Big\downarrow {\scriptstyle \mathrm{Sh}_w} \\ H^i(I_w, k) \otimes_k H^j(I_w, k) & \xrightarrow{\cup} & H^{i+j}(I_w, k) \end{array} \tag{3.6}$$

for any $w \in \widetilde{W}$, where the bottom row is the usual cup product on the cohomology algebra $H^*(I_w, k)$. In particular, we see that the cup product (3.4) is anticommutative.

4. Representing cohomological operations on resolutions

4.1. The Shapiro isomorphism. The Shapiro isomorphism (3.2) also holds for nontrivial coefficients provided we choose once and for all, as we will do in the following, a representative $\dot{w} \in N(T)$ for each $w \in \widetilde{W}$. Compact induction is an exact functor

$$\begin{aligned} \mathrm{ind}_I^G : \mathrm{Mod}(I) &\longrightarrow \mathrm{Mod}(G) \\ Y &\longmapsto \mathrm{ind}_I^G(Y) \ . \end{aligned}$$

Moreover, as before we have the decomposition $\mathrm{ind}_I^G(Y) = \oplus_{w \in \widetilde{W}} \mathrm{ind}_I^{IwI}(Y)$ and the isomorphism

$$\begin{aligned} \mathrm{ind}_I^{IwI}(Y) &\xrightarrow{\cong} \mathrm{ind}_{I_{\dot{w}}}^I(\dot{w}_* \,\mathrm{res}_{I_{\dot{w}^{-1}}}^I(Y)) \\ f &\longmapsto \phi_f(a) := f(a\dot{w}) \end{aligned}$$

as I-representations. On cohomology we obtain the commutative diagram

$$\begin{array}{ccccc} H^*(I, \mathrm{ind}_I^{IwI}(Y)) & \xrightarrow{\cong} & H^*(I, \mathrm{ind}_{I_w}^I(\dot{w}_* \,\mathrm{res}_{I_{w^{-1}}}^I(Y)) & \xrightarrow{\cong} & H^*(I_w, \dot{w}_* Y) \\ & {\scriptstyle \mathrm{res}} \searrow & & \nearrow {\scriptstyle H^*(I_w, \mathrm{ev}_{\dot{w}})} & \\ & & H^*(I_w, \mathrm{ind}_I^{IwI}(Y)) & & \end{array} \tag{4.1}$$

in which $\mathrm{ev}_{\dot{w}}$ now denotes the evaluation map in $\dot{w}$ and in which the composite map in the top row is an isomorphism denoted by $\mathrm{Sh}_{\dot{w}}$.

To lift this to the level of complexes we first make the following observation.

LEMMA 4.1. *If $\mathcal{J}$ is an injective object in* $\mathrm{Mod}(I)$ *then* $\mathrm{ind}_I^{IwI}(\mathcal{J})$, *for any* $w \in \widetilde{W}$, *is an injective object in* $\mathrm{Mod}(I)$ *as well.*

PROOF. We use the isomorphism $\mathrm{ind}_I^{IwI}(\mathcal{J}) \cong \mathrm{ind}_{I_w}^I(\dot{w}_* \operatorname{res}^I_{I_{w^{-1}}}(\mathcal{J}))$. As recalled before, the restriction functor to open subgroups preserves injective objects. The functor $\dot{w}_* : \mathrm{Mod}(I_{w^{-1}}) \xrightarrow{\simeq} \mathrm{Mod}(I_w)$ is an equivalence of categories and hence preserves injective objects. Finally, the induction functor from open subgroups of finite index also preserves injective objects (cf. [**32**] I.5.9.b)). □

Let now $k \xrightarrow{\sim} \mathcal{I}^\bullet$ and $k \xrightarrow{\sim} \mathcal{J}^\bullet$ be any two injective resolutions in $\mathrm{Mod}(I)$ of the trivial representation. By Lemma 4.1 then $\mathbf{X}(w) \xrightarrow{\sim} \mathrm{ind}_I^{IwI}(\mathcal{J}^\bullet)$ is an injective resolution in $\mathrm{Mod}(I)$ as well. Hence

$$H^*(I, \mathbf{X}(w)) = \mathrm{Hom}_{K(I)}(\mathcal{I}^\bullet, \mathrm{ind}_I^{IwI}(\mathcal{J}^\bullet)[*]) \ ,$$

i.e., any cohomology class in $H^*(I, \mathbf{X}(w))$ is of the form $[\alpha^\bullet]$ for some homomorphism of complexes $\alpha^\bullet : \mathcal{I}^\bullet \longrightarrow \mathrm{ind}_I^{IwI}(\mathcal{J}^\bullet)[*]$ in $\mathrm{Mod}(I)$ (which is unique up to homotopy). Composition with the evaluation map gives rise to the homomorphism of injective complexes

$$\mathrm{Sh}_{\dot{w}}(\alpha^\bullet) := \mathrm{ev}_{\dot{w}} \circ \alpha^\bullet : \mathcal{I}^\bullet \longrightarrow \dot{w}_* \mathcal{J}^\bullet[*]$$

in $\mathrm{Mod}(I_w)$ whose cohomology class is

$$\mathrm{Sh}_w([\alpha^\bullet]) \in H^*(I_w, k) = \mathrm{Hom}_{K(I_w)}(\mathcal{I}^\bullet, \dot{w}_* \mathcal{J}^\bullet[*]) \ .$$

In fact it will be more convenient later on to use the following modified version of the Shapiro isomorphism. For this we assume that $k \xrightarrow{\sim} \mathcal{J}^\bullet$ is actually an injective resolution in $\mathrm{Mod}(G)$ (and hence in $\mathrm{Mod}(I)$). Then the composite map

$$\mathrm{Sh}'_{\dot{w}}(\alpha^\bullet) : \mathcal{I}^\bullet \xrightarrow{\mathrm{Sh}_{\dot{w}}(\alpha^\bullet)} \dot{w}_* \mathcal{J}^\bullet[*] \xrightarrow{y \mapsto \dot{w} y} \mathcal{J}^\bullet[*] \tag{4.2}$$

is defined and is also a homomorphism of injective complexes in $\mathrm{Mod}(I_w)$ representing the same cohomology class as $\mathrm{Sh}_{\dot{w}}(\alpha^\bullet)$ but viewed in the group $H^*(I_w, k) = \mathrm{Hom}_{K(I_w)}(\mathcal{I}^\bullet, \mathcal{J}^\bullet[*])$, i.e., we have

$$[\mathrm{Sh}'_{\dot{w}}(\alpha^\bullet)] = [\mathrm{Sh}_{\dot{w}}(\alpha^\bullet)] = \mathrm{Sh}_w([\alpha^\bullet]) \ . \tag{4.3}$$

The homomorphism $\alpha^\bullet$ can be reconstructed from $\mathrm{Sh}'_{\dot{w}}(\alpha^\bullet)$ by the formula

$$\begin{aligned} \alpha^\bullet(x)(a\dot{w}b) &= b^{-1}(({}^{a^{-1}}(\alpha^\bullet(x)))(\dot{w})) = b^{-1}(\alpha^\bullet(a^{-1}x)(\dot{w})) \\ &= (a\dot{w}b)^{-1} a(\mathrm{Sh}'_{\dot{w}}(\alpha^\bullet)(a^{-1}x)) \end{aligned} \tag{4.4}$$

for any $x \in \mathcal{I}^\bullet$ and any $a, b \in I$.

4.2. The Yoneda product. Here we consider an injective resolution $\mathbf{X} \xrightarrow{\sim} \mathcal{I}^\bullet$ of our G-representation $\mathbf{X}$ in $\mathrm{Mod}(G)$. Then

$$E^* = \mathrm{Ext}^*_{\mathrm{Mod}(G)}(\mathbf{X}, \mathbf{X}) = \mathrm{Hom}_{D(G)}(\mathbf{X}, \mathbf{X}[*]) = \mathrm{Hom}_{K(I)}(\mathcal{I}^\bullet, \mathcal{I}^\bullet[*]) \ ,$$

and the Yoneda product is the obvious composition of homomorphisms of complexes (cf. [**8**] Cor. I.6.5). We recall, though, that our convention is to consider the opposite of this composition. For our purposes it is crucial to replace $\mathcal{I}^\bullet$ by a quasi-isomorphic complex constructed as follows.

We begin with an injective resolution $k \xrightarrow{\sim} \mathcal{J}^\bullet$ in $\mathrm{Mod}(I)$ of the trivial representation. Then $\mathbf{X} \xrightarrow{\sim} \mathrm{ind}_I^G(\mathcal{J}^\bullet) = \oplus_{w \in \widetilde{W}} \mathrm{ind}_I^{IwI}(\mathcal{J}^\bullet)$ is a resolution in $\mathrm{Mod}(G)$. By Lemma 4.1 each term $\mathrm{ind}_I^{IwI}(\mathcal{J}^\bullet)$ is injective in $\mathrm{Mod}(I)$. Since the cohomology functor $H^*(I,-)$ commutes with arbitrary sums in $\mathrm{Mod}(I)$ it follows that each term $\mathrm{ind}_I^G(\mathcal{J}^\bullet)$ is an $H^0(I,-)$-acyclic object in $\mathrm{Mod}(I)$. But by Frobenius reciprocity we have the isomorphism $\mathrm{Hom}_{\mathrm{Mod}(G)}(\mathrm{ind}_I^G(1),-) \cong H^0(I,-)$ of left exact functors on $\mathrm{Mod}(G)$. We conclude that $\mathbf{X} \xrightarrow{\sim} \mathrm{ind}_I^G(\mathcal{J}^\bullet)$ is a resolution of $\mathbf{X}$ in $\mathrm{Mod}(G)$ by objects which are acyclic for $\mathrm{Hom}_{\mathrm{Mod}(G)}(\mathbf{X},-)$. It follows that

$$
\begin{aligned}
(4.5)\qquad \mathrm{Ext}^*_{\mathrm{Mod}(G)}(\mathbf{X},\mathbf{X}) &= h^*(\mathrm{Hom}_{\mathrm{Mod}(G)}(\mathbf{X}, \mathrm{ind}_I^G(\mathcal{J}^\bullet))) \\
&= h^*(\mathrm{ind}_I^G(\mathcal{J}^\bullet)^I) \\
&= \oplus_{w\in\widetilde{W}}\, h^*(\mathrm{ind}_I^{IwI}(\mathcal{J}^\bullet)^I) \cong \oplus_{w\in\widetilde{W}}\, h^*(\mathrm{ind}_{I_w}^I(\dot{w}_*\,\mathrm{res}^I_{I_{w^{-1}}}(\mathcal{J}^\bullet))^I) \\
&\cong \oplus_{w\in\widetilde{W}}\, h^*((\mathcal{J}^\bullet)^{I_{w^{-1}}})\ .
\end{aligned}
$$

In order to lift these equalities to the level of complexes we consider the commutative diagram

$$
\begin{array}{ccc}
\mathrm{Ext}^*_{\mathrm{Mod}(G)}(\mathbf{X},\mathbf{X}) & & \\
\| & & \\
\mathrm{Hom}_{D(G)}(\mathrm{ind}_I^G(\mathcal{J}^\bullet), \mathrm{ind}_I^G(\mathcal{J}^\bullet)[*]) & = & h^*(\mathrm{Hom}_{\mathrm{Mod}(G)}(\mathbf{X}, \mathrm{ind}_I^G(\mathcal{J}^\bullet))) \\
\uparrow & & \uparrow \\
\mathrm{Hom}_{K(G)}(\mathrm{ind}_I^G(\mathcal{J}^\bullet), \mathrm{ind}_I^G(\mathcal{J}^\bullet)[*]) & & \cong \Big|\ \text{Frobenius reciprocity} \\
\text{Frobenius reciprocity}\ \Big\uparrow \cong & & \\
\mathrm{Hom}_{K(I)}(\mathcal{J}^\bullet, \mathrm{ind}_I^G(\mathcal{J}^\bullet)[*]) & \longrightarrow & H^*(I,\mathbf{X})) \\
\uparrow & & \| \\
\oplus_{w\in\widetilde{W}}\, \mathrm{Hom}_{K(I)}(\mathcal{J}^\bullet, \mathrm{ind}_I^{IwI}(\mathcal{J}^\bullet)[*]) & \xrightarrow{\ \cong\ } & \oplus_{w\in\widetilde{W}} H^*(I,\mathbf{X}(w)).
\end{array}
$$

The isomorphism in the bottom row is a consequence of Lemma 4.1. The computation (4.5) shows that the composite map in the first column is an isomorphism.

We point out that these two ways of representing E^* by homomorphisms of complexes, through $\mathcal{I}^\bullet$ and through $\mathrm{ind}_I^G(\mathcal{J}^\bullet)$, are related by the unique (up to homotopy) homomorphism of complexes in $\mathrm{Mod}(G)$ which makes the diagram

$$
\begin{array}{ccc}
 & & \mathrm{ind}_I^G(\mathcal{J}^\bullet) \\
 & \overset{\sim}{\nearrow} & \downarrow \\
\mathbf{X} & & \\
 & \underset{\sim}{\searrow} & \\
 & & \mathcal{I}^\bullet
\end{array}
$$

commutative.

Consider any classes $[\alpha^\bullet] \in H^i(I,\mathbf{X}(v))$ and $[\beta^\bullet] \in H^j(I,\mathbf{X}(w))$ represented by homomorphisms of complexes

$$
\begin{aligned}
\alpha^\bullet &: \mathcal{J}^\bullet \longrightarrow \mathrm{ind}_I^{IvI}(\mathcal{J}^\bullet)[i] \subseteq \mathrm{ind}_I^G(\mathcal{J}^\bullet)[i] \quad \text{and} \\
\beta^\bullet &: \mathcal{J}^\bullet \longrightarrow \mathrm{ind}_I^{IwI}(\mathcal{J}^\bullet)[j] \subseteq \mathrm{ind}_I^G(\mathcal{J}^\bullet)[j]\ ,
\end{aligned}
$$

respectively. According to the above diagram these induce, by Frobenius reciprocity, homomorphisms of complexes $\tilde{\alpha}^\bullet : \mathrm{ind}_I^G(\mathcal{J}^\bullet) \longrightarrow \mathrm{ind}_I^G(\mathcal{J}^\bullet)[i]$ and $\tilde{\beta}^\bullet : \mathrm{ind}_I^G(\mathcal{J}^\bullet) \longrightarrow \mathrm{ind}_I^G(\mathcal{J}^\bullet)[j]$ which represent our original classes when viewed in $\mathrm{Ext}^i_{\mathrm{Mod}(G)}(\mathbf{X},\mathbf{X})$ and $\mathrm{Ext}^j_{\mathrm{Mod}(G)}(\mathbf{X},\mathbf{X})$, respectively. The Yoneda product of the latter is represented by the composite $\tilde{\beta}^\bullet[i]\circ\tilde{\alpha}^\bullet$, which we may write as $\tilde{\beta}^\bullet[i]\circ\tilde{\alpha}^\bullet = \tilde{\gamma}^\bullet$ for a homomorphism of complexes $\gamma^\bullet : \mathcal{J}^\bullet \longrightarrow \mathrm{ind}_I^G(\mathcal{J}^\bullet)[i+j]$. By writing out the Frobenius reciprocity isomorphism we see that

$$\tilde{\beta}^\bullet : \mathrm{ind}_I^G(\mathcal{J}^\bullet) \longrightarrow \mathrm{ind}_I^G(\mathcal{J}^\bullet)[j] \qquad (4.6)$$
$$f \longmapsto \sum_{g\in G/I} g\beta^\bullet(f(g)) \ .$$

We deduce that, if f has support in the subset $S\subseteq G/I$, then $\tilde{\beta}^\bullet(f)=\sum_{g\in S} g\beta^\bullet(f(g))$ has support in $S\cdot IwI$. Applying this to functions in the image of $\alpha^\bullet$, which are supported in IvI, we obtain that $\gamma^\bullet$, in fact, is a homomorphism of complexes

$$\gamma^\bullet : \mathcal{J}^\bullet \longrightarrow \mathrm{ind}_I^{IvI\cdot IwI}(\mathcal{J}^\bullet)[i+j] \ .$$

This shows that, if $\cdot$ denotes the multiplication on $H^*(I,\mathbf{X})$ induced by the opposite of the Yoneda product, then we have $[\alpha^\bullet]\cdot[\beta^\bullet] = (-1)^{ij}[\gamma^\bullet]$ and hence

$$H^i(I,\mathbf{X}(v))\cdot H^j(I,\mathbf{X}(w)) \subseteq H^{i+j}(I,\mathrm{ind}_I^{IvI\cdot IwI}(1)) \ . \qquad (4.7)$$

4.3. The cup product. Let U be any profinite group. It is well known that, under the identification $H^*(U,k) = \mathrm{Ext}^*_{\mathrm{Mod}(U)}(k,k)$, the cup product pairing

$$H^i(U,k)\times H^j(U,k) \xrightarrow{\cup} H^{i+j}(U,k)$$

coincides with the Yoneda composition product

$$\mathrm{Ext}^i_{\mathrm{Mod}(U)}(k,k)\times \mathrm{Ext}^j_{\mathrm{Mod}(U)}(k,k) \xrightarrow{\circ} \mathrm{Ext}^{i+j}_{\mathrm{Mod}(U)}(k,k) \ .$$

For discrete groups this is, for example, explained in [**4**] V§4. The argument there uses projective resolutions and therefore cannot be generalized directly to profinite groups. Instead one may use the axiomatic approach in [**14**] Chap. IV (see also p. 136).

We will use this in the following way. Let $k \xrightarrow{\sim} \mathcal{I}^\bullet$, $k \xrightarrow{\sim} \mathcal{J}^\bullet$, and $k \xrightarrow{\sim} \mathcal{K}^\bullet$ be three injective resolutions in $\mathrm{Mod}(U)$. Any two cohomology classes $\alpha \in H^i(U,k)$ and $\beta \in H^j(U,k)$ can be represented by homomorphisms of complexes $\alpha^\bullet : \mathcal{J}^\bullet \to \mathcal{K}^\bullet[i]$ and $\beta : \mathcal{I}^\bullet \to \mathcal{J}^\bullet[j]$, respectively. Then $\alpha\cup\beta \in H^{i+j}(U,k)$ is represented by the composite $\alpha^\bullet[j]\circ\beta^\bullet : \mathcal{I}^\bullet \to \mathcal{K}^\bullet[i+j]$.

4.4. Conjugation. The cohomology of profinite groups is functorial in pairs (ξ, f) where $\xi : V' \to V$ is a continuous homomorphism of profinite groups and $f : M \to M'$ is a k-linear map between an M in $\mathrm{Mod}(V)$ and an M' in $\mathrm{Mod}(V')$ such that

$$f(\xi(g)m) = gf(m) \qquad \text{for any } g\in V \text{ and } m \in M.$$

One method to construct the corresponding map on cohomology $(\xi,f)^* : H^i(V,M) \to H^i(V',M')$ proceeds as follows. We pick injective resolutions $M \xrightarrow{\sim} \mathcal{I}^\bullet_M$ in $\mathrm{Mod}(V)$ and $M' \xrightarrow{\sim} \mathcal{I}^\bullet_{M'}$ in $\mathrm{Mod}(V')$. Via ξ, we view $M \xrightarrow{\sim} \mathcal{I}^\bullet_M$ as a resolution in $\mathrm{Mod}(V')$ and we see that f extends to a homomorphism of complexes $f^\bullet : \mathcal{I}^\bullet_M \to \mathcal{I}^\bullet_{M'}$ such that

$$f^i(\xi(g)x) = gf^i(x) \qquad \text{for any } i\geq 0,\ g\in V, \text{ and } x \in \mathcal{I}^i_M. \qquad (4.8)$$

Then

$$H^i(V,M) = \mathrm{Hom}_{K(V)}(k,\mathcal{I}_M^\bullet[i]) \xrightarrow{(\xi,f)^*} \mathrm{Hom}_{K(V')}(k,\mathcal{I}_{M'}^\bullet[i]) = H^i(V',M')$$

sends $\alpha^\bullet$ to $f^\bullet[i]\circ\alpha^\bullet$.

We are primarily interested in the case where $M=k$ and $M'=k$ are the trivial representations, $f=\mathrm{id}_k$, and ξ is an isomorphism. We simply write $\xi^* := (\xi,\mathrm{id}_k)^*$ in this case. We may then take $\mathcal{I}_{M'}^\bullet := \xi^*\mathcal{I}_k^\bullet$ to be $\mathcal{I}_k^\bullet$ but with V' acting through ξ and $f^\bullet := \mathrm{id}_{\mathcal{I}_k^\bullet}$. Let $k \xrightarrow{\sim} \mathcal{I}^\bullet$ be another injective resolution in $\mathrm{Mod}(V)$. We have the commutative diagram

$$\begin{array}{ccc}
\mathrm{Hom}_{K(V)}(k,\mathcal{I}_k^\bullet[i]) & \xrightarrow{\alpha^\bullet\mapsto\alpha^\bullet} & \mathrm{Hom}_{K(V')}(k,\xi^*\mathcal{I}_k^\bullet[i]) \\
\cong\uparrow & & \uparrow\cong \\
\mathrm{Hom}_{K(V)}(\mathcal{I}^\bullet,\mathcal{I}_k^\bullet[i]) & \xrightarrow{\alpha^\bullet\mapsto\alpha^\bullet} & \mathrm{Hom}_{K(V')}(\xi^*\mathcal{I}^\bullet,\xi^*\mathcal{I}_k^\bullet[i]).
\end{array}$$

In other words, if the cohomology class $\alpha \in H^i(V,k)$ is represented by the homomorphism of complexes $\alpha^\bullet : \mathcal{I}^\bullet \to \mathcal{I}_k^\bullet[i]$, then its image $\xi^*\alpha \in H^i(V',k)$ is represented by

$$\xi^*\alpha^\bullet := \alpha^\bullet : \xi^*\mathcal{I}^\bullet \to \xi^*\mathcal{I}_k^\bullet[i] \tag{4.9}$$

(viewed as a V'-equivariant homomorphism via ξ).

A specific instance of this situation is the following. Assume that the profinite group V is a subgroup of some topological group H, let $h\in H$ be a fixed element, $V' := hVh^{-1}$, and $\xi : V' \to V$ be the isomorphism given by conjugation by h^{-1}. We then write

$$h_* = (h^{-1})^* := \xi^* : H^i(V,k) \longrightarrow H^i(hVh^{-1},k)$$

for the map on cohomology and $h_*\alpha^\bullet$ for the representing homomorphisms. We suppose now that V is open in H, in which case there is the following alternative description. We choose injective resolutions $k \xrightarrow{\sim} \mathcal{I}^\bullet$ and $k \xrightarrow{\sim} \mathcal{J}^\bullet$ in $\mathrm{Mod}(H)$. Then they are also injective resolutions in $\mathrm{Mod}(V)$ and $\mathrm{Mod}(V')$, so that we may take $\mathcal{I}_k^\bullet := \mathcal{J}^\bullet$. The map $f^\bullet : \mathcal{J}^\bullet \xrightarrow{h\cdot} \mathcal{J}^\bullet$ satisfies the condition (4.8), and we obtain

$$\begin{aligned}
H^i(V,k) = \mathrm{Hom}_{K(V)}(k,\mathcal{J}^\bullet[i]) &\xrightarrow{h_*} \mathrm{Hom}_{K(hVh^{-1})}(k,\mathcal{J}^\bullet[i]) = H^i(hVh^{-1},k) \\
\alpha^\bullet &\longmapsto h\alpha^\bullet .
\end{aligned}$$

This time one checks that the diagram

$$\begin{array}{ccc}
\mathrm{Hom}_{K(V)}(k,\mathcal{J}^\bullet[i]) & \xrightarrow{\alpha^\bullet\mapsto h\alpha^\bullet} & \mathrm{Hom}_{K(hVh^{-1})}(k,\mathcal{J}^\bullet[i]) \\
\cong\uparrow & & \uparrow\cong \\
\mathrm{Hom}_{K(V)}(\mathcal{I}^\bullet,\mathcal{J}^\bullet[i]) & \xrightarrow{\alpha^\bullet\mapsto h\alpha^\bullet h^{-1}} & \mathrm{Hom}_{K(hVh^{-1})}(\mathcal{I}^\bullet,\mathcal{J}^\bullet[i])
\end{array}$$

is commutative. We conclude that, if the cohomology class $\alpha \in H^i(V,k)$ is represented by the homomorphism of complexes $\alpha^\bullet : \mathcal{I}^\bullet \to \mathcal{J}^\bullet[i]$, then its image $h_*\alpha \in H^i(hVh^{-1},k)$ is represented by

$$h_*\alpha^\bullet := h\alpha^\bullet h^{-1} : \mathcal{I}^\bullet \to \mathcal{J}^\bullet[i] . \tag{4.10}$$

4.5. The corestriction. Let U be a profinite group with open subgroup $V \subseteq U$ and let M be in $\mathrm{Mod}(U)$. In this situation we have the corestriction map $\mathrm{cores}_U^V : H^*(V,M) \to H^*(U,M)$. It can be constructed as follows (cf. [**17**] I§5.4). Let $M \xrightarrow{\sim} \mathcal{I}_M^\bullet$ be an injective resolution in $\mathrm{Mod}(U)$. Then

$$\begin{aligned} \mathrm{cores}_U^V : H^i(V,M) = \mathrm{Hom}_{K(V)}(k, \mathcal{I}_M^\bullet[i]) &\longrightarrow \mathrm{Hom}_{K(U)}(k, \mathcal{I}_M^\bullet[i]) = H(U,M) \\ \alpha^\bullet &\longmapsto \sum_{g \in U/V} g\alpha^\bullet . \end{aligned}$$

For a variant of this, which we will need, let $k \xrightarrow{\sim} \mathcal{I}^\bullet$ be an injective resolution in $\mathrm{Mod}(U)$. One easily checks that the diagram

$$\begin{array}{ccc} \mathrm{Hom}_{K(V)}(k, \mathcal{I}_M^\bullet[i]) & \xrightarrow{\alpha^\bullet \mapsto \sum_{g\in U/V} g\alpha^\bullet} & \mathrm{Hom}_{K(U)}(k, \mathcal{I}_M^\bullet[i]) \\ \cong \uparrow & & \uparrow \cong \\ \mathrm{Hom}_{K(V)}(\mathcal{I}^\bullet, \mathcal{I}_M^\bullet[i]) & \xrightarrow{\alpha^\bullet \mapsto \sum_{g\in U/V} g\alpha^\bullet g^{-1}} & \mathrm{Hom}_{K(U)}(\mathcal{I}^\bullet, \mathcal{I}_M^\bullet[i]). \end{array}$$

is commutative. This means that, if the cohomology class $\alpha \in H^i(V,M)$ is represented by the homomorphism of complexes $\alpha^\bullet : \mathcal{I}^\bullet \to \mathcal{I}_M^\bullet[i]$, then its image $\mathrm{cores}_U^V(\alpha) \in H^i(U,M)$ is represented by

$$\sum_{g \in U/V} g\alpha^\bullet g^{-1} : \mathcal{I}^\bullet \to \mathcal{I}_M^\bullet[i] . \tag{4.11}$$

4.6. Basic properties. For later reference we record from [**17**] Prop. 1.5.4 that, on cohomology, restriction as well as corestriction commute with conjugation, and from [**17**] Prop. 1.5.3(iv) that the projection formulas

$$\begin{aligned} \mathrm{cores}_U^V(\alpha \cup \mathrm{res}_V^U(\beta)) &= \mathrm{cores}_U^V(\alpha) \cup \beta \quad \text{and} \\ \mathrm{cores}_U^V(\mathrm{res}_V^U(\beta) \cup \alpha) &= \beta \cup \mathrm{cores}_U^V(\alpha) \end{aligned}$$

hold when V is an open subgroup of the profinite group U and $\alpha \in H^*(V,M)$, $\beta \in H^*(U,M)$.

5. The product in E^*

5.1. A technical formula relating the Yoneda and cup products. We fix classes $[\alpha^\bullet] \in H^i(I, \mathbf{X}(v))$ and $[\beta^\bullet] \in H^j(I, \mathbf{X}(w))$ represented by homomorphisms of complexes

$$\alpha^\bullet : \mathcal{J}^\bullet \longrightarrow \mathrm{ind}_I^{IvI}(\mathcal{J}^\bullet)[i] \quad \text{and} \quad \beta^\bullet : \mathcal{J}^\bullet \longrightarrow \mathrm{ind}_I^{IwI}(\mathcal{J}^\bullet)[j] ,$$

respectively. Here we always take $k \xrightarrow{\sim} \mathcal{I}^\bullet$ and $k \xrightarrow{\sim} \mathcal{J}^\bullet$ to be injective resolutions in $\mathrm{Mod}(G)$ (and hence in $\mathrm{Mod}(I)$). By (4.6) and (4.7) their Yoneda product $[\gamma^\bullet] := (-1)^{ij}[\alpha^\bullet] \cdot [\beta^\bullet]$ is represented by the homomorphism

$$\begin{aligned} \gamma^\bullet : \mathcal{J}^\bullet &\longrightarrow \mathrm{ind}_I^{IvI \cdot IwI}(\mathcal{J}^\bullet)[i+j] \\ x &\longmapsto \sum_{g \in IvI/I} g\beta^\bullet[i](\alpha^\bullet(x)(g)) . \end{aligned}$$

In fact, we introduce, for any $u \in \widetilde{W}$ such that $IuI \subseteq IvI \cdot IwI$, the homomorphism

$$\gamma_u^\bullet(-) := \gamma^\bullet(-)_{|IuI} : \mathcal{J}^\bullet \longrightarrow \mathrm{ind}_I^{IuI}(\mathcal{J}^\bullet)[i+j] \ .$$

Then

$$(-1)^{ij}[\alpha^\bullet] \cdot [\beta^\bullet] = \sum_{IuI \subseteq IvI \cdot IwI} [\gamma_u^\bullet] \ . \tag{5.1}$$

Our goal here is to give a formula for the class $[\mathrm{Sh}'_{\dot{u}}(\gamma_u^\bullet)] = \mathrm{Sh}_u([\gamma_u^\bullet]) \in H^{*+i+j}(I_u, k)$ (cf. (4.3)) in terms of group cohomological operations. We fix throughout a $u \in \widetilde{W}$ such that $IuI \subseteq IvI \cdot IwI$.

REMARK 5.1. The map

$$\{a \in I/I_v : v^{-1}au \in IwI\} \xrightarrow{\simeq} I_{v^{-1}} \backslash (v^{-1}Iu \cap IwI)$$
$$a \longmapsto v^{-1}a^{-1}\dot{u}$$

is a well defined bijection.

PROOF. The map is well defined since $v^{-1}I_v = I_{v^{-1}}v^{-1}$. It is obviously surjective. For injectivity suppose that $I_{v^{-1}}v^{-1}a\dot{u} = I_{v^{-1}}v^{-1}b\dot{u}$ for some $a, b \in I$. Then $v^{-1}I_v a^{-1} = v^{-1}I_v b^{-1}$ and hence $aI_v = bI_v$. □

Using the above formula for $\gamma^\bullet$ and Remark 5.1 we compute

$$\begin{aligned} \mathrm{Sh}'_{\dot{u}}(\gamma_u^\bullet)(x) &= \dot{u}(\gamma_u^\bullet(x)(\dot{u})) = \dot{u}(\gamma^\bullet(x)(\dot{u})) \\ &= \dot{u}\Big(\Big(\sum_{a \in I/I_v} av\beta^\bullet[i](\alpha^\bullet(x)(av))\Big)(\dot{u})\Big) \\ &= \sum_{a \in I/I_v} \dot{u}(\beta^\bullet[i](\alpha^\bullet(x)(av))(v^{-1}a^{-1}\dot{u})) \\ &= \sum_{a \in I/I_v, v^{-1}a^{-1}u \in IwI} \dot{u}(\beta^\bullet[i](\alpha^\bullet(x)(av))(v^{-1}a^{-1}\dot{u})) \\ &= \sum_{h \in I_{v^{-1}} \backslash (v^{-1}Iu \cap IwI)} \dot{u}(\beta^\bullet[i](\alpha^\bullet(x)(\dot{u}h^{-1}))(h)) \ . \end{aligned} \tag{5.2}$$

We fix, for the moment, an element $h \in v^{-1}Iu \cap IwI$ written as $h = c\dot{w}d = \dot{v}^{-1}a^{-1}\dot{u}$ with $a, c, d \in I$ and put

$$\begin{aligned} \Gamma^\bullet_{\dot{u},h}(x) &:= \dot{u}(\beta^\bullet[i](\alpha^\bullet(x)(\dot{u}h^{-1}))(h)) = \dot{u}(\beta^\bullet[i](\alpha^\bullet(x)(a\dot{v}))(c\dot{w}d)) \\ &= \dot{u}h^{-1}({}^{c_*}\,\mathrm{Sh}'_{\dot{w}}(\beta^\bullet[i])(\alpha^\bullet(x)(a\dot{v}))) \\ &= \dot{u}h^{-1}({}^{c_*}\,\mathrm{Sh}'_{\dot{w}}(\beta^\bullet[i])(\dot{v}^{-1}a^{-1}({}^{a_*}\,\mathrm{Sh}'_{\dot{v}}(\alpha^\bullet)(x)))) \\ &= \dot{u}h^{-1}\dot{v}^{-1}a^{-1}({}^{(a\dot{v})_* c_*}\,\mathrm{Sh}'_{\dot{w}}(\beta^\bullet[i])({}^{a_*}\,\mathrm{Sh}'_{\dot{v}}(\alpha^\bullet)(x))) \\ &= {}^{(a\dot{v}c)_*}\,\mathrm{Sh}'_{\dot{w}}(\beta^\bullet[i])({}^{a_*}\,\mathrm{Sh}'_{\dot{v}}(\alpha^\bullet)(x)) \ . \end{aligned}$$

In the second and third line we have used the reconstruction formula (4.4) for $\beta^\bullet$ and $\alpha^\bullet$, respectively, as well as (4.10).

LEMMA 5.2. i. *In the commutative diagram of surjective projection maps*

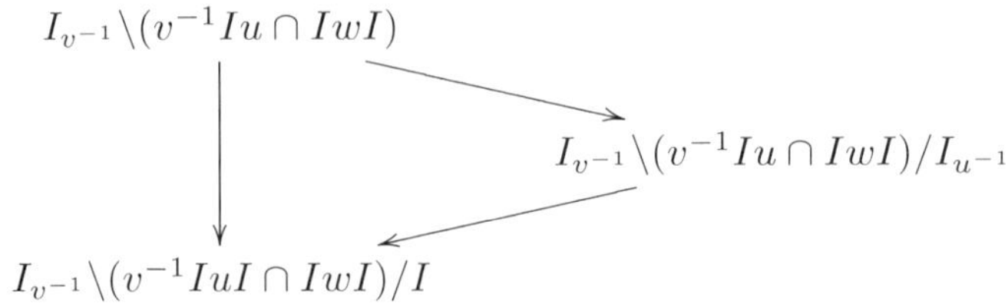

the lower oblique arrow is bijective.

ii. *Let* $h \in v^{-1}Iu \cap IwI$.
The map $b \mapsto I_{v^{-1}}hb$ *from the set* $(I_{u^{-1}} \cap h^{-1}Ih)\backslash I_{u^{-1}}$ *to the fiber of the projection map* $I_{v^{-1}}\backslash(v^{-1}Iu \cap IwI) \twoheadrightarrow I_{v^{-1}}\backslash(v^{-1}Iu \cap IwI)/I_{u^{-1}}$ *in the point* $I_{v^{-1}}hI_{u^{-1}}$ *is a bijection.*

PROOF. i. Let $I_{v^{-1}}hI = I_{v^{-1}}h'I$ with $h = \dot{v}^{-1}a^{-1}\dot{u}$, $h' = \dot{v}^{-1}a'^{-1}\dot{u}$, and $a, a' \in I$. Then $I_v a^{-1}\dot{u}I = I_v a'^{-1}\dot{u}I$ and hence $a'^{-1} = Aa^{-1}\dot{u}B\dot{u}^{-1}$ for some $A \in I_v$ and $B \in I$. It follows that $B \in I_{u^{-1}}$ and $h' = \dot{v}^{-1}Aa^{-1}\dot{u}B\dot{u}^{-1}\dot{u} = (\dot{v}^{-1}A\dot{v})hB \in I_{v^{-1}}hI_{u^{-1}}$.

ii. The equality $I_{v^{-1}}hb = I_{v^{-1}}h$ for some $b \in I_{u^{-1}}$ is equivalent to

$$b \in h^{-1}I_{v^{-1}}h = h^{-1}Ih \cap h^{-1}v^{-1}Ivh = h^{-1}Ih \cap u^{-1}Iu \ .$$

But the latter is equivalent to $b \in I \cap u^{-1}Iu \cap h^{-1}Ih = I_{u^{-1}} \cap h^{-1}Ih$. □

Coming back to $\Gamma^{\bullet}_{\dot{u},h}$ we note that, for $b \in I_{u^{-1}}$, we have $hb = cw(db) = \dot{v}^{-1}(\dot{u}b^{-1}\dot{u}^{-1}a)^{-1}\dot{u}$, where $c, db, \dot{u}b^{-1}\dot{u}^{-1}a \in I$. It follows that

$$\Gamma^{\bullet}_{\dot{u},hb}(x) = \dot{u}b^{-1}\dot{u}^{-1}\Gamma^{\bullet}_{\dot{u},h}(\dot{u}b\dot{u}^{-1}x) \ . \tag{5.3}$$

By inserting (5.3) into (5.2) and by using Lemma 5.2 we obtain

$$\begin{aligned} \mathrm{Sh}'_{\dot{u}}(\gamma^{\bullet}_{u})(x) &= \sum_{h \in I_{v^{-1}}\backslash(v^{-1}Iu \cap IwI)} \Gamma^{\bullet}_{\dot{u},h}(x) \\ &= \sum_{h \in I_{v^{-1}}\backslash(v^{-1}Iu \cap IwI)/I_{u^{-1}}} \ \sum_{b \in (I_{u^{-1}} \cap h^{-1}Ih)\backslash I_{u^{-1}}} \dot{u}b^{-1}\dot{u}^{-1}\Gamma^{\bullet}_{\dot{u},h}(\dot{u}b\dot{u}^{-1}x) \\ &= \sum_{h \in I_{v^{-1}}\backslash(v^{-1}Iu \cap IwI)/I_{u^{-1}}} \ \sum_{b \in (I_u \cap \dot{u}h^{-1}Ih\dot{u}^{-1})\backslash I_u} b^{-1}\Gamma^{\bullet}_{\dot{u},h}(bx) \ . \end{aligned} \tag{5.4}$$

Above and in the following every summation over h is understood to be over a chosen set of representatives in G of the respective double cosets. It also follows from (5.3) that

$$b\Gamma^{\bullet}_{\dot{u},h}(-) = \Gamma^{\bullet}_{\dot{u},h}(b-) \qquad \text{for any } b \in I_u \cap \dot{u}h^{-1}Ih\dot{u}^{-1}.$$

This says that $\Gamma^\bullet_{\dot u,h} : \mathcal{J}^\bullet \longrightarrow \mathcal{J}^\bullet[*+i+j]$ is a homomorphism of injective complexes in $\mathrm{Mod}(I_u \cap \dot u h^{-1} I h \dot u^{-1})$ and therefore defines a cohomology class $[\Gamma^\bullet_{\dot u,h}] \in H^{*+i+j}(I_u \cap \dot u h^{-1} I h \dot u^{-1}, k)$. By (4.11) the equality (5.4) then gives rise on cohomology to the equality

$$(5.5)\quad \mathrm{Sh}_u([\gamma_u^\bullet]) = [\mathrm{Sh}'_{\dot u}(\gamma_u^\bullet)] = \sum_{h \in I_{v^{-1}} \backslash (v^{-1} I u \cap I w I)/I_{u^{-1}}} \mathrm{cores}_{I_u}^{I_u \cap \dot u h^{-1} I h \dot u^{-1}} ([\Gamma^\bullet_{\dot u,h}])$$

in $H^{*+i+j}(I_u, k)$.

We recall that $h = c\dot w d = \dot v^{-1} a^{-1} \dot u$ with $a, c, d \in I$ and

$$\Gamma^\bullet_{\dot u,h}(x) = {}^{(a\dot v c)_*}\,\mathrm{Sh}'_{\dot w}(\beta^\bullet[i])({}^{a_*}\,\mathrm{Sh}'_{\dot v}(\alpha^\bullet)(x)) .$$

Note that both groups $aI_v a^{-1} = I \cap \dot u h^{-1} I h \dot u^{-1}$ and $(a\dot v c) I_w (a\dot v c)^{-1} = uIu^{-1} \cap \dot u h^{-1} I h \dot u^{-1}$ contain $I_u \cap \dot u h^{-1} I h \dot u^{-1}$; in fact the latter is the intersection of the former two. Therefore the above identity should, more precisely, be written as

$$\Gamma^\bullet_{\dot u,h} = \mathrm{res}^{uIu^{-1} \cap \dot u h^{-1} I h \dot u^{-1}}_{I_u \cap \dot u h^{-1} I h \dot u^{-1}} \left({}^{(a\dot v c)_*}\,\mathrm{Sh}'_{\dot w}(\beta^\bullet[i])\right) \circ \mathrm{res}^{I \cap \dot u h^{-1} I h \dot u^{-1}}_{I_u \cap \dot u h^{-1} I h \dot u^{-1}} \left({}^{a_*}\,\mathrm{Sh}'_{\dot v}(\alpha^\bullet)\right).$$

Using Subsection 4.3 as well as (4.3) we deduce that on cohomology classes we have the equality

$$(5.6)\quad [\Gamma^\bullet_{\dot u,h}] = \mathrm{res}^{uIu^{-1} \cap \dot u h^{-1} I h \dot u^{-1}}_{I_u \cap \dot u h^{-1} I h \dot u^{-1}} \left((a\dot v c)_*\,\mathrm{Sh}_w([\beta^\bullet])\right) \cup \mathrm{res}^{I \cap \dot u h^{-1} I h \dot u^{-1}}_{I_u \cap \dot u h^{-1} I h \dot u^{-1}} \left(a_*\,\mathrm{Sh}_v([\alpha^\bullet])\right)$$

in $H^{*+i+j}(I_u \cap \dot u h^{-1} I h \dot u^{-1}, k)$.

PROPOSITION 5.3. *For any cohomology classes $\alpha \in H^i(I, \mathbf{X}(v))$ and $\beta \in H^j(I, \mathbf{X}(w))$ we have $\alpha \cdot \beta = \sum_{u \in \widetilde W, IuI \subseteq IvI \cdot IwI} \gamma_u$ with uniquely determined $\gamma_u \in H^{i+j}(I, \mathbf{X}(u))$ and*

$$\mathrm{Sh}_u(\gamma_u) = \sum_{h \in I_{v^{-1}} \backslash (v^{-1} I u \cap I w I)/I_{u^{-1}}} \mathrm{cores}_{I_u}^{I_u \cap \dot u h^{-1} I h \dot u^{-1}} (\tilde\Gamma_{u,h})$$

with

$$\tilde\Gamma_{u,h} := \mathrm{res}^{I \cap \dot u h^{-1} I h \dot u^{-1}}_{I_u \cap \dot u h^{-1} I h \dot u^{-1}} \left(a_*\,\mathrm{Sh}_v(\alpha)\right) \cup \mathrm{res}^{uIu^{-1} \cap \dot u h^{-1} I h \dot u^{-1}}_{I_u \cap \dot u h^{-1} I h \dot u^{-1}} \left((a\dot v c)_*\,\mathrm{Sh}_w(\beta)\right),$$

where $h = c\dot w d = \dot v^{-1} a^{-1} \dot u$ with $a, c, d \in I$.

PROOF. Insert (5.6) into (5.5) and use the anticommutativity of the cup product together with (5.1). □

REMARK 5.4. Let $u, v, w \in \widetilde W$ such that $IuI \subseteq IvI \cdot IwI$. Suppose that $I_{v^{-1}} \backslash (v^{-1} I u \cap I w I)/I_{u^{-1}}$ contains a single element $I_{v^{-1}} h I_{u^{-1}}$ and that $I_u \subset \dot u h^{-1} I h \dot u^{-1}$. Then for any cohomology classes $\alpha \in H^i(I, \mathbf{X}(v))$ and $\beta \in H^j(I, \mathbf{X}(w))$ the component γ_u of $\alpha \cdot \beta$ in $H^{i+j}(I, \mathbf{X}(u))$ is such that

$$\mathrm{Sh}_u(\gamma_u) = \mathrm{res}^{I \cap \dot u h^{-1} I h \dot u^{-1}}_{I_u} \left(a_*\,\mathrm{Sh}_v(\alpha)\right) \cup \mathrm{res}^{uIu^{-1} \cap \dot u h^{-1} I h \dot u^{-1}}_{I_u} \left((a\dot v c)_*\,\mathrm{Sh}_w(\beta)\right)$$

with a and c as in the proposition. If $i = 0$ and $\alpha = \tau_v$ (resp. $j = 0$ and $\beta = \tau_w$) it is simply equal to $\mathrm{res}^{uIu^{-1} \cap \dot u h^{-1} I h \dot u^{-1}}_{I_u}\left((a\dot v c)_* \mathrm{Sh}_w(\beta)\right)$ (resp. $\mathrm{res}^{I \cap \dot u h^{-1} I h \dot u^{-1}}_{I_u}\left(a_* \mathrm{Sh}_v(\alpha)\right)$) because τ_v (resp. τ_w) corresponds to the constant function equal to 1 in $H^0(I_v, k)$ (resp. $H^0(I_w, k)$). Therefore for general α and β in the context of this remark, the components of $\alpha \cdot \beta$ and of $\alpha \cdot \tau_w \cup \tau_v \cdot \beta$ in $H^{i+j}(I, \mathbf{X}(u))$ coincide.

COROLLARY 5.5. *Let $v, w \in \widetilde{W}$ such that $\ell(vw) = \ell(v)+\ell(w)$. For any cohomology classes $\alpha \in H^i(I, \mathbf{X}(v))$ and $\beta \in H^j(I, \mathbf{X}(w))$ we have $\alpha \cdot \beta \in H^{i+j}(I, \mathbf{X}(vw))$, and*

$$(5.7) \qquad \alpha \cdot \beta = (\alpha \cdot \tau_w) \cup (\tau_v \cdot \beta) \ ,$$

where we use the cup product in the sense of subsection 3.3; moreover

$$(5.8) \quad \mathrm{Sh}_{vw}(\alpha \cdot \tau_w) = \mathrm{res}^{I_v}_{I_{vw}} \left(\mathrm{Sh}_v(\alpha)\right) \text{ and } \mathrm{Sh}_{vw}(\tau_v \cdot \beta) = \mathrm{res}^{vI_wv^{-1}}_{I_{vw}} \left(v_* \mathrm{Sh}_w(\beta)\right).$$

PROOF. We note before starting the proof that, for $x \in \widetilde{W}$, we have $\mathrm{Sh}_x(\tau_x) = 1 \in H^0(I_x, k) = k$. If $\ell(vw) = \ell(v) + \ell(w)$, then $IvI \cdot IwI = IvwI$ (cf. Cor. 2.5.ii) and there is only $u = vw$ to consider in Prop. 5.3. We have $\alpha \cdot \tau_w \in H^i(I, \mathbf{X}(vw))$ and $\tau_v \cdot \beta \in H^j(I, \mathbf{X}(vw))$ and $\alpha \cdot \beta \in H^{i+j}(I, \mathbf{X}(vw))$ by Prop. 5.3.

From Lemma 5.2.i we know that the projection map

$$I_{v^{-1}} \backslash (v^{-1} Ivw \cap IwI)/I_{(vw)^{-1}} \xrightarrow{\sim} I_{v^{-1}} \backslash (v^{-1} IvwI \cap IwI)/I$$

is a bijection. On the other hand we have the inclusions

$$IwI \subseteq I_{v^{-1}} wI \subseteq v^{-1} IvwI \cap IwI \ ,$$

the left one coming from (2.7) and the right one being trivial. It follows that, in fact,

$$I_{v^{-1}} wI = v^{-1} IvwI \cap IwI \ .$$

Furthermore, it is straightforward to check

$$I_{v^{-1}} wI_{(vw)^{-1}} \subseteq v^{-1} Ivw \cap IwI \subseteq I_{v^{-1}} wI \ .$$

We claim that the left inclusion actually is an equality. Let $g \in v^{-1} Ivw \cap IwI$ be an arbitrary element. Using the second inclusion above we may find a $g_0 \in I_{v^{-1}} w \subseteq v^{-1} Ivw \cap IwI$ and a $g_1 \in I$ such that $g = g_0 g_1$. The above bijectivity then implies that necessarily $g_1 \in I_{(vw)^{-1}}$. We conclude that $g \in I_{v^{-1}} wI_{(vw)^{-1}}$. This proves that indeed

$$v^{-1} Ivw \cap IwI = I_{v^{-1}} wI_{(vw)^{-1}} \ .$$

Hence it suffices to consider $h = \dot{w}$. We have $I_u \cap \dot{u}h^{-1} Ih\dot{u}^{-1} = I_u \cap I_v = I_{vw}$ by (2.6). Using Remark 5.4, we have proved (5.7). Moreover, we have $I \cap \dot{u}h^{-1} Ih\dot{u}^{-1} = I_v$ and $uIu^{-1} \cap \dot{u}h^{-1} Ih\dot{u}^{-1} = vI_w v^{-1}$ for obvious reasons. Furthermore, we may take $c = 1$ and $a \in T^1$ such that $\dot{u} = a\dot{v}\dot{w}$. Note that a is contained in both $vI_w v^{-1}$ and I_v so its acts trivially on the cohomology spaces $H^i(I_v, k)$ and $H^j(vI_w v^{-1}, k)$. Therefore in this case the formula of Proposition 5.3 gives:

$$\mathrm{Sh}_{vw}(\alpha \cdot \tau_w) = \mathrm{res}^{I_v}_{I_{vw}} \left(\mathrm{Sh}_v(\alpha)\right) \quad \text{and} \quad \mathrm{Sh}_{vw}(\tau_v \cdot \beta) = \mathrm{res}^{vI_wv^{-1}}_{I_{vw}} \left(\dot{v}_* \mathrm{Sh}_w(\beta)\right)$$

and

$$\mathrm{Sh}_{vw}(\alpha \cdot \beta) = \mathrm{Sh}_{vw}(\alpha \cdot \tau_w) \cup \mathrm{Sh}_{vw}(\tau_v \cdot \beta) \ .$$

∎

5.2. Explicit left action of H on the Ext-algebra. Here we draw from Prop. 5.3 the formula for the explicit left action of H on E^*. The proposition and its proof use notation introduced in §2.1.6. See in particular Remark 2.6.

PROPOSITION 5.6. *Let $\beta \in H^j(I, \mathbf{X}(w))$ with $w \in \widetilde{W}$ and $j \geq 0$. For $\omega \in \widetilde{\Omega}$, we have $\tau_\omega \cdot \beta \in H^j(I, \mathbf{X}(\omega w))$ and*

$$\mathrm{Sh}_{\omega w}(\tau_\omega \cdot \beta) = \omega_* \,\mathrm{Sh}_w(\beta)\ . \tag{5.9}$$

For $s = s_{(\alpha,\mathfrak{h})} \in S_{aff}$ we have:
- *either $\ell(\tilde{s}w) = \ell(w) + 1$ and $\tau_{\tilde{s}} \cdot \beta \in H^j(I, \mathbf{X}(\tilde{s}w))$ with*

$$\mathrm{Sh}_{\tilde{s}w}(\tau_{\tilde{s}} \cdot \beta) = \mathrm{res}^{\tilde{s}I_w\tilde{s}^{-1}}_{I_{sw}} \left(\tilde{s}_* \,\mathrm{Sh}_w(\beta)\right)\ , \tag{5.10}$$

- *or $\ell(\tilde{s}w) = \ell(w) - 1$ and*

$$\tau_{\tilde{s}} \cdot \beta = \gamma_{\tilde{s}w} + \sum_{t \in \check{\alpha}([\mathbb{F}_q^\times])} \gamma_{\bar{t}w} \in H^j(I, \mathbf{X}(\tilde{s}w)) \oplus \bigoplus_{t \in \check{\alpha}([\mathbb{F}_q^\times])} H^j(I, \mathbf{X}(\bar{t}w)) \tag{5.11}$$

with

$$\mathrm{Sh}_{\tilde{s}w}(\gamma_{\tilde{s}w}) = \mathrm{cores}^{\tilde{s}I_w\tilde{s}^{-1}}_{I_{\tilde{s}w}} \left(\tilde{s}_* \,\mathrm{Sh}_w(\beta)\right) \ \text{and} \tag{5.12}$$

$$\mathrm{Sh}_{\bar{t}w}(\gamma_{\bar{t}w}) = \sum_{z \in \mathbb{F}_q^\times,\ \check{\alpha}([z]) = t} (n_s t^{-1} x_\alpha(\pi^{\mathfrak{h}}[z]) n_s^{-1})_* \,\mathrm{Sh}_w(\beta)\ . \tag{5.13}$$

(Note that the statements above in the case $\ell(\tilde{s}w) = \ell(w) + 1$ are true for any lift $\tilde{s}$ for s in $\widetilde{W}$ whereas (5.11) and the subsequent formulas are valid only for the specific choice of $\tilde{s}$ made in (2.15).)

PROOF. Except for the case of $\ell(\tilde{s}w) = \ell(w) - 1$, the statements follow easily from Corollary 5.5, see in particular formula (5.8). So we consider the remaining case $\ell(\tilde{s}w) = \ell(w) - 1$ and recall that there is $(\alpha, \mathfrak{h}) \in \Pi_{aff}$ such that $s = s_{(\alpha,\mathfrak{h})}$ and that $\tilde{s} = n_s T^1$ was defined in (2.15). Using the notation from §2.1.6 (see in particular (2.18)), we have

$$n_s I n_s^{-1} I = I \,\dot\cup \dot{\bigcup}_{z \in \mathbb{F}_q^\times} x_\alpha(\pi^{\mathfrak{h}}[z]) \check{\alpha}([z]) n_s^{-1} I \subset I \,\dot\cup \dot{\bigcup}_{t \in \check{\alpha}([\mathbb{F}_q^\times])} I t n_s^{-1} I$$

and hence, using Cor. 2.5.ii,

$$\begin{aligned} n_s I w I = n_s I n_s^{-1} I n_s \dot{w} I &= I n_s \dot{w} I \,\dot\cup \dot{\bigcup}_{z \in \mathbb{F}_q^\times} x_\alpha(\pi^{\mathfrak{h}}[z]) \check{\alpha}([z]) n_s^{-1} I n_s \dot{w} I \\ &\subset I n_s \dot{w} I \,\dot\cup \dot{\bigcup}_{t \in \check{\alpha}([\mathbb{F}_q^\times])} I t \dot{w} I\ . \end{aligned} \tag{5.14}$$

This proves (5.11) using Proposition 5.3. It remains to compute $\gamma_{\tilde{s}w}$ and $\gamma_{\bar{t}w}$ for $t \in \check{\alpha}([\mathbb{F}_q^\times])$.

Let $u := \tilde{s}w$. From (2.7) we deduce that $\tilde{s}^{-1} I \tilde{s} w \subset I_{s^{-1}} w I$. Therefore $\tilde{s}^{-1} I \tilde{s} w I \cap I w I = I_{s^{-1}} w I$. Lemma 5.2.i implies that $I_{s^{-1}} \backslash (\tilde{s}^{-1} I u \cap I w I) / I_{u^{-1}}$ is made of the single double coset $I_{s^{-1}} w I_{u^{-1}}$. We have $I_u = I_{\tilde{s}w}$ and

$$\begin{aligned} I_u \cap u w^{-1} I w u^{-1} &= I_{\tilde{s}w} \cap \tilde{s} I \tilde{s}^{-1} = \tilde{s}(w I w^{-1} \cap \tilde{s}^{-1} I \tilde{s} \cap I)\tilde{s}^{-1} \\ &= \tilde{s}(I_w \cap \tilde{s}^{-1} I \tilde{s})\tilde{s}^{-1} = \tilde{s} I_w \tilde{s}^{-1}\ , \end{aligned}$$

where the last equality is justified by (2.6). Furthermore

$$u I u^{-1} \cap u w^{-1} I w u^{-1} = \tilde{s}(w I w^{-1} \cap I)\tilde{s}^{-1} = \tilde{s} I_w \tilde{s}^{-1}\ .$$

So Proposition 5.3 says that the component $\gamma_{\tilde{s}w}$ in $H^{i+j}(I, \mathbf{X}(\tilde{s}w))$ of $\tau_{\tilde{s}} \cdot \beta$ is given by

$$\mathrm{Sh}_{\tilde{s}w}(\gamma_{\tilde{s}w}) = \mathrm{cores}^{\tilde{s}I_w\tilde{s}^{-1}}_{I_{\tilde{s}w}}\left(\mathrm{res}^{\tilde{s}I_w\tilde{s}^{-1}}_{\tilde{s}I_w\tilde{s}^{-1}}(\tilde{s}_* \,\mathrm{Sh}_w(\beta)\right) = \mathrm{cores}^{\tilde{s}I_w\tilde{s}^{-1}}_{I_{\tilde{s}w}}\left(\tilde{s}_* \,\mathrm{Sh}_w(\beta)\right) .$$

Let $t \in \check{\alpha}([\mathbb{F}_q^\times])$ and $u_t := \bar{t}w$. We pick $\dot{u}_t := t\dot{w} \in N(T)$. We have $n_s^{-1}Iu_tI \cap IwI = n_s^{-1}(ItwI \cap n_sIwI)$. From (5.14) we obtain that

$$\begin{aligned} I\bar{t}wI \cap n_sIwI &= \dot{\bigcup}_{z\in\mathbb{F}_q^\times,\, \check{\alpha}([z])=t} x_\alpha(\pi^{\mathfrak{h}}[z])tn_s^{-1}In_s\dot{w}I \\ &= \dot{\bigcup}_{z\in\mathbb{F}_q^\times,\, \check{\alpha}([z])=t} I_s x_\alpha(\pi^{\mathfrak{h}}[z])t\dot{w}I \ . \end{aligned}$$

The second equality comes from the fact that t and $x_\alpha(\pi^{\mathfrak{h}}[z])$ normalize I_s (see Cor. 2.5.iii and Lemma 2.1 for the latter) and from (2.7). Therefore

$$n_s^{-1}Iu_tI \cap IwI = \dot{\bigcup}_{z\in\mathbb{F}_q^\times,\, \check{\alpha}([z])=t} I_{s^{-1}}n_s^{-1}x_\alpha(\pi^{\mathfrak{h}}[z])t\dot{w}I \ .$$

Let $z \in \mathbb{F}_q^\times$ such that $\check{\alpha}([z]) = t$ and $h_{t,z} := n_s^{-1}x_\alpha(\pi^{\mathfrak{h}}[z])\dot{u}_t$. It lies in $n_s^{-1}I\dot{u}_t \cap IwI$. Using Lemma 5.2 and the above equalities, we obtain that $I_{s^{-1}}\backslash(n_s^{-1}I\dot{u}_t \cap IwI)/I_{u_t^{-1}}$ is made of the (distinct) double cosets $I_{s^{-1}}h_{t,z}I_{u_t^{-1}}$ where $z \in \mathbb{F}_q^\times$ is such that $\check{\alpha}([z]) = t$. (By Remark 2.7, there is one or two such double cosets.) Furthermore, we have $I_{u_t} = I_w$ and $\dot{u}_th_{t,z}^{-1}Ih_{t,z}\dot{u}_t^{-1} = x_\alpha(\pi^{\mathfrak{h}}[z])^{-1}n_sIn_s^{-1}x_\alpha(\pi^{\mathfrak{h}}[z])$. Therefore

$$\begin{aligned} I_{u_t} \cap \dot{u}_th_{t,z}^{-1}Ih_{t,z}\dot{u}^{-1} &= I \cap wIw^{-1} \cap x_\alpha(\pi^{\mathfrak{h}}[z])^{-1}n_sIn_s^{-1}x_\alpha(\pi^{\mathfrak{h}}[z]) \\ &= x_\alpha(\pi^{\mathfrak{h}}[z])^{-1}I_sx_\alpha(\pi^{\mathfrak{h}}[z]) \cap wIw^{-1} = I_s \cap wIw^{-1} \\ &= I_s \cap I_w = I_w = I_{u_t} \ , \end{aligned}$$

where the third equality uses Cor. 2.5.iii and the fifth equality uses (2.6). Now to apply the formula of Prop. 5.3, we need to find $a_{t,z}$ and $c_{t,z}$ in I such that $h_{t,z} = n_s^{-1}a_{t,z}^{-1}\dot{u}_t \in c_{t,z}\dot{w}I$ where $z \in \mathbb{F}_q^\times$ is such that $\check{\alpha}([z]) = t$. Before giving them explicitly, first notice that $a_{t,z}n_sc_{t,z}$ lies in $\bar{t}wIw^{-1}$ thus it normalizes wIw^{-1} and it also lies in In_sI thus normalizes I_s by Corollary 2.5.iii. By (2.6) we have $I_w = wIw^{-1} \cap I_s$ hence $a_{t,z}n_sc_{t,z}$ normalizes I_w and it follows that $(a_{t,z}n_sc_{t,z})I_w(a_{t,z}n_sc_{t,z})^{-1} = u_tIu_t^{-1} \cap u_th_{t,z}^{-1}Ih_{t,z}u_t{}^{-1}$ coincides with $I_w = I_{u_t}$. Therefore, we have

$$\mathrm{Sh}_{u_t}(\gamma_{u_t}) = \sum_{z\in\mathbb{F}_q^\times,\, \check{\alpha}([z])=t} (a_{t,z}n_sc_{t,z})_* \,\mathrm{Sh}_w(\beta) \ .$$

By the above definitions we have $a_{t,z} = x_\alpha(\pi^{\mathfrak{h}}[z])^{-1}$. To find a suitable element $c_{t,z}$, notice that

$$\begin{aligned} n_sx_\alpha(\pi^{\mathfrak{h}}[z^{-1}])h_{t,z}\dot{w}^{-1}n_s^{-1} &= n_sx_\alpha(\pi^{\mathfrak{h}}[z^{-1}])n_s^{-1}x_\alpha(\pi^{\mathfrak{h}}[z])tn_s^{-1} \\ &= x_\alpha(\pi^{\mathfrak{h}}[-z]) \in \mathfrak{U}_{(\alpha,\mathfrak{h})} \end{aligned}$$

by (2.17). By (2.1), we have $\dot{w}^{-1}n_s^{-1}\mathfrak{U}_{(\alpha,\mathfrak{h})}n_s\dot{w} = \mathfrak{U}_{(s\hat{w})^{-1}(\alpha,\mathfrak{h})}$ where $\hat{w}$ denotes the image of w in W. By (2.4), $(s\hat{w})^{-1}(\alpha,\mathfrak{h})$ lies in Φ^+_{aff} and from Lemma 2.1.ii we deduce that $\mathfrak{U}_{(s\hat{w})^{-1}(\alpha,\mathfrak{h})}$ is contained in I. Therefore we have $\dot{w}^{-1}x_\alpha(\pi^{\mathfrak{h}}[z^{-1}])h_{t,z} \in I$ and we may pick $c_{t,z} := x_\alpha(\pi^{\mathfrak{h}}[z^{-1}])^{-1}$. Lastly using (2.17), we see that with this choice we have $a_{t,z}n_sc_{t,z} = n_st^{-1}x_\alpha(\pi^{\mathfrak{h}}[z])n_s^{-1}$, which concludes the proof of (5.13). □

5.3. Appendix. In [**23**] §4.1 a groupoid cohomology class is a G-equivariant function

$$f : (G \times G)/(I \times I) \longrightarrow \oplus_{(g_1,g_2)\in G^2/I^2} H^*(g_1 I g_1^{-1} \cap g_2 I g_2^{-1}, k)$$

such that

- f is supported on finitely many G-orbits, and
- $f(g_1 I, g_2 I) \in H^*(g_1 I g_1^{-1} \cap g_2 I g_2^{-1}, k)$ for any $(g_1, g_2) \in G^2$.

The product of two such functions f and $\tilde{f}$ is defined as follows. We use the maps

$$\begin{aligned} \iota_{\mu,\nu} : G \times G \times G &\longrightarrow G \times G \\ (g_1, g_2, g_3) &\longmapsto (g_\mu, g_\nu) , \end{aligned}$$

for $1 \le \mu < \nu \le 3$, in order to first introduce the pulled back functions

$$(\iota_{1,2}^* f)(g_1, g_2, g_3) := \mathrm{res}^{g_1 I g_1^{-1} \cap g_2 I g_2^{-1}}_{g_1 I g_1^{-1} \cap g_2 I g_2^{-1} \cap g_3 I g_3^{-1}} f(g_1, g_2)$$

and

$$(\iota_{2,3}^* f)(g_1, g_2, g_3) := \mathrm{res}^{g_2 I g_2^{-1} \cap g_3 I g_3^{-1}}_{g_1 I g_1^{-1} \cap g_2 I g_2^{-1} \cap g_3 I g_3^{-1}} f(g_2, g_3)$$

(with value in $H^*(g_1 I g_1^{-1} \cap g_2 I g_2^{-1} \cap g_3 I g_3^{-1}, k)$) on G^3/I^3. These functions are no longer supported on finitely many G-orbits. Nevertheless we consider their cup product

$$F(g_1, g_2, g_3) := (\iota_{1,2}^* f)(g_1, g_2, g_3) \cup (\iota_{2,3}^* f)(g_1, g_2, g_3) .$$

One can check (in fact, it follows from the subsequent computations) that its push forward

$$(\iota_{1,3*} F)(g_1, g_3) := \sum_{g \in (g_1 I g_1^{-1} \cap g_3 I g_3^{-1}) \backslash G/I} \mathrm{cores}^{g_1 I g_1^{-1} \cap g I g^{-1} \cap g_3 I g_3^{-1}}_{g_1 I g_1^{-1} \cap g_3 I g_3^{-1}} F(g_1, g, g_3)$$

is well defined and again is a groupoid cohomology class, which is defined in [**23**] Prop. 7 to be the product $f \cdot \tilde{f}$.

We now fix two "ordinary" cohomology classes $\alpha \in H^i(I, \mathbf{X}(v))$ and $\beta \in H^j(I, \mathbf{X}(w))$ and introduce the corresponding groupoid cohomology classes f_α and f_β supported on $G(1,v)I^2$ and $G(1,w)I^2$ by

$$f_\alpha(1, v) := \mathrm{Sh}_v(\alpha) \in H^i(I_v, k) \quad \text{and} \quad f_\beta(1, w) := \mathrm{Sh}_w(\beta) \in H^j(I_w, k) ,$$

respectively. In the following we compute the product $f_\alpha \cdot f_\beta$. By the G-equivariance it suffices to compute the classes

$$(f_\alpha \cdot f_\beta)(1, u) \in H^{i+j}(I_u, k) \qquad \text{for } u \in W.$$

We have the three injective maps between double coset spaces

$$\begin{aligned} D_v : I_v \backslash G/I &\hookrightarrow G \backslash G^3/I^3 \\ h &\mapsto (1, v, h), \\ D_w : I_w \backslash G/I &\hookrightarrow G \backslash G^3/I^3 \\ h &\mapsto (h, 1, w), \text{ and} \\ D_u : I_u \backslash G/I &\hookrightarrow G \backslash G^3/I^3 \\ h &\mapsto (1, h, u). \end{aligned}$$

The functions $\iota_{1,2}^* f_\alpha$ and $\iota_{2,3}^* f_\beta$ are supported on the G-orbits in $\mathrm{im}(D_v)$ and $\mathrm{im}(D_w)$, respectively. Hence $F := \iota_{1,2}^* f_\alpha \cup \iota_{2,3}^* f_\beta$ is supported on the G-orbits in $\mathrm{im}(D_v) \cap \mathrm{im}(D_w)$. Moreover, we have

$$(5.15) \qquad (f_\alpha \cdot f_\beta)(1,u) = (\iota_{1,3*}F)(1,u) = \sum_{h \in I_u \backslash G/I} \mathrm{cores}_{I_u}^{I_u \cap hIh^{-1}} F(1,h,u) \ .$$

Of course, on the right hand side only those h can occur for which we have $D_u(h) \in \mathrm{im}(D_v) \cap \mathrm{im}(D_w)$.

LEMMA 5.7. $D_u^{-1}(\mathrm{im}(D_v) \cap \mathrm{im}(D_w)) = I_u \backslash (uIw^{-1}I \cap IvI)/I$.

PROOF. Let $h = \dot{u}x_0\dot{w}^{-1}x_1 = x_2\dot{v}x_3$ with $x_j \in I$. Then

$$\begin{aligned} D_u(h) &= G(1, x_2\dot{v}x_3, \dot{u})I^3 = Gx_2(1, \dot{v}, x_2^{-1}\dot{u})(x_2^{-1}, x_3, 1)I^3 \\ &= G(1, \dot{v}, x_2^{-1}\dot{u})I^3 \in \mathrm{im}(D_v) \end{aligned}$$

and

$$\begin{aligned} D_u(h) &= G(1, \dot{u}x_0\dot{w}^{-1}x_1, \dot{u})I^3 = G\dot{u}x_0\dot{w}^{-1}(\dot{w}x_0^{-1}\dot{u}^{-1}, 1, \dot{w})(1, x_1, x_0^{-1})I^3 \\ &= G(\dot{w}x_0^{-1}\dot{u}^{-1}, 1, \dot{w})I^3 \in \mathrm{im}(D_w) \ . \end{aligned}$$

On the other hand, let $h \in G$ be such that $D_u(h) = G(1,h,u)I^3 \in \mathrm{im}(D_v) \cap \mathrm{im}(D_w)$. Then $(1,h,\dot{u}) \in G(1,v,\tilde{h})I^3 \cap G(\bar{h},1,w)I^3$ for some $\tilde{h}, \bar{h} \in G$. Write

$$(1,h,\dot{u}) = (gx_1, g\dot{v}x_2, g\tilde{h}x_3) = (g'\bar{h}y_1, g'y_2, g'\dot{w}y_3)$$

with $g, g' \in G$ and $x_j, y_j \in I$. We see that $1 = gx_1$ and $h = g\dot{v}x_2 = x_1^{-1}\dot{v}x_2 \in IvI$ as well as $\dot{u} = g'\dot{w}y_3$ and $h = g'y_2 = \dot{u}y_3^{-1}\dot{w}^{-1}y_2 \in uIw^{-1}I$. □

We deduce that (5.15) simplifies to

$$(f_\alpha \cdot f_\beta)(1,u) = \sum_{h \in I_u \backslash (uIw^{-1}I \cap IvI)/I} \mathrm{cores}_{I_u}^{I_u \cap hIh^{-1}} F(1,h,u) \ .$$

Recall that

$$\begin{aligned} F(1,h,u) &= (\iota_{1,2}^* f_\alpha)(1,h,u) \cup (\iota_{2,3}^* f_\beta)(1,h,u) \\ &= \mathrm{res}_{I_u \cap hIh^{-1}}^{I \cap hIh^{-1}} f_\alpha(1,h) \cup \mathrm{res}_{I_u \cap hIh^{-1}}^{uIu^{-1} \cap hIh^{-1}} f_\beta(h,u) \ . \end{aligned}$$

We write

$$h = \dot{u}A\dot{w}^{-1}B = C_h\dot{v}D \qquad \text{with } A, B, C_h, D \in I$$

and put $x_h := \dot{u}A\dot{w}^{-1}$. Then

$$(1,h) = C_h(1,\dot{v})(C_h^{-1}, D) \quad \text{and} \quad (h,\dot{u}) = x_h(1,\dot{w})(B, A^{-1})$$

and hence, by G-equivariance,

$$f_\alpha(1,h) = C_{h*}\,\mathrm{Sh}_v(\alpha) \quad \text{and} \quad f_\beta(h,u) = x_{h*}\,\mathrm{Sh}_w(\beta) \ .$$

Inserting this into the above formulas we arrive at

$$(5.16) \quad (f_\alpha \cdot f_\beta)(1,u) = \sum_{h \in I_u \backslash (uIw^{-1}I \cap IvI)/I} \mathrm{cores}_{I_u}^{I_u \cap hIh^{-1}} \Big(\mathrm{res}_{I_u \cap hIh^{-1}}^{I \cap hIh^{-1}} C_{h*}\,\mathrm{Sh}_v(\alpha) \cup \mathrm{res}_{I_u \cap hIh^{-1}}^{uIu^{-1} \cap hIh^{-1}} x_{h*}\,\mathrm{Sh}_w(\beta) \Big) \ .$$

REMARK 5.8. 1. $hIh^{-1} \cap uIu^{-1} = C_h vIv^{-1} C_h^{-1} \cap uIu^{-1}$.
2. $C_h I_v C_h^{-1} = I \cap hIh^{-1} = I \cap x_h I x_h^{-1}$.

3. $uIu^{-1} \cap hIh^{-1} = x_h I_w x_h^{-1} = x_h I x_h^{-1} \cap x_h w I w^{-1} x_h^{-1} = x_h I x_h^{-1} \cap uIu^{-1}$.
4. The coset $C_h I_v$ only depends on the coset hI (for $c, c' \in I$, we have $cvI = c'vI \iff c^{-1}c' \in I_v$).
5. The coset $x_h I_w$ only depends on the coset hI.
6. $x_h I x_h^{-1} uI = x_h IwI$.

LEMMA 5.9. i. *The projection map*

$$I_u\backslash(uIw^{-1} \cap IvI)/I_w \xrightarrow{\simeq} I_u\backslash(uIw^{-1}I \cap IvI)/I$$

is bijective.

ii. *The map*

$$\begin{aligned} I_{v^{-1}}\backslash(v^{-1}Iu \cap IwI)/I_{u^{-1}} &\xrightarrow{\simeq} I_u\backslash(uIw^{-1} \cap IvI)/I_w \\ h = cwd &\longmapsto \dot{u}h^{-1}c \end{aligned}$$

is bijective.

PROOF. i. Replace in Lemma 5.2.i the elements v^{-1}, u, w with u, w^{-1}, v.

ii. First of all we note that the coset cI_w only depends on w (and h). We therefore wrote $cwd = h$ by a slight abuse of notation. The map is well defined since $I_u uh^{-1}c = I_u ud^{-1}\dot{w}^{-1} \subseteq uIw^{-1}$ and, if $h = \dot{v}^{-1}a\dot{u}$, then $\dot{u}h^{-1}c = a^{-1}\dot{v}c \in IvI$. The bijectivity follows by checking that the map $h' = CvD \mapsto Dh'^{-1}u$ is a well defined inverse. □

We note that if $h = c\dot{w}d = \dot{v}^{-1}a^{-1}\dot{u}$ then $h' := \dot{u}h^{-1}c = \dot{u}d^{-1}\dot{w}^{-1} = a\dot{v}c$ and $x_{h'} = a\dot{v}c$ and $C_{h'} = a$. Hence, rewriting the right hand sum in (5.16) by using the composite bijection in Lemma 5.9, we obtain

$$\begin{aligned} (f_\alpha \cdot f_\beta)(1, u) = \sum_{h \in I_{v^{-1}}\backslash(v^{-1}Iu \cap IwI)/I_{u^{-1}}} & \operatorname{cores}_{I_u}^{I_u \cap \dot{u}h^{-1}Ih\dot{u}^{-1}} \Big(\\ & \operatorname{res}^{I \cap \dot{u}h^{-1}Ih\dot{u}^{-1}}_{I_u \cap \dot{u}h^{-1}Ih\dot{u}^{-1}} a_* \operatorname{Sh}_v(\alpha) \cup \operatorname{res}^{uIu^{-1} \cap \dot{u}h^{-1}Ih\dot{u}^{-1}}_{I_u \cap \dot{u}h^{-1}Ih\dot{u}^{-1}} (a\dot{v}c)_* \operatorname{Sh}_w(\beta) \Big) , \end{aligned}$$

where $h = c\dot{w}d = \dot{v}^{-1}a^{-1}\dot{u}$ with $a, c, d \in I$. By comparing this equality with the equality in Prop. 5.3 we deduce the following result.

PROPOSITION 5.10. *Let $\alpha \in H^i(I, \mathbf{X}(v))$, $\beta \in H^j(I, \mathbf{X}(w))$, and $\alpha \cdot \beta = \sum_u \gamma_u$ with $\gamma_u \in H^{i+j}(I, \mathbf{X}(u))$ as in 5.3; then $f_\alpha \cdot f_\beta = \sum_u f_{\gamma_u}$.*

6. An involutive anti-automorphism of the algebra E^*

For $w \in \widetilde{W}$, we have $I_{w^{-1}} = w^{-1}I_w w$ and a linear isomorphism

$$(w^{-1})_* : H^i(I_w, k) \xrightarrow{\cong} H^i(I_{w^{-1}}, k) ,$$

for all $i \geq 0$. Recall that conjugation by an element in $T^1 \subset I_w$ is a trivial operator on $H^i(I_w, k)$ and therefore the conjugation above is well defined and does not depend on the chosen lift for w^{-1} in $N(T)$. Via the Shapiro isomorphism (3.2), this induces the linear isomorphism $\mathcal{J}_w$:

(6.1)
$$\begin{array}{ccc} H^i(I, \mathbf{X}(w)) & \xrightarrow[\cong]{\mathcal{J}_w} & H^i(I, \mathbf{X}(w^{-1})) \\ \Big\downarrow \operatorname{Sh}_w & & \cong \Big\downarrow \operatorname{Sh}_{w^{-1}} \\ H^i(I_w, k) & \xrightarrow[\cong]{(w^{-1})_*} & H^i(I_{w^{-1}}, k) \end{array}$$

Summing over all $w \in \widetilde{W}$, the maps $(\mathcal{J}_w)_{w\in\widetilde{W}}$ induce a linear isomorphism

$$\mathcal{J} : H^i(I,\mathbf{X}) \xrightarrow{\cong} H^i(I,\mathbf{X}) \ .$$

PROPOSITION 6.1. *The map $\mathcal{J}$ defines an involutive anti-automorphism of the graded* Ext-*algebra E^*, namely*

$$\mathcal{J}(\alpha\cdot\beta) = (-1)^{ij}\mathcal{J}(\beta)\cdot\mathcal{J}(\alpha)$$

where $\alpha \in H^i(I,\mathbf{X})$ and $\beta \in H^j(I,\mathbf{X})$ for all $i,j \geq 0$. Restricted to $H^0(I,\mathbf{X})$ it yields the anti-involution

$$\tau_g \mapsto \tau_{g^{-1}} \text{ for any } g \in G$$

of the algebra H.

PROOF. First note that, for $w \in \widetilde{W}$, the element $\tau_w \in H^0(I,\mathbf{X}(w)) = \mathbf{X}(w)^I$ corresponds to $1 \in H^0(I_w,k) = k$. Therefore, $\mathcal{J}(\tau_w) = \tau_{w^{-1}}$. Now we turn to the proof of the first statement of the proposition. Let $\alpha \in H^i(I,\mathbf{X}(v))$ and $\beta \in H^j(I,\mathbf{X}(w))$. On the one hand, recall that we have

$$\alpha\cdot\beta = \sum_{u\in\widetilde{W},IuI\subseteq IvI\cdot IwI} \gamma_u$$

with $\gamma_u \in H^{i+j}(I,\mathbf{X}(u))$ as in Proposition 5.3 given by

$$\mathrm{Sh}_u(\gamma_u) = (-1)^{ij} \sum_{h\in I_{v^{-1}}\backslash(v^{-1}IuI\cap IwI)/I_{u^{-1}}} \mathrm{cores}_{I_u}^{I_u\cap\dot{u}h^{-1}Ih\dot{u}^{-1}}\big(\Gamma_{u,h}\big)$$

where

$$\Gamma_{u,h} := \mathrm{res}^{uIu^{-1}\cap\dot{u}h^{-1}Ih\dot{u}^{-1}}_{I_u\cap\dot{u}h^{-1}Ih\dot{u}^{-1}}\big((a\dot{v}c)_*(\mathrm{Sh}_w(\beta))\big) \cup \mathrm{res}^{I\cap\dot{u}h^{-1}Ih\dot{u}^{-1}}_{I_u\cap\dot{u}h^{-1}Ih\dot{u}^{-1}}\big(a_*(\mathrm{Sh}_v(\alpha))\big)$$

and $h = c\dot{w}d = \dot{v}^{-1}a^{-1}\dot{u}$ with $a,c,d \in I$.

We compute $\mathcal{J}(\beta)\cdot\mathcal{J}(\alpha)$. Recall that $\mathrm{Sh}_{v^{-1}}(\mathcal{J}(\alpha)) = (v^{-1})_*\,\mathrm{Sh}_v(\alpha)$ and $\mathrm{Sh}_{w^{-1}}(\mathcal{J}(\beta)) = (w^{-1})_*\,\mathrm{Sh}_w(\beta)$. Note that the map $u \mapsto u^{-1}$ yields a bijection between the set of $u \in \widetilde{W}$ such that $IuI \subseteq IvI\cdot IwI$ and the set of $u' \in \widetilde{W}$ such that $Iu'I \subseteq Iw^{-1}I\cdot Iv^{-1}I$. Therefore, by Proposition 5.3, we have $\mathcal{J}(\beta)\cdot\mathcal{J}(\alpha) = \sum_{u\in\widetilde{W},IuI\subseteq IvI\cdot IwI}\delta_{u^{-1}}$ with $\delta_{u^{-1}} \in H^{i+j}(I,\mathbf{X}(u^{-1}))$ given by

$$\mathrm{Sh}_{u^{-1}}(\delta_{u^{-1}}) = (-1)^{ij} \sum_{h'\in I_w\backslash(wIu^{-1}\cap Iv^{-1}I)/I_u} \mathrm{cores}_{I_{u^{-1}}}^{I_{u^{-1}}\cap\dot{u}^{-1}h'^{-1}Ih'\dot{u}}\big(\Delta_{u^{-1},h'}\big)$$

where

$$\begin{aligned}\Delta_{u^{-1},h'} &= \mathrm{res}^{u^{-1}Iu\cap\dot{u}^{-1}h'^{-1}Ih'\dot{u}}_{I_{u^{-1}}\cap\dot{u}^{-1}h'^{-1}Ih'\dot{u}}\big((a'\dot{w^{-1}}c')_*(\mathrm{Sh}_{v^{-1}}(\mathcal{J}(\alpha)))\big)\\ &\qquad\qquad \cup \mathrm{res}^{I\cap\dot{u}^{-1}h'^{-1}Ih'\dot{u}}_{I_{u^{-1}}\cap\dot{u}^{-1}h'^{-1}Ih'\dot{u}}\big(a'_*(\mathrm{Sh}_{w^{-1}}(\mathcal{J}(\beta)))\big)\\ &= \mathrm{res}^{u^{-1}Iu\cap\dot{u}^{-1}h'^{-1}Ih'\dot{u}}_{I_{u^{-1}}\cap\dot{u}^{-1}h'^{-1}Ih'\dot{u}}\big((a'\dot{w^{-1}}c'\dot{v^{-1}})_*(\mathrm{Sh}_v(\alpha))\big)\\ &\qquad\qquad \cup \mathrm{res}^{I\cap\dot{u}^{-1}h'^{-1}Ih'\dot{u}}_{I_{u^{-1}}\cap\dot{u}^{-1}h'^{-1}Ih'\dot{u}}\big((a'\dot{w^{-1}})_*(\mathrm{Sh}_w(\beta))\big)\end{aligned}$$

and $h' = c'\dot{v^{-1}}d' = (\dot{w^{-1}})^{-1}a'^{-1}\dot{u^{-1}}$ with $a',c',d' \in I$.

Now we compute $\mathcal{J}(\mathcal{J}(\beta)\cdot\mathcal{J}(\alpha))$. Recall that corestriction commutes with conjugation (cf. §4.6). We have:

$$\begin{aligned}\mathrm{Sh}_u(\mathcal{J}(\delta_{u^{-1}})) &= u_*\,\mathrm{Sh}_{u^{-1}}(\delta_{u^{-1}})\\ &= (-1)^{ij}\sum_{h'\in I_w\backslash(wIu^{-1}\cap Iv^{-1}I)/I_u}\mathrm{cores}_{I_u}^{I_u\cap h'^{-1}Ih'}u_*\big(\Delta_{u^{-1},h'}\big)\end{aligned}$$

To an element $h'\in wIu^{-1}\cap Iv^{-1}I$ written in the form $h'=c'\dot{v}^{-1}d'=(\dot{w}^{-1})^{-1}a'^{-1}\dot{u}^{-1}$ as above, we attach the double coset $I_{v^{-1}}hI_{u^{-1}}$ where $h:=c'^{-1}h'\dot{u}=\dot{v}^{-1}d'\dot{u}\in v^{-1}Iu\cap IwI$. This is well defined because d' is defined up to multiplication on the left by an element in I_v. It is easy to see that this yields a map $I_w\backslash(wIu^{-1}\cap Iv^{-1}I)/I_u\to I_{v^{-1}}\backslash(v^{-1}Iu\cap IwI)/I_{u^{-1}}$. One can check that the map in the opposite direction induced by attaching to $h=c\dot{w}d\in v^{-1}Iu\cap IwI$ the double coset $I_wh'I_u$ where $h'=c^{-1}h\dot{u}^{-1}=\dot{w}d\dot{u}^{-1}$ is well defined. Therefore, these maps are bijective. Note that for h' and h corresponding to each other as above, we have $I_u\cap h'^{-1}Ih'=I_u\cap\dot{u}h^{-1}Ih\dot{u}^{-1}$ and $h=c\dot{w}d=\dot{v}^{-1}a^{-1}\dot{u}$ with $a^{-1}\in T^1d'$, $c^{-1}\in T^1c'$ and $d^{-1}\in a'T^1$, therefore we compute

$$\begin{aligned}u_*(\Delta_{u^{-1},h'}) &= \dot{u}_*\,\mathrm{res}^{u^{-1}Iu\cap h^{-1}Ih}_{I_{u^{-1}}\cap h^{-1}Ih}\big((a'\dot{w}^{-1}c'\dot{v}^{-1})_*(\mathrm{Sh}_v(\alpha))\big)\\ &\qquad\cup\mathrm{res}^{I\cap h^{-1}Ih}_{I_{u^{-1}}\cap h^{-1}Ih}\big((a'\dot{w}^{-1})_*(\mathrm{Sh}_w(\beta))\big)\\ &= \mathrm{res}^{I\cap\dot{u}h^{-1}Ih\dot{u}^{-1}}_{I_{u^{-1}}\cap h^{-1}Ih}\big((\dot{u}a'\dot{w}^{-1}c'\dot{v}^{-1})_*(\mathrm{Sh}_v(\alpha))\big)\\ &\qquad\cup\mathrm{res}^{\dot{u}I\dot{u}^{-1}\cap\dot{u}h^{-1}Ih\dot{u}^{-1}}_{I_u\cap\dot{u}h^{-1}Ih\dot{u}^{-1}}\big((\dot{u}a'\dot{w}^{-1})_*(\mathrm{Sh}_w(\beta))\big)\\ &= \mathrm{res}^{I\cap\dot{u}h^{-1}Ih\dot{u}^{-1}}_{I_{u^{-1}}\cap h^{-1}Ih}\big((d'^{-1})_*(\mathrm{Sh}_v(\alpha))\big)\\ &\qquad\cup\mathrm{res}^{\dot{u}I\dot{u}^{-1}\cap\dot{u}h^{-1}Ih\dot{u}^{-1}}_{I_u\cap\dot{u}h^{-1}Ih\dot{u}^{-1}}\big((\dot{u}a'\dot{w}^{-1})_*(\mathrm{Sh}_w(\beta))\big)\\ &= \mathrm{res}^{I\cap\dot{u}h^{-1}Ih\dot{u}^{-1}}_{I_{u^{-1}}\cap h^{-1}Ih}\big(a_*(\mathrm{Sh}_v(\alpha))\big)\\ &\qquad\cup\mathrm{res}^{\dot{u}I\dot{u}^{-1}\cap\dot{u}h^{-1}Ih\dot{u}^{-1}}_{I_u\cap\dot{u}h^{-1}Ih\dot{u}^{-1}}\big((a\dot{v}c)_*(\mathrm{Sh}_w(\beta))\big)\\ &= (-1)^{ij}\Gamma_{u,h}\end{aligned}$$

and $\mathrm{Sh}_u(\mathcal{J}(\delta_{u^{-1}}))=(-1)^{ij}\,\mathrm{Sh}_u(\gamma_u)$. We proved $\mathcal{J}(\mathcal{J}(\beta)\cdot\mathcal{J}(\alpha))=(-1)^{ij}\alpha\cdot\beta$. □

REMARK 6.2. By (3.6) our cup product commutes with the Shapiro isomorphism. It also commutes with conjugation of the group ([**17**] Prop. 1.5.3(i)). Therefore the anti-automorphism $\mathcal{J}$ respects the cup product.

REMARK 6.3. We want to show that the anti-involution of H induced by $\mathcal{J}$ preserves the ideal $\mathfrak{J}$ of §2.3. Consider the space $\mathbb{Z}[G/I]$ of finitely supported functions $G/I\to\mathbb{Z}$. The ring of its G-equivariant $\mathbb{Z}$-endomorphisms is isomorphic to the convolution ring $\mathbb{Z}[I\backslash G/I]$ where the product is given by $f\star f'(\_)=\sum_{x\in G/I}f(x^{-1}\_)f'(x)$. The opposite ring is denoted by $H_{\mathbb{Z}}$. One easily checks that the map $f\mapsto[g\mapsto f(g^{-1})]$ defines an anti-involution $j_{\mathbb{Z}}$ of the convolution ring $\mathbb{Z}[I\backslash G/I]$. It induces an anti-involution of the k-algebra $H=H_{\mathbb{Z}}\otimes k$ which coincides with $\mathcal{J}$. A basis $(z_{\{\lambda\}})_{\{\lambda\}\in\widetilde{\Lambda}/W_0}$ of the center of the ring $H_{\mathbb{Z}}$ is described in [**33**]. It is indexed by the set of W_0-orbits in the preimage $\widetilde{\Lambda}=T/T^1$ of $\Lambda=T/T^0$ in $\widetilde{W}$ (see §2.1.4). From [**19**] Lemma 3.4 we deduce that $j_{\mathbb{Z}}(z_{\{\lambda\}})=z_{\{\lambda^{-1}\}}$ for any $\lambda\in\widetilde{\Lambda}$ where $\{\lambda^{-1}\}$ denotes the W_0-orbit of λ^{-1}.

Recall that the map ν defined in section 2.1.1 induces an isomorphism $\Lambda \cong X_*(T)$. As in [**19**] 1.2.6, notice that the map

$$\begin{aligned} X_*(T) &\longrightarrow T/T^1 \\ \xi &\longmapsto \xi(\pi^{-1}) \bmod T^1 \end{aligned}$$

composed with ν splits the exact sequence

$$0 \longrightarrow T^0/T^1 \longrightarrow \widetilde{\Lambda} \longrightarrow \Lambda \longrightarrow 1$$

and we may see Λ as a subgroup of $\widetilde{\Lambda}$ which is preserved by the action of W_0. The set Λ/W_0 of all W_0-orbits of in Λ contains the set $(\Lambda/W_0)'$ of orbits of elements with nonzero length (when seen in W). Note that via the map ν, it is indexed by the set $X_*^{dom}(T) \setminus (-X_*^{dom}(T))$. By the above remarks, the $\mathbb{Z}$-linear subspace of $H_{\mathbb{Z}}$ with basis $(z_{\{\lambda\}})_{\{\lambda\}\in(\Lambda/W_0)'}$ is preserved by $j_{\mathbb{Z}}$. Therefore, its image $\mathfrak{J}$ in H (as defined in [**19**] 5.2) is preserved by $\mathcal{J}$. Note in passing that the algebra $\mathcal{Z}^0(H)$ (as introduced in §2.3) has basis $(z_{\{\lambda\}})_{\{\lambda\}\in\Lambda/W_0}$.

We deduce from this that if M is a left (resp. right) supersingular module in the sense of §2.3, namely if any element of M is annihilated by a power of $\mathfrak{J}$, then $M^{\mathcal{J}}$ (resp. ${}^{\mathcal{J}}M$) is a right (resp. left) supersingular module.

Alternatively, if $\mathbf{G}$ is semisimple, one can argue that $\mathcal{J}$ preserves supersingular modules as follows. First of all it suffices to show this for cyclic supersingular modules and such modules have finite length. (This is because H is finitely generated over $\mathcal{Z}^0(H)$ by [**19**] Prop. 2.5ii and since $\mathbf{G}$ is semisimple, $\mathfrak{J}$ has finite codimension in $\mathcal{Z}^0(H)$ and therefore in H.) It then further suffices to do this after a suitable extension of the coefficient field. By the equivalence in the proof of Lemma 2.13 this finally reduces us to quotients of $H \otimes_{H_{aff}} \chi$ (or $\chi \otimes_{H_{aff}} H$) where χ is a supersingular character of H_{aff}. It is easy to see that the composite $\chi \circ \mathcal{J} : H_{aff} \to k$ is also a supersingular character. The right (resp. left) H-module $(H \otimes_{H_{aff}} \chi)^{\mathcal{J}}$ (resp. ${}^{\mathcal{J}}(\chi \otimes_{H_{aff}} H)$) is generated as an H-module by $1 \otimes 1$ which supports the character $\chi \circ \mathcal{J}$ of H_{aff}. Therefore by Lemma 2.13 it is annihilated by $\mathfrak{J}$ and supersingular as an H-module.

7. Dualities

7.1. Finite and twisted duals. Given a vector space Y, we denote by $Y^\vee$ the dual space $Y^\vee := \mathrm{Hom}_k(Y, k)$ of Y. If Y is a left, resp. right, module over H, then $Y^\vee$ is naturally a right, resp. left, module over H. Recall that H is endowed with an anti-involution respecting the product and given by the map $\mathcal{J}$ (see Proposition 6.1). We may twist the action of H on a left, resp. right, module Y by $\mathcal{J}$ and thus obtain the right, resp. left module $Y^{\mathcal{J}}$, resp. ${}^{\mathcal{J}}Y$, with the twisted action of H given by $(y, h) \mapsto \mathcal{J}(h)y$, resp. $(h, y) \mapsto y\mathcal{J}(h)$. If Y is an H-bimodule, then we may define the twisted H-bimodule ${}^{\mathcal{J}}Y^{\mathcal{J}}$ the obvious way.

REMARK 7.1. For a left, resp right, resp. bi-, H-module, the identity map yields an isomorphism of right, resp. left, resp. bi-, H-modules

$$({}^{\mathcal{J}}Y)^\vee = (Y^\vee)^{\mathcal{J}}, \quad \text{resp. } (Y^{\mathcal{J}})^\vee = {}^{\mathcal{J}}(Y^\vee), \quad \text{resp. } ({}^{\mathcal{J}}Y^{\mathcal{J}})^\vee = {}^{\mathcal{J}}(Y^\vee)^{\mathcal{J}}.$$

Since $\mathbf{X}$ is a right H-module, the space $\mathbf{X}^\vee$ is naturally a left H-module via $(h, \varphi) \mapsto \varphi(\_h)$. It is also endowed with a left action of G which commutes with the action of H via $(g, \varphi) \mapsto \varphi(g^{-1}\_)$. It is however not a smooth representation of G. Since $\mathbf{X}$ decomposes into $\oplus_{w\in\widetilde{W}}\mathbf{X}(w)$ as a vector space, $\mathbf{X}^\vee$ identifies with

$\prod_{w\in\widetilde{W}} \mathbf{X}(\omega)^\vee$ which contains $\oplus_{w\in\widetilde{W}}\mathbf{X}(w)^\vee$. We denote by $\mathbf{X}^{\vee,f}$ the image of the latter in $\mathbf{X}^\vee$. It is stable under the action of G on $\mathbf{X}^\vee$, and $\mathbf{X}^{\vee,f}$ is a smooth representation of G. Moreover it follows from Cor. 2.5.ii and (2.20) that $\mathbf{X}^{\vee,f}$ is an H-submodule of $\mathbf{X}^\vee$.

More generally, for Y a vector space which decomposes into a direct sum $Y = \oplus_{w\in\widetilde{W}}Y_w$, we denote by $Y^{\vee,f}$ the so-called finite dual of Y which is defined to be the image in $Y^\vee = \prod_{w\in\widetilde{W}} Y_w^\vee$ of $\oplus_{w\in\widetilde{W}}Y_w^\vee$.

For $g \in G$ denote by ev_g the evaluation map $\mathbf{X} \to k, f \mapsto f(g)$. This is an element in $\mathbf{X}^{\vee,f}$. For $g_0, g \in G$ and $f \in \mathbf{X}$ we have $({}^{g_0}\mathrm{ev}_g)(f) = \mathrm{ev}_g({}^{g_0^{-1}}f) = ({}^{g_0^{-1}}f)(g) = f(g_0 g) = \mathrm{ev}_{g_0 g}(f)$. In particular, $\mathrm{ev}_1 \in \mathbf{X}^{\vee,f}$ is fixed under the action of I and there is a well defined morphism of smooth representations of G:

$$\text{(7.1)}\qquad \begin{aligned} \mathbf{ev} : \mathbf{X} &\longrightarrow \mathbf{X}^{\vee,f} \\ \mathrm{char}_{gI} = {}^g\,\mathrm{char}_I &\longmapsto \mathrm{ev}_g = {}^g\,\mathrm{ev}_1 \qquad \text{for any } g \in G. \end{aligned}$$

It is clearly a bijection. The basis $(\mathrm{ev}_g)_{g\in G/I}$ of $\mathbf{X}^{\vee,f}$ is dual to the basis $(\mathrm{char}_{gI})_{g\in G/I}$ of $\mathbf{X}$. For $w \in \widetilde{W}$, the space $\mathbf{X}(w)$ corresponds to $\mathbf{X}(w)^\vee$ under the isomorphism $\mathbf{ev}$.

LEMMA 7.2. *The map* $\mathbf{ev}$ *induces an isomorphism of right* H*-modules* $\mathbf{X} \xrightarrow{\cong} (\mathbf{X}^{\vee,f})^{\mathfrak{J}}$.

PROOF. We only need to show that the composite map $\mathbf{X} \xrightarrow{\mathbf{ev}} (\mathbf{X}^{\vee,f})^{\mathfrak{J}} \xrightarrow{\subseteq} (\mathbf{X}^\vee)^{\mathfrak{J}}$ is right H-equivariant. Since $\mathbf{X}$ and $(\mathbf{X}^\vee)^{\mathfrak{J}}$ are (G,H)-bimodules and since $\mathbf{X}$ is generated by char_I under the action of G, it is enough to prove that $\mathbf{ev}((\mathrm{char}_I)\tau) = \mathfrak{J}(\tau)\mathbf{ev}(\mathrm{char}_I)$ for any $\tau \in H$, or equivalently, that

$$\mathbf{ev}(\mathrm{char}_{IwI}) = \tau_{w^{-1}}\mathbf{ev}(\mathrm{char}_I)$$

for any $w \in \widetilde{W}$. Decompose IwI into simple cosets $IwI = \sqcup_x xwI$ for $x \in (I \cap wIw^{-1})\backslash I$. Then on the one hand, $\mathbf{ev}(\mathrm{char}_{IwI}) = \sum_x \mathrm{ev}_{xw}$. For $g \in G$, it sends the function char_{gI} onto 1 if and only if $g \in IwI$ and to 0 otherwise. On the other hand, we have $(\tau_{w^{-1}}\,\mathrm{ev}_1)(\mathrm{char}_{gI}) = \mathrm{ev}_1(\mathrm{char}_{gI}\mathrm{char}_{Iw^{-1}I}) = \mathrm{ev}_1(\mathrm{char}_{gIw^{-1}I})$. It is equal to 1 if and only if $g^{-1} \in Iw^{-1}I$ and to 0 otherwise. This proves the lemma. □

7.2. Duality between E^i and E^{d-i} when I is a Poincaré group. In this section we always **assume** that the pro-p Iwahori group I is torsion free. This forces the field $\mathfrak{F}$ to be a finite extension of $\mathbb{Q}_p$ with $p \geq 5$. Then I is a Poincaré group of dimension d where d is the dimension of G as a p-adic Lie group: According to [**15**] Thm. V.2.2.8 and [**28**] the group I has finite cohomological dimension; then [**15**] Thm. V.2.5.8 implies that I is a Poincaré group of dimension d. Any open subgroup of a Poincaré group is a Poincaré group of the same dimension (cf. [**26**] Cor. I.4.5). This applies to our groups I_w for any $w \in \widetilde{W}$. It follows that

$$E^* = H^*(I, \mathbf{X}) = 0 \qquad \text{for } * > d$$

and that

$$\text{(7.2)}\qquad H^d(I, \mathbf{X}(w)) \cong H^d(I_w, k) \text{ is one dimensional for any } w \in \widetilde{W}.$$

REMARK 7.3. Let L be a proper open subgroup of I. By [**26**] Chap. 1 Prop. 30(4) and Exercise 5) respectively, $\mathrm{cores}_I^L : H^d(L,k) \to H^d(I,k)$ is a linear isomorphism while $\mathrm{res}_L^I : H^d(I,k) \to H^d(L,k)$ is the zero map.

Let $\mathcal{S} \in \mathbf{X}^\vee$ be the linear map given by

$$\mathcal{S} := \sum_{g \in G/I} \mathrm{ev}_g \ . \tag{7.3}$$

It is easy to check that $\mathcal{S} : \mathbf{X} \to k$ is G-equivariant when k is endowed with the trivial action of G. We denote by $\mathcal{S}^i := H^i(I, \mathcal{S})$ the maps induced on cohomology.

REMARK 7.4. We decompose $\mathcal{S} = \sum_{w \in \widetilde{W}} \mathcal{S}_w$ where $\mathcal{S}_w = \sum_{g \in IwI/I} \mathrm{ev}_g$. Each summand $\mathcal{S}_w : \mathbf{X} \to k$ is I-equivariant and $\mathcal{S}_w|_{\mathbf{X}(v)} = 0$ if $v \neq w \in \widetilde{W}$ and the following diagram is commutative:

$$\begin{array}{ccc} & & H^i(I, \mathbf{X}(w)) \\ & \swarrow^{\mathrm{Sh}_w} & \\ H^i(I_w, k) & & \Big\downarrow{\scriptstyle H^i(I, \mathcal{S}_w)} \\ & \searrow_{\mathrm{cores}_I^{I_w}} & \\ & & H^i(I, k) \end{array}$$

PROOF. We contemplate the larger diagram

$$\begin{array}{ccccc} H^i(I, \mathbf{X}(w)) & & \xrightarrow{\quad H^i(I, \mathcal{S}_w) \quad} & & H^i(I, k) \\ & \nwarrow^{\mathrm{cores}_I^{I_w}} & & \nwarrow^{\mathrm{cores}_I^{I_w}} & \\ \Big\downarrow{\scriptstyle \mathrm{Sh}_w} & & H^i(I_w, \mathbf{X}(w)) & \xrightarrow{H^i(I_w, \mathcal{S}_w)} & H^i(I_w, k) \\ & \nearrow^{?} & & \nearrow_{=} & \\ H^i(I_w, k). & & & & \end{array}$$

Here the map ? is induced by the map between coefficients which sends $a \in k$ to $a \, \mathrm{char}_{wI}$. The parallelogram is commutative since the corestriction is functorial in the coefficients. The right lower triangle is commutative since $\mathcal{S}_w(a \, \mathrm{char}_{wI}) = a$. The left triangle is commutative since the composite of the upwards pointing arrows is the inverse of the Shapiro isomorphism by (3.3). □

LEMMA 7.5. *For $0 \leq i \leq d$, the bilinear map defined by the composite*

$$H^i(I, \mathbf{X}) \otimes_k H^{d-i}(I, \mathbf{X}) \xrightarrow{\cup} H^d(I, \mathbf{X}) \xrightarrow{\mathcal{S}^d} H^d(I, k) \cong k$$

is nondegenerate.

PROOF. Let $w \in \widetilde{W}$. We consider the diagram

$$\begin{array}{ccccc} H^i(I, \mathbf{X}(w)) \otimes_k H^{d-i}(I, \mathbf{X}(w)) & \xrightarrow{\cup} & H^d(I, \mathbf{X}(w)) & \xrightarrow{H^d(I, \mathcal{S}_w)} & H^d(I, k) \\ \cong \Big\downarrow {\scriptstyle \mathrm{Sh}_w \otimes \mathrm{Sh}_w} & & \cong \Big\downarrow {\scriptstyle \mathrm{Sh}_w} & & \Big\| \\ H^i(I_w, k) \otimes_k H^{d-i}(I_w, k) & \xrightarrow{\cup} & H^d(I_w, k) & \xrightarrow[\cong]{\mathrm{cores}_I^{I_w}} & H^d(I, k), \end{array}$$

where the lower right corestriction map is an isomorphism by Remark 7.3. The left square is commutative by (3.6) and the right one by Remark 7.4. The lower pairing is nondegenerate since I_w is a Poincaré group of dimension d. Therefore, the top horizontal composite induces a perfect pairing. Using (3.5), this proves the lemma. □

7.2.1. *Congruence subgroups.* We consider a smooth affine group scheme $\mathcal{G} = \mathrm{Spec}(A)$ over $\mathfrak{O}$ of dimension δ. In particular, A is an $\mathfrak{O}$-algebra via a homomorphism $\alpha : \mathfrak{O} \to A$. A point $s \in \mathcal{G}(\mathfrak{O})$ is an $\mathfrak{O}$-algebra homomorphism $s : A \to \mathfrak{O}$; it necessarily satisfies $s \circ \alpha = \mathrm{id}$. The reduction map is

$$\begin{aligned}\mathcal{G}(\mathfrak{O}) &\longrightarrow \mathcal{G}(\mathfrak{O}/\pi\mathfrak{O})\\ s &\longmapsto \bar{s} := [A \xrightarrow{s} \mathfrak{O} \xrightarrow{\mathrm{pr}} \mathfrak{O}/\pi\mathfrak{O}]\ .\end{aligned}$$

Let $\epsilon : A \to \mathfrak{O}$ denote the unit element in $\mathcal{G}(\mathfrak{O})$; then $\bar{\epsilon}$ is the unit element in $\mathcal{G}(\mathfrak{O}/\pi\mathfrak{O}$.

Let $\mathfrak{p} := \ker(\epsilon)$. The formal completion $\widehat{\mathcal{G}}$ of $\mathcal{G}$ in the unit section is the formal group scheme $\widehat{\mathcal{G}} := \mathrm{Spf}(\widehat{A}^{\mathfrak{p}})$ where $\widehat{A}^{\mathfrak{p}}$ is the $\mathfrak{p}$-adic completion of A. By our smoothness assumption $\widehat{A}^{\mathfrak{p}} = \mathfrak{O}[[X_1, \ldots, X_\delta]]$ is a formal power series ring in δ many variables $X_1, \ldots, X_\delta$. A point in $\widehat{\mathcal{G}}(\mathfrak{O})$ is a point $s : A \to \mathfrak{O}$ in $\mathcal{G}(\mathfrak{O})$ which extends to a continuous homomorphism $s : \widehat{A}^{\mathfrak{p}} \to \mathfrak{O}$, i.e., which satisfies $s(\mathfrak{p}) \subseteq \mathfrak{M}$, or equivalently, $\bar{s}(\mathfrak{p}) = 0$. One checks that $\bar{s}(\mathfrak{p}) = 0$ if and only if $\bar{s} = \bar{\epsilon}$. This shows that

$$\widehat{\mathcal{G}}(\mathfrak{O}) = \ker\left(\mathcal{G}(\mathfrak{O}) \xrightarrow{\text{reduction}} \mathcal{G}(\mathfrak{O}/\pi\mathfrak{O})\right)\ .$$

On the other hand we have the bijection

$$\begin{aligned}\xi : \widehat{\mathcal{G}}(\mathfrak{O}) &\xrightarrow{\simeq} \mathfrak{M}^\delta\\ s &\longmapsto (s(X_1), \ldots s(X_\delta))\ .\end{aligned}$$

We see that $\widehat{\mathcal{G}}(\mathfrak{O})$ is a standard formal group in the sense of [**27**] II Chap. IV §8. We then have in $\widehat{\mathcal{G}}(\mathfrak{O})$ the descending sequence of normal subgroups

$$\widehat{\mathcal{G}}_m(\mathfrak{O}) := \xi^{-1}((\pi^m\mathfrak{O})^\delta) \qquad \text{for } m \geq 1$$

(loc. cit. II Chap. IV §9). It is clear that

$$\widehat{\mathcal{G}}_m(\mathfrak{O}) = \ker\left(\mathcal{G}(\mathfrak{O}) \xrightarrow{\text{reduction}} \mathcal{G}(\mathfrak{O}/\pi^m\mathfrak{O})\right)\ .$$

PROPOSITION 7.6. *Suppose that $\mathfrak{O} = \mathbb{Z}_p$; then $\widehat{\mathcal{G}}_m(\mathbb{Z}_p)$, for any $m \geq 1$ if $p \neq 2$, resp. $m \geq 2$ if $p = 2$, is a uniform pro-p group.*

PROOF. By [**7**] §13.2 (the discussion before Lemma 13.21) and Exercise 5 the group $\widehat{\mathcal{G}}_m(\mathfrak{O})$ is standard in the sense of loc. cit. Def. 8.22. Hence it is uniform by loc. cit. Thm. 8.31. □

To treat the general case we observe that the Weil restriction $\mathcal{G}_0 := \mathrm{Res}_{\mathfrak{O}/\mathbb{Z}_p}(\mathcal{G})$ is a smooth affine group scheme over $\mathbb{Z}_p$ (cf. [**2**] §7.6 Thm. 4 and Prop. 5). Let $e(\mathfrak{F}/\mathbb{Q}_p)$ denote the ramification index of the extension $\mathfrak{F}/\mathbb{Q}_p$. By the definition of the Weil restriction we have $\mathcal{G}_0(\mathbb{Z}_p) = \mathcal{G}(\mathfrak{O})$ and

$$\mathcal{G}_0(\mathbb{Z}_p/p^m\mathbb{Z}_p) = \mathcal{G}(\mathfrak{O}/p^m\mathfrak{O}) = \mathcal{G}(\mathfrak{O}/\pi^{me(\mathfrak{F}/\mathbb{Q}_p)}\mathfrak{O}) \quad \text{for } m \geq 1.$$

We therefore obtain the following consequence of the above proposition.

COROLLARY 7.7. *Let $m = je(\mathfrak{F}/\mathbb{Q}_p)$ with $j \geq 1$ if $p \neq 2$, resp. $j \geq 2$ if $p = 2$. Then $\widehat{\mathcal{G}}_m(\mathfrak{O})$ is a uniform pro-p group.*

7.2.2. *Bruhat-Tits group schemes.* We fix a facet F in the standard apartment $\mathscr{A}$. Let $\mathbf{G}_F$ denote the Bruhat-Tits group scheme over $\mathfrak{O}$ corresponding to F (cf. [**31**]). It is affine smooth with general fiber $\mathbf{G}$, and $\mathbf{G}_F(\mathfrak{O})$ is the pointwise stabilizer in G of the preimage of F in the extended building (denoted by $\mathscr{G}_{\mathrm{pr}^{-1}(F)}$ in [**31**] 3.4.1). Its neutral component is denoted by $\mathbf{G}_F^\circ$. The group of points $K_F := \mathbf{G}_F^\circ(\mathfrak{O})$ is the parahoric subgroup associated with the facet F. We introduce the descending sequence of normal congruence subgroups

$$K_{F,m} := \ker\Big(\mathbf{G}_F^\circ(\mathfrak{O}) \xrightarrow{\text{reduction}} \mathbf{G}_F^\circ(\mathfrak{O}/\pi^m\mathfrak{O})\Big) \quad \text{for } m \geq 1$$

in K_F. Let $\mathcal{P}_F^\dagger$ denote the stabilizer of F in G. It follows from [**31**] 3.4.3 (or [**6**] 4.6.17) that each $K_{F,m}$, in fact, is a normal subgroup of $\mathcal{P}_F^\dagger$. Note that, given F, any open subgroup of G contains some $K_{F,m}$.

COROLLARY 7.8. *For any $m = je(\mathfrak{F}/\mathbb{Q}_p)$ with $j \geq 2$ the group $K_{F,m}$ is a uniform pro-p group.*

PROOF. Apply Cor. 7.7 with $\mathcal{G} := \mathbf{G}_F^\circ$. □

In the following we will determine the groups $K_{F,m}$ in terms of the root subgroups $\mathfrak{U}_\alpha$ and the torus $\mathbf{T}$. Let $\mathscr{T}$ over $\mathfrak{O}$ denote the neutral component of the Neron model of $\mathbf{T}$. We have $\mathscr{T}(\mathfrak{O}) = T^0$, and we put

$$T^m := \ker\Big(\mathscr{T}(\mathfrak{O}) \xrightarrow{\text{reduction}} \mathscr{T}(\mathfrak{O}/\pi^m\mathfrak{O})\Big) \quad \text{for } m \geq 1.$$

By [**6**] 5.2.2-4 the group scheme $\mathbf{G}_F^\circ$ possesses, for each root $\alpha \in \Phi$, a smooth closed $\mathfrak{O}$-subgroup scheme $\mathscr{U}_{\alpha,F}$ such that

$$\mathscr{U}_{\alpha,F}(\mathfrak{O}) = \mathfrak{U}_{\alpha,f_F(\alpha)} \,. \tag{7.4}$$

Moreover the product map induces an open immersion of $\mathfrak{O}$-schemes

$$\prod_{\alpha\in\Phi^-} \mathscr{U}_{\alpha,F} \times \mathscr{T} \times \prod_{\alpha\in\Phi^+} \mathscr{U}_{\alpha,F} \hookrightarrow \mathbf{G}_F^\circ \,. \tag{7.5}$$

PROPOSITION 7.9. *For any $m \geq 1$ the map* (7.5) *induces the equality*

$$\prod_{\alpha\in\Phi^-} \mathfrak{U}_{\alpha,f_F(\alpha)+m} \times T^m \times \prod_{\alpha\in\Phi^+} \mathfrak{U}_{\alpha,f_F(\alpha)+m} = K_{F,m} \,.$$

PROOF. Let $\mathscr{Y}$ denote the left hand side of the open immersion (7.5). Because of (7.4) the left hand side of our assertion is equal to the subset of all points in $\mathscr{Y}(\mathfrak{O})$ which reduce to the unit element modulo π^m and hence is contained in $K_{F,m}$. On the other hand it follows from [**25**] Prop. I.2.2 and (7.4) that any point in $\mathbf{G}_F^\circ(\mathfrak{O})$, which reduces to a point of the unipotent radical of its special fiber, already lies in $\mathscr{Y}(\mathfrak{O})$. It follows that $K_{F,m}$ corresponds to points in $\mathscr{Y}(\mathfrak{O})$ which reduce to the unit element modulo π^m and hence is contained in the left hand side of the assertion. □

REMARK 7.10. In Chap. I of [**25**] certain pro-p subgroups $U_F^{(e)} \subseteq G$ for $e \geq 0$ were introduced and studied. If $F = x$ is a hyperspecial vertex then $K_{x,m} = U_x^{(m+1)}$. On the other hand, if $F = D$ is a chamber then $K_{D,m} = U_D^{(m)} T^m$.

COROLLARY 7.11. *Suppose that m is large enough so that $K_{F,m}$ is uniform. Then the Frattini quotient $(K_{F,m})_\Phi$ of $K_{F,m}$ satisfies*

$$\prod_{\alpha\in\Phi^-} \frac{\mathfrak{U}_{\alpha,f_F(\alpha)+m}}{\mathfrak{U}^p_{\alpha,f_F(\alpha)+m}} \times \frac{T^m}{(T^m)^p} \times \prod_{\alpha\in\Phi^+} \frac{\mathfrak{U}_{\alpha,f_F(\alpha)+m}}{\mathfrak{U}^p_{\alpha,f_F(\alpha)+m}} \xrightarrow{\sim} (K_{F,m})_\Phi \ .$$

PROOF. As a consequence of Prop. 7.9 the map in the assertion, which is given by multiplication, exists and is a surjection of $\mathbb{F}_p$-vector spaces. But both sides have the same dimension d. Hence the map is an isomorphism. □

In the case where the facet is a vertex x in the closure of our fixed chamber C we also introduce the notation

$$\begin{aligned}
&\Phi_x := \{(\alpha,\mathfrak{h}) \in \Phi_{aff} : (\alpha,\mathfrak{h})(x) = 0\},\ \Phi_x^\pm := \Phi_x \cap \Phi_{aff}^\pm, \\
&\Pi_x := \Phi_x \cap \Pi_{aff},\ S_x := \{s \in S_{aff} : s(x) = x\}, \\
&W_x := \text{subgroup of } W_{aff} \text{ generated by all } s_{(\alpha;\mathfrak{h})} \text{ such that } (\alpha,\mathfrak{h}) \in \Phi_x\}.
\end{aligned}$$

The pair (W_x, S_x) is a Coxeter system with finite group W_x (cf. [**20**] §4.3 and the references therein).

For any such vertex we have the inclusions $K_{x,1} \subseteq I \subseteq J \subseteq K_x$.

LEMMA 7.12. *The parahoric subgroup K_x is the disjoint union of the double cosets JwJ for all $w \in W_x$.*

PROOF. See [**20**] Lemma 4.9. □

7.2.3. *Triviality of actions on the top cohomology.* We recall from section 2.1.3 that I and J are normal subgroups of $\mathcal{P}_C^\dagger$ and that $\mathcal{P}_C^\dagger = \bigcup_{\omega\in\Omega} \omega J$. In the case where the root system is irreducible the following result was shown in [**13**] Thm. 7.1. The first part of our proof is essentially a repetition of his arguments.

LEMMA 7.13. *For $g \in \mathcal{P}_C^\dagger$, the endomorphism g_* on the one dimensional k-vector space $H^d(I,k)$ is the identity.*

PROOF. As noted above, g normalizes each subgroup $K_{C,m}$. Moreover, by Lemma 2.1.ii and Prop. 7.9, $K_{C,m}$ is contained in I. Hence the same argument as at the beginning of the proof of Lemma 7.15 reduces us to showing that the endomorphism g_* on the one dimensional k-vector space $H^d(K_{C,m},k)$ is the identity. Using Cor. 7.8 we may, by choosing m large enough, assume that $K_{C,m}$ is a uniform pro-p group.

Then, by [**15**] V.2.2.6.3 and V.2.2.7.2, the one dimensional k-vector space $H^d(K_{C,m},k)$ is the maximal exterior power (via the cup product) of the d dimensional k-vector space $H^1(K_{C,m},k)$. Conjugation commuting with the cup product, we see that the endomorphism g_* on $H^d(K_{C,m},k)$ is the determinant of g_* on $H^1(K_{C,m},k)$. We have

$$H^1(K_{C,m},k) = \operatorname{Hom}_{\mathbb{F}_p}((K_{C,m})_\Phi, k)$$

where $(K_{C,m})_\Phi$ is the Frattini quotient of the group $K_{C,m}$. This further reduces us to showing that the conjugation by g on $(K_{C,m})_\Phi$ has trivial determinant. In Cor. 7.11 we computed this Frattini quotient to be

$$(K_{C,m})_\Phi = \prod_{\alpha\in\Phi^-} \frac{\mathfrak{U}_{\alpha,f_C(\alpha)+m}}{\mathfrak{U}^p_{\alpha,f_C(\alpha)+m}} \times \frac{T^m}{(T^m)^p} \times \prod_{\alpha\in\Phi^+} \frac{\mathfrak{U}_{\alpha,f_C(\alpha)+m}}{\mathfrak{U}^p_{\alpha,f_C(\alpha)+m}} \ .$$

using Lemma 2.1.i this simplifies to

$$(7.6) \qquad (K_{C,m})_\Phi = \prod_{\alpha\in\Phi^-} \frac{\mathfrak{U}_{\alpha,m+1}}{\mathfrak{U}^p_{\alpha,m+1}} \times \frac{T^m}{(T^m)^p} \times \prod_{\alpha\in\Phi^+} \frac{\mathfrak{U}_{\alpha,m}}{\mathfrak{U}^p_{\alpha,m}} .$$

Recall that, for any $\alpha \in \Phi$, we have the additive isomorphism $x_\alpha : \mathfrak{F} \xrightarrow{\cong} \mathfrak{U}_\alpha$ defined in (2.14) by $x_\alpha(u) := \varphi_\alpha((\begin{smallmatrix} 1 & u \\ 0 & 1 \end{smallmatrix}))$. Put $s_0 := \left(\begin{smallmatrix} 0 & 1 \\ -1 & 0 \end{smallmatrix}\right) \in \mathrm{SL}_2(\mathfrak{F})$ and $n_\alpha := \varphi_\alpha(s_0)$. We observe that

- $n_\alpha = n_{s_{(\alpha,0)}}$, and
- $x_{-\alpha}(u) = n_\alpha x_\alpha(u) n_\alpha^{-1} = \varphi_\alpha((\begin{smallmatrix} 1 & 0 \\ -u & 1 \end{smallmatrix}))$ for any $u \in \mathfrak{F}$.

By [**31**] 1.1 and 1.4 the map x_α restricts, for any $r \in \mathbb{Z}$, to an isomorphism $\pi^r\mathfrak{O} \xrightarrow{\cong} \mathfrak{U}_{\alpha,r}$. This implies that all the $\mathbb{F}_p$-vector spaces $\frac{\mathfrak{U}_{\alpha,m}}{\mathfrak{U}^p_{\alpha,m}}$ and $\frac{\mathfrak{U}_{-\alpha,m+1}}{\mathfrak{U}^p_{-\alpha,m+1}}$, for $\alpha \in \Phi^+$, have the same dimension equal to $[\mathfrak{F} : \mathbb{Q}_p]$.

First let $g \in T^0$. Obviously g centralizes T^m. It acts on $\mathfrak{U}_\alpha$, resp. $\mathfrak{U}_{-\alpha}$, via α, resp. $-\alpha$. Therefore on the right hand side of (7.6) the conjugation by g visibly has trivial determinant. Since the conjugation action of I on $H^d(I,k)$ is trivial we obtain our assertion for any $g \in J$.

For the rest of the proof we fix an $\omega \in \Omega$. It remains to establish our assertion for the elements $g \in \omega J$. In fact, by the above observation, it suffices to do this for one specific $\ddot{\omega} \in \omega J$, which we choose as follows. We write the image of ω in W as a reduced product $s_{\alpha_1} \cdots s_{\alpha_\ell}$ of simple reflections and put

$$w_\omega := n_{\alpha_1} \cdots n_{\alpha_\ell} \in K_{x_0} .$$

Then $t := \omega w_\omega^{-1} \in T$, and we now define

$$\ddot{\omega} := t w_\omega .$$

Let $\Phi = \Phi_1 \dot{\cup} \ldots \dot{\cup} \Phi_r$ be the decomposition into orbits of (the image in W of) ω and put

$$\Theta_i := \Big(\prod_{\alpha\in\Phi^-\cap\Phi_i} \frac{\mathfrak{U}_{\alpha,m+1}}{\mathfrak{U}^p_{\alpha,m+1}} \Big) \times \Big(\prod_{\alpha\in\Phi^+\cap\Phi_i} \frac{\mathfrak{U}_{\alpha,m}}{\mathfrak{U}^p_{\alpha,m}} \Big) .$$

The Chevalley basis $(x_\alpha)_{\alpha\in\Phi}$ has the following property (cf. [**6**] 3.2):

(7.7) For any $\alpha \in \Phi$ there exists $\epsilon_{\alpha,\beta} \in \{\pm 1\}$ such that

$$x_{s_\beta(\alpha)}(u) = n_\beta x_\alpha(\epsilon_{\alpha,\beta} u) n_\beta^{-1} \quad \text{for any } u \in \mathfrak{F}.$$

This implies that the conjugation by w_ω preserves each Θ_i. The conjugation t_* preserves each root subgroup. Since $\ddot{\omega}_*$ preserves $(K_{C,m})_\Phi$ it follows that $\ddot{\omega}_*$ preserves each Θ_i. Setting $\Theta_0 := \frac{T^m}{(T^m)^p}$ we conclude that

$$(K_{C,m})_\Phi = \Theta_0 \times \Theta_1 \times \cdots \times \Theta_r$$

is an $\ddot{\omega}_*$-invariant decomposition. We will determine the determinant of $\ddot{\omega}_*$ on each factor.

The $\ddot{\omega}_*$-action on

$$\Theta_0 = \frac{X_*(T)}{pX_*(T)} \otimes_{\mathbb{F}_p} \frac{1+\pi^m\mathfrak{O}}{(1+\pi^m\mathfrak{O})^p}$$

is through the product $s_{\alpha_1} \cdots s_{\alpha_\ell} \in W$ acting on the left factor. Note that the dimension of the $\mathbb{F}_p$-vector space $\frac{1+\pi^m\mathfrak{O}}{(1+\pi^m\mathfrak{O})^p}$ is equal to $[\mathfrak{F} : \mathbb{Q}_p]$. Let $Q^\vee \subseteq X_*(T)$

denote the coroot lattice. By [**3**] VI.1.9 Prop. 27 the action of a simple reflection s_β on the quotient $X_*(T)/Q^\vee$ is trivial. Therefore in the exact sequence

$$0 \longrightarrow \mathrm{Tor}_1^{\mathbb{Z}}(X_*(T)/Q^\vee, \mathbb{F}_p) \longrightarrow Q^\vee \otimes_{\mathbb{Z}} \mathbb{F}_p \longrightarrow X_*(T) \otimes_{\mathbb{Z}} \mathbb{F}_p \longrightarrow (X_*(T)/Q^\vee) \otimes_{\mathbb{Z}} \mathbb{F}_p \longrightarrow 0$$

the action of s_β on the two outer terms is trivial. Hence the determinant of s_β on $X_*(T)/pX_*(T)$ is equal to its determinant on $Q^\vee/pQ^\vee$. If $p \neq 2$ then s_β is a reflection on the $\mathbb{F}_p$-vector space $Q^\vee/pQ^\vee$ and therefore has determinant -1. For $p = 2$, as $s_\beta^2 = \mathrm{id}$, the determinant is $-1 = 1$ as well. We deduce that

(7.8) the determinant of $\ddot{\omega}_*$ on Θ_0 is equal to $(-1)^{\ell[\mathfrak{F}:\mathbb{Q}_p]}$.

Next we consider Θ_i for some $1 \leq i \leq r$. Let c_i denote its cardinality, and fix some root $\beta \in \Phi_i$. We distinguish two cases.

Case 1: $-\Phi_i \cap \Phi_i = \emptyset$. Note that $\Phi_{i'} := -\Phi_i$ is the orbit of $-\beta$. By interchanging i and i' we may assume that $\beta \in \Phi^+$. We then have for completely formal reasons that

$$\text{determinant of } \ddot{\omega}_* \text{ on } \Theta_i = (-1)^{(c_i-1)[\mathfrak{F}:\mathbb{Q}_p]} \cdot \text{determinant of } \ddot{\omega}_*^{c_i} \text{ on } \frac{\mathcal{U}_{\beta,m}}{\mathcal{U}_{\beta,m}^p}$$

and correspondingly that

$$\text{determinant of } \ddot{\omega}_* \text{ on } \Theta_{i'} = (-1)^{(c_i-1)[\mathfrak{F}:\mathbb{Q}_p]} \cdot \text{determinant of } \ddot{\omega}_*^{c_i} \text{ on } \frac{\mathcal{U}_{-\beta,m+1}}{\mathcal{U}_{-\beta,m+1}^p}.$$

We have $\ddot{\omega}^{c_i} = w_\omega^{c_i} t_i$ for some $t_i \in T$. The determinants of t_{i*} on $\frac{\mathcal{U}_{\beta,m}}{\mathcal{U}_{\beta,m}^p}$ and $\frac{\mathcal{U}_{-\beta,m+1}}{\mathcal{U}_{-\beta,m+1}^p}$ (we must have $\beta(t_i) \in \mathfrak{O}^\times$) are inverse to each other. On the other hand let w_i denote the image of $w_\omega^{c_i}$ in W, which fixes β. As a consequence of the property (7.7) of the Chevalley basis, which we have recalled above, there are signs $\epsilon_{\pm\beta,w_i} \in \{\pm 1\}$ such that

$$x_{\pm\beta}(u) = w_\omega^{c_i} x_{\pm\beta}(\epsilon_{\pm\beta,w_i} u) w_\omega^{-c_i} \qquad \text{for any } u \in \mathfrak{F}.$$

Altogether we obtain that

$$\text{determinant of } \ddot{\omega}_* \text{ on } \Theta_i \times \Theta_{i'} = (\epsilon_{\beta,w_i}\epsilon_{-\beta,w_i})^{[\mathfrak{F}:\mathbb{Q}_p]} .$$

But we have $\epsilon_{\beta,w_i} = \epsilon_{-\beta,w_i}$. This follows by a straightforward induction from the fact that $\epsilon_{-\beta,\alpha} = \epsilon_{\beta,\alpha}$ (cf. [**30**] Lemma 9.2.2(ii)). Hence in the present case we deduce that

(7.9) the determinant of $\ddot{\omega}_*$ on $\Theta_i \times \Theta_{i'}$ is equal to 1.

Case 2: $-\Phi_i = \Phi_i$. Again we may assume that $\beta \in \Phi^+$. Then c_i is even, and $\ddot{\omega}_*^{c_i/2}$ preserves the product $\frac{\mathcal{U}_{\beta,m}}{\mathcal{U}_{\beta,m}^p} \times \frac{\mathcal{U}_{-\beta,m+1}}{\mathcal{U}_{-\beta,m+1}^p}$. The formal argument now says that

$$\text{determinant of } \ddot{\omega}_* \text{ on } \Theta_i = \text{determinant of } \ddot{\omega}_*^{c_i/2} \text{ on } \frac{\mathcal{U}_{\beta,m}}{\mathcal{U}_{\beta,m}^p} \times \frac{\mathcal{U}_{-\beta,m+1}}{\mathcal{U}_{-\beta,m+1}^p}.$$

This time we write $\ddot{\omega}^{c_i/2} = w_\omega^{c_i/2} t_i$ for some $t_i \in T$ and we let w_i denote the image of $w_\omega^{c_i/2}$ in W, which maps β to $-\beta$. We have $\beta(t_i) \in \pi\mathfrak{O}^\times$. Using the Chevalley basis we compute the above right hand determinant as being equal to

$$(-1)^{[\mathfrak{F}:\mathbb{Q}_p]}(\beta(t_i)\epsilon_{\beta,w_i})^{[\mathfrak{F}:\mathbb{Q}_p]}(\beta(t_i)^{-1}\epsilon_{-\beta,w_i})^{[\mathfrak{F}:\mathbb{Q}_p]} = (-1)^{[\mathfrak{F}:\mathbb{Q}_p]} .$$

Hence in this case we deduce that

(7.10) $\qquad$ the determinant of $\ddot{\omega}_*$ on Θ_i is equal to $(-1)^{[\mathfrak{F}:\mathbb{Q}_p]}$.

Combining (7.8), (7.9), and (7.10) we have established at this point that

(7.11) $\qquad$ the determinant of $\ddot{\omega}_*$ on $(K_{C,m})_\Phi$ is equal to $(-1)^{(N+\ell)[\mathfrak{F}:\mathbb{Q}_p]}$

where N is the number of ω-orbits $\Phi_i = -\Phi_i$ and, we repeat, ℓ is the length of w_ω. It remains to show that the sum $N + \ell$ always is even. This will be done in the subsequent lemma in a more general situation. □

LEMMA 7.14. *Let $w \in W$ be any element and let $N(w)$ be the number of $w^{\mathbb{Z}}$-orbits $\Psi \subseteq \Phi$ with the property that $-\Psi = \Psi$; then the number $N(w) + \ell(w)$ is even.*

PROOF. We have to show that $(-1)^{N(w)} = (-1)^{\ell(w)}$ holds true for any $w \in W$.

Step 1: The map $w \mapsto (-1)^{\ell(w)}$ is a homomorphism. This is immediate from the fact that $(-1)^{\ell(w)}$ is equal to the determinant of w in the reflection representation of W (cf. [**9**]).

Step 2: The map $w \mapsto (-1)^{N(w)}$ is a homomorphism. For this let $\mathbb{Z}[S]$ denote the free abelian group on a set S. We consider the obvious action of W on $\mathbb{Z}[\Phi]$. If $\Phi = \Phi_1 \dot\cup \ldots \dot\cup \Phi_r$ is the decomposition into $w^{\mathbb{Z}}$-orbits, then

$$\mathbb{Z}[\Phi] = \mathbb{Z}[\Phi_1] \oplus \ldots \oplus \mathbb{Z}[\Phi_r]$$

is a w-invariant decomposition. Obviously $\det(w|\mathbb{Z}[\Phi_i]) = (-1)^{|\Phi_i|-1}$. If $-\Phi_i \neq \Phi_i$ then $-\Phi_i = \Phi_j$ for some $j \neq i$. If $-\Phi_i = \Phi_i$ then $|\Phi|$ is even. It follows that $\det(w|\mathbb{Z}[\Phi]) = (-1)^{N(w)}$.

Step 3: If $w = s$ is a reflection at some $\alpha \in \Phi$ then $N(s) = 1$ and $(-1)^{\ell(s)} = -1$. □

LEMMA 7.15. *Let L be an open subgroup of I. For $g \in K_x$ normalizing L, the endomorphism g_* on the one dimensional k-vector space $H^d(L,k)$ is the identity.*

PROOF. We choose $m \geq 1$ large enough so that $K_{x,m}$ is contained in L. Recall that $\mathrm{cores}_L^{K_{x,m}} : H^d(K_{x,m},k) \xrightarrow{\cong} H^d(L,k)$ is an isomorphism by Remark 7.3. Since corestriction commutes with conjugation (§4.6), the following diagram commutes:

$$\begin{array}{ccc} H^d(L,k) & \xrightarrow{\;g_*\;} & H^d(L,k) \\ \cong \big\uparrow \mathrm{cores}_L^{K_{x,m}} & & \mathrm{cores}_L^{K_{x,m}} \big\uparrow \cong \\ H^d(K_{x,m},k) & \xrightarrow{\;g_*\;} & H^d(K_{x,m},k) . \end{array}$$

Therefore it is enough to prove the assertion for $L = K_{x,m}$. In this case the group K_x acts by conjugation on the one dimensional space $H^d(K_{x,m},k)$. This action is given by a character $\xi : K_x \to k^\times$. Its kernel $\Xi := \ker(\xi)$ is a normal subgroup of K_x.

First of all we recall again that the corestriction map commutes with conjugation, that $\mathrm{cores}_I^{K_{x,m}} : H^d(K_{x,m},k) \xrightarrow{\cong} H^d(I,k)$ is an isomorphism, and that conjugation by g_*, for $g \in I$, induces the identity on the cohomology $H^*(I,k)$.

Therefore we have the commutative diagram

$$\begin{array}{ccc} & & H^d(K_{x,m},k) \\ & \swarrow^{\mathrm{cores}_I^{K_{x,m}}}_{\cong} & \\ H^d(I,k) & & \uparrow g_* \\ & \nwarrow^{\cong}_{\mathrm{cores}_I^{K_{x,m}}} & \\ & & H^d(K_{x,m},k)\ . \end{array}$$

This shows that $I \subseteq \Xi$. Since $J = IT^0$ we deduce from Lemma 7.13 that even $J \subseteq \Xi$. Since (W_x, S_x) is a Coxeter system with finite group W_x, we may consider its (unique) longest element w_x. The normal subgroup Ξ of K_x then must contain $Jw_xJw_x^{-1}J$. For any $s \in S_x$, we have $\ell(sw_x) = \ell(w) - 1$. By (2.20) we have

$$JsJ \cdot Jw_xJ = Jsw_xJ \,\dot\cup\, Jw_xJ$$

and hence $\dot{s}J\dot{w}_x \cap Jw_xJ \neq \emptyset$. It follows that

$$\dot{s} \in Jw_xJw_x^{-1}J \subseteq \Xi$$

and therefore $\Xi = K_x$ by Lemma 7.12. □

PROPOSITION 7.16. *For $g \in G$, and $v, w \in \widetilde{W}$ such that $\ell(vw) = \ell(v) + \ell(w)$, the following diagrams of one dimensional k-vector spaces are commutative:*

$$\begin{array}{ccc} & & H^d(I_g,k) \\ & \swarrow^{\mathrm{cores}_I^{I_g}}_{\cong} & \\ H^d(I,k) & & \uparrow g_* \\ & \nwarrow^{\cong}_{\mathrm{cores}_I^{I_{g^{-1}}}} & \\ & & H^d(I_{g^{-1}},k) \end{array} \qquad \textit{and} \qquad \begin{array}{ccc} & & H^d(I_{vw},k)\ . \\ & \swarrow^{\mathrm{cores}_I^{I_{vw}}}_{\cong} & \\ H^d(I,k) & & \uparrow v_* \\ & \nwarrow^{\cong}_{\mathrm{cores}_I^{v^{-1}I_{vw}v}} & \\ & & H^d(v^{-1}I_{vw}v,k). \end{array}$$

PROOF. We prove the commutativity of the left diagram. We will see along the way that the commutativity of the right one follows.

Step 1: We claim that it suffices to establish the commutativity of the left diagram for elements $w \in \widetilde{W}$.

Let $g \in G$ and $h_1, h_2 \in I$. We have the commutative diagram

$$\begin{array}{ccccccc} H^d(I_{(h_1gh_2)^{-1}},k) & \xrightarrow{h_{2*}} & H^d(I_{g^{-1}},k) & \xrightarrow{g_*} & H^d(I_g,k) & \xrightarrow{h_{1*}} & H^d(I_{h_1gh_2},k) \\ {\scriptstyle \mathrm{cores}_I^{I_{(h_1gh_2)^{-1}}}}\downarrow\cong & & {\scriptstyle \mathrm{cores}_I^{I_{g^{-1}}}}\downarrow\cong & & {\scriptstyle \mathrm{cores}_I^{I_g}}\downarrow\cong & & {\scriptstyle \mathrm{cores}_I^{I_{h_1gh_2}}}\downarrow\cong \\ H^d(I,k) & \xrightarrow{=} & H^d(I,k) & \xrightarrow{?} & H^d(I,k) & \xrightarrow{=} & H^d(I,k), \end{array}$$

where the equality signs in the lower row use the facts that corestriction commutes with conjugation (cf. §4.6) and that the conjugation by an element in I is trivial on $H^d(I,k)$. This shows that the commutativity of the left diagram in the assertion only depends on the double coset IgI.

Step 2: Let $v, w \in \widetilde{W}$ be two elements for which the left diagram commutes and such that $\ell(vw) = \ell(v) + \ell(w)$; we claim that then the left diagram for vw as well as the right diagram commute.

This is straightforward from the following commutative diagram (cf. Lemma 2.2):

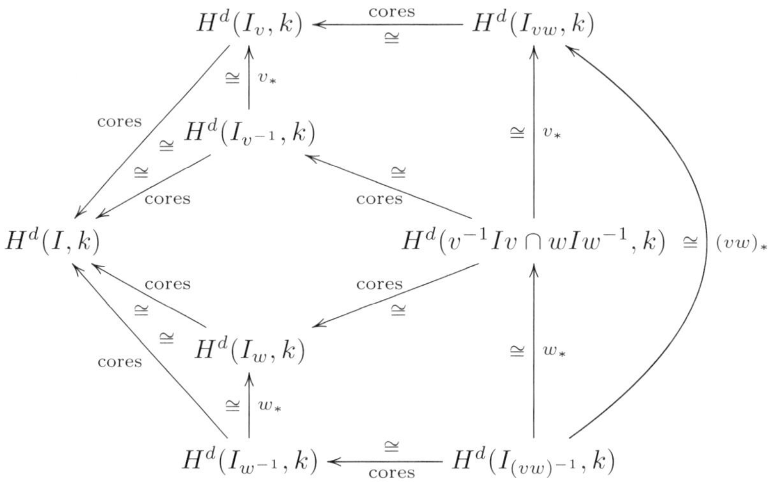

At this point we are reduced to establishing the commutativity of the left diagram in our assertion in the following two cases:

(A) $g = \dot{\omega}$ for $\omega \in \widetilde{\Omega}$. In that case, the claim is given by Lemma 7.13.

(B) $g = n_s$ for $s = s_{(\beta,\mathfrak{h})}$ where $(\beta, \mathfrak{h}) \in \Pi_{aff}$ is any simple affine root.

Step 3: It remains to treat case (B). We pick a vertex x in the closure of the chamber C such that $s(x) = x$. Then g lies in K_x and therefore normalizes $K_{x,1}$. Hence $K_{x,1} \subseteq I_g \subseteq I$, and we have the commutative diagram

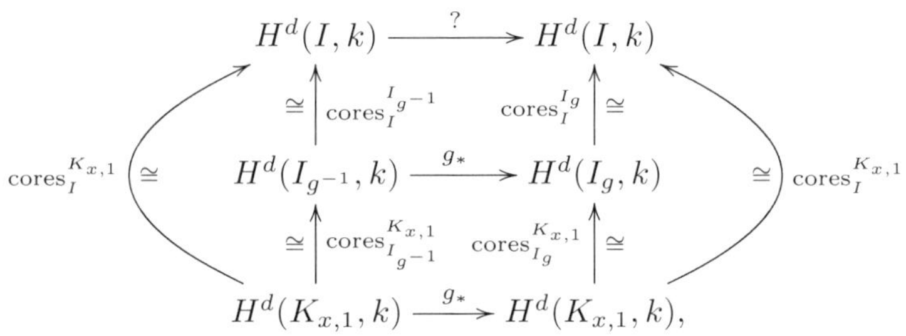

using again that corestriction commutes with conjugation. This reduces us to showing that the endomorphism g_* on $H^d(K_{x,1}, k)$ is the identity. This claim is given by Lemma 7.15. □

COROLLARY 7.17. *We have* $\mathcal{S}^d \circ \mathcal{J} = \mathcal{S}^d$ *on* $E^d = H^d(I, \mathbf{X})$.

PROOF. Let $w \in \widetilde{W}$ and $\alpha \in H^d(I, \mathbf{X}(w))$. Recalling that $\mathcal{J}(\alpha) \in H^d(I, \mathbf{X}(w^{-1}))$ satisfies $\mathrm{Sh}_{w^{-1}}(\mathcal{J}(\alpha)) = (w^{-1})_* \mathrm{Sh}_w(\alpha)$, we have

$$\begin{aligned}
\mathcal{S}^d \circ \mathcal{J}(\alpha) &= H^d(I, \mathcal{S}_{w^{-1}}) \circ \mathcal{J}(\alpha) = H^d(I, \mathcal{S}_{w^{-1}}) \circ \mathrm{Sh}_{w^{-1}}^{-1} \circ (w^{-1})_*(\mathrm{Sh}_w(\alpha)) \\
&= \mathrm{cores}_I^{I_{w^{-1}}} \circ (w^{-1})_*(\mathrm{Sh}_w(\alpha)) && \text{by Remark 7.4} \\
&= \mathrm{cores}_I^{I_w}(\mathrm{Sh}_w(\alpha)) && \text{by Prop. 7.16} \\
&= H^d(I, \mathcal{S}_w)(\alpha) = \mathcal{S}^d(\alpha) && \text{by Remark 7.4.}
\end{aligned}$$

□

7.2.4. *The duality theorem.* We fix an isomorphism $\eta : H^d(I,k) \longrightarrow k$. By Lemma 7.5 the map

$$\begin{aligned}\Delta^i : E^i = H^i(I,\mathbf{X}) &\longrightarrow H^{d-i}(I,\mathbf{X})^\vee = (E^{d-i})^\vee \\ \alpha &\longmapsto l_\alpha(\beta) := \eta \circ \mathcal{S}^d(\alpha \cup \beta)\end{aligned} \tag{7.12}$$

is a linear injection with image $(E^{d-i})^{\vee,f}$. The space E^{d-i} is naturally a bimodule under H. As in §7.1, we consider the twisted H-bimodule ${}^{\mathcal{J}}(E^{d-i})^{\mathcal{J}}$ namely the space E^{d-i} with the action of H on $\beta \in E^{d-i}$ given by

$$(\tau, \beta, \tau') \mapsto \mathcal{J}(\tau') \cdot \beta \cdot \mathcal{J}(\tau) \quad \text{for } \tau, \tau' \in H.$$

PROPOSITION 7.18. *The map* (7.12) *induces an injective morphism of H-bimodules*

$$E^i \longrightarrow ({}^{\mathcal{J}}(E^{d-i})^{\mathcal{J}})^\vee$$

with image $({}^{\mathcal{J}}(E^{d-i})^{\mathcal{J}})^{\vee,f}$.

PROOF. The fact that the map is injective with image $(E^{d-i})^{\vee,f}$ comes directly from Lemma 7.5 and its proof.

We prove that for all $\alpha \in H^i(I,\mathbf{X})$ and $\beta \in H^{d-i}(I,\mathbf{X})$ and all $\tau, \tau' \in H$, we have $l_{\tau\cdot\alpha\cdot\tau'}(\beta) = l_\alpha(\mathcal{J}(\tau)\cdot\beta\cdot\mathcal{J}(\tau'))$ namely

$$\mathcal{S}^d(\tau\cdot\alpha\cdot\tau' \cup \beta) = \mathcal{S}^d(\alpha \cup \mathcal{J}(\tau)\cdot\beta\cdot\mathcal{J}(\tau'))\ . \tag{7.13}$$

We first show that

$$\mathcal{S}^d(\alpha\cdot\tau' \cup \beta) = \mathcal{S}^d(\alpha \cup \beta\cdot\mathcal{J}(\tau'))\ . \tag{7.14}$$

The right action of H on $H^*(I,\mathbf{X})$ being through the coefficients it is enough to prove that for any $a, b \in \mathbf{X}$ and $\tau' \in H$ we have

$$\mathcal{S}((a\cdot\tau')b - a(b\cdot\mathcal{J}(\tau'))) = 0\ .$$

We check this equality for $a = \mathrm{char}_{xI}$ and $b = \mathrm{char}_{yI}$ with $x, y \in G$ and $\tau' = \tau_g$ with $g \in G$. We then have

$$(a\cdot\tau')b - a(b\cdot\mathcal{J}(\tau')) = \mathrm{char}_{xIgI\cap yI} - \mathrm{char}_{xI\cap yIg^{-1}I}\ .$$

It lies in the kernel of $\mathcal{S}$ if and only if $xIgI \cap yI/I$ and $xI \cap yIg^{-1}I/I$ have the same cardinality. But observe that $xIgI \cap yI$ is equal to yI if $x^{-1}y \in IgI$ and is empty otherwise, while $xI \cap yIg^{-1}I$ is equal to xI if $y^{-1}x \in Ig^{-1}I$ and is empty otherwise. This proves the equality and (7.14) follows.

That

$$\mathcal{S}^d(\tau\cdot\alpha \cup \beta) = \mathcal{S}^d(\alpha \cup \mathcal{J}(\tau)\cdot\beta)\ . \tag{7.15}$$

holds true as well follows now from the following computation: We may assume that $\alpha \in E^i$, $\beta \in E^j$, and $\gamma \in E^m$. We then compute

$$\begin{aligned}
\mathcal{S}^d(\tau \cdot \alpha \cup \beta) &= \mathcal{S}^d(\mathcal{J}(\tau \cdot \alpha \cup \beta)) && \text{by Cor. 7.17}\\
&= \mathcal{S}^d(\mathcal{J}(\tau \cdot \alpha) \cup \mathcal{J}(\beta)) && \text{by Remark 6.2}\\
&= \mathcal{S}^d(\mathcal{J}(\alpha) \cdot \mathcal{J}(\tau) \cup \mathcal{J}(\beta)) && \text{by Prop. 6.1}\\
&= \mathcal{S}^d(\mathcal{J}(\alpha) \cup \mathcal{J}(\beta) \cdot \tau) && \text{by (7.14)}\\
&= \mathcal{S}^d(\mathcal{J}(\alpha) \cup \mathcal{J}(\mathcal{J}(\tau) \cdot \beta)) && \text{by Prop. 6.1}\\
&= \mathcal{S}^d(\mathcal{J}(\alpha \cup \mathcal{J}(\tau) \cdot \beta)) && \text{by Remark 6.2}\\
&= \mathcal{S}^d(\alpha \cup \mathcal{J}(\tau) \cdot \beta) && \text{by Cor. 7.17.}
\end{aligned}$$

□

COROLLARY 7.19. *For any $i \in \{0, \cdots, d\}$ the space $(E^i)^{\vee,f}$ is a sub-H-bimodule of $(E^i)^\vee$.*

8. The structure of E^d

In this section we still **assume**, as in §7.2, that the pro-p Iwahori group I is torsion free. In this section we describe the top cohomology space $E^d = H^d(I, \mathbf{X})$ as an H-bimodule.

REMARK 8.1. The space E^d as a right H-module had already been computed in [**24**] §5.1.

Recall that the pairing $E^d \times E^0 \to k$, $(\alpha, \beta) \mapsto \eta \circ \mathcal{S}^d(\alpha \cup \beta)$ induces the isomorphism of H-bimodules

$$\Delta^d : \quad E^d \xrightarrow{\cong} ({}^{\mathcal{J}}E^{0\,\mathcal{J}})^{\vee,f} \tag{8.1}$$

of Proposition 7.18 when $i = d$. We denote by $(\phi_w)_{w \in \widetilde{W}}$ the basis of E^d obtained by dualizing the basis of $(\tau_w)_{w \in \widetilde{W}}$ of E^0. For each $w \in \widetilde{W}$, the element ϕ_w is the only one in $H^d(I, \mathbf{X}(w))$ satisfying $\eta(\mathcal{S}^d(\phi_w)) = 1$ (see (3.5)). Now $\mathcal{J}(\phi_w) \in H^d(I, \mathbf{X}(w^{-1}))$ and $\eta(\mathcal{S}^d(\mathcal{J}(\phi_w)))$ and $\eta(\mathcal{S}^d(\phi_{w^{-1}}))$ are both equal to 1 by Corollary 7.17. Therefore,

$$\mathcal{J}(\phi_w) = \phi_{w^{-1}} \ . \tag{8.2}$$

We now describe the explicit action of H on the elements $(\phi_w)_{w \in \widetilde{W}}$ in E^d. For any $s \in S_{aff}$ recall that we introduced the idempotent element θ_s in (2.27):

$$\theta_s := -|\mu_{\check{\alpha}}| \sum_{t \in \check{\alpha}([\mathbb{F}_q^\times])} \tau_t \ \in H \ .$$

PROPOSITION 8.2. *Let $w \in \widetilde{W}$, $\omega \in \widetilde{\Omega}$ and $s \in S_{aff}$. We have the formulas:*

$$\phi_w \cdot \tau_\omega = \phi_{w\omega} \ , \ \tau_\omega \cdot \phi_w = \phi_{\omega w}, \tag{8.3}$$

(8.4)

$$\phi_w \cdot \tau_{n_s} = \begin{cases} \phi_{w\tilde{s}} + |\mu_{\check{\alpha}}| \sum_{t \in \check{\alpha}([\mathbb{F}_q^\times])} \phi_{wt} = \phi_{w\tilde{s}} - \phi_w \cdot \theta_s & \text{if } \ell(w\tilde{s}) = \ell(w) - 1, \\ 0 & \text{if } \ell(w\tilde{s}) = \ell(w) + 1, \end{cases}$$

(8.5)
$$\tau_{n_s}\cdot\phi_w=\begin{cases}\phi_{\tilde{s}w}+|\mu_{\check{\alpha}}|\sum_{t\in\check{\alpha}([\mathbb{F}_q^\times])}\phi_{\bar{t}w}=\phi_{\tilde{s}w}-\theta_s\cdot\phi_w & \text{if } \ell(\tilde{s}w)=\ell(w)-1,\\ 0 & \text{if } \ell(\tilde{s}w)=\ell(w)+1.\end{cases}$$

PROOF. Before proving the proposition, we recall that the quadratic relations in H are given by $\tau_{n_s}^2=-\theta_s\tau_{n_s}=\tau_{n_s}\theta_s$ for any $s\in S_{aff}$. Let $w\in\widetilde{W}$, $s\in S_{aff}$, and $\omega\in\Omega$. We study the right action of τ_ω and of τ_{n_s} on $\phi_w\in E^d$. Recall that $\phi_w\cdot\tau_\omega(\_)=\phi_w(\_\tau_{\omega^{-1}})$ and $\phi_w\cdot\tau_{n_s}(\_)=\phi_w(\_\tau_{n_s^{-1}})$ (see (7.13)). Let $w'\in\widetilde{W}$. Below we use the braid relation (2.26) repeatedly.

- We have $\phi_w\cdot\tau_\omega(\tau_{w'})=\phi_w(\tau_{w'\omega^{-1}})$. It is nonzero if and only if $w'=w\omega$ in which case $\phi_w\cdot\tau_\omega(\tau_{w'})=1$.
- If $\ell(w'\tilde{s})=\ell(w')+1$ then $\phi_w\cdot\tau_{n_s}(\tau_{w'})=\phi_w(\tau_{w'}\tau_{n_s^{-1}})=\phi_w(\tau_{w'\tilde{s}^{-1}})$ and it is nonzero if and only if IwI is contained in $Iw'\tilde{s}^{-1}I$ which is equivalent to $w\tilde{s}=w'$ and in which case $\phi_w.\tau_{n_s}(\tau_{w'})=1$.
- If $\ell(w'\tilde{s})=\ell(w')-1$ then $\tau_{w'}=\tau_{w'\tilde{s}}\tau_{\tilde{s}^{-1}}$ and $\phi_w\cdot\tau_{n_s}(\tau_{w'})=\phi_w(\tau_{w'\tilde{s}}\tau_{n_s^{-1}}\tau_{n_s^{-1}})=\phi_w(\tau_{w'\tilde{s}}\tau_{n_s}^2\tau_{n_s^2})=-\phi_w(\tau_{w'}\theta_s)$ which is nonzero if and only if there is $t\in\check{\alpha}([\mathbb{F}_q^\times])$ such that $w'\bar{t}=w$ and in which case $\phi_w\cdot\tau_{n_s}(\tau_{w'})=|\mu_{\check{\alpha}}|$.

From the discussion above, we immediately deduce that $\phi_w\cdot\tau_\omega=\phi_{w\omega}$. Now we compute $\phi_w\cdot\tau_{n_s}$. Suppose that $\ell(w\tilde{s})=\ell(w)+1$, then for any $w'\in\widetilde{W}$ we have $\phi_w\cdot\tau_{n_s}(\tau_{w'})=0$, so $\phi_w\cdot\tau_{n_s}=0$. Suppose that $\ell(w\tilde{s})=\ell(w)-1$, then for any $w'\in\widetilde{W}$ we have $\phi_w\cdot\tau_{n_s}(\tau_{w'})=0$ except if $w'=w\tilde{s}$ in which cases $\phi_w\cdot\tau_{n_s}(\tau_{w'})=1$, or if there is $t\in\check{\alpha}([\mathbb{F}_q^\times])$ such that $w'=w\bar{t}^{-1}$ in which case $\phi_w\cdot\tau_{\tilde{s}}(\tau_{w'})=|\mu_{\check{\alpha}}|$, so $\phi_w\cdot\tau_{n_s}=\phi_{w\tilde{s}}+|\mu_{\check{\alpha}}|\sum_{t\in\check{\alpha}([\mathbb{F}_q^\times])}\phi_{w\bar{t}}$.

Since (8.1) is an isomorphism of right H-modules, the calculation above gives the right action of H on the ϕ's. The formulas for the left action follow using (8.2), Proposition 6.1 and

$$\begin{aligned}\tau_x\cdot\phi_w&=\mathcal{J}(\mathcal{J}(\tau_x\cdot\phi_w))=\mathcal{J}(\mathcal{J}(\phi_w)\cdot\mathcal{J}(\tau_x))\\&=\mathcal{J}(\phi_{w^{-1}}\cdot\tau_{x^{-1}}).\end{aligned}$$

□

REMARK 8.3. i. The formula for the right action coincides with the one given in [**24**] §5.1 p.9.

ii. Recall that $\iota(\tau_{n_s})=-\tau_{n_s}-\theta_s$ where ι is the involutive automorphism of H defined in (2.29). Formulas (8.4) and (8.5) can be given in the form:

$$\phi_w\cdot\iota(\tau_{n_s})=\begin{cases}-\phi_{w\tilde{s}} & \text{if } \ell(w\tilde{s})=\ell(w)-1\\ -\phi_w\cdot\theta_s & \text{if } \ell(w\tilde{s})=\ell(w)+1\end{cases}$$

and

$$\iota(\tau_{n_s})\cdot\phi_w=\begin{cases}-\phi_{\tilde{s}w} & \text{if } \ell(\tilde{s}w)=\ell(w)-1\\ -\theta_s\cdot\phi_w & \text{if } \ell(\tilde{s}w)=\ell(w)+1.\end{cases}$$

Recall that we defined in (7.3) the G-equivariant map $\mathcal{S}=\sum_{g\in G/I}\mathrm{ev}_g:\mathbf{X}\longrightarrow k$, where k is endowed with the trivial action of G. Note that the restriction of $\mathcal{S}$ to $H=\mathbf{X}^I$ coincides with the trivial character $\chi_{triv}:H\to k$ as defined in §2.2.2. This is because for $w\in\widetilde{W}$ we have $\mathcal{S}(\tau_w)=\mathcal{S}(\mathrm{char}_{IwI})=|IwI/I|=|I/I_w|=q^{\ell(w)}.1_k$ by Corollary 2.5.i. Since $\mathbf{X}$ is generated by char_I as a representation of G, it follows that $\mathcal{S}$ is a morphism of (G, H) modules $\mathbf{X} \to k_{triv}$ where k_{triv} denotes the (G, H)-bimodule k with the trivial action of G and the action of H via χ_{triv}.

The cohomology group $H^d(I, k_{triv}) = \mathrm{Ext}^d_{\mathrm{Mod}(G)}(\mathbf{X}, k_{triv})$ is naturally an H-bimodule where the left action comes from the right action on $\mathbf{X}$ and the right action comes from the trivial action on k_{triv}.

PROPOSITION 8.4. i. *The left as well as the right H-action on $H^d(I, k_{triv})$ are trivial, i.e., are through χ_{triv}.*
ii. *The map $\mathcal{S}^d$ induced on cohomology by $\mathcal{S}$ yields an exact sequence of H-bimodules*

$$0 \longrightarrow \ker(\mathcal{S}^d) \longrightarrow E^d \longrightarrow H^d(I, k_{triv}) \longrightarrow 0 \ . \tag{8.6}$$

PROOF. i. The triviality of the right action is obvious. For the left action we first consider the following diagram

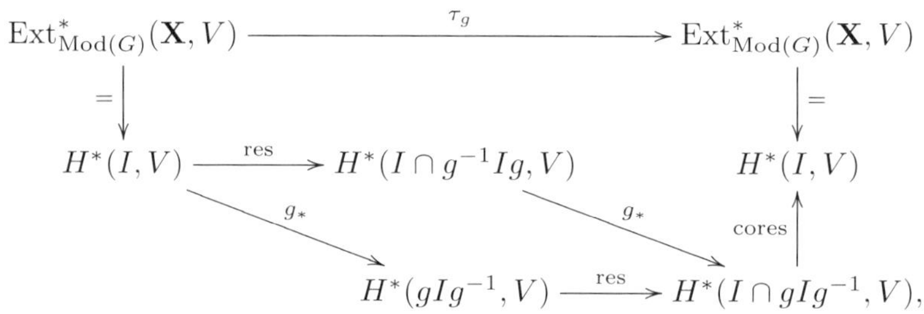

where V is an arbitrary object in $\mathrm{Mod}(G)$ and where the upper horizontal arrow is the action of $\tau_g \in H$ induced by its right action on $\mathbf{X}$. For its commutativity it suffices, by using an injective resolution of V, to consider the case $* = 0$, i.e., the diagram

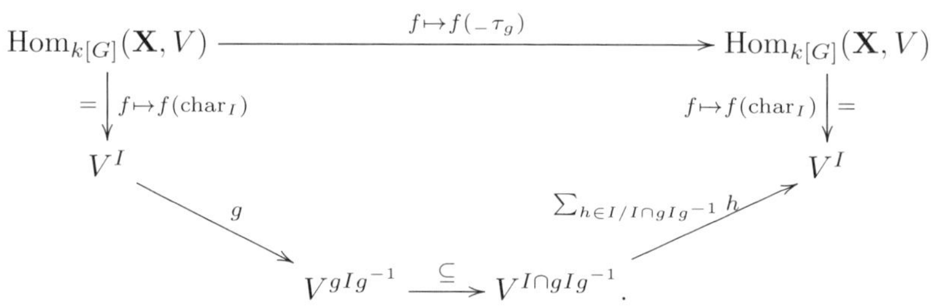

Its commutativity is verified easily by direct inspection. So our assertion reduces to the claim that the composed map

$$H^d(I, k) \xrightarrow{\mathrm{res}} H^d(I_{g^{-1}}, k) \xrightarrow{g_*} H^d(I_g, k) \xrightarrow{\mathrm{cores}} H^d(I, k)$$

coincides with the multiplication by $\chi_{triv}(\tau_g)$. Suppose that $g \in IwI$ with $w \in \widetilde{W}$. By Cor. 2.5.i we have $I_{g^{-1}} \subsetneqq I$ if and only if $\ell(w) > 0$. In this case the left restriction map above is the zero map by Remark 7.3 and $\chi_{triv}(\tau_w) = 0$. If $\ell(w) = 0$ then $\chi_{triv}(\tau_w) = 1$ and the above composed map simply is the map $H^d(I, k) \xrightarrow{w_*} H^d(I, k)$, which is the identity by Lemma 7.13.

ii. The map $\mathcal{S}^d : \mathrm{Ext}^d_{\mathrm{Mod}(G)}(\mathbf{X}, \mathbf{X}) \longrightarrow \mathrm{Ext}^d_{\mathrm{Mod}(G)}(\mathbf{X}, k_{triv})$ is surjective, for example, by Remark 7.4. It is right H-equivariant because $\mathcal{S}$ is right H-equivariant, and because the right action of H on the cohomology spaces is through the coefficients. Finally, it is left H-equivariant since the left actions are functorially induced by the right H-action on $\mathbf{X}$. $\square$

REMARK 8.5. Using directly the Hecke operators $\mathrm{cores} \circ g_* \circ \mathrm{res}$ in order to define the left H-action on $H^d(I, k_{triv})$ the Prop. 8.4.i is proved independently in [**13**] Thm. 7.1 in case the root system is irreducible.

The kernel $\ker(\chi_{triv})$ of χ_{triv} is a sub-H-bimodule of $E^0 = H$. Passing to duals, we have the restriction map

$$(E^0)^\vee \longrightarrow \ker(\chi_{triv})^\vee \ . \tag{8.7}$$

Its kernel is a one dimensional vector space isomorphic to χ_{triv} as an H-bimodule. It is generated by the linear map

$$\phi : \tau_\omega \mapsto 1, \text{ for } \omega \in \widetilde{\Omega} \text{ and } \tau_w \mapsto 0, \text{ for } w \in \widetilde{W} \text{ with length} > 0. \tag{8.8}$$

Assume that Ω is finite. Then ϕ lies in $(E^0)^{\vee,f}$ and we have a short exact sequence of H-bimodules

$$0 \longrightarrow k\phi \longrightarrow (E^0)^{\vee,f} \longrightarrow \ker(\chi_{triv})^{\vee,f} \longrightarrow 0 \tag{8.9}$$

where $\ker(\chi_{triv})^{\vee,f}$ denotes the image of $(E^0)^{\vee,f} \subset (E^0)^\vee$ by (8.7).

PROPOSITION 8.6. *If Ω is finite and $|\Omega|$ is invertible in k, then we have a decomposition of $(E^0)^{\vee,f}$ into a direct sum of H-bimodules*

$$(E^0)^{\vee,f} \cong \chi_{triv} \oplus \ker(\chi_{triv})^{\vee,f}$$

where χ_{triv} is supported by the element ϕ defined in (8.8) *and* $\ker(\chi_{triv})^{\vee,f}$ *is the image of $(E^0)^{\vee,f}$ by the restriction map* (8.7). *Via the isomorphism* (8.1), *this corresponds to the decomposition*

$$E^d \cong H^d(I, k_{triv}) \oplus \ker(\mathcal{S}^d). \tag{8.10}$$

PROOF. Note that the image of ϕ in E^d is $\sum_{\omega \in \widetilde{\Omega}} \phi_\omega$. Combining (8.6), (8.9) and (8.1), we obtain a diagram of morphisms of H-bimodules

$$\begin{array}{ccccccccc}
0 & \longrightarrow & {}^{\mathfrak{J}}(k\phi)^{\mathfrak{J}} & \longrightarrow & ({}^{\mathfrak{J}}E^{0\,\mathfrak{J}})^{\vee,f} & \longrightarrow & ({}^{\mathfrak{J}}\ker(\chi_{triv})^{\mathfrak{J}})^{\vee,f} & \longrightarrow & 0 \\
 & & \Big\downarrow \partial & & \Delta^d \Big\uparrow \cong & & \varepsilon \Big\uparrow & & \\
0 & \longleftarrow & H^d(I, k_{triv}) & \longleftarrow & E^d & \longleftarrow & \ker(\mathcal{S}^d) & \longleftarrow & 0
\end{array}$$

where ∂ and ε are such that the diagram commutes. The map ∂ is a map between one dimensional vector spaces, and it is not trivial since

$$\partial(\phi) = \mathcal{S}^d(\sum_{\omega \in \widetilde{\Omega}} \phi_\omega) = |\widetilde{\Omega}| \eta^{-1}(1) = -|\Omega| \eta^{-1}(1) \neq 0 \ .$$

This implies that the second short exact sequence splits. By a standard argument, ε is also an isomorphism and the first exact sequence also splits. $\square$

Recall that we defined in (2.38) a decreasing filtration of H as an H-bimodule. For the sake of homogeneity of notations, we denote it here by $(F^n E^0)_{n \geq 0}$ and recall that

$$F^n E^0 = \oplus_{\ell(w) \geq n} k\tau_w \ . \tag{8.11}$$

For $n \geq 1$, define $(F^n E^0)^{\vee,f}$ to be the H-bimodule image of $(E^0)^{\vee,f}$ by the surjective restriction map $(E^0)^\vee \to (F^n E^0)^\vee$. Since Ω is assumed to be finite, the kernel of $(E^0)^{\vee,f} \twoheadrightarrow (F^n E^0)^{\vee,f}$ coincides with the dual space $(E^0/F^n E^0)^\vee$. Furthermore $\ker(\chi_{triv})^{\vee,f}$ is the increasing union of the subspaces $(\ker(\chi_{triv})/F^n E^0)^\vee$ of all linear maps φ in $\ker(\chi_{triv})^\vee$ which are trivial on $F^n E^0$ for some $n \geq 1$. The supersingular H-modules were defined in §2.3.

COROLLARY 8.7. *If* **G** *is semisimple simply connected with irreducible root system, then the H-module $\ker(\mathcal{S}^d)$ is a union of H-bimodules which are (finite length) supersingular on the right and on the left.*

PROOF. It suffices to prove, for $n \geq 1$ that the finite dimensional space ${}^{\mathcal{J}}((\ker(\chi_{triv})/F^n E^0)^\vee)^{\mathcal{J}}$ is a supersingular H-module on the left and on the right. Since $\Omega = \{1\}$, we have

$$\ker(\chi_{triv}) = (1 - e_1)F^0 E^0 + F^1 E^0 \ ,$$

and $\ker(\chi_{triv})/F^n E^0$, by Lemma 2.14.ii, is annihilated by the action of $\mathfrak{J}^n$ on the left (resp. right). Therefore, $(\ker(\chi_{triv})/F^n E^0)^\vee$ is annihilated by the action of $\mathfrak{J}^n$ on the right (resp. left) and therefore supersingular. Due to Remark 6.3 this remains the case after twisting by $\mathcal{J}$. □

References

[1] J. N. Bernstein, *Le "centre" de Bernstein* (French), Representations of reductive groups over a local field, Travaux en Cours, Hermann, Paris, 1984, pp. 1–32. Edited by P. Deligne. MR771671

[2] S. Bosch, W. Lütkebohmert, and M. Raynaud, *Néron models*, Ergebnisse der Mathematik und ihrer Grenzgebiete (3) [Results in Mathematics and Related Areas (3)], vol. 21, Springer-Verlag, Berlin, 1990. MR1045822

[3] N. Bourbaki, *Lie groups and Lie algebras. Chapters 4–6*, Elements of Mathematics (Berlin), Springer-Verlag, Berlin, 2002. Translated from the 1968 French original by Andrew Pressley. MR1890629

[4] K. S. Brown, *Cohomology of groups*, Graduate Texts in Mathematics, vol. 87, Springer-Verlag, New York-Berlin, 1982. MR672956

[5] F. Bruhat and J. Tits, *Groupes réductifs sur un corps local* (French), Inst. Hautes Études Sci. Publ. Math. **41** (1972), 5–251. MR0327923

[6] F. Bruhat and J. Tits, *Groupes réductifs sur un corps local. II. Schémas en groupes. Existence d'une donnée radicielle valuée* (French), Inst. Hautes Études Sci. Publ. Math. **60** (1984), 197–376. MR756316

[7] J. D. Dixon, M. P. F. du Sautoy, A. Mann, and D. Segal, *Analytic pro-p groups*, 2nd ed., Cambridge Studies in Advanced Mathematics, vol. 61, Cambridge University Press, Cambridge, 1999. MR1720368

[8] R. Hartshorne, *Residues and duality*, Lecture notes of a seminar on the work of A. Grothendieck, given at Harvard 1963/64. With an appendix by P. Deligne. Lecture Notes in Mathematics, No. 20, Springer-Verlag, Berlin-New York, 1966. MR0222093

[9] J. E. Humphreys, *Reflection groups and Coxeter groups*, Cambridge Studies in Advanced Mathematics, vol. 29, Cambridge University Press, Cambridge, 1990. MR1066460

[10] N. Iwahori and H. Matsumoto, *On some Bruhat decomposition and the structure of the Hecke rings of* $\mathfrak{p}$*-adic Chevalley groups*, Inst. Hautes Études Sci. Publ. Math. **25** (1965), 5–48. MR0185016

[11] J. C. Jantzen, *Representations of algebraic groups*, 2nd ed., Mathematical Surveys and Monographs, vol. 107, American Mathematical Society, Providence, RI, 2003. MR2015057

[12] K. Kozioł, *Pro-p-Iwahori invariants for* SL_2 *and L-packets of Hecke modules*, Int. Math. Res. Not. IMRN **4** (2016), 1090–1125, DOI 10.1093/imrn/rnv162. MR3493443

[13] K. Koziol, *Hecke module structure on first and top pro-p-Iwahori Cohomology.* arXiv:1708.03013 (2017)

[14] S. Lang, *Topics in cohomology of groups*, Lecture Notes in Mathematics, vol. 1625, Springer-Verlag, Berlin, 1996. Translated from the 1967 French original by the author; Chapter X based on letters written by John Tate. MR1439508

[15] M. Lazard, *Groupes analytiques p-adiques* (French), Inst. Hautes Études Sci. Publ. Math. **26** (1965), 389–603. MR0209286

[16] G. Lusztig, *Affine Hecke algebras and their graded version*, J. Amer. Math. Soc. **2** (1989), no. 3, 599–635, DOI 10.2307/1990945. MR991016

[17] J. Neukirch, A. Schmidt, and K. Wingberg, *Cohomology of number fields*, Grundlehren der Mathematischen Wissenschaften [Fundamental Principles of Mathematical Sciences], vol. 323, Springer-Verlag, Berlin, 2000. MR1737196

[18] R. Ollivier, *Le foncteur des invariants sous l'action du pro-p-Iwahori de* $\mathrm{GL}_2(F)$ (French, with English summary), J. Reine Angew. Math. **635** (2009), 149–185, DOI 10.1515/CRELLE.2009.078. MR2572257

[19] R. Ollivier, *Compatibility between Satake and Bernstein isomorphisms in characteristic p*, Algebra Number Theory **8** (2014), no. 5, 1071–1111, DOI 10.2140/ant.2014.8.1071. MR3263136

[20] R. Ollivier and P. Schneider, *Pro-p Iwahori-Hecke algebras are Gorenstein*, J. Inst. Math. Jussieu **13** (2014), no. 4, 753–809, DOI 10.1017/S1474748013000303. MR3249689

[21] R. Ollivier and P. Schneider, *A canonical torsion theory for pro-p Iwahori-Hecke modules*, Adv. Math. **327** (2018), 52–127, DOI 10.1016/j.aim.2017.06.013. MR3761991

[22] V. Paškūnas, *The image of Colmez's Montreal functor*, Publ. Math. Inst. Hautes Études Sci. **118** (2013), 1–191, DOI 10.1007/s10240-013-0049-y. MR3150248

[23] N. Ronchetti, *Satake map for the mod p derived Hecke algebra*. Preprint 2016

[24] P. Schneider, *Smooth representations and Hecke modules in characteristic p*, Pacific J. Math. **279** (2015), no. 1-2, 447–464, DOI 10.2140/pjm.2015.279.447. MR3437786

[25] P. Schneider and U. Stuhler, *Representation theory and sheaves on the Bruhat-Tits building*, Inst. Hautes Études Sci. Publ. Math. **85** (1997), 97–191. MR1471867

[26] J.-P. Serre, *Cohomologie galoisienne* (French), With a contribution by Jean-Louis Verdier. Lecture Notes in Mathematics, No. 5. Troisième édition, vol. 1965, Springer-Verlag, Berlin-New York, 1965. MR0201444

[27] J.-P. Serre, *Lie algebras and Lie groups*, 2nd ed., Lecture Notes in Mathematics, vol. 1500, Springer-Verlag, Berlin, 1992. 1964 lectures given at Harvard University. MR1176100

[28] J.-P. Serre, *Sur la dimension cohomologique des groupes profinis* (French), Topology **3** (1965), 413–420, DOI 10.1016/0040-9383(65)90006-6. MR0180619

[29] J.-P. Serre, *Local fields*, Graduate Texts in Mathematics, vol. 67, Springer-Verlag, New York-Berlin, 1979. Translated from the French by Marvin Jay Greenberg. MR554237

[30] T. A. Springer, *Linear algebraic groups*, 2nd ed., Progress in Mathematics, vol. 9, Birkhäuser Boston, Inc., Boston, MA, 1998. MR1642713

[31] J. Tits, *Reductive groups over local fields*, Automorphic forms, representations and L-functions (Proc. Sympos. Pure Math., Oregon State Univ., Corvallis, Ore., 1977), Proc. Sympos. Pure Math., XXXIII, Amer. Math. Soc., Providence, R.I., 1979, pp. 29–69. MR546588

[32] M.-F. Vignéras, *Représentations l-modulaires d'un groupe réductif p-adique avec* $l \neq p$ (French, with English summary), Progress in Mathematics, vol. 137, Birkhäuser Boston, Inc., Boston, MA, 1996. MR1395151

[33] M.-F. Vignéras, *Erratum: "Pro-p-Iwahori Hecke ring and supersingular* $\overline{\mathbf{F}}_p$*-representations" [Math. Ann.* **331** *(2005), no. 3, 523–556; MR2122539]*, Math. Ann. **333** (2005), no. 3, 699–701, DOI 10.1007/s00208-005-0679-6. MR2198804

[34] M.-F. Vignéras, *The pro-p-Iwahori Hecke algebra of a reductive p-adic group III (spherical Hecke algebras and supersingular modules)*, J. Inst. Math. Jussieu **16** (2017), no. 3, 571–608, DOI 10.1017/S1474748015000146. MR3646282

Mathematics Department, University of British Columbia, 1984 Mathematics Road, Vancouver, BC V6T 1Z2, Canada
Email address: ollivier@math.ubc.ca
URL: http://www.math.ubc.ca/~ollivier

Mathematisches Institut, Universität Münster, Einsteinstr. 62, 48291 Münster, Germany
Email address: pschnei@uni-muenster.de
URL: http://www.uni-muenster.de/math/u/schneider/

Proceedings of Symposia in Pure Mathematics
Volume **101**, 2019
https://doi.org/10.1090/pspum/101/01799

Affine Hecke algebras and the conjectures of Hiraga, Ichino and Ikeda on the Plancherel density

Eric Opdam

Dedicated to Joseph Bernstein

ABSTRACT. Hiraga, Ichino and Ikeda have conjectured an explicit expression for the Plancherel density of the group of points of a reductive group defined over a local field F, in terms of local Langlands parameters. In these lectures we shall present a proof of these conjectures for Lusztig's class of representations of unipotent reduction if F is p-adic and G is of adjoint type and splits over an unramified extension of F. This is based on the author's paper [Spectral transfer morphisms for unipotent affine Hecke algebras, Selecta Math. (N.S.) **22** (2016), no. 4, 2143–2207].

More generally for G connected reductive (still assumed to be split over an unramified extension of F), we shall show that the requirement of compatibility with the conjectures of Hiraga, Ichino and Ikeda essentially determines the Langlands parameterisation for tempered representations of unipotent reduction. We shall show that there exist parameterisations for which the conjectures of Hiraga, Ichino and Ikeda hold up to rational constant factors. The main technical tool is that of spectral transfer maps between normalised affine Hecke algebras used in *op. cit.*

CONTENTS

2010 *Mathematics Subject Classification.* Primary 20C08; Secondary 22D25, 43A30.

Key words and phrases. Cuspidal unipotent representation, formal degree, discrete unramified Langlands parameter.

It is a pleasure to thank Maarten Solleveld for many useful comments. It is also a pleasure to acknowledge the excellent comments and suggestions of the participants of the workshop "Representation Theory of Reductive Groups Over Local Fields and Applications to Automorphic forms" at the Weizmann institute, and of the workshop "Representation theory of p-adic groups" at IISER, Pune.

1. Introduction

Let F be local field of characteristic 0, let $\Gamma_F := \mathrm{Gal}(\overline{F}/F)$ be the absolute Galois group of F, and let G be a the group of points of a connected reductive algebraic group defined over F. Let $G^\vee$ denote the Langlands dual group of G (a complex Lie group with root system dual to that of G), and let ${}^LG := G^\vee \rtimes \Gamma_F$ be the Galois form of the L-group of G. The Langlands group L_F of F is defined to be W_F (the Weil group of F) if F is archimedean, and $W_F \times \mathrm{SL}_2(\mathbb{C})$ otherwise. Let ψ be a fixed additive character of F. To a finite dimensional complex representation V of ${}^LG$ one attaches epsilon factors $\epsilon(s, V, \psi)$ and L-functions $L(s, V)$, where $s \in \mathbb{C}$ is a complex variable (see [**Tate**]). A Langlands parameter for G is a homomorphism $\varphi : L_F \to {}^LG$ some natural conditions (cf. Section 2.3). With all this in place, the adjoint γ-factor of a Langlands parameter φ of G is defined as

$$\gamma(s, \mathrm{Ad} \circ \varphi, \psi) := \frac{\epsilon(s, \mathrm{Ad} \circ \varphi, \psi) L(1 - s, \mathrm{Ad} \circ \varphi)}{L(s, \mathrm{Ad} \circ \varphi)},$$

where Ad is the adjoint representation of ${}^LG$ on $\mathrm{Lie}(G^\vee)/\mathrm{Lie}(Z^\vee)$, with $Z^\vee$ the center of $G^\vee$. Let $\widehat{G}$ be the space of equivalence classes of irreducible unitary representations π of G, equipped with the Fell topology. We will denote by Θ_π the distribution character of π. Then Harish-Chandra's Plancherel formula states the existence of a unique positive measure ν_{Pl} on $\widehat{G}$, such that $f(e) = \int_{\widehat{G}} \Theta_\pi(f) d\nu_{Pl}(\pi)$, for all $f \in C_c^\infty(G)$. The measure $d\nu_{Pl}$ is called the Plancherel measure of G. If π is a discrete series representation of G, its formal degree is equal to $\mathrm{fdeg}(\pi) := \nu_{Pl}(\pi) > 0$. For more general tempered representations $d\nu_{Pl}$ is described in terms of a density function. Hiraga, Ichino, and Ikeda formulated two conjectures [**HII**, **HIIcor**] expressing the Plancherel density at a tempered representation π in terms of the conjectural enhanced Langlands parameter attached to π. For an essentially discrete series representation π_ρ in an L-packet $\Pi_\varphi(G)$ attached to a (discrete) Langlands parameter $\varphi : L_F \to {}^LG$, enhanced with a local system $\rho \in \mathrm{Irr}(\mathcal{S}_\varphi, \chi_G)$ on the $G^\vee$-orbit of φ (see Section 2.5 for unexplained notation and more details), the conjecture reduces to the equality

$$\mathrm{fdeg}(\pi_\rho) = \frac{\dim(\rho)}{|\mathcal{S}_\varphi^\natural|}|\gamma(0, \mathrm{Ad}\circ\varphi, \psi)|.$$

Hiraga, Ichino, and Ikeda proved the conjecture for $F = \mathbb{R}$ [**HII**, **HIIcor**]. For F non-archimedean, it has been proved in several cases but not in general (see Section 2.7 for an overview of known results). From now on we assume that F is a non-archimedean field and that G splits over an unramified extension of F. When G is absolutely almost simple of adjoint type, the conjecture above and its extension to general tempered representations are known to hold for representations of G with unipotent reduction from previous works [**R1**], [**R3**], [**HO1**], [**FO**], [**Opd5**], [**Fe2**], [**FOS**]. The proof in [**Opd5**] uses the Lusztig classification (composed with the Iwahori-Matsumoto involution in order to map tempered representations to bounded parameters) as a Langlands parameterisation. The main goal of the present manuscript is to extend this result to a general connected reductive group G (still assumed to be split over an unramified extension of F). This is achieved in the main result Theorem 4.5.1.

The irreducible representations with unipotent reduction (a terminology introduced by Moeglin and Waldspurger), called "unipotent" by Lusztig, are the representations of G which admit non-zero invariant vectors by the pro-p unipotent radical of a parahoric subgroup of G. In particular, they are depth-zero representations. They are expected to correspond to unramified Langlands parameters. From their definition it follows that the category of representations with unipotent reduction of G is Morita equivalent to the module category of a direct sum of affine Hecke algebras $\mathcal{H}_t(G)$, where t runs over the set of equivalence classes of unipotent types of G. Let G^* be the quasi-split group in the inner class of G, and let $\mathbb{I}$ denote an Iwahori subgroup of G^*. Let $\mathcal{H}_{\mathbb{I}}(G^*)$ be the Iwahori Hecke algebra of G^*.

One of the main ingredients of the proof of Theorem 4.5.1 is the notion of spectral transfer morphism between normalised affine Hecke algebras [**Opd4**], [**Opd5**], which allows us to construct a bijection between the set $\widehat{G}_{uni}^{\mathrm{temp}}$ of equivalence classes of tempered irreducible representations of G with unipotent reduction and the set $\Phi_{nr}^{\mathrm{temp}}(G)$ of $G^\vee$-conjugacy classes of unramified bounded enhanced Langlands parameters for G. The construction of such a bijection has some interest in its own right. A key point is the fact that a spectral transfer morphism $\Psi : \mathcal{H}_t(G) \rightsquigarrow \mathcal{H}_{\mathbb{I}}(G^*)$ from the Hecke algebra $\mathcal{H}_t(G)$ of a unipotent type t of G to the Iwahori Hecke algebra $\mathcal{H}_{\mathbb{I}}(G^*)$ defines Langlands parameters $\pi \to \varphi_\pi \in \Phi_{nr}^{\mathrm{temp}}(G)$ for the tempered representations covered by t such that the conjectures of Hiraga, Ichino and Ikeda hold (up to rational constant factors independent of the cardinality q of the residue field of F). This is explained in Corollary 4.3.5, which is itself based on the Iwahori-spherical case Theorem 3.8.1 and Theorem 3.8.2. Conversely, a Langlands parameterisation such that the conjectures of Hiraga, Ichino and Ikeda hold (up to rational constant factors independent of the cardinality q of the residue field of F) and satisfies a certain algebraic condition (see Theorem 4.5.1(a)) defines such STMs uniquely.

It is remarkable that for tempered representations of unipotent reduction the conditions imposed on a Langlands parameterisation by the conjectures of Hiraga, Ichino and Ikeda determine it up to twisting by certain diagram automorphism. This can be viewed as a generalisation and strengthening of the principle expressed by Mark Reeder [**R3**] for discrete series L-packets, saying that "alleged L-packets can only be convicted upon circumstantial evidence, of which the formal degrees are one piece".

2. The conjecture of Hiraga, Ichino and Ikeda

Let F be a local field of characteristic 0, and let $\mathbf{G}$ be a connected reductive group defined over F. The group $G = \mathbf{G}(F)$ of F-points of $\mathbf{G}$ is a separable locally compact topological group which is unimodular. Let μ_G denote a Haar measure on G. Let $C^*(G)$ be the group C^*-algebra of G, i.e. the C^*-envelope of the Banach algebra $L^1(G, \mu_G)$ with respect to convolution. By famous results of Harish-Chandra [**HC2**] (if F is archimedean) and by Bernstein [**Be**] (if F is non-archimedean) we know that $C^*(G)$ is liminal, hence of Type I. Let $\hat{G}$ denote the space of equivalence classes of irreducible unitary representations of G, equipped with the Fell topology. For each $\pi \in \hat{G}$ we choose a representative denoted by (V_π, π). The abstract Plancherel formula for separable locally compact unimodular topological groups of Type I asserts that:

THEOREM 2.0.1. *There exists a unique positive measure ν_{Pl} (called the Plancherel measure of G) on $\hat{G}$ such that:*

$$L^2(G, \mu_G) \simeq \int_{\pi \in \hat{G}}^{\oplus} H(\pi) d\nu_{Pl}(\pi) \tag{2.1}$$

where $H(\pi) := V_\pi^* \hat{\otimes} V_\pi$ denotes the algebra of Hilbert-Schmidt operators on V_π.

Much of study of harmonic analysis on reductive groups is devoted to making the abstract Plancherel formula in this context explicit. This is a problem with many different facets, some of which are poorly understood or even unsolved even after more than 70 years into the subject. One part of this is conceptual. The tremendous success of the approach of Langlands towards harmonic analysis on reductive groups points out that number theory and algebraic geometry are inherent parts of this endeavour. An explicit Plancherel formula has to reflect the deep number theoretical problems which are conspiring in the background. There are formidable technical obstacles as well, stemming from the fact that one is forced to deal with representation theory on rather general topological vector spaces even if one's goal is the study unitary representations.

Harish-Chandra made deep contributions to our understanding of the structure of the explicit Plancherel formula ([**HC3**], [**Wal**]). He discovered that support of the Plancherel measure is not all of $\hat{G}$, except if G happens to be built from anisotropic and commutative almost factors. In general, the support of the Plancherel measure is the set of so-called irreducible *tempered* representations of G. A connected component (Harish-Chandra series) of this set consists of the irreducible summands of the representations obtained by unitary parabolic induction of a discrete series representation of a Levi subgroup of G. There are countably many Harish-Chandra series.

Assuming the local Langlands correspondence for tempered representations, Hiraga, Ichino and Ikeda conjectured an explicit formula for the Plancherel measure ν_{Pl} of G. The appeal of these conjectures is that they formulate the answers in terms of a natural number theoretical invariant which is associated with an irreducible representation π in the conjectural local Langlands correspondence, the so-called adjoint gamma factor.

2.1. The decomposition of the trace. The regular representation $L^2(G, \mu_G)$ corresponds to the semi-finite positive trace $\tau_G(f) := f(e)$ on $C^*(G)$, and in particular $C^*(G)$ is a Hilbert algebra. The dense subalgebra $C_c^\infty(G) \subset C^*(G)$ has the special property that for all $\pi \in \hat{G}$ and $f \in C_c^\infty(G)$ the operator $\pi(f) \in \mathcal{B}(V_\pi)$ is of trace class. This defines a distribution Θ_π defined by:

$$\Theta_\pi(f) := \mathrm{Tr}_{V_\pi}(\pi(f)) \tag{2.2}$$

This is called Harish-Chandra's distributional character of π.

The positive trace τ_G is defined on the dense subalgebra $C_c^\infty(G) \subset C^*(G)$, and the Plancherel measure ν_{Pl} is completely determined by the decomposition of τ_G as a superposition of the distributional characters Θ_π with $\pi \in \hat{G}$:

COROLLARY 2.1.1. *The Plancherel measure ν_{Pl} is the unique positive measure on $\hat{G}$ such that for all $f \in C_c^\infty(G)$:*

$$f(e) = \int_{\pi \in \hat{G}} \Theta_\pi(f) d\nu_{Pl}(\pi) \tag{2.3}$$

Some remarks are in order:

(1) The measure ν_{Pl} depends on the normalization of the Haar measure. If we replace μ_G by $a\mu_G$ (for $a > 0$) then ν_{Pl} is replaced by $a^{-1}\nu_{Pl}$.
(2) The definition of the distributional character Θ_π can be extended naturally to the class of admissible representations of G. An irreducible admissible representation (V_π, π) is tempered iff the distribution Θ_π is tempered, by which we mean that Θ_π extends continuously to the Harish-Chandra Schwartz algebra $\mathcal{C}(G) \supset C_c^\infty(G)$ of G. In turn this is equivalent to the requirement that for every standard parabolic subgroup $P \subset G$, the exponents $\chi \in \mathrm{Exp}(\pi_P)$ of the Jacquet module $(V_{\pi,P}, \pi_P)$ satisfy the Casselman conditions $\mathrm{Re}(\chi) \in \overline{{}^+\mathfrak{a}_P^{G,*}}$ (see [**Wal**]). Here $\overline{{}^+\mathfrak{a}_P^{G,*}}$ denotes the closed convex cone inside the real span $\mathfrak{a}_M^{G,*}$ of the set of G-roots $\Sigma(A_M, G)$ of the connected split center A_M of the standard Levi-factor M of P spanned by the set of roots associated to the unipotent radical of P.
(3) By a deep result of Harish-Chandra, the support of $d\nu_{Pl}$ is contained in the set $\hat{G}^{\mathrm{temp}}$ of equivalence classes of tempered representations of G. This was explained more conceptually by Joseph Bernstein [**Be1**].
(4) We call an admissible irreducible representation (V_π, π) of G a discrete series representation if the matrix coefficients of π are in $L^2(G, \mu_G)$. Let $\hat{G}^{\mathrm{disc}} \subset \hat{G}$ denote the subset of equivalence classes of discrete series representations. It is well known by Casselman's results (see [**Wal**]) that $\pi \in \hat{G}^{\mathrm{disc}}$ iff $\mathrm{Re}(\chi) \in {}^+\mathfrak{a}_P^{G,*}$ for all standard parabolic subgroups $P \subset G$ and all $\chi \in \mathrm{Exp}(\pi_P)$. The set $\hat{G}^{\mathrm{disc}}$ is not empty iff the center $Z(G)$ is anisotropic. By a well known characterisation of Dixmier we have $\pi \in \hat{G}^{\mathrm{disc}}$ iff $\nu_{Pl}(\{\pi\}) > 0$.

DEFINITION 2.1.2. If $\pi \in \hat{G}^{\mathrm{disc}}$ then we define the formal degree of π as $\mathrm{fdeg}(\pi) := \nu_{Pl}(\{\pi\}) > 0$.

(5) Let $A \subset G$ be the maximal split component of $Z(G)$. Then G has discrete series only if A is trivial. More generally, we call an irreducible representation π a discrete series modulo the center if π is tempered and its matrix coefficients are $L^2(G/A)$. If π is a discrete series representation modulo the center then one can show that there exists a constant $\mathrm{fdeg}(\pi) > 0$ such that for all $v, w \in V_\pi$:

$$\int_{G/A} |(\pi(g)v, w)|^2 d\mu_G(g) = \mathrm{fdeg}(\pi)^{-1} \|v\|^2 \|w\|^2 \tag{2.4}$$

This generalizes the notion of formal degree for a discrete series representation as in (3).

2.2. Normalization of Haar measure. As we have seen above, we need to fix the Haar measure μ_G in order to fix ν_{Pl} uniquely. The Haar measure depends on the choice of an additive character ψ of F. The construction is explained in [**GG**] (also see [**G**, Section 4], and the discussion in [**HIIcor**] of the differences between these two constructions). The corresponding Haar measure will be denoted by μ_G^ψ if we need to stress the dependence on the choice of ψ.

LEMMA 2.2.1. (a) *Suppose that F is non-archimedean and that $\mathbf{G}$ is split over an unramified extension. Let ψ_0 be an additive character with conductor $\mathfrak{p} \subset \mathfrak{o}$. Let q be the cardinality of the residue field $\mathfrak{o}/\mathfrak{p}$. Then for any parahoric subgroup $\mathbb{P} \subset G$ with reductive quotient $\overline{\mathbb{P}}$ we have*

$$\mathrm{Vol}(\mathbb{P}, \mu_G^{\psi_0}) = q^{-\dim(\overline{\mathbb{P}})/2} |\overline{\mathbb{P}}| \tag{2.5}$$

(This is the normalization of Haar measure used in [**DeRe**].*)*

(b) *Suppose that $F = \mathbb{R}$, and that $\psi_0(x) = \exp(2\pi\sqrt{-1}x)$. Assume that G has discrete series representations. By Harish-Chandra's criterion this is equivalent to the existence of an anisotropic maximal torus $T \subset G$ of G. Let μ_G^0 be the Haar measure on G defined by the volume form on $\mathfrak{g} = \mathrm{Lie}(G)$ corresponding to invariant norm $\|x\|^2 = -B(x, \theta(x))$, where $x \in \mathfrak{g}$, θ denotes the Cartan involution, and B a nondegenerate bilinear form on $\mathfrak{g}$ as in* [**HC2**]. *Let μ_T^0 be the Haar measure on T defined similarly. We denote by $\Sigma^\vee$ the root system of $\mathfrak{g}^\vee := Lie(G^\vee)$, with $|\Sigma^\vee| = 2N$ and $\dim(T) = l$. Then:*

$$\mu_G^{\psi_0} = 2^N (2\pi)^l \mathrm{Vol}(T, \mu_T^0)^{-1} \prod_{\alpha^\vee \in \Sigma_+^\vee} (\alpha^\vee, \alpha^\vee) \mu_G^0 \tag{2.6}$$

PROOF. Part (a) follows from comparing [**HII**, (1.1) and further] with [**DeRe**]. Part (b) is an easy consequence of the computation in [**HII**, §2]. □

In the sequel we will use $\mu_G := \mu_G^{\psi_0}$ as the standard normalization of the Haar measure on G.

2.3. Local Langlands parameters. Let $\Gamma := \mathrm{Gal}(\overline{F}/F)$ denote the absolute Galois group of F. Choose a Borel subgroup B and maximal torus $T \subset B$ of $G' := \mathbf{G}_{\overline{F}}$, and let $\beta(G) = (X^*(T), \Delta, X_*(T), \Delta^\vee)$ be the corresponding based root datum. Choose a pinning $(G', B, T, \{x_\alpha\}_{\alpha \in \Delta})$, which induces a splitting of the exact sequence:

$$1 \to \mathrm{Int}(G') \to \mathrm{Aut}(G') \to \mathrm{Aut}(\beta(G')) \to 1 \tag{2.7}$$

The action of Γ on $G(\overline{F})$ induces an action of Γ on $\beta(G')$, and via the above splitting this gives rise to an (algebraic) action a of Γ on G'. There is a unique split F-structure on G' which fixes the chosen splitting, and clearly this commutes with a. Therefore the composition of these two actions defines a quasi-split F-structure $\mathbf{G}^*$ on G'. We denote by $G^* = \mathbf{G}^*(F)$ the corresponding group of points.

Let $G^\vee$ be the connected complex reductive group with $\beta(G^\vee) := \beta(G')^\vee$. Choose a pinning $(G^\vee, B^\vee, T^\vee, \{y_{\alpha^\vee}\}_{\alpha^\vee \in \Delta^\vee})$ of $G^\vee$. The action of Γ on $\beta(G^\vee)$ induced by the action on $\beta(G')$ gives rise to an (algebraic) action of Γ on $G^\vee$. We define ${}^L G := G^\vee \rtimes \Gamma$, the Langlands dual group of $\mathbf{G}$.

The Langlands group L_F of F is defined as follows:

$$L_F = \begin{cases} W_F \text{ if } F \text{ is archimedean,} \\ W_F \times \mathrm{SL}_2(\mathbb{C}) \text{ else.} \end{cases} \tag{2.8}$$

Here W_F denotes the Weil group of F (see [**Tate**]). A Langlands parameter is a homomorphism $\varphi : L_F \to {}^L G$ such that

(1) φ is continuous.
(2) In the non-archimedean case, $\varphi|_{\mathrm{SL}_2(\mathbb{C})}$ is algebraic.
(3) $\mathrm{pr}_2 \circ \varphi|_{W_F} \to \Gamma$ is the canonical homomorphism $W_F \to \Gamma$.
(4) $\varphi(W_F)$ is semisimple.
(5) If $\mathrm{Im}(\varphi)$ is contained in the Levi-subgroup of a parabolic subgroup P of ${}^L G$ then P is G-relevant (in the sense of [**Bo**, §8]).

DEFINITION 2.3.1. We call φ *discrete* if $C_{G^\vee}(\varphi)$ is finite. We call φ *essentially discrete* if $\mathrm{Im}(\varphi)$ is not contained in the Levi subgroup of a proper relevant parabolic subgroup of ${}^L G$.

LEMMA 2.3.2. *Discrete parameters exist iff the connected center of G is anisotropic and F is non-archimedean, or else if G admits an anisotropic maximal F-torus. In this situation φ is discrete iff φ is essentially discrete.*

PROOF. It is easy to see that the connected center of G is F-anisotropic if and only if the center ${}^L Z := Z(G^\vee)^\Gamma$ is finite. The group $C := C_{G^\vee}(\varphi) \subset G^\vee$ is reductive since $\varphi(W_F)$ is semisimple and $\varphi(\mathrm{SL}_2(\mathbb{C}))$ is reductive. Hence C is finite iff C does not contain a nontrivial torus. By the above remark this can happen only if the connected center of G is anisotropic. In this case [**Bo**, Proposition 3.5, 3.6] implies that C does not contain a nontrivial torus iff $\mathrm{Im}(\varphi)$ is not contained in any proper Levi subgroup of ${}^L G$. By definition this is equivalent to saying that $\mathrm{Im}(\varphi)$ is not contained in a Levi subgroup of ${}^L G$ of any proper relevant parabolic subgroup of ${}^L G$. In the non-archimedean case we can define a discrete character φ_0 which is trivial on W_F and corresponds to the regular unipotent orbit on $\mathrm{SL}_2(\mathbb{C})$ (the principal parameter). In the case $F = \mathbb{C}$ there are no discrete parameters. If $F = \mathbb{R}$ and φ is discrete then $\varphi(\mathbb{C}^\times)$ (with $\mathbb{C}^\times = W_\mathbb{C} \subset W_\mathbb{R}$) must contain regular semisimple elements. Thus $\varphi(\mathbb{C}^\times)$ is contained in a unique maximal torus $T^\vee$ of $G^\vee$ which must be θ-stable (with θ the automorphism corresponding to the nontrivial element of $\mathrm{Gal}(\mathbb{C}/\mathbb{R})$). Since $\theta\varphi(z)\theta^{-1} = \varphi(\overline{z})$, the discreteness of φ implies that $\varphi(\overline{z}) = \varphi(z)^{-1}$, and thus that θ restricted to $T^\vee$ is sending t to t^{-1}. This implies that G has an anisotropic F-torus (this is clear if $G = G^*$ is quasi-split, and it is well known that this condition is independent of the inner form [**S**, Corollary 2.9]). Conversely, when this condition holds it is easy to write down discrete parameters. □

DEFINITION 2.3.3. We call a Langlands parameter φ tempered if $\varphi(W_F)$ is bounded.

It is not difficult to see that φ is tempered if φ is discrete.

DEFINITION 2.3.4. Two Langlands parameters φ, φ' are called equivalent iff they are in the same orbit of $\mathrm{Int}(G^\vee)$ (acting on ${}^LG$). The set of equivalence classes of Langlands parameters of G is denoted by $\Phi(G)$ and the subset of equivalence classes of tempered Langlands parameters of G is denoted by $\Phi^{\mathrm{temp}}(G)$.

2.4. L-functions and ϵ factors. We associate an L-function and an ϵ-factor to a representation V of L_F in the usual way (see [**Tate**]). The L-function $L(s, V)$ is a meromorphic function of a parameter $s \in \mathbb{C}$ which only depends on the semisimplification of V, and satisfies by inductivity (i.e. if $F' \subset F$ then $L(s, V) = L(s, \mathrm{Ind}_{L_F}^{L_{F'}}(V)))$ and additivity. It is known these properties determine $L(s, V)$ completely if $L(s, \chi)$ is known for all characters χ of L_F.

In the archimedean case the L-functions assigned to characters are as follows:

(a) If $F = \mathbb{R}$, $L_F^{ab} \simeq \mathbb{R}^\times$. A character has a unique representation of form $\chi(x) = x^{-n}|x|^{s_0}$ with $n \in \{0, 1\}$; then $L(s, \chi) = \pi^{-(s+s_0)/2}\Gamma((s + s_0)/2)$.

(b) If $F = \mathbb{C}$, $L_F^{ab} \simeq \mathbb{C}^\times$. A character has a unique representation of form $\chi(z) = \sigma(z)^{-n}\|z\|^{s_0}$ with $n \in \mathbb{Z}_{\geq 0}$ and $\sigma \in \Gamma(\mathbb{C}/\mathbb{R})$; then $L(s, \chi) = 2(2\pi)^{-(s+s_0)}\Gamma(s + s_0)$.

For the non-archimedean case, let Fr denote the Frobenius automorphism of the maximal unramified extension F_{ur} of F. We choose once and for all an extension of Fr to $\overline{F}$, defining an element of W_F which we will also denote by $\mathrm{Fr} \in W_F$. Define

$$\tilde{\mathrm{Fr}} = (\mathrm{Fr}, \begin{pmatrix} v^{-1} & 0 \\ 0 & v \end{pmatrix}) \in L_F, \text{ with } v = q^{1/2} \tag{2.9}$$

Now we define for a representation (V, φ) of L_F:

$$\begin{aligned} L(s, V) &= \det(1 - q^{-s}\varphi(\tilde{\mathrm{Fr}})|_{V_N^{I_F}})^{-1} \\ &= \prod_{n \geq 0} \det(1 - q^{-s-n/2}\varphi(\mathrm{Fr})|_{V_n^{I_F}})^{-1} \end{aligned}$$

where $I_F \subset W_F$ denotes the inertia subgroup $I_F := \Gamma(\overline{F}/F_{ur})$, and $V_N^{I_F}$ the space of highest weight vectors in the $(W_F/I_F) \times \mathrm{SL}_2(\mathbb{C})$-module V^{I_F}. In the second line we decomposed V as $V \simeq \oplus_{n\geq 0} V_n \otimes \mathrm{Sym}^n(\mathbb{C}^2)$ for certain representations V_n of W_F.

The ϵ-factors depend on the choice of the additive character ψ of F. It is known that ϵ is also additive, and inductive for virtual representations of degree 0. In the non-archimedean case we have $\epsilon(s, V, \psi) := \omega(V, \psi)q^{a(V)(1/2-s)}$ where $a(V)$ is the Artin conductor of V [**GR**, Section 2]. Here $\omega(V, \psi) \in \mathbb{C}^\times$ is independent of s. In the archimedean case we have $\epsilon(s, \chi, \psi) = c_\psi.(\sqrt{-1})^n$ where χ is a character of W_F expressed as above (see the discussion of the L-functions in the archimedean case).

Given a Langlands parameter $\varphi : L_F \to {}^LG$, we define the adjoint γ factor of φ as follows. Let Ad denote the adjoint representation of ${}^LG$ on $\mathrm{Lie}(G^\vee)/\mathrm{Lie}({}^LZ)$.

$$\gamma(s, \mathrm{Ad} \circ \varphi, \psi) := \frac{\epsilon(s, \mathrm{Ad} \circ \varphi, \psi)L(1 - s, \mathrm{Ad} \circ \varphi)}{L(s, \mathrm{Ad} \circ \varphi)} \tag{2.10}$$

It is not difficult to show that [**HII**, Lemma 1.2]:

PROPOSITION 2.4.1. *If φ is tempered then $\gamma(s, \mathrm{Ad} \circ \varphi, \psi)$ is regular at $s = 0$. Moreover φ is tempered and essentially discrete iff $\gamma(s, \mathrm{Ad} \circ \varphi, \psi)$ is nonzero at $s = 0$.*

2.5. A conjectural tempered local Langlands correspondence. The conjecture on Plancherel densities of Hiraga, Ichino and Ikeda presupposes the existence of a local Langlands correspondence for tempered representations. A satisfactory formulation of a refined local Langlands conjecture in full detail (including certain desired properties of transfer factors) at this level of generality seems not to be known. We refer the reader to [**Ar1**], [**Vog**], [**ABV**], [**Ar2**], [**HII**], [**HS**] and [**Kal**] for more background, discussion and overview of known results supporting various forms of the conjecture.

In this section we would like to formulate a more crude version of the local Langlands conjecture for tempered representations covering the aspects which are relevant to our goals. We mainly follow [**Ar2**, Section 3], [**HII**, Section 1], [**HS**, Section 9].

Put $\Pi(G)$ for the set of admissible irreducible representations of G. The local Langlands conjecture predicts that there exists a partition

$$\Pi(G) = \sqcup_{[\varphi]} \Pi_\varphi(G) \tag{2.11}$$

where the disjoint union is over the set of equivalence classes $[\varphi]$ of local Langlands parameters $\varphi : L_F \to {}^L G$. The sets $\Pi_\varphi(G)$ are called L-packets. Some of the fundamental expected properties of this conjectural partitioning are:

(i) Π_φ is a non-empty finite set.
(ii) Π_φ contains tempered characters iff φ is tempered. In this case all members of Π_φ are tempered.
(iii) Π_φ contains characters which are discrete modulo center iff φ is essentially discrete. In this case all members of Π_φ are discrete modulo center.
(iv) Suppose that F is p-adic. Then Π_φ contains a character which is generic and supercuspidal iff $\varphi|_{W_F}$ is discrete. In this case all members of Π_φ are supercuspidal.

Let us now look in more detail into the conjectural parameterisation of the L-packets Π_φ for φ tempered, following [**HII**].

Let A denote the (group of points of) the maximal split torus of $Z(G)$. Let $(G_{ad})^\vee = G^\vee_{sc}$ be the simply connected cover of the derived group of $G^\vee$. Let $G^{\vee,\natural}$ denote the dual group of G/A, and let $G^\vee_{ad} = G^\vee_{sc}/Z(G^\vee_{sc})$ be the dual group of the simply connected cover of the derived group of G.

We have homomorphisms $G^\vee_{sc} \xrightarrow{\beta} G^\vee_{ad} \xleftarrow{\alpha} G^{\vee,\natural}$. Given a tempered Langlands parameter $\varphi : L_F \to {}^L G$ for G, we define $S^\natural_\varphi := \{s \in G^{\vee,\natural} \mid \mathrm{Ad}(s) \circ \varphi = \varphi\}$ and $S_\varphi := \beta^{-1}\alpha(S^\natural_\varphi)$. Next we define $\mathcal{S}^\natural_\varphi = \pi_0(S^\natural_\varphi)$ and $\mathcal{S}_\varphi = \pi_0(S_\varphi)$.

Recall that G is (the group of F-points of) an inner form the quasi-split F-group G^*, which defines a class in $H^1(F, \mathbf{G}^*_{ad})$. Kottwitz constructed a canonical map $H^1(F, \mathbf{G}^*_{ad}) \to ({}^L Z_{sc})^*$ (with ${}^L Z_{sc} = Z(G^\vee_{sc})^\Gamma$) which is bijective in the p-adic case. Let $\chi_G \in ({}^L Z_{sc})^*$ be the character that corresponds to G. We have $\chi_{G^*} = \mathrm{triv}$. By [**Ar2**, §3], [**HS**, Lemma 9.1] we have:

LEMMA 2.5.1. *The kernel* $\mathrm{Ker}(\chi_G)$ *contains ${}^L Z_{sc} \cap S^0_\varphi$, hence χ_G descends to $\mathrm{Im}({}^L Z_{sc} \to \mathcal{S}_\varphi) \subset \mathcal{S}_\varphi$ (also called χ_G).*

Following [**Ar2**], we choose an extension of χ_G to $\mathrm{Im}(Z_{sc} \to \mathcal{S}_\varphi)$ (also denoted by χ_G) such that $\chi_{G^*} = \mathrm{triv}$ and define: $\Pi(\mathcal{S}_\varphi, \chi_G) = \{\rho \in \mathrm{Irr}(\mathcal{S}_\varphi) \mid \rho|_{\mathrm{Im}(Z_{sc} \to \mathcal{S}_\varphi)} = n\chi_G\}$.

From [**Ar2**, Section 3], [**HII**] and [**HS**, Section 9] we distill the following crude form of the local Langlands conjecture:

CONJECTURE 2.5.2. *There exists a bijection $\rho \to \pi_\rho$ between $\Pi(\mathcal{S}_\varphi, \chi_G)$ and $\Pi_\varphi(G)$ such that for all tempered local Langlands parameters φ for G,*

$$\Theta_\varphi := \sum_{\rho \in \Pi(\mathcal{S}_\varphi, \chi_G)} \dim(\rho) \Theta_{\pi_\rho} \tag{2.12}$$

is a stable character of G. Here Θ_{π_ρ} denotes the distributional character of G corresponding to the tempered irreducible representation π_ρ. Any stable linear combination of characters from $\Pi(\mathcal{S}_\varphi, \chi_G)$ is a multiple of Θ_φ.

DEFINITION 2.5.3. We define $\tilde{\Phi}^{\mathrm{temp}}(G) = \{(\varphi, \rho) \mid \varphi \in \Phi^{\mathrm{temp}}(G) \text{ and } \rho \in \Pi(\mathcal{S}_\varphi, \chi_G)\}$. Suppose that for all $\varphi \in \Phi^{\mathrm{temp}}(G)$ a parameterisation $\rho \to \pi_\rho$ of $\Pi_\varphi(G)$ as in Conjecture 2.5.2 exists. The corresponding bijection $\hat{G}^{\mathrm{temp}} \to \tilde{\Phi}^{\mathrm{temp}}(G)$, $\pi \to (\varphi_\pi, \rho_\pi)$ such that for each $\varphi \in \Phi^{\mathrm{temp}}(G)$, $\Pi_\varphi(G) = \{\pi \in \tilde{\Phi}^{\mathrm{temp}}(G) \mid \varphi_\pi = \varphi\}$ and such that the bijection $\Pi_\varphi(G) \to \Pi(\mathcal{S}_\varphi, \chi_G)$, $\pi \to \rho_\pi$ is the inverse of the bijection $\rho \to \pi_\rho$ in Conjecture 2.5.2, is called an enhancement of the Langlands parameterisation $\pi \to \varphi_\pi$.

2.6. The conjectures of Hiraga, Ichino and Ikeda. We now have everything in place in order to formulate the conjectures of Hiraga, Ichino and Ikeda [**HII**]. Suppose that we have given an enhanced Langlands parameterisation $\hat{G}^{\mathrm{temp}} \to \tilde{\Phi}^{\mathrm{temp}}(G)$.

CONJECTURE 2.6.1 (Conjecture 1.4 of [**HII**]). *Let $\varphi : L_F \to {}^L G$ be a discrete Langlands parameter for G, let $\rho \in \Pi(\mathcal{S}_\varphi, \chi_G)$ and let $\pi_\rho \in \Pi_\varphi$ be the tempered essentially discrete series representation corresponding to (φ, ρ). Then*

$$\mathrm{fdeg}(\pi_\rho) = \frac{\dim(\rho)}{|\mathcal{S}_\varphi^\natural|} |\gamma(0, \mathrm{Ad} \circ \varphi, \psi)| \tag{2.13}$$

For general tempered representations [**HII**] formulate a conjecture expressing the Plancherel density. This amplification is based on Harish-Chandra's Plancherel Theorem ([**HC3**], [**Wal**]) and Langlands' conjecture on the Plancherel measure [**L**, Appendix II], [**Sha**].

CONJECTURE 2.6.2 (Conjecture 1.5 of [**HII**]). *Let $P = MN \subset G$ be a semi-standard F parabolic subgroup. Let $\mathcal{O}$ be an orbit of tempered essentially discrete series characters of M. Let $d\pi$ denote the Haar measure on $\mathcal{O}$, normalised as in* [**Wal**, pages 239 and 302]. *For $\pi \in \mathcal{O}$ we put*

$$d\nu(\pi) = \frac{\dim(\rho)}{|\mathcal{S}_{\varphi_M}^\natural|} |\gamma(0, r_M \circ \varphi, \psi)| d\pi \tag{2.14}$$

where r_M denotes the adjoint representation of ${}^L M$ on $\mathrm{Lie}(G^\vee)/\mathrm{Lie}({}^L Z_M)$. Then the Plancherel density at $\mathrm{Ind}_P^G(\pi)$ is $c_M d\nu(\pi)$ for some explicit constants $c_M \in \mathbb{R}_+$ independent of F and $\mathcal{O}$.

2.7. Known results and further comments. Conjecture 2.6.1 is reduced to the case of generic tempered representations by Shahidi's paper [**Sha**], if one knows the stability of Θ_φ in Conjecture 2.5.2.

The Conjecture 2.6.1 is known for $F = \mathbb{R}$ [**HII**, Section 3]. For F non-archimedean Conjecture 2.6.1 is known in the following cases:

(a) G an inner form of GL_n ([**SZ**], [**Z**], [**HII**]).
(b) G an inner form of SL_n ([**HS**], [**HII**]).
(c) G arbitrary, π the Steinberg representation, φ the principal parameter (due essentially to Borel, [**Bo1**], [**HII**]).
(d) G split exceptional of adjoint type, π discrete series of unipotent reduction (due to Reeder, [**R3**]).
(e) G arbitrary, π depth 0 supercuspidal representation (tame regular semi-simple case) (DeBacker and Reeder [**DeRe**], [**HII**]).
(f) Ichino, Lapid and Mao proved Conjecture 2.6.1 for odd orthogonal groups.
(g) Beuzart-Plessis proved Conjecture 2.6.1 for unitary groups.
(h) For supercuspidal representations of unipotent reduction of connected semisimple p-adic groups which split over an unramified extension [**FOS**, Theorem 1.3].

The main result we will discuss in these lectures is and extension to general connected reductive G of the following result: [1]

THEOREM 2.7.1 ([**R3**], [**Opd4**], [**Opd5**], [**Fe2**]). *Let G be absolutely almost simple of adjoint type over a non-archimedean field F such that G splits over an unramified extension of F. Then Conjectures* 2.6.1, 2.6.2 *hold for representations of unipotent reduction, when we use Lusztig's classification* [**Lus4**], [**Lus5**] *as a Langlands parameterisation.*

The proof of Theorem 2.7.1 and its extension is based on two techniques for affine Hecke algebras:

(1) Spectral transfer maps between Hecke algebras [**Opd4**], [**Opd5**], [**FO**], [**FOS**]. We use these tools to deal with the q-rational factors of the formal degree.
(2) Dirac induction for affine Hecke algebras [**COT**], [**CO**]. This tool is useful to determine the precise rational constant factors of the formal degree.

In fact, by the theory of types and Theorem 4.5.1 I expect that these techniques may reduce the general case of Conjectures 2.6.1 and 2.6.2 to the case of generic supercuspidal representations.

The "converse results" Theorem 4.2.3, Theorem 4.5.1 are interesting in their own right, and are closely related to the theory of spectral transfer maps between normalised affine Hecke algebras [**Opd5**].

3. The Plancherel formula for affine Hecke algebras

Let F be a nonarchimedean local field with residue field of cardinality q from here onwards.

---

[1]The Lusztig parameterisation should be twisted by the Iwahori-Matsumoto involution in order to map tempered representations to bounded parameters (cf. [**AMS**, Text below Theorem 2]). Here and elsewhere we will tacitly assume this modification.

3.1. The Bernstein center. Let $\mathcal{C}(G)$ be the abelian category of smooth representations of G, and let $\Pi(G)$ denote the space of classes of irreducible objects of $\mathcal{C}(G)$. Let $\mathcal{B}(G)$ be the set of inertial equivalence classes of cuspidal pairs (M,τ) (with M a Levi subgroup, and τ a supercuspidal representation of M). For $s\in\mathcal{B}(G)$, let Ω_s be the corresponding set of cuspidal pairs in the class s, modulo G-conjugacy (an affine variety). We have a central character map

$$\mathrm{cc}:\Pi(G)\to\sqcup_{s\in\mathcal{B}(G)}\Omega_s \tag{3.1}$$

such that $\mathrm{cc}(\pi)=(M,\tau)$ if π is a subquotient of $i_P^G(\tau)$, where $P=MU$ is a parabolic subgroup with Levi factor M. Let $\mathcal{O}(\Omega_s)$ be the ring of regular functions on Ω_s, and put $\mathfrak{z}(G)=\prod_{s\in\mathcal{B}(G)}\mathcal{O}(\Omega_s)$. We put $Z_{\mathcal{B}}(G)=\mathrm{End}(\mathrm{Id}_{\mathcal{C}(G)})$, the Bernstein center. As is well known, we can interpret $Z_{\mathcal{B}}(G)$ as the set of G-invariant distributions z on G such that $z\mathcal{H}(G)\subset\mathcal{H}(G)$ and $\mathcal{H}(G)z\subset\mathcal{H}(G)$, where $\mathcal{H}(G)$ denotes the Hecke algebra of G. The famous theorem of Bernstein and Deligne states:

THEOREM 3.1.1 ([**BeDe**], Theorem 2.13). *There is a unique algebra isomorphism (the Fourier transform)*

$$Z_{\mathcal{B}}(G)\to\mathfrak{z}(G) \tag{3.2}$$

$$z\to\hat{z} \tag{3.3}$$

characterised by the property that for all $\pi\in\Pi(G)$, one has $z_\pi=\hat{z}(\mathrm{cc}(\pi))\mathrm{Id}_{V_\pi}$.

As an immediate consequence one obtains the Bernstein decomposition of $\mathcal{C}(G)$:

COROLLARY 3.1.2. *We have a family of orthogonal idempotents $e_s\in Z_{\mathcal{B}}(G)$ (with $s\in\mathcal{B}(G)$) such that $\widehat{e_s}$ is the characteristic function of Ω_s. We have corresponding decompositions*

$$\mathcal{C}(G)=\prod_{s\in\mathcal{B}(G)}\mathcal{C}(G)_s,\quad \Pi(G)=\sqcup_{s\in\mathcal{B}(G)}\Pi(G)_s \tag{3.4}$$

where $\Pi(G)_s$ is the set of irreducible objects of $\mathcal{C}(G)_s$.

3.2. Types, Hecke algebras and Plancherel measure. A *type* is a pair $t=(J,\rho)$ such that:

(1) $J\subset G$ is a compact open subgroup of G.
(2) ρ is a finite dimensional irreducible representation of J.
(3) Let $e_t\in\mathcal{H}(G)$ be the idempotent given by

$$e_t(x)=\begin{cases}\frac{\mathrm{Tr}(\rho(x^{-1}))}{\mathrm{Vol}(J)} & \text{if } x\in J\\ 0 & \text{else.}\end{cases} \tag{3.5}$$

Let $\mathcal{C}^t(G)$ be the full subcategory of $\mathcal{C}(G)$ consisting of representations (π,V_π) such that $\mathcal{H}(G)(e_tV_\pi)=V_\pi$, and let $\mathcal{H}_t=e_t\mathcal{H}(G)e_t$. Then the functor $m_t:\mathcal{C}^t(G)\to\mathcal{H}_t-\mathrm{mod}$ given by $V_\pi\to e_t(V_\pi)$ is an equivalence of categories.

THEOREM 3.2.1 (Bushnell and Kutzko [**BK**]). *Let t be a type. Then $\mathcal{C}^t(G)=\prod_{s\in\mathcal{B}^t(G)}\mathcal{C}(G)_s$ where $\mathcal{B}^t(G)$ is a finite set.*

As a consequence of Theorem 3.1.1 we see:

COROLLARY 3.2.2. *Let t be a type, and let $Z(\mathcal{H}_t)$ denote the center of $\mathcal{H}_t$. There exists a unique isomorphism:*

$$\beta^t : \Omega^t := \sqcup_{s\in\mathcal{B}^t(G)}\Omega_s \to \mathrm{Spec}(Z(\mathcal{H}_t)) \tag{3.6}$$

such that for all $\pi \in \Pi(G)^t := \sqcup_{s\in\mathcal{B}^t(G)}\Pi(G)_s$ we have: $\mathrm{cc}^t(m_t(\pi)) = \beta^t(cc(\pi))$. Here cc^t denotes the central character map of the algebra $\mathcal{H}_t$.

THEOREM 3.2.3 (Yu, J-L Kim). *If p is sufficiently large then for all $s \in \mathcal{B}(G)$ we can find a type t such that $\mathcal{C}^t(G) = \mathcal{C}(G)_s$.*

EXAMPLE 3.2.4.
(1) The archetypical example is that of the Borel component: Let $\mathbb{B} \subset G$ be the Iwahori subgroup, then $(\mathbb{B}, \mathrm{triv})$ is a type. More generally, if $\mathbb{B}_m$ is the m-th filtration subgroup of $\mathbb{B}$ in the Moy-Prasad filtration, then $(\mathbb{B}_m, \mathrm{triv})$ is a type.
(2) (Moy-Prasad, Morris, Lusztig) Let $t = (\mathbb{P}, \sigma)$ with $\mathbb{P} \subset G$ a parahoric subgroup, and σ a cuspidal unipotent representation of the reductive quotient $\overline{\mathbb{P}}(\mathbb{F}_q)$. We refer to such t as a "unipotent type", and to the objects in the associated categories $\mathcal{C}^t(G)$ as "representations of unipotent reduction".
(3) Let $x \in B(G)$ be a point in the building of G, and let $r \geq 0$. Let $G_{x,r,+}$ denote the corresponding Moy-Prasad subgroups. Then $(G_{x,r,+}, \mathrm{triv})$ is a type (Bestvina-Savin).

The Hecke algebra $\mathcal{H}_t = e_t\mathcal{H}(G)e_t$ of a type t inherits a $*$ (an anti-linear anti-involution) and trace τ from $\mathcal{H}(G)$, defined by

(1) $f^*(g) := \overline{f(g^{-1})}$ for $f \in \mathcal{H}_t$.
(2) $\tau(f) := f(1)$ (observe that for the unit $e_t \in \mathcal{H}_t$ we have $\tau(e_t) = \frac{\dim(\rho)}{\mathrm{Vol}(J)}$).

The Hermitian form $(x, y) := \tau(x^*y)$ is positive definite, and defines a Hilbert space completion $L^2(\mathcal{H}_t)$ of $\mathcal{H}_t$. This turns $\mathcal{H}_t$ into a Hilbert algebra with trace τ. It is well known that the irreducible representations of $\mathcal{H}_t$ are finite dimensional, hence this Hilbert algebra has a type I C^*-algebra envelop. As a consequence of Dixmier's central decomposition theorem for Type I C^*-algebras we conclude:

COROLLARY 3.2.5. *Let $\hat{\mathcal{H}}_t = \{[\pi] \mid \pi$ is an irrducible $*$-unitary $\mathcal{H}_t$-mod$\}$. There exists a unique positive measure $\nu_{\mathcal{H}_t}$ on $\hat{\mathcal{H}}_t$ such that*

$$\tau = \int_{\pi\in\hat{\mathcal{H}}_t} \chi_\pi d\nu_{\mathcal{H}_t}(\pi) \tag{3.7}$$

The support $\hat{\mathcal{H}}_t^{\mathrm{temp}} := \mathrm{Supp}(\nu_{\mathcal{H}_t}) \subset \hat{\mathcal{H}}_t$ is called the tempered dual of $\mathcal{H}_t$.

THEOREM 3.2.6 ([**BKH**]). *The functor m_t defines a homeomorphism $\hat{m}_t^{\mathrm{temp}}$: $\hat{G}^{t,\mathrm{temp}} := \Pi(G)^t \cap \hat{G}^{\mathrm{temp}} \to \hat{\mathcal{H}}_t^{\mathrm{temp}}$ such that $(\hat{m}_t^{\mathrm{temp}})_*(\nu_{Pl}|_{\hat{G}^{t,\mathrm{temp}}}) = \nu_{\mathcal{H}_t}$.*

Therefore we can compute ν_{Pl} by computing $\nu_{\mathcal{H}_t}$ for the Hecke algebras $\mathcal{H}_t$ of a collection of types t such that the open closed sets $\hat{G}^{t,\mathrm{temp}}$ cover $\hat{G}^{\mathrm{temp}}$. In this sense the measures $\nu_{\mathcal{H}_t}$ are the building blocks of the Plancherel measure of G.

3.3. Affine Hecke algebras as Hilbert algebras. The algebras $\mathcal{H}_t$ are slight generalisations of affine Hecke algebras. In Lusztig's case of unipotent types ([**Lus4**], [**Lus5**]; also see [**Mo**], [**MP1**], [**MP2**]) the associated "unipotent Hecke algebras" are precisely (extended) affine Hecke algebras.

Let us therefore review the theory of affine Hecke algebras and the spectral decomposition of the corresponding Hilbert algebra.

Let $W = W^a \rtimes \Omega$ be an extended affine Weyl group. By this we mean that we have an affine Coxeter group W^a with set of simple reflections S say, and a group of *special automorphisms* $\Omega \subset \mathrm{Out}_S(W^a, S)$. (A diagram automorphism $\omega \in \mathrm{Out}_S(W^a, S)$ is called special if its restriction to the canonical normal subgroup $Q \subset W^a$ consisting of the elements with a finite conjugacy class, equals the restriction to Q of an inner automorphism of W^a.)

Let $\Lambda = \mathbb{Z}[v_s^{\pm 1} \mid s \in S;\ v_s = v_{s'} \text{ if } s \sim_W s']$ (the v_s are commuting indeterminates), and put $\mathtt{Q} = \{\mathtt{v} \in \mathrm{Hom}_{alg}(\Lambda, \mathbb{C}^\times) \mid \mathtt{v}_s = \mathtt{v}(v_s) \in \mathbb{R}_+ \forall s \in S\} \simeq \mathbb{R}_+^N$ where N denotes the number of conjugacy classes of affine reflections in W. We have a length function l on W^a relative to the set S of simple reflections, which we extend to W by giving elements of Ω length 0.

Definition 3.3.1 (Coxeter presentation of the Hecke algebra). Let $\mathcal{H}_\Lambda$ be the free Λ-algebra with basis $\{N_w\}_{w \in W}$ such that

(i) $N_u N_v = N_{uv}$ for $u, v \in W$ such that $l(uv) = l(u) + l(v)$.
(ii) For all $s \in S$ we have $(N_s - v_s)(N_s + v_s^{-1}) = 0$.

Given $\mathtt{v} \in \mathtt{Q}$ we define $\mathcal{H}_{\mathtt{v}} = \mathcal{H}_\Lambda \otimes \mathbb{C}_{\mathtt{v}}$. Let $d \in \Lambda$ be positive on $\mathtt{Q}$ (or on some subset of $\mathtt{Q}$ which contains $\mathtt{v}$). Then we define a Hilbert algebra structure on $\mathcal{H}_{\mathtt{v}}$ by defining a $*$-operator and a *positive trace* τ as we did before with $\mathcal{H}(G)$:

(i) $\tau(N_w) = \delta_{w,e} d(\mathtt{v})$.
(ii) $N_w^* = N_{w^{-1}}$.

We recall that the positivity of τ means that the Hermitian form $(x, y) = \tau(x^* y)$ on $\mathcal{H}_{\mathtt{v}}$ is positive definite. This elementary fact is crucial in all that follows.

Next we would like to express the Hilbert algebra structure in terms of the Bernstein presentation of $\mathcal{H}_{\mathtt{v}}$. Let $X \subset W$ be the canonical normal subgroup of W consisting of the elements which have finitely many conjugates. This is the translation subgroup of W, and we can choose a splitting of $W/X \simeq W_0$ by choosing a special point $0 \in C$, where $C \subset V = \mathbb{R} \otimes X$ denotes the alcove.

The length $l(w)$ of $w \in W$ can be interpreted more geometrically as the number of affine reflection hyperplanes of W^a separating the fundamental alcove $C \subset V = \mathbb{R} \otimes Q$ and $w(C)$. In particular we have $l(x) = 2\rho(x)$ for $x \in X^+$, the dominant cone in X. By the defining relations of $\mathcal{H}_\Lambda$ this implies that the elements N_x with $x \in X^+$ form a commutative monoid of invertible elements of $\mathcal{H}_\Lambda$. Bernstein and Zelevinski turned this into a very important alternative presentation of $\mathcal{H}_\Lambda$, (see [**Lus3**] for further background).

Lemma 3.3.2. *There exists a unique homomorphism $X \ni x \to \theta_x \in \mathcal{H}_\Lambda$ such that for $x \in X^+$ one has $\theta_x = N_x$. Let $A \subset \mathcal{H}_\Lambda$ be the commutative subalgebra $\Lambda[\theta_x \mid x \in X] \subset \mathcal{H}_\Lambda$ generated by the θ_x. Let $\mathcal{H}_0 = \mathcal{H}_\Lambda(W_0, S_0) \subset \mathcal{H}_\Lambda$ be the finite type Hecke subalgebra associated to the isotropy group (W_0, S_0) of the chosen special point $0 \in C$. Then the multiplication map $A \otimes \mathcal{H}_0 \to \mathcal{H}_\Lambda$ is an isomorphism of $(A, \mathcal{H}_0)$-bimodules, and $\mathcal{H}_0 \otimes A \to \mathcal{H}_\Lambda$ is an isomorphism of $(\mathcal{H}_0, A)$-bimodules,* *and the algebra structure of $\mathcal{H}_\Lambda$ is determined by the Bernstein relation:*

$$\theta_x N_s - N_s\theta_{s(x)} = \left((v_s - v_s^{-1}) + (v_{s'} - v_{s'}^{-1})\theta_{-\alpha}\right)\frac{\theta_x - \theta_{s(x)}}{1-\theta_{-2\alpha}} \tag{3.8}$$

where $s = s_{\alpha^\vee} \in S_0$ for some simple root α_0, and $s' \in S$ is such that $s' \sim_W s_{\alpha^\vee+1}$.

REMARK 3.3.3. (a) Let $\mathcal{R} = (\Delta_0, X, \Delta_0^\vee, Y)$ the based root datum associated with $W = W_0 \rtimes X$, where S_0 corresponds to Δ_0. For $\alpha \in \Delta_0$ one can put $q_\alpha^+ = v_s v_{s'}$ and $q_\alpha^- = v_s/v_{s'}$. The parameters $q_\alpha^\pm$ will be more convenient than the v_s in the spectral theory of $\mathcal{H}_v$.

(b) If $v_{s'} \neq v_s$ then $\alpha^\vee \in 2Y$. In irreducible cases this happens only if $\mathcal{R}$ is of type $C_n^{(1)}$. That means that Δ_0 is the basis $\{e_1 - e_2, \dots, e_{n-1} - e_n, e_n\}$ of the irreducible root system of type B_n, and $X = \mathbb{Z}^n$ is the root lattice of this root system. The affine Dynkin diagram of W is the untwisted affine extension of the Dynkin diagram of $\Delta_0^\vee$ of type C_n. Note that $q_\alpha^- = 1$ unless we are in the $C_n^{(1)}$ case, and $\alpha = e_n$.

COROLLARY 3.3.4. *The center $Z := Z(\mathcal{H}_\Lambda)$ is equal to A^{W_0}, which is naturally isomorphic to the ring $\Lambda[\theta_x \mid x \in X]^{W_0}$. Let T denote the algebraic torus with character lattice $\mathbb{Z}^{S/\sim} \times X$, viewed as a split torus over Λ via $v_s^n \to (ne_s, 0)$. Then $Z(\mathcal{H}_\Lambda) = \mathbb{C}[W_0\backslash\backslash T]$.*

DEFINITION 3.3.5. We have $T = T_0 \times \mathrm{Spec}(\Lambda)$ where T_0 is algebraic torus over $\mathbb{C}$ with character lattice X. If $\mathtt{v} \in \mathtt{Q}$ then we will write $T^{\mathtt{v}}$ for the fibre of T above $\mathtt{v}$, i.e. the spectrum of $\mathbb{C}_{\mathtt{v}}[X]$ where $\mathbb{C}_{\mathtt{v}}$ denotes the residue field of Λ at $\mathtt{v}$.

3.4. A formula for the trace of an affine Hecke algebra. We will now write the trace τ of $\mathcal{H}_{\mathtt{v}}$ (for some $\mathtt{v} \in \mathtt{Q}$) in terms of the Bernstein presentation of $\mathcal{H}_{\mathtt{v}}$. First we introduce "intertwining elements" $R_s \in \mathcal{H}_\Lambda$ for every $s = s_\alpha \in S_0$ by:

$$R_s = v_s\left((1-\theta_{-2\alpha})N_s - ((v_s - v_s^{-1}) + (v_{s'} - v_{s'}^{-1})\theta_{-\alpha})\right) \tag{3.9}$$

These elements satisfy:

(i) For all $x \in X$ and $s = s_\alpha \in S_0$ we have: $R_s\theta_x = \theta_{s(x)}R_s$.

(ii) $R_s^2 = v_s^2\Delta_{2\alpha}\Delta_{-2\alpha}c_\alpha c_{-\alpha}$, where $\Delta_{\pm 2\alpha} = (1-\theta_{\pm 2\alpha})$ and where the rational functions c_α are the famous *Harish Chandra c-functions* in the present context: $c_\alpha = (\Delta_{-2\alpha})^{-1}(1+\theta_{-\alpha}/q_\alpha^-)(1-\theta_{-\alpha}/q_\alpha^+)$.

COROLLARY 3.4.1. *We have $\mathcal{H}_\Lambda \otimes_Z Z' \simeq A'\#W_0$, where Z' is the localization of Z on the open subset of $W_0\backslash T$ which is the intersection of the open subsets $U_\alpha \subset T$ ($\alpha \in \Sigma_0$) where c_α is an invertible regular function (the complement of the union of the hyperplanes of the form $\alpha(t) = \pm q_\alpha^\pm$ and $\alpha(t) = \pm 1$).*

DEFINITION 3.4.2. Define

$$\mu(t) = \frac{d}{q(w_0)}\frac{1}{\prod_{\alpha\in\Sigma_0^+} c_\alpha(t)c_{-\alpha}(t)} = \frac{1}{q(w_0)}\frac{d}{c(t)c(t^{-1})}, \tag{3.10}$$

the μ-function of $\mathcal{H}_\Lambda$ with normalising factor $d \in \Lambda$.

THEOREM 3.4.3 ([**Opd1**]). *Let $t \in T^{\mathtt{v}}$ where $\mathtt{v}$ is such that $q_\alpha^\pm > 0$ for all α. Via the nondegenerate symmetric bilinear form $\langle x, y\rangle := \tau(xy)$ we view $\mathcal{H}_{\mathtt{v}}$ as a subspace of $\mathcal{H}_{\mathtt{v}}^*$, and equip $\mathcal{H}_{\mathtt{v}}^*$ with the weak topology. Then $\mathcal{E}_t := \sum_{x\in X} t(-x)\theta_x \in \mathcal{H}_{\mathtt{v}}^*$*

is convergent in $\mathcal{H}_{\mathrm{v}}^*$ *if for all* $\alpha \in \Delta_0$*, we have* $|\alpha(t)| < \min\{(q_\alpha^-)^{-1}, (q_\alpha^+)^{-1}\}$*. Moreover,*

$$\mathcal{E}_t = \frac{E_t}{q(w_0)\Delta(t)}\mu(t) \tag{3.11}$$

where $T^{\mathrm{v}} \ni t \to E_t \in \mathcal{H}_{\mathrm{v}}^*$ *is a certain regular family of matrix coefficients of minimal principal series at* t *such that* $E_t(1) = q(w_0)\Delta(t)$ *and such that for all* $a, b \in A$*,* $E_t(ahb) = a(t)b(t)E_t(h)$*.*

COROLLARY 3.4.4 ([**Opd1**]). *We have the following disintegration of* τ *on* $\mathcal{H}_{\mathrm{v}}$*:*

$$\tau = \int_{t_0 T_u^{\mathrm{v}}} \frac{E_t}{q(w_0)\Delta(t)}\mu(t)dt \tag{3.12}$$

where T_u^{v} *denotes the compact form of* T^{v}*, and* $t_0 \in T_v^{\mathrm{v}}$ *is a real base point such that the inequality* $|\alpha(t_0)| < \min\{(q_\alpha^-)^{-1}, (q_\alpha^+)^{-1}\}$ *holds.*

PROOF. Immediate from Theorem 3.4.3 by the Fourier inversion formula on T_u^{v}. □

3.5. Spectral decomposition of τ**.** The disintegration of τ given in Corollary 3.4.4 is not yet a spectral decomposition because the matrix coefficients E_t are neither tempered on $t_0 T_u^{\mathrm{v}}$, nor tracial (i.e. they do not vanish on commutators). To arrive at the spectral decomposition of τ several steps of refinement are necessary. The first step uses "residue distributions" for integrals in the form (3.12), and symmetrizing the resulting distributions on T over W_0. This leads to a decomposition of the form [**Opd2**]:

$$\tau = \int_{W_0 t \in W_0 \backslash T^{\mathrm{v}}} \chi_{W_0 t} d\nu(W_0 t) \tag{3.13}$$

where

(i) ν denotes the spectral measure of the decomposition of $\tau|_{Z_{\mathrm{v}}}$, where $Z_{\mathrm{v}} \subset \mathcal{H}_{\mathrm{v}}$ denotes the center. We remark that Z_{v} is invariant for $*$, and the restriction of $*$ and τ to Z_{v} equips it with the structure of a commutative Hilbert algebra.

(ii) The support of ν is denoted by $W_0 \backslash T^{\mathrm{v},\mathrm{temp}}$. For each $W_0 t$ in $W_0 \backslash T^{\mathrm{v},\mathrm{temp}}$, $\chi_{W_0 t}$ is a tempered positive trace of $\mathcal{H}_{\mathrm{v}}$, with central character $W_0 t$.

(iii) We have $T^{\mathrm{v},\mathrm{temp}} = \cup_{L \text{ residual coset}} L^{\mathrm{temp}}$, where a coset $L \subset T^{\mathrm{v}}$ of a subtorus is called *residual* if

$$\#\{\alpha \in \Sigma_0 \mid \alpha|_L = \pm q_\alpha^{\pm}\} - \#\{\alpha \in \Sigma_0 \mid \alpha|_L = \pm 1\} = \mathrm{codim}(L) \tag{3.14}$$

Furthermore if $L \subset T$ a residual coset, we define its tempered part $L^{\mathrm{temp}} \subset L$ as follows. Let $\Sigma_L \subset \Sigma_0$ be the parabolic subsystem of the roots which are constant on L. Let $T^L \subset T^{\mathrm{v}}$ be the identity component of the simultaneous kernel of the $\alpha \in \Sigma_L$. Let $T_L \subset T^{\mathrm{v}}$ be the subtorus associated with the subspace $\mathbb{C}\Sigma_L^\vee \subset \mathfrak{t} = \mathrm{Lie}(T^{\mathrm{v}})$. If $r_L \in T_L \cap L$ then $L = r_L T^L$. Now put $L^{\mathrm{temp}} = r_L T_u^L$ (this does not depend on the choice $r_L \in T_L \cap L$).

The computation of the spectral decomposition of τ as trace on $\mathcal{H}^{\mathrm{v}}$ now reduces to the problem of computing the measure ν explicitly, and for each $W_0 t \in W_0 \backslash T^{\mathrm{v},\mathrm{temp}}$, decomposing $\chi_{W_0 t}$ as a (positive) superposition of irreducible tempered characters (with central character $W_0 t$).

In the special case of the discrete series of $\mathcal{H}^{\mathrm{v}}$ we see that these correspond to the W_0-orbits of *residual points*. This case is the basic building block for the spectral decomposition:

THEOREM 3.5.1 ([**Opd2**]). (i) *An orbit $W_0 r \subset W_0\backslash T^{\mathrm{v}}$ is the central character of a discrete series representation π of $\mathcal{H}^{\mathrm{v}}$ if and only if r is a residual point (a residual coset of dimension 0).*

(ii) *If $W_0 r \subset T^{\mathrm{v}}$ is an orbit of residual points then $\nu(\{W_0 r\}) = c\mu^{\{r\}}(r) := c\frac{d(\mathrm{v})}{q(w_0)}m_r(\mathrm{v})$ (the residue of μ at $W_0 r$), where $c \in \mathbb{Q}^\times$ and where the regularisation $\mu^{\{r\}}$ of μ at r is defined by:*

$$\mu^{\{r\}} = \frac{d(\mathrm{v})}{q(w_0)} \prod_{\alpha\in\Sigma_0}{}' \frac{(1-\alpha^{-2})}{(1+\alpha^{-1}/q_\alpha^-)(1-\alpha^{-1}/q_\alpha^+)} = \frac{d(\mathrm{v})}{q(w_0)} m_r(\mathrm{v}) \tag{3.15}$$

where the symbol $\prod'$ means that all irreducible factors of the numerator and the denominator which become identically 0 upon evaluation at $r \in T^{\mathrm{v}}$ are omitted.

(iii) *We have $\chi_{W_0 r} = \sum_{\delta\ ds, \mathrm{cc}(\delta)=W_0 r} d_{\mathcal{H},\delta}(\mathrm{v})\chi_\delta$ where $d_{\mathcal{H},\delta}(\mathrm{v}) > 0$.*

(iv) *(Scaling invariance.) Define $\mathrm{v}(\epsilon)$ by $\mathrm{v}_s(\epsilon) = \mathrm{v}_s^\epsilon$ for $\epsilon \in \mathbb{R}_+$. Every orbit of residual points $W_0 r \in T^{\mathrm{v}}$ has a unique extension to a real analytic ϵ-family of orbits of residual points $W_0\tilde{r}$ such that $W_0\tilde{r}(\mathrm{v}(1)) = W_0 r$. A discrete series character δ with $\mathrm{cc}(\delta) = W_0 r$ has a unique extension to a continuous ϵ-family of discrete series characters $\tilde{\delta}$ of $\mathcal{H}_{\mathrm{v}(\epsilon)}$, and we have $\mathrm{cc}(\tilde{\delta}(\epsilon)) = W_0\tilde{r}(\mathrm{v}(\epsilon))$ for all $\epsilon > 0$. This yields for all $\epsilon > 0$ a canonical bijection between $\{\delta$ ds of $\mathcal{H}_{\mathrm{v}} \mid \mathrm{cc}(\delta) = W_0 r\}$ and $\{\delta'$ ds of $\mathcal{H}_{\mathrm{v}(\epsilon)} \mid \mathrm{cc}(\delta') = W_0 r(\epsilon)\}$. Then $d_{\mathcal{H},\tilde{\delta}(\epsilon)}(\mathrm{v}(\epsilon))$ is independent of $\epsilon > 0$.*

(v) *If δ is a discrete series representation of $\mathcal{H}_{\mathrm{v}}$ then $\mathrm{fdeg}(\delta) = d_{\mathcal{H},\delta}\frac{d(\mathrm{v})}{q(w_0)}|m_r(\mathrm{v})|$ where $d_{\mathcal{H},\delta} \in \mathbb{R}_+$ as defined in (iii) and m_r as defined in (ii) (we will see below that in fact $d_{\mathcal{H},\delta} \in \mathbb{Q}_+$).*

The non-discrete contributions to the spectral decomposition of τ can be obtained from the discrete summand of the spectral decomposition of the corresponding traces of "parabolic subalgebras" by a process of unitary parabolic induction, analogous to Harish-Chandra's theory of the Plancherel decomposition for reductive groups. More precisely we have [**Opd2**]:

THEOREM 3.5.2. *Let $L = r_L T^L \subset T^{\mathrm{v}}$ be a residual coset, such that $\Sigma_L \subset \Sigma_0$ is a standard parabolic subsystem. Let $\mathcal{H}^{P(L)} \subset \mathcal{H}$ be the subalgebra corresponding to the based root datum $P(L) := (\Delta_L, X, \Delta_L^\vee, Y)$. Let X_L be the character lattice of T_L, and $Y_L \subset Y$ its dual. Let $\mathcal{H}_{P(L)}$ be the Hecke algebra with the semisimple based root datum $P_{ss}(L) = (\Delta_L, X_L, \Delta_L^\vee, Y_L)$ and Hecke parameters $q_\alpha^\pm$ obtained by restriction from Δ_0 to Δ_L. Given $t^L \in T^L$ there exists a homomorphism $\phi_{t^L} : \mathcal{H}^{P(L)} \to \mathcal{H}_{P(L)}$ defined by (in the Bernstein presentation) $N_w \to N_w$ for all $w \in W_L$, and $\theta_x \to x(t^L)\theta_{\mathrm{pr}(x)}$ where $\mathrm{pr} : X \to X_L$ is the canonical projection.*

(i) *Let $t = r_L t^L \in L^{\mathrm{temp}}$ be a generic point, i.e. c_α defines a regular and invertible germ at t for all $\alpha \in \Sigma_0\backslash\Sigma_L$. Then in (3.13) we have:*

$$|W_0/W_L|\chi_{W_0 t} = \sum_{\delta\in\hat{\mathcal{H}}_{P(L),ds}, \mathrm{cc}(\delta)=W_0 r_L} \mathrm{Ind}_{\mathcal{H}^L}^{\mathcal{H}}(\chi_{L,W_L r_L} \circ \phi_{t^L}) \tag{3.16}$$

(ii) *We have* $\nu = \sum_{L \text{ residual coset}} \nu_L$ *where* ν_L *is the push forward of a measure on* L^{temp} *given by* $d\nu_L(t) = \mu^L(t)dt = c_L \mu_{\mathcal{H}_L}^{\{r_L\}}(r_L) m^L(t) dt^L$ *with* $c_L \in \mathbb{Q}_+$ *and*

$$m^L(t) = \frac{1}{q(w^L)} \prod_{\alpha \in \Sigma_+^L := \Sigma_{0,+} \setminus \Sigma_{L,+}} \frac{1}{c_\alpha(t) c_\alpha(t^{-1})} \tag{3.17}$$

COROLLARY 3.5.3. *The explicit spectral decomposition of the trace* τ *on* $\mathcal{H}_{\mathbf{v}}$ *reduces, by Theorem 3.5.2, to the classification of the discrete series of the standard parabolic semisimple subquotient Hecke algebras* $\mathcal{H}_P$ *of* $\mathcal{H}$*, and the computation of their formal degree. This reduces further to the classification of the set of* W_0*-orbits of residual points* $\{W_0 r\}$*, and of the finite set of discrete series characters* δ *with* $\mathrm{cc}(\delta) = W_0 r$ *(which has been carried out in* [**OS2**]*), and the computation of the constants* $d_{\mathcal{H}_P,\delta} \in \mathbb{R}_+$ *(carried out in* [**CO**]*).*

REMARK 3.5.4. In the context of Hecke algebras of a type of a reductive group over a non-archimedean local field F, changing the base field to an unramified extension of F of degree n corresponds to the scaling $\mathbf{v} \to \mathbf{v}(n)$. This explains the importance of the scaling invariance properties.

3.6. Residual cosets and their properties. Given the importance of residual subspaces for the spectral decomposition of τ we discuss some of their properties [**Opd2, Opd3, Opd4**] and [**OS2**].

THEOREM 3.6.1. *For every coset of a subtorus* $L \subset T^{\mathbf{v}}$ *we have*

$$\#\{\alpha \in \Sigma_0 \mid \alpha|_L = \pm q_\alpha^{\pm}\} - \#\{\alpha \in \Sigma_0 \mid \alpha|_L = \pm 1\} \leq \mathrm{codim}(L) \tag{3.18}$$

PROPOSITION 3.6.2. *Let* $L \subset T$ *be a residual coset of the subtorus* T^L*, with* $\mathfrak{t}^L := \mathrm{Lie}(T^L) = \Sigma_L^{\perp}$*. Let* $T_L \subset T$ *be the subtorus such that* $\mathfrak{t}_L := \mathrm{Lie}(T_L) = \mathbb{R}\Sigma_L^\vee$*. Then* $T = T_L T^L$ *and* $T^L \cap T_L = K_L$ *is a finite abelian group. Moreover* $L \cap T_L = K_L r_L$ *for a residual point* $r_L \in T_L$ *of* $\mathcal{H}_L$*.*

COROLLARY 3.6.3. *There exists only finitely many residual cosets* $L \subset T^{\mathbf{v}}$*.*

COROLLARY 3.6.4. *If* $L, M \subset T$ *are residual cosets then* $L^{\text{temp}} \subset M^{\text{temp}}$ *if and only if* $L = M$*.*

COROLLARY 3.6.5. *The measure* ν_L *defined in 3.5.2 is smooth on* L^{temp}*.*

THEOREM 3.6.6. *For* $L \subset T$ *residual, put* $S_L := W_0 \backslash W_0 L^{\text{temp}} \subset \mathrm{supp}(\nu) = S$ *with* $S = S(\mathcal{H}) := \mathrm{cc}(\hat{\mathcal{H}}_{\mathbf{v}}^{\text{temp}}) \subset W_0 \backslash T^{\mathbf{v}}$*. The sets* $S_L \subset S$ *are the connected components of* S*.*

COROLLARY 3.6.7. *For every connected component* $C \subset \hat{\mathcal{H}}_{\mathbf{v}}^{\text{temp}}$ *there exists a residual coset* L *such that* $\mathrm{cc}(C) = S_L$*. Then* $(\nu_{\mathcal{H}})|_C = c_C i_*(\mathrm{cc}|_{C^{reg}})^*(\nu_L)$ *for some constant* $c_C > 0$*. Here* $C^{reg} \subset C$ *is open and dense, and has a unique structure of a smooth manifold such that* $\mathrm{cc}|_{C^{reg}}$ *is a smooth finite covering map to* S_L^{reg}*, where* $S_L^{reg} \subset S_L$ *is the image of the largest stratum with respect to the action of* W_0 *on* $W_0 L^{\text{temp}}$*, and* $i : C^{reg} \to C$ *denotes the embedding.*

THEOREM 3.6.8 ([**OS2**]). *Let* $L^{\mathbf{v}} \subset T^{\mathbf{v}}$ *be a residual coset. Then there exists a residual coset* $L_\Lambda \subset T = T_\Lambda$ *defined over* Λ *such that* $L^{\mathbf{v}} = \{\mathbf{v}\} \times_{\mathrm{Spec}(\Lambda)} L_\Lambda$*.*

THEOREM 3.6.9 ([**OS2**], [**CO**]). *Let $Q(\Lambda)$ denote the quotient field of Λ, and let $r_\Lambda \in T_\Lambda$ be a residual point defined over Λ. Then $m_{r_\Lambda} \in Q(\Lambda)$ is regular on $\mathtt{Q}$, and if $\mathtt{v} \in \mathtt{Q}$ then $r_\Lambda^{\mathtt{v}} \in T^{\mathtt{v}}$ is residual if and only if $m_{r_\Lambda}(\mathtt{v}) \neq 0$. The set $\{\mathtt{v} \in \mathtt{Q} \mid m_{r_\Lambda}(\mathtt{v}) = 0\}$ is a union of finitely many hyperplanes in the real vector group $\mathtt{Q}$.*

3.7. Deformation of discrete series and the computation of $d_{\mathcal{H},\delta}$. In this section we review a deformation principle in the parameters $\mathtt{v} \in \mathtt{Q}$ for the discrete series characters δ of an affine Hecke algebra $\mathcal{H}_{\mathtt{v}}$. As we will see, this leads to an important tool to compute the rational constants $d_{\mathcal{H}_{\mathtt{v}},\delta}$ for unequal parameter Hecke algebras $\mathcal{H}_{\mathtt{v}}$ (which are abundant among unipotent Hecke algebras).

THEOREM 3.7.1 ([**OS2**]). *Assume that $\mathcal{H}$ is a semisimple affine Hecke algebra. Let $\mathcal{S} = \{\sum_{w\in W} c_w N_w \mid \forall N \in \mathbb{N} : W \ni w \to l(w)^N |c_w| \text{ is a bounded function}\}$ denote the Schwartz completion of $\mathcal{H}$. Note that this nuclear Frechet space is independent of the Hecke parameter v. Suppose that δ is a discrete series character of $\mathcal{H}_{\mathtt{v}}$ with $cc(\delta) = W_0 r$. There exists an (analytic) open neighbourhood $\mathtt{U} \subset \mathtt{Q}$ of $\mathtt{v}$, a* unique *W_0-orbit $W_0 r_\Lambda$ of residual points defined over Λ, and a* unique *continuous family $\mathtt{U} \ni \mathtt{v}' \to \tilde{\delta}^{\mathtt{v}'} \in \mathcal{S}$ of discrete series characters $\tilde{\delta}^{\mathtt{v}'}$ of $\mathcal{H}_{\mathtt{v}'}$ such that $\tilde{\delta}^{\mathtt{v}} = \delta$.*

We now review a remarkable rationality property of the formal degree of $\tilde{\delta}$ as in Theorem 3.7.1. According to [**CO**] there exists an orthonormal set $\mathcal{B}_{gm}$ of elliptic virtual characters of the affine Weyl group $W = X \rtimes W_0$ (cf. [**CO**, 2A1]) (with respect to the Euler-Poincaré pairing) which naturally parameterises the generic families of discrete series characters. More precisely, to each $b \in \mathcal{B}_{gm}$ we assign (based in part on the previous subsection):

(i) An orbit of generic residual points $W_0 r_b \in W_0\backslash T(\Lambda)$.
(ii) The open set $\mathtt{Q}_b^{reg} = \{\mathtt{v} \in \mathtt{Q} \mid m_b(\mathtt{v}) := m_{W_0 r_b}(\mathtt{v}) \neq 0\}$ (the complement of finitely many hyperplanes of $\mathtt{Q}$).
(iii) A continuous family $\mathtt{Q}_b^{reg} \ni \mathtt{v} \to \mathrm{Ind}_D(b,\mathtt{v})$ of virtual characters of $\mathcal{H}_{\mathtt{v}}$ (the "Dirac induction" of b, cf. [**CO**, 2B4]) with the properties that:
  (a) For each $b \in \mathcal{B}_{gm}$ there exists a locally constant function $\epsilon(b,\mathtt{v}) \in \{\pm 1\}$ on $\mathtt{Q}_b^{reg}$ such that for all $\mathtt{v} \in \mathtt{Q}_b^{reg}$, $\epsilon(b,\mathtt{v})\mathrm{Ind}_D(b,\mathtt{v})$ is an irreducible discrete series character of $\mathcal{H}_{\mathtt{v}}$.
  (b) We have $cc(\mathrm{Ind}_D(b,\mathtt{v})) = W_0 r_b$, for all $\mathtt{v} \in \mathtt{Q}_b^{reg}$.
  (c) For all $\mathtt{v} \in \mathtt{Q}$, the set of irreducible discrete series character of $\mathcal{H}_{\mathtt{v}}$ is equal to $\{\epsilon(b,\mathtt{v})\mathrm{Ind}_D(b,\mathtt{v}) \mid b \in \mathcal{B}_{gm}^{\mathtt{v}}\}$, where $\mathcal{B}_{gm}^{\mathtt{v}} := \{b \in Bc_{gm} \mid \mathtt{v} \in \mathtt{Q}_b^{reg}\}$.
  (d) Let $[\pi]$ denote the elliptic class of a virtual character π of W. Then we have $[\lim_{\epsilon\to 0} \mathrm{Ind}_D(b,\mathtt{v}^\epsilon)] = b$.

REMARK 3.7.2. By the Langlands classification, virtual elliptic characters of $\mathcal{H}_{\mathtt{v}}$ can be written as linear combinations of *tempered* characters of $\mathcal{H}$. Hence it makes sense to view fdeg as a linear function on the space of virtual elliptic characters of $\mathcal{H}_{\mathtt{v}}$.

THEOREM 3.7.3 ([**CO**]). *Assume that τ is normalised by $\tau(1) = 1$. For all $b \in \mathcal{B}_{gm}$ there exist a constant $d_b \in \mathbb{Q}_+$ such that for all $\mathtt{v} \in \mathtt{Q}_b^{reg}$:* $\mathrm{fdeg}(\mathrm{Ind}_D(b,\mathtt{v})) = d_b m_b(\mathtt{v})$.

Hence for each $b \in \mathcal{B}_{gm}$ the function $\mathrm{fdeg}(\mathrm{Ind}_D(b,\mathtt{v}))$ is a rational function of $\mathtt{v}$ which is regular on $\mathtt{Q}$. This rationality is remarkable, because the family of characters $\mathtt{Q}_b^{reg} \ni \mathtt{v} \to \mathrm{Ind}_D(b, \mathtt{v})$ does not extend continuously to $\mathtt{Q}$. The rationality is very powerful to compute the rational constants $d_{\mathcal{H}_{\mathtt{v}},\delta}$ for the discrete series characters δ of $\mathcal{H}_{\mathtt{v}}$ in the unequal parameters cases, because we see that it is enough to compute the single constant d_b in each generic family. These "generic constants" d_b are known for all irreducible root data and are quite simple (for example, for the classical Hecke algebras of the type $\mathrm{C}_n[m_-, m_+](q^\beta)$ we have $d_b = 1$ for all b, (cf. [**CO**], [**CK**], [**Opd5**])).

3.8. Central characters and Langlands parameters. The orbits $S_L = W_0 \backslash W_0 L^{\mathrm{temp}}$ of tempered residual cosets for affine Hecke algebras can be viewed as "parameter deformations of unramified Langlands parameters". This is a crucial point in order to be able to cast the results on spectral decompositions of traces of affine Hecke algebras as discussed above in terms of adjoint gamma factors.

The basic result is the following ([**KL**], [**HO1**], [**Opd2**], [**Opd5**]):

THEOREM 3.8.1. *Consider the special case of the Iwahori Hecke algebra* $\mathcal{H}_{\mathbb{I},\mathtt{v}}$ *of (the group of points of) an unramified connected reductive group* G *over* F. *By* [**Bo**, Lemma 6.5, Proposition 6.7] *we have* $\mathrm{Spec}(Z(\mathcal{H}_{\mathbb{I},\mathtt{v}})) \simeq (G^\vee\theta)_{ss} = W_{\mathbb{I},0}\backslash T_{\mathbb{I},\mathtt{v}}$ *where* $W_{\mathbb{I},0} := W^\theta$ *is the* F*-Weyl group of* G, *and where the torus* $T_{\mathbb{I},\mathtt{v}}$ *can be identified with the quotient* $T_{\mathbb{I},\mathtt{v}} = T^\vee/(1-\theta)T^\vee$ *of the maximal torus* $T^\vee$ *of* $G^\vee$. *Let*

$$S_{\mathbb{I},\mathtt{v}} := S(\mathcal{H}_{\mathbb{I},\mathtt{v}}) = \bigsqcup_{L \subset T_{\mathbb{I},\mathtt{v}} \ residual} S_L \subset W_{\mathbb{I},0}\backslash T_{\mathbb{I},\mathtt{v}} = (G^\vee\theta)_{ss}$$

denote the central support of the tempered spectrum of $\mathcal{H}_{\mathbb{I},\mathtt{v}}$, *and let* $\Phi_{nr}^{\mathrm{temp}}(G)$ *denote the set of equivalence classes of unramified tempered Langlands parameters for* G. *The map*

$$\gamma^{\mathbb{I}} : \Phi_{nr}^{\mathrm{temp}}(G) \to S_{\mathbb{I},\mathtt{v}} \subset W_{\mathbb{I},0}\backslash T_{\mathbb{I},\mathtt{v}} \tag{3.19}$$

$$[\varphi] \to \overline{W_{\mathbb{I},0}\varphi(\tilde{\mathrm{Fr}})} \tag{3.20}$$

is a bijection.

Using this fact it is not difficult to translate the results on the spectral decomposition of τ in this special case using adjoint γ-factors, a remark that essentially goes back to [**HII**]. In fact, using the work of Reeder [**R5**] and results from [**CO**] one can deduce:

THEOREM 3.8.2 ([**HII**], [**HO1**], [**Opd2**], [**KL**], [**R3**], [**R5**], [**CO**]). *Suppose* **G** *is unramified over* F. *There exists an enhanced Langlands parameterisation of the tempered Iwahori-spherical representations of the packets* $\Pi_\varphi(G)$ *such that the conjectures* 2.6.1 *and* 2.6.2 *hold true for Iwahori-spherical representations.*

PROOF. Based on the results of [**R3**] this was shown in [**HII**, §3.4] for Iwahori spherical discrete series of a group G which is split of adjoint type. The results of [**R3**] have been extended to Iwahori-spherical representations of general semisimple unramified groups G in [**CO**, Proposition 4.9] using [**R5**], [**Opd2**], [**HO1**], and Theorem 3.2.6. Applying the same proof as in [**HII**, §3.4] shows the result for the discrete series in the general unramified semisimple case. By Theorems 3.5.1, 3.5.2 this proves the required results for Iwahori spherical tempered representations of an arbitrary unramified group G. □

4. Lusztig's representations of unipotent reduction and spectral transfer maps. Main result.

Let $\mathbf{G}$ be a connected reductive group over F which is split over an unramified extension of F. Our main theorem is a slight sharpening and extension of the main result of [**Opd5**].

In the formulation of the main result, the action of the group $X_{wur}(G)$ of weakly unramified characters of G plays an important role. Let F_{ur}/F be an unramified extension of F over which G splits, and let $\mathrm{Fr} \in \mathrm{Gal}(F_{ur}/F)$ denote the geometric Frobenius map. We also denote by Fr the corresponding automorphism on $G(F_{ur})$, and by Fr^* an inner twist of Fr which defines an F-quasi-split structure (denoted by $\mathbf{G}^*$, with group of points G^*) on $\mathbf{G}$. We denote by θ the action of Fr on $G^\vee$, to that ${}^LG = G^\vee \rtimes \langle\theta\rangle$. Let $\Omega = \mathrm{Hom}(Z(G^\vee), \mathbb{C}^\times)$. A complex character χ of G is called *weakly unramified* if χ is trivial on the kernel G_1 of the Kottwitz homomorphism $w_G : G \to \mathrm{Hom}(Z(G^\vee), \mathbb{C}^\times)^\theta = \Omega^\theta$ (cf. [**Kot1, PR**]). We denote by $X_{wur}(G) = (\Omega^\theta)^* = Z(G^\vee)/(1-\theta)Z(G^\vee)$ the diagonalizable group of weakly unramified characters. This is the group of characters π_α of G attached to the set of unramified Langlands parameters $H^1_{nr}(W_F, Z(G^\vee))$ as constructed in [**Bo**, 10.2]. We note that if $\mathbb{I}_1 \subset \mathbf{G}(F_{ur})_1$ is a Fr-stable Iwahori subgroup (such exist by [**Tits**, 1.10.3]), then there is a canonical isomorphism $\Omega = N_{\mathbf{G}(F_{ur})}(\mathbb{I}_1)/\mathbb{I}_1$. Put $\mathbb{I} = \mathbb{I}_1^{\mathrm{Fr}}$, then $\Omega^\theta = N_G(\mathbb{I})/\mathbb{I}$.

Denote by $X_{wur}^{\mathrm{temp}}(G) \subset X_{wur}(G)$ the subgroup of tempered weakly unramified characters. Tensoring by (tempered) weakly unramified characters defines a natural action of $X_{wur}^{\mathrm{temp}}(G)$ on the set of (tempered) irreducible characters of G of unipotent reduction which is Plancherel density preserving. There is also a natural action of $X_{wur}(G)$ on $\Phi_{nr}(G)$ as follows. If $\omega \in \Omega^* = Z(G^\vee)$ represents a weakly unramified character $[\omega] \in X_{wur}(G) = (\Omega^\theta)^* = \Omega^*/(1-\theta)\Omega^*$ and $\varphi : \mathbb{Z} \times \mathrm{SL}_2(\mathbb{C}) \to {}^LG$ represents a class $[\varphi] \in \Phi_{nr}(G)$, then we define $[\omega].[\varphi] = [\varphi']$ where $\varphi'(\mathrm{Fr}) = \omega\varphi(\mathrm{Fr})$ while φ' coincides with φ on $\mathrm{SL}_2(\mathbb{C})$. One easily verifies that φ' defines an unramified Langlands parameter of G, that $[\varphi']$ is independent of the lift ω of $[\omega]$, and that this defines by restriction an action of $X_{wur}^{\mathrm{temp}}(G)$ on $\Phi_{nr}^{\mathrm{temp}}(G)$ which preserves the adjoint γ-factors. The following lemma is obvious:

LEMMA 4.0.1. *The bijection $\gamma^{\mathbb{I}}$ of Theorem 3.8.1 is equivariant with respect to the action of $X_{wur}^{\mathrm{temp}}(G)$ on $\Phi_{nr}^{\mathrm{temp}}(G)$ and on the set $S_{\mathbb{I},\mathrm{v}}$ of central characters of the tempered irreducible $\mathcal{H}_{\mathbb{I},\mathrm{v}}(G^*)$-modules corresponding to the Iwahori-spherical tempered irreducible representations of unipotent reduction of G^*.*

The *unramified* characters of G are the complex (quasi-)characters of G which are trivial on the intersection G^1 of the kernels of the compositions $\mathrm{Val}_F \circ \chi$ with $\chi \in X^*(G)$. It is clear that G/G^1 can be identified with a sublattice of the dual of $X^*(G)$, thus $X_{nr}(G)$ is the group of complex points of an algebraic torus. By the functoriality of Kottwitz's map w_G it follows easily that $G_{der} \subset G_1 \subset G^1$, so we have a natural embedding $X_{nr}(G) \subset X_{wur}(G)$. If $A \subset Z(G)$ is the group of F-points of the maximal F-split subtorus of the center $Z(G)$ of G then the restriction map gives a canonical embedding $X^*(G) \subset X^*(A)$ with finite cokernel. It follows that there is a canonical surjection with finite kernel $X_{nr}(G) \to X_{nr}(A)$ which we will denote by m. Again by the functoriality of Kottwitz's map we also see easily that there exits a canonical surjection with finite kernel $X_{wur}(G) \to X_{wur}(A) = X_{nr}(A)$. Therefore $X_{nr}(G) \subset X_{wur}(G)$ is the identity component. An irreducible representation of unipotent reduction π canonically defines an unramified character $z_\pi \in X_{nr}(A)$ since the scalar action of A on V_π is clearly unramified. For $\omega \in X_{wur}(G)$ we have $z_{\omega.\pi} = m(\omega)z_\pi$.

We mention in passing that $X_{wur}(G)$ and $X_{nr}(G)$ are not sensitive to inner twists; in particular we have canonical isomorphisms $X_{wur}(G) = X_{wur}(G^*)$ and $X_{nr}(G) = X_{nr}(G^*)$.

Our main result Theorem 4.5.1 deals with existence and uniqueness of *a parameterisation* (the precise meaning of the set $\hat{G}^{\mathrm{temp}}_{uni}$ is explained in Section 4.1):

$$\varphi : \hat{G}^{\mathrm{temp}}_{uni} \to \Phi^{\mathrm{temp}}_{nr},\ \pi \to \varphi_\pi$$

such that the conjectures 2.6.1 and 2.6.2 hold. By a parameterisation we mean:

Definition 4.0.2. A map $\hat{G}^{\mathrm{temp}}_{uni} \to \Phi^{\mathrm{temp}}_{nr}$, $\pi \to \varphi_\pi$ is called "a parameterisation" only if it satisfies following properties:

(i) The map $\pi \to \varphi_\pi$ is equivariant for the actions of the group of tempered weakly unramified characters $X^{\mathrm{temp}}_{wur}(G)$ as defined above (cf. [**Bo**, §10.2; 10.3(2)]).
(ii) We can express the character by which the center $Z(G)$ acts on V_π in terms of φ_π, as in [**Bo**, 10.1; 10.3(1)].
(iii) Compatibility with unitary parabolic induction as in [**Bo**, §10.3(3), 11.3, 11.7]. In particular, an irreducible tempered representation is a direct summand of an induced tempered representation $i_P^G(\delta)$ for some relevant parabolic $P = MN$ if and only if φ_π is equivalent to a parameter $\varphi_\delta^M \in \Phi^{\mathrm{temp}}_{nr}(M)$ for M (considering ${}^LM$ as a subgroup of ${}^LG$), where φ^M is a parameterisation for M.
(iv) The parameterisation is compatible with restriction of scalars [**Bo**, §10.1], and taking products of reductive groups.
(v) Let $\eta : H \to G$ is an F-morphism of connected reductive groups with commutative kernel and cokernel, and let $\varphi \in \Phi^{\mathrm{temp}}_{nr}(G)$. Given $\pi \in \Pi_\varphi(G)^{\mathrm{temp}}$, the pull-back of π to H is a finite direct sum of tempered irreducible representations in $\Pi^{\mathrm{temp}}_{{}^L\eta\circ\varphi}(H)$, where ${}^L\eta : {}^LG \to {}^LH$ denotes the natural map (cf. [**Bo**, §10.3(5)]). It follows in particular that a representation $\pi \in \hat{G}^{\mathrm{temp}}_{uni}$ factorizes through a representation of G/A (with A as above) if and only if $\mathrm{Im}(\varphi_\pi) \subset {}^LG^\natural$ (the L-group of G/A). (We use here that G maps surjectively to $(\mathbf{G}/\mathbf{A})(F)$.)
(vi) For Iwahori-spherical representations of unramified connected reductive groups, the correspondence equals that of Theorem 3.8.2 (compare with [**Bo**, 10.4]).

4.1. Unipotent types and unipotent affine Hecke algebras. Let $\Lambda_0 = \mathbb{C}[v^{\pm}]$, with v a formal variable. Let $\mathtt{v} \in \mathbb{R}_+$ be such that $\mathtt{v}^2 = \mathtt{q} = |\mathfrak{O}/\mathfrak{p}|$. We remark that there are no "bad primes" for representations of unipotent reduction [**Lus1**], and we may and will often replace $\mathtt{v}$ by the indeterminate v in the theory. For example, we can view the Hecke algebra over $\mathbb{C}$ with parameter $\mathtt{v}$ as specialization of the corresponding generic Hecke algebra over Λ_0.

We assume from now on that $\mathbf{G}$ is connected reductive over F and splits over an unramified extension of F. Let $F_{ur} \supset F$ be a maximal unramified extension of F. Let $\mathbf{T} \subset \mathbf{G}$ be a maximally split F-torus which is F_{ur}-split. In the apartment $\mathcal{A}_1$ of $\mathbf{T}(F_{ur})$ we can choose an Fr-stable alcove, by [**Tits**, 1.10.3]. Let $\mathbb{I}_1 \subset G(F_{ur})$ be the corresponding Fr-stable Iwahori subgroup, and let $\mathbb{I} = \mathbb{I}_1^{\mathrm{Fr}}$ be the corresponding Iwahori subgroup of G.

Steinberg's vanishing theorem $H^1(F_{ur}, \mathbf{G}) = 1$ implies that:

$$H^1(F, \mathbf{G}) = H^1(\mathrm{Gal}(F_{ur}/F), \mathbf{G}(F_{ur}))$$

. Kottwitz's Theorem expresses this in terms of the center of the Langlands dual group:

$$H^1(F, \mathbf{G}) = H^1(\mathrm{Fr}, \mathbf{G}(F_{ur})) = [\Omega/(1-\theta)\Omega]_{tor} = \mathrm{Irr}(\pi_0(Z(G^\vee)^\theta))$$

with $\Omega = X_*(T)/Q$, where Q denotes the root lattice of the dual group $G^\vee$. [2] The inner forms of G are parametrised by $H^1(F, \mathbf{G}_{ad}) = (Z(G^\vee_{sc})^\theta)^* = \Omega_{sc}/(1-\theta)\Omega_{sc}$ with $\Omega_{sc} = P/Q$, where P is the weight lattice of Q. Given $\omega \in H^1(F, \mathbf{G}_{ad})$ we may choose a representative $u \in N_{G_{ad}(F_{ad})}(\mathbb{I}_1)$ whose image in $H^1(F, \mathbf{G}_{ad}) = \Omega_{sc}/(1-\theta)\Omega_{sc}$ is ω. Then the inner twist $\mathrm{Fr}_u := \mathrm{Ad}(u) \circ \mathrm{Fr}^*$ of Fr^* defines an inner form of G^* corresponding to ω, which we will often denoted by G^u.

Let $\mathbb{P} \subset G$ be a parahoric subgroup. There exists an Fr-stable parahoric $\mathbf{P} \subset \mathbf{G}(F_{ur})$ such that $\mathbb{P} = \mathbf{P}^{\mathrm{Fr}}$. We put as before

$$\mathrm{Vol}(\mathbb{P}) := \mathrm{v}^{-\dim(\overline{\mathbb{P}})}|\overline{\mathbb{P}}|$$

This is the specialization at $v = \mathrm{v}$ of a Laurent polynomial $\mathrm{Vol}(\mathbb{P}) \in \Lambda_0$ in v. Let σ be a cuspidal unipotent representation of $\overline{\mathbb{P}}$, lifted to $\mathbb{P}$. Let $\tilde{\mathbb{P}} := N_G(\mathbb{P})$, and choose an extension of σ to a representation $\tilde{\sigma}$ of $\tilde{\mathbb{P}} \subset G$ (such extensions exist [**Lus4**]). Then $t = (\mathbb{P}, \sigma)$ is a type (see Theorem 4.1.1) for a finite set of Bernstein components of representations of unipotent reduction of G. Notice that t and the extension $\tilde{t} = (\tilde{\mathbb{P}}, \tilde{\sigma})$ are determined by data (the local Tits index of G (with trivial action of the inertia group), a facet of the apartment of T, a cuspidal unipotent representation σ of the corresponding parahoric subgroup $\mathbb{P}$, and an extension to its normalizer in G) which are independent of the base field F of G. (We use here that the classification of cuspidal unipotent characters of finite groups of Lie type is independent of the field of definition [**Lus1**]). We write $t = (\mathbb{P}, \sigma)$ to refer to this "abstract" unipotent type (in which the base field F is undetermined, and the cardinality of its residue field considered as indeterminate v), while we often write $t_{\mathrm{v}} = (\mathbb{P}_{\mathrm{v}}, \sigma_{\mathrm{v}})$ if we want to refer to the "concrete" type of G "specialised at $v = \mathrm{v}$". Similarly for $\tilde{t}$. Such "families of unipotent types" t (with varying base field F) have explicit meaning on the Langlands dual side, as we will see shortly.

THEOREM 4.1.1 ([**MP1**], [**MP2**], [**Mo**], [**Lus4**], [**Lus5**]). *Let $t = (\mathbb{P}, \sigma)$ and $\tilde{t} = (\tilde{\mathbb{P}}, \tilde{\sigma})$ be as above, and let $\Omega^{\mathbb{P},\theta} = \tilde{\mathbb{P}}/\mathbb{P} \subset \Omega^\theta$ be the stabilizer of $\mathbb{P}$ (see [**Lus4**, 1.16]).*

Then t_{v} is a type for G with Hecke algebra $\mathcal{H}_{t,\mathrm{v}}$ which is of the form $\mathcal{H}_{t,\mathrm{v}} = \mathcal{H}^a_{t,\mathrm{v}} \rtimes \Omega^{\mathbb{P},\theta}$, where $\mathcal{H}^a_{t,\mathrm{v}}$ is the specialization at $v = \mathrm{v}$ of a generic (unextended) affine Hecke algebra $\mathcal{H}^a_t$ defined over Λ_0 (depending on t only), on which $\Omega^{\mathbb{P},\theta}$ acts via diagram automorphisms. If $\Omega_1^{\mathbb{P},\theta} \subset \Omega^{\mathbb{P},\theta}$ denotes the subgroup which acts trivially on $\mathcal{H}^a_t$, and $\Omega_2^{\mathbb{P},\theta} = \Omega^{\mathbb{P},\theta}/\Omega_1^{\mathbb{P},\theta}$ then

$$\mathcal{H}_t = \mathcal{H}^e_{\tilde{t}} \otimes \mathbb{C}[\Omega_1^{\mathbb{P},\theta}]; \quad \mathcal{H}^e_{\tilde{t}} = \mathcal{H}^a_t \rtimes \Omega_2^{\mathbb{P},\theta} \tag{4.1}$$

where $\mathcal{H}^e_{\tilde{t}}$ is an extended affine Hecke algebra. The set of extensions $\tilde{t}$ of t is a torsor for the quotient $(\Omega^{\mathbb{P},\theta})^*$ of the group $X_{wur}(G) = (\Omega^\theta)^*$, and $\mathcal{H}^e_{\tilde{t}}$ corresponds

[2] From now on we will call roots of $(G^\vee, T^\vee)$ "roots", and roots of $(\mathbf{G}, \mathbf{T})$ "coroots". We apologize for the inconvenience this may cause.

to the orbit of $\tilde{t}$ *under the subgroup* $(\Omega^{\mathbb{P},\theta}/\Omega_1^{\mathbb{P},\theta})^*$. *Thus the summands* $\mathcal{H}_{\tilde{t}}^e$ *of* $\mathcal{H}_t$ *form a torsor for the quotient* $(\Omega_1^{\mathbb{P},\theta})^*$, *and* $X_{wur}(G) = (\Omega^\theta)^*$ *acts on the center of* $\mathcal{H}_t$.

Let $\mathrm{Vol}(\mathbb{P}) \in \Lambda_0$ *be the Laurent polynomial defined above, let* $\deg(\sigma) \in \Lambda_0$ *be the polynomials such that* $\deg(\sigma)(\mathrm{v}) = \deg(\sigma_{\mathrm{v}})$, *then the trace* τ *of* $\mathcal{H}_{\tilde{t}}^e$ *is normalised by* $\tau(1) = d^t = |\Omega_1^{\mathbb{P},\theta}|^{-1}\frac{\deg(\sigma)}{\mathrm{Vol}(\mathbb{P}^{\mathrm{Fr}_u}}$. *This turns the summands* $\mathcal{H}_{\tilde{t}}^e$ *into normalised affine Hecke algebras in the sense of* [**Opd5**].

PROOF. We refer the reader to [**Opd5**, Section 2.4, Theorem 2.8], where this is worked out in detail, and to the discussion in [**Lus4**]. □

DEFINITION 4.1.2 (Lusztig [**Lus4**]). The category $\mathcal{C}_{uni}(G)$ of smooth representations of G of unipotent reduction is the direct product over all conjugacy classes of unipotent types $t = (\mathbb{P}, \sigma)$ of the abelian subcategories $\mathcal{C}^t(G)$ of $\mathcal{C}(G)$, where $\mathcal{C}^t(G)$ is Morita equivalent to the Hecke algebra $\mathcal{H}_{t,\mathrm{v}}$ of t. Since $\mathcal{H}_t$ is a direct sum (4.1) of (mutually isomorphic) extended affine Hecke algebras $\mathcal{H}_{\tilde{t}}^e$ parameterised by the set of characters of $\Omega_1^{\mathbb{P},\theta}$, each subcategory $\mathcal{C}^t(G)$ decomposes as a product over a finite set of Bernstein components $\mathcal{C}^{\tilde{t},e}(G)$ parameterised by the set of extensions $\tilde{t}$ of t to the inverse image of $\Omega_1^{\mathbb{P},\theta} \subset \Omega^{\mathbb{P},\theta}$ in $\tilde{\mathbb{P}}$. This is a torsor for $(\Omega_1^{\mathbb{P},\theta})^*$, such that for each extended type $\tilde{t}$, the map $(\pi, V_\pi) \to \mathrm{Hom}_{N_G(\mathbb{P}^{\mathrm{Fr}_u})}(\tilde{\sigma}_{\mathrm{v}}, V_\pi|_{N_G(\mathbb{P}^{\mathrm{Fr}_u})})$ is a Morita equivalence from $\mathcal{C}^{\tilde{t},e}(G)$ to $\mathcal{H}_{\tilde{t},\mathrm{v}}^e$.

Using his arithmetic-geometric diagram correspondences Lusztig constructed [**Lus4**], [**Lus5**] a parameterisation of the irreducible objects of $\mathcal{C}_{uni}(G)$ if G is simple of adjoint type. In particular:

THEOREM 4.1.3 (Lusztig). *Let* $G = G^u$ *be (the group of points of) a simple group of adjoint type defined over* F. *There exists a partitioning*

$$\mathrm{Irr}_{uni}^{\mathrm{temp}}(G) = \bigsqcup_{\varphi \in \Phi_{nr}^{\mathrm{temp}}} \Pi_\varphi \tag{4.2}$$

such that for all $\varphi \in \Phi_{nr}^{\mathrm{temp}}$, *there is a bijection between* Π_φ *and* $\Pi(\mathcal{S}_\varphi, \chi_G)$ *(where* $\chi_G \in \Omega = (Z_{sc}^\theta)^*$ *and* χ_G *and* $\Pi(\mathcal{S}_\varphi, \chi_G)$ *as in Conjecture* 2.5.2*). This map* $\pi \to \varphi_\pi$ *can be taken* $X_{wur}(G)$*-equivariantly.*

For $G = G^*$ and $t_{\mathbb{I}} = (\mathbb{I}, \mathrm{triv})$, $\mathcal{C}^{t_{\mathbb{I}}}(G)$ is the Bernstein component of the minimal unramified principal series. Indeed, by Borel's classical result this abelian subcategory of $\mathcal{C}(G)$ is equal to the category of smooth representations of G which are generated by their Iwahori-fixed vectors. Then the equivalence of Definition 4.1.2 is the classical equivalence between $\mathcal{C}^{t_{\mathbb{I}}}(G)$ and the module category of the (extended) Iwahori-spherical Hecke algebra $\mathcal{H}_{t_{\mathbb{I}},\mathrm{v}}^e = \mathcal{H}_{\mathbb{I},\mathrm{v}}$. Via this equivalence, the restriction of the correspondence of Theorem 4.1.3 to $\mathcal{C}^{t_{\mathbb{I}}}(G)$ becomes the classical Kazhdan-Lusztig parameterisation [**KL**]. Theorem 3.8.2 (and its proof) shows that in this special case, this parameterisation restricted to tempered representations satisfies the conjectures 2.6.1 and 2.6.2, and that this can in fact be extended to general reductive groups.

The basic problem one is facing when trying to extend this result to all tempered representations of $\mathcal{C}_{uni}(G^u)$ (with G^u an arbitrary inner twist of G), is how one should parameterise the tempered irreducible representations of the affine Hecke algebras of the form $\mathcal{H}_{\tilde{t}}^e$ (with $t = (\mathbb{P}, \sigma)$ a unipotent type of G^u, and $\tilde{t}$ an extension) in terms of tempered unramified Langlands parameters for G^u. Lusztig [**Lus4**], [**Lus5**] does this via his theory of local systems on $G^\vee$-orbits of Langlands parameters, and the remarkable isomorphisms between arithmetic diagrams (related to the affine diagram of $\mathcal{H}^e_{\tilde{t}}$) and geometric diagrams related to graded affine Hecke algebras associated with a local system. We follow a different approach in which the conjectures 2.6.1 and 2.6.2 play a central role. In view of Theorem 3.8.1, it is clear that one would like to map the spectrum of the center $Z(\mathcal{H}^e_{\tilde{t},\mathrm{v}})$ of $\mathcal{H}^e_{\tilde{t},\mathrm{v}}$ to the spectrum of the center of $Z(\mathcal{H}_{\mathbb{I},\mathrm{v}})$, so that $S(\mathcal{H}^e_{\tilde{t},\mathrm{v}})$ maps to $S(\mathcal{H}_{\mathbb{I},\mathrm{v}})$, and such that the Plancherel measures (see Theorem 3.5.1 and Theorem 3.5.2) up to constant factors correspond. We call such a map a *spectral transfer map* (STM) $\mathcal{H}^e_{\tilde{t},\mathrm{v}} \rightsquigarrow \mathcal{H}_{\mathbb{I},\mathrm{v}}$. Such maps turns out to exist and turn out to be *essentially unique*. Moreover, these STMs are essentially "the same" as the maps implicit in Lusztig's arithmetic-geometric correspondences. Seeing this through (cf. [**Opd5**], [**FO**], [**FOS**]) recovers Lusztig parameterisation $\pi \to \varphi_\pi$, proving at the same time the conjectures 2.6.1 and 2.6.2 for this parameterisation. In our Main Theorem 4.5.1 we have further extended this to general connected reductive group.

4.2. Langlands parameters and residual cosets. Let us first consider the case of cuspidal representations of unipotent reduction.

Let $t = (\mathbb{P}, \sigma)$ be a unipotent type of G, with $\mathbb{P} \subset G$ a maximal parahoric subgroup. Assume that G has anisotropic center. Then $\mathcal{H}_t = \Lambda_0[\Omega^{\mathbb{P},\theta}]$ has rank 0, and its trace is normalised by $\tau(1) = d^t = \frac{\deg(\sigma)}{\mathrm{vol}(\mathbb{P})}$. Let π be a supercuspidal unipotent character of G^u belonging to the finite set $\mathcal{C}^t(G)$. Then, in view of Theorems 3.5.1, 3.8.1 and 3.8.2 we need to find a Λ_0-valued residual point $r \in T_{\mathbb{I}}(\Lambda_0)$ such that:

$$\mathrm{fdeg}(\pi) = c\mu^{\{r\}}(r) = \frac{c}{q(w_0)\mathrm{Vol}(\mathbb{I})} m_r(v) \tag{4.3}$$

for some constant factor $c \in \mathbb{R}_+$, where

$$\mathrm{fdeg}(\pi) = \frac{\deg(\sigma)}{|\Omega^{\mathbb{P},\theta}|\mathrm{Vol}(\mathbb{P})} \tag{4.4}$$

Recall that by Theorem 3.8.1, an orbit of residual points $W_0 r \in W_0\backslash T_{\mathbb{I}}$ corresponds to a unique orbit of discrete unramified Langlands parameters $\varphi \in \Phi^{\mathrm{temp}}_{nr}(G^*)$ such that $W_0 r = [\varphi(\tilde{\mathrm{Fr}})] \in W_0\backslash T_{\mathbb{I}}$. By Theorem 3.8.2 we then have, in terms of φ, for some constant factor c':

$$\mathrm{fdeg}(\pi) = c\mu^{\{r\}}(r) = c'\gamma(0, \mathrm{Ad} \circ \varphi, \psi) \tag{4.5}$$

viewed as identity of rational functions in v (where $v^2 = q$).

EXAMPLE 4.2.1. We begin with a very basic example. Let $G = \mathrm{PGL}_{m+1}(F)$. The only inner form of G which has a cuspidal type in this case is the anisotropic inner form $G^u = \mathbb{D}^\times/F^\times$ with $\mathbb{D}$ the tame division algebra of degree $m+1$ over F. The unique parahoric subgroup of G^u is $\mathbb{P}^u = G^u_1 = \cap_{\chi \in X_{wur}(G^u)}\mathrm{Ker}(\chi)$, which has a unique unipotent cuspidal $\sigma = \mathrm{triv}$. This gives a maximal unipotent type $t = (\mathbb{P}^u, \sigma)$ whose extensions $\tilde{t}$ to G^u are given by $X_{wur}(G^u) = \Omega^*$. Let us call the corresponding cuspidal unipotent characters of G^u: π_χ (with $\chi \in X_{wur}(G^u)$). Now G has essentially only one unramified discrete Langlands parameter, the regular parameter φ_0, up to the action of $X_{wur}(G) = \Omega = C_{m+1}$. We have $\varphi_0|_{W_F} = 1$, and

$\varphi_0(\begin{pmatrix} 1 & 1 \\ 0 & 1 \end{pmatrix})$ is a regular unipotent element. Hence $W_0 r = [\varphi_0(\tilde{\mathrm{Fr}})] \in W_0\backslash T_{\mathbb{I}}$, and the Ω^*-orbit $\Omega^* r$ is defined by the equations $\alpha_i(r) = \mathsf{q}$ for $i = 1 \ldots m$. We check simply from (4.3) and (4.4) that

$$\text{(4.6)} \qquad \mathrm{fdeg}(\pi_\chi) = \mu^r(r) = \frac{1}{(m+1)[m+1]_q}, \quad \text{with } [n]_q := \frac{v^{m+1} - v^{-(m+1)}}{v - v^{-1}}.$$

EXAMPLE 4.2.2. Another example which is known for a long time [**R1**]. Let $G = \mathbf{G}_2(F)$. Let π be the cuspidal unipotent character of G which is compactly induced from the cuspidal unipotent representation $\mathbf{G}_2(\mathbb{F}_\mathsf{q})$ which is denoted by $\mathbf{G}_2[1]$, inflated to $\mathbb{P} = \mathbf{G_2}(\mathcal{O}_F)$. Then

$$\text{(4.7)} \qquad \mathrm{fdeg}(\pi) = \frac{1}{6}\mu^r(r) = \frac{1}{6(v + v^{-1})^2(v^2 + 1 + v^{-2})}.$$

where $W_0 r = [\varphi_{sub}(\tilde{\mathrm{Fr}})] \in W_0\backslash T_{\mathbb{I}}$ is the Λ_0-point associated with the real discrete unramified Langlands parameter φ_{sub}, associated with the subregular unipotent orbit of $\mathbf{G_2}$. This uniquely determines $W_0 r$ and thus φ_{sub}. Indeed, there is a cuspidal local system supported by $G^\vee \dot{\varphi}_{sub}$, and Lusztig maps π to this cuspidal local system.

It turns out that this always works, at least in the case of G being absolutely simple and of adjoint type:

THEOREM 4.2.3 (Reeder [**R1**], [**R3**] (split exceptional groups), [**HO1**], [**FO**], [**Fe2**], [**FOS**]). *Let G be a simple group of adjoint type defined over F and split over an unramified extension of F. Let π be a supercuspidal representation of G of unipotent reduction. Let $\mathfrak{t} = (\mathbb{P}, \sigma)$ be a unipotent type for G such that π belongs to $\mathcal{C}^{\mathfrak{t}}(G)$. Then $\mathbb{P} \subset G$ is a maximal parahoric subgroup. Conversely, if $\mathbb{P} \subset G$ is maximal then $\mathcal{C}^{\mathfrak{t}}(G)$ consists of supercuspidal unipotent representations. In this situation there exists a* unique *$X_{wur}(G)$-orbit of discrete unramified Langlands parameters $[\varphi] \in \Phi_{nr}^{\mathrm{temp}}(G)$ such that:*

$$\text{(4.8)} \qquad \mathrm{fdeg}(\pi) = c\gamma(0, \mathrm{Ad} \circ \varphi, \psi)$$

The collection of classes of discrete unramified Langlands parameters $[\varphi] \in \Phi_{nr}^{\mathrm{temp}}(G)$ thus obtained is exactly equal to the set of classes of discrete unramified Langlands parameters which support a cuspidal local system.

Recall that $\mathcal{H}_{\mathfrak{t}}^{\mathfrak{e}} = \Lambda_0$, and that the trace of this Hecke algebra is normalised by $\tau^{\mathfrak{e}}(1) = \frac{\deg(\sigma)}{|\Omega^{\mathbb{P},\theta}|\mathrm{Vol}(\mathbb{P})}$. In the terminology of STMs we view a Λ_0-valued point $r : \mathrm{Spec}(\Lambda_0) \to T_{\mathbb{I}}(\Lambda_0)$ such that (4.8) holds as realizing a cuspidal STM $\mathcal{H}_{\mathfrak{t}}^{\mathfrak{e}} \rightsquigarrow \mathcal{H}_{\mathbb{I}}$. In the next section we will discuss the notion of STM in the general case.

The proof of Theorem 4.2.3 is difficult. In the exceptional cases it reduces to explicit case by case computations due to [**HO1**], [**R3**], [**Fe2**]. In the classical case, see [**FO**]. This is based on the existence an explicit set of generators between the unipotent normalised affine Hecke algebras of type C_n, and the observation that for classical cases other than $\mathrm{PGL}_{m+1}(F)$, the expressions $\frac{\deg(\sigma)}{\mathrm{Vol}(\mathbb{P})}$ do not contain odd cyclotomic factors in the numerator or the denominator. This property turns out to eliminate most of the discrete unramified Langlands parameters φ for which an identity of the form (4.8) could hold. Analysing the remaining cases carefully using the STMs whose *existence* was established directly in [**Opd5**], [**Fe2**] completes the proof.

4.3. Spectral transfer maps. Let us now introduce the notion of STM formally. Assume we are given two normalised affine Hecke algebras $\mathcal{H}_1, \mathcal{H}_2$ defined over Λ_0. For $i = 1, 2$ we have the torus T_i defined over Λ_0 associated with the character lattice X_i of the root datum of $\mathcal{H}_1$. We have the Weyl groups $W_{0,i}$ acting on T_i, and the μ-function μ_i, a rational function on T_i.

Given a residual coset $L \subset T_2$ defined over Λ_0, we know that $L = r_L T^L$ for some residual point $r_L \in T_L$, by Proposition 3.6.2. Let $L_n = L/K_L^n$ be the smallest quotient of L such that for each $K_L^n t \in L_n$, the orbit $W_{0,2}K_L^n t = W_{0,2}t$ is a single W_0-orbit in T_2. Then $K_L^n \subset K_L = T_L \cap T^L$ is the finite abelian subgroup $K_L^n = N_{W(\Sigma_L)}(L)/Z_{W(\Sigma_L)}(L) \subset K_L$. Then μ_2^L (see Theorem 3.5.2) is K_L^n invariant, and descends to a μ-function $\mu_2^{L_n}$.

Definition 4.3.1. A spectral transfer map $\mathcal{H}_1 \rightsquigarrow \mathcal{H}_2$ between the normalised affine Hecke algebras over Λ_0 is an equivalence class of morphisms $\Psi : T_1 \to L_n$ of torsors of algebraic tori defined over Λ_0, where $L \subset T_2$ is a residual coset for $\mathcal{H}_2$, such that

(1) Ψ is surjective with finite fibres.
(2) For all $w_1 \in W_{0,1}$ there exists a $w_2 \in N_{W_{0,2}}(L)$ such that $\Psi \circ w_1 = w_2 \circ \Psi$.
(3) $\Psi^*(\mu_2^{L_n}) = D\mu_1$ for some constant $D \in \mathbb{Q}^\times$.

We call Ψ and Ψ' equivalent if $\Psi' = w \circ \Psi$ for some $w \in W_{0,2}$.

Remark 4.3.2. (i) In the special case $\dim(T_1) = 0$ this recovers the notion cuspidal STMs as discussed above for unipotent affine Hecke algebras $\mathcal{H}_{\tilde{t}}^e$.
(ii) The property (2) in Definition 4.3.1 is almost always superfluous, except in degenerate cases [**Opd4**, cf. Proposition 5.4]. For unipotent affine Hecke algebras this property is always automatic.
(iii) Note that Ψ defines a morphism $\Psi_Z : \mathrm{Spec}(Z_1) \to \mathrm{Spec}(Z_2)$ between the spectra of the centers of the $\mathcal{H}_i$.
(iv) Suppose that $L_1 \subset T_1$ is a residual coset for $\mathcal{H}_1$, and let $\mathrm{Im}(\Psi) = L_n$. Then $\Psi_Z(W_{1,0}L_1) = N_{W_{0,2}}(L)(L_2)$ for some residual coset $L_2 \subset L$, and all residual cosets $L_2 \subset L$ are the image under Ψ of a residual coset of $\mathcal{H}_1$ in this sense.
(v) We can compose STMs. This is useful as we may generate in this way all STMs between the unipotent affine Hecke algebras of the various groups G^u from a small number of generators.

The most important property of STMs with respect to the spectral decomposition of affine Hecke algebras is the following result [**Opd4**, Theorem 6.1]:

Theorem 4.3.3. *Let $\Psi : \mathcal{H}_1 \rightsquigarrow \mathcal{H}_2$ be an STM. Suppose that $C_{1,\mathrm{v}} \subset \hat{\mathcal{H}}_{1,\mathrm{v}}^{\mathrm{tem}}$ is a component of the space of irreducible tempered characters of $\mathcal{H}_{1,\mathrm{v}}$. Let $cc_{i,\mathrm{v}} : \hat{\mathcal{H}}_i^{\mathrm{temp,v}} \to \mathrm{Spec}(Z_{i,\mathrm{v}})$ be the central character map of $\mathcal{H}_{i,\mathrm{v}}$. Then $cc_1(C_{1,\mathrm{v}}) = S_{L_{1,\mathrm{v}}}$ where $L_1 \subset T_1$ is a residual coset. There exists a unique orbit $W_{0,2}L_2 \subset T_2$ of residual cosets of $\mathcal{H}_2$ such that $\Psi_Z(S_{L_{1,\mathrm{v}}}) = S_{L_{2,\mathrm{v}}}$. If $C_{2,\mathrm{v}} \subset \hat{\mathcal{H}}_{2,\mathrm{v}}^{\mathrm{tem}}$ is such that*

$cc_2(C_{2,\mathrm{v}}) = S_{L_{2,\mathrm{v}}}$ *then consider the fibred product* C_{12} *defined by the diagram*

$$
\begin{array}{ccc}
C_{12} & \xrightarrow{P_2} & C_2 \\
{\scriptstyle P_1}\downarrow & & \downarrow{\scriptstyle cc_2} \\
C_1 & \xrightarrow[\Psi_Z \circ cc_1]{} & S_{L_2}
\end{array}
\tag{4.9}
$$

Then there exist constants $r_i \in \mathbb{Q}_+$ *and a family of positive measures* ν_{v} *on* $C_{12,\mathrm{v}}$ *(for all* $\mathrm{v} > 0$*) such that for* $i = 1, 2$ *we have* $r_i.P_{i,*}(\nu_{\mathrm{v}}) = \nu_{i\mathrm{v}}|_{C_{1,\mathrm{v}}}$.

This result states that an STM $\Psi : \mathcal{H}_1 \rightsquigarrow \mathcal{H}_2$ defines a correspondence between the tempered spectrum of $\mathcal{H}_1$ and of $\mathcal{H}_2$ which respects the connected components in these tempered irreducible spectra and, up to constant rational factors only depending on the components, the Plancherel measures $\nu_{i,\mathrm{v}}$ of $\mathcal{H}_i$.

Let $\mathbf{G}$ be a connected reductive group defined over F which splits over an unramified extension of F. Let $t = (\mathbb{P}, \sigma)$ be a unipotent type of G, i.e. $\mathbb{P} \subset G$ is a parahoric subgroup, and σ is a cuspidal unipotent representation of $\mathbb{P}$. Suppose in Theorem 4.3.3 that $\mathcal{H}_1 = \mathcal{H}_t(G)$ (a finite direct sum of extended affine Hecke algebras of the form $\mathcal{H}_{\mathfrak{t}}^{\mathfrak{e}}$), and that $\mathcal{H}_2 = \mathcal{H}_{\mathbb{I}}(G^*)$, the Iwahori Hecke algebra of a quasi-split inner form G^* of G (with the obvious extension of the notion of an STM on a finite direct sum of extended affine Hecke algebras such as $\mathcal{H}_t(G)$, by allowing T_1 to be disjoint union of algebraic tori over Λ_0).

Let $M \subset G$ be a Levi subgroup such that the set Ω^t of G-conjugacy classes of cuspidal pairs which belong in the inertial classes covered by t are of the form $[(M, \delta)]$. By Corollary 3.2.2 the diagonalizable group $T_{1,\mathrm{v}}$ over $\mathbb{C}$ can be identified with the space $\Omega^t(M)$ of M-conjugacy classes of such cuspidal pairs. The natural action (by taking tensor products) of $X_{wur}(M)$ on $\Omega^t(M)$ turns $T_{1,\mathrm{v}}$ into a $X_{wur}(M)$-space, and in fact each component of $T_{1,\mathrm{v}}$ is a quotient of $X_{wur}(M)$ with finite kernel (because $X_{nr}(M) \subset X_{wur}(M)$ is the identity component, and the components of $\Omega^t(M)$ are by definition already homogeneous for $X_{nr}(M)$ with finite kernel).

REMARK 4.3.4. Recall that (cf. Theorem 3.8.1) the semisimple conjugacy classes of the set $G^\vee\theta \subset {}^LG$ are in natural bijection with the set of W^θ-orbits in the complex torus $T_{\mathbb{I}} = \mathrm{Hom}(X_*(T)^\theta, \mathbb{C}^\times)$. Observe that

$$X_{wur}(M) = \mathrm{Hom}((X_*(T)/Q_M)^\theta, \mathbb{C}^\times) \to \mathrm{Hom}(X_*(T)^\theta, \mathbb{C}^\times) = T_2 = T_{\mathbb{I}}.$$

Hence $X_{wur}(M)$ also acts naturally on $T_{\mathbb{I}}$ (and faithfully, in fact).

COROLLARY 4.3.5. *Let* G, $\mathcal{H}_1$ *and* $\mathcal{H}_2$ *be as above. Suppose that* $\Psi : \mathcal{H}_t(G) \rightsquigarrow \mathcal{H}_{\mathbb{I}}(G^*)$ *is a* $X_{wur}(G)$*-equivariant STM. Let* $\psi : \hat{G}^{t,temp} \to \Phi_{nr}^{\mathrm{temp}}(G)$ *be the map* $(\gamma^{\mathbb{I}})^{-1} \circ \Psi_Z \circ \beta^t \circ cc$ *(cf. (3.1), Corollary 3.2.2, Theorem 3.8.1 and Definition 4.3.1).*

(1) *Via the map* $\pi \to \psi_\pi$ *on* $\hat{G}^{t,\mathrm{temp}}$*, the restriction of the Plancherel measure of* G *to* $\hat{G}^{t,\mathrm{temp}}$ *is expressed as in conjectures 2.6.1 and 2.6.2, up to rational constant factors.*

(2) ψ *satisfies the conditions (i), (ii), and (vi) of a parameterisation if and only if the following additional compatibility conditions hold:*
   (a) $\Psi(\eta.t) = \eta.\Psi(t)$ *for all* $\eta \in X_{wur}(M)$, $t \in T_{1,\mathrm{v}}$.

(b) *If* $t = (\mathbb{I}, \mathrm{triv})$, *with* $\mathbb{I} \subset G^*$ *the Iwahori subgroup of a quasi-split group* G^* *which is an inner twist of* G. *Then we require that* $\Psi : \mathcal{H}_{\mathbb{I}} \rightsquigarrow \mathcal{H}_{\mathbb{I}}$ *is the identity.*

PROOF. By Theorem 3.2.6 $\hat{m}_t$ defines a Plancherel measure preserving homeomorphism from $\hat{G}^{t,\mathrm{temp}}$ to $\hat{\mathcal{H}}_t(G)$. By Theorem 3.5.1, Theorem 3.5.2 a component C_1 in the tempered irreducible spectrum of $\mathcal{H}_t(G)$ is defined by unitary parabolic induction of discrete series characters modulo center of Levi-subalgebras. The image $cc^t(C_1)$ under the central character map cc^t is the image $S_{L_1} \subset \mathrm{mSpec}(Z(\mathcal{H}_t(G)))$ of the tempered form of a residual coset L_1^{temp}. Moreover the Plancherel measure of $\mathcal{H}_t(G)$ on C_1 is given, up to a constant factor, by the pull back $cc^{t,*}(\mu_t^{L_1}|_{S_{L_1}})$ on L_1 to C_1 (using that cc^t is a smooth finite covering map on a dense open subset of C_1). Hence modulo constant factors, the Plancherel density on C_1 is a function of the central character only.

By Theorem 4.3.3 $\Psi_Z(S_{L_1}) = S_{L_2} \subset W_{\mathbb{I},0}\backslash T_{\mathbb{I}}$ for a unique residual coset $L_2 \subset L = \mathrm{Im}(\Psi)$, and up to rational constant factors $\Psi_Z^*(\mu_{\mathbb{I},L_2}) = c\mu_t^{L_1}$. Hence Theorem 3.8.1 and Theorem 3.8.2 imply that in this way we can express the Plancherel density at $\pi \in \hat{G}^{t,\mathrm{remp}}$ up to locally constant rational factors by the appropriate adjoint γ-factors in ψ_π, proving (1).

For (2): It is easy to see that ψ satisfies (i), (ii) iff Ψ satisfies the stated compatibility condition (a). Clearly (vi) makes sense only in the case of the Borel component of G^*. In this case, (2)(b) forces (vi) by insisting that Ψ is identical. $\square$

REMARK 4.3.6. Without this condition (2)(b) it would be allowed in the case $t = (\mathbb{I}, \mathrm{triv})$ that our map $\pi \to \psi_\pi$ is that of Theorem 3.8.2 (as required in (vi)) but twisted by an STM $\Psi : \mathcal{H}_{\mathbb{I}} \rightsquigarrow \mathcal{H}_{\mathbb{I}}$ satisfying (2)(a). Such STMs are given by the action of $X_{wur}(G) = (\Omega^\theta)^*$. For general components of $\hat{G}^t$ in general I do not know of a preferred choice for Ψ within its $X_{wur}(G)$-orbit.

4.4. Lusztig's geometric-arithmetic correspondences and STMs. In this subsection we will assume that $\mathbf{G}$ is absolutely simple and adjoint (we will reduce the general case to this case), and that $\mathbf{G}$ is split over an unramified extension of F.

The following result was essentially proven in [**Opd4**, Theorem 3.4]:

THEOREM 4.4.1. *Let* $t = (\mathbb{P}, \sigma)$ *be a unipotent type for* $G = \mathbf{G}(F)$, *and let* $\mathcal{H}_{\mathbb{I}}(G^*)$ *denote the Iwahori Hecke algebra of the quasi-split inner twist* G^* *of* G. *Let* $\tilde{t} = (\tilde{\mathbb{P}}, \tilde{\sigma})$ *be an extension of* t *to* $\tilde{\mathbb{P}} = N_G(\mathbb{P})$.

(a) *There exist STMs* $\Phi_{\tilde{t}} : \mathcal{H}_{\tilde{t}}^e \rightsquigarrow \mathcal{H}_{\mathbb{I}}(G^*)$.

(b) *If* $\Phi'_{\tilde{t}} : \mathcal{H}_{\tilde{t}}^e \rightsquigarrow \mathcal{H}_{\mathbb{I}}(G^*)$ *is also an STM, then there exists an* $\omega \in X_{wur}(G) = X_{wur}(G^*)$ *and a spectral automorphism* α *of* $\mathcal{H}_t$ *such that* $\Phi'_{\tilde{t}} = m_\omega \circ \Phi_{\tilde{t}} \circ \alpha$. *Here* $m_\omega : \mathcal{H}_{\mathbb{I}}(G^*) \rightsquigarrow \mathcal{H}_{\mathbb{I}}(G^*)$ *denotes the STM given by multiplication with* ω.

(c) *There exists a* $\Phi_t : \mathcal{H}_t^e \rightsquigarrow \mathcal{H}_{\mathbb{I}}(G^*)$ *which is* $X_{wur}(G)$*-equivariant and which satisfies the conditions (2)(a) and (b) of Corollary* 4.3.5. *Such* Φ_t *is unique up to the action of* $X_{wur}(G)$.

(d) *The parameterisation* $\varphi_t : \hat{G}^t \to \Phi_{nr}^{\mathrm{temp}}(G)$ *associated to* Φ_t *as in Corollary* 4.3.5 *is, up to a twist by* $X_{wur}(G)$, *the same as Lusztig's parameterisation* $\varphi_{Lu} : \hat{G}^t \to \Phi_{nr}^{\mathrm{temp}}(G)$ *of* [**Lus4**], [**Lus5**].

In this subsection I will sketch the proof of the Theorem, which is quite involved. For details the reader is referred to [**Opd5**], [**Opd4**], [**FO**], [**Fe2**] and [**FOS**].

First the existence (a) is proved in [**Opd5**] and [**Fe2**].

In the exceptional cases the existence is shown by constructing STMs associated to Lusztig's arithmetic-geometric correspondences, in a way we will explain below. Given a arithmetic-geometric correspondence of diagrams as in Lusztig, it is not difficult to find the candidate map $\Phi_{\tilde{t},T} : T_{\tilde{t}} \to L_n$ underlying the alleged STM $\Phi_{\tilde{t}} : \mathcal{H}^e_{\tilde{t}} \rightsquigarrow \mathcal{H}_{\mathbb{I}}(G^*)$. To verify the main property (3) of Definition 4.3.1 one needs to do a rather cumbersome computation.

In the classical cases this is not practical. Fortunately the required STMs can be obtained from a small set of generators of STMs between certain unipotent affine Hecke algebras of the form $\mathrm{C}_d[m_-, m_+](q^\beta)$ (see [**Opd5**, 3.2.6; 3.2.7]). There are three kind of generating STMs: The translation STMs which decrease one of $m_\pm$ by 1 (if this parameter is in $\mathbb{Z}_+ + 1/2$) or by 2 (if this parameter is in $\mathbb{Z}_+$), increasing the rank accordingly, and do not change β; the spectral isomorphisms, which interchange m_- and m_+, or give one of these a minus sign, and the extraspecial STMs. In this case $m_\pm \in \mathbb{Z} \pm 1/4$ and $\beta = 2$, while the target Hecke algebra is of the form $C_n[\delta_-, \delta_+](q)$ with $\delta_\pm \in \{0, 1\}$. The latter STMs correspond to Hecke algebras of types of inner forms G^u of G^* of even orthogonal or symplectic groups where $u \in \Omega/(1-\theta)\Omega$ has maximal order.

Let us now explain the construction of STMs associated with a correspondence of diagrams as in Lusztig in the exceptional cases. For more details, see [**Opd5**].

First one needs to establish the cuspidal case of Theorem 4.4.1 in the exceptional case. This is an explicit case by case verification based on the classification of the cuspidal unipotent characters of finite groups of Lie type [**Lus1**], the classification of the residual points for $\mathcal{H}_{\mathbb{I}}$, and the computation of the formal degree (up to constant factors) for discrete series representations of $\mathcal{H}_{\mathbb{I}}$ supported by the corresponding central characters [**HO1**], [**OS2**]. This was carried out in [**Opd5**, 3.2.2].

Next we discuss the diagram of an STM $\Phi : \mathcal{H}^e_{\tilde{t}} \rightsquigarrow \mathcal{H}_{\mathbb{I}}(G^*)$. Consider the underlying map $\Phi : T_{\tilde{t}} \to L_n$ with $L \subset T = T_{\mathbb{I}} = T^\vee/(1-\theta)T^\vee$ a residual coset. Such a map is determined by assigning weights to the nodes of the Kac diagram $D(\mathfrak{g}^\vee, \theta)$ [**R4**] of ${}^LG = G^\vee \rtimes \langle\theta\rangle$ according to the following steps (i)-(vii).

(i) Lift $\tilde{\Phi}$ to an affine linear map $\mathfrak{t}_{\tilde{t}} \to \mathfrak{t} = \mathfrak{t}_{\mathbb{I}}$ and use the action of the dual affine Weyl group $W^{a,\vee}_{\mathbb{I}} = Q^\vee \rtimes W_{\mathbb{I},0}$ to replace $\tilde{\Phi}$ by the lift $\tilde{\Phi}'$ of a map equivalent to Φ, such that $P_L := \mathrm{Im}(\tilde{\Phi}'_{\mathrm{v}=1})$ meets the closure $\overline{C^\vee}$ of the dual alcove $C^\vee$ in a facet of dimension equal to $\dim(T_{\tilde{t}})$. We assume from now on that $\Phi = \Phi'$ has this property.

(ii) We have $\mathrm{Im}(\Phi) = L_n$ for a unique residual coset $L_\Phi = r_L T^L \subset T_{\mathbb{I}}$, where $r_L = s_L c_L \in T_L$ is a residual point. Then P_L is a lift of $L_{\Phi,\mathrm{v}=1}$.

(iii) Let us denote the spectral diagram [**OS2**, Section 8], [**Opd4**, 2.3] of $\mathcal{H}^e_{\tilde{t}}$ by Σ^{spec}_t. Let $(W^{a,\vee}_t, S^{a,\vee}_t)$ be the associated affine Coxeter group. It turns out that property (3) of Definition 4.3.1 implies that there exists an isomorphism of affine Coxeter groups $(W^{a,\vee}_t, S^{a,\vee}_t) \to W^*_L := N_{W^{a,\vee}_{\mathbb{I}}}(P_L)/C_{W^{a,\vee}_{\mathbb{I}}}(P_L)$. Here $W^{a,\vee}_{\mathbb{I}}$ is the affine Weyl group associated to the Kac diagram $D(\mathfrak{g}^\vee, \theta)$ acting on $\mathfrak{t}_{\mathbb{I}} = \mathfrak{t}^\theta$ (see [**R4**]).

Indeed, using the theory of intertwiners Harish-Chandra proved that the reflection in a hyperplane on which the Plancherel density of a generalised principal series vanishes $i^{G^*}_{P=LN}(\delta)$ is an element of $N_{W_{\mathbb{I}}}(L, \delta)$. Apply this to the generalised

principal series of G^* supported by L_Φ, then we see that such hyperplane reflection is in $N_{W_{\mathbb{I}}}(L_\Phi)$. On the other hand, by (3) of Definition 4.3.1 all reflection hyperplanes of $W_t^{a,\vee}$ correspond to such zeroes of the Plancherel density μ^{L_Φ} via Φ_T. This implies that the simple affine reflections in P_L associated with the faces of P_L are restrictions to P_L of elements of the normalizer of P_L in $W_{\mathbb{I}}^{a,\vee}$.
(iv) By (iii), the material in [**Lus4**, Section 2] applies. We see that if P_L corresponds to the set $J \subset I$ (with I an index set for the set of nodes of the Kac diagram $D(\mathfrak{g}^\vee, \theta)$), then $J \subset I$ is an *excellent* subset, and the set of affine simple reflections of $W^* \simeq (W_t^{a,\vee}, S_t^{a,\vee})$ is in natural bijection with $K = I\backslash J$. Hence the set I decomposes in a subset J correspond to the affine simple roots constant which are constant on L, and $K = I\backslash J$, the which is in canonical bijection with the set $\tilde{K}$ of simple reflections of Σ_t^{spec} via the isomorphism $(W_t^{a,\vee}, S_t^{a,\vee}) \to W_L^*$ induced by Φ.
(v) Assign weights to the nodes of I: If $i \in I$ let $a_i^\vee$ be the corresponding affine simple Kac root, and let $Da_i^\vee$ be its gradient, viewed as character of $T_{\mathbb{I}}$ (defined over Λ_0).

We define $w_i(s) := D(a_i^\vee)(\Phi(s))$, viewed as a function of $s \in T_{\tilde{t},v}$ (the real vector group $T_{\tilde{t},v} \subset T_{\tilde{t}}$). We note that w_i is independent of $s \in T_{\tilde{t},v}$ if and only if $i \in J$, and that $\prod_{i\in I} w_i^{n_i} = 1$, where n_i are such that $\sum_{i\in I} n_i a_i^\vee = 1$.

Observe that for all $j \in J$, $w_j = v^{\mathsf{d}_j}$ for certain $\mathsf{d}_j \in \mathbb{Z}_{\geq 0}$, and these coordinates determine the residue point $r_L \in T_L = T_J$ modulo the action of $(\Omega_{J,ad})^*$.

We identify $\tilde{K}$ and K via the isomorphism of (iv). Suppose that $\sum_{k\in K} \tilde{n}_k b_k^\vee = 1$ is the affine relation for the affine simple roots of Σ_t^{Spec}. It turns out that $\mathsf{z}_k := n_k/\tilde{n}_k \in \mathbb{N}$ for all $k \in \tilde{K} = K$. We then have for all $k \in K$ that $w_k(s) = \zeta_k v^{\mathsf{c}_k}(Db_k^\vee(s))^{1/\mathsf{z}_k}$, where $b_k^\vee$ is the affine root of $(W_t^{a,\vee}, S_t^{a,\vee})$ corresponding to k via Φ. Moreover, if k_0 denotes the unique affine extension root of Σ_t^{Spec} then ζ_{k_0} is a primitive n_{k_0}-th root of unity, $\zeta_k = 1$ if $k \neq k_0$, and $\mathsf{c}_k \in \mathbb{Z}$ can be computed by $Da_k^\vee(c_L) = v^{c_k}$.
(vi) We call the diagram $D(\mathfrak{g}^\vee, \theta)$ so obtained, with the weights w_i attached, the diagram $D(\Phi)$ of the STM Φ. Note that Φ is determined by $D(\Phi)$, since $\tilde{\Phi}$ is.
(vii) Given such a diagram, to check that the map it defines is an STM comes down to checking first of all that the constant weights w_j on J define a residue point of T_J, and secondly that (3) of Definition 4.3.1 holds. The latter verification is straightforward but cumbersome (a lot of cancellations are taking place).

We claim that the diagram $D(\Phi)$ without the weights w_i, but with the subsets J and K remembered, is the "geometric diagram" which Lusztig [**Lus4**], [**Lus5**] attaches to the arithmetic diagram of $\mathcal{H}_{\tilde{t}}^{\mathfrak{e}}$.

Conversely, given the "geometric diagram" Lusztig attaches to the arithmetic diagram of $\mathcal{H}_{\tilde{t}}^{\mathfrak{e}}$. Now one needs to assign the appropriate weights w_i to the nodes of the diagram so that it becomes the diagram of an STM. This is done as follows. From $(\mathbb{P}, \sigma)$ we obtain a corresponding cuspidal pair (M, δ) with a Levi subgroup $M \subset G$ and a supercuspidal representations of unipotent reduction δ of M/A_M (with A_M the maximal split torus in $Z(M)$). Note that M/A_M is again of adjoint type. Thus by the cuspidal case of Theorem 4.4.1 (which we have established already for exceptional groups) we obtain a *unique* $X_{wur}(M)$-orbit of central characters $W_M r_M$ with $r_M \in T_M$ a residue points for M/A_M associated to δ such that (4.3) holds (for (M, δ) instead of (G, π)). Let $L = r_M T^M \subset T_{\mathbb{I}}$ be the corresponding residual coset. Using $W_{\mathbb{I}}$ we may assume that L is in the position as in (a),(b). We observe that $X_{wur}(M)/(T_M \cap T^M)$ is a quotient of $X_{wur}(G)$. Thus (M, δ)

determines up to the action of $W_{\mathbb{I}}$ a unique $X_{wur}(G)$-orbit of residual subspaces, and we have a representative $L = r_M T^M \subset T_{\mathbb{I}}$ of this orbit in the position described in (b) above. We now assign weights w_i to the nodes of $D(\mathfrak{g}^\vee, \theta)$, exactly as has been described above in (a)-(g), see [**Opd5**, 3.2.4; 3.2.5]. This defines a map $\Phi : T_{\tilde{t}} \to L_n$. Finally one needs to verify the property (3) of Definition 4.3.1 for this map, which amounts to case by case computations [**Opd5**].

At this stage we have established the existence of enough STMs, Theorem 4.4.1(a).

Next we now look at the essential uniqueness statement Theorem 4.4.1(b) for $\Phi : \mathcal{H}^e_{\tilde{t}} \rightsquigarrow \mathcal{H}_{\mathbb{I}}(G^*)$ This reduces to the cuspidal case using STM diagrams as follows: By Theorem [**Opd4**, Proposition 7.13] if suffices to show that one can find a $\omega \in X_{wur}(G)$ such that $\Phi'_{\tilde{t}}$ and $m_\omega \circ \Phi_{\tilde{t}}$ have the same image (in the sense of [**Opd5**, Definition 5.10]). Assume that the existence and uniqueness property (b) for the cuspidal case has been solved already, and let $\mathcal{H}^e_{\tilde{t}}$ be given. Let (M, δ) be the cuspidal pair associated to t by [**Mo**], then $\mathbb{P}_M = M \cap \mathbb{P}$ is a maximal parahoric subgroup of M, and $\sigma_M = \mathbb{P}|_{\mathbb{P}_M}$ is a cuspidal unipotent character for $\mathbb{P}_M$. By the already established cuspidal case of Theorem 4.4.1(a),(b), equation (4.3) determines a unique $X_{wur}(M)$-orbit of W_M-orbits $W_M r_M$ of residual points $r_M = s_M c_M$ for $\mathcal{H}_{\mathbb{I}_M}(M^*)$. Let $\tilde{J}$ be the maximal proper subset of the spectral diagram $D(\mathfrak{m}, \theta_M)$ of $\mathcal{H}_{\mathbb{I}_M}(M^*)$ which corresponds to r_M, i.e. the set of affine simple roots of $D(\mathfrak{m}, \theta_M)$ which are trivial on s_M. Then it turns out that in all cases, $\tilde{J}$ fits as excellent subdiagram of $D(\mathfrak{g}^\vee, \theta)$, with image $J \subset D(\mathfrak{g}^\vee, \theta)$ which is *unique* up to the action of $X_{wur}(G)$. This determines T_L and the image $W_0 r_L \in T_L \subset T_{\mathbb{I}}$ of $W_M r_M \in T_{\mathbb{I}_M}$, up to the action of $X_{wur}(G)$. Hence the image $L = r_L L$ of Φ is uniquely determined by $t = (\mathbb{P}, \sigma)$ up to the action of $X_{wur}(G)$. By [**Opd4**, Proposition 7.13] this proves the desired uniqueness.

The above reasoning reduces the general uniqueness statement Theorem 4.4.1(b) to the existence and uniqueness 4.4.1(a), (b) for the cuspidal case. This existence and uniqueness for the cuspidal case was shown in general in [**FO**] (using the general existence statement Theorem 4.4.1(a)), thereby finishing the proof of Theorem 4.4.1(b).

The existence and essential uniqueness of a $X_{wur}(G)$-equivariant STM, and its compatibility to Lusztig's parameterisation (Theorem 4.4.1(c), (d)) also reduces to the cuspidal case, by using the same construction of "inducing higher rank STMs from cuspidal ones" as discussed above. In the cuspidal case, these statements follow from a case by case verification [**FO**], [**FOS**]. This finishes the discussion of the proof of Theorem 4.4.1.

4.5. Main Theorem. The following Theorem is the main result. It is a generalisation of results of [**R3**], [**Opd5**], [**FO**], [**FOS**].

THEOREM 4.5.1. *Let $G = \mathbf{G}(F)$ be the group of points of a connected reductive group $\mathbf{G}$ defined over a non-archimedean local field F which splits over an unramified extension of F. Let $\hat{G}^{\text{temp}}_{uni} = \sqcup_{s \in \mathcal{B}_{uni}(G)} \hat{G}^{\text{temp}}_s$ denote the set of equivalence classes of irreducible tempered representations of G of unipotent reduction. Here $\mathcal{B}_{uni}(G) \subset \mathcal{B}(G)$ is the finite subset of unipotent inertial equivalence classes, and for each $s \in \mathcal{B}_{uni}(G)$, $\hat{G}^{\text{temp}}_s \subset \Pi(G)_s$ is the subsets of equivalence classes of tempered irreducible representations in $\mathcal{C}(G)_s$. For a unipotent type $t = (\mathbb{P}, \sigma)$, $\mathcal{B}^t(G)$ is an $X_{wur}(G)$-orbit in $\mathcal{B}_{uni}(G)$. For $s \in \mathcal{B}^t(G)$ choose a corresponding extension $\tilde{t}(s)$*

of t to $N_G(\mathbb{P})$, and extend this choice $X_{wur}(G)$-equivariantly. Let $\mathcal{H}_t$ be the Hecke algebra of the type t, then $\mathcal{H}_t = \oplus_{s\in\mathcal{B}^t(G)}\mathcal{H}^{\mathrm{e}}_{\tilde{t}(s)}$. Put $\mathcal{H}_{uni}(G) := \oplus_{s\in\mathcal{B}_{uni}(G)}\mathcal{H}^{\mathrm{e}}_{\tilde{t}(s)}$. Let $\mathcal{H}_{\mathbb{I}} = \mathcal{H}_{\mathbb{I}}(G^*)$ be the Iwahori Hecke algebra of the quasi-split group G^* in the inner class of G.

(a) *There exists a parameterisation $\varphi : \hat{G}^{\mathrm{temp}}_{uni} \to \Phi^{\mathrm{temp}}_{nr}(G)$, $\pi \to \varphi_\pi$ such that:*
  (1) *Conjectures* 2.6.1 *and* 2.6.2 *hold, up to constant factors independent of q.*
  (2) *For every unipotent type t of G there exists a morphism Φ^t_Z : $\mathrm{Spec}(Z(\mathcal{H}_t)) \to \mathrm{Spec}(Z(\mathcal{H}_{\mathbb{I}}))$ such that the map $\gamma^{\mathbb{I}} \circ \varphi : \hat{G}^{t,\mathrm{temp}} \to S_{\mathbb{I}} \subset \mathrm{Spec}(Z(\mathcal{H}_{\mathbb{I}}))$ factorises as $\gamma^{\mathbb{I}} \circ \varphi = \Phi^t_Z \circ \beta^t \circ \mathrm{cc}$.*

  The morphism Φ^t_Z is associated to an STM $\Phi^t : \mathcal{H}_t \rightsquigarrow \mathcal{H}_{\mathbb{I}}$ satisfying the conditions (2)(a) and (b) of Corollary 4.3.5.

(b) *Such a parameterisation is unique up to automorphisms $\alpha \in \mathrm{Aut}(\mathcal{H}_{uni})$ such that $\varphi' := \varphi \circ \alpha^*$ is again a parameterisation.*

(c) *If G is of adjoint type then Lusztig's enhanced parameterisation* [**Lus4**, **Lus5**] *$\tilde{\varphi}_{Lu} : \hat{G}^{\mathrm{temp}}_{uni} \to \tilde{\Phi}^{\mathrm{temp}}_{nr}(G)$ satisfies the conjectures* 2.6.1 *and* 2.6.2 *of Hiraga, Ichino and Ikeda* [**HII**]*, and moreover satisfies property (a)(2) above. (For general unramified G: see Theorem* 3.8.2.*)*

Proof. (c) For G simple and of adjoint type, Conjecture 2.6.1 for discrete series representations is [**Opd5**, Theorem 4.11]. The proof is based on Theorem 4.4.1 (to show the validity modulo rational constant factors), and the precise computation of the rational constant $d_{\mathcal{H},\delta}$ for the discrete series of affine Hecke algebras based on Theorem 3.7.3. Obviously this implies the result for all semisimple groups G of adjoint type. If G is of adjoint type and $M \subset G$ is a Levi subgroup with maximal split central torus $A_M \subset Z_M$, then M/A_M is also of adjoint type. Hence Conjecture 2.6.1 also holds for the discrete series of M modulo the center and of unipotent reduction. Thus Conjecture 2.6.2 follows from the Plancherel formula of Harish-Chandra [**Wal**, Théorème VIII.1.1]. By Theorem 4.4.1 (see also [**Opd5**, Theorem 3.4]), Lusztig's parameterisation $\varphi_{Lu} : \hat{G}^{\mathrm{temp}}_{uni}(G) \to \Phi^{\mathrm{temp}}_{nr}(G)$ of the irreducible tempered representations of unipotent reduction gives rise to a spectral transfer map $\Phi_{uni} : \mathcal{H}_{uni}(G) \rightsquigarrow \mathcal{H}_{\mathbb{I}}(G^*)$. By [**Opd4**, Theorem 6.1] and [**BKH**] this implies that, up to constants independent of q, the map $\gamma^{\mathbb{I}} \circ \varphi_{Lu} : \hat{G}^{\mathrm{temp}}_{uni} \to S_{\mathbb{I}} \subset \mathrm{mSpec}(Z(\mathcal{H}_{\mathbb{I}}(G^*)_{\mathrm{v}}))$ also satisfies (a)(2).

(a) and (b): For an absolutely semisimple group G of adjoint type, (c) implies the existence of a parameterisation $\pi \to \varphi_\pi$ of tempered representations of unipotent reduction satisfying (a)(1) and (2). Let us also prove the uniqueness property (b) for such G. Given a parameterisation satisfying (a)(1) and (2) it follows from Definition 4.0.2(iv) of the parameterisation and [**Bo**, Proposition 8.4], that we may assume that G is absolutely simple. For all unipotent types t of G, the morphism Φ^t_Z defines a spectral transfer map $\Phi^t : \mathcal{H}_t \rightsquigarrow \mathcal{H}_{\mathbb{I}}$ by definition. Indeed, by Definition 4.0.2 (i), (iii) and (v) of a parameterisation, the morphism Φ^t_Z comes from an affine morphism of algebraic tori $\Phi^t : T^t = X_{nr}(M) \to L'$ where L' is an intermediate quotient $L \to L' \to L_n$ of a residual coset $L \subset T_{\mathbb{I},\mathrm{v}}$ (indeed, the points of L' are in bijection with the orbit of $X_{nr}(M)$ twists of δ, which is a finite quotient of $X_{nr}(M)$, and L_n corresponds by definition to the central character map for representations of M of this type, hence is a quotient of L'). By (a)(1), [**BKH**], Theorems 3.5.1 and

3.5.2, and [**Opd4**, Definition 5.1] it follows that Φ^t defines a spectral transfer map $\mathcal{H}_t \rightsquigarrow \mathcal{H}_{\mathbb{I}}$. By Definition 4.0.2(i) and by Theorem 4.4.1 or [**Opd5**, Theorem 3.4] the required uniqueness property follows. Hence we have established the existence and uniqueness of parameterisations satisfying (a)(1) and (2) for G absolutely simple of adjoint type.

Let us now show (a) and (b) when G is connected reductive with anisotropic center. We write $\mathbf{G} = \mathbf{Z}^0\mathcal{D}(\mathbf{G})$ with $\mathcal{D}(\mathbf{G})$ the derived subgroup (which is a connected semisimple group) and $\mathbf{Z}^0$ an anisotropic torus. Then the quotient $\mathbf{Z} = \mathbf{G}/\mathcal{D}(\mathbf{G})$ is an isogenous quotient of $\mathbf{Z}^0$, in particular an anisotropic torus. Let $\mathbf{G}_{ad}$ be the adjoint quotient of $\mathbf{G}$, a connected semisimple group of adjoint type. Now consider the isogeny $\psi : \mathbf{G} \to \mathbf{Z} \times \mathbf{G}_{ad}$, and the corresponding dual isogeny $\psi^\vee : Z^\vee \times G^\vee_{sc} \to G^\vee$. Let $T^\vee$ be a maximal torus of $G^\vee$, and $T^\vee_{sc}$ its inverse image in $G^\vee_{sc}$. The kernel ψ can be expressed as $\mathrm{Hom}(K^*_G, F^\times_{ur})^{\mathrm{Fr}}$, where $K_G = X^*(T^\vee_{sc})/X^*(T^\vee)$.

We have a surjection with finite kernel ${}^L(Z^\vee \times G^\vee_{sc}) \to {}^LG$. Now $\Phi^{\mathrm{temp}}_{nr}(Z \times G_{ad}) = \Phi^{\mathrm{temp}}_{nr}(Z) \times \Phi^{\mathrm{temp}}_{nr}(G_{ad}) \simeq \Phi^{\mathrm{temp}}_{nr}(G_{ad})$, since for the anisotropic torus Z, $\Phi^{\mathrm{temp}}_{nr}(Z) = \{[\varphi_{triv}]\}$. The corresponding map $\Phi^{\mathrm{temp}}_{nr}(Z \times G_{ad}) \to \Phi^{\mathrm{temp}}_{nr}(G)$ can be identified, via these isomorphisms, with the natural map $\Phi^{\mathrm{temp}}_{nr}(G_{ad}) \to \Phi^{\mathrm{temp}}_{nr}(G)$. Via the isomorphism of Theorem 3.8.1 this map comes from the covering map $T^\vee_{sc}/(1-\theta)T^\vee_{sc} \to T^\vee/(1-\theta)T^\vee$ applied to $S_{sc,\mathbb{I},\mathbf{v}}$. Therefore this map is surjective, with fibres that are orbits for the canonical image $I_G \subset (\Omega^\theta_{sc})^*$ of the finite abelian group $(K^\theta_G)^*$ in $X_{wur}(G_{ad}) = (\Omega^\theta_{sc})^*$. Then I_G is the kernel of the canonical surjection $(\Omega^\theta_{sc})^* \to (\Omega^\theta_G)^*$, and we have an identification $\Phi^{\mathrm{temp}}_{nr}(G) = I_G \backslash \Phi^{\mathrm{temp}}_{nr}(G_{ad})$.

We propose the following parameterisation $\hat{G}^{\mathrm{temp}}_{uni} \to \Phi^{\mathrm{temp}}_{nr}(G)$. Given an irreducible tempered representation (π, V_π) of unipotent reduction of G. The kernel $\mathrm{Hom}(K^*_G, F^\times_{nr})^{\mathrm{Fr}}$ of the isogeny $G \to Z \times G_{ad}$ acts trivially in V_π, hence (π, V_π) descends to a representation of the image of G in $Z \times G_{ad}$. Since the image in Z acts trivially on V_π, we can extend to a representation $\overline{\pi}$ of $Z \times \mathrm{Im}(G)$ (with trivial action of the Z-factor) where $\mathrm{Im}(G) \subset G_{ad}$. We *claim* that $\overline{\pi}$ is a summand of the restriction of a representation $\tilde{\pi}$ of unipotent reduction of G_{ad}, and that this uniquely determines the Lusztig-Langlands parameter $\varphi_{\tilde{\pi}}$ of $\tilde{\pi}$ up to the action of I_G. We define, in view of the above identification, $\varphi_\pi := I_G\varphi_{\pi_{ad}}$. In order that this definition makes sense we need to verify the above claim.

Let $t = (\mathbb{P}, \delta)$ be a "unipotent type" for the unipotent Bernstein component of G, where $\mathbb{P}$ is a parahoric subgroup, δ a cuspidal unipotent of $\mathbb{P}$, and $\tilde{\delta}$ an extension of δ to $N_G\mathbb{P}$. The Hecke algebra $\mathcal{H}_t$ of t is a direct sum of extended affine Hecke algebras $\mathcal{H}^e_{\tilde{t}}$. We have (see [**PR**]) $\mathbb{P} = G_1 \cap N_G\mathbb{P}$, and this group is self-normalizing within G_1, the kernel of the Kottwitz homomorphism w_G. The set of Bernstein components described by t corresponds bijectively to the set of extensions $\tilde{t} = (N_G\mathbb{P}, \tilde{\delta}')$ of δ to $N_G\mathbb{P}$. This is a torsor for the character group $(\Omega^{\mathbb{P},\theta}_G)^*$ of the abelian group $\Omega^{\mathbb{P},\theta}_G$, the subgroup of $\Omega^\theta_G = G/G_1$ fixing $\mathbb{P}$ under conjugation. The subalgebra $\mathcal{H}^a_t$ of the extended affine Hecke algebra $\mathcal{H}^e_{\tilde{t}}$ of functions with support in G_1 is the (unextended) underlying affine Hecke algebra, by the arguments of Lusztig [**Lus4**, 1.10–1.20]. The Hecke algebra $\mathcal{H}_t$ of the type t is isomorphic to $\mathcal{H}^a_t\#(\Omega^{\mathbb{P},\theta}_G)^*$. This algebra can be written as a direct sum of the various extended affine Hecke algebras $\mathcal{H}^e_{\tilde{t}}$ associated with the extensions $\tilde{t}$. This is a direct sum of mutually isomorphic extended affine Hecke algebras $\mathcal{H}^e_{\tilde{t}}$ whose set of summands form a torsor for the group of characters of the subgroup $(\Omega^{\mathbb{P},\theta}_G)_1$ of $\Omega^{\mathbb{P},\theta}_G$ of elements acting trivially on $\mathcal{H}_t^a$. In any case, one obtains a canonical action of $(\Omega_G^{\mathbb{P},\theta})^*$, and thus of $(\Omega_G^\theta)^*$, on the disjoint union of the centers of the summands $\mathcal{H}_{\tilde{t}}^e$ of $\mathcal{H}_t$.

There is a bijection between the set of parahoric subgroups of G and of $Z \times G_{ad}$ defined as follows. First choose a maximally F-split maximal F-torus T of G which splits over F_{ur}. Let $\mathcal{A}$ be the apartment of G associated with T, which can be embedded in the apartment of $\mathbf{G}(F_{ur})$ as the set of Fr-invariant elements. By [**PR**, Definition 1; Remark 16], [**Opd5**, Corollary 2.2], the stabilizer of a facet f in $\mathcal{A}$ in $G_1 = \mathbf{G}(F_{ur})^{\mathrm{Fr}}$ is also the pointwise fixator of f in G_1, since $\mathbf{G}(F_{ur})$ is generated by $\mathbf{G}_{der}(F_{ur})$ and elements acting trivially on $\mathcal{A}$. Hence by [**PR**] it follows that the parahoric subgroup $\mathbb{P}_f$ of G is also equal to the normalizer $\mathbb{P}_f = N_{G_1}(\mathbb{P}_f \cap G_{der})$. Indeed, $\subset$ follows since $\mathbb{P}_f$ is self-normalizing in G_1, while $\supset$ follows since the right hand side is stabilizing, hence pointwise fixing f. By the functoriality of Kottwitz's map w_G we have a homomorphism $G_1 \to (Z \times G_{ad})_1$. The bijection between the parahoric subgroups is defined as follows: Given $\mathbb{P}$ of G, we have $\mathbb{P}' := N_{(Z\times G_{ad})_1}(\psi(\mathbb{P}))$ is a parahoric group of $Z \times G_{ad}$, as follows from the discussion above and the remark that $(Z \times G_{ad})_{der} \subset \psi(G)$; conversely if $\mathbb{P}' \subset Z \times G_{ad}$ is parahoric then $\mathbb{P} := \psi^{-1}(\mathbb{P}') \cap G_1$ is parahoric in G. These maps are inverse to each other. Moreover, the cuspidal unipotent representations of $\mathbb{P}$ and $\mathbb{P}'$ correspond bijectively to each other, since it is known that the set of cuspidal unipotent characters of a connected reductive group $\mathcal{G}$ over a finite field is independent of the type of $\mathcal{G}$ within its isogeny class (cf. [**Lus1**]). It follows that the type $t_1 := (\mathbb{P}, \delta)$ of G corresponds to a unique type $\tilde{t}_1 = (\mathbb{P}', \delta')$ of $Z \times G_{ad}$, and the affine Hecke algebras $\mathcal{H}'_{t_1}$ and $\mathcal{H}'_{\tilde{t}_1}$ are isomorphic unextended affine Hecke algebras via ψ, with the same normalization of their respective traces.

Now let us return to the verification of the above claim. Consider an irreducible tempered representation (π, V_π) of G of unipotent reduction, and let $\tilde{t} = (N_G\mathbb{P}, \tilde{\delta})$ be an extension of a unipotent type t associated to the Bernstein component to which π belongs. The induction of $\overline{\pi}$ to $Z \times G_{ad}$ is a unitary tempered representation of $Z \times G_{ad}$ with finitely many irreducible summands $\tilde{\pi}$, and this gives an obvious construction of tempered irreducible representations $\tilde{\pi}$ which contain $\overline{\pi}$ when restricted to the image of G in $Z \times G_{ad}$. By restriction to $\psi : G_1 \to (Z \times G_{ad})_1$ it follows that an extension $\tilde{t}$ describing the Bernstein component of $\tilde{\pi}$ must restrict to $\tilde{t}_1 = (\mathbb{P}', \delta')$ as described above.

Hence $\tilde{t}$ is of the form $\tilde{t} = (N_{Z\times G_{ad}}\mathbb{P}', \tilde{\delta}')$. As before, the algebra $\mathcal{H}_{\tilde{t}_1}$ is a direct sum of isomorphic extended affine Hecke algebras. Via ψ we can view $\mathcal{H}_{t_1}$ as a subalgebra of $\mathcal{H}_{\tilde{t}_1}$, and by the Bernstein description of the center of an affine Hecke algebra it follows that $Z(\mathcal{H}_{t_1}) \subset Z(\mathcal{H}_{\tilde{t}_1})$ is a subalgebra of finite index. The kernel of the corresponding surjective homomorphism $\mathrm{Spec}(Z(\mathcal{H}_{\tilde{t}_1})) \to \mathrm{Spec}(Z(\mathcal{H}_{t_1}))$ is the image of $I_G \subset (\Omega_{sc}^\theta)^*$ under the surjection $(\Omega_{sc}^\theta)^* \to (\Omega_{sc}^{\mathbb{P},\theta})^*$, and the natural action by the group $(\Omega_{sc}^\theta)^*$ on the spectrum of $\mathrm{Spec}Z(\mathcal{H}_{\tilde{t}_1})$ corresponds to the natural action of $(\Omega_G^\theta)^*$ on $\mathrm{Spec}(Z(\mathcal{H}_{\tilde{t}_1}))$. Clearly $cc(\tilde{\pi})$ belongs to the I_G-orbit of central characters in the fibre above $cc(\pi)$ under this map.

Now recall that for the representation $\tilde{\pi}$ of G_{ad} (or $Z \times G_{ad}$) the Langlands parameter $\varphi_{\tilde{\pi}}$ is defined by $\varphi_{\tilde{\pi}} := (\gamma^{\mathbb{I}})^{-1}(\Phi_Z(cc^t(m_t(\tilde{\pi}))))$, where $\Phi : \mathcal{H}_{uni}(G_{ad}) \rightsquigarrow \mathcal{H}_{\mathbb{I}}(G_{ad}^*)$ is the spectral transfer map of [**Opd5**, Theorem 3.4]. We just explained above that $cc^t(m_t(\tilde{\pi}))$ belongs to a single I_G-orbit, and then the $X_{wur}(G_{ad}) := (\Omega_{sc}^\theta)^*$-equivariance of Φ and of $\gamma^{\mathbb{I}}$ (by Lemma 4.0.1) finally establishes the claim.

Hence for G connected reductive with anisotropic kernel, we have now shown existence of a Langlands correspondence $\varphi : \hat{G}^{\text{temp}}_{uni} \to \Phi^{\text{temp}}_{nr}(G)$ which obviously satisfies (a)(1) (we reduced this to the case G_{ad} where we know (a)(1) already, via spectral transfer maps which preserve Plancherel densities up to constants by [**Opd4**]) and (a)(2) (since spectral transfer maps yield such morphisms by definition). The uniqueness property (b) follows from the case of G_{ad} if we can show that there always exists a lift of a $X_{wur}(G) = I_G\backslash X_{wur}(G_{ad})$-equivariant parameterisation $\varphi : \hat{G}^{\text{temp}}_{uni} \to \Phi^{\text{temp}}_{nr}(G)$ to a $X_{wur}(G_{ad})$-equivariant parameterisation $\varphi_{ad} : (\hat{G}_{ad})^{\text{temp}}_{uni} \to \Phi^{\text{temp}}_{nr}(G_{ad})$. For this we need to invoke a stronger form of the uniqueness property of STMs $\Phi^t : \mathcal{H}_t \to \mathcal{H}_{\mathbb{I},\text{v}}(G_{ad})$ (where t is an (extended) unipotent type which represents a Bernstein component of representations of unipotent reduction for an inner form of G_{ad}) as explained in [**Opd5**, 3.2.8] and [**Opd4**, Proposition 7.13]: If $\Phi^t_1 : \mathcal{H}_t \to \mathcal{H}_{\mathbb{I},\text{v}}(G_{ad})$ is another such STM, then there exists an isomorphism $\alpha : \mathcal{H}_t \to \mathcal{H}_t$, $w \in W_{\mathbb{I},0}$, and $\omega \in X_{wur}(G_{ad}) = (\Omega^{\theta}_{sc})^*$ such that $\Phi^t_1 = \omega \circ w \circ \Phi^t \circ \alpha$. It follows that any two matching I_G-orbits of connected components under φ must also correspond under any equivariant correspondence $\tilde{\varphi}_{ad}$ (which we know exists) up to the action of $X_{wur}(G_{ad})$. In this way we can compose $\tilde{\varphi}_{ad}$ with an element of $X_{wur}(G_{ad})$ to obtain an I_G-equivariant lifting of φ defined on this I_G-orbit of components. Clearly we can do this for any I_G-orbit, thus defining the desired equivariant lifting φ_{ad} of φ.

Finally let us consider the general case. Let G be connected reductive and let A be the maximal F-split torus in the center of G. Then $H = G/A$ is the group of F-points of the quotient $\mathbf{G}/\mathbf{A}$, whose center is anisotropic. In particular the functoriality of the Kottwitz homomorphism implies that G_1 maps surjectively to H_1.

Given a tempered irreducible representation π of unipotent reduction of G, choose a $x \in X_{wur}(G)$ such that $x^{-1} \otimes \pi$ is trivial on A. Hence $x^{-1} \otimes \pi$ descends to a tempered irreducible representation of unipotent reduction $\overline{x^{-1} \otimes \pi}$ of H. Let $\varphi_{\overline{x^{-1}\otimes\pi}} \in \Phi^{\text{temp}}_{nr}(H)$. Let $p_G(\varphi_{\overline{x^{-1}\otimes\pi}})$ in $\Phi^{\text{temp}}_{nr}(G)$ be the image of $\varphi_{\overline{x^{-1}\otimes\pi}}$ under the canonical map ${}^LH \to {}^LG$. Now define $\varphi_\pi := x.p_G(\varphi_{\overline{x^{-1}\otimes\pi}}) \in \Phi^{\text{temp}}_{nr}(G)$. In order that this makes sense, we need to show the independence of the choice of $x \in X_{wur}(G)$ such that $x^{-1} \otimes \pi$ is trivial on A. Suppose that $z \in X_{wur}(G)$ such that $z|_A = 1$. Then z restricts to $\overline{z}$, a character of H. Since G_1 surjects onto H_1, we see that $\overline{z} \in X_{wur}(H)$. Hence if we would replace x by zx then we would get $\varphi'_\pi := (zx).p_G((\varphi_{\overline{(z^{-1}x^{-1})\otimes\pi}}))$. The equivariance for $X_{wur}(H)$ of the parameterisation of H and of p_G (which is obvious) implies that $\varphi'_\pi = \varphi_\pi$. Hence our definition is sound, $X_{wur}(G)$-equivariant by construction, and is the only possible extension of a given equivariant parameterisation for H. The properties (a)(1), (a)(2) and (b) easily follow.

This finishes the proof of the existence and essential uniqueness of $X_{wur}(G)$-equivariant STM's $\Phi^t_Z : \text{Spec}(Z(\mathcal{H}_t)) \to \text{Spec}(Z(\mathcal{H}_{\mathbb{I}}))$ in the general case for all unipotent types t, satisfying the conditions (2)(a) and (b) of Corollary 4.3.5. We finally need to show that the corresponding map $\hat{G}^{\text{temp}}_{uni} \to \Phi^{\text{temp}}_{nr}$, $\pi \to \varphi_\pi$ defines a parameterisation in the sense of Definition 4.0.2. Of course, in conditions (iii) and (v) where two such maps play a role we need to allow for the fact that the maps are only uniquely determined up to certain spectral automorphisms of $\mathcal{H}_{uni}$, which means that we may need to choose the relevant morphisms Φ^t_Z judiciously.

The conditions (i), (ii) and (vi) hold because Φ^t_Z satisfies conditions (2)(a) and (b) of Corollary 4.3.5. Furthermore condition (iv) follows easily from [**Bo**, Proposition 8.4].

Let us show that condition (iii) (the compatibility with unitary parabolic induction) holds. Let $\pi \in \hat{G}^{\mathrm{temp}}_{uni}$ be a tempered representation of unipotent reduction of G, and suppose that π is a summand of $i^G_P(\pi')$ (unitary induction), for an F-parabolic subgroup $P = HU$ with H a standard F-Levi subgroup of G, and $\pi' \in \hat{H}^{\mathrm{temp}}_{uni}$. Let $\varphi^H_{\pi'} \in \Phi^{\mathrm{temp}}_{nr}(H)$ be the parameter for π', defined by a morphism $\Phi^{t_H}_Z$ as in (a) (thus originating from an STM $\Phi^{t_H} : \mathcal{H}_{t_H} \rightsquigarrow \mathcal{H}_{\mathbb{I}}(H^*)$ which satisfies the conditions (2)(a) and (b) of Corollary 4.3.5) for H). Then we want to show that we can find such a morphism $\Phi^{t_G}_Z$ as in (a) for G, such that it attaches the parameter $\varphi^H_{\pi'} \in \Phi^{temp}_{nr}(H) \subset \Phi^{temp}_{nr}(G)$ to π. In order to do so, suppose that $\pi \in \hat{G}^{\mathrm{temp}}_{t_G}$ for some unipotent type t_G for G (such type exists by [**Mo**],[**MP1**]). Then t_G is a G-cover of a cuspidal unipotent type t_M for a cuspidal pair (M, τ) in the cuspidal support of π ([**BK**], [**Mo**], [**MP2**]), where we choose $M \subset G$ to be a standard parabolic subgroup. Since the cuspidal support of π' equals that of π, we see that we may assume $M \subset H$, and that we have an intermediate H-cover t_H of t_M such that $\pi' \in \Pi(H)^{t_H}$. In this situation we have ([**BK**], [**Mo**])) an injective homomorphism $j : \mathcal{H}_{t_H} \to \mathcal{H}_{t_G}$ (a "parabolic subalgebra" in the sense of [**Opd2**]) inducing a canonical embedding of commutative algebras $Z(\mathcal{H}_{t_G}) \subset Z(\mathcal{H}_{t_H})$. Similarly we have $\mathcal{H}_{\mathbb{I}}(H^*) \subset \mathcal{H}_{\mathbb{I}}(G^*)$ (likewise a parabolic subalgebra), inducing a canonical embedding $Z(\mathcal{H}_{\mathbb{I}}(G^*)) \subset Z(\mathcal{H}_{\mathbb{I}}(H^*))$. The morphisms defined by induction on the spaces of central characters of these Hecke algebras are given by these inclusion maps.

It follows from [**Opd5**, 3.1.1] that $\Phi^{t_H} : \mathcal{H}_{t_H} \to \mathcal{H}_{\mathbb{I}}(H^*)$ is induced form a cuspidal STM $\Phi'_M : \mathcal{H}_{t_M} \to \mathcal{H}_{\mathbb{I}}(M^*)_M$. Similarly Φ_{t_G} (whose existence we have established in the first part of the proof) is induced from a cuspidal STM $\Phi_M : \mathcal{H}_{t_M} \to \mathcal{H}_{\mathbb{I}}(M^*)_M$. But it is clear that Φ^{t_G} also defines an STM $\mathcal{H}_{t_H} \rightsquigarrow \mathcal{H}_{\mathbb{I}}(H^*)$ satisfying the conditions (2)(a) and (b) of Corollary 4.3.5 for H (defined by the same map as Φ^{t_G} on $T_{t_H} = T_{t_G}$; indeed this clearly defines a spectral preserving map for the parabolic subalgebra $\mathcal{H}_{t_H}$ if it does so for $\mathcal{H}_{t_G}$). In particular, we may just as well choose $\Phi'_M = \Phi_M$. The resulting commuting diagram

$$\begin{array}{ccc} \mathrm{Spec}(Z(\mathcal{H}_{t_H})) & \longrightarrow & \mathrm{Spec}(Z(\mathcal{H}_{t_G})) \\ {\scriptstyle \Phi^{t_H}_Z}\downarrow & & \downarrow{\scriptstyle \Phi^{t_G}_Z} \\ \mathrm{Spec}(Z(\mathcal{H}_{\mathbb{I}}(H^*))) & \longrightarrow & \mathrm{Spec}(Z(\mathcal{H}_{\mathbb{I}}(G^*))) \end{array}$$

now easily shows the desired compatibility with unitary induction for the associated maps φ^H and φ^G.

The proof of (v) is comparable to that of (iv). Suppose that $\eta : H \to G$ is an F-morphism with abelian kernel and co-kernel. We want to show that if $\varphi \in \Phi^{\mathrm{temp}}_{nr}(G)$ is given and $\pi \in \Pi_\varphi(G)^{\mathrm{temp}}$, the pull-back of π to H is a finite direct sum of tempered irreducible representations in $\Pi^{\mathrm{temp}}_{{}^L\eta\circ\varphi}(H)$, where ${}^L\eta : {}^LG \to {}^LH$ denotes the natural map.

We choose maximally F-split tori T_H and T_G such that $\eta(T_H) \subset T_G$. The induced map by η on the (absolute) root data of (H, T_H) and (G, T_G) is an isomorphism on the root systems, but the lattices may differ. This defines a bijective correspondence between the sets of standard F-Levi subgroups of H and G.

Let M_G and M_H be matching standard Levi subgroups of G and H respectively. Then $\eta^M := \eta|_{M_H} : M_H \to M_G$ also has abelian kernel and co-kernel. The pull-back of a cuspidal character of M_G of unipotent reduction defines a cuspidal character of unipotent reduction of M_H, and conversely, every cuspidal character of unipotent reduction of M_H is obtained like this. Suppose that we have such matching unipotent inertial equivalence classes $\mathfrak{s}_H$ and $\mathfrak{s}_G$, and let t_H and t_G be unipotent types for H and G respectively covering the corresponding Bernstein components. Let $\Phi^{t_H} : \mathcal{H}_{t_H} \rightsquigarrow \mathcal{H}_{\mathbb{I}}(H^*)$ and $\Phi^{t_G} : \mathcal{H}_{t_G} \rightsquigarrow \mathcal{H}_{\mathbb{I}}(G^*)$ be corresponding STMs which satisfy the conditions (2)(a) and (b) of Corollary 4.3.5 (the existence and essential uniqueness of such STMs was proved in the first part of the proof). By [**Opd5**, 3.1.1] these STMs are induced from cuspidal unipotent STMs $\Phi_{M_H} : \mathcal{H}_{t_{M_H}} \to \mathcal{H}_{\mathbb{I}}(M_H^*)_{M_H}$ and $\Phi_{M_G} : \mathcal{H}_{t_{M_G}} \to \mathcal{H}_{\mathbb{I}}(M_G^*)_{M_H}$ of the reductive groups $(M_H)_{ssa} := M_H/A_H^M$ and $(M_G)_{ssa} = M_G/A_G^M$ respectively, which are almost direct products of a semisimple and an anisotropic group (here A_G^M denotes the connected split center of M_G, and likewise A_H^M is the connected split center of M_H). Hence the latter Hecke algebras are all semisimple. By the equivariance property (2)(a) of Corollary 4.3.5, the STMs Φ^{t_H} and Φ^{t_G} are completely determined by Φ_{M_H} and Φ_{M_G} respectively. Note that η induces a spectral isomorphism $\Phi^t_\eta : \mathcal{H}_{t_{M_H}} \rightsquigarrow \mathcal{H}_{t_{M_G}}$ (since these Hecke algebras have rank zero, if an STM between them exists it is an isomorphism) and an embedding (in particular an STM–see [**Opd4**, 7.1.3]) $\Phi^{\mathbb{I}}_\eta : \mathcal{H}(M_H^*)_{M_H} \rightsquigarrow \mathcal{H}_{t_{M_G}}(M_G^*)_{M_G}$.

Because the root systems of G and H are isomorphic via η, it is clear that $\Phi'_{M_G} = \Phi^{\mathbb{I}}_\eta \circ \Phi_{M_H} \circ (\Phi^t_\eta)^{-1}$ also defines an STM $\mathcal{H}_{t_{M_G}} \to \mathcal{H}_{\mathbb{I}}(M_G^*)_{M_H}$ that can be induced to an STM $(\Phi^{t_G})' : \mathcal{H}_{t_G} \rightsquigarrow \mathcal{H}_{\mathbb{I}}(G^*)$. (Indeed, the μ-functions of the $\mathcal{H}_{\mathbb{I}}(G^*)$ and $\mathcal{H}_{\mathbb{I}}(H^*)$ are given by "the same" formula.) Thus we may and will from now on assume that $\Psi^{t_G} = (\Phi^{t_G})'$ is constructed like this. It follows easily that we have a commutative diagram of morphisms

$$\begin{array}{ccc}
\mathrm{Spec}(Z(\mathcal{H}_{t_G})) & \xrightarrow{(\eta_Z^t)^*} & \mathrm{Spec}(Z(\mathcal{H}_{t_H})) \\
\Phi_Z^{t_G}\Big\downarrow & & \Big\downarrow\Phi_Z^{t_H} \\
\mathrm{Spec}(Z(\mathcal{H}_{\mathbb{I}}(G^*))) & \xrightarrow{(\eta_Z^{\mathbb{I}})^*} & \mathrm{Spec}(Z(\mathcal{H}_{\mathbb{I}}(H^*)))
\end{array}$$

where η_Z^t and $\eta_Z^{\mathbb{I}}$ denote the natural morphisms which η induces on the centers of the unipotent affine Hecke algebras involved. By comparing the central characters of a tempered representation of $\mathcal{H}_{t_G}$ and a summand of its pull back via η the desired property (v) now follows in a straightforward fashion. □

REMARK 4.5.2. In [**FOS**, Theorem 1.3] the conjecture 2.6.1 (including the rational constants) was proved for supercuspidal representations of unipotent reduction of semisimple groups. We hope that this can be extended to prove the conjectures 2.6.1 and 2.6.2 of [**HII**] for all tempered representations of unipotent reduction of a general connected reductive group over F (split over an unramified extension), using [**CO**] (in particular Theorem 3.7.3) and drawing on ideas from [**R5**].

References

[Ar1] J. Arthur, *Unipotent automorphic representations: conjectures*, Astérisque **171-172** (1989), 13–71. Orbites unipotentes et représentations, II. MR1021499

[Ar2] J. Arthur, *A note on L-packets*, Pure Appl. Math. Q. **2** (2006), no. 1, Special Issue: In honor of John H. Coates., 199–217, DOI 10.4310/PAMQ.2006.v2.n1.a9. MR2217572

[ABV] J. Adams, D. Barbasch, and D. A. Vogan Jr., *The Langlands classification and irreducible characters for real reductive groups*, Progress in Mathematics, vol. 104, Birkhäuser Boston, Inc., Boston, MA, 1992. MR1162533

[AMS] A-M. Aubert, A. Moussaoui, and M. Solleveld, "Affine Hecke algebras for Langlands parameters", math.RT:1701.03593 (2017).

[Be] J. N. Bernstein, "All reductive p-adic groups are tame", Functional Anal. Appl. **8** (1974), 91–93.

[BeDe] J. N. Bernstein, *Le "centre" de Bernstein* (French), Representations of reductive groups over a local field, Travaux en Cours, Hermann, Paris, 1984, pp. 1–32. Edited by P. Deligne. MR771671

[Be1] J. N. Bernstein, *On the support of Plancherel measure*, J. Geom. Phys. **5** (1988), no. 4, 663–710 (1989), DOI 10.1016/0393-0440(88)90024-1. MR1075727

[Bo] A. Borel, *Automorphic L-functions*, Automorphic forms, representations and L-functions (Proc. Sympos. Pure Math., Oregon State Univ., Corvallis, Ore., 1977), Proc. Sympos. Pure Math., XXXIII, Amer. Math. Soc., Providence, R.I., 1979, pp. 27–61. MR546608

[Bo1] A. Borel, *Admissible representations of a semi-simple group over a local field with vectors fixed under an Iwahori subgroup*, Invent. Math. **35** (1976), 233–259, DOI 10.1007/BF01390139. MR0444849

[BT] F. Bruhat and J. Tits, *Groupes réductifs sur un corps local. II. Schémas en groupes. Existence d'une donnée radicielle valuée* (French), Inst. Hautes Études Sci. Publ. Math. **60** (1984), 197–376. MR756316

[BK] C. J. Bushnell and P. C. Kutzko, *Smooth representations of reductive p-adic groups: structure theory via types*, Proc. London Math. Soc. (3) **77** (1998), no. 3, 582–634, DOI 10.1112/S0024611598000574. MR1643417

[BKH] C. J. Bushnell, G. Henniart, and P. C. Kutzko, *Types and explicit Plancherel formulæfor reductive p-adic groups*, On certain L-functions, Clay Math. Proc., vol. 13, Amer. Math. Soc., Providence, RI, 2011, pp. 55–80. MR2767510

[Car] R. W. Carter, *Finite groups of Lie type*, Pure and Applied Mathematics (New York), John Wiley & Sons, Inc., New York, 1985. Conjugacy classes and complex characters; A Wiley-Interscience Publication. MR794307

[CK] D. Ciubotaru and S. Kato, *Tempered modules in exotic Deligne-Langlands correspondence*, Adv. Math. **226** (2011), no. 2, 1538–1590, DOI 10.1016/j.aim.2010.08.013. MR2737793

[COT] D. Ciubotaru, E. M. Opdam, and P. E. Trapa, *Algebraic and analytic Dirac induction for graded affine Hecke algebras*, J. Inst. Math. Jussieu **13** (2014), no. 3, 447–486, DOI 10.1017/S147474801300008X. MR3211796

[CO] D. Ciubotaru and E. Opdam, *A uniform classification of discrete series representations of affine Hecke algebras*, Algebra Number Theory **11** (2017), no. 5, 1089–1134, DOI 10.2140/ant.2017.11.1089. MR3671432

[DeRe] S. DeBacker and M. Reeder, *Depth-zero supercuspidal L-packets and their stability*, Ann. of Math. (2) **169** (2009), no. 3, 795–901, DOI 10.4007/annals.2009.169.795. MR2480618

[DeLu] P. Deligne and G. Lusztig, *Representations of reductive groups over finite fields*, Ann. of Math. (2) **103** (1976), no. 1, 103–161, DOI 10.2307/1971021. MR0393266

[Fe1] Y. Feng, "On Cuspidal Unipotent Representations", Thesis 2015, University of Amsterdam.

[Fe2] Y. Feng, "Notes on Spectral Transfer Maps for Affine Hecke Algebras", in preparation.

[FO] Y. Feng and E. M. Opdam, "On a uniqueness property of supercuspidal unipotent representations", arXiv:1504.03458[math.RT].

[FOS] Y. Feng, E. M. Opdam, and M. Solleveld, "Supercuspidal unipotent representations and L-packets", arXiv:1805.01888.

[GK] S. S. Gelbart and A. W. Knapp, *L-indistinguishability and R groups for the special linear group*, Adv. in Math. **43** (1982), no. 2, 101–121, DOI 10.1016/0001-8708(82)90030-5. MR644669

[G] B. H. Gross, *On the motive of a reductive group*, Invent. Math. **130** (1997), no. 2, 287–313, DOI 10.1007/s002220050186. MR1474159

[GG] B. H. Gross and W. T. Gan, *Haar measure and the Artin conductor*, Trans. Amer. Math. Soc. **351** (1999), no. 4, 1691–1704, DOI 10.1090/S0002-9947-99-02095-4. MR1458303

[GR] B. H. Gross and M. Reeder, *Arithmetic invariants of discrete Langlands parameters*, Duke Math. J. **154** (2010), no. 3, 431–508, DOI 10.1215/00127094-2010-043. MR2730575

[PR] G. Pappas and M. Rapoport, *Twisted loop groups and their affine flag varieties*, Adv. Math. **219** (2008), no. 1, 118–198, DOI 10.1016/j.aim.2008.04.006. With an appendix by T. Haines and Rapoport. MR2435422

[HC1] Harish-Chandra, *Harmonic analysis on real reductive groups. I. The theory of the constant term*, J. Functional Analysis **19** (1975), 104–204. MR0399356

[HC2] Harish-Chandra, *Representations of a semisimple Lie group on a Banach space. I*, Trans. Amer. Math. Soc. **75** (1953), 185–243, DOI 10.2307/1990730. MR0056610

[HC3] Harish-Chandra, *Harmonic analysis on real reductive groups. III. The Maass-Selberg relations and the Plancherel formula*, Ann. of Math. (2) **104** (1976), no. 1, 117–201, DOI 10.2307/1971058. MR0439994

[HT] M. Harris and R. Taylor, *The geometry and cohomology of some simple Shimura varieties*, Annals of Mathematics Studies, vol. 151, Princeton University Press, Princeton, NJ, 2001. With an appendix by Vladimir G. Berkovich. MR1876802

[HO1] G. J. Heckman and E. M. Opdam, *Harmonic analysis for affine Hecke algebras*, Current developments in mathematics, 1996 (Cambridge, MA), Int. Press, Boston, MA, 1997, pp. 37–60. MR1724944

[HO] G. J. Heckman and E. M. Opdam, *Yang's system of particles and Hecke algebras*, Ann. of Math. (2) **145** (1997), no. 1, 139–173, DOI 10.2307/2951825. MR1432038

[H1] G. Henniart, *Une preuve simple des conjectures de Langlands pour* $\mathrm{GL}(n)$ *sur un corps p-adique* (French, with English summary), Invent. Math. **139** (2000), no. 2, 439–455, DOI 10.1007/s002220050012. MR1738446

[H2] G. Henniart, *Caractérisation de la correspondance de Langlands locale par les facteurs* ϵ *de paires* (French, with English and French summaries), Invent. Math. **113** (1993), no. 2, 339–350, DOI 10.1007/BF01244309. MR1228128

[H3] G. Henniart, *Une caractérisation de la correspondance de Langlands locale pour* $\mathrm{GL}(n)$ (French, with English and French summaries), Bull. Soc. Math. France **130** (2002), no. 4, 587–602. MR1947454

[HII] K. Hiraga, A. Ichino, and T. Ikeda, *Formal degrees and adjoint* γ*-factors*, J. Amer. Math. Soc. **21** (2008), no. 1, 283–304, DOI 10.1090/S0894-0347-07-00567-X. MR2350057

[HIIcor] K. Hiraga, A. Ichino, and T. Ikeda, *Correction to: "Formal degrees and adjoint* γ*-factors" [J. Amer. Math. Soc.* **21** *(2008), no. 1, 283–304; MR2350057]*, J. Amer. Math. Soc. **21** (2008), no. 4, 1211–1213, DOI 10.1090/S0894-0347-08-00605-X. MR2425185

[HS] K. Hiraga and H. Saito, *On L-packets for inner forms of* SL_n, Mem. Amer. Math. Soc. **215** (2012), no. 1013, vi+97, DOI 10.1090/S0065-9266-2011-00642-8. MR2918491

[Kac] V. G. Kac, *Infinite-dimensional Lie algebras*, 3rd ed., Cambridge University Press, Cambridge, 1990. MR1104219

[Kal] T. Kaletha, *The local Langlands conjectures for non-quasi-split groups*, Families of automorphic forms and the trace formula, Simons Symp., Springer, [Cham], 2016, pp. 217–257. MR3675168

[L] R. P. Langlands, *On the functional equations satisfied by Eisenstein series*, Lecture Notes in Mathematics, Vol. 544, Springer-Verlag, Berlin-New York, 1976. MR0579181

[KL] D. Kazhdan and G. Lusztig, *Proof of the Deligne-Langlands conjecture for Hecke algebras*, Invent. Math. **87** (1987), no. 1, 153–215, DOI 10.1007/BF01389157. MR862716

[Kot] R. E. Kottwitz, *Stable trace formula: elliptic singular terms*, Math. Ann. **275** (1986), no. 3, 365–399, DOI 10.1007/BF01458611. MR858284

[Kot1] R. E. Kottwitz, *Isocrystals with additional structure. II*, Compositio Math. **109** (1997), no. 3, 255–339, DOI 10.1023/A:1000102604688. MR1485921

[Lus1] G. Lusztig, *Characters of reductive groups over a finite field*, Annals of Mathematics Studies, vol. 107, Princeton University Press, Princeton, NJ, 1984. MR742472

[Lus2] G. Lusztig, *Intersection cohomology complexes on a reductive group*, Invent. Math. **75** (1984), no. 2, 205–272, DOI 10.1007/BF01388564. MR732546

[Lus3] G. Lusztig, *Affine Hecke algebras and their graded version*, J. Amer. Math. Soc. **2** (1989), no. 3, 599–635, DOI 10.2307/1990945. MR991016

[Lus4] G. Lusztig, *Classification of unipotent representations of simple p-adic groups*, Internat. Math. Res. Notices **11** (1995), 517–589, DOI 10.1155/S1073792895000353. MR1369407

[Lus5] G. Lusztig, *Classification of unipotent representations of simple p-adic groups. II*, Represent. Theory **6** (2002), 243–289, DOI 10.1090/S1088-4165-02-00173-5. MR1927955

[Mo] L. Morris, *Level zero* **G**-*types*, Compositio Math. **118** (1999), no. 2, 135–157, DOI 10.1023/A:1001019027614. MR1713308

[MP1] A. Moy and G. Prasad, *Unrefined minimal K-types for p-adic groups*, Invent. Math. **116** (1994), no. 1-3, 393–408, DOI 10.1007/BF01231566. MR1253198

[MP2] A. Moy and G. Prasad, *Jacquet functors and unrefined minimal K-types*, Comment. Math. Helv. **71** (1996), no. 1, 98–121, DOI 10.1007/BF02566411. MR1371680

[Opd1] E. M. Opdam, *A generating function for the trace of the Iwahori-Hecke algebra*, Studies in memory of Issai Schur (Chevaleret/Rehovot, 2000), Progr. Math., vol. 210, Birkhäuser Boston, Boston, MA, 2003, pp. 301–323. MR1985730

[Opd2] E. M. Opdam, *On the spectral decomposition of affine Hecke algebras*, J. Inst. Math. Jussieu **3** (2004), no. 4, 531–648, DOI 10.1017/S1474748004000155. MR2094450

[Opd3] E. Opdam, *The central support of the Plancherel measure of an affine Hecke algebra* (English, with English and Russian summaries), Mosc. Math. J. **7** (2007), no. 4, 723–741, 767–768. MR2372211

[Opd4] E. Opdam, *Spectral correspondences for affine Hecke algebras*, Adv. Math. **286** (2016), 912–957, DOI 10.1016/j.aim.2015.09.019. MR3415700

[Opd5] E. Opdam, *Spectral transfer morphisms for unipotent affine Hecke algebras*, Selecta Math. (N.S.) **22** (2016), no. 4, 2143–2207, DOI 10.1007/s00029-016-0273-7. MR3573955

[OS1] E. Opdam and M. Solleveld, *Homological algebra for affine Hecke algebras*, Adv. Math. **220** (2009), no. 5, 1549–1601, DOI 10.1016/j.aim.2008.11.002. MR2493620

[OS2] E. Opdam and M. Solleveld, *Discrete series characters for affine Hecke algebras and their formal degrees*, Acta Math. **205** (2010), no. 1, 105–187, DOI 10.1007/s11511-010-0052-9. MR2736154

[PR] G. Pappas and M. Rapoport, *Twisted loop groups and their affine flag varieties*, Adv. Math. **219** (2008), no. 1, 118–198, DOI 10.1016/j.aim.2008.04.006. With an appendix by T. Haines and Rapoport. MR2435422

[R1] M. Reeder, *On the Iwahori-spherical discrete series for p-adic Chevalley groups; formal degrees and L-packets*, Ann. Sci. École Norm. Sup. (4) **27** (1994), no. 4, 463–491. MR1290396

[R2] M. Reeder, *Whittaker models and unipotent representations of p-adic groups*, Math. Ann. **308** (1997), no. 4, 587–592, DOI 10.1007/s002080050091. MR1464911

[R3] M. Reeder, *Formal degrees and L-packets of unipotent discrete series representations of exceptional p-adic groups*, J. Reine Angew. Math. **520** (2000), 37–93, DOI 10.1515/crll.2000.023. With an appendix by Frank Lübeck. MR1748271

[R4] M. Reeder, *Torsion automorphisms of simple Lie algebras*, Enseign. Math. (2) **56** (2010), no. 1-2, 3–47, DOI 10.4171/LEM/56-1-1. MR2674853

[R5] M. Reeder, *Isogenies of Hecke algebras and a Langlands correspondence for ramified principal series representations*, Represent. Theory **6** (2002), 101–126, DOI 10.1090/S1088-4165-02-00167-X. MR1915088

[Sha] F. Shahidi, *A proof of Langlands' conjecture on Plancherel measures; complementary series for p-adic groups*, Ann. of Math. (2) **132** (1990), no. 2, 273–330, DOI 10.2307/1971524. MR1070599

[S] D. Shelstad, *Characters and inner forms of a quasi-split group over* **R**, Compositio Math. **39** (1979), no. 1, 11–45. MR539000

[SZ] A. J. Silberger and E.-W. Zink, "The formal degree of discrete series representations of central simple algebras over p-adic fields", Max-Planck-Institut für Mathematik, 1996.

[Slo] K. Slooten, *Generalized Springer correspondence and Green functions for type B/C graded Hecke algebras*, Adv. Math. **203** (2006), no. 1, 34–108, DOI 10.1016/j.aim.2005.04.003. MR2231067

[Spr] T. A. Springer, *Linear algebraic groups*, 2nd ed., Progress in Mathematics, vol. 9, Birkhäuser Boston, Inc., Boston, MA, 1998. MR1642713

[Tate] J. Tate, *Number theoretic background*, Automorphic forms, representations and L-functions (Proc. Sympos. Pure Math., Oregon State Univ., Corvallis, Ore., 1977), Proc.

Sympos. Pure Math., XXXIII, Amer. Math. Soc., Providence, R.I., 1979, pp. 3–26. MR546607

[Tits] J. Tits, *Reductive groups over local fields*, Automorphic forms, representations and L-functions (Proc. Sympos. Pure Math., Oregon State Univ., Corvallis, Ore., 1977), Proc. Sympos. Pure Math., XXXIII, Amer. Math. Soc., Providence, R.I., 1979, pp. 29–69. MR546588

[Vog] D. A. Vogan Jr., *The local Langlands conjecture*, Representation theory of groups and algebras, Contemp. Math., vol. 145, Amer. Math. Soc., Providence, RI, 1993, pp. 305–379, DOI 10.1090/conm/145/1216197. MR1216197

[Wal] J.-L. Waldspurger, *Représentations de réduction unipotente pour* $\mathrm{SO}(2n+1)$*: quelques conséquences d'un article de Lusztig* (French), Contributions to automorphic forms, geometry, and number theory, Johns Hopkins Univ. Press, Baltimore, MD, 2004, pp. 803–910. MR2058629

[Z] E.-W. Zink, "Comparison of GL_N and division algebra representations II", Max-Planck-Institut für Mathematik, 1993.

Korteweg-de Vries Institute for Mathematics, Universiteit van Amsterdam, Science Park 105-107, 1098 XG Amsterdam, The Netherlands

Email address: `e.m.opdam@uva.nl`

Proceedings of Symposia in Pure Mathematics
Volume **101**, 2019
https://doi.org/10.1090/pspum/101/01800

Limiting cycles and periods of Maass forms

Andre Reznikov

To Joseph with gratitude.

ABSTRACT. We study (generalized) periods of Maass forms along non-closed geodesics having a closed geodesic as the limit set.

1. Introduction

The aim of this note is to introduce a new kind of a period for Maass forms. For outsiders of the automorphic world, we recall that in the theory of automorphic functions eigenfunctions of the Laplacian on a finite volume hyperbolic Riemann surface are called Maass forms (after H. Maass [**Ma**] who realized their importance in Number Theory). There are 3 types of closed cycles naturally appearing in the theory of automorphic functions on the group $G = \mathrm{PGL}_2(\mathbb{R})$ with respect to a lattice $\Gamma \subset G$: 1) closed horocycles which are associated to closed orbits in the automorphic space $X = \Gamma \setminus G$ of a unipotent subgroup $N \subset G$; 2) closed geodesics and geodesic rays starting and ending in a cusp (both of these types are associated to closed orbits on X of the diagonal subgroup $A \subset G$); and 3) closed geodesics circles which are associated to orbits on X of a maximal compact subgroup $K \subset G$. Periods of Maass forms along all these cycles play a crucial role in Number Theory.

We propose to consider one more period along a special type of a *non-closed* orbit of the subgroup A, i.e., along a *non-closed* special geodesic on the corresponding Riemann surface. These geodesics will have closed geodesics as their limit sets. Our justification for introducing such cycles is that (generalized) periods of Maass forms along these geodesics satisfy nice analytic properties.

Periods (with characters) of Hecke-Maass forms along "classical" cycles lead to Fourier coefficients of cusp forms and to L-functions (e.g., the Hecke L-function given by a period along the geodesic ray connecting two cusps and a special value of a quadratic base change L-function from a theorem of J.-L. Waldspurger [**W**] appearing as the square of a period along a closed geodesic), and hence play an important role in Number Theory. Admittedly, we do not know yet what is the arithmetic meaning of these new periods (although their residues are connected to periods along closed geodesics and hence to special values of L-functions).

2010 *Mathematics Subject Classification.* Primary 54C40, 14E20; Secondary 46E25, 20C20.

Key words and phrases. Automorphic forms, periods, analytic continuation.

The research was partially supported by the ERC grant 291612 and by the ISF grant 533/14.

1.1. Limiting cycles. Although our methods do *not* use arithmetic, we describe the phenomenon in the simplest (and probably in the most interesting) case of the arithmetic lattice $\Gamma = \mathrm{PGL}_2(\mathbb{Z}) \subset G$. Denote by $Y = X/K = \Gamma \setminus \mathcal{H}$ the corresponding Riemann surface (here $K = PO(2)$ and $\mathcal{H} = G/K$ denotes the hyperbolic upper half plane). We denote by $\mathcal{F} \subset \mathcal{H}$ a (Dirichlet) fundamental domain of Γ. Let $\ell \subset Y$ be a closed geodesic. Lifting ℓ to the upper half plane $\mathcal{H}$, we obtain a geodesic $\tilde{\ell} \subset \mathcal{H}$ connecting points α, $\overline{\alpha}$ on the absolute $\mathbb{P}^1(\mathbb{R})$ of $\mathcal{H}$ (for the lattice $\mathrm{PGL}_2(\mathbb{Z})$, α and $\overline{\alpha}$ are two conjugate real quadratic irrationalities).

We consider the vertical geodesic $\tilde{L} = \tilde{L}_\alpha \subset \mathcal{H}$ connecting the cusp of Γ at ∞ and α (correspondingly, $L_{\bar{\alpha}}$ connects ∞ and $\bar{\alpha}$).

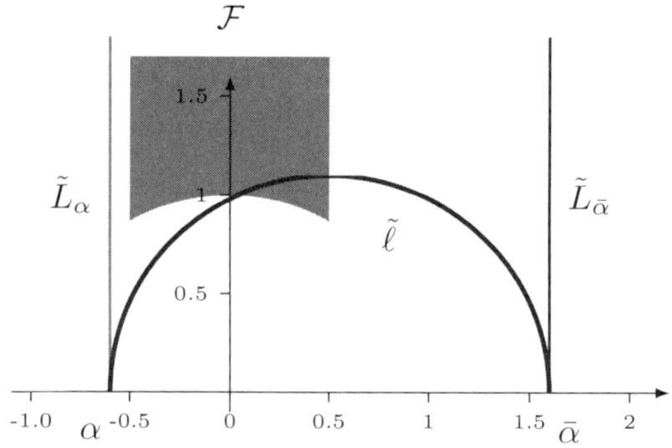

FIGURE 1

On Y the corresponding geodesic $L \subset Y$ becomes a geodesic winding around the closed geodesic ℓ in one direction and escaping into the cusp in the opposite direction. Clearly, ℓ is in the closure of L (in fact, $\overline{L} = L \cup \ell$). In general, L is a piecewise smooth geodesic in the case it passes through a conical point on Y (i.e., through a fixed point of an elliptic element in Γ).

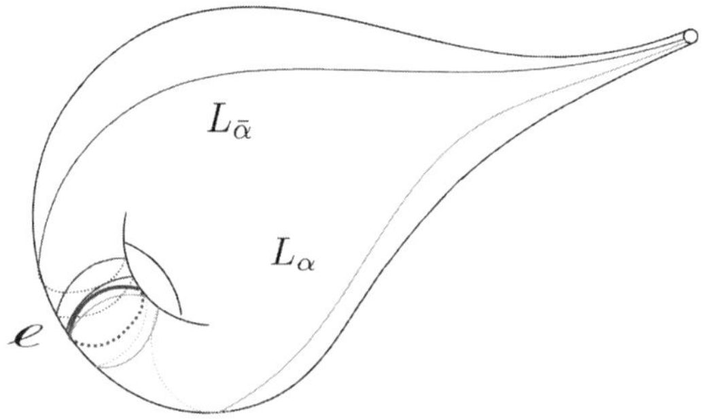

FIGURE 2

1.2. Limiting periods of Maass forms. Let $\varphi = \varphi_\lambda$ be a cuspidal Maass form on Y with the eigenvalue $\mu = \frac{1-\lambda^2}{4}$ of the Laplacian Δ on Y, where $\lambda \in i\mathbb{R} \cup (0,1)$. For $s \in \mathbb{C}$, we consider a twisted period of φ along L, namely,

$$I_s(\varphi) = I_{s,L_\alpha}(\varphi_\lambda) = \int_L \varphi(l)|l|^s dl = \int_{\tilde{L}} \varphi(\tilde{l})|\tilde{l}|^s d\tilde{l}, \tag{1.1}$$

where we identify L with the subgroup $A_{\alpha,\infty} = \mathrm{Stab}_G(\tilde{L})$ (e.g., by choosing a point on $\tilde{L}$), and, using this identification, consider multiplicative characters $|\cdot|^s$ on $\tilde{L} \simeq \mathbb{R}_+^\times$, and an $A_{\alpha,\infty}$ invariant measure $d\tilde{l}$ on $\tilde{L}$. We can write the same integral in "additive" coordinates on L using the natural geodesic parameter $t \in \mathbb{R}$ on L (i.e., $t = \ln a$) and the corresponding measure dt given by the length. We have then $I_s(\varphi) = \int_L \varphi(t) e^{st} dt$.

It is easy to see that the integral $I_s(\varphi)$ is absolutely convergent for $\Re(s) > 1$ (e.g., using the fact that φ is a bounded function with an exponential decay in cusps).

Our main result is the following

THEOREM 1.1. *The integral $I_s(\varphi)$ is absolutely convergent for $\Re(s) > 1$, and has meromorphic continuation to $\mathbb{C}$.*

Moreover, we will identify possible poles and residues of $I_s(\varphi)$ in terms of periods of φ along the closed geodesic $\ell \subset Y$.

One can give a reformulation of Theorem 1.1 in terms of Fourier coefficients of φ. Let

$$\varphi_\lambda(x+iy) = \sum_{n \neq 0} a_n \cdot K_{\lambda,n}(y) e^{2\pi i n x} \tag{1.2}$$

be the Fourier expansion of φ_λ at ∞. Here we denote by $K_{\lambda,n}$ a multiple of the classical K-Bessel function which we will normalize by the corresponding matrix coefficient (see (2.2)). Coefficients $a_n = a_n(\varphi_\lambda)$, up to a simple twist, coincide with classical Fourier coefficients of φ_λ, and, in particular, satisfy the asymptotic $\sum_{|n| \leq T} |a_n|^2 \sim C \cdot T$ with some constant $C = C(\varphi_\lambda) > 0$ (of course, this depends on normalization of functions $K_{\lambda,n}$).

By writing the function φ on $\tilde{L}$ through its Fourier expansion at the cusp, we obtain a Dirichlet series expression for the integral $I_s(\varphi)$. This leads to the following, more "classical", form of Theorem 1.1.

THEOREM 1.2. *The twisted Dirichlet series*

$$D_{s,\alpha}(\varphi) = \sum_{n \neq 0} \frac{a_n}{|n|^s} \cdot e^{2\pi i n \alpha} \tag{1.3}$$

is absolutely convergent for $\Re(s) > 1$, and has meromorphic continuation to $\mathbb{C}$.

Remarks. 1. We compute poles and residues of $D_{s,\alpha}$ (which are given in terms of twisted periods of φ along ℓ). As a result, one can see that the combination $D_{s,\alpha} - D_{s,\overline{\alpha}}$ is holomorphic (or what is the same, the combination $I_{s,L_\alpha}(\varphi_\lambda) + I_{s,\overline{L_{\overline{\alpha}}}}(\varphi_\lambda)$ is holomorphic where $L_{\overline{\alpha}}$ is the cycle connecting $\overline{\alpha}$ to the cusp at ∞ and "$\overline{L}$" stands for the cycle with the reversed orientation).

Our method also naturally treats the signed sum

$$D^{\pm}_{s,\alpha}(\varphi) = \sum_{n \neq 0} \frac{a_n}{|n|^s} \cdot e^{2\pi i n \alpha} \mathrm{sign}(n) , \tag{1.4}$$

and as a result the sum $D^{>}_{s,\alpha}(\varphi) = \sum_{n>0} \frac{a_n}{|n|^s} \cdot e^{2\pi i n \alpha}$.

Our proof also gives a polynomial in s and λ bound for $D_{s,\alpha}(\varphi_\lambda)$.

2. Dirichlet series for the "additively twisted" L-function

$$D_{s,a}(\varphi) = \sum_{n \geq 1} \frac{a_n(\varphi)}{n^s} e^{2\pi i n a}, \quad a \in \mathbb{R}$$

were studied extensively starting with Hecke for $a \in \mathbb{Q}$ (this leads to the usual Hecke L-function and hence such a series has holomorphic continuation to $\mathbb{C}$). Properties of $D_{s,a}$ for an irrational a were studied, among other places, in [**C**] and [**MS**], and in [**Mi**] for $GL(n)$, and it is certainly expected that for a general a, the corresponding series does not have analytic continuation to $\mathbb{C}$.

3. The proof we give does not use arithmetic. It is interesting to see if for an arithmetic Γ and Hecke-Maass forms, the above Dirichlet series has some arithmetical meaning (from what we have seen above, residues of this series do have a connection to special values of L-functions via the theorem of Waldspurger [**W**]).

We note that the cycle over which we integrate has clear arithmetic description (for a congruence subgroup of $\Gamma = \mathrm{PGL}_2(\mathbb{Z})$). Let E be the quadratic field corresponding to the geodesic ℓ (in particular, the primitive element γ_ℓ corresponds to a unit in an order in E). We have the natural imbedding $E^\times \hookrightarrow \mathrm{GL}_2(\mathbb{Q})$ which gives rise to the real tori $T_E(\mathbb{R}) = E^\times(\mathbb{R}) \hookrightarrow \mathrm{GL}_2(\mathbb{R})$. In the quotient space, this gives the geodesic ℓ (or rather a collection of geodesics corresponding to the class group for the relevant order). The corresponding tori $T_E(\mathbb{R})$ has α and $\bar{\alpha}$ as fixed points on $\mathbb{P}^1(\mathbb{R})$. In the light of the Waldspurger theorem [**W**] which deals with the period squared $\left|\int_\ell \varphi\right|^2$, it is natural to consider the imbedding $T_E(\mathbb{R}) \times T_E(\mathbb{R}) \hookrightarrow \mathrm{GL}_2(\mathbb{R}) \times \mathrm{GL}_2(\mathbb{R})$ coming from two conjugated imbeddings $E \hookrightarrow \mathbb{R}$. Note that pointwise these two imbeddings of $T_E(\mathbb{R})$ coincide with one fixing the pair $\alpha, \bar{\alpha}$ and another the pair $\bar{\alpha}, \alpha$. In the quotient space $X \times X$, the corresponding cycle $\ell \times \ell$ is compact. Our period L_α or rather the product $L_\alpha \times L_{\bar{\alpha}} \subset X \times X$ could be described in similar terms. Consider another tori $T'_E \subset \mathrm{GL}_2(E)$ which fixes ∞ and α on $\mathbb{P}^1(E)$. T'_E is defined over E while T_E is defined over $\mathbb{Q}$ (both are conjugate in $\mathrm{GL}_2(E)$). $T'_E(\mathbb{R})$ has two conjugated imbeddings into $\mathrm{GL}_2(\mathbb{R})$. Hence we obtain the imbedding $T'_E(\mathbb{R}) \times T'_E(\mathbb{R}) \hookrightarrow \mathrm{GL}_2(\mathbb{R}) \times \mathrm{GL}_2(\mathbb{R})$ with one component fixing the pair ∞, α and another $\infty, \bar{\alpha}$. In the quotient space, this becomes our cycle $L_\alpha \times L_{\bar{\alpha}}$. The corresponding cycle is not compact since there are no points in T'_E which are defined over $\mathbb{Q}$ (unlike for the tori T_E). This shows that the product of integrals $I_s(\varphi) \cdot I_{-s}(\bar{\varphi})$ is given by an integral over a cycle in $L_\alpha \times L_{\bar{\alpha}}$ which is defined over E while the Waldspurger cycle $\ell \times \ell$ is defined over $\mathbb{Q}$.

4. For the group $\mathrm{PSL}_2(\mathbb{Z})$, integrals along limiting cycles L_α were considered in [**MM**] from a different point of view. For $x, y \in \tilde{L}_\alpha$, denote by $\tilde{L}_{x,y} \subset \tilde{L}_\alpha$ the segment between points x and y. It is shown in [**MM**] that for any fixed $x \in \tilde{L}_\alpha$, the mean value of the integral $I(\varphi, \alpha, x) = \lim_{y \to \alpha} |\tilde{L}_{x,y}|^{-1} \int_{\tilde{L}_{x,y}} \varphi(l)dl$ converges to the period integral $\int_\ell \varphi(l)dl$ (and in particular is independent of x) for α which is a fixed point of a hyperbolic element in $\mathrm{PSL}_2(\mathbb{Z})$, and vanishes for other irrational α. For α which is a fixed point of a hyperbolic element in $\mathrm{PSL}_2(\mathbb{Z})$, we can deduce this result from the existence of the simple pole of $I_s(\varphi)$ at $s = 1$ with the residue given by the period along the closed geodesic ℓ. In fact, we can treat also twisted integrals $\lim_{y \to \alpha} |\tilde{L}_{x,y}|^{-1} \int_{\tilde{L}_{x,y}} \varphi(l)\chi(l)dl$ where, as in (1.1), we view L_α as an orbit of A and $\chi : A \to \mathbb{C}^\times$ a character trivial on $\Gamma_\ell = A \cap \Gamma$. Looking at poles of $I_s(\varphi)$

at appropriate s, we see that the limit is given by the twisted period $\int_\ell \varphi(l)\chi(l)dl$ along the corresponding closed geodesic ℓ.

5. We also have analogous results for integrals over limiting geodesics winding down from one closed geodesic to another closed geodesic (such a geodesic has two rotational numbers, for a general discussion, see [**P**]), although these integrals are not easily expressible in terms of a Fourier coefficient (and in fact, such integrals could be treated for co-compact lattices as well).

We note that there are infinitely many non-homotopic geodesic having two closed geodesics (or a closed geodesic and a cusp) as their limit set. Our result is valid for any such a geodesic.

We also note that non-cuspidal Maass (e.g., Eisenstein series) forms could be treated by our method as well.

Acknowledgments. It is a pleasure to thank Leonid Polterovich for an enlightening discussion of the underlying geometry relevant to this paper.

The paper is a byproduct of a joint project with Joseph Bernstein. It is a great pleasure to thank Joseph for teaching the author a lot.

2. Proof

We now present the proof of Theorem 1.2. Theorem 1.1 follows easily from Theorem 1.2 by writing φ on L_α through its Fourier expansion (see (2.1)).

Our proof is based on standard methods in representation theory of automorphic functions (and, in particular, we use co-invariants of representation of $\mathrm{PGL}_2(\mathbb{R})$ with respect to some cyclic subgroups coming from Γ).

2.1. Representation theoretic setup. Let π_λ be an automorphic representation associated to a cuspidal Maass form φ_λ (this means that we have an abstract unitary representation $(\pi_\lambda, L_{\pi_\lambda})$ of G and an isometry $\nu : L_{\pi_\lambda} \to L^2(X)$ such that $\nu(e_0) = \varphi_\lambda$ for a K-fixed vector $e_0 \in L_{\pi_\lambda}$ of norm one). We will work with the space of smooth vectors $V_\lambda = L^\infty_{\pi_\lambda} \subseteq L_{\pi_\lambda}$ and will assume for simplicity that $\lambda \in i\mathbb{R}$ (i.e., π_λ is a principal series representation of G). Frobenius reciprocity (e.g., see [**BR2**]) gives the corresponding functional $I = I_{\varphi_\lambda} \in \mathrm{Hom}_\Gamma(V_\lambda, \mathbb{C})$, i.e., a Γ-invariant functional $I : V_\lambda \to \mathbb{C}$ such that $\varphi_v(g) := \nu(v)(g) = I(\pi_\lambda(g)v)$ for any smooth vector $v \in V_\lambda$. The functional I belongs to some Sobolev class of functionals on V_λ. In [**BR1**] it is shown that the functional I has finite $-\left(\frac{1}{2}+\varepsilon\right)$-Sobolev L^2-norm for any $\varepsilon > 0$ (results for the Hölder class were obtained simultaneously in [**O**] and [**S**] and for the Besov scale in [**BR2**]). The space V_λ has natural realization in the space of smooth functions $C^\infty_\lambda(\mathbb{R})$ on $\mathbb{R}$ with some condition on the decay at infinity (e.g., functions have asymptotic $|x|^{-1+\lambda}$ as $|x| \to \infty$). We call this model the linear model of π_λ. Note that the representation theoretic Sobolev norm constructed in [**BR1**] and [**BR2**] does not coincide with the standard Sobolev norm on $C^\infty(\mathbb{R})$. However, for vectors with a *fixed* compact support in the linear model $V_\lambda \simeq C^\infty_\lambda(\mathbb{R})$, these two norms are equivalent. This is discussed at length in [**BR2**].

We can view the Frobenius functional I as a distribution on $\mathbb{R}$. The element $\left(\begin{smallmatrix}1&1\\0&1\end{smallmatrix}\right) \in \Gamma$ generates the cuspidal subgroup $\Gamma_\infty = \Gamma \cap N$, and acts by the usual translation $\pi_\lambda\left(\begin{smallmatrix}1&1\\0&1\end{smallmatrix}\right) f(x) = f(x-1)$ in the linear model of V_λ. Hence the Γ-invariant distribution I has the periodic Fourier expansion

$$I = \sum_{0 \neq n \in \mathbb{Z}} a_n W_{n,\lambda}, \tag{2.1}$$

where $W_{n,\lambda} : V_\lambda \to \mathbb{C}$ is a Whittaker functional, i.e., a (unique up to a multiple constant) functional satisfying $W_{n,\lambda}(\pi_\lambda\left(\begin{smallmatrix}1 & t\\ 0 & 1\end{smallmatrix}\right)v) = e^{2\pi int}W_{n,\lambda}(v)$. We normalize Whittaker functionals by their norm on the space of Γ_∞-co-invariants $(V_\lambda)_{\Gamma_\infty} \simeq C^\infty([0,1])$ in the following way. We view $W_{n,\lambda}$ as a generalized vector in $\pi_{-\lambda}$. We choose the "base" Whittaker functional as given by the kernel $W_{1,\lambda}(x) = e^{2\pi ix}$ and denote by $W_{n,\lambda}(x) = |n|^{-\frac{1}{2}} \cdot \pi_{-\lambda}\begin{pmatrix} n^{-\frac{1}{2}} & 0\\ 0 & n^{\frac{1}{2}}\end{pmatrix} W_{1,\lambda} = |n|^{-\frac{\lambda}{2}} \cdot e^{2\pi inx}$. Hence up to a factor of the absolute norm one, $W_{n,\lambda}$ is given by the exponent $e^{2\pi inx}$.

Note that with such a normalization of functionals $W_{n,\lambda}$, coefficients a_n in (2.1) do not depend on the change of the sign of λ (but do depend on a choice of $W_{1,\lambda}$).

Coefficients a_n in (2.1) will be consistent with the Fourier coefficients of φ_λ in (1.2) once we normalize K-Bessel functions in (1.2) by the corresponding matrix coefficient, i.e., by the following integral representation (compare to the classical normalization in [**M**])

$$
\begin{aligned}
K_{\lambda,n}(y) &= \left\langle \pi_\lambda \begin{pmatrix} y^{\frac{1}{2}} & 0\\ 0 & y^{-\frac{1}{2}}\end{pmatrix} e_0,\, W_{n,\lambda}\right\rangle \\
&= |n|^{-\lambda/2}|y|^{\frac{\lambda-1}{2}} \int_{\mathbb{R}} |1+(y^{-1}x)^2|^{\frac{\lambda-1}{2}} e^{-2\pi inx}dx\ , \qquad (2.2)
\end{aligned}
$$

where $e_0(x) = |1+x^2|^{\frac{\lambda-1}{2}}$ is a K-fixed vector in the linear model of V_λ.

From (2.1) we see that the Dirichlet series $D_{s,\alpha}$, up to a ratio of Γ-functions, is given by the pairing of I with a (generalized) vector $d_{s,\alpha}$ given by the kernel $d_{s,\alpha}(x) = |x-\alpha|^{-1-\frac{\lambda}{2}+s}$ in the linear model of the representation π_λ. We have for $\Re(s) \gg 1$,

$$
\begin{aligned}
\langle I, d_{s,\alpha}\rangle &= \sum_{0\neq n\in\mathbb{Z}} a_n \cdot W_{n,\lambda}(d_{s,\alpha}) = \gamma(s-\tfrac{1}{2}\lambda)\cdot \sum_{n\neq 0} a_n\cdot |n|^{-s}\cdot e^{2\pi in\alpha}\\
&= \gamma(s-\tfrac{1}{2}\lambda)\cdot D_{s,\alpha(\varphi)}\ ,
\end{aligned}
$$

where $\gamma(s)$ is given by the Fourier transform relation $\mathcal{F}(|x|^{s-1})[\xi] = \gamma(s)\cdot|\xi|^{-s}$ with $\gamma(s) = \pi^{-\frac{s}{2}}\Gamma\left(\frac{s}{2}\right)/\pi^{-\frac{1-s}{2}}\Gamma\left(\frac{1-s}{2}\right)$.

We remark that it is more natural to shift s by $\frac{1}{2}$ and consider the vector (which we will denote by the same letter) $d_{s,\alpha}(x) = |x-\alpha|^{-\frac{1}{2}-\frac{\lambda}{2}+s}$.

The vector $d_{s,\alpha}$ is not a smooth vector, and we have to make sense out of the pairing $\langle I, d_{s,\alpha}\rangle$. We will show that in fact this pairing is well-defined for $\Re(s) > 1$ (since the corresponding vector belongs to an appropriate Sobolev space); and, moreover, the corresponding pairing has meromorphic continuation (because $d_{s,\alpha}$ belongs to an appropriate Sobolev space after the projection to co-invariants of V_λ with respect to the cyclic subgroup $\Gamma_\ell \subset \Gamma$ corresponding to the geodesic ℓ).

2.2. Co-invariants. Our aim is to analytically continue the pairing $\langle I, d_{s,\alpha}\rangle$, where $d_{s,\alpha}(x) = |x-\alpha|^{-\frac{1}{2}-\frac{\lambda}{2}+s}$. It is easy to see that $d_{s,\alpha}$ is an eigenvector of the action of the torus $A_{\alpha,\infty} \subset G$. Namely, we have $\pi_\lambda(a_{\alpha,\infty})d_{s,\alpha} = |a_{\alpha,\infty}|^{-2s}\cdot d_{s,\alpha}$. Here $|a_{\alpha,\infty}|^{-2s}$ denotes the natural character $\chi_{-2s} : A_{\alpha,\infty} \to \mathbb{C}^\times$, $s\in\mathbb{C}$, induced by the conjugation of $A_{\alpha,\infty}$ to the standard torus $A_{0,\infty} = A = \left\{\left(\begin{smallmatrix} a & 0\\ 0 & a^{-1}\end{smallmatrix}\right) \mid a \in \mathbb{R}^\times\right\}$ and the corresponding (quasi-) character $\chi_\sigma : A_{0,\infty} \to \mathbb{C}^\times$ is given by $\chi_\sigma\left(\begin{smallmatrix} a & 0\\ 0 & a^{-1}\end{smallmatrix}\right) = |a|^\sigma$ for $\sigma\in\mathbb{C}$.

We now use the fact that $\mathrm{Stab}_\Gamma(\infty) = \Gamma_\infty \neq \{e\}$ and that $\mathrm{Stab}_\Gamma(\alpha) = \Gamma_\ell \neq \{e\}$. Note that $\mathrm{Stab}_\Gamma((\alpha,\infty)) = \Gamma\cap A_{\alpha,\infty} = \{e\}$.

The generalized vector $d_{\alpha,s}$ is smooth outside of ∞ and α. We use partition of unity to separate these singularities. Let u_0 be a real valued function supported near α and u_∞ the complimentary function (i.e., $1 \geq u(x) \geq 0$, $\operatorname{supp} u \subseteq [\alpha - 1, \alpha + 1]$, $u\big|_{[\alpha-1/2,\alpha+1/2]} \equiv 1$ and $u_0 + u_\infty \equiv 1$). Denote by $d^0_{\alpha,s}(x) = u_0(x)d_{\alpha,s}(x)$, $d^\infty_{\alpha,s}(x) = u_\infty(x)d_{\alpha,s}(x)$, $d_{\alpha,s}(x) = d^0_{\alpha,s}(x) + d^\infty_{\alpha,s}(x)$.

2.2.1. **Claim I.** $\langle I, d^\infty_{\alpha,s}\rangle$ is holomorphic.

This follows from the fact that in the space of co-invariants $(V_\lambda)_{\Gamma_\infty} = V_\lambda/\mathrm{Span}\{(1-n)f\}_{n\in\Gamma_\infty, f\in V_\lambda}$, the vector $d^\infty_{\alpha,S}$ is differentiable N-times for any $N \geq 1$ (and, in particular, it has a representative in V_λ which belongs to the $\left(\frac{1}{2}+\varepsilon\right)$-Sobolev space of V_λ for any $\varepsilon > 0$).

2.2.2. **Claim II.** $\langle I, d^0_{\alpha,s}\rangle$ is well-defined for $\Re(s) > \frac{1}{2}$, and has the meromorphic continuation to $\mathbb{C}$.

The function $d^0_{\alpha,s}(x) = |x-\alpha|^{-\frac{1}{2}-\frac{\lambda}{2}+s}u_0(x)$ has a polynomial singularity at $x = \alpha$ which belongs to a $\left(\frac{1}{2}+\varepsilon\right)$-Sobolev space for $\Re(s) > \frac{1}{2}+\varepsilon$. Hence the value of $\langle I, d^0_{\alpha,s}\rangle$ is well-defined and is holomorphic in s in this domain.

Consider a non-trivial element $\gamma_\ell \in \Gamma_\ell = \Gamma \cap A_{\alpha,\overline{\alpha}}$ (e.g., a primitive hyperbolic element in Γ which corresponds to the geodesic ℓ). The element γ_ℓ is conjugated to a diagonal element $a_\ell = \begin{pmatrix} q_\ell^{\frac{1}{2}} & 0 \\ 0 & q_\ell^{-\frac{1}{2}} \end{pmatrix} \in A$ for some $q_\ell > 1$ (q_ℓ is easily given in terms of length of ℓ). We denote by $A_\ell = \langle a_\ell\rangle \subset A$ the corresponding subgroup. The space of co-invariants $(V_\lambda)_{\Gamma_\ell}$ is naturally isomorphic to the space of (smooth) functions on $A_\ell \backslash A \simeq \langle q\rangle \backslash \mathbb{R}^\times$. Up to a conjugation (by an element g_ℓ conjugating γ_ℓ to a_ℓ), our function $d^0_{\alpha,s}$ becomes the function

$$\tilde{d}^0_{\beta,s}(x) = \pi_\lambda(g_\ell)d^0_{\alpha,s} = |x|^{-\frac{1}{2}-\frac{\gamma}{2}+s}|x-\beta|^{-\frac{1}{2}-\frac{\lambda}{2}-s}\tilde{u}_0(x))$$

with $\tilde{u}_0$ supported in an interval which includes 0, but not β (here $\beta = g_\ell(\infty)$ and $0 = g_\ell(\alpha)$), and $\tilde{u}_0 \equiv 1$ in some interval which includes 0. Hence $\tilde{d}^0_{\beta,s}$ has the only singularity at $x = 0$.

To describe the image of $\tilde{d}^0_{\beta,s}$ in the space of co-invariants $(V_\lambda)_{\langle a_\ell\rangle}$ which is isomorphic to the space $C^\infty(\langle q\rangle \setminus \mathbb{R}^\times)$, we compute (multiplicative) Fourier coefficients $b^+_j(s) = \left\langle \tilde{d}^0_{\beta,s}, |x|^{-\frac{1}{2}-\frac{\lambda}{2}+s_j}\right\rangle_{V_\lambda \simeq L^2(\mathbb{R})}$ where $s_j \in i\mathbb{R}$ are defined by the condition $\left(\pi_\lambda\begin{pmatrix} q_\ell{}^{\frac{1}{2}} & 0 \\ 0 & q_\ell^{-\frac{1}{2}} \end{pmatrix}\right)|x|^{-\frac{1}{2}-\frac{\lambda}{2}+s_j} = |x|^{-\frac{1}{2}-\frac{\lambda}{2}+s_j}$ (namely, $|x|^{-\frac{1}{2}-\frac{\lambda}{2}+s_j}$ are A-equivariant functionals on V_λ which are trivial on $A_\ell = \langle a_\ell\rangle$), i.e., $s_j = \frac{2\pi i}{\ln q_\ell}\cdot j$, $j \in \mathbb{Z}$.

In fact, since $A \simeq \mathbb{R}^\times$ has two connected components, we also have to consider "odd" multiplicative Fourier coefficients given by odd characters of A, i.e., $b^-_j(s) = \left\langle \tilde{d}^0_{\beta,s}, |x|^{-\frac{1}{2}-\frac{\lambda}{2}+s_j}\operatorname{sign}(x)\right\rangle_{V_\lambda \simeq L^2(\mathbb{R})}$.

We claim that coefficients $b^\pm_j(s)$ are meromorphic in s, and for a fixed s outside of poles, are rapidly decreasing as $|j| \to \infty$. This would imply that the vector $d^0_{\alpha,s}$ has a smooth representative in co-invariants (outside of its poles) and hence could be paired with I.

To show that coefficients $b_j^{\pm}(s)$ are meromorphic, we consider first N-terms of the Taylor expansion of the *smooth* function $|x-\beta|^{-\frac{1}{2}-\frac{\lambda}{2}-s}$ at $x=0$. We have

$$F_{0,\beta,s}(x) = |x-\beta|^{-\frac{1}{2}-\frac{\lambda}{2}-s} = a_0(\lambda,s) + a_1(\lambda,s)x + \cdots + O_{s,\lambda,\beta}(|x|^N)\ ,$$

with functions $a_k(\lambda,s)$ which are holomorphic in s, and for fixed s and λ, are rapidly decreasing in k. Hence it is enough to compute scalar products in $V_\lambda \subset L^2(\mathbb{R})$ of the form $\left\langle |x|^{-\frac{1}{2}-\frac{\lambda}{2}+s+n}\tilde{u}_0(x),\, |x|^{-\frac{1}{2}-\frac{\lambda}{2}+s_j}\operatorname{sign}^{\varepsilon}(x)\right\rangle$, where $n \geq 0$, $\varepsilon = 0,1$. These are well-defined for $\Re(s) \gg 0$ (since $\tilde{u}_0$ is supported near 0) and have meromorphic continuation to $\mathbb{C}$ with possible poles appearing when

$$-\frac{1}{2} - \frac{\lambda}{2} + s + n - \frac{1}{2} + \frac{\lambda}{2} - s_j = -1 + n + s - s_j = k$$

is a *negative* integer, e.g., $s = s_j$ and $n = 0$ (see the classic [**GS**]). The residue at such a point is given by the coefficients $\varrho^{\Gamma_\ell}_{s_j,\varepsilon} = \langle I, d^{\Gamma_\ell}_{s_j,\varepsilon}\rangle_{(V_\lambda)_{\Gamma_\ell}}$, where $d^{\Gamma_\ell}_{s_j,\varepsilon}$ is the $A_{\alpha,\overline{\alpha}}$ -equivariant functional $d^{\Gamma_\ell}_{s_j,\varepsilon} : V_\lambda \to \mathbb{C}$, i.e., $\pi_{-\lambda}(a)d^{\Gamma_\ell}_{s_j,\varepsilon} = |a|^{s_j}\operatorname{sign}^{\varepsilon}(a)d^{\Gamma_\ell}_{s_j,\varepsilon}$, for $a \in A_{\alpha,\overline{\alpha}}$, and with the natural meaning of the character $|a|^s\operatorname{sign}^{\varepsilon}(a)$ (recall that $A_{\alpha,\overline{\alpha}} = A_\ell$ is the subgroup which is conjugate to the diagonal subgroup $A = A_{0,\infty}$ and such that $\Gamma \cap A_\ell = \Gamma_\ell$). Coefficients $\varrho^{\Gamma_\ell}_{s_j,\varepsilon}$ are nothing else but (appropriately normalized) twisted (by the character $|a|^{s_j}\operatorname{sign}^{\varepsilon}(a)$) periods of φ along the geodesic ℓ. In fact, since we deal with the group $\mathrm{PGL}_2(\mathbb{R})$ and the corresponding diagonal subgroup A has two connected components, orbits of A on X are coming in pairs of connected orbits or one connected orbit with an automorphism (which could be trivial). Periods $\varrho^{\Gamma_\ell}_{s_j,\varepsilon}$ reflect this structure (this is discussed in detail in [**R2**]).

We note that it is easy to see that coefficients $b_j^{\pm}(s)$ are polynomial in s (and in λ) and this gives polynomial bounds for the analytically continued series $D_{s,\alpha}(\varphi_\lambda)$ with possible poles at $s = s_j - n$ for integer $n \geq 0$ and $s_j = \frac{2\pi i}{\ln q_\ell}\cdot j$, $j \in \mathbb{Z}$.

Remark. For another example of an analytic continuation of a pairing $\langle I, v_s\rangle$ for a singular vector $v_s \in V_\lambda$, see [**R1**]. In that case, singularities of the test vector v_s are at two rational points on $\mathbb{P}^1(\mathbb{R})$, and the analytic continuation is achieved with the help of Hecke operators.

References

[BR1] J. Bernstein and A. Reznikov, *Sobolev norms of automorphic functionals and Fourier coefficients of cusp forms* (English, with English and French summaries), C. R. Acad. Sci. Paris Sér. I Math. **327** (1998), no. 2, 111–116, DOI 10.1016/S0764-4442(98)80001-3. MR1645080

[BR2] J. Bernstein and A. Reznikov, *Sobolev norms of automorphic functionals*, Int. Math. Res. Not. **40** (2002), 2155–2174, DOI 10.1155/S1073792802101139. MR1930758

[C] F. Chamizo, *Automorphic forms and differentiability properties*, Trans. Amer. Math. Soc. **356** (2004), no. 5, 1909–1935, DOI 10.1090/S0002-9947-03-03349-X. MR2031046

[GS] I. M. Gel'fand and G. E. Shilov, *Generalized functions. Vol. I: Properties and operations*, Translated by Eugene Saletan, Academic Press, New York-London, 1964. MR0166596

[Ma] H. Maass, *Über eine neue Art von nichtanalytischen automorphen Funktionen und die Bestimmung Dirichletscher Reihen durch Funktionalgleichungen* (German), Math. Ann. **121** (1949), 141–183, DOI 10.1007/BF01329622. MR0031519

[M] W. Magnus et al., *Formulas and Theorems for the Special Functions*, Springer, 1966.

[MM] Y. I. Manin and M. Marcolli, *Continued fractions, modular symbols, and noncommutative geometry*, Selecta Math. (N.S.) **8** (2002), no. 3, 475–521, DOI 10.1007/s00029-002-8113-3. MR1931172

[Mi] S. D. Miller, *Cancellation in additively twisted sums on* GL(n), Amer. J. Math. **128** (2006), no. 3, 699–729. MR2230922

[MS] S. D. Miller and W. Schmid, *The highly oscillatory behavior of automorphic distributions for* SL(2), Lett. Math. Phys. **69** (2004), 265–286, DOI 10.1007/s11005-004-0470-8. MR2104447

[O] J.-P. Otal, *Sur les fonctions propres du laplacien du disque hyperbolique* (French, with English and French summaries), C. R. Acad. Sci. Paris Sér. I Math. **327** (1998), no. 2, 161–166, DOI 10.1016/S0764-4442(98)80081-5. MR1645108

[P] L. V. Polterovich, *Geodesics on a two-dimensional torus with two rotation numbers* (Russian), Izv. Akad. Nauk SSSR Ser. Mat. **52** (1988), no. 4, 774–787, 895, DOI 10.1070/IM1989v033n01ABEH000816; English transl., Math. USSR-Izv. **33** (1989), no. 1, 101–114. MR966984

[R1] A. Reznikov, *Torus period coefficients on PGL(2) and Dirichlet series*, J. Number Theory **133** (2013), no. 1, 328–342, DOI 10.1016/j.jnt.2012.07.013. MR2981414

[R2] A. Reznikov, *A uniform bound for geodesic periods of eigenfunctions on hyperbolic surfaces*, Forum Math. **27** (2015), no. 3, 1569–1590, DOI 10.1515/forum-2012-0185. MR3341487

[S] W. Schmid, *Automorphic distributions for* $SL(2,\mathbb{R})$, Conférence Moshé Flato 1999, Vol. I (Dijon), Math. Phys. Stud., vol. 21, Kluwer Acad. Publ., Dordrecht, 2000, pp. 345–387. MR1805897

[W] J.-L. Waldspurger, *Sur les valeurs de certaines fonctions L automorphes en leur centre de symétrie* (French), Compositio Math. **54** (1985), no. 2, 173–242. MR783511

Department of Mathematics, Bar-Ilan University, Ramat-Gan 52900, Israel
Email address: `reznikov@math.biu.ac.il`

Proceedings of Symposia in Pure Mathematics
Volume **101**, 2019
https://doi.org/10.1090/pspum/101/01801

On vector-valued twisted conjugation invariant functions on a group; with an appendix by Stephen Donkin

Liang Xiao and Xinwen Zhu

To Joseph Bernstein with deep admiration

Abstract. We study the space of vector-valued (twisted) conjugation invariant functions on a connected reductive group.

Contents

1. Introduction

We first describe the problems we are studying in this article. Let us fix an algebraically closed field k throughout this article. Let G be a connected simply-connected semisimple group over k, and let $k[G]$ denote its ring of regular functions. Consider the conjugation action of G on itself, and let $\boldsymbol{J} = k[G]^{c(G)}$ denote the space of conjugation invariant regular functions on G. More generally, for an algebraic representation V of G, let

$$\boldsymbol{J}(V) := (k[G] \otimes V)^{c(G)} = \{f : G \to V \mid f(gxg^{-1}) = g \cdot f(x), \quad x, g \in G\}$$

denote the $\boldsymbol{J}$-module of V-valued conjugation invariant (algebraic) functions on G, where G acts on $k[G] \otimes V$ diagonally. There are also similarly defined rings of invariants $\boldsymbol{J}_0$ and $\boldsymbol{J}_+$ and the corresponding modules $\boldsymbol{J}_0(V)$ and $\boldsymbol{J}_+(V)$, when G is replaced by its asymptotic cone As_G and the corresponding Vinberg monoid V_G. Our first result is as follows.

L.X. was partially supported by Simons Collaboration grant #278433 and NSF grant DMS–1502147. X.Z. was partially supported by NSF grant DMS–1602092 and the Sloan Fellowship.

THEOREM 1.0.1. *Assume that V admits a good filtration. Then the module $\boldsymbol{J}(V)$ (resp. $\boldsymbol{J}_+(V)$, resp. $\boldsymbol{J}_0(V)$) is finite free over $\boldsymbol{J}$ (resp. $\boldsymbol{J}_+$, resp. $\boldsymbol{J}_0$) of rank* $\dim V(0)$, *where $V(0)$ denotes the zero weight space of V (with respect to a maximal torus of G). In addition, if* $\operatorname{char} k = 0$, *there are natural bases of $\boldsymbol{J}(V)$, $\boldsymbol{J}_+(V)$, and $\boldsymbol{J}_0(V)$ determined by certain bases of V.*

We refer to § 4.4 for detailed discussions regarding to the last statement. This theorem resembles the classical result of Kostant ([**Ko63**]): Assume $\operatorname{char} k = 0$, and let $\mathfrak{g}$ be the Lie algebra of G. Then $(k[\mathfrak{g}] \otimes V)^G$ is a graded free $k[\mathfrak{g}]^G$-module with a basis given by harmonic polynomials. Recall that the study of $(k[\mathfrak{g}] \otimes V)^G$ leads to a natural filtration on the weight spaces (the Kostant–Brylinski filtration), which can be studied either via the geometry of the flag variety (as in [**Br89**]) or via the geometric Satake correspondence (as in [**Gi95**]). Likewise, our proof of the above theorem leads us to define a multi-filtration on the weight spaces (see § 3.3), which can also be studied either via the geometry of the flag variety (§ 3.4) or via the geometric Satake correspondence (§ 3.5). In particular, we obtain a multivariable analogue of the weight multiplicity $P_{\mu,\nu}(q)$ of a representation (see Remark 3.3.3). Note that Kostant's theorem has a very simple proof ([**BL96**]). It would be interesting to see whether there is a similar argument in the group case (at least when $\operatorname{char} k = 0$).

Note that finite freeness of $\boldsymbol{J}(V)$ over $\boldsymbol{J}$ was previously known to Richardson (see [**Ric79**]) when $\operatorname{char} k = 0$ and Donkin (see [**Do88b**] and the appendix) in positive characteristic, under the same assumption on V, but by a different method[1].

Now let V^* be the dual representation of V, which is also assumed to admit a good filtration. Then there is a natural $\boldsymbol{J}$-bilinear pairing

$$\boldsymbol{J}(V) \otimes \boldsymbol{J}(V^*) \to \boldsymbol{J} \tag{1.0.1}$$

induced by pairing V-valued functions with V^*-valued functions. One can show that this pairing is a certain deformation of the natural pairing between $V(0)$ and $V^*(0)$ (see Remark 6.1.3 (2)). Our main result (Theorem 6.1.2) calculates the determinant of this pairing as an element in $\boldsymbol{J}$ up to a unit (or more precisely as a divisor on $\operatorname{Spec} \boldsymbol{J}$.) We choose a maximal torus T of G. Let $\Phi(G,T)$ denote the root system of G. For a root α of G, let e^α denote the corresponding character function on T. For a weight λ of T, let $V(\lambda)$ denote the corresponding weight space. Using the classical Chevalley isomorphism, we identify $\boldsymbol{J} = k[G]^{c(G)}$ with the Weyl group invariant functions on T.

THEOREM 1.0.2. *Assume that* $\operatorname{char} k > 2$. *Then the determinant of the pairing* (1.0.1) *is*

$$c \prod_{\alpha \in \Phi(G,T)} (e^\alpha - 1)^{\zeta_\alpha},$$

where c is a non-zero constant and $\zeta_\alpha = \sum_{n \geq 1} \dim V(n\alpha)$.

As mentioned above, when $\operatorname{char} k = 0$, it is possible to construct bases of $\boldsymbol{J}(V)$ and $\boldsymbol{J}(V^*)$ so the pairing (1.0.1) can be represented by a matrix. We refer to § 6.4 for some examples where the matrices are calculated explicitly. Our main interest in these matrices (and their determinants) lies in the fact that under the Satake

[1] For example, the method in *loc. cit.* requires the flatness of the Chevalley map $G \to G /\!\!/ c(G)$ as an input, whereas our approach could deduce the flatness as a corollary (see Corollary 4.3.3).

isomorphism they exactly correspond to the intersection matrices for certain cycles on the mod p fibers of some Shimura varieties; we refer to [**XZ17**] for more details.

In fact, in this article, we will consider a more general situation. Assume that τ is an automorphism of the algebraic group G. We consider the τ-twisted conjugation action of G on itself

$$c_\tau(h)(g) = hg\tau(h)^{-1}, \quad \text{for } g, h \in G. \tag{1.0.2}$$

This is equivalent to considering the usual conjugation action of G on the coset $G\tau$ inside the semidirect product $G\rtimes\langle\tau\rangle$. We can similarly define $\boldsymbol{J}(V) := (k[G]\otimes V)^{c_\tau(G)}$ as the space of V-valued functions f on G satisfying $f(gx\tau(g)^{-1}) = g\cdot f(x)$, $x, g \in G$, and in particular $\boldsymbol{J} := k[G]^{c_\tau(G)}$. We prove the above mentioned results for general τ (see Theorem 4.3.2 and Theorem 6.1.2 for the general statements). This generality is needed in [**XZ17**] for applications, but will cause some additional complications for this article. Readers who are happy with $\tau = \mathrm{id}$ can skip § 5 and parts of § 4 and § 6.

We expect that results in this article can be generalized to the case of symmetric pairs. In addition, by reformulating the argument purely algebraically, it is likely that they also extend to quantum groups.

We briefly describe the remaining parts of the note. In § 2, we discuss multi-filtered vector spaces and the corresponding Rees construction in some generality. Unlike the classical situation, the dimension of the associated graded vector space might be larger than the dimension of the original vector space. Whether the two dimensions are equal is a subtle question and is closely related to the flatness of certain Rees modules. We give some sufficient conditions for the equality, which might be of independent interest.

In § 3.1, we define the canonical filtration on a G-representations and study the associated Rees modules. We apply these discussions to the Vinberg monoid in § 3.2.

In § 5, we discuss twisted conjugacy classes of G for the action (1.0.2). In particular, we discuss the notion of τ-regular conjugacy class, and study the (twisted) Chevalley map and the Grothendieck–Springer resolution, generalizing some well-known results for $\tau = \mathrm{id}$.

In § 6.3, we study $\boldsymbol{J}(V \otimes V^*)$. It is naturally the space of endomorphisms of a vector bundle $\widetilde{V}$ on the quotient stack $[G/c_\tau(G)]$ for the τ-twisted conjugation action of G on itself, and therefore is a $\boldsymbol{J}$-algebra. This is non-commutative in general. But we will study a commutative subalgebra generated by a "tautological" element in $\gamma_{\mathrm{taut}} \in \boldsymbol{J}(V \otimes V^*)$.

Acknowledgments. Some of the work was finished when both authors visited Beijing International Center of Mathematical Research, we thank the staff members for the hospitality. We thank Xuhua He, George Lusztig, and Jun Yu for useful discussions, Stephen Donkin for useful comments and the referee for carefully reading the early version of this article.

Notations and conventions. Throughout the note, let k denote an algebraically closed field. By a variety over k, we mean a separated, integral scheme of finite type over k. If X is an affine algebraic variety over k, let $k[X]$ denote the ring of regular functions on X. If $\mathfrak{X}$ is an algebraic stack of finite presentation over k, let $\mathrm{Coh}(\mathfrak{X})$ denote the category of coherent sheaves on $\mathfrak{X}$.

For an algebraic group H, let $\mathbb{B}H = [\operatorname{Spec} k/H]$ denote the classifying stack. We identify $\operatorname{Coh}(\mathbb{B}H)$ with the category $\operatorname{Rep}^f(H)$ of finite dimensional representations of H, and use them interchangeably. In particular, the trivial representation, sometimes denoted by $\mathbf{1}$ corresponds to the structure sheaf of $\mathbb{B}H$. If X is an H-space and V is an H-representation, let $V_{[X/H]} := X \times^H V$ be the locally free sheaf on $[X/H]$ via the usual associated construction. Alternatively, it is the pullback of V (as coherent sheaf on $\mathbb{B}H$) along the natural projection $[X/H] \to \mathbb{B}H$.

For a reductive group H acting on an affine variety X over k, we write $X/\!\!/H := \operatorname{Spec} k[X]^H$ for the GIT quotient. There is a natural morphism of stacks $[X/H] \to X/\!\!/H$.

The ind-completion $\operatorname{Rep}(H)$ of $\operatorname{Rep}^f(H)$ is the category of algebraic representations of H. If $H_1 \subset H$ is a closed subgroup, and V an H_1-representation, let

$$\operatorname{ind}_{H_1}^H V := (k[H] \otimes V)^{H_1} = \Gamma(H/H_1, V_{H/H_1})$$

be the induced H-representation. It is the *right* adjoint of the restriction functor $\operatorname{Res}_{H_1}^H$ from H-representations to H_1-representations. If no confusion will likely arise, for an H-representation V, we write V for $\operatorname{Res}_{H_1}^H V$ for simplicity.

For an algebraic group H, let $\operatorname{Dist}(H)$ denote the Hopf algebra of invariant distributions on H. Sometimes, $\operatorname{Dist}(H)$ is also called the hyperalgebra of H.

Let X be a set (or scheme) with an automorphism τ. We denote by X^τ the subset (subscheme) of fixed points. If X is an abelian group, let X_τ denote its group of coinvariants.

Throughout this note, $\mathbb{N}$ stands for the set of *nonnegative* integers.

2. Filtered vector spaces and Rees modules

2.1. Filtered vector spaces. We need to first discuss (multi)-filtered vector spaces and (multi)-graded modules. It seems best to start with the following setup. Let $(S, \leq)$ be a partially ordered set (or sometimes called a poset for simplicity). As usual, we write $s_1 < s_2$ if $s_1 \leq s_2$ but $s_1 \neq s_2$. Recall that S is called *directed* if for $s, s' \in S$, there exists $s'' \in S$ such that $s \leq s''$ and $s' \leq s''$. We say a partially ordered set *poly-directed* if $S = \sqcup_i S_i$ is a disjoint union of directed poset S_i, and if $s \in S_i$ and $s' \in S_j$ are incomparable if $i \neq j$.

Definition 2.1.1. Assume that S is poly-directed. We define an *S-filtered vector space* (or a vector space with an *S-filtration*) to be a vector space M, equipped with a collection of subspace $\{\operatorname{fil}_s M, s \in S\}$ such that $\operatorname{fil}_s M \subset \operatorname{fil}_{s'} M$ if $s \leq s'$, and that

$$M = \bigoplus_{S_i} \bigcup_{s \in S_i} \operatorname{fil}_s M. \tag{2.1.1}$$

Let $\operatorname{Vect}^{S-\mathrm{fil}}$ denote the category of S-filtered vector spaces, with morphisms given by k-linear maps preserving the filtrations. Similarly, we define an *S-graded vector space* to be a vector space M equipped with a direct sum decomposition $M = \oplus_{s \in S} M_s$ indexed by S. Let $\operatorname{Vect}^{S-\mathrm{gr}}$ denote the corresponding category. There is a natural functor

$$\operatorname{gr} : \operatorname{Vect}^{S-\mathrm{fil}} \to \operatorname{Vect}^{S-\mathrm{gr}}, \quad M \mapsto \operatorname{gr}_S M = \bigoplus_{s \in S} \operatorname{gr}_s M, \quad \operatorname{gr}_s M = \operatorname{fil}_s M / \sum_{s' < s} \operatorname{fil}_{s'} M.$$

In the above definition, if s is *minimal* in S, i.e. the set $\{s' \in S \mid s' < s\}$ is empty, then we set $\sum_{s'<s} \mathrm{fil}_{s'} M = 0$ so $\mathrm{gr}_s M = \mathrm{fil}_s M$. If S is clear from the context, we write $\mathrm{gr} M$ for $\mathrm{gr}_S M$ for simplicity.

We say a nonzero element $m \in \mathrm{gr} M$ is of *homogeneous* of degree s if $m \in \mathrm{gr}_s M$. For a homogeneous element m of degree s, a *lifting* of m is an element $\widetilde{m} \in \mathrm{fil}_s M$ whose projection to $\mathrm{gr}_s M$ is m.

EXAMPLE 2.1.2. (1) If $S = \mathbb{N}$ equipped with the natural order, an S-filtered vector space is a vector space M equipped with an increasing filtration $\mathrm{fil}_0 M \subset \mathrm{fil}_1 M \subset \cdots$ in the usual sense such that $M = \cup_s \mathrm{fil}_s M$. In this case, for every non-zero element m, there is a unique s such that $m \in \mathrm{fil}_s M - \mathrm{fil}_{s-1} M$ and its projection to $\mathrm{gr}_s M$ is called the *symbol* of m, denoted by $\overline{m}$.

(2) Assume $S = \mathbb{N}^r$ with the partial order $(s_1, \dots, s_r) \le (s_1', \dots, s_r')$ if $s_j \le s_j'$ for every j. Let M be a vector space equipped with r independent increasing filtrations $\mathrm{fil}_0^j M \subset \mathrm{fil}_1^j M \subset \cdots$ such that $\bigcup_s \mathrm{fil}_s^j M = M$, for each $j = 1, \dots, r$. Then one can define an S-filtration of M as follows: for $s = (s_1, \dots, s_r)$, let $\mathrm{fil}_s M = \cap_j \mathrm{fil}_{s_j}^j M$.

The partially ordered sets we encounter in this article will satisfy the following descending chain condition

$$\text{(DCC)} \qquad \text{Every descending chain } s_0 > s_1 > \cdots \text{ is finite.}$$

It is clear that the condition (DCC) passes to subsets of a partially order subset.

The following lemma is usually referred to as the *graded Nakayama lemma.*

LEMMA 2.1.3. *Let S be a poly-directed poset satisfying* (DCC). *Let M be an S-filtered vector space.*

(1) *If $\mathrm{gr} M = 0$, then $M = 0$.*
(2) *Let $\{m_i\}$ be a set of homogeneous elements that span $\mathrm{gr} M$, and let $\{\widetilde{m_i}\}$ be a set of liftings. Then $\{\widetilde{m_i}\}$ span M. In particular, $\dim M \le \dim \mathrm{gr} M$.*

PROOF. (1) Assume $M \neq 0$. Then there should exist some $s_1 \in S$ such that $\mathrm{fil}_{s_1} M \neq 0$. Since $\mathrm{gr} M = 0$, there should be another $s_2 < s_1$ such $\mathrm{fil}_{s_2} M \neq 0$. In this way, one could produce a descending chain $s_1 > s_2 > \cdots$ of infinite length. Contradiction.

(2) We may assume that S is directed. Let $\widetilde{M} \subset M$ be the subspace spanned by $\{\widetilde{m_i}\}$, and let $M' = M/\widetilde{M}$, equipped with an S-filtration defined as $\mathrm{fil}_s M' = \mathrm{Im}(\mathrm{fil}_s M \to M')$. By construction, $\mathrm{gr}_S M' = 0$. Therefore $M' = 0$ by Part (1). □

REMARK 2.1.4. In the classical situation, i.e. Example 2.1.2 (1), it is always true that $\dim \mathrm{gr} M = \dim M$. One can also show that this is the case if S is as in Example 2.1.2 (2) with $r = 2$. However, the inequality $\dim \mathrm{gr} M \ge \dim M$ could be strict in general. For example, take S as in Example 2.1.2 (2) with $r = 3$ and let M be a 2-dimensional vector space equipped with three filtrations $0 = \mathrm{fil}_0^j M \subset \mathrm{fil}_1^j M \subset \mathrm{fil}_2^j M = M$, $j = 1, 2, 3$ such that $\mathrm{fil}_1^1 M, \mathrm{fil}_1^2 M, \mathrm{fil}_1^3 M$ are three different lines in M. Then if we define the S-filtration on M as in Example 2.1.2 (2), the graded spaces $\mathrm{gr}_{(1,2,2)} M, \mathrm{gr}_{(2,1,2)} M, \mathrm{gr}_{(2,2,1)} M$ are all nontrivial. It then follows that $\dim \mathrm{gr} M = 3 > 2 = \dim M$.

As we shall see soon, whether $\dim M = \dim \mathrm{gr} M$ is closely related to the flatness of certain Rees module. So it would be desirable to have some criteria for the equality. Here is one which is a direct consequence of the above lemma.

COROLLARY 2.1.5. *Let* (S, M) *be as in Lemma* 2.1.3. *If*

(*) *there is a basis* $\{m_i\}$ *of* $\mathrm{gr}M$ *consisting of homogeneous elements (such a basis is called a* homogeneous basis*), and for each* i *a lifting* $\widetilde{m_i}$ *of* m_i*, such that* $\{\widetilde{m_i}\}$ *form a basis of* M*,*

then $\dim M = \dim \mathrm{gr}M$.

Conversely, if $\dim M = \dim \mathrm{gr}M < \infty$*, then for every homogeneous basis of* $\mathrm{gr}M$*, any set of liftings of elements in this basis to elements in* M *form a basis of* M.

A particular situation where the above criterion is applicable is as follows.

LEMMA 2.1.6. *Consider an* S*-filtered module* M *as in Example* 2.1.2 *(2). Assume that there is a basis* $\mathbb{B} = \{m_i\}$ *of* M *such that for each* j*, the corresponding symbols* $\{\overline{m_i}^j\}$ *for the filtration* $\{\mathrm{fil}^j\}$ *form a basis of* $\mathrm{gr}^j M = \oplus_d \mathrm{fil}^j_d M/\mathrm{fil}^j_{d-1} M$. *Then* $\dim M = \dim \mathrm{gr}M$.

PROOF. For $m_i \in \mathbb{B}$, define its multidegree to be $\underline{d}(m_i) := (d_1(m_i), \dots, d_r(m_i)) \in \mathbb{N}^r$ such that with respect to the jth filtration $\mathrm{fil}^j_\bullet M$, $m_i \in \mathrm{fil}^j_{d_j(m_i)} M - \mathrm{fil}^j_{d_j(m_i)-1} M$. Given $\underline{d} = (d_1, \dots, d_r)$, let

$$\mathbb{B}_{\underline{d}} = \{m_i \in \mathbb{B} \mid d_j(m_i) = d_j\}, \quad \mathbb{B}_{\leq \underline{d}} = \sqcup_{\underline{d}' \leq \underline{d}} \mathbb{B}_{\underline{d}'}.$$

Then it is easy to show that $\mathbb{B}_{\leq \underline{d}}$ form a basis of $\mathrm{fil}_{\underline{d}} M$. Indeed, let $m \in \mathrm{fil}_{\underline{d}} M$, and write $m = \sum a_i m_i$ in terms of the basis $\mathbb{B}$. If not every element in $\{\underline{d}(m_i) \mid a_i \neq 0\}$ is $\leq \underline{d}$, we can find some $1 \leq j \leq r$ such that $d'_j := \max\{d_j(m_i) \mid a_i \neq 0\} > d_j$. Then projecting to $\mathrm{gr}^j_{d'_j} M$ gives $0 = \sum a_i \overline{m_i}^j$ for the sum over those m_i with $d_j(m_i) = d'_j$. This contradicts our assumption.

It follows that the projection of elements in $\mathbb{B}_{\underline{d}}$ in $\mathrm{gr}_{\underline{d}} M$ form a basis of $\mathrm{gr}_{\underline{d}} M$. The lemma follows from Corollary 2.1.5. □

We will also use of the following lemma.

LEMMA 2.1.7. *Let* $(S, M, \{\mathrm{fil}_s M\}_{s \in S})$ *be an* S*-filtered vector as above (in particular* S *is poly-directed), and assume that* S *satisfies* (DCC). *Let* $S' \subset S$ *be a subset. Assume for every* $s \in S$*, the set* $\{s' \in S' \mid s \leq s'\}$ *is nonempty and has a unique least element, denoted by* s^h. *Then* S' *with the induced partial order is also poly-directed, and also satisfies (DCC). In addition,* $\dim \mathrm{gr}_{S'} M \leq \dim \mathrm{gr}_S M$*, where* $\mathrm{gr}_{S'} M$ *denotes the associated graded of* M *with respect to the sub-filtration* $\{\mathrm{fil}_s M\}_{s \in S'}$. *In particular, if* $\dim M = \dim \mathrm{gr}_S M$*, then* $\dim M = \dim \mathrm{gr}_{S'} M$.

PROOF. The statement for S' is clear. We prove the inequality. Then last equality follows from it by Lemma 2.1.3.

If $s \in S'$, the s-graded piece in $\mathrm{gr}_S M$ is denote by $\mathrm{gr}_s M$ while the s-graded piece in $\mathrm{gr}_{S'} M$ is denoted by $\mathrm{gr}'_s M$. For $t \in S'$, let $S_t = \{s \in S \mid s^h = t\} \subset S$. Note that this is a partially ordered set with the greatest element t and $S = \sqcup_{t \in S'} S_t$. We can define a filtration on $\mathrm{gr}'_t M$ labeled by S_t as $\mathrm{fil}_s \mathrm{gr}'_t M = \mathrm{Im}(\mathrm{fil}_s M \to \mathrm{gr}'_t M)$. Note that $\mathrm{fil}_s \mathrm{gr}'_t M \subset \mathrm{fil}_{s'} \mathrm{gr}'_t M$ if $s \leq s'$. Let

$$\mathrm{gr}_s \mathrm{gr}'_t M = \mathrm{fil}_s \mathrm{gr}'_t M \Big/ \sum_{s' < s,\, s' \in S_t} \mathrm{fil}_{s'} \mathrm{gr}'_t M.$$

Note that S_t also satisfies (DCC), so by Lemma 2.1.3,

$$\sum_{s\in S_t} \dim \mathrm{gr}_s \mathrm{gr}'_t M \geq \dim \mathrm{gr}'_t M.$$

On the other hand, for a fixed $s \in S_t$, if $s' < s$ and $s'^h \neq t$, then $s'^h < t$ and therefore the image of $\mathrm{fil}_{s'} M \to \mathrm{gr}'_t M$ is zero. It follows that $\mathrm{gr}_s \mathrm{gr}'_t M$ is a quotient of $\mathrm{gr}_s M = \mathrm{fil}_s M / \sum_{s'<s} \mathrm{fil}_{s'} M$. Therefore, it follows that for every $t \in S'$,

$$\sum_{s\in S_t} \dim \mathrm{gr}_s M \geq \sum_{s\in S_t} \dim \mathrm{gr}_s \mathrm{gr}'_t M \geq \dim \mathrm{gr}'_t M,$$

giving $\dim \mathrm{gr}_S M \geq \dim \mathrm{gr}_{S'} M$. □

2.1.8. *Monoid and partially ordered set.* The partially ordered sets we encounter in this paper will mostly arise in the following way. Let Γ be a commutative monoid, and let S be a non-empty set with a Γ-action (where multiplication and the action map will be written additively). Consider the following condition

(Can) The only invertible element in Γ is the unit element (such a monoid is called *sharp*), and the action of Γ on S is free, i.e. for $s \in S$ and $\gamma, \gamma' \in \Gamma$, the identity $s + \gamma = s + \gamma'$ implies $\gamma = \gamma'$.

This condition in particular implies that Γ is *integral*, i.e. the map from Γ to its group completion Γ^{gp} is injective. In addition, there is a well-defined partial order on S defined by $s \leq s'$ if $s' = s + \gamma$ for some $\gamma \in \Gamma$. Note that since $s_1 \leq s_2$ and $s_2 \leq s_1$ will imply $s_1 = s_2$, this is indeed a partial order.

LEMMA 2.1.9. *The set S equipped with the above partial order is poly-directed.*

PROOF. The only property of this partial order we need is the following: if $s \leq s_1$ and $s \leq s_2$, then there is s' such that $s_1 \leq s'$ and $s_2 \leq s'$. Indeed, by definition $s_1 = s + \gamma_1$ and $s = s + \gamma_2$. Then we can choose $s' = s + \gamma_1 + \gamma_2$. We show that any partially ordered set S satisfying this property is poly-directed.

Consider the equivalence relation on S generated by the partial order. It is enough to show that every equivalence class equipped with the induced partial order is directed. But if s, s' are in one equivalence class, then there exists a chain of elements $s = s_0, s_1, \ldots, s_r = s' \in S$ such that either $s_i \leq s_{i+1} \geq s_{i+2}$ or $s_i \geq s_{i+1} \leq s_{i+2}$. By replacing the second relation by $s_i \leq s'_{i+1} \geq s_{i+2}$ repeatedly, one find s'' such that $s \leq s'' \geq s'$. □

2.2. Rees construction. Let Γ be a commutative monoid acting on a set S. Let $R = k[\Gamma]$ be the monoid algebra. For $\gamma \in \Gamma$, the corresponding element in R is denoted by e^γ. Let $I_1 \subset R$ be the ideal generated by $e^\gamma - 1$ for all $\gamma \in \Gamma$, and let $I_0 \subset R$ be the ideal spanned by e^γ for all $0 \neq \gamma \in \Gamma$.

We define an *S-graded R-module* to be a k-vector space N with a direct sum decomposition $N = \oplus_{s\in S} N_s$, such that $e^\gamma N_s \subset N_{s+\gamma}$. They naturally form a category, with morphisms given by R-module homomorphisms preserving the grading, denoted by $R\text{–}\mathrm{Mod}^{S-\mathrm{gr}}$. There is a natural functor

$$R\text{–}\mathrm{Mod}^{S-\mathrm{gr}} \to \mathrm{Vect}^{S-\mathrm{gr}}, \quad N \mapsto N/I_0 N,$$

where $(N/I_0N)_s = N_s / \sum_{\gamma'+s'=s} e^{\gamma'} N_{s'}$. Given a homogeneous element $n \in N/I_0N$ of degree s, *a homogeneous lifting* of n to N is an element $\widetilde{n} \in N_s$ which projects to n.

If (Γ, S) satisfies (Can), there is also a natural functor

$$R\text{–Mod}^{S-\mathrm{gr}} \to \mathrm{Vect}^{S-\mathrm{fil}}, \quad N \mapsto N/I_1N,$$

where we define the S-filtration by $\mathrm{fil}_s(N/I_1N) = \mathrm{Im}(N_s \to N \to N/I_1N)$. The following lemma is another version of the graded Nakayama lemma.

LEMMA 2.2.1. *Let* (Γ, S) *be as above and assume that* (Can) *and* (DCC) *hold for* (Γ, S). *Let* $R = k[\Gamma]$. *Let* N *be an* S*-graded* R*-modules.*

(1) *If* $N/I_0N = 0$, *then* $N = 0$.
(2) *Let* $\{n_i\}$ *be a set of homogeneous elements that span* N/I_0N, *and let* $\{\widetilde{n_i}\}$ *be a set of homogeneous liftings. Then* $\{\widetilde{n_i}\}$ *generates* N *as an* R*-module.*
(3) *Assume that in addition, the images of* $\{\widetilde{n_i}\}$ *in* N/I_1N *form a basis of* N/I_1N. *Then* $N \cong \oplus_i R\widetilde{n_i}$. *In particular,* N *is a free* R*-module.*

PROOF. The proof of Part (1) and (2) is the same as the proof of Lemma 2.1.3. For Part (3), let $\sum r_i\widetilde{n_i} = 0$ be a homogeneous linear relation in N, and write $r_i = a_ie^{\gamma_i}$. Then in N/I_1N, there is a linear relation $\sum a_i\widetilde{n_i} = 0$, which implies all $a_i = 0$. □

We continue to assume that (Γ, S) satisfies (Can). Then the functor $N \mapsto N/I_1N$ as above admits a right adjoint

$$\mathrm{Vect}^{S-\mathrm{fil}} \to R\text{–Mod}^{S-\mathrm{gr}}, \quad M \mapsto R_SM = \bigoplus_{s\in S} \mathrm{fil}_s M,$$

where we define the multiplication $e^\gamma : \mathrm{fil}_s M \to \mathrm{fil}_{s+\gamma} M$ (i.e. the R-module structure) to be the natural inclusion. The functor $M \to R_SM$ sometimes is called the *Rees construction* and R_SM is called the *Rees module*. It is convenient to introduce the formal symbols e^s, $s \in S$ and write elements in the degree s summand in R_SM as me^s for $m \in \mathrm{fil}_s M$. Then the R-module structure on R_SM is given by $e^\gamma \cdot me^s = me^{s+\gamma}$.

We need some criteria of flatness of R_SM over R. Note that if it is the case, then by (2.2.1) $\dim \mathrm{gr}M$ should be equal to $\dim M$ (if both are finite). It turns out that under some mild assumptions on M, this condition is also sufficient. First, note that

$$R_SM/I_0R_SM \cong \mathrm{gr}M, \quad R_SM/I_1R_SM \cong M, \tag{2.2.1}$$

where the second isomorphism makes use of (2.1.1).

LEMMA 2.2.2. *Let* $(\Gamma, S, M, \{\mathrm{fil}_s M\}_{s\in S})$ *be as above and assume that Conditions* (Can) *and* (DCC) *hold for* (Γ, S). *Let* R_SM *be the corresponding Rees module. Assume that Condition* (*) *in Corollary* 2.1.5 *holds. Then* R_SM *is free over* R *with basis* $\{\widetilde{m_i e^{s_i}}\}$. *In particular, if* $\dim \mathrm{gr}M = \dim M < \infty$, *then* R_SM *is free over* R.

PROOF. This follows from Lemma 2.2.1 and (2.2.1). □

2.2.3. *Rees algebra.* Let $(\Gamma, S, M, \{\mathrm{fil}_s M\}_{s\in S})$ be as before, where (Γ, S) satisfy condition (Can). If S itself is a commutative monoid such that the action of Γ on S is induced by a monoid homomorphism $f : \Gamma \to S$ (the action of S on itself is the natural translation), and if M is a (not necessarily commutative) k-algebra, then it makes sense to assume that the filtration $\{\mathrm{fil}_s M\}_{s\in S}$ satisfies the additional condition

(Alg) $k \subset \mathrm{fil}_0 M$ and $\mathrm{fil}_s M \cdot \mathrm{fil}_{s'} M \subset \mathrm{fil}_{s+s'} M$.

In this case, $R_S M$ is naturally a (not necessarily commutative) algebra over R, with the multiplication given by $me^s \cdot m'e^{s'} = mm'e^{s+s'}$, and the map $R \to R_S M$ is given by $e^\gamma \mapsto 1 \cdot e^{f(\gamma)}$. We call it the *Rees algebra* of M.

If in addition, M has a co-algebra structure such that each $\mathrm{fil}_s M$ is a sub-coalgebra, then $R_S M$ is also a coalgebra.

3. Filtration on representations

In this section, let G be a connected reductive group over k. We discuss the canonical filtration on G-representations and the associated Rees modules in § 3.1, and apply discussions to the Vinberg monoid in § 3.2. In § 3.3–3.5, we define and study a multi-filtration on the weight spaces of a representation of G. Just as the Kostant–Brylinski filtration plays an important role in the study of vector-valued invariant functions on $\mathfrak{g}$, this filtration is important for the study of vector-valued invariant functions on the group. We will make use of the following notations and conventions throughout this section.

We fix a maximal torus T of G, contained in a Borel subgroup B. Let U denote the unipotent radical of B. Let $\Phi = \Phi(G,T)$ denote the root system, and $\Delta \subset \Phi$ the set of simple roots with respect to B. For every $\alpha \in \Phi$, we fix an isomorphism $x_\alpha : \mathbb{G}_a \simeq U_\alpha$, where U_α is the root subgroup corresponding to α. The tuple $(G, B, T, \{x_\alpha\}_{\alpha\in\Delta})$ forms a pinning of G.

The Lie algebra $\mathfrak{u}_\alpha$ of U_α is spanned over k by $E_\alpha := \frac{d}{dy_\alpha}$ (here we regard $y_\alpha := x_\alpha^{-1}$ as a coordinate function on U_α). More generally, the algebra of invariant distributions $\mathrm{Dist}(U_\alpha)$ of U_α is spanned over k by $\{E_\alpha^{(n)},\ n \geq 0\}$, where $E_\alpha^{(n)}(y_\alpha^j) = \binom{j}{n} y_\alpha^{j-n}$. In particular, $E_\alpha^{(n)}(y_\alpha^j) = 0$ if $n > j$. Informally, we may think $E_\alpha^{(n)} = E_\alpha^n/n!$.

Let B^- be the opposite Borel with respect to (T, B) and U^- its unipotent radical. Let $N = N_G(T)$ be the normalizer of T in G, and let $W = N/T$ denote the (absolute) Weyl group. Let $w_0 \in W$ be the longest Weyl group element. Let $\mu \mapsto \mu^* := -w_0(\mu)$ be the involution on the character lattice $\mathbb{X}^\bullet(T)$, which preserves the set of dominant weights $\mathbb{X}^\bullet(T)^+$ (with respect to B). When we regard a weight $\nu \in \mathbb{X}^\bullet(T)$ as a regular function on T, we write it as e^ν.

Let Z_G denote the scheme-theoretic center of G. Let T_{ad} be the adjoint torus of G, i.e. the quotient of T by Z_G. Its character lattice $\mathbb{X}^\bullet(T_{\mathrm{ad}})$ is the subgroup of $\mathbb{X}^\bullet(T)$ generated by roots. We write $\mathbb{X}^\bullet(T_{\mathrm{ad}})_{\mathrm{pos}}$ for the monoid of nonnegative integer linear combinations of simple roots in Δ. We consider the partial order $\preceq$ on $\mathbb{X}^\bullet(T)$ induced by the action of $\mathbb{X}^\bullet(T_{\mathrm{ad}})_{\mathrm{pos}}$ in the sense of § 2.1.8, i.e. $\lambda_1 \preceq \lambda_2$ if and only if $\lambda_2 - \lambda_1$ is a nonnegative integral linear combination of simple roots of G.

For a root α, let G_α be the rank one subgroup of G generated by $T, U_\alpha, U_{-\alpha}$. Let $B_\alpha = TU_\alpha$ and $B_\alpha^- = TU_{-\alpha}$ be the pair of opposite Borel subgroups of G_α. We similarly have the partial order $\preceq_\alpha$ on $\mathbb{X}^\bullet(T)$ induced by the action of $\mathbb{Z}_{\geq 0}\alpha$. We say λ is *α-dominant* if $\langle \lambda, \alpha^\vee\rangle \geq 0$, where $\alpha^\vee$ is the coroot corresponding to α. Note that if $0 \preceq_\alpha \lambda$, then λ is α-dominant.

For a weight $\nu \in \mathbb{X}^\bullet(T)$, let k_ν denote the corresponding one-dimensional T-module. For a representation V of T and $\nu \in \mathbb{X}^\bullet(T)$, we write $V(\nu)$ for the ν-weight space, so $V(\nu) \cong \mathrm{Hom}_T(k_\nu, V) \otimes k_\nu$.

3.1. The canonical filtration on G-modules. We first review Weyl and Schur modules. Via inflation, the T-module k_ν can be regarded as a representation of B or B^-. Let

$$\mathtt{S}_\nu := \mathrm{ind}_{B^-}^G k_\nu \cong \mathrm{ind}_B^G k_{w_0(\nu)}$$

be the *Schur module* of highest weight ν, and let

$$\mathtt{W}_\nu := \mathtt{S}_{\nu^*}^*$$

denote the *Weyl module* of highest weight ν. More geometrically, we write

$$\mathcal{O}_{G/B}(\nu) = G \times^B k_\nu, \quad \mathcal{O}_{G/B^-}(\nu) = G \times^{B^-} k_\nu$$

to denote the line bundle on the flag variety. Then

$$\mathtt{S}_\nu = \Gamma(G/B^-, \mathcal{O}(\nu)) = \Gamma(G/B, \mathcal{O}(w_0(\nu))).$$

It is known that $\mathtt{S}_\nu = \mathtt{W}_\nu = 0$ unless ν is dominant.

We call a dominant weight $\omega \in \mathbb{X}^\bullet(T)^+$ *minuscule* if all weights in $\mathtt{S}_\omega$ form a single orbit under the action of the Weyl group W. Note that in this case, the multiplicity of each weight space is one-dimension and $\mathtt{S}_\omega \cong \mathtt{W}_\omega$. The set of minuscule weights is denoted by $\mathtt{Min} \subset \mathbb{X}^\bullet(T)^+$. Note that in our convention, the zero weight is minuscule.

LEMMA 3.1.1. *Let $\mathbb{X}^\bullet(T)^+_{\mathrm{pos}} \subset \mathbb{X}^\bullet(T)$ be the submonoid generated by $\mathbb{X}^\bullet(T_{\mathrm{ad}})_{\mathrm{pos}}$ and $\mathbb{X}^\bullet(T)^+$. Then under the natural action of the monoid $\mathbb{X}^\bullet(T_{\mathrm{ad}})_{\mathrm{pos}}$,*

$$\mathbb{X}^\bullet(T)^+_{\mathrm{pos}} = \bigsqcup_{\omega \in \mathtt{Min}} \big(\omega + \mathbb{X}^\bullet(T_{\mathrm{ad}})_{\mathrm{pos}}\big).$$

In particular, the pair $(\Gamma, S) = (\mathbb{X}^\bullet(T_{\mathrm{ad}})_{\mathrm{pos}}, \mathbb{X}^\bullet(T)^+_{\mathrm{pos}})$ satisfies Conditions (DCC) *and* (Can) *in the previous section.*

PROOF. It is known that the set $\mathtt{Min}$ gives a collection of coset representatives of the quotient $\mathbb{X}^\bullet(T)/\mathbb{X}^\bullet(T_{\mathrm{ad}})$ so that every $\lambda \in \mathbb{X}^\bullet(T)$ can be uniquely written as $\lambda = \gamma_\lambda + \omega_\lambda$ with $\gamma_\lambda \in \mathbb{X}^\bullet(T_{\mathrm{ad}})$ and $\omega_\lambda \in \mathtt{Min}$. In addition, if $\lambda \in \mathbb{X}^\bullet(T)^+$, then $\gamma_\lambda \in \mathbb{X}^\bullet(T_{\mathrm{ad}})_{\mathrm{pos}}$. The lemma then clearly follows. □

For a representation V of G, we define an $\mathbb{X}^\bullet(T)^+_{\mathrm{pos}}$-filtration on V, called *the canonical filtration* of V,[2] as follows. For $\lambda \in \mathbb{X}^\bullet(T)$, we denote by $V_{\preceq\lambda}$ the *maximal* subrepresentation of G such that $V_{\preceq\lambda}(\nu) \neq 0$ implies $\nu \preceq \lambda$. Clearly, $V_{\preceq\lambda} \neq 0$ only if $\lambda \in \mathbb{X}^\bullet(T)^+_{\mathrm{pos}}$. Moreover, the functor $V \mapsto V_{\preceq\lambda}$ is left exact. Therefore, we obtained the quadruple

$$(\Gamma, S, M, \{\mathrm{fil}_s M\}_{s\in S}) = \big(\mathbb{X}^\bullet(T_{\mathrm{ad}})_{\mathrm{pos}}, \mathbb{X}^\bullet(T)^+_{\mathrm{pos}}, V, \{V_{\preceq\lambda}\}_{\lambda\in\mathbb{X}^\bullet(T)^+_{\mathrm{pos}}}\big).$$

Let $R = k[\mathbb{X}^\bullet(T_{\mathrm{ad}})_{\mathrm{pos}}]$, and let $R_{\mathbb{X}^\bullet(T)^+_{\mathrm{pos}}} V$ be the associated Rees module, which is an $\mathbb{X}^\bullet(T)^+_{\mathrm{pos}}$-graded R-module. Note that the functor

$$G\text{–Mod} \to R\text{–Mod}^{\mathbb{X}^\bullet(T)^+_{\mathrm{pos}}-\mathrm{gr}}, \quad V \mapsto R_{\mathbb{X}^\bullet(T)^+_{\mathrm{pos}}} V$$

is left exact.

[2]This is closely related, but not the same as the notion of canonical filtration in [**Ma90**, §3]. In particular, our definition is independent of any choice.

There is an important class of G-modules, whose Rees module associated to the canonical filtration is R-flat. Recall that a *good filtration* of a representation V of G is a filtration of V by G-submodules (in the classical sense as in Example 2.1.2 (1)) whose associated graded are Schur modules. We recall some important properties of this class of representations.

THEOREM 3.1.2. (1) *If V admits a good filtration, then its restriction to every Levi subgroup $M \subset G$ also admits a good filtration (as an M-module).*

(2) *The tensor product of two G-modules that admit a good filtration also admits a good filtration.*

(3) *The following are equivalent.*
- *V admits a good filtration.*
- $\mathrm{Ext}^i(\mathtt{W}_{\nu^*}, V) = \mathrm{H}^i(G, \mathtt{S}_\nu \otimes V) = 0$ *for every dominant weight ν and every $i > 0$.*
- $\mathrm{Ext}^1(\mathtt{W}_{\nu^*}, V) = \mathrm{H}^1(G, \mathtt{S}_\nu \otimes V) = 0$ *for every dominant weight ν.*

(4) *Regard $k[G]$ as a $G\times G$-bimodule via left and right translations. Then $k[G]$ admits a good filtration. For every dominant weight ν, $\mathtt{S}_\nu \otimes \mathtt{S}_{\nu^*}$ appears in the composition factors of any good filtration of $k[G]$ exactly once.*

Part (1) and (2) are due to Mathieu [**Ma90**] (and were already obtained earlier by Donkin [**Do85**] in most cases) and (3) is due to Donkin [**Do81**]. Part (4) is due to Donkin [**Do88a**] and independently Koppinen [**Ko84**]. It follows easily from Part (3) that the number of factors in the successive quotients of a good filtration of V that are isomorphic to $\mathtt{S}_\nu$ is equal to $\dim \mathrm{Hom}_G(\mathtt{W}_\nu, V)$, and therefore is independent of the choice of the good filtration. In particular, if V is finite dimensional

$$\dim V = \sum_\nu \dim \mathrm{Hom}_G(\mathtt{W}_\nu, V) \cdot \dim \mathtt{S}_\nu. \tag{3.1.1}$$

The main result of this subsection is as follows.

PROPOSITION 3.1.3. *Let V be a (finite dimensional) G-module that admit a good filtration. Then the Rees module $R_{\mathbb{X}^\bullet(T)^+_{\mathrm{pos}}} V$ associated to the canonical filtration is a (finite) flat R-module.*

PROOF. As every G-module V that admits a good filtration is the union of finite dimensional G-submodules V_i that admit a good filtration, and as $R_{\mathbb{X}^\bullet(T)^+_{\mathrm{pos}}} V$ is the union of $R_{\mathbb{X}^\bullet(T)^+_{\mathrm{pos}}} V_i$, we may assume that V is finite dimensional. Now by Lemma 2.2.2, it suffices to show that $\dim V = \dim \mathrm{gr} V$. But this follows from (3.1.1) and the following lemma. □

LEMMA 3.1.4. *Let V be a representation of G that admits a good filtration. Let λ be a weight.*

(1) *Both $V_{\preceq\lambda} \subset V$ and $V/V_{\preceq\lambda}$ admit good filtrations.*

(2) *Let $V_{\prec\lambda} = \sum_{\lambda' \preceq \lambda} V_{\preceq\lambda'} \subset V_{\preceq\lambda}$. Then $V_{\prec\lambda}$ admits a good filtration, and $V_{\preceq\lambda}/V_{\prec\lambda}$ is isomorphic to $\mathrm{Hom}_G(\mathtt{W}_\lambda, V) \otimes \mathtt{S}_\lambda$.*

PROOF. This is a special case of [**Do85**, 12.1.6]. We include a proof for completeness. (1) By Theorem 3.1.2 (3), it is enough to show that for every dominant ν, the map

$$\mathrm{Hom}_G(\mathtt{W}_\nu, V) \to \mathrm{Hom}_G(\mathtt{W}_\nu, V/V_{\preceq\lambda}) \tag{3.1.2}$$

is surjective. If $\nu \preceq \lambda$, then $\mathrm{Hom}_G(\mathtt{W}_\nu, V/V_{\preceq\lambda}) = \mathrm{Hom}_G(\mathtt{W}_\nu, (V/V_{\preceq\lambda})_{\preceq\lambda}) = 0$, since $(V/V_{\preceq\lambda})_{\preceq\lambda}$ is evidently zero. Assume $\nu \not\preceq \lambda$. By Frobenius reciprocity, the map (3.1.2) may be identified with the map $\mathrm{Hom}_B(k_\nu, V) \to \mathrm{Hom}_B(k_\nu, V/V_{\preceq\lambda})$. Let $L \subset V(\nu)$ be a line, such that B acts on its image in $V/V_{\preceq\lambda}$ via the character $B \to T \xrightarrow{\nu} \mathbb{G}_m$. Then B acts on L by the same way (as any weights $\succeq \nu$ does not appear in $V_{\preceq\lambda}$). Therefore (3.1.2) is an isomorphism in this case.

(2) The same argument as in the proof of Part (1) shows that

- both $V_{\prec\lambda}$ and $V_{\preceq\lambda}/V_{\prec\lambda}$ admit good filtrations; and that
- the natural map $\mathrm{Hom}_G(\mathtt{W}_\nu, V_{\preceq\lambda}) \to \mathrm{Hom}_G(\mathtt{W}_\nu, V_{\preceq\lambda}/V_{\prec\lambda})$ is zero unless $\nu = \lambda$, in which case, the map is an isomorphism.

Now the isomorphism $V_{\preceq\lambda}/V_{\prec\lambda} \cong \mathrm{Hom}_G(\mathtt{W}_\lambda, V) \otimes \mathtt{S}_\lambda$ follows by combining the isomorphisms

- $V_{\preceq\lambda}/V_{\prec\lambda} \cong (V_{\preceq\lambda}/V_{\prec\lambda})(\lambda) \otimes \mathtt{S}_\lambda$;
- $(V_{\preceq\lambda}/V_{\prec\lambda})(\lambda) \cong \mathrm{Hom}_G(\mathtt{W}_\lambda, V_{\preceq\lambda}/V_{\prec\lambda}) \cong \mathrm{Hom}_G(\mathtt{W}_\lambda, V_{\preceq\lambda}) \cong \mathrm{Hom}_G(\mathtt{W}_\lambda, V)$.

□

We end this subsection with the following results.

LEMMA 3.1.5. *Let V be a G-module that admits a good filtration, and let $\lambda \in \mathbb{X}^\bullet(T)^+_{\mathrm{pos}}$. Then the following sequence is exact*

(3.1.3)
$$0 \to V_{\preceq\lambda-\alpha_1-\cdots-\alpha_r} \to \cdots \to \bigoplus_{i<j} V_{\preceq\lambda-\alpha_i-\alpha_j} \to \bigoplus_i V_{\preceq\lambda-\alpha_i} \to V_{\preceq\lambda} \to V_{\preceq\lambda}/V_{\prec\lambda} \to 0.$$

PROOF. Note that $(e^{\alpha_1}, \ldots, e^{\alpha_r})$ form a regular sequence in R generating the ideal $I_0 \subset R$, and since $R_{\mathbb{X}^\bullet(T)^+_{\mathrm{pos}}} V$ is R-flat, they form a regular sequence in $R_{\mathbb{X}^\bullet(T)^+_{\mathrm{pos}}} V$. The lemma follows from taking the λ-graded piece of the corresponding Koszul complex. □

COROLLARY 3.1.6. *In the exact sequence* (3.1.3), *both the kernel and the image of $\bigoplus_i V_{\preceq\lambda-\alpha_i} \to V_{\preceq\lambda}$ admit good filtrations.*

PROOF. By Theorem 3.1.2 (4) and Lemma 3.1.4, each term in (3.1.3) admits a good filtration. The corollary follows from the criteria for existence of good filtrations in Theorem 3.1.2 (3). □

3.2. Vinberg monoid via the canonical filtration. We apply the previous discussion to the $G \times G$-modules $k[G]$, with the $G \times G$-module structure given by left and right translation of G on itself. We regard a pair of weights (ν_1, ν_2) as a weight of $T \times T$. Then for $\nu \in \mathbb{X}^\bullet(T)$, by regarding $k[G]$ as a $G \times G$-representation, we can define

$$\mathrm{fil}_\nu k[G] := k[G]_{\preceq(\nu^*,\nu)}. \tag{3.2.1}$$

LEMMA 3.2.1. *The above $\mathbb{X}^\bullet(T)^+_{\mathrm{pos}}$ filtration $\{\mathrm{fil}_\mu\, k[G]\}_{\mu\in\mathbb{X}^\bullet(T)^+_{\mathrm{pos}}}$ of $k[G]$ satisfies the following properties.*

(1) $\mathrm{fil}_\nu\, k[G] \subset \mathrm{fil}_{\nu+\lambda}\, k[G]$ *if* $\lambda \in \mathbb{X}^\bullet(T_{\mathrm{ad}})_{\mathrm{pos}}$.
(2) $\mathrm{fil}_\nu\, k[G] \cdot \mathrm{fil}_{\nu'}\, k[G] \subset \mathrm{fil}_{\nu+\nu'}\, k[G]$;
(3) *Each* $\mathrm{fil}_\nu\, k[G]$ *is a sub-coalgebra of* $k[G]$.
(4) $\mathrm{gr}\, k[G] = \oplus_{\nu\in\mathbb{X}^\bullet(T)^+} \mathtt{S}_{\nu^*} \otimes \mathtt{S}_\nu$;

PROOF. Properties (1)–(3) are clear. Property (4) follows from Theorem 3.1.2 (4) and Proposition 3.1.3. □

Write $T_{\mathrm{ad}}^+ = \operatorname{Spec} R$. This is a natural monoid (in fact, the affine space with coordinate function indexed by simple roots, and equipped with coordinate multiplication), containing the adjoint torus T_{ad} of G as the open subset of the group of invertible elements. In particular, R has a coalgebra structure. To avoid possible confusion of notations in later discussions, we write $\bar{e}^\alpha$ (instead of e^α) for the coordinate function on T_{ad}^+ corresponding the simple root α. The comultiplication sends $\bar{e}^\alpha$ to $\bar{e}^\alpha \otimes \bar{e}^\alpha$.

According to the discussion of Rees algebra in § 2.2.3, $R_{\mathbb{X}^\bullet(T)^+_{\mathrm{pos}}} k[G]$ is an R-algebra, and the map $R \to R_{\mathbb{X}^\bullet(T)^+_{\mathrm{pos}}} k[G]$ is also a coalgebra homomorphism. Then

$$V_G := \operatorname{Spec} R_{\mathbb{X}^\bullet(T)^+_{\mathrm{pos}}} k[G]$$

is a monoid, which is usually called the *Vinberg monoid* (at least when G is semisimple and simply-connected, see Remark 3.2.3 below). In addition, it is equipped with a monoid homomorphism

$$\mathfrak{d} : V_G \to T_{\mathrm{ad}}^+, \tag{3.2.2}$$

usually called the *abelianization map*. We pick two distinguished representatives for the open and closed T_{ad}-orbit on T_{ad}^+: $\mathbf{1} = (1, \ldots, 1), \mathbf{0} = (0, \ldots, 0) \in \mathbb{A}^r \cong T_{\mathrm{ad}}^+$. Then by (2.2.1),

$$\mathfrak{d}^{-1}(\mathbf{1}) \cong \operatorname{Spec}(k[G]) = G, \quad \mathfrak{d}^{-1}(\mathbf{0}) \cong \operatorname{Spec}(\operatorname{gr} k[G]) =: \mathrm{As}_G\,.$$

The affine scheme As_G is usually called the *asymptotic cone* of G. We will make use of the following basic facts.

PROPOSITION 3.2.2. (1) *Let $G \times^{Z_G} T$ be the quotient of $G \times T$ by the action of Z_G given by $z \cdot (g, t) = (zg, zt)$. The affine monoid V_G contains the open affine scheme $G \times^{Z_G} T$ as the group of invertible elements, such that the abelianization map* (3.2.2) *extends the natural group homomorphism $G \times^{Z_G} T \to T_{\mathrm{ad}}$, $(g, t) \mapsto (t \mod Z_G)$. In particular, there is a natural $G \times G \times T$-action on V_G, where $G \times G$ acts on V_G by left and right translation and T acts on V_G by multiplication.*
(2) *The map $\mathfrak{d}$ is faithfully flat.*

PROOF. Part (1) is clear. Part (2) follows from Proposition 3.1.3. □

REMARK 3.2.3. The original construction of the Vinberg monoid as in [**Vi95**] (when $\operatorname{char} k = 0$) and in [**Rit01**] (in general) is different. However, it is easy to see that V_G is uniquely characterized by the properties in Proposition 3.2.2 and the fact that $\mathfrak{d}^{-1}(\mathbf{0}) = \mathrm{As}_G$. Since the Vinberg monoid as constructed in *loc. cit.* also satisfies these properties, it must be isomorphic to V_G introduced here.

Let V_T be the closure of $T \times^{Z_G} T \subset G \times^{Z_G} T$ in V_G. The image of the $\mathbb{X}^\bullet(T)^+_{\mathrm{pos}}$-filtration on $k[G]$ under the map $k[G] \to k[T]$ defines an $\mathbb{X}^\bullet(T)^+_{\mathrm{pos}}$-filtration on $k[T]$ given by

$$\mathrm{fil}_\nu k[T] = \bigoplus_{\lambda_{\mathrm{dom}} \preceq \nu} k \cdot e^\lambda, \tag{3.2.3}$$

where for a weight λ, λ_{dom} denotes the unique dominant element in the Weyl group orbit $W\lambda$ of λ. Then the embedding $T\times^{Z_G}T\to V_T$ is given by

$$k[V_T]=\bigoplus_{(\lambda,\nu),\ \nu-\lambda_{\mathrm{dom}}\in\mathbb{X}^\bullet(T_{\mathrm{ad}})_{\mathrm{pos}}} k(e_1^\lambda\otimes e_2^\nu)\subset\bigoplus_{(\lambda,\nu),\ \nu-\lambda\in\mathbb{X}^\bullet(T_{\mathrm{ad}})} k(e_1^\lambda\otimes e_2^\nu)=k[T\times^{Z_G}T],$$

where e_i^μ are the corresponding character functions on the ith factor of $T\times^{Z_G}T$, and the map $V_T\to T_{\mathrm{ad}}^+$ is given by $\bar{e}^\lambda\mapsto 1\otimes e_2^\lambda$ for $\lambda\in\mathbb{X}^\bullet(T_{\mathrm{ad}})_{\mathrm{pos}}$.

3.3. The filtration on the weight spaces. To prepare our study of vector-valued conjugation invariant functions $\boldsymbol{J}_G(V)$, we need to introduce a different filtration on each weight space of a representation V of G. This is not directly related to the filtration we discussed in the previous subsections. Fix a simple root α of G. Let V be a representation of G and let ν be a weight of T. Define a filtration on $V(\nu)$ as follows

$$\mathrm{fil}_i^\alpha V(\nu):=V(\nu)\cap(\mathrm{Res}^G_{G_\alpha}V)_{\preceq_\alpha\nu+i\alpha}. \tag{3.3.1}$$

There are two equivalent descriptions of the filtration. First, we claim that

$$\mathrm{fil}_i^\alpha V(\nu)=\ker\Big(\bigoplus_{j\geq1}E_\alpha^{(i+j)}:V(\nu)\to\bigoplus_{j\geq 1}V(\nu+(i+j)\alpha)\Big), \tag{3.3.2}$$

Indeed, let v be a vector in the right hand side of (3.3.2). Then $\mathrm{Dist}(G_\alpha)v$ is a G_α-module whose weights $\preceq_\alpha \nu+i\alpha$. Conversely, if $v\in V(\nu)\cap V_{\preceq_\alpha\nu+i\alpha}$, then clearly $E_\alpha^{(i+j)}v=0$ for every $j\geq1$. The claim follows.

In addition, recall that if V is a finite dimensional representation of G_α whose weights $\preceq_\alpha\lambda$, then all of its weights $\succeq_\alpha s_\alpha(\lambda)$. It follows that we can also define the filtration as

$$\mathrm{fil}_i^\alpha V(\nu)=\ker\Big(\bigoplus_{j\geq1}F_\alpha^{(\langle\nu,\alpha^\vee\rangle+i+j)}:V(\nu)\to\bigoplus_{j\geq1}V\big(\nu-(\langle\nu,\alpha^\vee\rangle+i+j)\alpha\big)\Big). \tag{3.3.3}$$

Next we define the multi-filtration we need. Let T_{sc} denote the preimage of T in the simply-connected cover G_{sc} of G, and $\Gamma=S=\mathbb{X}^\bullet(T_{\mathrm{sc}})^+$ be the monoid of dominant weights, which acts on itself by translations. We identify

$$\mathbb{X}^\bullet(T_{\mathrm{sc}})^+\cong\mathbb{N}^\Delta,\quad \mu\mapsto(\langle\mu,\alpha^\vee\rangle)_{\alpha\in\Delta}. \tag{3.3.4}$$

Under this identification, the partial order on $\mathbb{X}^\bullet(T_{\mathrm{sc}})^+$ induced by the translation action (as in § 2.1.8) is just the standard partial order on $\mathbb{N}^\Delta$, as in Example 2.1.2 (2). (Note that this is different from the restriction to $\mathbb{X}^\bullet(T_{\mathrm{sc}})^+$ of partial order $\preceq$ on $\mathbb{X}^\bullet(T_{\mathrm{sc}})$ induced by the action of $\mathbb{X}^\bullet(T_{\mathrm{ad}})_{\mathrm{pos}}$.) We define an $\mathbb{X}^\bullet(T_{\mathrm{sc}})^+$-filtration on $V(\nu)$ as in Example 2.1.2 (2), i.e.

$$\mathrm{fil}_\lambda V(\nu)=\bigcap_{\alpha\in\Delta}\mathrm{fil}^\alpha_{\langle\lambda,\alpha^\vee\rangle}V(\nu).$$

We obtain a quadruple $(\Gamma,S,M,\{\mathrm{fil}_sM\}_{s\in S})=\big(\mathbb{X}^\bullet(T_{\mathrm{sc}})^+,\mathbb{X}^\bullet(T_{\mathrm{sc}})^+,V(\nu),\{\mathrm{fil}_\lambda V(\nu)\}_{\lambda\in\mathbb{X}^\bullet(T_{\mathrm{sc}})^+}\big)$. The main result of this section is

THEOREM 3.3.1. *Assume that V admits a good filtration, then* $\dim\mathrm{gr}V(\nu)=\dim V(\nu)$.

We prove this theorem here, assuming two ingredients that will be established in § 3.4–§ 3.5.

PROOF. We first reduce to the case when V is a Schur module. Indeed, suppose we have a short exact sequence of G-modules

$$0 \to V' \to V \to V'' \to 0.$$

We deduce easily from (3.3.2) or (3.3.3) an exact sequence

$$0 \to \mathrm{fil}_\lambda V'(\nu) \to \mathrm{fil}_\lambda V(\nu) \to \mathrm{fil}_\lambda V''(\nu).$$

In addition, if V' admits a good filtration, by Theorem 3.1.2 and Proposition 3.4.1 below, this sequence is also exact on the right.

It follows that it is enough to prove the theorem for a Schur module V. In this case, by Lemma 2.1.6, it is a consequence of the following proposition. □

PROPOSITION 3.3.2. *Let V be a Schur module. Then there exists a basis $\{v_j\}$ of $V(\nu)$ such that for every simple root α, the corresponding symbols $\{\overline{v_j}\}$ for the filtration fil^α form a basis of the associated graded $\mathrm{gr}^\alpha V(\nu) = \oplus_i \mathrm{fil}^\alpha_i V(\nu)/\mathrm{fil}^\alpha_{i-1} V(\nu)$.*

There are several ways to construct such a basis. For example, the canonical basis constructed by Lusztig and Kashiwara, or the semi-canonical basis constructed by Lusztig satisfies the required properties in the proposition (see [**Lu00**, Theorem 3.1 and Corollary 3.9]). At the end of § 3.5, we will give an alternative construction of a basis with the needed properties using the MV basis from the geometric Satake correspondence.[3]

REMARK 3.3.3. For $\mu, \nu \in \mathbb{X}^\bullet(T)$ with μ dominant, we define

$$P_{\mu,\nu}(q) := \sum_{\lambda \in \mathbb{X}^\bullet(T_{\mathrm{sc}})^+} \dim \mathrm{gr}_\lambda \mathsf{S}_\mu(\nu) q^\lambda \in \mathbb{Z}[\mathbb{X}^\bullet(T_{\mathrm{sc}})^+]$$

as an element in the monoid algebra for $\mathbb{X}^\bullet(T_{\mathrm{sc}})^+$. (Here we use q^λ instead of e^λ to denote the element in $\mathbb{Z}[\mathbb{X}^\bullet(T_{\mathrm{sc}})^+]$ corresponding to λ.) If we identify $\mathbb{X}^\bullet(T_{\mathrm{sc}})^+$ with $\mathbb{N}^l$ as above, it becomes to a polynomial of l-variables

$$P_{\mu,\nu}(q_1, \ldots, q_l) = \sum \dim \mathrm{gr}_{(s_1,\ldots,s_l)} \mathsf{S}_\mu(\nu) q_1^{s_1} \cdots q_l^{s_l}.$$

Then Theorem 3.3.1 implies that $P_{\mu,\nu}(1) = \dim \mathsf{S}_\mu(\nu)$. This may be viewed as a multivariable analogue of the Lusztig–Kato polynomials (which give the q-analogue of the weight multiplicities [**Lu83**]).

3.4. The filtration via Borel–Weil. We will explain how to obtain the filtration defined in § 3.3 using the geometry of flag varieties.

Recall that by Frobenius reciprocity, there is a canonical morphism of B^--modules $\mathsf{S}_\mu \to k_\mu$, and a canonical morphism of B-modules $\mathsf{S}_\mu \to k_{w_0(\mu)}$.

Let μ, ν be two dominant weights. Let V be a representation of G. The map $\mathsf{S}_{\mu^*} \otimes \mathsf{S}_\nu \otimes V \to k_{-\mu} \otimes k_\nu \otimes V$ induces a natural map

$$\ell_{\mu,\nu} : (\mathsf{S}_{\mu^*} \otimes \mathsf{S}_\nu \otimes V)^G \to (k_{-\mu} \otimes k_\nu \otimes V)^T \cong V(\mu - \nu).$$

Note that the collection of the maps $\{\ell_{\mu,\nu}\}$ has the following property. For a dominant weight η, there is a canonical G-equivariant map in $\mathrm{Hom}_G(\mathtt{W}_\eta, \mathsf{S}_\eta)$, which gives a G-invariant element $\sigma_\eta \in \mathsf{S}_{\eta^*} \otimes \mathsf{S}_\eta$. Multiplying by σ_η induces

$$(\mathsf{S}_{\mu^*} \otimes \mathsf{S}_\nu \otimes V)^G \to (\mathsf{S}_{\mu^*+\eta^*} \otimes \mathsf{S}_{\nu+\eta} \otimes V)^G.$$

[3]This is the set of basis we use in [**XZ17**].

We denote this map still by σ_η. Then it is clear that

$$\ell_{\mu+\eta,\nu+\eta}\circ\sigma_\eta=\ell_{\mu,\nu}. \tag{3.4.1}$$

This subsection is devoted to proving the following.

PROPOSITION 3.4.1. *The map $\ell_{\mu,\nu}$ above induces an isomorphism*

$$(\mathtt{S}_{\mu^*}\otimes\mathtt{S}_\nu\otimes V)^G\cong \mathrm{fil}_\nu V(\mu-\nu)\subset V(\mu-\nu).$$

Here by abuse of notations, the image of ν under the map $\mathbb{X}^\bullet(T)\to\mathbb{X}^\bullet(T_{\mathrm{sc}})$ is still denoted by ν.

We will later in § 3.4.4 deduce this proposition from the following natural exact sequence of B-modules

$$0\to\mathtt{S}_\nu\to\mathrm{ind}_T^B k_\nu\to\bigoplus_\alpha \mathrm{ind}_{B_\alpha}^B \mathtt{M}^\alpha_{s_\alpha(\nu)-\alpha}. \tag{3.4.2}$$

Here, for a weight $\lambda\in\mathbb{X}^\bullet(T)$,

$$\mathtt{M}^\alpha_\lambda:=\mathrm{Dist}(G_\alpha)\otimes_{\mathrm{Dist}(B_\alpha)}k_\lambda$$

is the *restricted Verma module* of G_α of highest weight λ. The sequence (3.4.2) is in fact the first two terms of the (restricted) dual BGG complex. Since this sequence in the above form (and in characteristic $p>0$) might be not familiar to some readers, we give a self-contained construction.

3.4.2. *Case of* SL_2. First, we review some facts about representations of $G=\mathrm{SL}_2$. Let B (resp. B^-) be the subgroup of upper (resp. lower) triangular matrices in SL_2, and let $T=B\cap B^-\cong\mathbb{G}_m$ be the group of diagonal matrices. We identify $G/B^-=\mathbb{P}^1$ in the way such that

- $j:\mathbb{A}^1\to\mathbb{P}^1$ corresponds to the open B-orbit and $i:\{\infty\}\to\mathbb{P}^1$ corresponds to the closed B-orbit,
- $0\in\mathbb{A}^1$ is fixed by B^-.

For $n\in\mathbb{N}$, consider the exact sequence of sheaves

$$0\to\mathcal{O}(n)\to j_*j^*\mathcal{O}(n)\to j_*j^*\mathcal{O}(n)/\mathcal{O}(n)\to 0 \tag{3.4.3}$$

on $\mathbb{P}^1$. Since the sequence is B-equivariant, we may also regard it as an exact sequence on $[B\backslash G/B^-]$ via descent.

We fix two nonzero sections t_0 and t_∞ of $\mathcal{O}(1)$ that vanish at 0 and ∞ respectively and view $x=t_0/t_\infty$ as a coordinate function on $\mathbb{A}^1$. Then

$$\Gamma(\mathbb{A}^1,\mathcal{O}(n))=k[x]t_\infty^n\cong k[x]=:\mathtt{M}_n^\vee$$

is a $(\mathrm{Dist}(G),B)$-module with highest weight n, on which e acts as $\frac{d}{dx}$, h acts as $n-2x\frac{d}{dx}$ and f acts as $x(n-x\frac{d}{dx})$. Note that as a $\mathrm{Dist}(G)$-module, it is isomorphic to the restricted dual Verma module, and as a B-module, it is isomorphic to the induced representation $\mathrm{ind}_T^B k_n$. The subspace

$$\Gamma(\mathbb{P}^1,\mathcal{O}(n))\cong\{f(x)\in k[x]\mid \deg f\le n\}:=\mathtt{S}_n$$

is the Schur module for SL_2 of highest weight n. The section of the quotient sheaf is

$$\Gamma(\mathbb{P}^1,j_*j^*\mathcal{O}(n)/\mathcal{O}(n))\cong x^{n+1}k[x]=:\mathtt{M}_{-n-2},$$

which as a $\mathrm{Dist}(G)$-module is isomorphic to the restricted Verma module for SL_2 of highest weight $-n-2$.

For a B-module V, write $V = \oplus V(j)$ for the weight decomposition with respect to T, then

$$(V \otimes \mathtt{M}_n^\vee)^B = \Big\{ \sum_i (-1)^i e^{(i)} v \otimes x^i \mid v \in V(-n) \Big\} \cong V(-n).$$

It follows that the following diagram is commutative with horizontal sequences exact

$$\begin{array}{ccccccc} 0 & \longrightarrow & (V \otimes \mathtt{S}_n)^B & \longrightarrow & (V \otimes \mathtt{M}_n^\vee)^B & \longrightarrow & (V \otimes \mathtt{M}_{-n-2})^B \\ & & \| & & \downarrow{\cong} & & \downarrow \\ 0 & \longrightarrow & (V \otimes \mathtt{S}_n)^B & \longrightarrow & V(-n) & \xrightarrow{\oplus_{i\geq 1}(-1)^{n+i} e^{(n+i)}} & \bigoplus_{i\geq 1} V(n+2i), \end{array} \tag{3.4.4}$$

where the right vertical map is the inclusion

$$(V \otimes \mathtt{M}_{-n-2})^B \hookrightarrow (V \otimes \mathtt{M}_{-n-2})^T \cong \bigoplus_{i\geq 1} V(n+2i).$$

Note that the above discussions may equally apply to G_α, with n replaced by a weight ν of T such that $\langle \nu, \alpha^\vee \rangle \geq 0$. The corresponding $(\mathrm{Dist}(G_\alpha), B_\alpha)$-modules $\mathtt{M}_n^\vee$, $\mathtt{S}_n$, and $\mathtt{M}_{-n-2}$ will be denoted by $\mathtt{M}_\nu^{\alpha,\vee}$, $\mathtt{S}_\nu^\alpha$, and $\mathtt{M}^\alpha_{s_\alpha(\nu)-\alpha}$, respectively.

3.4.3. *General case.* Recall that the B-orbits on G/B^- are parameterized by the Weyl group W. For $w \in W$, let C_w denote the corresponding B-orbit through $\dot{w} \in G/B^-$, where $\dot{w}$ is any lifting of w to $N_G(T)$. In particular, C_e is open and isomorphic to B/T, and C_{s_α}'s are of codimension one, where s_α is the simple reflection corresponding to the simple root α. In addition, the natural map $B \times^{B_\alpha} (G_\alpha/B_\alpha^-) \to G/B^-$ is an open embedding, with the image $C_{\leq s_\alpha} = C_e \sqcup C_{s_\alpha}$. Let $C_\Delta = \cup_{\alpha\in\Delta} C_{\leq s_\alpha}$ be the open subset of G/B^- complement to the union of B-orbits of codimension at least two. The inclusion $j : C_e \to C_\Delta$ is open and the inclusion $i : \sqcup C_\alpha \to C_\Delta$ is closed. For simplicity, the restriction of $\mathcal{O}_{G/B^-}(\nu)$ to C_Δ is denoted by $\mathcal{O}(\nu)$. Consider the following exact sequence of B-equivariant quasi-coherent sheaves on C_Δ

$$0 \to \mathcal{O}(\nu) \to j_* j^* \mathcal{O}(\nu) \to j_* j^* \mathcal{O}(\nu)/\mathcal{O}(\nu) \to 0. \tag{3.4.5}$$

Note that $C_e \cong B/T$, and $\mathcal{O}(\nu)|_{C_e} \cong B \times^T k_\nu$. In addition, the restriction of the map $j_* j^* \mathcal{O}(\nu) \to j_* j^* \mathcal{O}(\nu)/\mathcal{O}(\nu)$ to $C_{\leq s_\alpha}$ is the pullback of (3.4.3) (with n replaced by ν) along the natural projection $C_{\leq s_\alpha} \cong B \times^{B_\alpha} (G_\alpha/B_\alpha^-) \to [B_\alpha \backslash G_\alpha/B_\alpha^-]$. It follows from these observations and the previous discussions about SL_2 that (3.4.2) is obtained by taking the global sections of (3.4.5).

3.4.4. *Proof of Proposition* 3.4.1. We may assume that V is finite dimensional. Using Frobenius reciprocity, we have

$$(\mathtt{S}_{\mu^*} \otimes \mathtt{S}_\nu \otimes V)^G = \mathrm{Hom}_G(\mathtt{W}_{\nu^*}, \mathtt{S}_{\mu^*} \otimes V) = \mathrm{Hom}_B(\mathtt{W}_{\nu^*}, k_{-\mu} \otimes V) = \mathrm{Hom}_B(k, \mathtt{S}_\nu \otimes k_{-\mu} \otimes V).$$

Then by (3.4.2), there is an exact sequence

$$0 \to \mathrm{Hom}_B(k, \mathtt{S}_\nu \otimes k_{-\mu} \otimes V) \to (k_\nu \otimes k_{-\mu} \otimes V)^T \to \bigoplus_\alpha (\mathtt{M}^\alpha_{s_\alpha(\nu)-\alpha} \otimes k_{-\mu} \otimes V)^{B_\alpha}.$$

It is easy to check that the resulting map

$$(\mathtt{S}_{\mu^*} \otimes \mathtt{S}_\nu \otimes V)^G = \mathrm{Hom}_B(k, \mathtt{S}_\nu \otimes k_{-\mu} \otimes V) \to (k_\nu \otimes k_{-\mu} \otimes V)^T$$

is $\ell_{\mu,\nu}$. In addition, using (3.4.4) and (3.3.2), we see that the kernel of the map

$$V(\mu-\nu) = (k_\nu \otimes k_{-\mu} \otimes V)^T \to (\mathtt{M}^\alpha_{s_\alpha(\nu)-\alpha} \otimes k_{-\mu} \otimes V)^{B_\alpha}$$

is exactly $\mathrm{fil}^\alpha_{\langle \nu, \alpha^\vee\rangle} V(\mu-\nu)$. Proposition 3.4.1 now follows from the definition of the $\mathbb{X}^\bullet(T)^+$-filtration on $V(\mu-\nu)$.

3.5. The filtration via geometric Satake correspondence. We give a geometric construction of the above filtrations via the geometric Satake correspondence and in particular give a proof of Proposition 3.3.2.

We refer to [**Gi95, MV07**] for the geometric Satake correspondence (see also [**Zhu17, BR17**] for an exposition). Let $\hat{G}$ be the Langlands dual group of G over $\mathbb{C}$. Let $\mathrm{Gr}_{\hat{G}} = L\hat{G}/L^+\hat{G}$ denote its affine Grassmannian over $\mathbb{C}$, equipped with the analytic topology. Here for an affine variety Z over $\mathbb{C}$, let LZ (resp. L^+Z) denote its loop (resp. jet) space as usual, so $LZ(\mathbb{C}) = Z(\mathbb{C}((t)))$ and $L^+Z(\mathbb{C}) = Z(\mathbb{C}[[t]])$. Let $\mathrm{P}_{L^+\hat{G}}(\mathrm{Gr}_{\hat{G}})$ be the category of $L^+\hat{G}$-equivariant perverse sheaves on $\mathrm{Gr}_{\hat{G}}$, with k-coefficients as in [**MV07**]. It is known from *loc. cit.* that this is an abelian tensor category. Recall that the geometric Satake correspondence is a natural equivalence of abelian tensor categories

$$\mathrm{Sat} : \mathrm{Rep}^f(G) \to \mathrm{P}_{L^+\hat{G}}(\mathrm{Gr}_{\hat{G}}),$$

such that its composition with hypercohomology functor $\mathrm{H}^*(\mathrm{Gr}_{\hat{G}}, -)$ is isomorphic to the forgetful functor from $\mathrm{Rep}^f(G)$ to the category of finite dimensional k-vector spaces. Let us also recall from *loc. cit.* the dictionary between some geometry and representation theory under this equivalence. For a weight $\nu \in \mathbb{X}^\bullet(T) = \mathbb{X}_\bullet(\hat{T})$, let t^ν denote the $\hat{T}$-fixed point of $\mathrm{Gr}_{\hat{G}}$ corresponding to ν as usual, i.e. it is the image of $t \in \mathbb{C}((t))^\times = L\mathbb{G}_m(\mathbb{C})$ under the map $L\mathbb{G}_m \xrightarrow{\nu} L\hat{T} \to L\hat{G} \to \mathrm{Gr}_{\hat{G}}$.

(1) For each $\lambda \in \mathbb{X}^\bullet(T)^+$, let $\mathring{\mathrm{Gr}}_{\hat{G},\lambda}$ denote the $L^+\hat{G}$-orbit through t^λ, and $\mathrm{Gr}_{\hat{G},\lambda}$ its closure. They are $\langle 2\check{\rho}, \lambda\rangle$-dimensional, where $2\check{\rho}$ is the sum of positive coroots of G. Let $i_\lambda : \mathrm{Gr}_{\hat{G},\lambda} \to \mathrm{Gr}_{\hat{G}}$ denote the closed embedding and $\mathring{i}_\lambda : \mathring{\mathrm{Gr}}_{\hat{G},\lambda} \to \mathrm{Gr}_{\hat{G}}$ the locally closed embedding. Let $k[\langle 2\check{\rho}, \lambda\rangle]$ denote the constant sheaf on $\mathring{\mathrm{Gr}}_{\hat{G},\lambda}$, shifted to degree $-\langle 2\check{\rho}, \lambda\rangle$. Then

$$\mathrm{Sat}(\mathtt{W}_\lambda) \simeq {}^p\mathrm{H}^0(\mathring{i}_\lambda)_! k[\langle 2\check{\rho}, \lambda\rangle], \quad \mathrm{Sat}(\mathtt{S}_\lambda) \simeq {}^p\mathrm{H}^0(\mathring{i}_\lambda)_* k[\langle 2\check{\rho}, \lambda\rangle].$$

(2) We fix a Borel subgroup $\hat{B}$ of $\hat{G}$, and a maximal torus $\hat{T} \subset \hat{B}$. Let $\hat{U} \subset \hat{B}$ denote the unipotent radical. Let S_ν be the $L\hat{U}$-orbit through t^ν. Then the functor

$$F_\nu := \mathrm{H}^{\langle 2\check{\rho}, \nu\rangle}_{S_\nu}(\mathrm{Gr}_{\hat{G}}, -) : D^b_c(\mathrm{Gr}_{\hat{G}}, k) \to \mathrm{Vect}_k \tag{3.5.1}$$

is exact when restricted to $\mathrm{P}_{L^+\hat{G}}(\mathrm{Gr}_{\hat{G}})$, which corresponds to the weight functor $V \mapsto V(\nu)$ under the geometric Satake correspondence.

(3) For a simple root α of G, let $\hat{P}_\alpha$ be the standard parabolic subgroup whose Levi quotient $\hat{G}_\alpha$ is the Langlands dual group of G_α. There is the

following diagram

$$\begin{array}{ccc} & \mathrm{Gr}_{\hat{P}_\alpha} & \\ {}^{q_\alpha}\swarrow & & \searrow^{i_\alpha} \\ \mathrm{Gr}_{\hat{G}_\alpha} & & \mathrm{Gr}_{\hat{G}}. \end{array}$$

The morphism i_α is a locally closed embedding and hence we also regard S_ν as a subscheme of $\mathrm{Gr}_{\hat{P}_\alpha}$. The connected components of the affine Grassmannian $\mathrm{Gr}_{\hat{G}_\alpha}$ are parameterized by $\mathbb{X}^\bullet(T)/\mathbb{Z}\alpha$. For $\theta \in \mathbb{X}^\bullet(T)/\mathbb{Z}\alpha$, let $\mathrm{Gr}^\theta_{\hat{G}_\alpha}$ be the corresponding component, and $\mathrm{Gr}^\theta_{\hat{P}_\alpha} = q_\alpha^{-1}(\mathrm{Gr}^\theta_{\hat{G}_\alpha})$. The restrictions of i_α and q_α to $\mathrm{Gr}^\theta_{\hat{P}_\alpha}$ are denoted by i^θ_α and q^θ_α respectively. Recall that $\langle 2\check{\rho}, \alpha\rangle = \langle \check{\alpha}, \alpha\rangle$ and therefore $\langle 2\check{\rho} - \check{\alpha}, \theta\rangle$ makes sense, which we denote by l_θ. Then there is a perverse exact functor

$$\mathrm{CT}_\alpha := \bigoplus_\theta (q^\theta_\alpha)_*(i^\theta_\alpha)^![l_\theta] : \mathrm{P}_{L^+\hat{G}}(\mathrm{Gr}_{\hat{G}}) \to \mathrm{P}_{L^+\hat{G}_\alpha}(\mathrm{Gr}_{\hat{G}_\alpha})$$

which corresponds under the geometric Satake correspondence to the restriction functor from G-representations to G_α-representations. For a weight ν, let $S^\alpha_\nu \subset \mathrm{Gr}_{\hat{G}_\alpha}$ be the $L\hat{U}_\alpha$-orbit through t^ν, and $F^\alpha_\nu = \mathrm{H}^{\langle\check{\alpha},\nu\rangle}_{S^\alpha_\nu}(\mathrm{Gr}_{\hat{G}_\alpha}, -)$ the corresponding weight functor on $\mathrm{P}_{L^+\hat{G}_\alpha}(\mathrm{Gr}_{\hat{G}_\alpha})$. Since $q_\alpha^{-1}(S^\alpha_\nu) = S_\nu$, the proper base change theorem implies that there is a canonical isomorphism $F_\nu \cong F^\alpha_\nu \circ \mathrm{CT}_\alpha$, which corresponds to the natural identification $V(\nu) = (\mathrm{Res}^G_{G_\alpha} V)(\nu)$ under the geometric Satake correspondence.

Now we give a geometric construction of the filtration defined in § 3.3. First, we define a filtration on S_ν by open subsets

$$S^{>\nu+i\alpha}_\nu := q_\alpha^{-1}(S^\alpha_\nu \cap (\mathrm{Gr}_{\hat{G}_\alpha} - \mathrm{Gr}_{\hat{G}_\alpha,\nu+i\alpha})).$$

Proposition 3.5.1. *Let V be a finite dimensional representation of G. Then we have an exact sequence*

$$0 \to \mathrm{fil}^\alpha_i V(\nu) \to \mathrm{H}^{\langle 2\check{\rho},\nu\rangle}_{S_\nu}(\mathrm{Gr}_{\hat{G}}, \mathrm{Sat}(V)) \to \mathrm{H}^{\langle 2\check{\rho},\nu\rangle}_{S^{>\nu+i\alpha}_\nu}(\mathrm{Gr}_{\hat{G}}, \mathrm{Sat}(V)).$$

Remark 3.5.2. It will follow from the arguments in § 3.5.4 that if V is a Schur module (or more generally V admits a good filtration), the above sequence is also surjective at the right.

Proof. For an α-dominant weight ν of G, we denote the closed embedding $\mathrm{Gr}_{\hat{G}_\alpha,\nu} \to \mathrm{Gr}_{\hat{G}_\alpha}$ by i^α_ν, and the complementary open embedding by j^α_ν. Then for $\mathcal{A} \in \mathrm{P}_{L^+\hat{G}}(\mathrm{Gr}_{\hat{G}})$, there is the following distinguished triangle of sheaves on $\mathrm{Gr}_{\hat{G}_\alpha}$

$$(i^\alpha_{\nu+i\alpha})_*(i^\alpha_{\nu+i\alpha})^! \,\mathrm{CT}_\alpha(\mathcal{A}) \to \mathrm{CT}_\alpha(\mathcal{A}) \to (j^\alpha_{\nu+i\alpha})_*(j^\alpha_{\nu+i\alpha})^* \,\mathrm{CT}_\alpha(\mathcal{A}) \to .$$

Applying the functor F^α_ν as in (3.5.1), and noticing by the proper base change,

$$F^\alpha_\nu((j^\alpha_{\nu+i\alpha})_*(j^\alpha_{\nu+i\alpha})^* \,\mathrm{CT}_\alpha(\mathcal{A})) = \mathrm{H}^{\langle 2\check{\rho},\nu\rangle}_{S^{>\nu+i\alpha}_\nu}(\mathrm{Gr}_{\hat{G}}, \mathcal{A})),$$

we obtain

$$0 \to F^\alpha_\nu(i^\alpha_{\nu+i\alpha})_* {}^p\mathrm{H}^0(i^\alpha_{\nu+i\alpha})^! \,\mathrm{CT}_\alpha(\mathcal{A}) \to \mathrm{H}^{\langle 2\check{\rho},\nu\rangle}_{S_\nu}(\mathrm{Gr}_{\hat{G}}, \mathcal{A}) \to \mathrm{H}^{\langle 2\check{\rho},\nu\rangle}_{S^{>\nu+i\alpha}_\nu}(\mathrm{Gr}_{\hat{G}}, \mathcal{A}).$$

The injectivity at the left follows from the fact that $(j^\alpha_{\nu+i\alpha})_*(j^\alpha_{\nu+i\alpha})^*\,\mathrm{CT}_\alpha(\mathcal{A})$ lives in perverse cohomological degree ≥ 0 and the exactness of F^α_ν. Now let $\mathcal{A} = \mathrm{Sat}(V)$. To conclude the proof, we apply the following lemma (which follows from the description of the Weyl modules under the geometric Satake correspondence as in (1) above) and the definition of the filtration (3.3.1) to replace $F^\alpha_\nu(i^\alpha_{\nu+i\alpha})_*{}^p\,\mathrm{H}^0(i^\alpha_{\nu+i\alpha})^!\,\mathrm{CT}_\alpha(\mathrm{Sat}(V))$ by $\mathrm{fil}^\alpha_i V(\nu)$. □

LEMMA 3.5.3. *Let λ be a dominant weight of G. Under the geometric Satake correspondence, the left exact functor $V \mapsto V_{\preceq\lambda}$ corresponds to $(i_\lambda)_*{}^p\,\mathrm{H}^0(i_\lambda)^!$: $\mathrm{P}_{L^+\hat{G}}(\mathrm{Gr}_{\hat{G}}) \to \mathrm{P}_{L^+\hat{G}}(\mathrm{Gr}_{\hat{G}})$.*

3.5.4. *Proof of Proposition* 3.3.2. We construct the needed basis explicitly via MV cycles. Let V be a Schur module so $\mathrm{Sat}(V) = {}^p\,\mathrm{H}^0(\mathring{i}_\mu)_*k[\langle 2\check{\rho},\mu\rangle]$ for some dominant weight μ of G. In this case, the natural map

$$
{}^p\,\mathrm{H}^0(\mathring{i}_\mu)_*k[\langle 2\check{\rho},\mu\rangle] \to (\mathring{i}_\mu)_*k[\langle 2\check{\rho},\mu\rangle]
$$

induces the following commutative diagram
(3.5.2)

$$
\begin{array}{ccc}
\mathrm{H}^{\langle 2\check{\rho},\nu\rangle}_{S_\nu}(\mathrm{Gr}_{\hat{G}}, {}^p\,\mathrm{H}^0(\mathring{i}_\mu)_*k[\langle 2\check{\rho},\mu\rangle]) & \longrightarrow & \mathrm{H}^{\langle 2\check{\rho},\nu\rangle}_{S_\nu^{>\nu+i\alpha}}(\mathrm{Gr}_{\hat{G}}, {}^p\,\mathrm{H}^0(\mathring{i}_\mu)_*k[\langle 2\check{\rho},\mu\rangle]) \\
\downarrow & & \downarrow \\
\mathrm{H}^{\langle 2\check{\rho},\nu+\mu\rangle}_{S_\nu\cap\mathring{\mathrm{Gr}}_{\hat{G},\mu}}(\mathring{\mathrm{Gr}}_{\hat{G},\mu}, k) & \longrightarrow & \mathrm{H}^{\langle 2\check{\rho},\nu+\mu\rangle}_{S_\nu^{>\nu+i\alpha}\cap\mathring{\mathrm{Gr}}_{\hat{G},\mu}}(\mathring{\mathrm{Gr}}_{\hat{G},\mu}, k).
\end{array}
$$

The two vertical arrows are isomorphisms by (the proof of) [**MV07**, Proposition 3.10], and therefore we identify their sources and targets. The groups in the bottom row are canonically isomorphic to the top Borel–Moore homology of $S_\nu\cap\mathring{\mathrm{Gr}}_{\hat{G},\mu}$ and of $S_\nu^{>\nu+i\alpha}\cap\mathring{\mathrm{Gr}}_{\hat{G},\mu}$ respectively, and therefore have a basis given by the fundamental classes of their irreducible components. Such a fundamental class in the lower left corner maps to zero under the horizontal arrow if and only if the corresponding irreducible component does not intersect with $S_\nu^{>\nu+i\alpha}$.

Now we take the basis of $V(\nu) = F_\nu(\mathrm{Sat}(V))$ given by the aforementioned fundamental classes $\{v_j\}$ in $\mathrm{H}^{\langle 2\check{\rho},\nu+\mu\rangle}_{S_\nu\cap\mathring{\mathrm{Gr}}_{\hat{G},\mu}}(\mathring{\mathrm{Gr}}_{\hat{G},\mu}, k)$, i.e. the MV basis. The discussion above says that, for every $\alpha \in \Delta$ and every $i \in \mathbb{Z}_{\geq 0}$, those v_j's which lie in the kernel of the top horizontal arrow of (3.5.2), or equivalently in $\mathrm{fil}^\alpha_i V(\nu)$ by Proposition 3.5.1, in fact span $\mathrm{fil}^\alpha_i V(\nu)$. Therefore, their symbols $\overline{v_j}$ for the filtration fil^α form a basis of the associated graded $\mathrm{gr}^\alpha V(\nu)$. This concludes the proof of Proposition 3.3.2.

4. Vector-valued twisted conjugation invariant functions

In this section, we start to study the space of vector-valued (twisted) conjugation invariant functions on a group.

We will continue making use of the conventions and notations in § 3. In particular, we have a pinned reductive group $(G, B, T, \{x_\alpha\}_{\alpha\in\Delta})$ over k. Let σ be an automorphism of G preserving (B, T), and hence the root system $\Phi(G, T)$. We define the σ-action on the character group $\mathbb{X}^\bullet(T)$ by

$$
\sigma(\alpha)(t) = \alpha(\sigma^{-1}(t)), \quad \text{for } \alpha \in \mathbb{X}^\bullet(T),\ t \in T.
$$

so that $\sigma(U_\alpha) = U_{\sigma(\alpha)}$. For a G-representation V, let σV denote the representation

$$G \xrightarrow{\sigma^{-1}} G \to \mathrm{GL}(V).$$

We identify elements in σV with $\{\sigma v \,|\, v \in V\}$ and thus $g(\sigma v) = \sigma(\sigma^{-1}(g)v)$ for $g \in G$. This way, we have $\sigma \mathtt{S}_\mu = \mathtt{S}_{\sigma(\mu)}$ and $\sigma \mathtt{W}_\mu = \mathtt{W}_{\sigma(\mu)}$ for $\mu \in \mathbb{X}^\bullet(T)^+$.

4.1. Vector-valued twisted conjugation invariant functions on a group. Let H be a linear algebraic group over k equipped with an automorphism τ. We still denote the τ-twisted conjugation action of H on itself (as defined in (1.0.2)) by c_τ, i.e

$$c_\tau(h)(g) = hg\tau(h)^{-1}, \quad h, g \in H.$$

Let $\boldsymbol{J}_H = k[H]^{c_\tau(H)}$ denote the space of τ-twisted conjugation invariant functions on H, i.e. those $f \in k[H]$ satisfying $f(hg\tau(h)^{-1}) = f(g)$ for $h, g \in H$. More generally, for a representation V of H, we will denote by

$$\boldsymbol{J}_H(V) = (k[H] \otimes V)^{c_\tau(H)}$$

the space of vector-valued τ-twisted conjugation invariant functions on H. Equivalently, let $[H/c_\tau(H)]$ (or sometimes $[H\tau/H]$) denote the quotient stack of H by the τ-twisted conjugation action of H on itself, and let $\widetilde{V} := V_{[H/c_\tau(H)]}$ denote the corresponding vector bundle on $[H/c_\tau(H)]$ (see § 1 for the notations and conventions). Then

$$\boldsymbol{J}_H = \Gamma([H\tau/H], \mathcal{O}), \quad \boldsymbol{J}_H(V) = \Gamma([H\tau/H], \widetilde{V}).$$

Moreover, $\boldsymbol{J}_H(V)$ is naturally a $\boldsymbol{J}_H$-module.

REMARK 4.1.1. Note that the spaces $\boldsymbol{J}_H(V)$ depend only on the image of τ in the outer automorphism group $\mathrm{Out}(H)$ of H. Indeed, if $\tau_2 = c(h) \circ \tau_1$, where $c(h) : H \to H, \ \ g \mapsto hgh^{-1}$ is the inner automorphism of H induced by some $h \in H$, then the map $H \to H, \ x \mapsto xh$ is an isomorphism intertwining the action c_{τ_1} and c_{τ_2}, and therefore induces isomorphisms between the spaces of vector-valued twisted conjugation invariant functions.

We discuss a few basic properties of these spaces. First, if $H = H_1 \times H_2$ and $V = V_1 \boxtimes V_2$ is the exterior tensor product of two representations of H_1 and H_2 respectively, then clearly $\boldsymbol{J}_H(V)$ is compatible with tensor product

$$\boldsymbol{J}_H(V) = \boldsymbol{J}_{H_1}(V_1) \otimes_k \boldsymbol{J}_{H_2}(V_2). \tag{4.1.1}$$

Next, if K is another linear algebraic group equipped with an automorphism τ and there is a τ-equivariant homomorphism $K \to H$, we have a natural homomorphism

$$\mathrm{Res}_V^\tau : \boldsymbol{J}_H(V) \to \boldsymbol{J}_K(V)$$

by restricting (a.k.a. pulling back) of the V-valued functions on H to K; this is compatible with the $\boldsymbol{J}_H$-module and $\boldsymbol{J}_K$-module structure through the natural restriction map $\mathrm{Res}_1^\tau : \boldsymbol{J}_H \to \boldsymbol{J}_K$, and in particular induces the map of $\boldsymbol{J}_K$-modules

$$\mathrm{Res}_V^\tau \otimes 1 : \boldsymbol{J}_H(V) \otimes_{\boldsymbol{J}_H} \boldsymbol{J}_K \to \boldsymbol{J}_K(V).$$

Next we discuss the compatibility of $\boldsymbol{J}_H(V)$ with tensor induction. Let H_0 be a linear algebraic group equipped with an automorphism τ_0. Suppose we have an embedding $\langle \tau_0 \rangle \subset \langle \tau \rangle$ of cyclic groups of finite index so that $\tau_0 = \tau^d$. Then we can form the tensor induction $H = \mathrm{Ind}_{\langle \tau_0 \rangle}^{\langle \tau \rangle} H_0$, which is the group of τ_0-equivariant maps from $\langle \tau \rangle$ to H, where τ_0 acts on $\langle \tau \rangle$ by translation. In addition, τ acts on H since it acts on $\langle\tau\rangle$ by translation. Since $\{\mathrm{id}, \tau, \dots, \tau^{d-1}\}$ forms a set of representatives of $\langle\tau\rangle/\langle\tau_0\rangle$, we may explicitly identify H with the product $\prod_{i=0}^{d-1} H_0$, on which τ acts by $(h_0, \dots, h_{d-1}) \mapsto (\tau_0(h_{d-1}), h_0, \dots, h_{d-2})$.

We have two natural maps from H_0 to H, given by embedding i_0 into the 0th factor of H and by the diagonal embedding Δ, i.e.

$$i_0(h) = (h, 1, \dots, 1), \quad \Delta(h) = (h, \dots, h).$$

It is straightforward to check that these two embeddings satisfy the following relation: for $g, h \in H_0$,

$$\begin{aligned} c_\tau(\Delta(h))(i_0(g)) &= c_\tau(h, \dots, h)(g, 1, \dots, 1) \\ &= (hg\tau_0(h)^{-1}, hh^{-1}, \dots, hh^{-1}) = i_0\big(c_{\tau_0}(h)(g)\big), \end{aligned}$$

i.e. i_0 intertwines the τ_0-twisted conjugation action of H_0 on the source, and the τ-twisted conjugation action of $\Delta(H_0)$ on the target:

$$\begin{array}{ccc} H_0 & \xrightarrow{\ i_0\ } & H \\ c_{\tau_0} \circlearrowright & & \circlearrowright c_\tau \\ H_0 & \xrightarrow{\ \Delta\ } & H. \end{array}$$

From this, we naturally deduce a morphism of stacks

$$(i_0/\Delta) : [H_0/c_{\tau_0}(H_0)] \to [H/c_\tau(H)]. \tag{4.1.2}$$

If $K_0 \subset H_0$ is a τ_0-stable subgroup, then $K = \mathrm{Ind}_{\langle\tau_0\rangle}^{\langle\tau\rangle} K_0$ is a τ-stable subgroup of H, and we obtain a commutative diagram

$$\begin{array}{ccc} [K_0/c_{\tau_0}(K_0)] & \longrightarrow & [K/c_\tau(K)] \\ \downarrow & & \downarrow \\ [H_0/c_{\tau_0}(H_0)] & \longrightarrow & [H/c_\tau(H)]. \end{array}$$

LEMMA 4.1.2. (1) *The morphism (i_0/Δ) in (4.1.2) is an isomorphism of stacks.*

(2) *If V is an H-module, regarded as a representation of H_0 via the* diagonal *embedding Δ, then we may form $\boldsymbol{J}_{H_0}(V)$ and $\boldsymbol{J}_{K_0}(V)$ with respect to the τ_0-twisted conjugation action, and $\boldsymbol{J}_H(V)$ and $\boldsymbol{J}_K(V)$ with respect to the τ-twisted conjugation action. Then we have a natural commutative diagram,*

$$\begin{array}{ccc} \boldsymbol{J}_H(V) \otimes_{\boldsymbol{J}_H} \boldsymbol{J}_K & \xrightarrow{\mathrm{Res}_V^\tau \otimes 1} & \boldsymbol{J}_K(V) \\ \Big\downarrow {\scriptstyle (i_0/\Delta)^*} & & \Big\downarrow {\scriptstyle (i_0/\Delta)^*} \\ \boldsymbol{J}_{H_0}(V) \otimes_{\boldsymbol{J}_{H_0}} \boldsymbol{J}_{K_0} & \xrightarrow{\mathrm{Res}_V^{\tau_0} \otimes 1} & \boldsymbol{J}_{K_0}(V) \end{array}$$

where the two vertical morphisms are isomorphisms.

PROOF. (1) The group $\{1\} \times H_0^{d-1}$ is a coset representative of $H/\Delta(H_0)$. It is enough to show that the following map is an isomorphism

$$\begin{aligned} c_\tau(-, i_0(-)) : (\{1\} \times H_0^{d-1}) \times H_0 &\longrightarrow H = H_0^d \\ ((h_1, \ldots, h_{d-1}), g) &\longmapsto c_\tau(1, h_1, \ldots, h_{d-1})(i_0(g)) \\ &= \big(g\tau_0(h_{d-1}^{-1}), h_1, h_2h_1^{-1}, \ldots, h_{d-1}h_{d-2}^{-1}\big). \end{aligned}$$

But this is clear, and the inverse map is given by

$$(g_0, \ldots, g_{d-1}) \mapsto \big(g_1, g_2g_1, \ldots, g_{d-1}g_{d-2}\cdots g_1), g_0\tau_0(g_{d-1}g_{d-2}\cdots g_1)\big).$$

So (4.1.2) is an isomorphism.

(2) follows from (1) immediately. In an explicit form, the inverse of the pull back $(i_0/\Delta)^*$ is given by sending $\bar{f} \in \boldsymbol{J}_{H_0}(V)$ to a τ-twisted conjugation invariant function $f : H \to V$ defined by the following formula: for $g_0, \ldots, g_{d-1} \in H_0$

$$f(g_0, \ldots, g_{d-1}) = (g_1^{-1}g_2^{-1}\cdots g_{d-1}^{-1}, \ldots, g_{d-2}^{-1}g_{d-1}^{-1}, g_{d-1}^{-1}, 1) \cdot \bar{f}(g_{d-1}g_{d-2}\cdots g_0). \quad \square$$

Finally, let us discuss the compatibility of $\boldsymbol{J}_H(V)$ with central homomorphisms. Let $F \subset H$ denote a τ-stable central subgroup of H. We assume that F is of multiplicative type, i.e. if $\Lambda = \operatorname{Hom}(F, \mathbb{G}_m)$ denote its character group, then $F = \operatorname{Spec} k[\Lambda]$. The exact sequence

$$1 \to \Lambda^\tau \to \Lambda \xrightarrow{1-\tau} \Lambda \to \Lambda_\tau \to 1$$

induces

$$1 \to F^\tau \to F \xrightarrow{1-\tau} F \to F_\tau \to 1.$$

Let $H' = H/(1-\tau)F$ and $H'' = H/F$. Then the kernel of the map $H' \to H''$ is F_τ. Left multiplication by F induces an F-action on $k[H]$ via $(z \cdot f)(h) = f(zh)$. Then we have the decomposition $k[H] = \oplus_{\psi\in\Lambda} k[H]_\psi$ according to the weights of F, where $k[H]_{\psi=\mathbf{1}} = k[H'']$ and each $k[H']_\psi$ is an invertible $k[H'']$-module. In addition, $\oplus_{\psi\in\Lambda^\tau} k[H]_\psi = k[H']$, and for given $\chi \in (1-\tau)\Lambda$, $\oplus_{\tau(\psi)-\psi=\chi} k[H]_{\tau(\psi)}$ is an invertible $k[H']$-module.

Let V be an H-module, decomposed as $\oplus_\chi V_\chi$ according to the weights of F. Then

$$\boldsymbol{J}_H(V) = \bigoplus_{\psi\in\Lambda} \big(k[H]_{\tau(\psi)} \otimes V_{\tau(\psi)-\psi}\big)^{c_\tau(H'')} = \bigoplus_{\chi\in(\tau-1)\Lambda} \Big(\bigoplus_{\tau(\psi)-\psi=\chi} k[H]_{\tau(\psi)} \otimes V_\chi\Big)^{c_\tau(H'')}.$$

Each direct summand $\big(\oplus_{\tau(\psi)-\psi=\chi} k[H]_{\tau(\psi)} \otimes V_\chi\big)^{c_\tau(H'')}$ is acted on by F_τ. In particular,

$$\boldsymbol{J}_H - k[H']^{c_\tau(H'')}$$

is acted by F_τ, and the $\boldsymbol{J}_H$-module structure on $\boldsymbol{J}_H(V)$ is compatible with this action. Note that if V is a representation of H'', then

$$\boldsymbol{J}_H(V)^{F_\tau} = \boldsymbol{J}_{H''}(V). \tag{4.1.3}$$

In particular, $\boldsymbol{J}_H^{F_\tau} = \boldsymbol{J}_{H''}$.

LEMMA 4.1.3. *Let V be a representation of H''. Assume that the above F_τ-action realizes* $\operatorname{Spec} \boldsymbol{J}_H$ *as an F_τ-torsor (in fpqc topology) over* $\operatorname{Spec} \boldsymbol{J}_{H''}$, *then the natural map*

$$\operatorname{Res}_V^\tau \otimes 1 : \boldsymbol{J}_{H''}(V) \otimes_{\boldsymbol{J}_{H''}} \boldsymbol{J}_H \to \boldsymbol{J}_H(V)$$

is an isomorphism.

PROOF. Note that the action of F_τ on $\boldsymbol{J}_H(V)$ equips $\boldsymbol{J}_H(V)$ with a descent datum for the map $\boldsymbol{J}_{H''} \to \boldsymbol{J}_H$. The lemma now follows from (4.1.3) and faithfully flat descent. □

4.2. Vector-valued twisted conjugation invariant functions on the Vinberg monoid. Now let $H = G$ be the reductive group as fixed at the beginning of the section. Let τ be an automorphism of G. Let V be a representation of G. Sometimes we write $\boldsymbol{J}(V) = \boldsymbol{J}_G(V)$ and $\boldsymbol{J} = \boldsymbol{J}_G$ for simplicity. We now define their analogues $\boldsymbol{J}_+(V)$ (resp. $\boldsymbol{J}_0(V)$) of vector-valued twisted conjugation invariant functions on V_G (resp. As_G).

We consider the G-action on V_G via the twisted diagonal embedding $G \to G \times G$, $g \mapsto (g, \tau(g))$ so that its restriction to $\mathfrak{d}^{-1}(\mathbf{1}) = G$ is the τ-twisted conjugation action of G on itself. For this reason, we also denote this action of G on V_G and on As_G by c_τ. Let $[V_G/c_\tau(G)]$ and $[\mathrm{As}_G/c_\tau(G)]$ denote the stack quotients. The map $\mathfrak{d}$ induces

$$[\mathfrak{d}] : [V_G/c_\tau(G)] \to T^+_{\mathrm{ad}}, \quad [\mathfrak{d}]^{-1}(\mathbf{1}) = [G/c_\tau(G)], \quad [\mathfrak{d}]^{-1}(\mathbf{0}) = [\mathrm{As}_G/c_\tau(G)].$$

Write $\widetilde{V}_+ := V_{[V_G/c_\tau(G)]}$ (resp. $\widetilde{V}_0 = V_{[\mathrm{As}_G\,/c_\tau(G)]}$) for the vector bundle on $[V_G/c_\tau(G)]$ (resp. $[\mathrm{As}_G\,/c_\tau(G)]$) corresponding to V. Then clearly $\widetilde{V} := \widetilde{V}_+|_{[\mathfrak{d}]^{-1}(\mathbf{1})}$ and $\widetilde{V}_0 = \widetilde{V}_+|_{[\mathfrak{d}]^{-1}(\mathbf{0})}$. Now we can define

$$\boldsymbol{J}_+ = \Gamma([V_G/c_\tau(G)], \mathcal{O}), \quad \boldsymbol{J}_+(V) = \Gamma([V_G/c_\tau(G)], \widetilde{V}_+),$$

and

$$\boldsymbol{J}_0 = \Gamma([\mathrm{As}_G/c_\tau(G)], \mathcal{O}), \quad \boldsymbol{J}_0(V) = \Gamma([\mathrm{As}_G/c_\tau(G)], \widetilde{V}_0).$$

Then $\boldsymbol{J}_+(V)$ (resp. $\boldsymbol{J}_0(V)$) is a $\boldsymbol{J}_+$ (resp. $\boldsymbol{J}_0$)-module.

Remark 4.1.1 also applies to the study of the spaces $\boldsymbol{J}_+(V)$ and $\boldsymbol{J}_0(V)$. Recall that a choice of a pinning of G defines a section of the projection $\mathrm{Aut}(G) \to \mathrm{Out}(G)$. Therefore, to study $\boldsymbol{J}_*(V)$ for $* = +, 0, \emptyset$, without loss of generality we may and will assume that $\tau = \sigma$ is an automorphism preserving the pinning $(G, B, T, \{x_\alpha\}_{\alpha\in\Delta})$ we fix at the beginning of this section.

We explain the relations between $\boldsymbol{J}_+(V), \boldsymbol{J}_0(V)$ and $\boldsymbol{J}(V)$. Let

$$V_G /\!\!/ c_\sigma(G) = \mathrm{Spec}\, \boldsymbol{J}_+$$

be the GIT quotient. Then $[\mathfrak{d}]$ factors as

$$[V_G/c_\sigma(G)] \to V_G /\!\!/ c_\sigma(G) \xrightarrow{\bar{\mathfrak{d}}} T^+_{\mathrm{ad}}.$$

LEMMA 4.2.1. (1) *For any representation* V, $\boldsymbol{J}_+(V) \otimes_{\boldsymbol{J}_+} k[\bar{\mathfrak{d}}^{-1}(\mathbf{1})] \cong \boldsymbol{J}(V)$. *In particular,* $k[\bar{\mathfrak{d}}^{-1}(\mathbf{1})] \cong \boldsymbol{J}$.

(2) *If* V *admits a good filtration, then* $\boldsymbol{J}_+(V) \otimes_{\boldsymbol{J}_+} k[\bar{\mathfrak{d}}^{-1}(\mathbf{0})] \cong \boldsymbol{J}_0(V)$. *In particular,* $k[\bar{\mathfrak{d}}^{-1}(\mathbf{0})] \cong \boldsymbol{J}_0$.

PROOF. Recall that there is an $\mathbb{X}^\bullet(T)^+_{\mathrm{pos}}$-filtration on $k[G]$ by (3.2.1). Then

$$\boldsymbol{J}_+(V) \otimes_{\boldsymbol{J}_+} k[\bar{\mathfrak{d}}^{-1}(\mathbf{1})] = \bigoplus_{\omega \in \mathsf{Min}} \varinjlim_{\mu \in \omega + \mathbb{X}^\bullet(T)^+_{\mathrm{pos}}} \left(\mathrm{fil}_\mu k[G] \otimes V\right)^{c_\sigma(G)}.$$

Since taking G-invariants commutes with taking direct limits, Part (1) follows.

To prove Part (2), first notice that for $\mu \in \mathbb{X}^\bullet(T)^+_{\mathrm{pos}}$, there is the following commutative diagram with all rows exact.
(4.2.1)

$$\begin{array}{ccccccc} \bigoplus_i \left(\mathrm{fil}_{\mu-\alpha_i} k[G] \otimes V\right)^{c_\sigma(G)} & \to & (\mathrm{fil}_\mu k[G] \otimes V)^{c_\sigma(G)} & \to & \left(\boldsymbol{J}_+(V) \otimes_{\boldsymbol{J}_+} k[\bar{\mathfrak{d}}^{-1}(\mathbf{0})]\right)_\mu & \to 0 \\ \downarrow & & \| & & \downarrow & \\ 0 \to \left(\sum_i \mathrm{fil}_{\mu-\alpha_i} k[G] \otimes V\right)^{c_\sigma(G)} & \to & (\mathrm{fil}_\mu k[G] \otimes V)^{c_\sigma(G)} & \longrightarrow & (\mathrm{gr}_\mu k[G] \otimes V)^{c_\sigma(G)} & \longrightarrow 0. \end{array}$$

Here $(\boldsymbol{J}_+(V) \otimes_{\boldsymbol{J}_+} k[\bar{\mathfrak{d}}^{-1}(\mathbf{0})])_\mu$ denotes the μ-graded piece of $\boldsymbol{J}_+(V) \otimes_{\boldsymbol{J}_+} k[\bar{\mathfrak{d}}^{-1}(\mathbf{0})]$, and $\sum_i \mathrm{fil}_{\mu-\alpha_i} k[G]$ denotes the image of $\bigoplus_i \mathrm{fil}_{\mu-\alpha_i} k[G] \to \mathrm{fil}_\mu k[G]$. The second row is clearly left exact, and the right exactness follows from the fact that $\sum_i \mathrm{fil}_{\mu-\alpha_i} k[G]$ has a good filtration as a $G \times G$-module, by Theorem 3.1.2, Corollary 3.1.6. By the same reasoning, the left vertical map is also surjective. It follows that the right vertical map is an isomorphism. Part (2) follows. □

REMARK 4.2.2. If V is the trivial representation, one can directly show the surjectivity of the map

$$(\mathrm{fil}_\mu k[G])^{c_\sigma(G)} \to (\mathrm{gr}_\mu k[G])^{c_\sigma(G)}$$

by constructing a splitting. Namely, note that

$$\dim(\mathrm{gr}_\mu k[G])^{c_\sigma(G)} = \begin{cases} 0 & \sigma(\mu) \neq \mu \\ 1 & \sigma(\mu) = \mu. \end{cases}$$

In the latter case, the composition

$$(\mathtt{W}^*_\mu \otimes \mathtt{W}_\mu)^{c_\sigma(G)} \to \mathrm{fil}_\mu k[G]^{c_\sigma(G)} \to (\mathtt{S}_{\mu^*} \otimes \mathtt{S}_\mu)^{c_\sigma(G)}$$

is an isomorphism, giving the splitting.

In addition, by applying Lemma 2.2.2 to the case $M = k[V_G]^{c_\sigma(G)}$, we conclude that $\bar{\mathfrak{d}} : V_G /\!\!/ c_\sigma(G) \to T^+_{\mathrm{ad}}$ is flat.

We finish this subsection with the following *twisted Chevalley isomorphism*. Recall that we denote by $W = N(T)/T$ the Weyl group. Let $W_0 = W^\sigma$ and let N_0 be the preimage of W_0 in N. It acts on T via the twisted conjugation c_σ and it also acts on V_T via c_σ.

PROPOSITION 4.2.3. *The restriction of a function on V_G to V_T induces an isomorphism*

$$\mathrm{Res}^\sigma_{+,\mathbf{1}} : k[V_G]^{c_\sigma(G)} \xrightarrow{\cong} k[V_T]^{c_\sigma(N_0)}.$$

In particular, restricting a σ-conjugation invariant function on G to T induces the twisted Chevalley isomorphism

$$\mathrm{Res}^\sigma_{\mathbf{1}} : \boldsymbol{J} = k[G]^{c_\sigma(G)} \xrightarrow{\cong} k[T]^{c_\sigma(N_0)}.$$

The last statement was also proved in [**Sp06**, Theorem 1], essentially by the same argument given below.

PROOF. Recall the $\mathbb{X}^\bullet(T)^+_{\mathrm{pos}}$-filtration on $k[T]$ defined by (3.2.3), which is the image of the $\mathbb{X}^\bullet(T)^+_{\mathrm{pos}}$-filtration on $k[G]$ defined by (3.2.1). Then

$$(\mathrm{fil}_\nu k[T])^{c_\sigma(N_0)} := \bigoplus_{\substack{\lambda \in \mathbb{X}^\bullet(T)^{+,\sigma} \\ \mu - \lambda \in \mathbb{X}^\bullet(T_{\mathrm{ad}})_{\mathrm{pos}}}} \Big(k \cdot \sum_{\nu \in W_0 \lambda} e^\nu\Big),$$

By Remark 4.2.2, it is straightforward to see that $\mathrm{gr}k[G]^{c_\sigma(G)} \cong \mathrm{gr}k[T]^{c_\sigma(N_0)}$. Therefore, $\mathrm{Res}^\sigma_{+,1}$ is an isomorphism. □

4.3. Freeness. Here are the main results of this section. We define a number

$$r_V := \dim V|_{T^\sigma}(0).$$

ASSUMPTION 4.3.1. In this subsection, let G be a simply-connected semisimple group over k and V a G-representation that admits a good filtration.

THEOREM 4.3.2. *Keep Assumption* 4.3.1. *Then* $\boldsymbol{J}_+(V)$ *(resp.* $\boldsymbol{J}_0(V)$*, resp.* $\boldsymbol{J}(V)$*) is a free* $\boldsymbol{J}_+$*-module (resp.* $\boldsymbol{J}_0$*-module, resp.* $\boldsymbol{J}$*-module) of rank* r_V*.*

COROLLARY 4.3.3. *Keep Assumption* 4.3.1. *The morphisms*

$$\chi_+ : V_G \to V_G/\!\!/c_\sigma(G), \quad \chi_0 : \mathrm{As}_G \to \mathrm{As}_G/\!\!/c_\sigma(G), \quad \chi : G \to G/\!\!/c_\sigma(G)$$

are faithfully flat.

We call these morphisms the *(twisted) Chevalley maps*. When $\sigma = \mathrm{id}$, this corollary is also proved by [**Bou15**] by a different method.

PROOF. Let Reg denote the regular representation of G, i.e. the representation of G on $k[G]$ induced by left multiplication, which as mentioned before admits a good filtration. Since for any affine variety $X = \mathrm{Spec}\, R$ equipped with a G-action, $(R \otimes \mathrm{Reg})^G = R$, the statement follows from Theorem 4.3.2. □

REMARK 4.3.4. The Chevalley map $\chi : G \to G/\!\!/c_\sigma(G)$ is not flat in general, and therefore Theorem 4.3.2 cannot hold for arbitrary reductive group. We refer to [**Ric79**, Proposition 4.1] for a discussion of this point when $\sigma = \mathrm{id}$.

Nevertheless, there is a sufficient condition for the flatness of $\boldsymbol{J}(V)$ over $\boldsymbol{J}$ when G is a general connected reductive group. Let G_{sc} be the simply-connected cover of the derived subgroup of G, and let F be the kernel of the central isogeny $1 \to F \to G' := G_{\mathrm{sc}} \times Z_G \to G \to 1$. It is known that σ lifts to a unique automorphism of G_{sc} (e.g. see [**St68**, §9.16]). We decompose $V = \oplus_\psi V_\psi \otimes k_\psi$ according to the central character for the action of Z_G on V so that each V_ψ is a G_{sc}-module. Then by (4.1.1)

$$\boldsymbol{J}_{G'}(V) = \bigoplus_\psi \boldsymbol{J}_{G_{\mathrm{sc}}}(V_\psi) \otimes \boldsymbol{J}_{Z_G}(k_\psi) = \bigoplus_{\psi|_{Z_G^\sigma} = 1} \boldsymbol{J}_{G_{\mathrm{sc}}}(V_\psi) \otimes \boldsymbol{J}_{Z_G}(k_\psi),$$

which is free over $G'/\!\!/c_\sigma(G') \cong G_s/\!\!/c_\sigma(G_s) \times (Z_G)_\sigma$ of rank r_V by Theorem 4.3.2. It follows from Lemma 4.1.3 that if the action of F_σ on $G'/\!\!/c_\sigma(G')$ is free, then $\boldsymbol{J}_G(V)$ is finite projective of rank r_V over $G/\!\!/c_\sigma(G)$. In particular, if the map $F_\sigma \to (Z_G)_\sigma$ is injective, then $J_G(V)$ is finite projective over J_G. For example, this is the case if $G_{\mathrm{sc}} = G_{\mathrm{der}}$ and $\sigma = \mathrm{id}$.

The rest of this subsection is devoted to the proof of Theorem 4.3.2. We will first prove the statement for $\boldsymbol{J}_0(V)$, and then deduce from it the statement for $\boldsymbol{J}_+(V)$ and $\boldsymbol{J}(V)$. Note that by Lemma 3.2.1 (4),

$$\boldsymbol{J}_0(V) = \Gamma([\mathrm{As}_G / c_\sigma(G)], \widetilde{V}_0) = \bigoplus_{\nu \in \mathbb{X}^\bullet(T)^+} (\mathsf{S}_{\sigma(\nu^*)} \otimes \mathsf{S}_\nu \otimes V)^G,$$

Let us fix $\xi \in \mathbb{X}^\bullet(T)$. Clearly, the part of Theorem 4.3.2 for $\boldsymbol{J}_0(V)$ will follow from the following refinement.

PROPOSITION 4.3.5. *Assume that V admits a good filtration. Then*

$$\boldsymbol{J}_0(V)_\xi := \bigoplus_{\sigma(\nu)-\nu=\xi} (\mathsf{S}_{\sigma(\nu^*)} \otimes \mathsf{S}_\nu \otimes V)^G$$

is a finite free $\boldsymbol{J}_0$-module of rank $= \dim V(\xi)$. Moreover, we may choose a basis $\{e_i\}$ of $\boldsymbol{J}_0(V)_\xi$ as a $\boldsymbol{J}_0$-module such that each $e_i \in (\mathrm{gr}_{\nu_i} k[G] \otimes V)^{c_\sigma(G)}$ for some $\nu_i \in \mathbb{X}^\bullet(T)^+$ satisfying $\sigma(\nu_i) - \nu_i = \xi$.

PROOF. Set

$$\mathbb{X}^\bullet(T)^{+,\sigma} = \{\nu \in \mathbb{X}^\bullet(T)^+ \mid \sigma(\nu) = \nu\} \text{ and } \mathbb{X}^\bullet(T)^+_\xi = \{\nu \in \mathbb{X}^\bullet(T)^+ \mid \sigma(\nu) - \nu = \xi\}.$$

Then $\mathbb{X}^\bullet(T)^{+,\sigma}$ naturally acts on $\mathbb{X}^\bullet(T)^+_\xi$. We shall apply the discussions from § 2 to the following tuple

$$(\Gamma, S, M, \{\mathrm{fil}_s M\}_{s\in S}) = \big(\mathbb{X}^\bullet(T)^{+,\sigma}, \mathbb{X}^\bullet(T)^+_\xi, V(\xi), \{\mathrm{fil}_\nu V(\xi)\}_{\nu\in\mathbb{X}^\bullet(T)^+_\xi}\big).$$

The following lemma implies that $(\mathbb{X}^\bullet(T)^{+,\sigma}, \mathbb{X}^\bullet(T)^+_\xi)$ satisfies (Can) and (DCC), and Lemma 2.1.7 is applicable (with $S' \subset S$ being $\mathbb{X}^\bullet(T)^+_\xi \subset \mathbb{X}^\bullet(T)^+ = \mathbb{X}^\bullet(T_{\mathrm{sc}})^+$).

LEMMA 4.3.6. *We equip $\mathbb{X}^\bullet(T)^+$ with a partial order by identifying $\mathbb{X}^\bullet(T)^+ \cong \mathbb{N}^\Delta$ as in (3.3.4), i.e. $\lambda_1 \geq \lambda_2$ if and only if $\langle \lambda_1, \check{\alpha}\rangle \geq \langle \lambda_2, \check{\alpha}\rangle$ for every simple coroot $\check{\alpha}$. Then for every $\nu_0 \in \mathbb{X}^\bullet(T)^+$, the set*

$$\{\nu \in \mathbb{X}^\bullet(T)^+ \mid \sigma(\nu) - \nu = \xi, \nu \geq \nu_0\}$$

has a unique minimal element, denoted by ν_0^h. In addition, in each σ-orbit of simple coroots, there is at least one $\check{\alpha}$ such that $\langle \nu_0^h, \check{\alpha}\rangle = \langle \nu_0, \check{\alpha}\rangle$.

Specializing this discussion to $\nu_0 = 0$, we deduce that $\mathbb{X}^\bullet(T)^+_\xi = \nu_0^h + \mathbb{X}^\bullet(T)^{+,\sigma}$.

PROOF. The last sentence of the lemma is a direct corollary of the existence and the properties of ν_0^h. We focus on constructing the needed element ν_0^h. Since G is a simply-connected semisimple group, we may take the set of fundamental weights $\{\omega_\alpha\}_{\alpha\in\Delta}$ such that $\langle \omega_\alpha, \check{\beta}\rangle = \delta_{\alpha\beta}$ for any pair $\alpha, \beta \in \Delta$. Then every weight can be written as $\xi = \sum \langle \xi, \check{\alpha}\rangle \omega_\alpha$.

If we write $\nu = \sum_\alpha \nu_\alpha \omega_\alpha$ with $\nu_\alpha \in \mathbb{Z}$, the equality $\sigma(\nu) - \nu = \xi$ is equivalent to the system of equations

$$\nu_\alpha - \nu_{\sigma(\alpha)} = \langle \xi, \sigma(\check{\alpha})\rangle, \quad \alpha \in \Delta,$$

with variables $\{\nu_\alpha\} \in \mathbb{Z}^\Delta$, and the condition $\nu \geq \nu_0$ is equivalent to that

$$\nu_\alpha \geq \langle \nu_0, \check{\alpha}\rangle \quad \text{for any} \quad \alpha \in \Delta. \tag{4.3.1}$$

For every σ-orbit $\mathcal{O} \subset \Delta$, let $\omega_{\mathcal{O}} = \sum_{\alpha\in\mathcal{O}} \omega_\alpha$. Then after modifying a solution ν by a multiple of $\omega_{\mathcal{O}}$ if necessary, we can always find ν satisfying the additional inequalities (4.3.1) and such that in each σ-orbit $\mathcal{O} \subset \Delta$, there is at least one $\alpha \in \mathcal{O}$ such that the equality in (4.3.1) holds. Then this is the desired ν_0^h. □

We now return to the proof of Proposition 4.3.5. By Proposition 3.4.1 and (3.4.1), we have an isomorphism

$$\boldsymbol{J}_0(V)_\xi = \bigoplus_{\nu\in\mathbb{X}^\bullet(T)^+_\xi} (\mathsf{S}_{\sigma(\nu^*)} \otimes \mathsf{S}_\nu \otimes V)^G \overset{\oplus_\nu \ell_{\sigma(\nu),\nu}}{\cong} \bigoplus_{\sigma(\nu)-\nu=\xi} \mathrm{fil}_\nu V(\xi) = R_{\mathbb{X}^\bullet(T)^+_\xi} V(\xi),$$

as modules over $\boldsymbol{J}_0 \cong k[\mathbb{X}^\bullet(T)^{+,\sigma}]$.

Therefore, by Lemma 2.2.2, the lemma will follow if we can show that

$$\dim \operatorname{gr}_{\mathbb{X}^\bullet(T)^+_\xi} V(\xi) = \dim V(\xi). \tag{4.3.2}$$

But by Lemma 4.3.6, we can apply Lemma 2.1.7 to deduce this equality from Theorem 3.3.1. □

By Proposition 4.3.5, we may choose a basis $\{e_i\}$ of $\boldsymbol{J}_0(V)_\xi$ as a $\boldsymbol{J}_0$-module such that each $e_i \in (\operatorname{gr}_{\nu_i} k[G] \otimes V)^{c_\sigma(G)}$ for some $\nu_i \in \mathbb{X}^\bullet(T)^+$. By the exactness of the bottom row of (4.2.1), we can lift each element in $\{e_i\}$ to $\boldsymbol{J}_+(V)$ so that $\widetilde{e_i} \in (\operatorname{fil}_{\nu_i} k[G] \otimes V)^{c_\sigma(G)}$. By Lemma 2.2.1 (2), the natural map

$$\bigoplus_i \boldsymbol{J}_+ \widetilde{e_i} \to \boldsymbol{J}_+(V)$$

is surjective. In particular, $\boldsymbol{J}_+(V)$ is finitely generated over $\boldsymbol{J}_+$. By Lemma 4.2.1 (1), $\boldsymbol{J}(V)$ is also finitely generated over $\boldsymbol{J}$.

Now Theorem 4.3.2 is reduced to show that for every point $x \in V_G /\!\!/ c_\sigma(G)$, the fiber of the module $\boldsymbol{J}_+(V)$ over x has dimension $\geq r_V$. By the semi-continuity, it is a consequence of the following.

LEMMA 4.3.7. *Over the generic point of $V_G /\!\!/ c_\sigma(G)$, the rank of $\boldsymbol{J}_+(V)$ is r_V.*

PROOF. Since the map $V_G \to T^+_{\mathrm{ad}}$ is T-equivariant, and this T-action commutes with the $G \times G$-action, it is enough to prove a similar statement for $\boldsymbol{J}_+(V)|_{\mathfrak{d}^{-1}(\mathbf{1})} = \boldsymbol{J}(V)$.

If V is a representation of G, the restriction from G to T gives a natural map

$$\operatorname{Res}^\sigma_V : \boldsymbol{J}(V) \to (k[T] \otimes V)^{c_\sigma(N_0)}, \tag{4.3.3}$$

compatible with the isomorphism $\boldsymbol{J} \cong k[T]^{c_\sigma(N_0)}$ from Proposition 4.2.3. We call $\operatorname{Res}^\sigma_V$ the *twisted Chevalley restriction homomorphism*. But unlike the case when V is the trivial representation, the map (4.3.3) in general fails to be an isomorphism and the failure is studied in details in [**Ba11, KNV11**] (when $\sigma = \mathrm{id}$). In the next section, we will also study in details (a variant of) this map. Currently, we just notice the following. By Remark 5.2.12 and Remark 5.2.15 below, there is an affine open N_0-stable subset

$$T^{\sigma\text{-reg}} = \left\{ t \in T \,\middle|\, I_t := \{g \in G \mid gt\sigma(g)^{-1} = t\} = T^\sigma \right\} \subset T,$$

an open subset $G^{\sigma\text{-rs}}$ of G, and an isomorphism $G \times^{N_0} T^{\sigma\text{-reg}} \xrightarrow{\cong} G^{\sigma\text{-rs}}$. It follows that

LEMMA 4.3.8. *The map (4.3.3), regarded as a map of coherent sheaves on $G/\!\!/c_\sigma(G) \cong T/\!\!/c_\sigma(N_0)$, restricts to an isomorphism over a (non-empty) open subset of $T^{\sigma\text{-reg}}/\!\!/c_\sigma(N_0)$.*

Note that the fiber of $(k[T] \otimes V)^{c_\sigma(N_0)}$ over any point of $T^{\sigma\text{-reg}}/\!\!/c_\sigma(N_0)$ is isomorphic to V^{T^σ}, whose dimension is $\dim V^{T^\sigma} = r_V$. Therefore, the generic fiber of $\boldsymbol{J}(V)$ over $\operatorname{Spec} \boldsymbol{J}$ is of dimension r_V. Again, since $\mathfrak{d} : V_G \to T^+_{\mathrm{ad}}$ is T-equivariant, Lemma 4.3.7 follows. □

4.4. Construction of the basis. In this subsection, we assume that char $k = 0$, and G is semisimple and simply-connected. Then every representation V of G admits a good filtration. The proof of Theorem 4.3.2 in fact gives a method to construct a basis of $\boldsymbol{J}(V)$, from certain basis of V.

For every weight $\xi \in (\sigma - 1)\mathbb{X}^\bullet(T)$, let $\mathbb{B}(\xi)$ be a basis of $V(\xi)$ satisfying the conditions in Proposition 3.3.2; for example, we may choose the MV basis as in § 3.5.

For each $\mathbf{b} \in \mathbb{B}(\xi)$ and each simple root α, let $\varepsilon_\alpha(\mathbf{b}) \geq 0$ be the integer such that $\mathbf{b} \in \mathrm{fil}^\alpha_{\varepsilon_\alpha(\mathbf{b})} V(\xi) - \mathrm{fil}^\alpha_{\varepsilon_\alpha(\mathbf{b})-1} V(\xi)$, i.e.

$$E_\alpha^{\varepsilon_\alpha(\mathbf{b})}(\mathbf{b}) \neq 0, \quad E_\alpha^{\varepsilon_\alpha(\mathbf{b})+1}(\mathbf{b}) = 0.$$

By Lemma 4.3.6, there is a unique minimal $\nu_{\mathbf{b}} \in \mathbb{X}^\bullet(T)^+$ such that $\sigma(\nu_{\mathbf{b}}) - \nu_{\mathbf{b}} = \xi$ and $\langle \nu_{\mathbf{b}}, \check{\alpha} \rangle \geq \varepsilon_\alpha(\mathbf{b})$ for every $\alpha \in \Delta$. Then by Proposition 3.4.1, we have an isomorphism

$$\ell_{\sigma(\nu_{\mathbf{b}}),\nu_{\mathbf{b}}} : (\mathtt{S}_{\sigma(\nu^*_{\mathbf{b}})} \otimes \mathtt{S}_{\nu_{\mathbf{b}}} \otimes V)^G \to \mathrm{fil}_{\nu_{\mathbf{b}}} V(\sigma(\nu_{\mathbf{b}}) - \nu_{\mathbf{b}}).$$

It follows that $\mathbf{b}$ comes from a unique element in $(\mathtt{S}_{\sigma(\nu^*_{\mathbf{b}})} \otimes \mathtt{S}_{\nu_{\mathbf{b}}} \otimes V)^G$, denoted by $f_{\mathbf{b},0}$.

REMARK 4.4.1. Recall that $\boldsymbol{J}_0 = k[\mathbb{X}^\bullet(T)^{+,\sigma}]$. For $\omega \in \mathbb{X}^\bullet(T)^{+,\sigma}$, let q^ω the corresponding element in $\boldsymbol{J}_0$. Let $x_1 \in \mathrm{Spec}\,\boldsymbol{J}_0$ be the point defined by $q^\omega(x_1) = 1$ for any ω. Then by the proof of Proposition 4.3.5, the fiber of $\boldsymbol{J}_0(V)$ at x_1 is canonically isomorphic to $\oplus_{\xi \in (\sigma-1)\mathbb{X}^\bullet(T)} V(\xi)$, and the restriction of $f_{\mathbf{b},0}$ to x_1 is just $\mathbf{b}$.

Since in characteristic zero, Schur and Weyl modules are isomorphic, there is a canonical G-equivariant map

$$\mathtt{S}_{\nu_{\mathbf{b}}} \otimes \mathtt{S}_{\sigma(\nu^*_{\mathbf{b}})} \cong \mathtt{W}_{\nu_{\mathbf{b}}} \otimes \mathtt{S}_{\sigma(\nu^*_{\mathbf{b}})} \to k[G\sigma] \tag{4.4.1}$$

given by taking (a twisted version of) matrix coefficients. Thus $f_{\mathbf{b},0}$ defines an element $f_{\mathbf{b}} \in (k[G\sigma] \otimes V)^G = \boldsymbol{J}(V)$. Explicitly, we write $f_{\mathbf{b},0}$ as $\sum_{i,j} \sigma e_i^* \otimes e_j \otimes v_{ij}$, where $\{e_i\}$ is a basis of $\mathtt{S}_{\nu_{\mathbf{b}}}$ and $\{e_i^*\}$ the dual basis, and $v_{ij} \in V$. Then

$$f_{\mathbf{b}}(g\sigma) = \sum_{i,j} \langle \sigma e_i^*, g\sigma e_j \rangle \cdot v_{ij}, \tag{4.4.2}$$

where $\langle \cdot, \cdot \rangle$ is the natural pairing between $\mathtt{S}_{\sigma(\nu_{\mathbf{b}})}$ and $\mathtt{S}_{\sigma(\nu^*_{\mathbf{b}})}$. In fact, (4.4.2) is valid as long as $\mathtt{S}_{\nu_{\mathbf{b}}} \cong \mathtt{W}_{\nu_{\mathbf{b}}}$ (even in positive characteristic).

PROPOSITION 4.4.2. *The collection* $\{f_{\mathbf{b}} \mid \mathbf{b} \in \sqcup_{\xi \in (\sigma-1)\mathbb{X}^\bullet(T)} \mathbb{B}(\xi)\}$ *forms a basis of* $\boldsymbol{J}(V)$ *as a* $\boldsymbol{J}$*-module.*

PROOF. This follows from the proof of the freeness of $\boldsymbol{J}(V)$ over $\boldsymbol{J}$. Indeed, since $\boldsymbol{J}_+(V) = \sum_\nu (\mathrm{fil}_\nu k[G] \otimes V)^{c_\sigma(G)}$ and the matrix coefficient map (4.4.1) lands in $(\mathrm{fil}_{\nu_{\mathbf{b}}} k[G] \otimes V)^{c_\sigma(G)}$, we may regard $f_{\mathbf{b}}$ as a homogeneous element in $\boldsymbol{J}_+(V)$, denoted as $f_{\mathbf{b},+}$. By definition, Lemma 2.2.2 and (4.3.2) from the proof of Proposition 4.3.5, the image of $\{f_{\mathbf{b},+}\}$ under the restriction $\boldsymbol{J}_+(V) \otimes_{\boldsymbol{J}_+} \boldsymbol{J}_0 \cong \boldsymbol{J}_0(V)$ forms a basis. The proof of Theorem 4.3.2 shows that $\{f_{\mathbf{b},+}\}$ forms a basis of $\boldsymbol{J}_+(V)$ over $\boldsymbol{J}_+$. Therefore, $\{f_{\mathbf{b}}\}$ forms a basis of $\boldsymbol{J}(V) = \boldsymbol{J}_+(V) \otimes_{\boldsymbol{J}_+} \boldsymbol{J}$ over $\boldsymbol{J}$. □

5. Chevalley groups with an automorphism

In this section, we establish a few results about Chevalley groups equipped with a pinned automorphism σ. Our conventions and notations are as in § 4. In particular, we have a pinned reductive group $(G, B, T, \{x_\alpha\}_{\alpha\in\Delta})$ over k. We further assume that σ is a finite order automorphism of G preserving the pinning, i.e. $\sigma \circ x_\alpha = x_{\sigma(\alpha)}$ for $\alpha \in \Delta$. Put $A = T/(\sigma - 1)T$. Let $W_0 = W^\sigma$, which acts on A, and let N_0 be the preimage of W_0 in N, which acts on T by twisted conjugation c_σ. Let $(\mathfrak{g}, \mathfrak{b}, \mathfrak{t}, \mathfrak{u})$ denote the Lie algebra of (G, B, T, U). Write G_{sc} for the simply-connected cover of G, and T_{sc} the maximal torus of G_{sc} which is the preimage of T.

5.1. Root datum with an automorphism. We start with some discussion of a version of folding of root systems. We also refer to [**Sp06**] for some related discussions.

For each σ-orbit $\mathcal{O} \subset \Phi(G, T)$, we write

$$\alpha_{\mathcal{O}} := \sum_{\gamma\in\mathcal{O}} \gamma,$$

which belongs to $\mathbb{X}^\bullet(A) \cong \mathbb{X}^\bullet(T)^\sigma \subset \mathbb{X}^\bullet(T)$. If we pick $\gamma \in \mathcal{O}$, then $\alpha_{\mathcal{O}} = \gamma + \sigma\gamma + \cdots + \sigma^{|\mathcal{O}|-1}\gamma$, where $|\mathcal{O}|$ denotes the cardinality of $\mathcal{O}$. Note that $\alpha_{\mathcal{O}}$ may be different from the image of γ under the usual norm map $\mathbb{X}^\bullet(T) \to \mathbb{X}^\bullet(A)$.

LEMMA 5.1.1. *The collection of $\alpha_{\mathcal{O}}$ for all σ-orbits $\mathcal{O} \subset \Phi(G,T)$, regarded as a subset of $\mathbb{X}^\bullet(A)$, has a structure of a root datum. Let G_σ denote the corresponding reductive group over k containing A as a maximal torus,[4] and let $\Phi(G_\sigma, A)$ denote the corresponding root system. Then the map $\mathcal{O} \mapsto \alpha_{\mathcal{O}}$ establishes a bijection between the set of σ-orbits in Δ and a set of simple roots in $\Phi(G_\sigma, A)$. With this choice of simple roots of $\Phi(G_\sigma, A)$, its subset of positive roots are $\Phi(G_\sigma, A)^+ = \{\alpha_{\mathcal{O}} \mid \mathcal{O} \subset \Phi(G,T)^+\}$.*

PROOF. Since σ lifts to a unique automorphism of the simply-connected cover of the derived group of G (e.g. see [**St68**, §9.16]), we may assume that G is semisimple and simply-connected. Then $(G,T) = \prod(G_i, T_i)$, where G_i is simple and simply-connected, and the action of σ permutes the simple factors. To prove the lemma, clearly we may assume that there are r simple factors, all of which are isomorphic to (G_0, T_0) and are cyclically permuted by σ, and that σ^r is an automorphism of $(\mathbb{X}^\bullet(T_0), \Phi(G_0, T_0))$ of order d. Then if $d = 1$, $G_\sigma = G_0$. If $d > 1$, then $\Phi(G,T)$ is of type A_n, D_n, or E_6, and one can check case by case that the reductive group G_σ is determined by the following table ($n \geq 1$).

G_0	SL_{2n+1}	SL_{2n+2}	$\mathrm{Spin}_{2n+2}, d=2$	E_6	$\mathrm{Spin}_8, d=3$
G_σ	SO_{2n+1}	Spin_{2n+3}	Sp_{2n}	F_4	G_2

The detailed calculations for the case $G_0 = \mathrm{SL}_{2n+1}, \mathrm{SL}_{2n+2}$ and $\mathrm{Spin}_{2n+2}, d = 2$ can be found in Example 5.1.5, Example 5.1.6 and Example 6.4.4 below. □

REMARK 5.1.2. Assume that G is semisimple and simply-connected. Note that the group of invariants G^σ of G under the σ-action is a connected semisimple group by [**St68**, Theorem 8.2] containing T^σ as a maximal torus (note however that the

[4] We choose the notation G_σ for the group because it has a maximal torus $A = T_\sigma$. This does not suggest that it relates to the σ-coinvariants of G, whatever it means.

group of invariants is denoted by G_σ in *loc. cit.*). The root system $\Phi(G^\sigma, T^\sigma)$ is isogenous to $\Phi(G_\sigma, T_\sigma)$ but in general is not isomorphic to it. For example, if $G = \mathrm{SL}_{2n}$ ($n > 1$) with σ nontrivial, then $G^\sigma = \mathrm{Sp}_{2n}$, whereas $G_\sigma = \mathrm{Spin}_{2n+1}$.

REMARK 5.1.3. Let $\hat{G}$ denote the (adjoint semisimple) Deligne–Lusztig dual group with dual torus $\hat{T}$ over a finite field κ (so that the absolute root datum of $(\hat{G}, \hat{T})$ is dual to the root datum $\Phi(G,T) \subset \mathbb{X}^\bullet(T)$ and the Weil descent datum defining $(\hat{G}, \hat{T})$ over κ induces the σ-action on $\Phi(G,T)$). Then the maximal split torus $\hat{A}$ of $\hat{G}$ is dual to A. For example, if $G = \mathrm{SL}_n$ with the non-trivial involution σ as in Example 5.1.5, Example 5.1.6 below, then $\hat{G}$ is isomorphic to the projectivel unitary group PU_n over κ. Then the dual root datum of $\Phi(G_\sigma, A)$ is equal to the sub-root system of the relative root system $\Phi_{\mathrm{rel}}(\hat{G}, \hat{A}) \subset \mathbb{X}^\bullet(\hat{A})$ consisting of those $\alpha^\vee \in \Phi_{\mathrm{rel}}(\hat{G}, \hat{A})$ such that $2\alpha^\vee \notin \Phi_{\mathrm{rel}}(\hat{G}, \hat{A})$.

It is easy to see (e.g. by a case-by-case inspection) that every root in $\Phi(G_\sigma, A)$ comes from one or two σ-orbits in $\Phi(G,T)$. In the latter case, the cardinality of one orbit is twice of the cardinality of the other.

DEFINITION 5.1.4. A root in $\Phi(G_\sigma, A)$ is called of *type* A if there is a unique σ-orbit $\mathcal{O} \subset \Phi(G,T)$ such that this root is $\alpha_{\mathcal{O}}$. In this case, the corresponding orbit $\mathcal{O}$ is also called of *type* A. Note that $\langle \alpha, \check{\beta} \rangle = 0$ for all pairs of distinct roots $\alpha, \beta \in \mathcal{O}$. If $\alpha_{\mathcal{O}}$ is a simple root in $\Phi(G_\sigma, A)$, then $\mathcal{O}$ is a σ-orbit of simple roots of $\Phi(G,T)$ and all vertices in the sub Dynkin diagram corresponding to $\mathcal{O}$ are isolated.

A root in $\Phi(G_\sigma, A)$ is called of *type* BC if there are two σ-orbits $\mathcal{O}^-$ and $\mathcal{O}^+$ such that this root is $\alpha_{\mathcal{O}^-} = \alpha_{\mathcal{O}^+}$, and that $|\mathcal{O}^-| = 2|\mathcal{O}^+|$. The orbit $\mathcal{O}^-$ (resp. $\mathcal{O}^+$) is called of *type* BC^- (resp. *type* BC^+). In this case, $\langle \sigma^{|\mathcal{O}^+|}\alpha, \check{\alpha} \rangle = 1$ for every $\alpha \in \mathcal{O}^-$, and $\beta := \alpha + \sigma^{|\mathcal{O}^+|}\alpha \in \mathcal{O}^+$. If $\alpha_{\mathcal{O}^-}$ is a simple root in $\Phi(G_\sigma, A)$, then $\mathcal{O}^-$ is a σ-orbit of simple roots in $\Phi(G,T)$, and the sub Dynkin diagram corresponding to $\mathcal{O}^-$ is a product of $|\mathcal{O}^+|$ copies of the root system A_2.

EXAMPLE 5.1.5. Let G be SL_{2r+1} ($r \geq 1$) with row and column indices in $\{-r, \dots, r\}$, the pinning (B, T, e) given by the group of standard upper triangular matrices, the subgroup of diagonal matrices, and $e = \sum_{i=-r}^{r-1} E_{i,i+1}$. Then the unique non-trivial pinned automorphism σ is given by

$$\sigma(X) = J\, {}^t X^{-1} J \quad \text{for } X \in \mathrm{SL}_{2r+1}, \tag{5.1.1}$$

where J is the anti-diagonal matrix, with entries $J_{i,-i} = (-1)^{r+i}$. Then

$$\Phi(G,T) = \{\varepsilon_i - \varepsilon_j \mid -r \leq i, j < r,\ i \neq j\},$$

where ε_i is the character of T given by evaluating at the (i,i)-entry. Since σ acts on $\Phi(G,T)$ by $\sigma(\varepsilon_i) = -\varepsilon_{-i}$, the σ-orbits on $\Phi(G,T)$ are

$$\mathcal{O}_{i,j} = \{\varepsilon_{-i} - \varepsilon_{-j},\ \varepsilon_j - \varepsilon_i\}, \quad \mathcal{O}_i^- = \{\varepsilon_{-i} - \varepsilon_0,\ \varepsilon_0 - \varepsilon_i\}, \quad \mathcal{O}_i^+ = \{\varepsilon_{-i} - \varepsilon_i\} \tag{5.1.2}$$

for $i \in \{\pm 1, \dots, \pm r\}$ and $j \in \{\pm 1, \dots, \pm(|i| - 1)\}$. They are of type A, BC^-, and BC^+ respectively.

As $A = T/(\sigma - 1)T$, its character group is

$$\mathbb{X}^\bullet(A) = \mathbb{X}^\bullet(T)^\sigma = \bigoplus_{i=1}^{r} \mathbb{Z}(\varepsilon_{-i} - \varepsilon_i).$$

The simple roots of $\Phi(G_\sigma, A)$ are $\varepsilon_{-i-1} - \varepsilon_{-i} + \varepsilon_i - \varepsilon_{i+1}$ for $i = 1, \dots, r-1$ and $\varepsilon_{-1} - \varepsilon_1$. The former ones are of type A (being equal to $\alpha_{\mathcal{O}_{i+1,i}}$) and the latter one is of type BC (being equal to $\alpha_{\mathcal{O}_1^-} = \alpha_{\mathcal{O}_1^+}$).

EXAMPLE 5.1.6. Let G be SL_{2r} ($r \geq 2$) with row and column indices in $\{-r, \dots, -1, 1, \dots, r\}$, the pinning (B, T, e) given by the group of standard upper triangular matrices, the subgroup of diagonal matrices, and $e = \sum_{i=-r}^{-2}(E_{i,i+1} + E_{-i-1,-i}) + E_{-1,1}$. Then the unique non-trivial pinned automorphism σ is given by

$$\sigma(X) = J\,{}^t X^{-1} J \quad \text{for } X \in \mathrm{SL}_{2r},$$

where J is the anti-diagonal matrix, with entries $J_{i,-i} = -J_{-i,i} = (-1)^{r+i}$ for $i = 1, \dots, r$. Then

$$\Phi(G, T) = \{\, \varepsilon_i - \varepsilon_j \mid -r \leq i, j \leq r,\, i \neq j,\, ij \neq 0\},$$

where ε_i is the character of T given by evaluating at the (i,i)-entry. Since σ acts on $\Phi(G,T)$ by $\sigma(\varepsilon_i) = -\varepsilon_{-i}$, the σ-orbits on $\Phi(G,T)$ are

$$\mathcal{O}_{i,j} = \{\varepsilon_{-i} - \varepsilon_{-j},\, \varepsilon_j - \varepsilon_i\}, \quad \mathcal{O}_i = \{\varepsilon_{-i} - \varepsilon_i\}$$

for $i \in \{\pm 1, \dots, \pm r\}$ and $j \in \{\pm 1, \dots, \pm(|i|-1)\}$. They are all of type A.

As $A = T/(\sigma - 1)T$, its character group is

$$\mathbb{X}^\bullet(A) = \mathbb{X}^\bullet(T)^\sigma = \Big(\bigoplus_{i=1}^r \mathbb{Z}(\varepsilon_{-i} - \varepsilon_i)\Big) + \mathbb{Z}(\varepsilon_{-r} + \cdots + \varepsilon_{-1}).$$

The simple roots of $\Phi(G_\sigma, A)$ are $\varepsilon_{-i-1} - \varepsilon_{-i} + \varepsilon_i - \varepsilon_{i+1}$ for $i = 1, \dots, r-1$ and $\varepsilon_{-1} - \varepsilon_1$.

LEMMA 5.1.7. *The action of the Weyl group $N_{G_\sigma}(A)/A$ on A is identified with the natural action of W_0 on A.*

PROOF. For a root $\alpha \in \Phi(G,T)$, let s_α denote the corresponding reflection acting on $\mathbb{X}^\bullet(T)$. Let $\alpha_{\mathcal{O}}$ be a simple root of $\Phi(G_\sigma, A)$ and $s_{\alpha_{\mathcal{O}}}$ the corresponding simple reflection. One checks easily that $s_{\alpha_{\mathcal{O}}} = \prod_{\gamma \in \mathcal{O}} s_\alpha \in W_0$ if $\alpha_{\mathcal{O}}$ is of type A, and $s_{\alpha_{\mathcal{O}}} = \prod_{\alpha \in \mathcal{O}^+} s_\alpha \in W_0$ if $\alpha_{\mathcal{O}}$ is of type BC. Therefore, $N_{G_\sigma}(A)/A \subset W_0$ as automorphisms of A. It remains to notice that they are isomorphic as abstract Coxeter groups, as can be checked easily from the table in Lemma 5.1.1. □

We discuss subgroups of G associated to σ-orbits of roots, which specialize to "root SL_2" when $\sigma = \mathrm{id}$. First, we have the following lemma.

LEMMA 5.1.8. *Let $\alpha \in \Phi(G,T)$. If the σ-orbit $\mathcal{O}$ containing α is of type* A *or* BC$^-$, *then $\sigma^{|\mathcal{O}|} \circ x_\alpha = x_\alpha$. If the σ-orbit $\mathcal{O}$ containing α is of type* BC$^+$, *then $\sigma^{|\mathcal{O}|} \circ x_\alpha = -x_\alpha$.*

PROOF. First, if α is a simple root, then $\sigma^{|\mathcal{O}|} \circ x_\alpha = x_\alpha$ since σ acts by pinned automorphisms. In general, every σ-orbit $\mathcal{O}$ of type A or of type BC$^-$ is conjugated to a σ-orbit of simple roots by an element $w \in W_0 = W^\sigma$. We can choose a lifting of w to a σ-invariant element $\dot{w}$ in N (see [**St68**, (5), p. 55]). For a root $\alpha \in \mathcal{O}$, write $\mathrm{Ad}_{\dot{w}}(E_\alpha) = cE_{w(\alpha)}$ for some invertible constant c. Then $\mathrm{ord}(|\mathcal{O}|) = \mathrm{ord}(|w(\mathcal{O})|)$ and

$$\mathrm{Ad}_{\dot{w}}(E_\alpha) = cE_{w(\alpha)} = \sigma^{|\mathcal{O}|}(cE_{w(\alpha)}) = \sigma^{|\mathcal{O}|}(\mathrm{Ad}_{\dot{w}}(E_\alpha)) = \mathrm{Ad}_{\dot{w}}(\sigma^{|\mathcal{O}|}(E_\alpha)).$$

Therefore, $\sigma^{|\mathcal{O}|} \circ x_\alpha = x_\alpha$.

Next, assume that α belongs to a σ-orbit $\mathcal{O}^+$ of type BC^+, then $\alpha = \beta + \sigma^{|\mathcal{O}^+|}(\beta)$ for some root β in a σ-orbit $\mathcal{O}^-$ of type BC^-, and $E_\alpha = c[E_\beta, E_{\sigma^{|\mathcal{O}^+|}(\beta)}]$ for some invertible constant c. Note that $\sigma^{|\mathcal{O}^+|}$ will send E_β to $c' E_{\sigma^{|\mathcal{O}^+|}(\beta)}$ and $E_{\sigma^{|\mathcal{O}^+|}(\beta)}$ to $c'^{-1} E_\beta$ for some invertible c'. Hence $\sigma^{|\mathcal{O}^+|}(E_\alpha) = -E_\alpha$, and therefore

$$\sigma^{|\mathcal{O}^+|} \circ x_\alpha = -x_\alpha.$$

□

Now, for a σ-orbit $\mathcal{O} \subset \Phi(G,T)^+$ of positive roots, let $G_{\mathcal{O}}$ be the subgroup of G generated by T and root subgroups $U_{\pm\alpha}$ for $\alpha \in \mathcal{O}$. Clearly, this is a σ-stable reductive subgroup of G. Let $U_{\mathcal{O}}$ (resp. $U_{\mathcal{O},-}$) be the subgroup of U generated by U_α for $\alpha \in \mathcal{O}$ (resp. $-\alpha \in \mathcal{O}$), and let $B_{\mathcal{O}} = U_{\mathcal{O}} T$. If $\mathcal{O}$ is a σ-orbit of simple roots, we also let $U^{\mathcal{O}}$ denote the subgroup of U generated by U_β, for those positive roots β that are not in the sub-root system spanned by $\mathcal{O}$. Then $B = U^{\mathcal{O}} B_{\mathcal{O}}$ is a semi-direct product decomposition, and $P_{\mathcal{O}} = U_{\mathcal{O}} G_{\mathcal{O}}$ is a σ-stable standard parabolic subgroup of G.

Note that $(G_{\mathcal{O}}, B_{\mathcal{O}}, T, \{x_\alpha\}_{\alpha \in \mathcal{O}})$ is a pinning of $G_{\mathcal{O}}$. When G is semisimple and simply-connected, there are essentially two cases for its derived subgroup $G_{\mathcal{O},\mathrm{der}}$ (which is always simply-connected). The following lemma also follows from Lemma 5.1.8 (and in fact is equivalent to Lemma 5.1.8).

LEMMA 5.1.9. *Assume that G is semisimple and simply-connected.*

(1) *If $\mathcal{O}$ is of type A, then $G_{\mathcal{O},\mathrm{der}} \cong \prod_{i=1}^{|\mathcal{O}|} \mathrm{SL}_2$ where the σ acts by permuting factors and preserves the pinning (up to possibly rescaling the x_α's).*

(2) *If $\mathcal{O}$ is of type BC^-, then $G_{\mathcal{O},\mathrm{der}} \cong \prod_{i=1}^{|\mathcal{O}|/2} \mathrm{SL}_3$ where the σ acts by permuting factors and $\sigma^{|\mathcal{O}|/2}$ acts on each factor SL_3 as in Example 5.1.5, which also preserves the pinning (up to possibly rescaling the x_α's).*

(3) *If $\mathcal{O}$ is of type BC^+, then $G_{\mathcal{O},\mathrm{der}} \cong \prod_{i=1}^{|\mathcal{O}|} \mathrm{SL}_2$, where the σ acts by*

$$(g_1, \ldots, g_{|\mathcal{O}|}) \longmapsto \left(\mathrm{Ad}_{\left(\begin{smallmatrix} 1 & 0 \\ 0 & -1 \end{smallmatrix}\right)}(g_{|\mathcal{O}|}), g_1, \ldots, g_{|\mathcal{O}|-1}\right). \tag{5.1.3}$$

Let $\mathbf{T} = B/U$ denote the abstract Cartan. Note that σ acts on $\mathbf{T}$, and $T \to B \to \mathbf{T}$ is a σ-equivariant isomorphism. Let $\mathbf{A} = \mathbf{T}/(1-\sigma)\mathbf{T}$, which is canonically isomorphic to A. By transport of structures, W_0 acts on $\mathbf{A}$, and for every root $\alpha_{\mathcal{O}} \in \Phi(G_\sigma, A)$, $e^{\alpha_{\mathcal{O}}}$ can be regarded as a regular function on $\mathbf{A}$. Let

$$q_B : B \to \mathbf{T} \to \mathbf{A} \tag{5.1.4}$$

denote the quotient map.

DEFINITION 5.1.10. Let $\mathcal{O} \subset \Phi(G, A)$ be a σ-orbit. We define a divisor

$$\mathbf{A}_{\mathcal{O}} = \begin{cases} \{t \in \mathbf{A} \mid e^{\alpha_{\mathcal{O}}}(t) = 1\} & \mathcal{O} \text{ is of type } \mathsf{A} \text{ or } \mathsf{BC}^-, \\ \{t \in \mathbf{A} \mid e^{\alpha_{\mathcal{O}}}(t) = -1\} & \mathcal{O} \text{ is of type } \mathsf{BC}^+. \end{cases}$$

Note that $\mathbf{A}_{\mathcal{O}} = \mathbf{A}_{-\mathcal{O}}$. Let

$$\mathring{\mathbf{A}} = \mathbf{A} - \cup_{\mathcal{O}} \mathbf{A}_{\mathcal{O}},$$

where the union is taken over all σ-orbits $\mathcal{O} \subset \Phi(G,T)$. This is a W_0-invariant open subset of $\mathbf{A}$. For a σ-orbit $\mathcal{O} \subset \Phi(G,T)$, let

$$\mathbf{A}^{[\mathcal{O}]} = \mathbf{A} - \bigcup_{\mathcal{O}' \neq \mathcal{O}} \mathbf{A}_{\mathcal{O}'}.$$

In particular, it is an open neighborhood of the generic points of $\mathbf{A}_{\mathcal{O}}$ in $\mathbf{A}$.

Let $\mathring{B}$ and $\mathring{T}$ be the preimage of $\mathring{\mathbf{A}}$ under the natural projections $q_B : B \to \mathbf{A}$ and $T \cong \mathbf{T} \to \mathbf{A}$. Similarly, let $B^{[\mathcal{O}]}$ and $B_{\mathcal{O}'}^{[\mathcal{O}]}$ be the preimage of $\mathbf{A}^{[\mathcal{O}]}$ under the projections $B \to \mathbf{A}$ and $B_{\mathcal{O}'} \to \mathbf{A}$. Finally, via the isomorphism $A \cong \mathbf{A}$, we have similarly defined spaces $A_{\mathcal{O}}, \mathring{A}$ and $A^{[\mathcal{O}]}$.

REMARK 5.1.11. Note that $A_{\mathcal{O}}$ may not be irreducible nor reduced in general. For example, if $G_\sigma = \mathrm{Sp}_{2n}$, then every long root in $\Phi(G_\sigma, A)$ is twice of a weight of A and therefore the corresponding divisor $A_{\mathcal{O}}$ consists of two connected components if char $k > 2$, and is non-reduced if char $k = 2$. On the other hand, if char $k > 2$, then $A_{\mathcal{O}}$ is always reduced.

LEMMA 5.1.12. *Assume that G is semisimple and simply-connected. Then $\mathbf{A}/\!/W_0$ is isomorphic to an affine space, and the natural morphism $\mathbf{A} \to \mathbf{A}/\!/W_0$ is a finite and flat W_0-cover, with branch loci $\cup_{\mathcal{O}} \mathbf{A}_{\mathcal{O}}$. In particular, the restriction $\mathring{\mathbf{A}} \to \mathring{\mathbf{A}}/\!/W_0$ is a finite étale Galois cover.*

PROOF. We may identify the action of W_0 on $\mathbf{A}$ with the action of $W_0 \cong N_{G_\sigma}(A)/A$ on A by Lemma 5.1.7. Note that by table in the proof of Lemma 5.1.1, G_σ is isomorphic to a product of simply-connected groups and odd orthogonal groups. Therefore, $k[\mathbf{A}]^{W_0}$ is a polynomial algebra (see [**St75**]), and $\mathbf{A} \to \mathbf{A}/\!/W_0$ is finite flat. The branch locus of the covering $\mathbf{A} \to \mathbf{A}/\!/W_0$ is the union of divisors $\{t \in \mathbf{A} \mid \check{\alpha}_{\mathcal{O}}(e^{\alpha_{\mathcal{O}}}(t)) = 1, \alpha_{\mathcal{O}} \in \Phi(G_\sigma, A)\}$. Here $\check{\alpha}_{\mathcal{O}}$ denotes the coroot of $\alpha_{\mathcal{O}}$, regarded as a homomorphism $\check{\alpha}_{\mathcal{O}} : \mathbb{G}_m \to A \cong \mathbf{A}$. Its construction is as follows. Let $\gamma \in \mathcal{O}$ if $\alpha_{\mathcal{O}}$ is of type A or $\gamma \in \mathcal{O}^+$ if $\alpha_{\mathcal{O}}$ is of type BC. Let $\check{\gamma}$ be the corresponding coroot, regarded as a homomorphism $\mathbb{G}_m \to T$. Then $\check{\alpha}_{\mathcal{O}}$ is the composition of this homomorphism with the projection $T \to A \cong \mathbf{A}$. Again, by checking the table in the proof of Lemma 5.1.1, one sees that $\ker \check{\alpha}_{\mathcal{O}}$ is trivial if $\alpha_{\mathcal{O}}$ is of type A and is $\{\pm 1\}$ if $\alpha_{\mathcal{O}}$ is of type BC. The lemma follows. □

5.2. Twisted conjugacy classes. In this subsection, we study σ-regular elements of G, generalizing some of the well-known results of Steinberg [**St65**] for $\sigma = \mathrm{id}$. Some results with restriction of the characteristic of k were also obtained by Mohrdieck [**Mo03**] before.

Let G be a reductive group over k. Recall the σ-twisted conjugation (or σ-conjugation for brevity)

$$c_\sigma : G \times G\sigma \to G\sigma, \quad (h, g\sigma) = hg\sigma(h)^{-1}\sigma =: c_\sigma(h)(g).$$

Since σ preserves (B,T), it acts on the set $\Delta \subset \Phi(G,T)$ of simple roots.

Let I be the *centralizer group scheme* for the action of G on itself by c_σ, i.e. it is the group scheme over G defined by the Cartesian diagram

$$\begin{array}{ccc} I & \longrightarrow & G \times G\sigma \\ \downarrow & & \downarrow{\scriptstyle c_\sigma \times \mathrm{pr}_2} \\ G & \longrightarrow & G \times G\sigma, \end{array}$$

where pr_2 denotes the projection to the second factor and $G \to G \times G\sigma$ denotes the "diagonal" embedding $g \mapsto (g, g\sigma)$. Its fiber over $g \in G$ is denoted by I_g.

DEFINITION 5.2.1. Let $G^{\sigma\text{-reg}}$ denote the *σ-regular locus* of G. Namely,

$$G^{\sigma\text{-reg}} = \{g \in G \mid \dim I_g = \dim T^\sigma =: r\} \subset G.$$

Let $B^{\sigma\text{-reg}} := G^{\sigma\text{-reg}} \cap B$, $T^{\sigma\text{-reg}} = G^{\sigma\text{-reg}} \cap T$ and $U^{\sigma\text{-reg}} = G^{\sigma\text{-reg}} \cap U$.

REMARK 5.2.2. (1) It will be clear from the following discussion that $\dim I_g \geq r$ for any $g \in G$. Therefore, $G^{\sigma\text{-reg}}$ is an open subset of G by semi-continuity.

(2) When $\mathrm{char}\, k = 0$, $\dim I_g = \mathrm{Lie}\, I_g = \ker(\mathrm{id} - \mathrm{Ad}_g\sigma : \mathfrak{g} \to \mathfrak{g})$. Complication arises in the positive characteristic case; for example, the center $\mathfrak{z}$ of $\mathfrak{g}$, i.e.

$$\mathfrak{z} := \bigcap_{\alpha \in \Phi(G,T)} \mathrm{Ker}(d\alpha : \mathfrak{t} \to k),$$

may not be trivial even if G is semisimple.

(3) For every automorphism τ of the algebraic group G, one can define the open subset of τ-regular elements $G^{\tau\text{-reg}} \subset G$ as those $g \in G$ such that I_g achieves the minimal dimension. But since every τ differs from some pinned automorphism σ by an inner automorphism, the study of $G^{\tau\text{-reg}}$ reduces to the study of $G^{\sigma\text{-reg}}$, by virtual of Remark 4.1.1.

We first study $U^{\sigma\text{-reg}}$. Let $V = [U, U] \cdot (\sigma - 1)(\prod_{\alpha \in \Delta} U_\alpha) \subset U$, and let

$$q_U : U \to W = U/V$$

denote the quotient map. The natural conjugation of T^σ on U induces an action of T^σ on W, and there is a unique T^σ-open orbit $\mathring{W} \subset W$. Explicitly, if we choose an element $\alpha \in \mathcal{O}$ for each σ-orbit of simple roots so that there is an isomorphism

$$\prod_{\text{chosen } \alpha} x_\alpha : \mathbb{G}_a^{|\Delta^\sigma|} \to U \to W$$

giving a basis of W (as a vector space over k), then $v \in \mathring{W}$ if and only if all the coefficients of v with respect to this basis are non-zero. Let $\mathring{U} = q_U^{-1}(\mathring{W}) \subset U$.

We say that an element $w \in W$ is a *σ-twisted Coxeter element* in W, if one can choose an element $\alpha \in \mathcal{O}$ for each σ-orbit $\mathcal{O}$ of simple roots and write w as the product of simple reflections corresponding to these chosen α's (in some order). Let $\dot{w}$ be a representative of w in N.

LEMMA 5.2.3. *Assume that G is simply-connected. Let w be a σ-twisted Coxeter element in W. Then for every $b \subset B$,*

$$\dim \ker(\mathrm{id} - \mathrm{Ad}_{b\dot{w}}\sigma : \mathfrak{g} \to \mathfrak{g}) \leq \dim \mathfrak{z}^\sigma + r.$$

PROOF. Given the properties of σ-twisted Coxeter elements presented in [**Sp74**, Theorem 7.6], the proof of [**St65**, Lemma 4.3] applies literally to our situation. □

Now we come to the first key result.

PROPOSITION 5.2.4. *For each σ-orbit of simple roots $\mathcal{O} \subset \Phi(G,T)$, choose one $\alpha \in \mathcal{O}$ and define*

$$x := \prod_\alpha x_\alpha^{-1}(1) \in U,$$

where the product is taken over the chosen simple roots in a fixed order. Then

(1) $(\mathrm{id} - \mathrm{Ad}_x\sigma)^{-1}(\mathfrak{u}) = \mathfrak{t}^\sigma + \mathfrak{u}$.
(2) *Let* $g \in G$. *If* $gx\sigma(g)^{-1} \in U$, *then* $g \in B$. *In addition,* $I_x \subset B$.
(3) $x \in U^{\sigma\text{-reg}}$; *in particular,* $U^{\sigma\text{-reg}}$ *is nonempty.*
(4) $\mathring{U} = U^{\sigma\text{-reg}}$ *is a single orbit under the* σ*-conjugation action of* $T^\sigma \cdot U$ *on* U. *In particular,* $I_u \subset B$ *and* $(\mathrm{id} - \mathrm{Ad}_u\sigma)^{-1}(\mathfrak{u}) = \mathfrak{t}^\sigma + \mathfrak{u}$ *for every* $u \in U^{\sigma\text{-reg}}$.

PROOF. Write $\mathfrak{g}_x$ for the Lie algebras of I_x, and similarly $\mathfrak{b}_x$ and $\mathfrak{u}_x$ for the Lie algebra of $I_x \cap B$ and $I_x \cap U$. We have the following lemma.

LEMMA 5.2.5. *We have* $\dim \mathfrak{u}_x = |\Delta^\sigma|$ *and* $\mathfrak{g}_x = \mathfrak{b}_x \subset \mathfrak{t}^\sigma + \mathfrak{u}$.

PROOF. The inclusion $\mathfrak{b}_x \subset \mathfrak{t}^\sigma + \mathfrak{u}$ is clear. We prove the two equalities in the statement. We first assume that G is semisimple and simply-connected. Then $r = \dim T^\sigma = |\Delta^\sigma|$. Let w_0 be the longest element in the Weyl group and $\dot{w}_0 \in N_0$ a representative of w_0. Then $\dot{w}_0 x \dot{w}_0 \in B\dot{w}B$ for some σ-twisted Coxeter element w. Then as in [**St65**, Theorem 4.6],

$$\dim \mathfrak{z}^\sigma + \dim \mathfrak{u}_x \leq \dim \mathfrak{b}_x \leq \dim \mathfrak{g}_x \leq \dim \mathfrak{z}^\sigma + r.$$

Therefore $\dim \mathfrak{u}_x \leq r$ and $\dim(I_x \cap U) \leq r$. On the other hand, U acts on fibers of q_U via σ-conjugation. Therefore, $\dim(I_x \cap U) \geq \dim W = r$. Putting them together gives $\dim \mathfrak{u}_x = \dim(I_x \cap U) = r$ and $\mathfrak{b}_x = \mathfrak{g}_x$. Now for a general reductive group G, let G_{sc} be the simply-connected cover of its derived group. Then σ lifts to a unique automorphism of G_{sc} (e.g. see [**St68**, §9.16]). Since the central isogeny $G_{\mathrm{sc}} \to G$ induces an isomorphism on unipotent subgroups, we have $\dim(I_x \cap U) = |\Delta^\sigma|$. In addition, since the kernels and the cokernels of the two maps $\mathfrak{b}_{\mathrm{sc}} \to \mathfrak{b}$ and $\mathfrak{g}_{\mathrm{sc}} \to \mathfrak{g}$ are equal, and since $(\mathfrak{b}_{\mathrm{sc}})_x = (\mathfrak{g}_{\mathrm{sc}})_x$, $\mathfrak{b}_x = \mathfrak{g}_x$ also holds for G. □

Now we prove the proposition. (1) First, note that $(\mathrm{id} - \mathrm{Ad}_x\sigma)(\mathfrak{u}) \subset \mathfrak{v}$, where $\mathfrak{v}$ is the Lie algebra of V. But since $\dim \mathfrak{u}_x = \dim \mathfrak{u} - \dim \mathfrak{v}$, we have $(\mathrm{id} - \mathrm{Ad}_x\sigma)(\mathfrak{u}) = \mathfrak{v}$. Next, if G is of adjoint type, then the composition of maps

$$\mathfrak{t}^\sigma \xrightarrow{\mathrm{id} - \mathrm{Ad}_x\sigma} \mathfrak{u} \to \mathfrak{u}/\mathfrak{v}$$

is an isomorphism. It follows that in this case $(\mathrm{id} - \mathrm{Ad}_x\sigma)(\mathfrak{t}^\sigma + \mathfrak{u}) = \mathfrak{u}$. Since $\mathfrak{g}_x = \mathfrak{b}_x \subset \mathfrak{t}^\sigma + \mathfrak{u}$, we see that $(\mathrm{id} - \mathrm{Ad}_x\sigma)^{-1}(\mathfrak{u}) = \mathfrak{t}^\sigma + \mathfrak{u}$. In general, let G_{ad} be the adjoint quotient of G. Since the central isogeny induces an isomorphism $\mathfrak{u} \cong \mathfrak{u}_{\mathrm{ad}}$, it is easy to see that $(\mathrm{id} - \mathrm{Ad}_x\sigma)^{-1}(\mathfrak{u}) \subset \mathfrak{b}$. In addition, if $Y \in \mathfrak{b}$ such that $Y - \mathrm{Ad}_x\sigma(Y) \in \mathfrak{u}$, then $(Y \mod \mathfrak{u}) \in \mathfrak{t}^\sigma$. Therefore, $(\mathrm{id} - \mathrm{Ad}_x\sigma)^{-1}(\mathfrak{u}) = \mathfrak{t}^\sigma + \mathfrak{u}$ holds in general.

(2) Write $g = u_1 n u_2$ under the Bruhat decomposition for some $n \in N, u_1$, $u_2 \in U$. Then $nu_2 x\sigma(u_2)^{-1} = u_1^{-1} u \sigma(u_1)\sigma(n)$ for some $u \in U$. It follows that $n = \sigma(n)$ by the uniqueness of the Bruhat decomposition. Then

$$nu_2 x\sigma(u_2)^{-1} n^{-1} = u_1^{-1} u \sigma(u_1) \in U.$$

Since $q_U(u_2 x \sigma(u_2)^{-1}) = q_U(x) \in \mathring{W}$, we must have $n \in T$ and therefore $g \in B$. It follows that $I_x^{\mathrm{red}} \subset B$. Together with $\mathfrak{g}_x = \mathfrak{b}_x$, we deduce that $I_x \subset B$.

(3) Let $b \in I_x \subset B$. We write $b = tu$ according to the decomposition $B = TU$. Then $tux = x\sigma(t)\sigma(u)$. By projecting along $B \to \mathbf{T}$, we see that $\sigma(t) = t$, and

thus $t^{-1}xtx^{-1} = ux\sigma(u)^{-1}x^{-1} \in V$. This shows that $t \in (\cap_{\alpha\in\Delta} \ker\alpha)^\sigma$, which has dimension $\dim T^\sigma - |\Delta^\sigma|$, and that $u \in I_x \cap U$. Therefore,

$$\dim I_x = (\dim T^\sigma - |\Delta^\sigma|) + \dim(I_x \cap U) = \dim T^\sigma,$$

i.e. x is σ-regular.

(4) Note that $q_U^{-1}(x) \subset U$ is a single orbit under the σ-conjugation action of U on itself. Indeed, since U is unipotent, the orbit through x is a closed subset of $q_U^{-1}(x)$. On the other hand, since $\dim(I_x \cap U) = \dim W$, the dimension of this orbit is equal to the dimension of $q_U^{-1}(x)$. Therefore, $q_U^{-1}(x)$ is an orbit. Now, let $u \in \mathring{U}$. After taking a conjugation by elements in T^σ, we may assume that $q_U(x) = q_U(u)$, and therefore x and u are σ-conjugate by an element in U. This shows that $\mathring{U} \subset U^{\sigma\text{-reg}}$, and $\mathring{U}$ is a single $T^\sigma \cdot U$-orbit. On the other hand, let $u \in U^{\sigma\text{-reg}}$. Since $\dim I_u = \dim T^\sigma$, the $T^\sigma \cdot U$-orbit through u is $\dim U$-dimensional, and therefore must meet $\mathring{U}$. This implies that $\mathring{U} = U^{\sigma\text{-reg}}$. The last statement follows from Part (1) and (2). □

LEMMA 5.2.6. *An element $u \in U$ is σ-regular if and only if the set $\mathcal{B}_u := \{gB \in G/B \mid g^{-1}u\sigma(g) \in B\}$ is finite.*

PROOF. Proposition 5.2.4 implies that if $u \in U^{\sigma\text{-reg}}$, then $\mathcal{B}_u$ consists of only one element. Now, let $u \in U - U^{\sigma\text{-reg}}$. Then after a σ-conjugation by an element in U, we may assume that there is a σ-orbit $\mathcal{O}$ of simple roots, such that $u \in U^{\mathcal{O}}$, where $U^{\mathcal{O}}$ is the subgroup of U introduced before Lemma 5.1.9. Then $\mathcal{B}_u$ contains a positive dimensional subvariety $(G_{\mathcal{O}}/B_{\mathcal{O}})^\sigma$. The lemma is proved. □

Next, we study $T^{\sigma\text{-reg}}$, and then $B^{\sigma\text{-reg}}$. Recall the map q_B as in (5.1.4) and Definition 5.1.10. For $t \in T(k)$, let $\Phi(G,T)_t \subset \Phi(G,T)$ be the smallest sub-root system containing those σ-orbits $\mathcal{O}$ such that $q_B(t\sigma) \in \mathbf{A}_{\mathcal{O}}(k)$. We allow $\Phi(G,T)_t = \emptyset$ if $t \in \mathring{T}$. Let $G_t \subset G$ denote the corresponding reductive subgroup, and $B_t = G_t \cap B$ a Borel subgroup of G_t. Its unipotent radical U_t is the subgroup generated by $\{U_\alpha\}$, for those positive roots $\alpha \in \Phi(G,T)_t$. We write $B_t = U_tT$. Let $\mathfrak{g}_t$ denote the Lie algebra of G_t.

LEMMA 5.2.7. *The sub-root system $\Phi(G,T)_t$ is exactly the union of σ-orbits of roots in $\Phi(G,T)$ of the following three types:*

(1) *$\mathcal{O}$, if $\mathcal{O}$ is of type* A *and $e^{\alpha_{\mathcal{O}}}(t) = 1$;*
(2) *$\mathcal{O}^- \cup \mathcal{O}^+$, if $\mathcal{O}^\pm$ is of type* $\mathsf{BC}^\pm$ *and $e^{\alpha_{\mathcal{O}}}(t) = 1$;*
(3) *$\mathcal{O}$, if $\mathcal{O}$ is of type* BC^+ *and $e^{\alpha_{\mathcal{O}}}(t) = -1$.*

PROOF. Indeed, the smallest sub-root system containing those σ-orbits $\mathcal{O}$ such that $q_B(t\sigma) \in \mathbf{A}_{\mathcal{O}}(k)$ would necessarily contain the these roots. Therefore, it is enough to show that they indeed form a sub-root system of $\Phi(G,T)$, or equivalently, if $\mathcal{O}_1, \mathcal{O}_2$ are two σ-orbits from the above types, and $\alpha_i \in \mathcal{O}_i$, then the σ-orbit $\mathcal{O}$ containing $s_{\alpha_i}(\alpha_j)$ would also be one of the above types. To check this, write $\beta = s_{\alpha_i}(\alpha_j)$, and $\beta_{\mathcal{O}}$ the sum over the σ-orbit of β. We may assume that $\beta \neq \pm\alpha_j$ so in particular α_i and α_j are in the same irreducible factor of the root system $\Phi(G,T)$. Assume that $\tau = \sigma^r$ fixes this irreducible factor Φ. If (Φ,τ) is as in Example 5.1.5, one checks directly that $\beta_{\mathcal{O}}$ also belongs to one of the above types. If (Φ,τ) is not as in Example 5.1.5, then it is readily to see that $\beta_{\mathcal{O}}$ is an integral linear combination of $\alpha_{\mathcal{O}_1}$ and $\alpha_{\mathcal{O}_2}$ and all $\alpha_{\mathcal{O}_1}$, $\alpha_{\mathcal{O}_2}$ and $\beta_{\mathcal{O}}$ are of type (1). The lemma is proven. □

LEMMA 5.2.8. *Let $\Delta_t \subset \Phi(G,T)_t$ be the set of simple roots (with respect to (B_t, T)). Then after possible rescaling x_a's, the automorphism $G_t \to G_t, g \mapsto t\sigma(g)t^{-1}$ preserves the pinning $(G_t, B_t, T, \{x_\alpha\}_{\alpha\in\Delta_t})$.*

PROOF. Note that if $\mathcal{O} \subset \Delta_t$ is a σ-orbit, then $e^{\alpha_{\mathcal{O}}}(t) = 1$ if $\mathcal{O}$ is of type A and BC^-, and $e^{\alpha_{\mathcal{O}}}(t) = -1$ if $\mathcal{O}$ is of type BC^+. Then the lemma follows from Lemma 5.1.9. □

LEMMA 5.2.9. *Let $b = ut \in U_tT$. Then $\mathrm{id} - \mathrm{Ad}_b\sigma : \mathfrak{g}/\mathfrak{g}_t \to \mathfrak{g}/\mathfrak{g}_t$ is an isomorphism. In particular,*

$$\ker(\mathrm{id} - \mathrm{Ad}_b\sigma : \mathfrak{g} \to \mathfrak{g}) = \ker(\mathrm{id} - \mathrm{Ad}_b\sigma : \mathfrak{g}_t \to \mathfrak{g}_t), \tag{5.2.1}$$

and $\dim I_b = \dim(I_b \cap G_t)$.

PROOF. For $i \in \mathbb{Z}$, let Φ_i denote the set of positive roots α of height i, i.e. $\langle\alpha, \check{\rho}\rangle = i$, where $\check{\rho}$ is the half of the sum of positive coroots. We choose the following basis of $\mathfrak{g}$ (in the given order): first a basis in $\mathfrak{g}_t$; then the standard root basis $\{E_\alpha\}$ associated to the roots in $\ldots, \Phi_i, \Phi_{i-1}, \ldots$ but not in $\Phi(G,T)_t$, with σ-orbits grouped together; and finally the standard root basis $\{E_\alpha\}$ associated to the corresponding negative roots in $\ldots, \Phi_{i-1}, \Phi_i, \ldots$ but not in $\Phi(G,T)_t$, with σ-orbits grouped together. With respect to this choice of basis, the linear operator $\mathrm{id} - \mathrm{Ad}_b\sigma$ is represented by a block upper triangular matrix M, where the first block corresponds to $\mathfrak{g}_t$, and other blocks correspond to σ-orbits of roots not in $\Phi(G,T)_t$.

Note that a diagonal block that corresponds to a σ-orbit $\mathcal{O}'$ not in $\Phi(G,T)_t$ is invertible due to: (i) its determinant is $\pm(e^{\alpha_{\mathcal{O}'}}(t) - 1)$ if $\mathcal{O}'$ is of type A or BC^- and $\pm(e^{\alpha_{\mathcal{O}'}}(t) + 1)$ if $\mathcal{O}'$ is of type BC^+, which follows from Lemma 5.1.8 and an easy computation; and (ii) $q_B(t) \notin \mathbf{A}_{\mathcal{O}'}$. The first claim of the lemma follows. Then clearly (5.2.1) holds, which in turn implies that $\mathrm{Lie}(I_b^{\mathrm{red}}) \subset \mathrm{Lie} I_b \subset \mathfrak{g}_t$. Therefore the neutral connected component of I_b^{red} is a closed subgroup of G_t. The lemma is proved. □

A very similar argument yields the following "σ-twisted" Jordan decomposition.

LEMMA 5.2.10. *Let $b = ut \in UT$. Then b is σ-conjugated by an element in U to an element $u't$ with $u' \in U_t$.*

PROOF. Let $u_1 = u$. By induction on i, one can show that we can σ-conjugate b by an element in U to u_it, where u_i is in the subgroup of U generated by U_t and U_α for $\alpha \in \Phi_i \cup \Phi_{i+1} \cup \cdots$. This is because if $t \notin \mathbf{A}_{\mathcal{O}'}$, then $\mathrm{id} - \mathrm{Ad}_t\sigma$ is invertible on the space $\oplus_{\alpha\in\mathcal{O}'} kE_\alpha$. □

LEMMA 5.2.11. *We have equalities $T^{\sigma\text{-reg}} = \mathring{T}$.*

PROOF. Note that for $t \in \mathring{T}$, $G_t = T$, and $I_t \cap T = T^\sigma$. It follows from Lemma 5.2.9 that $t \in T^{\sigma\text{-reg}}$.

Next we show that $T^{\sigma\text{-reg}} \subset \mathring{T}$, i.e. if $q_B(t\sigma) \in \mathbf{A}_{\mathcal{O}}(k)$ for some σ-orbit $\mathcal{O} \subset \Phi(G,T)$, then I_t contains a unipotent subgroup and therefore $\dim I_t > \dim T^\sigma$ (as clearly $T^\sigma \subset I_t$). If $\mathcal{O}$ is of type A or BC^+, pick $\alpha \in \mathcal{O}$ and write $\alpha_i = \sigma^i(\alpha)$ for $i = 1, \ldots, |\mathcal{O}|$. These α_i's are pairwise orthogonal and we may assume that $\sigma \circ x_{\alpha_i} = x_{\alpha_{i+1}}$ for $i = 1, \ldots, |\mathcal{O}| - 1$ and thus $\sigma \circ x_{|\mathcal{O}|}$ is equal to x_{α_1} if $\mathcal{O}$ is of type A and is equal to $-x_{\alpha_1}$ if $\mathcal{O}$ is of type BC^+. One checks that, for $c \in k$,

$$\prod_{i=1}^{|\mathcal{O}|} x_{\alpha_i}(e^{\alpha_2+\cdots+\alpha_i}(t)c)$$

form a unipotent subgroup of I_t.

It remains to consider the case $\mathcal{O}$ is of type BC^- and $\operatorname{char} k \neq 2$. Let $d = |\mathcal{O}|/2$. Pick $\alpha \in \mathcal{O}$ and write $\alpha_i = \sigma^i(\alpha)$ for $i = 1, \ldots, 2d$ and $\beta_i = \alpha_i + \alpha_{i+d}$ for $i = 1, \ldots, d$. We may assume that

- $\sigma \circ x_{\alpha_{2d}} = x_1$, $\sigma \circ x_{\alpha_i} = x_{\alpha_{i+1}}$ for $i = 1, \ldots, 2d-1$, and
- $x_{\alpha_i}(u)x_{\alpha_{d+i}}(v) = x_{\alpha_{d+i}}(v)x_{\alpha_i}(u)x_{\beta_i}(uv)$ for $i = 1, \ldots, d$.

This in particular implies that $\sigma \circ x_{\beta_i} = x_{\beta_{i+1}}$ for $i = 1, \ldots, d-1$, and $\sigma \circ x_{\beta_d} = -x_{\beta_1}$. Now one can check that for every $c \in k$,

$$\prod_{i=1}^{d} \Big(x_{\alpha_i}(e^{\alpha_2+\cdots+\alpha_i}(t)c) \cdot x_{\alpha_{d+i}}(e^{\alpha_2+\cdots+\alpha_{d+i}}(t)c) \\ \cdot x_{\beta_i}\big(-\tfrac{1}{2}e^{\beta_2+\cdots+\beta_i+\alpha_2+\cdots+\alpha_{d+1}}(t)c^2\big)\Big) \in I_t,$$

giving the desired unipotent subgroup. □

REMARK 5.2.12. We say an element $t \in T$ strongly σ-regular if $I_t = T^\sigma$. We claim that they form a non-empty affine open subset of $T^{\sigma\text{-reg}}$, sometimes denoted by $T^{\text{s.}\sigma\text{-reg}}$. Indeed, let $t \in \mathring{T}$, and let $g \in I_t(k)$. Using the Bruhat decomposition of $g = u_1 n u_2$ with $u_1, u_2 \in U$ and $n \in N$, it is easy to see that $g = n \in N^\sigma$, and $g \in T^\sigma$ if and only if no non-trivial elements in W_0 fixes the image of t under the projection $T \to A$. But the last condition clearly defines a non-empty affine open subset of T, verifying the claim. Note that if G is semisimple and simply-connected, then by Lemma 5.1.12, $T^{\text{s.}\sigma\text{-reg}} = T^{\sigma\text{-reg}} = \mathring{T}$.

To continue, we need the following lemma. We identify the tangent space $T_g G$ of G at g with $\mathfrak{g}$ via the right translation R_g by g. Let $(h, g\sigma) \in G \times G\sigma$ and $g' = c_\sigma(h)(g)$. A direct calculation shows the following.

LEMMA 5.2.13. *The differential of c_σ at $(h, g\sigma)$ is*

$$dc_\sigma : T_h G \oplus T_{g\sigma} G\sigma \to T_{g'} G\sigma, \quad (X, Y) \mapsto X - \operatorname{Ad}_{g'}(\sigma(X)) + \operatorname{Ad}_h(Y).$$

Recall that we denote by N_0 the preimage of $W_0 = W^\sigma \subset W$ in N. It acts on T via the twisted conjugation c_σ preserving $T^{\sigma\text{-reg}}$. Consider the map

$$G \times^{N_0} T^{\sigma\text{-reg}} \to G. \tag{5.2.2}$$

LEMMA 5.2.14. *The map (5.2.2) is an open embedding.*

Let $G^{\sigma\text{-rs}}$ denote the image of this map, called the *σ-regular semisimple locus* of G. Note that $G^{\sigma\text{-rs}} \cap T = T^{\sigma\text{-reg}}$ by Lemma 5.2.11.

PROOF. It follows from Lemma 5.2.13 that the map is étale. It remains to prove that it is also injective. Assume that $gt\sigma(g)^{-1} = t'$ with $t, t' \in T^{\sigma\text{-reg}} = \mathring{T}$. Using the Bruhat decomposition $g = u_1 n u_2$ as in Remark 5.2.12, one deduces that $u_1 = u_2 = 1$ and $t\sigma(n) = nt'$. The injectivity follows. □

REMARK 5.2.15. Note that N_0 also preserves $T^{\text{s.}\sigma\text{-reg}}$, and the map $G \times^{N_0} T^{\text{s.}\sigma\text{-reg}} \to G$ is open. The image is denoted by $G^{\text{s.}\sigma\text{-rs}}$, called the *strongly σ-regular semisimple locus* of G. Then $I|_{G^{\text{s.}\sigma\text{-rs}}}$ is smooth and fiberwise conjugate to T^σ in G. By Lemma 5.1.12, $G^{\text{s.}\sigma\text{-rs}} = G^{\sigma\text{-rs}}$ if G is semisimple and simply-connected.

REMARK 5.2.16. Note that σ is of finite order (since it preserves a pinning). If char k does not divide the order of σ, then $G^{\sigma\text{-rs}}\sigma$ is contained in the set of semisimple elements of the non-connected algebraic group $G \rtimes \langle\sigma\rangle$.

LEMMA 5.2.17. (1) *The restriction of c_σ to $U \times \mathring{T}\sigma \to \mathring{B}\sigma$ is an isomorphism.*

(2) *For a σ-orbit $\mathcal{O}$ of simple roots, let $U^{\mathcal{O}}$ be the subgroup of U introduced before Lemma 5.1.9. Then the restriction of c_σ to $U^{\mathcal{O}} \times B_{\mathcal{O}}^{[\mathcal{O}]}\sigma \to B^{[\mathcal{O}]}\sigma$ is an isomorphism.*

PROOF. We only prove (2) since the proof of (1) is similar (and simpler). Note that $U^{\mathcal{O}}$ is a normal subgroup in B and $B = U^{\mathcal{O}} B_{\mathcal{O}}$ as a semidirect product (of algebraic groups).

First, we claim that for any $b\sigma \in B_{\mathcal{O}}^{[\mathcal{O}]}\sigma$, the group

$$\{u \in U^{\mathcal{O}} \mid b\sigma(u)b^{-1} = u\}$$

is trivial. Indeed, let Φ_i be as in the proof of Lemma 5.2.9, and for $i \geq 1$, let U_i denote the group generated by the root groups U_α with $\alpha \in \Phi_i \cup \Phi_{i+1} \cup \cdots$. Then we obtain a filtration $U = U_1 \supset U_2 \supset \cdots$ by normal subgroups. Let $U_i^{\mathcal{O}} = U^{\mathcal{O}} \cap U_i$. Then $U_i^{\mathcal{O}}/U_{i+1}^{\mathcal{O}}$ is abelian, isomorphic to its Lie algebra. As argued in Lemma 5.2.9, $\{u \in U_i^{\mathcal{O}}/U_{i+1}^{\mathcal{O}} \mid b\sigma(u)b^{-1} = u\}$ is trivial because $q_B(b)$ does not lie in $\mathbf{A}_{\mathcal{O}'}$ for any σ-orbits $\mathcal{O}'$ that appear in $U^{\mathcal{O}}$. The claim follows by induction.

It follows that the map in Part (2) is a monomorphism, and that the map $\operatorname{Lie} U^{\mathcal{O}} \to \operatorname{Lie} U^{\mathcal{O}}$ given by $X \mapsto X - \operatorname{Ad}_b(\sigma(X))$ is an isomorphism. On the other hand, for any $b_1 \in U^{\mathcal{O}}$, and $Y \in \operatorname{Lie} B_{\mathcal{O}}$,

$$Y = \operatorname{Ad}_{b_1} Y \mod \operatorname{Lie} U^{\mathcal{O}}.$$

It follows from Lemma 5.2.13 that $U^{\mathcal{O}} \times B_{\mathcal{O}}^{[\mathcal{O}]}\sigma \to B^{[\mathcal{O}]}\sigma$ is étale and therefore is an open embedding. Note that the following diagram is commutative

$$\begin{array}{ccc} U^{\mathcal{O}} \times B_{\mathcal{O}}^{[\mathcal{O}]}\sigma & \xrightarrow{\quad c_\sigma \quad} & B^{[\mathcal{O}]}\sigma \\ & {\scriptstyle \mathrm{pr}_2}\searrow \quad \swarrow & \\ & B_{\mathcal{O}}^{[\mathcal{O}]}\sigma, & \end{array}$$

where $B^{[\mathcal{O}]}\sigma \to B_{\mathcal{O}}^{[\mathcal{O}]}\sigma$ is induced by the projection $B = U^{\mathcal{O}} B_{\mathcal{O}} \to B_{\mathcal{O}}$. In addition, for every point $b\sigma \in B_{\mathcal{O}}^{[\mathcal{O}]}\sigma$, the $U^{\mathcal{O}}$-orbit through this point is closed in $B^{[\mathcal{O}]}\sigma$ since $U^{\mathcal{O}}$ is unipotent. It follows that the $B_{\mathcal{O}}^{[\mathcal{O}]}\sigma$-morphism $c_\sigma : U^{\mathcal{O}} \times B_{\mathcal{O}}^{[\mathcal{O}]}\sigma \to B^{[\mathcal{O}]}\sigma$ is fiberwise open and closed. Therefore $U^{\mathcal{O}} \times B_{\mathcal{O}}^{[\mathcal{O}]}\sigma \to B^{[\mathcal{O}]}\sigma$ is open and surjective. Part (2) follows. □

We also need the following companion result. If $\mathcal{O}^+$ is a σ-orbit of type BC^+ such that $\alpha_{\mathcal{O}^+}$ is a simple root of $\Phi(G_\sigma, A)$, let $\mathcal{O}^-$ denote the corresponding σ-orbit of simple roots of type BC^- and we fix an order of roots in $\mathcal{O}^-$.

LEMMA 5.2.18. *Keep the notations as above. Then the restriction*

$$c_\sigma : \Big(U^{\mathcal{O}^-} \times \prod_{\alpha \in \mathcal{O}^-} U_\alpha\Big) \times B_{\mathcal{O}^+}^{[\mathcal{O}^+]}\sigma \to B^{[\mathcal{O}^+]}\sigma$$

is an isomorphism, where the product $\prod_{\alpha\in\mathcal{O}^-} U_\alpha$ *is taken with respect to a chosen order.*

PROOF. The same proof of Lemma 5.2.17 (2) for the σ-orbit $\mathcal{O}^-$ shows that the restriction of c_σ to $U^{\mathcal{O}^-} \times B_{\mathcal{O}^-}^{[\mathcal{O}^+]}\sigma \to B^{[\mathcal{O}^+]}\sigma$ is also an isomorphism. Therefore, it reduces to prove that

$$c_\sigma : \prod_{\alpha\in\mathcal{O}^-} U_\alpha \times B_{\mathcal{O}^+}^{[\mathcal{O}^+]}\sigma \to B_{\mathcal{O}^-}^{[\mathcal{O}^+]}\sigma$$

is an isomorphism. This follows from a simple explicit computation. Namely, we set $d = |\mathcal{O}^+|$, fix a root $\alpha \in \mathcal{O}^-$, and write $\alpha_i = \sigma^i(\alpha)$ for $1 \leq i \leq 2d$ and $\beta_i = \alpha_i + \sigma^d(\alpha_i)$. Then for $b = t\prod x_{\beta_i}(b_i)$ with $e^{\alpha_{\mathcal{O}}}(t) \neq 1$, and $u = \prod x_{\alpha_i}(a_i)$,

$$ub\sigma(u)^{-1} = t\prod x_{\alpha_i}(e^{-\alpha_i}(t)a_i - a_{i-1})\prod x_{\beta_i}(b_i'),$$

where $b_i' - b_i$ is a polynomial in $(a_1, \ldots, a_{2d})$ of degree two (with coefficients involving $e^{\alpha_i}(t)$'s). Since $e^{\alpha_{\mathcal{O}}}(t) - 1$ is invertible, for every $(c_1, \ldots, c_{2d})$, there is a unique $(a_1, \ldots, a_{2d})$ such that $c_i = e^{\alpha_i}(t)a_i - a_{i-1}$, the map in the lemma is an isomorphism. □

PROPOSITION 5.2.19. (1) *The map* $q_B : B^{\sigma\text{-reg}} \to \mathbf{A}$ *is surjective.*

(2) *Each fiber of* $q_B : B^{\sigma\text{-reg}} \to \mathbf{A}$ *is a single* B*-orbit for the* σ*-conjugation action.*
(3) *For every* $b \in B^{\sigma\text{-reg}}$, $(\mathrm{id} - \mathrm{Ad}_b\sigma)^{-1}(\mathfrak{u}) = \mathfrak{t}^\sigma + \mathfrak{u}$.
(4) *Assume that* G *is semisimple and simply-connected. Fix* $b \in B^{\sigma\text{-reg}}$*. Let* $g \in G$ *such that* $gb\sigma(g)^{-1} \in B$ *and* $q_B(gb\sigma(g)^{-1}) = q_B(b)$*. Then* $g \in B$*. In addition,* $I_b \subset B$*.*

PROOF. (1) For every $t \in T$, consider the automorphism of G_t given in Lemma 5.2.8, denoted by $t\sigma$. Applying Proposition 5.2.4 to $(G_t, t\sigma)$ gives an element $u \in U_t$ which is $t\sigma$-regular in G_t. Let $b = ut \in U_tT$. Then $I_b \cap G_t = \{g \in G_t \mid gut\sigma(g)^{-1}t^{-1} = g\}$ is the $t\sigma$-twisted centralizer of u in G_t. Therefore, $b = ut$ is also σ-regular in G_t by Lemma 5.2.9.

(2) If $b \in B^{\sigma\text{-reg}}$, then the dimension of the B-orbit through b under the σ-conjugation action of B on itself is equal to $\dim q_B^{-1}(q_B(b))$. This shows that any two σ-regular B-orbits in a fiber of q_B must meet since fibers of q_B are irreducible. Therefore, there is exact one B-orbit in each fiber of $q_B : B^{\sigma\text{-reg}} \to \mathbf{A}$.

(3) We may assume that $b = ut$, with $u \in U_t$ being $t\sigma$-regular in G_t. Then the claims follows from Lemma 5.2.9, and Proposition 5.2.4.

(4) We may assume that $b = ut$, with $u \in U_t$ being $t\sigma$-regular in G_t. Write an element $g \in I_b(k)$ as $g = u_1nu_2$ with $u_1, u_2 \in U$ and $n \in N$. Then $u_1nu_2b = b\sigma(u_1)\sigma(n)\sigma(u_2)$. It follows that $\sigma(n) = n$, so

$$nu_2b\sigma(u_2)^{-1}n^{-1} = u_1^{-1}b\sigma(u_1).$$

Taking projection to $\mathbf{A}$, we see that $w(q_B(t)) = q_B(t)$, where $w = n \mod T$. Since G is simply-connected, by Lemma 5.1.12 and Lemma 5.2.7, w must be in the Weyl group of G_t. Then by applying the same argument as in the proof of

Proposition 5.2.4 to G_t, we deduce that $w=1$. In particular, $I_b^{\mathrm{red}} \subset B$. On the other hand,

$$\ker(\mathrm{id} - \mathrm{Ad}_b\sigma : \mathfrak{g} \to \mathfrak{g}) = \ker(\mathrm{id} - \mathrm{Ad}_b\sigma : \mathfrak{g}_t \to \mathfrak{g}_t) = \ker(\mathrm{id} - \mathrm{Ad}_b\sigma : \mathfrak{b}\cap\mathfrak{g}_t \to \mathfrak{b}\cap\mathfrak{g}_t).$$

where the first equality follows from Lemma 5.2.9, and the second follows from Lemma 5.2.5. Putting these together, we see that $I_b \subset B$.

More generally let $g \in G$ such that $gb\sigma(g)^{-1} \in B$ and $q_B(gb\sigma(g)^{-1}) = q_B(b)$, there is some $b' \in B$ such that $gb\sigma(g)^{-1} = b'b\sigma(b')^{-1}$, by Part (2). Therefore $g^{-1}b' \in I_b \subset B$ and $g \in B$. □

COROLLARY 5.2.20. *Let $K = \ker q_B = (\sigma - 1)T \cdot U$, and let $\mathfrak{k}$ be its Lie algebra. Then for every $b \in B^{\sigma\text{-reg}}$, $(\mathrm{id} - \mathrm{Ad}_b\sigma)^{-1}(\mathfrak{k}) = \mathfrak{b}$.*

PROOF. Let $X \in \mathfrak{g}$ such that $X - \mathrm{Ad}_b\sigma(X) = (1-\sigma)H + \mathfrak{u}$, with $H \in \mathfrak{t}$. Then $(\mathrm{id} - \mathrm{Ad}_b\sigma)(X - H) \in \mathfrak{u}$. Therefore $X - H \in \mathfrak{t}^\sigma + \mathfrak{u}$ and $X \in \mathfrak{b}$. □

COROLLARY 5.2.21. *The codimension of $B - B^{\sigma\text{-reg}}$ is at least two.*

PROOF. This follows from Lemma 5.2.17 (1) and Proposition 5.2.19. □

COROLLARY 5.2.22. *An element $b \in B$ is σ-regular if and only if the set $\mathcal{B}_b = \{g \in G/B \mid g^{-1}b\sigma(b) \in B\}$ is finite.*

PROOF. This follows from Lemma 5.2.17 by applying Lemma 5.2.6 to $u \in U_t$. □

5.3. Twisted Grothendieck–Springer resolution. In this subsection, we assume that G is semisimple and simply-connected.

Note that $k[T]^{c_\sigma(N_0)} = k[A]^{W_0}$. Then the twisted Chevalley isomorphism (Proposition 4.2.3) implies that $G/\!\!/c_\sigma(G) \cong \mathbf{A}/\!\!/W_0 := \operatorname{Spec} k[\mathbf{A}]^{W_0}$. So we write the (twisted) Chevalley map as

$$\chi : G \to G/\!\!/c_\sigma(G) = \mathbf{A}/\!\!/W_0,$$

which we recall is faithfully flat by Corollary 4.3.3.

As in the untwisted case, there is the following commutative diagram

$$\begin{array}{ccc} \widetilde{G} & \xrightarrow{q} & \mathbf{A} \\ \downarrow & & \downarrow \\ G & \xrightarrow{\chi} & \mathbf{A}/\!\!/W_0, \end{array}$$

where the left vertical map

$$\widetilde{G} := \{(gB, x) \in G/B \times G \mid x \in gB\sigma(g)^{-1}\} \longrightarrow G \qquad (gB, x) \mapsto x,$$

is what we call the *(σ-twisted) Grothendieck–Springer resolution*. The map q is induced by

$$\widetilde{G} \cong G \times^B (B\sigma) \xrightarrow{q_B} G \times^B \mathbf{A} \to \mathbf{A}.$$

Together, these two maps induce a proper map $\pi : \widetilde{G} \to G \times_{\mathbf{A}/\!\!/W_0} \mathbf{A}$.

Let $\widetilde{G}^{\sigma\text{-reg}}$ be the preimage of $G^{\sigma\text{-reg}}$. We have the following proposition.

PROPOSITION 5.3.1. *The induced map $\widetilde{G}^{\sigma\text{-reg}} \to G^{\sigma\text{-reg}} \times_{\mathbf{A}/\!\!/W_0} \mathbf{A}$ is an isomorphism.*

PROOF. We start with the following special case.

LEMMA 5.3.2. *Let $\widetilde{G}^{\sigma\text{-rs}}$ be the preimage of $G^{\sigma\text{-rs}}$. Then the induced map*

$$\pi|_{\widetilde{G}^{\sigma\text{-rs}}} : \widetilde{G}^{\sigma\text{-rs}} \to G^{\sigma\text{-rs}} \times_{\mathring{\mathbf{A}}/\!/W_0} \mathring{\mathbf{A}} \tag{5.3.1}$$

is an isomorphism.

PROOF. By (5.2.2) and Lemma 5.2.11, we have $G^{\sigma\text{-rs}} \cong G \times^{N_0} \mathring{T}$. So we can write

$$\widetilde{G}^{\sigma\text{-rs}} = \{(gB, g't\sigma(g')^{-1}) \in G/B \times (G \times^{N_0} \mathring{T}) \mid g't\sigma(g')^{-1} \in gB\sigma(g)^{-1}\}.$$

The last condition is equivalent to $g^{-1}g't\sigma(g^{-1}g')^{-1} \in B$. By Proposition 5.2.19 (4), $g^{-1}g' \in B \times^T N_0$. From this, we deduce that $\widetilde{G}^{\sigma\text{-rs}} \cong G \times^T \mathring{T}$ and therefore (5.3.1) is an isomorphism. □

We return to the proof of Proposition 5.3.1. Since $\mathbf{A} \to \mathbf{A}/\!/W_0$ is finite and flat by Lemma 5.1.12, the fiber product $G^{\sigma\text{-rs}} \times_{\mathring{\mathbf{A}}/\!/W_0} \mathring{\mathbf{A}}$ is open and dense in $G \times_{\mathbf{A}/\!/W_0} \mathbf{A}$. But π is proper as G/B is, so π is surjective. In particular, $G \times_{\mathbf{A}/\!/W_0} \mathbf{A}$ is irreducible. Moreover, since $\mathbf{A}$ and G are smooth and $G \to \mathbf{A}/\!/W_0$ is flat, $G \times_{\mathbf{A}/\!/W_0} \mathbf{A}$ is a complete intersection (in particular Cohen–Macaulay), and it follows from the above lemma that $G \to \mathbf{A}/\!/W_0$ is generically smooth. Therefore, $G \times_{\mathbf{A}/\!/W_0} \mathbf{A}$ is also reduced. So $G \times_{\mathbf{A}/\!/W_0} \mathbf{A}$ is a closed subvariety of $G \times \mathbf{A}$.

Since $\widetilde{G} \to G$ is surjective, every element in G is σ-conjugate to an element in B. In particular, $\widetilde{G}^{\sigma\text{-reg}} = G \times^B B^{\sigma\text{-reg}}$. By Proposition 5.2.19 (4), the map $G \times^B B^{\sigma\text{-reg}} \to G \times \mathbf{A}$ is injective on points. By Lemma 5.2.13 and Corollary 5.2.20, the tangent map for $G \times^B B^{\sigma\text{-reg}} \to G \times \mathbf{A}$ is injective at every point, and therefore the morphism is unramified. It follows that $\widetilde{G}^{\sigma\text{-reg}} \to G^{\sigma\text{-reg}} \times \mathbf{A}$ is a closed embedding, with image (as topological space) $G^{\sigma\text{-reg}} \times_{\mathbf{A}/\!/W_0} \mathbf{A}$. But since $G^{\sigma\text{-reg}} \times_{\mathbf{A}/\!/W_0} \mathbf{A}$ is also reduced, the map in the proposition is indeed an isomorphism. □

Here are some standard corollaries.

COROLLARY 5.3.3. *The morphism $\chi : G^{\sigma\text{-reg}} \to \mathbf{A}/\!/W_0$ is smooth.*

PROOF. This follows from the smoothness of $\widetilde{G}^{\sigma\text{-reg}} \to \mathbf{A}$ and the flatness of $\mathbf{A} \to \mathbf{A}/\!/W_0$. □

COROLLARY 5.3.4. *The map π induces an isomorphism $k[G \times_{\mathbf{A}/\!/W_0} \mathbf{A}] \cong \Gamma(\widetilde{G}, \mathcal{O})$.*

PROOF. It is enough to show that $G \times_{\mathbf{A}/\!/W_0} \mathbf{A}$ is an integral normal scheme. Then it follows that π induces an isomorphism between rings of global regular functions, since it is proper birational.

We already seen that $G \times_{\mathbf{A}/\!/W_0} \mathbf{A}$ is integral, and a complete intersection. By Proposition 5.3.1, $G^{\sigma\text{-reg}} \times_{\mathbf{A}/\!/W_0} \mathbf{A}$ is smooth. By Corollary 5.2.21, the complement of $G^{\sigma\text{-reg}} \times_{\mathbf{A}/\!/W_0} \mathbf{A}$ in $G \times_{\mathbf{A}/\!/W_0} \mathbf{A}$ has codimension at least two. It follows from this lemma that $G \times_{\mathbf{A}/\!/W_0} \mathbf{A}$ is normal. □

COROLLARY 5.3.5.

(1) *For every $a \in \mathbf{A}/\!/W_0$, $\chi^{-1}(a) \cap G^{\sigma\text{-reg}}$, is a single G-orbit, and the codimension of the complement of this G-orbit in $\chi^{-1}(a)$ at least two.*
(2) *Each fiber of χ is a complete intersection and normal variety.*

PROOF. (1) Pick $\tilde{a} \in \mathbf{A}(k)$ that lifts $a \in \mathbf{A}/\!/W_0$. By Proposition 5.3.1, the fiber $\chi^{-1}(a) \cap G^{\sigma\text{-reg}}$ is isomorphic to the fiber of $q : \widetilde{G}^{\sigma\text{-reg}} \cong G \times^B B^{\sigma\text{-reg}} \to \mathbf{A}$ at $\tilde{a}$, which is clearly a single G-orbit by Proposition 5.2.19 (2). On the other hand, by Corollary 5.2.22, the fibers of $\widetilde{G} \to G$ over $G - G^{\sigma\text{-reg}}$ have positive dimension. Since $q_B^{-1}(\tilde{a}) - (q_B^{-1}(\tilde{a}) \cap B^{\sigma\text{-reg}})$ is a proper closed subset of $q_B^{-1}(\tilde{a})$, $q^{-1}(a) - (q^{-1}(a) \cap G^{\sigma\text{-reg}})$ has codimension at least two in $q^{-1}(a)$.

(2) Since $\chi : G \to \mathbf{A}/\!/W_0$ is flat, each of its fiber is a complete intersection and hence Cohen–Macaulay. By (1), each fiber of χ contains a (smooth) G-orbit whose complement has codimension at least two. So the fiber is also normal. □

6. The determinant of the pairing $\boldsymbol{J}(V) \otimes \boldsymbol{J}(V^*) \to \boldsymbol{J}$

ASSUMPTION 6.0.1. In this section, assume that char $k > 2$. Let G be a simply-connected semisimple group over k. Let V be a finite dimensional representation of G and V^* the dual representation. Assume that both V and V^* admit good filtrations.

We keep conventions and notations as in § 5. Then $\widetilde{V^*}$ is the dual of $\widetilde{V}$ as a vector bundle on $[G\sigma/G]$. The perfect pairing $\widetilde{V} \otimes \widetilde{V^*} \to \mathcal{O}_{[G\sigma/G]}$ induces a $\boldsymbol{J}$-bilinear pairing $\boldsymbol{J}(V) \otimes \boldsymbol{J}(V^*) \to \boldsymbol{J}$ of global sections, which however is not perfect in general. Our main result (Theorem 6.1.2) calculates the determinant of this pairing of two finite free $\boldsymbol{J}$-modules. A main intermediate step is to study the failure of the surjectivity of the twisted Chevalley restriction homomorphism (4.3.3).

6.1. Main results.

DEFINITION 6.1.1. For each σ-orbit $\mathcal{O} \subset \Phi(G,T)$, we choose $\alpha \in \mathcal{O}$ and view it as a character of T^σ by restriction, which is clearly independent of the choice of α. Define the number

$$\zeta_{\mathcal{O}} = \zeta_{\mathcal{O}}(V) := \sum_{n \geq 1} \dim V|_{T^\sigma}(n\alpha).$$

The main theorem of this section is the following.

THEOREM 6.1.2. *Keep Assumption 6.0.1. The determinant of the natural $\boldsymbol{J}$-bilinear pairing*[5]

$$\langle \cdot, \cdot \rangle_V : \boldsymbol{J}(V) \otimes \boldsymbol{J}(V^*) \to \boldsymbol{J} \tag{6.1.1}$$

is of the form

$$\big(\textit{some unit in } k\big) \cdot \prod_{\mathcal{O} \textit{ type } \mathsf{A} \textit{ or } \mathsf{BC}^-} (e^{\alpha_{\mathcal{O}}} - 1)^{\zeta_{\mathcal{O}}} \cdot \prod_{\mathcal{O} \textit{ type } \mathsf{BC}^+} (e^{\alpha_{\mathcal{O}}} + 1)^{\zeta_{\mathcal{O}}}. \tag{6.1.2}$$

REMARK 6.1.3. (1) Note that $\zeta_{\mathcal{O}}$ depends only on the W_0-orbits of $\mathcal{O}$, and therefore the product in (6.1.2) does belong to $\boldsymbol{J}$.

(2) When σ is trivial, the rank of $\boldsymbol{J}(V)$ as a $\boldsymbol{J}$-module is $r_V = \dim V(0)$ by Theorem 4.3.2. In this case, then $\zeta_{\mathcal{O}} = \zeta_\alpha$ can be alternatively computed via the multi-filtration on the weight spaces introduced in § 3.3 as

$$\zeta_\alpha = \sum_{\nu \in \mathbb{X}^\bullet(T)^+} \langle \nu, \check{\alpha} \rangle \cdot \dim \mathrm{gr}_\nu V(0).$$

[5] Taking the determinant makes sense because Theorem 4.3.2 shows that $\boldsymbol{J}(V)$ and $\boldsymbol{J}(V^*)$ are both free $\boldsymbol{J}$-modules of the same rank.

(3) Similarly we have the pairing $\boldsymbol{J}_+(V) \otimes \boldsymbol{J}_+(V^*) \to \boldsymbol{J}_+$. Its restriction to $\operatorname{Spec} \boldsymbol{J}$ is the pairing in the theorem, and restriction to the point $x_1 \in \operatorname{Spec} \boldsymbol{J}_0$ as defined in Remark 4.4.1 is the natural pairing between $(V|_{T^\sigma})(0)$ and $(V|_{T^\sigma})^*(0)$.

(4) When $\operatorname{char} k = 0$, we have explained in § 4.4 a construction of a basis of $\boldsymbol{J}(V)$ (resp. $\boldsymbol{J}(V^*)$) from a certain basis of V (resp. V^*) (e.g. the MV basis used in § 3.5). Then the $\langle \cdot, \cdot \rangle_V$ is represented by a square matrix. It seems to be an interesting (although probably difficult) question to calculate the entries of this matrix explicitly. See § 6.4 for some calculations and further discussions. We also refer to [**XZ17**] for the arithmetic and geometric meaning of this square matrix.

In what follows, we shall relate the pairing (6.1.1) to the twisted Chevalley restriction map (4.3.3). As we will show in Lemma 6.1.5 that Theorem 6.1.2 follows from Proposition 6.1.4 below. More precisely, we will not study the Chevalley restriction homomorphism (4.3.3) itself, but rather the induced $\boldsymbol{J}_T$-module homomorphism

$$\operatorname{Res}_V^\sigma \otimes 1 : \boldsymbol{J}_G(V) \otimes_{\boldsymbol{J}_G} \boldsymbol{J}_T \to \boldsymbol{J}_T(V), \tag{6.1.3}$$

where explicitly, we may write the target as

$$\boldsymbol{J}_T(V) = (k[T] \otimes V)^{c_\sigma(T)} \cong \bigoplus_{\xi \in (\sigma-1)\mathbb{X}^\bullet(T)} k[A]e^{\nu_\xi} \otimes V(\xi) \tag{6.1.4}$$

with $\nu_\xi \in \mathbb{X}^\bullet(T)$ some weight such that $\sigma(\nu_\xi) - \nu_\xi = \xi$. In particular, if $\sigma = \mathrm{id}$, $\boldsymbol{J}_T(V) \cong k[T] \otimes V(0)$.

We view $\operatorname{Res}_V^\sigma \otimes 1$ as a morphism of coherent sheaves over A. By Lemma 4.3.8 (which relies on Lemma 5.2.11), it is an isomorphism over $\mathring{A}$ (which is defined in Definition 5.1.10).

Now, let η be a generic point of the divisor $\bigcup_{\mathcal{O}} A_{\mathcal{O}}$ (which is reduced since we assumed that $\operatorname{char} k > 2$; see Remark 5.1.11). Then the complete local ring of A at η is isomorphic to $k(\eta)[[\varpi]]$, where $\varpi = e^{\alpha_{\mathcal{O}}} - 1$ or $e^{\alpha_{\mathcal{O}}} + 1$ for some σ-orbit $\mathcal{O} \subset \Phi(G,T)$. Note that $\boldsymbol{J}_G(V)$ is always a torsion free $\boldsymbol{J}_G$-module (even if V does not admit a good filtration). Therefore $\boldsymbol{J}_G(V) \otimes_{\boldsymbol{J}_G} k(\eta)[[\varpi]]$ is always free, and (6.1.3), base changed to $k(\eta)[[\varpi]]$, is a map Res_η of finite free $k(\eta)[[\varpi]]$-modules which becomes an isomorphism when further base changed to $k(\eta)((\varpi))$. (Such map is called a *modification* of vector bundles on $\operatorname{Spec} k(\eta)[[\varpi]]$ (in the sense as in [**XZ17**, § 3.1.3]). The top exterior power of this map Res_η is an element in $k(\eta)[[\varpi]] - \{0\}$, well-defined up to multiplying an element in $k[[\varpi]]^\times$, and therefore gives a well-defined element in

$$(k(\eta)[[\varpi]] - \{0\})/k(\eta)[[\varpi]]^\times \cong \mathbb{Z}_{\geq 0}.$$

We call this number *the length of the modification.*

Here is the main result regarding the map (6.1.3).

PROPOSITION 6.1.4. *For every G-module V with good filtration, the length of the modification* (6.1.3) *at every generic point of $A_{\mathcal{O}}$ is exactly $\zeta_{\mathcal{O}}$.*

This proposition will be proved in § 6.2. We note the following first.

LEMMA 6.1.5. *Proposition* 6.1.4 *implies Theorem* 6.1.2.

PROOF. Consider the following commutative diagram

$$\begin{array}{ccccc}
\boldsymbol{J}(V) & \times & \boldsymbol{J}(V^*) & \xrightarrow{\langle\cdot,\cdot\rangle_V} & \boldsymbol{J} \\
\downarrow & & \downarrow & & \downarrow \\
\boldsymbol{J}(V)\otimes_{\boldsymbol{J}}\boldsymbol{J}_T & \times & \boldsymbol{J}(V^*)\otimes_{\boldsymbol{J}}\boldsymbol{J}_T & \xrightarrow{\langle\cdot,\cdot\rangle_V} & \boldsymbol{J}_T \\
\downarrow{\scriptstyle \mathrm{Res}^\sigma_V\otimes 1} & & \downarrow{\scriptstyle \mathrm{Res}^\sigma_{V^*}\otimes 1} & & \| \\
\boldsymbol{J}_T(V) & \times & \boldsymbol{J}_T(V^*) & \xrightarrow{\langle\cdot,\cdot\rangle_V} & \boldsymbol{J}_T.
\end{array}$$

By Theorem 4.3.2, $\boldsymbol{J}(V)$ and $\boldsymbol{J}(V^*)$ are free $\boldsymbol{J}$-modules. So second row is simply a base change of the first row, and hence the matrices for the top two pairings are the same (when choosing compatible bases). Now the bottom row is a *perfect* $\boldsymbol{J}_T$-bilinear pairing (as can be easily seen from the explicit expression of $\boldsymbol{J}_T(V)$ in (6.1.4)). So the determinant of the middle row is the product of the determinant of the map $\mathrm{Res}^\sigma_V\otimes 1$ and the determinant of the map $\mathrm{Res}^\sigma_{V^*}\otimes 1$ (up to a unit). Thus, Proposition 6.1.4 would imply that the determinant of (6.1.1), as a divisor on A, is given by

$$\sum_{\mathcal{O}\subset\Phi(G,T)} (\zeta_{\mathcal{O}}(V)+\zeta_{\mathcal{O}}(V^*))\cdot A_{\mathcal{O}}.$$

By Lemma 6.1.6 below, this gives the same expression as the formula (6.1.2) up to a unit in $k[A]$. But note that the product in (6.1.1) is W_0-invariant, so is the determinant of (6.1.1). It follows that the ambiguous unit in $k[A]$ belongs to $(k[A]^{W_0})^\times$. As mentioned in the proof of Lemma 5.1.12 (see also [**Sp06**, Corollary 2]), $k[A]^{W_0}$ is a polynomial algebra and hence its units are just $k^\times$. Therefore, the determinant of (6.1.1) is given by (6.1.2) up to a unit in k. □

LEMMA 6.1.6. *For a representation V of G and a σ-orbit $\mathcal{O}\subset\Phi(G,T)$, we have $\zeta_{\mathcal{O}}(V)=\zeta_{\mathcal{O}}(V^*)$.*

PROOF. This follows from the following sequence of equalities

$$\dim V^*|_{T^\sigma}(n\alpha)=\dim V|_{T^\sigma}(-n\alpha)=\dim V|_{T^\sigma}(n\alpha),$$

where the first equality follows from the duality and the last equality follows from the fact that α and $-\alpha$ lie in the same W-orbit. □

6.2. Proof of Proposition 6.1.4. We will first reduce Proposition 6.1.4 to the cases of SL_2 and SL_3.

LEMMA 6.2.1. *Let V be a representation of G. The map $\boldsymbol{J}_G(V)\otimes_{\boldsymbol{J}_G}\boldsymbol{J}_T\to\boldsymbol{J}_B(V)$ is an isomorphism.*

PROOF. Recall that by Corollary 5.3.4, the pushforward of the structure sheaf along $\pi:[B\sigma/B]\to[G\sigma/G]\times_{\mathbf{A}/\!\!/W_0}\mathbf{A}$ is the structure sheaf. Then it follows from the projection formula that

$$\boldsymbol{J}_G(V)\otimes_{\boldsymbol{J}_G}\boldsymbol{J}_T=\Gamma([G\sigma/G]\otimes_{\mathbf{A}/\!\!/W_0}\mathbf{A},\widetilde{V}\boxtimes\mathcal{O})=\Gamma([B\sigma/B],\pi^*\widetilde{V})=\boldsymbol{J}_B(V)$$

is an isomorphism. □

Using the W_0-action on $\Phi(G_\sigma,A)$, it suffices to prove Proposition 6.1.4 for a σ-orbit $\mathcal{O}$ that is either a σ-orbit of *simple* roots (of type A or BC^-), or a σ-orbit of roots of type BC^+ that are sum of two roots from a σ-orbit $\mathcal{O}^-$ of simple roots.

In either case, let $G_{\mathcal{O}}$ be the subgroup of G generated by T and U_α with $\alpha \in \mathcal{O} \cup -\mathcal{O}$; and $B_{\mathcal{O}}$ the Borel subgroup generated by T and U_α with $\alpha \in \mathcal{O}$.

LEMMA 6.2.2. *The map* $\boldsymbol{J}_B(V) \to \boldsymbol{J}_{B_{\mathcal{O}}}(V)$ *is an isomorphism over* $\mathbf{A}^{[\mathcal{O}]}$.

PROOF. It follows from Lemma 5.2.17 and Lemma 5.2.18 that

$$[B^{[\mathcal{O}]}\sigma/B] \cong [B_{\mathcal{O}}^{[\mathcal{O}]}\sigma/B_{\mathcal{O}}].$$

So the map in the lemma induces an isomorphism when restricted to $\mathbf{A}^{[\mathcal{O}]}$. □

LEMMA 6.2.3. *To prove Proposition* 6.1.4, *it is enough to prove it for two cases:*

- $G = \mathrm{SL}_2$ *with* $\sigma = \mathrm{id}$*;*
- $G = \mathrm{SL}_3$ *with* σ *given as in Example* 5.1.5 *and* $\mathcal{O}$ *being a* σ*-orbit consisting of two roots.*

PROOF. By Lemma 6.2.1 and Lemma 6.2.2, we reduce Proposition 6.1.4 to prove that, for the σ-orbit $\mathcal{O}$ considered above and over every generic point η of $\mathbf{A}_{\mathcal{O}}$, the map $\boldsymbol{J}_{B_{\mathcal{O}}}(V) \to \boldsymbol{J}_T(V)$ is a modification of length $\zeta_{\mathcal{O}}$. (Note that $V|_{G_{\mathcal{O}}}$ also admits a good filtration by Theorem 3.1.2 (1).) We consider the central isogeny

$$G'_{\mathcal{O}} := G_{\mathcal{O},\mathrm{der}} \times Z_{G_{\mathcal{O}}} \to G_{\mathcal{O}}.$$

with kernel $F = G_{\mathcal{O},\mathrm{der}} \cap Z_{G_{\mathcal{O}}}$. Let T' denote the preimage of T and $A' = T'_\sigma$. Note that the right exact sequence $F_\sigma \to T'_\sigma \to T_\sigma \to 1$ is also exact on the left, so the induced map $A' \to A$ is an F_σ-torsor. Note that F_σ is indeed étale (given the classification of $G_{\mathcal{O},\mathrm{der}}$ by Lemma 5.1.9) since we assume $\mathrm{char}\, k \neq 2$. By Lemma 4.1.3 (applied to $B'_{\mathcal{O}} = B_{\mathcal{O},\mathrm{der}} \times Z_{G_{\mathcal{O}}}$ and $T_{\mathcal{O},\mathrm{der}} \times Z_{G_{\mathcal{O}}}$) and the fact that modifications commute with étale base change, the length of modification of $\boldsymbol{J}_{B_{\mathcal{O}}}(V) \to \boldsymbol{J}_T(V)$ at every generic point η of $\mathbf{A}_{\mathcal{O}}$ is the same as the length of modification of $\boldsymbol{J}_{B'_{\mathcal{O}}}(V) \to \boldsymbol{J}_{T'}(V)$ (or equivalently $\boldsymbol{J}_{G'_{\mathcal{O}}}(V) \otimes_{\boldsymbol{J}_{G'_{\mathcal{O}}}} \boldsymbol{J}_T \to \boldsymbol{J}_T(V)$ by Lemma 6.2.1) at any preimage of η under the map $\pi : \mathbf{A}' \to \mathbf{A}$.

We decompose $V = \oplus_\psi V_\psi \otimes k_\psi$ according to the central character for the action of $Z_{G_{\mathcal{O}}}$ on V so that each V_ψ is a $G_{\mathcal{O},\mathrm{der}}$-module. Then by (4.1.1)

$$\boldsymbol{J}_{G'_{\mathcal{O}}}(V) = \bigoplus_\psi \boldsymbol{J}_{G_{\mathcal{O},\mathrm{der}}}(V_\psi) \otimes \boldsymbol{J}_{Z_{G_{\mathcal{O}}}}(k_\psi) = \bigoplus_{\psi|_{Z^\sigma_{G_{\mathcal{O}}}} = \mathbf{1}} \boldsymbol{J}_{G_{\mathcal{O},\mathrm{der}}}(V_\psi) \otimes \boldsymbol{J}_{Z_{G_{\mathcal{O}}}}(k_\psi).$$

We write $A_{\mathrm{der}} = (T_{\mathcal{O},\mathrm{der}})_\sigma$, where $T_{\mathcal{O},\mathrm{der}} = T \cap G_{\mathcal{O},\mathrm{der}}$. Then $A' = A_{\mathrm{der}} \times (Z_{G_{\mathcal{O}}})_\sigma$, and one checks that $\pi : \mathbf{A}_{\mathrm{der},\mathcal{O}} \times (Z_{G_{\mathcal{O}}})_\sigma \to \mathbf{A}_{\mathcal{O}}$ is surjective. Clearly, the modification

$$\boldsymbol{J}_{G'_{\mathcal{O}}}(V) = \bigoplus_\psi \boldsymbol{J}_{G_{\mathcal{O},\mathrm{der}}}(V_\psi) \otimes \boldsymbol{J}_{Z_{G_{\mathcal{O}}}}(k_\psi) \to \boldsymbol{J}_{T'}(V) = \bigoplus_\psi \boldsymbol{J}_{T_{\mathcal{O},\mathrm{der}}}(V_\psi) \otimes \boldsymbol{J}_{Z_{G_{\mathcal{O}}}}(k_\psi)$$

only happens on the first factor. Since the second factor is of rank one over $\boldsymbol{J}_{Z_{G_{\mathcal{O}}}}$, and since

$$\dim(V|_{T^\sigma})(n\alpha) = \sum_{\psi|_{Z^\sigma_{G_{\mathcal{O}}}} = \mathbf{1}} (V_\psi|_{T^\sigma_{\mathcal{O},\mathrm{der}}})(n\alpha),$$

we see that to prove Proposition 6.1.4, it is enough to assume that $G = G_{\mathcal{O},\mathrm{der}}$.

By Lemma 5.1.9, there are three cases for $G_{\mathcal{O},\mathrm{der}}$. By Lemma 5.2.18, and Lemma 4.1.2, we reduce the proof of Proposition 6.1.4 to the case $G = \mathrm{SL}_2$ with $\sigma = \mathrm{id}$, and $G = \mathrm{SL}_3$ with σ given as in Example 5.1.5 and $\mathcal{O}$ being a σ-orbit consisting of two roots. This completes the proof of the lemma. □

Finally, we treat the above mentioned two cases.

LEMMA 6.2.4. *Proposition* 6.1.4 *holds for* $G = \mathrm{SL}_2$.

PROOF. It is enough to assume that V is the Schur module $\mathtt{S}_n$ (of dimension $n+1$), in which case we make explicit computations. We set

$$B = \left\{\left(\begin{smallmatrix} a & b \\ 0 & a^{-1}\end{smallmatrix}\right) \mid a \in k^\times,\ b \in k\right\}, \quad T = \left\{\left(\begin{smallmatrix} a & 0 \\ 0 & a^{-1}\end{smallmatrix}\right) \mid a \in k^\times\right\},$$
$$\text{and} \quad U = \left\{\left(\begin{smallmatrix} 1 & b \\ 0 & 1\end{smallmatrix}\right) \mid b \in k\right\}.$$

By Lemma 6.2.1, we need to show that

(1) $J_T(\mathtt{S}_n)$ is zero when n is odd, and
(2) the length of the modification of $J_B(\mathtt{S}_n) \to J_T(\mathtt{S}_n)$ at the generic point of T is exactly $n/2$ when n is even.

(1) is obvious by looking at the action of $\left(\begin{smallmatrix} -1 & 0 \\ 0 & -1\end{smallmatrix}\right) \in \mathrm{SL}_2$. For (2) (and hence n is even), we explicitly identify $\mathtt{S}_n \cong k[z]^{\deg \leq n}$ and $\mathcal{O}_B \cong k[x^{\pm 1}, y]$, where B acts on $\mathtt{S}_n$ by $\left(\begin{smallmatrix} a & b \\ 0 & a^{-1}\end{smallmatrix}\right)(h)(z) = a^n h\big(a^{-2}z - a^{-1}b\big)$ and the conjugation action of B on $\mathcal{O}_B$ is given by

$$\left(\begin{smallmatrix} a & b \\ 0 & a^{-1}\end{smallmatrix}\right)(h)(x,y) = h(x, a^{-2}y + a^{-1}b(x - x^{-1})).$$

From this, we see that $z(x - x^{-1}) + y$ is invariant under the U-action and explicitly

$$(\mathcal{O}_B \otimes \mathtt{S}_n)^U \cong \bigoplus_{i=0}^{n-1} k[x^{\pm 1}] \cdot \big(z(x - x^{-1}) + y\big)^i.$$

But an element $\left(\begin{smallmatrix} a & 0 \\ 0 & a^{-1}\end{smallmatrix}\right) \in T$ acts on $(z(x - x^{-1}) + y)^i$ by multiplication by a^{n-2i}. It follows that

$$(\mathcal{O}_B \otimes \mathtt{S}_n)^B \cong k[x^{\pm 1}] \cdot (z(x - x^{-1}) + y)^{n/2}.$$

Restricting to T, or equivalently setting $y = 0$, we see that the map $\boldsymbol{J}_B(V) \to \boldsymbol{J}_T(V)$ can be identified with the inclusion

$$k[x^{\pm 1}] \cdot (z(x - x^{-1}))^{n/2} \to k[x^{\pm 1}] \cdot z^{n/2}.$$

The length of the modification of the above map at each point of the divisor $x - x^{-1}$ is $n/2$. This completes the proof of the lemma. □

LEMMA 6.2.5. *Proposition* 6.1.4 *holds for* $G = \mathrm{SL}_3$ *with the automorphism* σ *given in Example* 5.1.5, *at the generic points of* $A_{\mathcal{O}}$ *for the* σ*-orbit* $\mathcal{O}$ *consisting of two roots.*

PROOF. Let $\alpha, \sigma\alpha, \beta = \alpha + \sigma\alpha$ denote the positive roots and let $\mathcal{O} = \{\alpha, \sigma\alpha\}$ so that $\alpha_{\mathcal{O}} = \beta$.

We consider the principal SL_2 of SL_3. Its image is exactly $H = \mathrm{SO}_3 = (\mathrm{SL}_3)^\sigma$, where SO_3 is the orthogonal group defined by $\left(\begin{smallmatrix} & & 1 \\ & -1 & \\ 1 & & \end{smallmatrix}\right)$, regarded as a symmetric bilinear form. The map $T^\sigma \to T_\sigma = A$ is the square mapping on $\mathbb{G}_m$. One checks immediately that when $\operatorname{char} k \neq 2$, $\boldsymbol{J}_T \to \boldsymbol{J}_{T^\sigma}$ is finite étale of degree two and the induced map

$$\boldsymbol{J}_T(V) \otimes_{\boldsymbol{J}_T} \boldsymbol{J}_{T^\sigma} \to \boldsymbol{J}_{T^\sigma}(V)$$

is an isomorphism.

The embedding $H \to G$ induces $\operatorname{Spec} \boldsymbol{J}_H = H /\!\!/ c(H) \to \operatorname{Spec} \boldsymbol{J}_G = G /\!\!/ c_\sigma(G)$, where c denotes the usual conjugation of H on H. Here $\boldsymbol{J}_H = k[y]$ (resp. $\boldsymbol{J}_G = k[x]$) is the space of conjugation invariant functions on SO_3 (resp. twisted conjugation invariant functions on SL_3), with the variable y (resp. x) representing the character of the standard representation of SO_3 (resp. the adjoint representation of SL_3). Note that $x=(y-1)^2$ when viewed as functions on T^σ. So $k[x,x^{-1}]\to k[y,(y-1)^{-1}]$ is finite étale of degree two.

The point $x = 4$ is the image of the divisor $A_{\mathcal{O}}$ under the isomorphism $A/\!\!/W_0 \cong \operatorname{Spec}\boldsymbol{J}_G$. Its preimages under the map $\operatorname{Spec}\boldsymbol{J}_H \to \operatorname{Spec}\boldsymbol{J}_G$ are points represented by $y = -1$ and $y = 3$, where $y = 3$ corresponds to the Weyl group orbit of the zero of $e^\beta - 1$ on T^σ. Let $u = x - 4$, and $v = y - 3$. Then the map $k[x] \to k[y]$ induces an isomorphism $k[[u]] \cong k[[v]]$. We claim that the natural map

$$\boldsymbol{J}_G(V) \otimes_{\boldsymbol{J}_G} k[[u]] \to \boldsymbol{J}_H(V) \otimes_{\boldsymbol{J}_H} k[[v]] \tag{6.2.1}$$

is an isomorphism.

By [**St65**, Theorem 6.11 c)] (or a direct computation), the preimage of $y = 3$ under the quotient map $\mathrm{SO}_3^{\mathrm{reg}} \to \operatorname{Spec}\boldsymbol{J}_H = \operatorname{Spec} k[y]$ is a single H-orbit of $g(0) = \begin{pmatrix} 1 & 1 & \frac{1}{2} \\ & 1 & 1 \\ & & 1 \end{pmatrix}$. One can easily check that the image of $g(0)$ in SL_3, denoted by $h(0)$, is also σ-regular. Using the smoothness of the Chevalley map $\mathrm{SO}_3^{\mathrm{reg}} \to \operatorname{Spec} k[y]$, we may lift $g(0)$ to $g : \operatorname{Spec} k[[v]] \to \mathrm{SO}_3^{\mathrm{reg}}$. Since $\mathrm{SL}_3^{\sigma\text{-reg}}$ is open and naturally $\operatorname{Spec} k[[u]] \cong \operatorname{Spec} k[[v]]$, we get a composition of maps

$$h : \operatorname{Spec} k[[u]] \cong \operatorname{Spec} k[[v]] \xrightarrow{g} \mathrm{SO}_3 \to \mathrm{SL}_3$$

whose image lands in $\mathrm{SL}_3^{\sigma\text{-reg}}$. Now, viewing g and h as $k[[u]]$-valued points on $\mathrm{SO}_3^{\mathrm{reg}}$ and $\mathrm{SL}_3^{\sigma\text{-reg}}$ respectively, the centralizer $Z_H(g)$ of g in SO_3 is naturally contained in the twisted centralizer I_h of h in SL_3. We claim that $Z_H(g) = I_h$. Indeed, this is true over the generic point of $\operatorname{Spec} k[[u]] \cong \operatorname{Spec} k[[v]]$ as both spaces are connected by Lemma 5.2.11. Over the special point, one can compute explicitly that

$$Z_H(g(0)) = I_{h(0)} = \left\{ \begin{pmatrix} 1 & x & \frac{1}{2}x^2 \\ & 1 & x \\ & & 1 \end{pmatrix} ;\ x \in k \right\}.$$

Thus $Z_H(g) = I_h$. Since a V-valued function is determined by its restriction to the regular locus, by Corollary 5.3.5 (1), we deduce that

$$\boldsymbol{J}_H(V) \otimes_{\boldsymbol{J}_H} k[[u]] = (V \otimes k[[u]])^{Z_H(g)} = (V \otimes k[[v]])^{I_h} = \boldsymbol{J}_G(V) \otimes_{\boldsymbol{J}_G} k[[v]].$$

This shows that (6.2.1) is an isomorphism.

Now we immediately reduce the lemma to the SL_2-case (Lemma 6.2.4) by the same argument as in the last two paragraphs of the proof of Lemma 6.2.3 (based on Lemma 5.2.18), provided that we can show that $V|_{\mathrm{SO}_3}$ admits a good filtration. It suffices to check this for Schur modules $\mathsf{S}_{a\omega_1+b\omega_2}$ with $a, b \geq 0$, where ω_1 and ω_2 are the fundamental weights of SL_3. We check by induction on $a + b$. This is true for ω_1 and ω_2 as they are standard representations of SO_3, and therefore true for $\mathsf{S}_{\omega_1}^{\otimes a} \otimes \mathsf{S}_{\omega_2}^{\otimes b}$ by Theorem 3.1.2 (2). By the Frobenius reciprocity, there is a unique zero (up to scalar) map $\mathsf{S}_{\omega_1}^{\otimes a} \otimes \mathsf{S}_{\omega_2}^{\otimes b} \to \mathsf{S}_{a\omega_1+b\omega_2}$. One easily identify the kernel V' as $(\mathsf{S}_{\omega_1}^{\otimes a} \otimes \mathsf{S}_{\omega_2}^{\otimes b})_{\prec a\omega_1+b\omega_2}$ in Lemma 3.1.4 (2), and therefore admits a good filtration (whose graded pieces are Schur modules with highest weights strictly lower than $a\omega_1 + b\omega_2$). So the restriction of V' to SO_3 admits a filtration by inductive hypothesis. Moreover, by Lemma 3.1.4 (2), $\mathsf{S}_{\omega_1}^{\otimes a} \otimes \mathsf{S}_{\omega_2}^{\otimes b} \to \mathsf{S}_{a\omega_1+b\omega_2}$ is surjective. By Theorem 3.1.2 (3), we conclude that $\mathsf{S}_{a\omega_1+b\omega_2}$ admits a good filtration when restricted to SO_3. □

REMARK 6.2.6. As in the proof of the lemma, we have an analogous map

$$\boldsymbol{J}_G(V) \otimes_{\boldsymbol{J}_G} k[[x-4]] \to \boldsymbol{J}_H(V) \otimes_{\boldsymbol{J}_H} k[[y+1]].$$

But it is not an isomorphism in general. The reason is that $y = -1$ lies on $\mathring{T}^\sigma /\!/ W_0$, so every regular element in $\mathrm{SO}_3^{\mathrm{reg}}$ that maps to the point $y = -1$ in $\operatorname{Spec} k[y]$ is conjugate to the *semisimple* matrix $\begin{pmatrix} -1 & 0 & 0 \\ & 1 & 0 \\ & & -1 \end{pmatrix}$; yet this matrix when viewed inside SL_3 is *not* σ-regular.

6.3. The ring $\operatorname{End}(\widetilde{V})$. Motivated by some applications in [**XZ17**], in this subsection we study the endomorphism ring $\operatorname{End}(\widetilde{V})$ of the vector bundle $\widetilde{V}$ on $[G\sigma/G]$. Here are some basic facts.

(1) The ring $\operatorname{End}(\widetilde{V})$ is a $\boldsymbol{J}$-algebra. In particular, if W is a representation of $G \rtimes \langle\sigma\rangle$, its character χ_W, restricted to $G\sigma$, defines a regular function on $[G\sigma/G]$ and therefore an element in $\operatorname{End}(\widetilde{V})$, still denoted by χ_W.

(2) The Chevalley restriction map is the pullback map $\operatorname{End}(\widetilde{V}) \to \operatorname{End}(\widetilde{V}|_{[T\sigma/T]})$ along $[T\sigma/T] \to [G\sigma/G]$. Note that

$$\operatorname{End}(\widetilde{V}|_{[T\sigma/T]}) \cong (k[T] \otimes \operatorname{End}_{T^\sigma} V)^{T/T^\sigma}.$$

So (6.1.3) gives an injective map

$$\operatorname{Res}^\sigma_{V\otimes V^*} \otimes 1 : \operatorname{End}(\widetilde{V}) \otimes_{\boldsymbol{J}_G} \boldsymbol{J}_T \to (k[T] \otimes \operatorname{End}_{T^\sigma} V)^{T/T^\sigma},$$

which is an isomorphism generically over $\boldsymbol{J}_T$. It follows that $\operatorname{End}(\widetilde{V})$ is commutative if and only if V is multiplicity free as a T^σ-module.

(3) Assume that V is the restriction of a representation $\rho : G \rtimes \langle\sigma\rangle \to \mathrm{GL}(V)$ to G. There is an element $\gamma_{\mathrm{taut}} \in \operatorname{End}(\widetilde{V})$, given by a tautological automorphism of $\widetilde{V}$. Namely, consider the trivial vector bundle $\mathcal{O}_{G\sigma} \otimes V$ over $G\sigma$. There is the tautological automorphism of this bundle whose restriction to the fiber over $g\sigma \in G\sigma$ is given by the automorphism $\rho(g\sigma)$ of V. This automorphism commutes with the G-equivariant structure on $\mathcal{O}_{G\sigma} \otimes V$, and therefore descends to the desired γ_{taut}. Note that

$$(\operatorname{Res}^\sigma_{V\otimes V^*} \otimes 1)(\gamma_{\mathrm{taut}}) = \sum_\mu e^{\sigma(\mu)} \otimes \sigma_\mu,$$

where the sum is taken over all weights in V and $\sigma_\mu : V(\mu) \to V(\sigma(\mu))$ is the natural isomorphism.

PROPOSITION 6.3.1 (Cayley–Hamilton). *Write $d = \dim V$. Define a degree d polynomial $f(x) \in \boldsymbol{J}[x]$ as*

$$f(x) = \sum (-1)^i \chi_{\wedge^{d-i} V} x^i.$$

Then there is an injective map of $\boldsymbol{J}$-algebras defined by

$$\boldsymbol{J}[x]/(f(x)) \to \operatorname{End}(\widetilde{V}), \quad x \mapsto \gamma_{\mathrm{taut}}. \tag{6.3.1}$$

In particular,

$$\sum (-1)^i \chi_{\wedge^{d-i} V} \gamma^i_{\mathrm{taut}} = 0 \tag{6.3.2}$$

as elements in $\operatorname{End}(\widetilde{V})$.

PROOF. For $G = \mathrm{GL}_n$, $\sigma = \mathrm{id}$ and $V = \mathrm{std}$ being the standard representation, the usual Cayley–Hamilton theorem gives (6.3.2). In general, the representation defines a homomorphism $G \rtimes \langle\sigma\rangle \to \mathrm{GL}(V)$, which induces $[G\sigma/G] \to [\mathrm{GL}(V)/\mathrm{GL}(V)]$. The pullback of the bundle $\widetilde{\mathrm{std}}$ on $[\mathrm{GL}(V)/\mathrm{GL}(V)]$ along this map is exactly $\widetilde{V}$, the pullback of the tautological automorphism γ_{taut} of $\widetilde{\mathrm{std}}$ is the tautological automorphism of $\widetilde{V}$, and the pullback of the regular function $\chi_{\wedge^i \mathrm{std}}$ is just $\chi_{\wedge^i V}$. Therefore, (6.3.2) holds in general. Or equivalently, (6.3.1) is well-defined. It remains to show that it is injective.

If $G = \mathrm{GL}_n, \sigma = \mathrm{id}$ and $V = \mathrm{std}$, by Lemma 6.3.4 below, (6.3.1) is an isomorphism. On the other hand, for a representation $G \rtimes \langle \sigma \rangle \to \mathrm{GL}(V)$, the map

$$\mathrm{End}(\widetilde{\mathrm{std}}) \otimes_{\boldsymbol{J}_{\mathrm{GL}(V)}} \boldsymbol{J}_G \to \mathrm{End}_{\mathrm{Coh}([G\sigma/G])}(\widetilde{V})$$

is injective. Therefore, the injectivity of (6.3.1) in the general case follows. □

REMARK 6.3.2. Note that the proof of the above proposition holds for *any* algebraic group G.

REMARK 6.3.3. We also briefly explain how to study $\mathrm{End}(\widetilde{V})$ if V is not a representation of $G \rtimes \langle \sigma \rangle$, but only a representation of a subgroup $G \rtimes \langle \tau \rangle$, where $\tau = \sigma^f$. In this case, we have the map

$$[G\sigma/G] \to [G\tau/G], \quad g\sigma \mapsto (g\sigma)^f = g\sigma(g)\sigma^2(g)\cdots\sigma^{f-1}(g)\tau.$$

Then we can pullback everything from $[G\tau/G]$. Note that for a representation W of $G \rtimes \langle \tau \rangle$, $W \otimes \sigma(W) \otimes \cdots \otimes \sigma^{f-1}(W)$ is a natural representation of $G \rtimes \langle \sigma \rangle$, usually called the tensor induction representation. Let us denote it by $\otimes_\tau^\sigma W$. Then the pullback of χ_W to $[G\sigma/G]$ is $\chi_{\otimes_\tau^\sigma W}$. So we have

$$\sum (-1)^i \chi_{\otimes_\tau^\sigma(\wedge^{d-i} V)} \gamma_{\mathrm{taut}}^i = 0.$$

As mentioned above, for a general V which is not multiplicity free as a T^σ-module, the ring $\mathrm{End}(\widetilde{V})$ is non-commutative. So $\mathrm{End}(\widetilde{V})$ is not generated by γ_{taut} as a $\boldsymbol{J}$-algebra. Even if $\mathrm{End}(\widetilde{V})$ is commutative, (6.3.1) will only be an isomorphism away from a divisor on $\mathrm{Spec}\, \boldsymbol{J}_G$. However, we can say more in the following special case.

LEMMA 6.3.4. *The map* (6.3.1) *is an isomorphism if*

- $G = \mathrm{GL}_n$ *or* SL_n *equipped with the trivial* σ*-action, and* V *is the standard representation or its dual.*
- $G = \mathrm{Sp}_{2n}$ *and* V *is the standard representation;*

PROOF. It is enough to show that $\boldsymbol{J}_G[x]/(f(x)) \otimes_{\boldsymbol{J}_G} \boldsymbol{J}_T = \boldsymbol{J}_T[x]/(f(x)) \to \mathrm{End}(\widetilde{V}) \otimes_{\boldsymbol{J}_G} \boldsymbol{J}_T$ is an isomorphism. For this, we calculate the composition (6.3.3)

$$\boldsymbol{J}_G[x]/(f(x)) \otimes_{\boldsymbol{J}_G} \boldsymbol{J}_T = \boldsymbol{J}_T[x]/(f(x)) \to \mathrm{End}(\widetilde{V}) \otimes_{\boldsymbol{J}_G} \boldsymbol{J}_T \to \boldsymbol{J}_T \otimes \mathrm{End}_T(V).$$

Under this map (6.3.3), the image of γ_{taut} is $\sum e^{\lambda_j} \otimes \mathrm{id}_{V(\lambda_j)}$, where the sum is taken over all weights $\{\lambda_j\}$ of V, and $\mathrm{id}_{V(\lambda_j)}$ is the identity map of $V(\lambda_j)$. Then the image of γ_{taut}^i is $\sum e^{i\lambda_j} \otimes \mathrm{id}_{V(\lambda_j)}$. Since V is minuscule, $\{\mathrm{id}_{V(\lambda_j)}\}$ form a basis of $\boldsymbol{J}_T \otimes \mathrm{End}_T(V)$ as a $\boldsymbol{J}_T$-module. By Vandermonde, the determinant of the composed map (6.3.3) is

$$\prod_{j<j'} (e^{\lambda_j} - e^{\lambda_{j'}}) = e^{\mu} \prod_{j<j'} (1 - e^{\lambda_j - \lambda_{j'}}),$$

from some $\mu \in \mathbb{X}^\bullet(T)$. For the cases in the lemma, the difference $\lambda_j - \lambda_{j'}$ is a root α, which is equal to the determinant (up to a unit in $\boldsymbol{J}_T$) of the second map in (6.3.3) by Proposition 6.1.4. (Note that Proposition 6.1.4 also holds for GL_n.) Therefore, the first map in (6.3.3) is an isomorphism. □

6.4. Some examples. In this subsection, we present some examples of the calculation of the determinant of (6.1.1).

EXAMPLE 6.4.1. Consider the case where $G = \prod_{i=1}^{d} \mathrm{SL}_2$ and σ permutes all d factors, and let $V = V_{a_1} \boxtimes \cdots \boxtimes V_{a_d}$ be the exterior tensor product representation, where V_a is the ath symmetric power representation of SL_2 (i.e. the representation of SL_2 on the space of homogeneous polynomials in (x, y) of degree a). By Lemma 4.1.2, this is equivalent to the case where $G = \mathrm{SL}_2$, $\sigma = \mathrm{id}$, and $V = V_{a_\bullet} := V_{a_1} \otimes \cdots \otimes V_{a_d}$ (so that $r_V = V_{a_\bullet}(0) \neq 0$). Now assume that $a_1 + \cdots + a_d$ is even. In this case, Theorem 6.1.2 says that the determinant of the matrix (6.1.1) is equal to

$$c \cdot (e^{\alpha} - 1)^{\zeta} (e^{-\alpha} - 1)^{\zeta}$$

for some $c \in k^{\times}$, where

$$\zeta = \sum_{n \geq 1} \dim V_{a_\bullet}(n) = \frac{\dim V_{a_\bullet} - \dim V_{a_\bullet}(0)}{2}.$$

In fact, the matrix for the pairing (6.1.1) (for some appropriate basis) can be obtained via a combinatorial description in terms of periodic meanders, which carries a quantum deformation. In the special case when $G = \mathrm{SL}_2$, d is even, and $a_1 = \cdots = a_d = 1$, this matrix (6.1.1) is the Gram matrix for the periodic meanders (see [**TX14**]).

EXAMPLE 6.4.2. Consider the case when $G = \mathrm{SL}_{2r+1}$ with the non-trivial pinned σ-action as explained in Example 5.1.5. We use freely the notation therein. In particular, $\mathbb{X}^\bullet(T)$ is generated by the characters $\varepsilon_{-r}, \ldots, \varepsilon_r$ with the relation that $\sum_{i=-r}^{r} \varepsilon_i = 0$, and σ acts by $\sigma(\varepsilon_i) = -\varepsilon_{-i}$ for each i. The Bruhat partial order on $\mathbb{X}^\bullet(T)$ is generated by $\varepsilon_i \succeq \varepsilon_{i+1}$ for $i = -r, \ldots, r-1$. For $A = T/(\sigma - 1)T$, $\mathbb{X}^\bullet(A) = \mathbb{X}^\bullet(T)^\sigma = \bigoplus_{i=1}^{r} \mathbb{Z} \cdot (\varepsilon_{-i} - \varepsilon_i)$. The absolute Weyl group $W \cong S_{2r+1}$ permutes $\varepsilon_{-r}, \ldots, \varepsilon_r$, and its σ-invariant elements are

$$W_0 = W^\sigma = \langle (i, -i), (ij)(-i, -j) \mid i, j = 1, \ldots, r \rangle = (\mathbb{Z}/2\mathbb{Z})^r \rtimes S_r.$$

Recall the σ-orbits of roots in (5.1.2); in particular, $\alpha_{\mathcal{O}_i^-} = \alpha_{\mathcal{O}_i^+} = \varepsilon_{-i} - \varepsilon_i$ for $i = 1, \ldots, r$. Write $\mathfrak{S}_i$ for the ith elementary symmetric power in

$$e^{\alpha_{\mathcal{O},1}} + e^{-\alpha_{\mathcal{O},1}}, \ldots, e^{\alpha_{\mathcal{O},r}} + e^{-\alpha_{\mathcal{O},r}},$$

then

$$\begin{aligned} \boldsymbol{J} := k[G\sigma]^G &\cong k[\mathbb{X}^\bullet(A)]^{W_0} \\ &\cong k[\mathfrak{S}_1, \ldots, \mathfrak{S}_r] \subset k[\mathbb{X}^\bullet(T)] = k[e^{\pm\varepsilon_{-r}}, \ldots, e^{\pm\varepsilon_{-1}}, e^{\pm\varepsilon_1}, \ldots, e^{\pm\varepsilon_r}]. \end{aligned}$$

When $r = 1$, $\boldsymbol{J} = k[e^{\alpha_{\mathcal{O},1}} + e^{-\alpha_{\mathcal{O},1}}]$.

We consider the standard representation $V = \mathsf{S}_{\varepsilon_{-r}} = \mathrm{std}$ of $G = \mathrm{SL}_{2r+1}$. Observe that the only nonzero weight space of V with weights in $(\sigma - 1)\mathbb{X}^\bullet(T)$ is $V(\varepsilon_0)$. We write

$$\varepsilon_0 = \sigma(\nu) - \nu, \quad \text{for} \quad \nu = \varepsilon_{-r} + \cdots + \varepsilon_{-1}.$$

The representation $\mathtt{S}_\nu \cong \mathtt{S}_{\sigma(\nu)^*} \cong \wedge^r\mathrm{std}$ is the rth wedge product of the standard representation. Then it is easy to see that

$$(\mathtt{S}_{\sigma(\nu^*)} \otimes \mathtt{S}_\nu \otimes V)^G = (\wedge^r\mathrm{std} \otimes \wedge^r\mathrm{std} \otimes \mathrm{std})^G \cong (\wedge^{2r+1}\mathrm{std})^G \cong k,$$

Let $\mathbf{b}$ denote the element on the left hand side that corresponds to $1 \in k$, or more precisely $e_{-r} \wedge \cdots \wedge e_r$. Then the recipe in (4.4.2) defines a function $f_{\mathbf{b}} \in \boldsymbol{J}(V)$ (note that in our case $\mathtt{W}_\nu \cong \mathtt{S}_\nu$). We will compute its restriction to $T\sigma$.

Let $e_{-r}, \dots, e_r$ denote the standard basis of the representation std. Then $\wedge^r\mathrm{std}$ has a basis $(e_I)_{I\subset\{-r,\dots,r\},|I|=r}$, where for $I = \{i_1, \dots, i_r\}$ with elements ordered increasingly, we write $e_I = e_{i_1} \wedge \cdots \wedge e_{i_r}$. The natural map

$$\sigma : \mathtt{S}_\nu = \wedge^r\mathrm{std} \to \mathtt{S}_{\sigma(\nu)} \cong (\wedge^r\mathrm{std})^*$$

is given by sending e_i to $(-1)^i e^*_{-i}$.[6] So $\sigma(e_I) = \mathrm{sgn}(I) \cdot e^*_{-I}$, where

$$-I = \{-i_r, \dots, -i_1\}$$

and $\mathrm{sgn}(I) = (-1)^{i_1+\cdots+i_r+r(r-1)/2}$. Using this notation, we write explicitly

$$\mathbf{b} = \sum_{|I|=r,|I'|=r} e_I \otimes e_{I'} \otimes v_{I,I'}, \quad f_{\mathbf{b}}(t\sigma) = \sum_{|I|=r,|I'|=r} \langle e_I, t\sigma e_{I'}\rangle \cdot v_{I,I'},$$

where $v_{I,I'} \in V = \mathrm{std}$, $t \in T$, and the sums run through all subsets I, I' of $\{-r, \dots, r\}$ of cardinality r. In the expression for $f_{\mathbf{b}}(t\sigma)$, the pairing $\langle e_I, t\sigma e_{I'}\rangle$ is nonzero precisely when $I' = -I$. In the expression for $\mathbf{b}$, the vector $v_{I,I'}$ is nonzero precisely when I and I' are disjoint. So we need only to discuss the case when there is $\underline{s} = \{s_1, \dots, s_r\} \in \{\pm 1\}^r$ such that $I = I_{\underline{s}} = \{s_1 \cdot 1, \dots, s_r \cdot r\}$ and $I' = -I$. In this case, $\mathrm{sgn}(-I_{\underline{s}}) = (-1)^{r(r+1)/2+r(r-1)/2} = (-1)^r$ is independent of $\underline{s}$. Moreover, we may take $v_{I,I'} = (-1)^r \,\mathrm{sgn}(\underline{s}) \cdot e_0$, where the sign $\mathrm{sgn}(\underline{s}) = (-1)^u$ with u equal to the number of -1's in $\underline{s}$. Thus,

$$\mathrm{Res}^\sigma_V(f_{\mathbf{b}}) = \sum_{\underline{s}\in\{\pm1\}^r} \mathrm{sgn}(\underline{s}) e^{-\varepsilon_{-s_1}-\varepsilon_{-2s_2}-\cdots-\varepsilon_{-rs_r}}(t) \cdot e_0 = \prod_{i=1}^r \big(e^{-\varepsilon_{-i}} - e^{-\varepsilon_i}\big) \cdot e_0 \in \boldsymbol{J}_T(V).$$

In the dual picture, we consider an element

$$\mathbf{b}^\vee \in (V_{\sigma(\nu)} \otimes V_{\nu^*} \otimes V^*)^G = \big((\wedge^r\mathrm{std}^*) \otimes (\wedge^r\mathrm{std}^*) \otimes \mathrm{std}^*\big)^G \cong (\wedge^{2r+1}\mathrm{std}^*)^G \cong k.$$

Explicitly, we write

$$\mathbf{b}^\vee = \sum_{|I|=r,|I'|=r} e_I \otimes e^*_{I'} \otimes v^\vee_{I,I'}, \quad f_{\mathbf{b}^\vee}(t\sigma) = \sum_{|I|=r,|I'|=r} \langle e_I, t\sigma e^*_{I'}\rangle \cdot e^\vee_{I,I'}.$$

Similar to above, we need only to discuss the case when $I = I_{\underline{s}}$ for some $\underline{s} \subset \{\pm 1\}^r$ and $I' = -I$. In this case, we may choose $v_{I,I'} = (-1)^r \,\mathrm{sgn}(\underline{s}) \cdot e^*_0$. Thus,

$$\mathrm{Res}^\sigma_{V^*}(f_{\mathbf{b}^\vee}) = \sum_{\underline{s}\in\{\pm1\}^r} \mathrm{sgn}(\underline{s}) e^{\varepsilon_{-s_1}+\varepsilon_{-2s_2}+\cdots+\varepsilon_{-rs_r}}(t) \cdot e^*_0 = \prod_{i=1}^r \big(e^{\varepsilon_{-i}} - e^{\varepsilon_i}\big) \cdot e^*_0 \in \boldsymbol{J}_T(V^*).$$

Now we compute $\langle f_{\mathbf{b}}, f_{\mathbf{b}^\vee}\rangle$ under the pairing $\boldsymbol{J}(V) \otimes_{\boldsymbol{J}} \boldsymbol{J}(V^*) \to \boldsymbol{J} \cong \boldsymbol{J}_T^{W_0}$:

$$\langle f_{\mathbf{b}}, f_{\mathbf{b}^\vee}\rangle = \prod_{i=1}^r \big(e^{-\varepsilon_{-i}} - e^{-\varepsilon_i}\big) \cdot \prod_{i=1}^r \big(e^{\varepsilon_{-i}} - e^{\varepsilon_i}\big) = \prod_{i=1}^r \big((e^{\alpha_{\mathcal{O}_i^-}} - 1)(e^{-\alpha_{\mathcal{O}_i^-}} - 1)\big).$$

[6] This is in fact $(-1)^r$ times the natural map given by (5.1.1).

This agrees with Theorem 6.1.2 by noting that $\zeta_{\mathcal{O}_i^-}(V)=1$ and $\zeta_{\mathcal{O}_i^+}(V)=\zeta_{\mathcal{O}_{i,j}}(V)=0$, for the σ-orbits in (5.1.2).

EXAMPLE 6.4.3. Consider SL_4 with row and column indices in $\{-2,-1,1,2\}$, the pinning (B,T,e) given by the subgroup of standard upper triangular matrices, the subgroup of diagonal matrices, and $e = E_{-2,-1} + E_{-1,1} + E_{1,2}$. The unique non-trivial pinned automorphism σ is given by $\sigma(X) = J\,{}^tX^{-1}J^{-1}$ for $X \in \mathrm{SL}_4$, with $J = \begin{pmatrix} 0&0&0&1\\ 0&0&-1&0\\ 0&1&0&0\\ -1&0&0&0 \end{pmatrix}$.

Let ε_i for $i = \pm1, \pm2$ be the character of T given by evaluating at the (i,i)-entry; so that $\mathbb{X}^\bullet(T)$ is generated by these ε_i with the relation $\varepsilon_{-2}+\varepsilon_{-1}+\varepsilon_1+\varepsilon_2 = 0$. Under the σ-action $\sigma(\varepsilon_i) = -\varepsilon_i$, the following are the σ-orbits of $\Phi(G,T)$:

$$\mathcal{O}_i=\{\varepsilon_{-i}-\varepsilon_i\},\quad \mathcal{O}_{\pm2,\pm1}=\{\varepsilon_{\mp2}-\varepsilon_{\mp1},\varepsilon_{\pm1}-\varepsilon_{\pm2}\},\quad \mathcal{O}_{\mp2,\pm1}=\{\varepsilon_{\pm2}-\varepsilon_{\mp1},\varepsilon_{\pm1}-\varepsilon_{\mp2}\}$$

with $i \in \{\pm1,\pm2\}$. They are all of type A. For $A = T/(\sigma-1)T$,

$$\mathbb{X}^\bullet(A) = \mathbb{X}^\bullet(T)^\sigma = \mathbb{Z}(\varepsilon_{-2}-\varepsilon_2) \oplus \mathbb{Z}(\varepsilon_{-1}+\varepsilon_{-2}).$$

(Note that $(\varepsilon_{-2}-\varepsilon_2) + (\varepsilon_{-1}-\varepsilon_1) = 2(\varepsilon_{-2}+\varepsilon_{-1})$.) The root system $\Phi(G_\sigma, A)$ consists of short roots $\alpha_{\mathcal{O}_i}$ and long roots $\alpha_{\mathcal{O}_{i,j}}$; the group $G_\sigma \cong \mathrm{Spin}_5$. The absolute Weyl group of $\Phi(G,T)$ is $W \cong S_4$ given by permuting the ε_i's, and its σ-invariant elements are $W_0 = W^\sigma = \langle(-1,1),(-2,2),(1,2)(-1,-2)\rangle = (\mathbb{Z}/2\mathbb{Z})^2 \rtimes S_2$. Write

$$\mathfrak{S}_1 = e^{\varepsilon_{-1}-\varepsilon_1} + e^{\varepsilon_1-\varepsilon_{-1}} + e^{\varepsilon_{-2}-\varepsilon_2} + e^{\varepsilon_2-\varepsilon_{-2}},$$
$$\mathfrak{S}_2 = e^{\varepsilon_{-2}+\varepsilon_{-1}} + e^{-\varepsilon_{-2}-\varepsilon_{-1}} + e^{\varepsilon_1+\varepsilon_2} + e^{-\varepsilon_1-\varepsilon_2}.$$

Then

$$\begin{aligned}\boldsymbol{J} := k[G\sigma]^G &\cong k[\mathbb{X}^\bullet(A)]^{W_0} \cong k[\mathfrak{S}_1,\mathfrak{S}_2]\\ &\subset k[\mathbb{X}^\bullet(T)] = k[e^{\pm\varepsilon_{-2}}, e^{\pm\varepsilon_{-1}}, e^{\pm\varepsilon_1}, e^{\pm\varepsilon_2}]/(e^{\varepsilon_{-2}+\varepsilon_{-1}+\varepsilon_1+\varepsilon_2}-1).\end{aligned}$$

Consider the minuscule representation $V = \mathsf{S}_{\varepsilon_{-2}+\varepsilon_{-1}}$ of SL_4; it is isomorphic to $\wedge^2\mathrm{std}$ for the standard representation std of SL_4. There are two nonzero weight spaces with weights in $(\sigma-1)\mathbb{X}^\bullet(T)$: $V(\lambda_1)$ and $V(\lambda_2)$ with $\lambda_i = \varepsilon_{-i}+\varepsilon_i$. They are both of one-dimensional. So $r_V = 2$. The minimal dominant weights ν_i (in the sense as in Lemma 4.3.6) satisfying $\sigma(\nu_i) - \nu_i = \lambda_i$ are

$$\nu_1 = \varepsilon_{-2}, \quad \nu_2 = -\varepsilon_2.$$

We remark here that for SL_n with $n \geq 5$ and the non-trivial automorphism σ, and for $V = \wedge^2\mathrm{std}$, we have $r_V = \lfloor \frac{n}{2} \rfloor$ and some (or rather most) of the minimal dominant weights ν above are no longer minuscule. So the computation will be much more involved.

Back to our case, $\mathsf{S}_{\nu_1} \cong \mathsf{S}_{\sigma(\nu_1)^*} \cong \mathrm{std}$ and $\mathsf{S}_{\nu_2} \cong \mathsf{S}_{\sigma(\nu_2)^*} \cong \mathrm{std}^*$. In what follows, we will describe nonzero elements

$$\mathbf{b}_1 \in (\mathsf{S}_{\sigma(\nu_1)^*} \otimes \mathsf{S}_{\nu_1} \otimes V)^G = (\mathrm{std} \otimes \mathrm{std} \otimes \wedge^2\mathrm{std})^G,$$
$$\mathbf{b}_2 \in (\mathsf{S}_{\sigma(\nu_2)^*} \otimes \mathsf{S}_{\nu_2} \otimes V)^G = (\mathrm{std}^* \otimes \mathrm{std}^* \otimes \wedge^2\mathrm{std})^G,$$

use the recipe in (4.4.2) to defines a function $f_{\mathbf{b}_1}, f_{\mathbf{b}_2} \in \boldsymbol{J}(V)$ (note that in our case $\mathsf{W}_{\nu_i} \cong \mathsf{S}_{\nu_i}$), and describe the restriction of $f_{\mathbf{b}_i}$ to $T\sigma$.

Write $\{e_{-2}, e_{-1}, e_1, e_2\}$ for the standard basis of std and $\{e_{-2}^*, e_{-1}^*, e_1^*, e_2^*\}$ the dual basis; and the isomorphism $\sigma : \mathrm{std} \cong \mathrm{std}^*$ is given by

$$e_{-2} \mapsto e_2^*, \quad e_{-1} \mapsto -e_1^*, \quad e_1 \mapsto e_{-1}^*, \quad e_2 \mapsto -e_{-2}^*.$$

With this notation, we write explicitly,

$$\mathbf{b}_1 = \sum_{i,j\in\{\pm1,\pm2\}} e_i \otimes e_j \otimes v_{i,j}, \quad f_{\mathbf{b}_1}(t\sigma) = \sum_{i,j\in\{\pm1,\pm2\}} \langle e_i, t\sigma e_j\rangle \cdot v_{i,j},$$

where $v_{i,j} \in V = \wedge^2\mathrm{std}$ and $t \in T$. In the expression for $f_{\mathbf{b}_1}(t\sigma)$, the pairing $\langle e_i, t\sigma e_j\rangle$ is nonzero precisely when $i = -j$. In the expression for $\mathbf{b}_1$, the vector $v_{i,j}$ is nonzero precisely when $i \neq j$. So we need only to discuss the case when $i = -j \in \{\pm1, \pm2\}$. In this case, we may choose $v_{-2,2} = -v_{2,-2} = -e_{-1} \wedge e_1$ and $v_{-1,1} = -v_{1,-1} = -e_{-2} \wedge e_2$. Thus,

$$\mathrm{Res}_V^\sigma(f_{\mathbf{b}_1}) = (e^{-\varepsilon_{-2}} + e^{-\varepsilon_2})e_{-1} \wedge e_1 - (e^{-\varepsilon_{-1}} + e^{-\varepsilon_1})e_{-2} \wedge e_2 \in \boldsymbol{J}_T(V).$$

In a similar way, we may compute (up to choosing a sign for $\mathbf{b}_2$)

$$\mathrm{Res}_V^\sigma(f_{\mathbf{b}_2}) = (e^{\varepsilon_{-1}} + e^{\varepsilon_1})e_{-1} \wedge e_1 - (e^{\varepsilon_{-2}} + e^{\varepsilon_2})e_{-2} \wedge e_2 \in \boldsymbol{J}_T(V).$$

On the dual side, we choose elements

$$\mathbf{b}_1^\vee \in (\mathsf{S}_{\sigma(\nu_1)} \otimes \mathsf{S}_{\nu_1^*} \otimes V^*)^G = (\mathrm{std}^* \otimes \mathrm{std}^* \otimes \wedge^2\mathrm{std}^*)^G,$$
$$\mathbf{b}_2^\vee \in (\mathsf{S}_{\sigma(\nu_2)} \otimes \mathsf{S}_{\nu_2^*} \otimes V^*)^G = (\mathrm{std} \otimes \mathrm{std} \otimes \wedge^2\mathrm{std}^*)^G,$$

and a computation similar to above gives the explicit formulas (up to choosing a sign for $\mathbf{b}_i^\vee$)

$$\mathrm{Res}_{V^*}^\sigma(f_{\mathbf{b}_1^\vee}) = (e^{\varepsilon_{-2}} + e^{\varepsilon_2})e_{-1}^* \wedge e_1^* - (e^{\varepsilon_{-1}} + e^{\varepsilon_1})e_{-2}^* \wedge e_2^* \in \boldsymbol{J}_T(V^*)$$
$$\mathrm{Res}_{V^*}^\sigma(f_{\mathbf{b}_2^\vee}) = (e^{-\varepsilon_{-1}} + e^{-\varepsilon_1})e_{-1}^* \wedge e_1^* - (e^{-\varepsilon_{-2}} + e^{-\varepsilon_2})e_{-2}^* \wedge e_2^* \in \boldsymbol{J}_T(V^*).$$

From this, we deduce that under the natural $\boldsymbol{J}$-linear pairing $\boldsymbol{J}(V)\times\boldsymbol{J}(V^*)\to\boldsymbol{J}$, we have

$$\begin{aligned}\langle f_{\mathbf{b}_1}, f_{\mathbf{b}_1^\vee}\rangle &= \langle f_{\mathbf{b}_2}, f_{\mathbf{b}_2^\vee}\rangle \\ &= (e^{\varepsilon_{-2}} + e^{\varepsilon_2})(e^{-\varepsilon_{-2}} + e^{-\varepsilon_2}) + (e^{\varepsilon_{-1}} + e^{\varepsilon_1})(e^{-\varepsilon_{-1}} + e^{-\varepsilon_1}) = \mathfrak{S}_1 + 4, \\ \langle f_{\mathbf{b}_1}, f_{\mathbf{b}_2^\vee}\rangle &= \langle f_{\mathbf{b}_2}, f_{\mathbf{b}_1^\vee}\rangle \\ &= 2(e^{-\varepsilon_{-1}} + e^{-\varepsilon_1})(e^{-\varepsilon_{-2}} + e^{-\varepsilon_2}) = 2(e^{\varepsilon_{-1}} + e^{\varepsilon_1})(e^{\varepsilon_{-2}} + e^{\varepsilon_2}) = 2\mathfrak{S}_2.\end{aligned}$$

(Note that $e^{\varepsilon_{-2}+\varepsilon_{-1}+\varepsilon_1+\varepsilon_2} = 1$.) We can compute the determinant of the pairing as

$$\begin{aligned}&\det\begin{pmatrix}\langle f_{\mathbf{b}_1}, f_{\mathbf{b}_1^\vee}\rangle & \langle f_{\mathbf{b}_1}, f_{\mathbf{b}_2^\vee}\rangle \\ \langle f_{\mathbf{b}_2}, f_{\mathbf{b}_1^\vee}\rangle & \langle f_{\mathbf{b}_2}, f_{\mathbf{b}_2^\vee}\rangle\end{pmatrix} \\ &= \big((e^{\varepsilon_{-2}} + e^{\varepsilon_2})(e^{-\varepsilon_{-2}} + e^{-\varepsilon_2}) + (e^{\varepsilon_{-1}} + e^{\varepsilon_1})(e^{-\varepsilon_{-1}} + e^{-\varepsilon_1})\big)^2 \\ &\qquad - 4(e^{\varepsilon_{-1}} + e^{\varepsilon_1})(e^{\varepsilon_{-2}} + e^{\varepsilon_2})(e^{\varepsilon_{-1}} + e^{\varepsilon_1})(e^{\varepsilon_{-2}} + e^{\varepsilon_2}) \\ &= \big((e^{\varepsilon_{-2}} + e^{\varepsilon_2})(e^{-\varepsilon_{-2}} + e^{-\varepsilon_2}) - (e^{\varepsilon_{-1}} + e^{\varepsilon_1})(e^{-\varepsilon_{-1}} + e^{-\varepsilon_1})\big)^2 \\ &= \big((e^{\varepsilon_{-2}-\varepsilon_2} - e^{\varepsilon_{-1}-\varepsilon_1})(1 - e^{\varepsilon_2-\varepsilon_{-2}+\varepsilon_1-\varepsilon_{-1}})\big)^2 \\ &= (e^{\alpha_{\mathcal{O}_{-2,-1}}} - 1)(e^{\alpha_{\mathcal{O}_{-2,1}}} - 1)(e^{\alpha_{\mathcal{O}_{2,-1}}} - 1)(e^{\alpha_{\mathcal{O}_{2,1}}} - 1).\end{aligned}$$

One can compare this with the computation that $\zeta_{\mathcal{O}_i}(V) = 0$ and

$$\zeta_{\mathcal{O}_{\pm2,\pm1}}(V) = \zeta_{\mathcal{O}_{\pm2,\mp1}}(V) = 1.$$

EXAMPLE 6.4.4. Assume that $\mathrm{char}\, k = 0$. Consider the case of $G = \mathrm{Spin}_{2r}$. More precisely, consider $Q := k^{\oplus 2r}$ with basis $e_{-r}, \dots, e_{-1}, e_1, \dots, e_r$ and a symmetric quadratic form $\langle e_i, e_{-j}\rangle = \delta_{ij}$. Let

$$Cl = Cl(Q) = Cl^{\mathrm{even}} \oplus Cl^{\mathrm{odd}} = \Big(\bigoplus_{n\geq 0} Q^{\otimes n}\Big) \Big/ \big(w \otimes w - \tfrac{1}{2}\langle w, w\rangle;\, w \in Q\big)$$

denote the associated ($\mathbb{Z}/2\mathbb{Z}$-graded) Clifford algebra. There is a natural (anti-commutative) involution $\bullet^t$ on Cl generated by sending $w \in Q$ to $-w$. The pin group Pin_{2r} is formed by all elements $x \in Cl$ such that $x \cdot x^t = 1$, and $xQx^t \subset Q$. The spin group $\mathrm{Spin}_{2r} := \mathrm{Pin}_{2r} \cap Cl^{\mathrm{even}}$ is the neutral connected component of Pin_{2r}. The maximal torus T of Spin_{2r} is given by the image of

$$\begin{aligned} \Gamma : \mathbb{G}_m^r &\longrightarrow \mathrm{Spin}_{2r} \\ (z_1, \dots, z_r) &\longmapsto \textstyle\prod_{i=1}^r \big(z_i e_i e_{-i} + z_i^{-1} e_i e_{-i}\big), \end{aligned}$$

where the kernel is $\{(z_1, \dots, z_r) \in \{\pm 1\}^r \mid z_1 \cdots z_r = 1\}$. Let $\frac{\varepsilon_1}{2}, \dots, \frac{\varepsilon_r}{2}$ denote the usual basis of the character group of $\mathbb{G}_m^r$; then

$$\mathbb{X}^\bullet(T) = \mathbb{Z} \cdot \tfrac{1}{2}(\varepsilon_1 + \cdots + \varepsilon_r) \oplus \bigoplus_{i=1}^{r-1} \mathbb{Z} \cdot \varepsilon_i.$$

A weight $\sum_{i=1}^r a_i \varepsilon_i$ is dominant if $a_1 \geq \cdots \geq a_{r-1} \geq |a_r|$.

We equip with Spin_{2r} with the outer automorphism σ, induced by the conjugation by $e_r + e_{-r}$. A simple computation shows that

$$(e_r + e_{-r})(z_i e_i e_{-i} + z_i^{-1} e_i e_{-i})(e_r + e_{-r}) = \begin{cases} z_i e_i e_{-i} + z_i^{-1} e_i e_{-i} & \text{if } i \neq r \\ z_r^{-1} e_r e_{-r} + z_r e_r e_{-r} & \text{if } i = r \end{cases}$$

So σ fixes $\varepsilon_1, \dots, \varepsilon_{r-1}$ and maps ε_r to $-\varepsilon_r$. So for $A = T/(\sigma - 1)T$, its character group $\mathbb{X}^\bullet(A) = \mathbb{X}^\bullet(T)^\sigma$ is the free abelian group with basis $\varepsilon_1, \dots, \varepsilon_{r-1}$.

The absolute Weyl group $W \cong H_r \rtimes S_r \subset \{\pm 1\}^r \times S_r$, where $H_r \subset \{\pm 1\}^r$ is the subgroup consisting of even number of -1's, S_r permutes $\varepsilon_1, \dots, \varepsilon_r$ and $(h_1, \dots, h_r) \in H_r$ sends ε_i to $h_i \varepsilon_i$ for $i = 1, \dots, r$. The σ-invariant elements of W are $W_0 := W^\sigma = \{\pm 1\}^{r-1} \rtimes S_{r-1}$, where S_{r-1} permutes $\varepsilon_1, \dots, \varepsilon_{r-1}$ and $(h_1, \dots, h_r) \in \{\pm 1\}^{r-1}$ sends ε_i to $h_i \varepsilon_i$ for each $i = 1, \dots, r-1$. It follows that the invariants of $\mathbb{C}[\mathbb{X}^\bullet(A)]$ under W_0 are

$$\boldsymbol{J} := \mathbb{C}[\mathbb{X}^\bullet(A)]^{W_0} = \mathbb{C}[\mathfrak{S}_1, \dots, \mathfrak{S}_{r-1}],$$

where $\mathfrak{S}_i$ for $i = 1, \dots, r-1$ is the ith elementary symmetric polynomial in $e^{\varepsilon_1} + e^{-\varepsilon_1}, \dots, e^{\varepsilon_{r-1}} + e^{-\varepsilon_{r-1}}$.

The root system $\Phi(G, T)$ consists of $\pm\varepsilon_i \pm \varepsilon_j$ and $\pm\varepsilon_i \mp \varepsilon_j$ for $i \in \{1, \dots, r\}$ and $j \in \{1, \dots, i-1\}$. The σ-orbits of these roots are

$$\mathcal{O}_{i,j}^{\pm} = \{\pm\varepsilon_i \pm \varepsilon_j\}, \quad \mathcal{O}_{j,i}^{\pm} = \{\pm\varepsilon_i \mp \varepsilon_j\}, \quad \mathcal{O}_i^{\pm} = \{\pm\varepsilon_i + \varepsilon_r, \pm\varepsilon_i - \varepsilon_r\},$$

with $i \in \{1, \dots, r\}, j \in \{1, \dots, i-1\}$; they are all of type A. Thus the root system $\Phi(G_\sigma, A)$ consists of short roots $\alpha_{\mathcal{O}_{i,j}^{\pm}} = \pm\varepsilon_i \pm \varepsilon_j$ and long roots $\alpha_{\mathcal{O}_i^{\pm}} = \pm 2\varepsilon_i$.

We shall consider $V = \mathsf{S}_{\varepsilon_1} \cong Q$, the vector representation of $G = \mathrm{Spin}_{2r}$. Its nonzero weight spaces with weights in $(\sigma - 1)\mathbb{X}^\bullet(T)$ are $V(-\varepsilon_r)$ and $V(\varepsilon_r)$. We write

$$\mp\varepsilon_r = \sigma(\nu_\pm) - \nu_\pm \quad \text{for} \quad \nu_\pm = \tfrac{1}{2}(\varepsilon_1 + \cdots + \varepsilon_{r-1} \pm \varepsilon_r).$$

(Be careful with the signs here.) So $\sigma(\nu_\pm) = \nu_\mp$. We shall see that

$$(\mathtt{S}_{\sigma(\nu_\pm^*)} \otimes \mathtt{S}_{\nu_\pm} \otimes V)^G \cong k. \tag{6.4.1}$$

Indeed, since V is minuscule, $\mathtt{S}_{\nu_\pm} \otimes V$ is a direct sum of $\mathtt{S}_{\nu_\pm+\tau}$ with all $\tau \in \{\pm\varepsilon_1, \dots, \pm\varepsilon_r\}$ such that $\nu_\pm + \tau$ is dominant. In particular, $\nu_\mp$ is among those weights $\nu_\pm + \tau$. Dually, we have

$$(\mathtt{S}_{\sigma(\nu_\pm)} \otimes \mathtt{S}_{\nu_\pm^*} \otimes V^*)^G \cong k. \tag{6.4.2}$$

LEMMA 6.4.5. *There exist bases $\mathbf{b}_\pm$ of* (6.4.1) *and bases $\mathbf{b}_\pm^\vee$ of* (6.4.2) *such that if we use $f_{\mathbf{b}_\pm} \in \boldsymbol{J}(V)$ and $f_{\mathbf{b}_\pm^\vee} \in \boldsymbol{J}(V^*)$ to denote the associated class functions via the recipe* (4.4.2), *then*
(6.4.3)

$$\mathfrak{M} := \begin{pmatrix} \langle f_{\mathbf{b}_+}, f_{\mathbf{b}_+^\vee}\rangle & \langle f_{\mathbf{b}_+}, f_{\mathbf{b}_-^\vee}\rangle \\ \langle f_{\mathbf{b}_-}, f_{\mathbf{b}_+^\vee}\rangle & \langle f_{\mathbf{b}_-}, f_{\mathbf{b}_-^\vee}\rangle \end{pmatrix} = \begin{pmatrix} \sum_{i\geq 0 \text{ even}} 2^{r-1-i}\mathfrak{S}_i & \sum_{i>0 \text{ odd}} 2^{r-1-i}\mathfrak{S}_i \\ \sum_{i>0 \text{ odd}} 2^{r-1-i}\mathfrak{S}_i & \sum_{i\geq 0 \text{ even}} 2^{r-1-i}\mathfrak{S}_i \end{pmatrix}.$$

The determinant of $\mathfrak{M}$ is

$$\prod_{i=1}^{r-1}(1 - e^{2\varepsilon_i})(1 - e^{2\varepsilon_{-i}}).$$

This agrees with our Theorem 6.1.2 because, for $i \neq j \in \{1, \dots, r-1\}$, one can check easily that $\zeta_{\mathcal{O}_i^\pm}(V) = 1$, $\zeta_{\mathcal{O}_{i,j}^\pm}(V) = 0$.

PROOF. Consider the (isotropic) subspace $U = \bigoplus_{i=1}^r ke_i$ of Q and its ($\mathbb{Z}/2\mathbb{Z}$-graded) wedge product space $\wedge^* U = (\wedge^* U)^{\mathrm{even}} \oplus (\wedge^* U)^{\mathrm{odd}}$. A basis of the latter is given by $e_I = e_{i_1} \wedge \cdots \wedge e_{i_m}$ for $I = \{i_1, \dots, i_m\}$ in which elements are ordered so that $i_1 > \cdots > i_m$. (Our unusual ordering is to avoid the appearance of unnecessary signs later.) The pin group Pin_{2r} acts on $\wedge^* U$ as, for $1 \leq i \leq m$,

$$\begin{aligned} e_i \bullet (u_1 \wedge \cdots \wedge u_m) &= e_i \wedge u_1 \wedge \cdots \wedge u_m, \\ e_{-i} \bullet (u_1 \wedge \cdots \wedge u_m) &= \textstyle\sum_{i=1}^m (-1)^{i-1} \langle e_{-i}, u_i\rangle \cdot u_1 \wedge \cdots \wedge \widehat{u_i} \wedge \cdots \wedge u_m. \end{aligned} \tag{6.4.4}$$

Restricting to Spin_{2r}, both $(\wedge^* U)^{\mathrm{even}}$ and $(\wedge^* U)^{\mathrm{odd}}$ are irreducible representations; when r is even, their highest weights are

$$\nu_+ = \frac{1}{2}(\varepsilon_1 + \cdots + \varepsilon_r) \text{ and } \nu_- = \frac{1}{2}(\varepsilon_1 + \cdots + \varepsilon_{r-1} - \varepsilon_r),$$

respectively; when r is odd, the highest weights are exchanged. In what follows, we assume that r is even. The other case can be treated similarly. We henceforth identify $\mathtt{S}_{\nu_+} \cong (\wedge^* U)^{\mathrm{even}}$ and $\mathtt{S}_{\nu_-} \cong (\wedge^* U)^{\mathrm{odd}}$. More generally, the weight of e_I (with $I = \{i_1, \dots, i_m\}$) is

$$\varepsilon_{i_1} + \cdots + \varepsilon_{i_m} - \nu_+.$$

Moreover, the action (6.4.4) also defines natural morphisms

$$V \otimes (\wedge^* U)^{\mathrm{even}} \to (\wedge^* U)^{\mathrm{odd}}, \quad V \otimes (\wedge^* U)^{\mathrm{odd}} \to (\wedge^* U)^{\mathrm{even}}.$$

The isomorphism $\sigma : \mathtt{S}_{\nu_+} \to \mathtt{S}_{\sigma(\nu_+)} = \mathtt{S}_{\nu_-}$ may be identified with the natural map

$$(\wedge^* U)^{\mathrm{even}} \to (\wedge^* U)^{\mathrm{odd}}, \quad x \mapsto (e_r + e_{-r}) \bullet x,$$

intertwining the natural action of Spin_{2r} on the source and the σ-twisted action of Spin_{2r} on the target. Explicitly, this map sends e_I to $e_{I \triangle \{r\}}$, where $I \triangle \{r\}$ is the symmetric difference.

With this discussion, we may express a basis element of

$$(\mathsf{S}_{\sigma(\nu_+)^*} \otimes \mathsf{S}_{\nu_+} \otimes V)^G = \big((\wedge^* U)^{\mathrm{odd},*} \otimes (\wedge^* U)^{\mathrm{even}} \otimes V\big)^G$$

and its class function as

$$\mathbf{b}_+ = \sum_{|I| \text{ odd}} \sum_{|I'| \text{ even}} e_I^* \otimes e_{I'} \otimes v_{I,I'}, \quad f_{\mathbf{b}_+}(t\sigma) = \sum_{|I| \text{ odd}} \sum_{|I'| \text{ even}} \langle e_I^* \otimes t\sigma e_{I'} \rangle \cdot v_{I,I'}$$

for $v_{I,I'} \in V$ and $t \in T$, where the sum is taken over all subsets $I, J \subset \{1, \ldots, r\}$ whose sizes are of the specified parity. In the expression for $f_{\mathbf{b}}(t\sigma)$, the pairing $\langle e_I^*, \sigma e_{I'} \rangle$ is nonzero precisely when $I' = I \triangle \{r\}$; in this case, we may rewrite $\{I, I'\}$ as $\{J, J \cup \{r\}\}$ for some $J \subset \{1, \ldots, r-1\}$, and we may take the vector $v_{I,I'}$ to be e_r if $r \in I$ and to be e_{-r} if $r \notin I$. Therefore, we have

$$\begin{aligned} f_{\mathbf{b}_+}(t\sigma) &= \sum_{|J| \text{ even}} \langle e^*_{J \cup \{r\}}, e^{\varepsilon_J + \varepsilon_r - \nu_+}(t) \sigma e_J \rangle \cdot e_r + \sum_{|J| \text{ odd}} \langle e_J^*, e^{\varepsilon_J - \nu_+}(t) e_{J \cup \{r\}} \rangle \cdot e_{-r} \\ &= \sum_{|J| \text{ even}} e^{\varepsilon_J + \varepsilon_r - \nu_+}(t) \cdot e_r + \sum_{|J| \text{ odd}} e^{\varepsilon_J - \nu_+}(t) \cdot e_{-r}, \end{aligned}$$

where the sums are taken over subsets $J \subset \{1, \ldots, r-1\}$ whose sizes are of the given parity, and for $J = \{i_1, \ldots, i_m\}$, $\varepsilon_J = \varepsilon_{i_1} + \cdots + \varepsilon_{i_m}$. To simplify notation, we put

$$C_+ = e^{-\nu_+} \cdot \sum_{|J| \text{ even}} e^{\varepsilon_J}, \quad C_- = e^{-\nu_+} \cdot \sum_{|J| \text{ odd}} e^{\varepsilon_J}$$

so that $\mathrm{Res}_V^\sigma(f_{\mathbf{b}_+}) = C_+ e^{\varepsilon_r} \cdot e_r + C_- \cdot e_{-r} \in \boldsymbol{J}_T(V)$.

Similarly, we may choose a basis element

$$\mathbf{b}_- \in (\mathsf{S}_{\sigma(\nu_-)^*} \otimes \mathsf{S}_{\nu_-} \otimes V)^G = \big((\wedge^* U)^{\mathrm{even},*} \otimes (\wedge^* U)^{\mathrm{odd}} \otimes V\big)^G,$$

with class function

$$\mathrm{Res}_V^\sigma(f_{\mathbf{b}_-}) = C_- e^{\varepsilon_r} \cdot e_r + C_+ \cdot e_{-r} \in \boldsymbol{J}_T(V).$$

On the dual side, we choose elements

$$\begin{aligned} \mathbf{b}_+^\vee &\in (\mathsf{S}_{\sigma(\nu_+)} \otimes \mathsf{S}_{\nu_+^*} \otimes V^*)^G = \big((\wedge^* U)^{\mathrm{odd}} \otimes (\wedge^* U)^{\mathrm{even},*} \otimes V^*\big)^G, \\ \mathbf{b}_-^\vee &\in (\mathsf{S}_{\sigma(\nu_-)} \otimes \mathsf{S}_{\nu_-^*} \otimes V)^G = \big((\wedge^* U)^{\mathrm{even}} \otimes (\wedge^* U)^{\mathrm{odd},*} \otimes V\big)^G. \end{aligned}$$

If we write $D_+ = e^{\nu_+} \cdot \sum_{|J| \text{ even}} e^{-\varepsilon_J}$, $D_- = e^{\nu_+} \cdot \sum_{|J| \text{ odd}} e^{-\varepsilon_J}$, the class functions $f_{\mathbf{b}_+^\vee}, f_{\mathbf{b}_-^\vee} \in \boldsymbol{J}(V^*)$ have restrictions

$$\begin{aligned} \mathrm{Res}_{V^*}^\sigma(f_{\mathbf{b}_+^\vee}) &= D_+ e^{-\varepsilon_r} \cdot e_r^* + D_- \cdot e_{-r}^*, \\ \mathrm{Res}_{V^*}^\sigma(f_{\mathbf{b}_-^\vee}) &= D_- e^{-\varepsilon_r} \cdot e_r^* + D_+ \cdot e_{-r}^* \quad \in \boldsymbol{J}_T(V^*). \end{aligned}$$

Combining all above, we deduce that

$$\mathfrak{M} = \begin{pmatrix} \langle f_{\mathbf{b}_+}, f_{\mathbf{b}_+^\vee} \rangle & \langle f_{\mathbf{b}_+}, f_{\mathbf{b}_-^\vee} \rangle \\ \langle f_{\mathbf{b}_-}, f_{\mathbf{b}_+^\vee} \rangle & \langle f_{\mathbf{b}_-}, f_{\mathbf{b}_-^\vee} \rangle \end{pmatrix} = \begin{pmatrix} C_+D_+ + C_-D_- & C_+D_- + C_-D_+ \\ C_+D_- + C_-D_+ & C_+D_+ + C_-D_- \end{pmatrix}.$$

Now the formula (6.4.3) follows from the following identity

$$C_+D_+ + C_-D_- = \sum_{i \geq 0 \text{ even}} 2^{r-1-i} \mathfrak{S}_i, \quad C_+D_- + C_-D_+ = \sum_{i \geq 0 \text{ odd}} 2^{r-1-i} \mathfrak{S}_i.$$

It is easier to compute the determinant of $\mathfrak{M}$ in the following way:

$$\begin{aligned}\det \mathfrak{M} &= (C_+D_+ + C_-D_-)^2 - (C_+D_- + C_-D_+)^2 \\ &= (C_+ + C_-)(D_+ + D_-)(C_+ - C_-)(D_+ - D_-).\end{aligned}$$

On the other hand, it is easy to see that

$$C_+ \pm C_- = e^{-\nu_+}\prod_{i=1}^{r-1}(1 \pm e^{\varepsilon_i}), \quad D_+ \pm D_- = e^{\nu_+}\prod_{i=1}^{r-1}(1 \pm e^{-\varepsilon_i}).$$

So

$$\det \mathfrak{M} = \prod_{i=1}^{m-1}\left((1 - e^{2\varepsilon_i})(1 - e^{-2\varepsilon_i})\right). \qquad \square$$

References

[Ba11] M. Balagović, *Chevalley restriction theorem for vector-valued functions on quantum groups*, Represent. Theory **15** (2011), 617–645, DOI 10.1090/S1088-4165-2011-00408-8. MR2833470

[BR17] P. Baumann, S. Riche, Notes on the geometric Satake equivalence, `arXiv:1703.07288`, to appear in Lecture Notes in Mathematics 2221.

[BL96] J. Bernstein and V. Lunts, *A simple proof of Kostant's theorem that $U(\mathfrak{g})$ is free over its center*, Amer. J. Math. **118** (1996), no. 5, 979–987. MR1408495

[Bou15] A. Bouthier, *Dimension des fibres de Springer affines pour les groupes* (French, with English and French summaries), Transform. Groups **20** (2015), no. 3, 615–663, DOI 10.1007/s00031-015-9326-9. MR3376144

[Br89] R. K. Brylinski, *Limits of weight spaces, Lusztig's q-analogs, and fiberings of adjoint orbits*, J. Amer. Math. Soc. **2** (1989), no. 3, 517–533, DOI 10.2307/1990941. MR984511

[Do81] S. Donkin, *A filtration for rational modules*, Math. Z. **177** (1981), no. 1, 1–8, DOI 10.1007/BF01214334. MR611465

[Do85] S. Donkin, *Rational representations of algebraic groups*, Lecture Notes in Mathematics, vol. 1140, Springer-Verlag, Berlin, 1985. Tensor products and filtration. MR804233

[Do88a] S. Donkin, *Skew modules for reductive groups*, J. Algebra **113** (1988), no. 2, 465–479, DOI 10.1016/0021-8693(88)90172-X. MR929773

[Do88b] S. Donkin, *On conjugating representations and adjoint representations of semisimple groups*, Invent. Math. **91** (1988), no. 1, 137–145, DOI 10.1007/BF01404916. MR918240

[Gi95] V. Ginzburg, Perverse sheaves on a loop group and Langlands' duality, `arXiv:alg-geom/9511007`.

[KNV11] S. Khoroshkin, M. Nazarov, and E. Vinberg, *A generalized Harish-Chandra isomorphism*, Adv. Math. **226** (2011), no. 2, 1168–1180, DOI 10.1016/j.aim.2010.08.001. MR2737780

[Ko84] M. Koppinen, *Good bimodule filtrations for coordinate rings*, J. London Math. Soc. (2) **30** (1984), no. 2, 244–250, DOI 10.1112/jlms/s2-30.2.244. MR771419

[Ko63] B. Kostant, *Lie group representations on polynomial rings*, Amer. J. Math. **85** (1963), 327–404, DOI 10.2307/2373130. MR0158024

[Lu83] G. Lusztig, *Singularities, character formulas, and a q-analog of weight multiplicities*, Analysis and topology on singular spaces, II, III (Luminy, 1981), Astérisque, vol. 101, Soc. Math. France, Paris, 1983, pp. 208–229. MR737932

[Lu00] G. Lusztig, *Semicanonical bases arising from enveloping algebras*, Adv. Math. **151** (2000), no. 2, 129–139, DOI 10.1006/aima.1999.1873. MR1758244

[Ma90] O. Mathieu, *Filtrations of G-modules*, Ann. Sci. École Norm. Sup. (4) **23** (1990), no. 4, 625–644. MR1072820

[MV07] I. Mirković and K. Vilonen, *Geometric Langlands duality and representations of algebraic groups over commutative rings*, Ann. of Math. (2) **166** (2007), no. 1, 95–143, DOI 10.4007/annals.2007.166.95. MR2342692

[Mo03] S. Mohrdieck, *Conjugacy classes of non-connected semisimple algebraic groups*, Transform. Groups **8** (2003), no. 4, 377–395, DOI 10.1007/s00031-003-0429-3. MR2015256

[Ric79] R. W. Richardson, *The conjugating representation of a semisimple group*, Invent. Math. **54** (1979), no. 3, 229–245, DOI 10.1007/BF01390231. MR553220

[Rit01] A. Rittatore, *Very flat reductive monoids* (English, with English and Spanish summaries), Publ. Mat. Urug. **9** (2001), 93–121 (2002). MR1935829

[Sp74] T. A. Springer, *Regular elements of finite reflection groups*, Invent. Math. **25** (1974), 159–198, DOI 10.1007/BF01390173. MR0354894

[Sp06] T. A. Springer, *Twisted conjugacy in simply connected groups*, Transform. Groups **11** (2006), no. 3, 539–545, DOI 10.1007/s00031-005-1113-6. MR2264465

[St65] R. Steinberg, *Regular elements of semisimple algebraic groups*, Inst. Hautes Études Sci. Publ. Math. **25** (1965), 49–80. MR0180554

[St68] R. Steinberg, *Endomorphisms of linear algebraic groups*, Memoirs of the American Mathematical Society, No. 80, American Mathematical Society, Providence, R.I., 1968. MR0230728

[St75] R. Steinberg, *On a theorem of Pittie*, Topology **14** (1975), 173–177, DOI 10.1016/0040-9383(75)90025-7. MR0372897

[TX14] Y. Tian and L. Xiao, Tate cycles on quaternionic Shimura varieties over finite fields, `arXiv:1410.2321`, to appear in Duke Math. Journal.

[Vi95] E. B. Vinberg, *On reductive algebraic semigroups*, Lie groups and Lie algebras: E. B. Dynkin's Seminar, Amer. Math. Soc. Transl. Ser. 2, vol. 169, Amer. Math. Soc., Providence, RI, 1995, pp. 145–182, DOI 10.1090/trans2/169/10. MR1364458

[XZ17] L. Xiao and X. Zhu, Cycles on Shimura varieties via geometric Satake, `arXiv:1707.05700`.

[Zhu17] X. Zhu, *An introduction to affine Grassmannians and the geometric Satake equivalence*, Geometry of moduli spaces and representation theory, IAS/Park City Math. Ser., vol. 24, Amer. Math. Soc., Providence, RI, 2017, pp. 59–154. MR3752460

Appendix: A remark on freeness

Stephen Donkin[7]

Let G be a semisimple, simply connected algebraic group over an algebraically closed field k. Then conjugation gives rise to a G-module structure on the coordinate algebra $k[G]$ and the algebra of class functions $C(G)$ is the invariant algebra. Let V be a finite dimensional rational G-module which admits a good filtration. In Theorem 1.0.1 it is shown that the space of invariants $(k[G] \otimes V)^G$ is free as a $C(G)$-module.

We here obtain this as a special case of Theorem B below. However, we would like to stress that this does not cover the freeness of $\boldsymbol{J}_0(V)$ and $\boldsymbol{J}_+(V)$ established in Theorem 1.0.1, nor does it cover the cases in which the action on the coordinate algebra is twisted via an automorphism, considered by the authors.

In [**3**] we considered the situation in which G acts rationally on a finitely generated k-algebra A, see [**3**], 1.5 Theorem and deduced freeness over the algebra of invariants, in some cases, in particular in the case $A = k[G]$ with the conjugation action. The argument we give below is essentially a re-run of the argument from [**3**], but where we now consider invariants of a finitely generated A-module and rational G-module. Our argument relies on flatness and we should mention that in the cases considered in § 2–§ 4 a more constructive approach is taken and flatness is obtained as a corollary of freeness.

Some General Recollections. By a G-algebra we mean a commutative k-algebra on which G acts rationally as k-algebra automorphisms. We will denote the action of an element g in G on an element a in a G-algebra A by $g \cdot a$.

[7]Department of Mathematics, University of York, York YO10 5DD. E-mail address: `stephen.donkin@york.ac.uk`

By an (A,G)-module we mean a k-space M which is an A-module and a rational G-module in such a way that $g(am) = (g \cdot a)gm$, for all $g \in G$, $a \in A$, $m \in M$. The space of invariants of a G-module M will be denoted M^G and C denotes A^G.

We here modify the arguments of [**3**], pp139–140, to give a generalization to (A,G)-modules of the main result of [**3**]. We begin by describing some properties of (A,G)-modules. These are no doubt well known but we include them for the convenience of the reader.

Recall that, from the Mumford Conjecture (proved by Haboush in [**5**]) if A is finitely generated then so is C. Moreover, if $\pi : A \to B$ is a surjection of G-algebras then for $b \in B^G$ we have $b^n = \pi(a)$ for some $a \in A^G$ and positive integer n (see [**7**], p54). In particular B^G is integral over $\pi(A^G)$ and so if A is finitely generated then B^G is a finitely generated A^G-module.

Now suppose that A is finitely generated and that M is a finitely generated (A,G)-module. We claim that M^G is finitely generated as a $C = A^G$-module. If this holds we say that M has the finite generation property. Note that a submodule of an (A,G)-module with the finite generation property has the finite generation property. Note also that an extension of (A,G)-modules with this property also has this property: Suppose $0 \to X \to Y \to Z \to 0$ is a short exact sequence of (A,G)-modules and X, Z have the property. Then we have a short exact sequence of C-modules

$$0 \to X^G \to Y^G \to (X + Y^G)/X \to 0.$$

Now $X + Y^G/X$ embeds in $(Y/X)^G$ and so is finitely generated as a C-module. Thus X^G and $(X+Y^G)/X$ are finitely generated C-modules so that Y^G is finitely generated too.

We consider the case in which M is generated by a non-zero invariant $m_0 \in M^G$. Thus $M = Am_0$ and M is isomorphic to A/I, where I is the annihilator of m_0. We have the natural map $\pi : A \to B = A/I$ and by the above B^G is integral over $B_0 = \pi(A^G)$. Hence B^G is finitely generated as a C-module and M has the finite generation property.

Now consider the general case. Let M be a finitely generated (A,G)-module and let X be a submodule maximal subject to the condition that it has the finite generation property. If $(M/X)^G = 0$ then $M^G = X^G$ so that M has the required property. Otherwise we choose $m_0 \in M \backslash X$ with $(m_0 + X) \in (M/X)^G$. Put $Y = X + Am_0$. Then Y/X is generated by an invariant so has the property. Hence X and Y/X have the finite generation property and therefore so has Y. But X is strictly contained in Y so we have a contradiction.

Thus we have the following.

(1) *If A is a finitely generated G-algebra and M is a finitely generated (A,G)-module then M^G is a finitely generated A^G-module.*

We need to improve (1) so that we can take invariants for the action of a subgroup of G. Now suppose that H is a Grosshans subgroup of G, i.e., a closed subgroup such that the algebra

$$k[H\backslash G] = \{f \in k[G] | f(hg) = f(g) \text{ for all } h \in H, g \in G\}$$

is finitely generated. (In fact we only need the case in which H is a maximal unipotent subgroup of G.). Suppose that A is a finitely generated G-algebra and M is a finitely generated (A,G)-module. We shall need that M^H is finitely generated as an A^H-module. By Frobenius reciprocity and the tensor identity one has

$M^H = (M \otimes k[H\backslash G])^G$ as k-spaces and $(A \otimes k[H\backslash G])^G = A^H$ so the argument should be to regard $M \otimes k[H\backslash G]$ as an $A \otimes k[H\backslash G]$-module and take invariants.

To see that this really works we write down explicitly the maps involved in identifying H-invariants and G-invariants. This is a slight extension of the context of [**8**], Theorem 4.

Let $\sigma : k[G] \to k[G]$ be the antipode, so $\sigma(f)(x) = f(x^{-1})$, for $f \in k[G]$, $x \in G$. Let A be a G-algebra and let M be an (A, G)-module. We choose a k-basis $(m_i)_{i \in I}$ of M. Let $f_{ij} \in k[G]$ be the corresponding coefficient functions so that

$$gm_i = \sum_{j \in I} f_{ji}(g) m_j$$

for $g \in G$, $i \in I$.

It is easy to check that there are inverse k-linear isomorphisms

$$\phi_M : (M \otimes k[H\backslash G])^G \to M^H \text{ and } \psi_M : M^H \to (M \otimes k[H\backslash G])^G$$

satisfying

$$\phi_M(\sum_{i \in I} m_i \otimes b_i) = \sum_{i \in I} b_i(1) m_i$$

for $\sum_{i \in I} m_i \otimes b_i \in (M \otimes k[H\backslash G])^G$ and

$$\psi_M(\sum_{\in I} \lambda_i m_i) = \sum_{i,j \in I} \lambda_i m_j \otimes \sigma(f_{ji})$$

for $\sum_{i \in I} \lambda_i m_i \in M^H$.

For $a \in A^H$ and $m \in M^H$ one checks that $\psi_M(am) = \psi_A(a)\psi_M(m)$. We regard $M \otimes k[H\backslash G]$ as an $(A \otimes k[H\backslash G], G)$-module in the natural way. By (1), $(M \otimes k[H\backslash G])^G$ is a finitely generated $(A \otimes k[H\backslash G])^G$-module, i.e.,

$$(M \otimes k[H\backslash G])^G = \sum_{i=1}^{n} (A \otimes k[H\backslash G])^G y_i$$

for some $y_1, \ldots, y_n \in (M \otimes k[H\backslash G])^G$. Let $x_i \in M^H$ be such that $\psi_M(x_i) = y_i$ for $1 \leq i \leq n$ and put $D = \sum_{i=1}^{n} A^H x_i \leq M^H$. Then

$$\psi_M(D) = \sum_{i=1}^{n} \psi_A(A^H)\psi_M(x_i) = \sum_{i=1}^{n} (A \otimes k[H\backslash G])^G y_i = (M \otimes k[H\backslash G])^G$$

and so $D = M^H$ and M^H is finitely generated.

To summarise, we have the following.

(2) *If A is a finitely generated G-algebra, H is a Grosshans subgroup of G and M is a finitely generated (A, G)-module then M^H is a finitely generated A^H-module.*

Freeness. We are now in a position to extend the main result of [**3**] to the context of (A, G)-modules. We adopt the notation of [**6**]. In particular we have the maximal torus T and Weyl group W. We attach a root system to G, with respect to T, and let B be the negative Borel subgroup and U its unipotent radical. We have the character group $X(T)$ with natural partial order $\leq$ and set of dominant weight $X^+(T)$. For $\lambda \in X^+(T)$ we write $L(\lambda)$ for the irreducible rational G-module with highest weight λ, write k_λ for the one dimensional B-module on which T acts via λ and write $\nabla(\lambda)$ for the induced module $\mathrm{ind}_B^G k_\lambda$. A subset π of $X^+(T)$ is said to be saturated if whenever $\lambda \in \pi$ and $\mu \in X^+(T)$ with $\mu \leq \lambda$ then $\mu \in \pi$. For

π a saturated subset of $X^+(T)$ and V a rational G-module the set of submodules of V which have all composition factors belonging to $\{L(\lambda)|\lambda \in \pi\}$ has a unique maximal element, which we denote $O_\pi(V)$.

Theorem 1.5 in [**3**], which we now extend to the context of (A, G)-modules, is obtained via three Propositions. The first, Proposition 1.2. and the third, Proposition 1.4, require no modification.

The modified Proposition 1.3 and its proof require only minimal changes, given the remarks on finite generation above, but we give it again for the sake of completeness, in the form of the following lemma.

LEMMA A. *Let A be a finitely generated G-algebra and let $C = A^G$. Suppose that M is a finitely generated (A, G)-module and that M has a good filtration. Then, for every finite saturated subset π of $X^+(T)$, the C-module $O_\pi(M)$ is a finitely generated.*

PROOF. Let λ be a maximal element of π and let $\pi' = \pi\backslash\{\lambda\}$. Then $O_\pi(M)/O_{\pi'}(M) \cong C \otimes O_\pi(M)^\lambda$, by [**3**], 1.2 Proposition. By induction on the size of π it therefore suffices to prove that $O_\pi(M)^\lambda$ is a finitely generated C-module. Moreover, multiplication by a coset representative of w_0 (the longest element of the Weyl group W) induces an isomorphism $O_\pi(M)^\lambda \to O_\pi(M)^{w_0\lambda}$, so it suffices to show that $O_\pi(M)^{w_0\lambda}$ is finitely generated.

Let $M_0 = M^U$ and $A_0 = A^U$. Since $w_0\lambda$ is a lowest weight of $O_\pi(M)$, we have $O_\pi(M)^{w_0\lambda} \leq M_0^{w_0\lambda}$. On the other hand $\bar{M}^{w_0\lambda} = 0$, by [**2**], (12.1.6) and (1.5.2), where $\bar{M} = (M/O_\pi(M))^U$, so that $O_\pi(M)^{w_0\lambda} = M_0^{w_0\lambda}$. Furthermore, M_0 is a T-module and $A_0^T = A^B = A^G$, by [**1**], (2.1) Theorem. Moreover, U is a Grosshans subgroup, e.g., by [**4**], so that A_0 is finitely generated and, by (2), M_0 is a finitely generated A_0-module. Hence we may (and do) replace M by M^U and G by T. So it suffices to prove that if A is finitely generated T-algebra and M is a finitely generated (A, T)-module then for any $\mu \in X(T)$ the weight space M^μ is a finitely generated A^T-module. We choose a finite dimensional T-invariant subspace of M which generates M, as an (A, T)-module. Then we have a surjective (A, T)-module map $A \otimes V \to M$ inducing a surjection on invariants (by complete reducibility of T-modules). Hence we may assume $M = A \otimes V$ and, by complete reducibility of T-modules again, that V is one dimensional, with weight τ say. Then $(A \otimes V)^\mu$ is isomorphic to $A^{\mu-\tau}$. Hence it suffices to prove that A^λ is a finitely generated A^T-module, for $\lambda \in X(T)$. So let $\chi = -\lambda$. Then $A \otimes k[\chi]$ is a finitely generated T-algebra and so $(A \otimes k[\chi])^T$ is finitely generated by $a_i \otimes \chi^{d_i}$ say, for $1 \leq i \leq n$, $d_i \geq 0$. Then A^λ is generated as a A^T-module by $\{a_i : 1 \leq i \leq n \text{ and } d_i = 1\}$. □

Now the proofs of [**3**] 1.5 Theorem and its Corollary go through in the context of (A, G)-modules with Lemma A replacing [**3**], Proposition 1.3, and we obtain the following.

THEOREM B. *Suppose that A is a finitely generated G-algebra whose algebra of invariants $C = A^G$ is a free polynomial algebra. Suppose that M is a finitely generated (A, G)-module such that M is flat as an A-module and has a good filtration (as a G-module).*

Let π be any finite saturated subset π of $X^+(T)$, let λ be a maximal element and put $\pi' = \pi\backslash\{\lambda\}$. Then, as a (C, G)-module, $O_\pi(M)/O_{\pi'}(M)$ is isomorphic to $E \otimes C$, where E is isomorphic to a direct sum of finitely many copies of $\nabla(\lambda)$.

Given a C-module N and a k-space of G-module E we write $|E| \otimes N$ for the vector space $E \otimes N$ viewed as a C-module with action $c(e \otimes n) = e \otimes cn$, for $c \in C$, $e \in E$, $n \in N$.

For a finite dimensional G-module E admitting a good filtration and $\lambda \in X^+(T)$, we write $(E : \nabla(\lambda))$ for the multiplicity of $\nabla(\lambda)$ as a section in a good filtration of E.

Corollary C. *Under the hypotheses of the Theorem, M has an ascending (C, G)-module filtration $0 = M_0, M_1, \ldots$, where $M_i/M_{i-1} \cong |E_i| \otimes C$, E_i is a finite direct sum of copies of $\nabla(\lambda_i)$ ($i \geq 1$) and $\lambda_1, \lambda_2, \ldots$ is a labelling of the elements of $X^+(T)$ such that $i < j$ whenever $\lambda_i < \lambda_j$. For a given labelling the multiplicity $(E_i : \nabla(\lambda_i))$ is independent of the choice of such a filtration.*

In particular M is a free C-module.

We specialize further to the situation of this paper, Theorem 1.0.1.

Corollary D. *Regard $k[G]$ as a G-module via the conjugating action and let V be a finite dimensional rational G-module which admits a good filtration. Then $(k[G] \otimes V)^G$ is free over $C(G)$.*

Note that $C(G)$ is a free polynomial algebra by [**10**], 6.1 Theorem. Moreover, $k[G] \otimes V$ is free, and hence flat over $k[G]$ and $k[G]$ is flat over $C(G)$, by [**9**], Proposition 2.3, so that $k[G] \otimes V$ is flat over $C(G)$. We take $A = k[G]$, considered as a G-algebra via the conjugating representation and $M = k[G] \otimes V$ and $\pi = \{0\}$. Then M has a good filtration, by [**6**], II, 4.20 Proposition and 4.21 Proposition so we may apply Corollary C.

References

[1] E. Cline, B. Parshall, L. Scott, and W. van der Kallen, *Rational and generic cohomology*, Invent. Math. **39** (1977), no. 2, 143–163, DOI 10.1007/BF01390106. MR0439856

[2] S. Donkin, *Rational representations of algebraic groups*, Lecture Notes in Mathematics, vol. 1140, Springer-Verlag, Berlin, 1985. Tensor products and filtration. MR804233

[3] S. Donkin, *On conjugating representations and adjoint representations of semisimple groups*, Invent. Math. **91** (1988), no. 1, 137–145, DOI 10.1007/BF01404916. MR918240

[4] F. Grosshans, *Observable groups and Hilbert's fourteenth problem*, Amer. J. Math. **95** (1973), 229–253, DOI 10.2307/2373655. MR0325628

[5] W. J. Haboush, *Reductive groups are geometrically reductive*, Ann. of Math. (2) **102** (1975), no. 1, 67–83, DOI 10.2307/1970974. MR0382294

[6] J. C. Jantzen, *Representations of algebraic groups*, 2nd ed., Mathematical Surveys and Monographs, vol. 107, American Mathematical Society, Providence, RI, 2003. MR2015057

[7] P. E. Newstead, *Introduction to moduli problems and orbit spaces*, Tata Institute of Fundamental Research Lectures on Mathematics and Physics, vol. 51, Tata Institute of Fundamental Research, Bombay; by the Narosa Publishing House, New Delhi, 1978. MR546290

[8] V. L. Popov, *Contraction of the actions of reductive algebraic groups*, Math. USSR, Sb. **58** (2), (1987), 311-335.

[9] R. W. Richardson, *The conjugating representation of a semisimple group*, Invent. Math. **54** (1979), no. 3, 229–245, DOI 10.1007/BF01390231. MR553220

[10] R. Steinberg, *Regular elements of semisimple algebraic groups*, Inst. Hautes Études Sci. Publ. Math. **25** (1965), 49–80. MR0180554

Department of Mathematics, University of Connecticut, Storrs, 341 Mansfield Road, Unit 1009, Storrs, Connecticut 06269
Email address: liang.xiao@uconn.edu

Department of Mathematics, Caltech, 1200 East California Boulevard, Pasadena, California 91125
Email address: xzhu@caltech.edu

Proceedings of Symposia in Pure Mathematics
Volume **101**, 2019
https://doi.org/10.1090/pspum/101/01802

Local theta correspondence and nilpotent invariants

Chen-Bo Zhu

Dedicated to Joseph Bernstein on the occasion of his seventy-second birthday

ABSTRACT. We consider two types of nilpotent invariants associated to smooth representations, namely generalized Whittaker models, and associated characters (in the case of a real reductive group). We survey some recent results on the behavior of these nilpotent invariants under local theta correspondence, and highlight the special role of a certain double fibration of moment maps.

1. Introduction

One of the most fruitful approaches in representation theory is to understand representations through their invariants. Among the most well-known invariants are matrix coefficients (and their growth), and the character of a representation, which are of analytic nature. In the case of a real reductive group, fundamental invariants of algebraic nature include the infinitesimal character and the restriction of the representation to a maximal compact subgroup. Each of the afore-mentioned invariants has played a prominent role in the most important tasks of representation theory, for example in the classification of representations [**28**, **53**].

In the current article, we will be concerned with smooth representations of a reductive group defined over a local field F of characteristic 0. By a smooth representation we mean a smooth representation in the usual sense for F non-Archimedean, namely it is locally constant, and a Casselman-Wallach representation for F Archimedean [**4**, **59**].

The invariants we will consider will be informally referred to as nilpotent invariants as they depend on various types of nilpotent orbits and thus reflect various kinds of singularities.

The first nilpotent invariant to be examined is the space of generalized Whittaker models [**39**, **57**, **63**]. As it is well-known, the study of Whittaker and generalized Whittaker models for representations evolved in connection with the theory of automorphic forms, and has found numerous applications. See for example [**22**, **23**, **40**, **48**, **57**, **63**]. We will recall the basic definitions regarding generalized Whittaker models in Section 3.1.

2010 *Mathematics Subject Classification.* Primary 22E46.
Key words and phrases. Dual pairs, local theta lifting, generalized Whittaker models, associated varieties, associated cycles, associated characters.

The second nilpotent invariant to be examined is defined by Vogan [**54**] for a Harish-Chandra $(\mathfrak{g}, K)$-module (and therefore for a Casselman-Wallach representation of a real reductive group G), and is called the associated character. This is an element of the Grothendieck group of $\mathbf{K}$-equivariant coherent sheaves on $\mathcal{N}(\mathfrak{p})$ (the nilpotent cone on $\mathfrak{p}$), and is a refinement of the associated cycle, which is a (finite) union of the closure of $\mathbf{K}$-orbits in $\mathcal{N}(\mathfrak{p})$ with multiplicities. Here $\mathfrak{g} = \mathfrak{k} \oplus \mathfrak{p}$ is the complexified Cartan decomposition, and $\mathbf{K}$ is the complexification of K. We will recall the basic definitions regarding the associated character in Section 4.1.

REMARK 1.1. Another well-known nilpotent invariant (which we will not examine) is the wave front cycle, defined by Harish-Chandra in the non-Archimedean case ([**16**]) and by Howe and Barbasch-Vogan in the Archimedean case ([**3, 19**]; see also [**46**]). In the non-Archimedean case, Mœglin and Waldspurger have established [**39**] that the wave front cycle controls the spaces of generalized Whittaker models. In the Archimedean case, Schmid and Vilonen have shown [**49**] that the wave front cycle and the associate cycle determine each other through the Kostant-Sekiguchi correspondence [**47**].

REMARK 1.2. While both of nilpotent invariants examined in this paper are defined for a smooth representation, it is generally very difficult to compute them, even when we have a precise description of the representation.

We now come to the other part of the title regarding local theta correspondence. Let $(G, G') \subseteq \mathrm{Sp}(W)$ be a reductive dual pair of type I in the sense of Howe [**18**]. By fixing a non-trivial unitary character ψ of F, we have the smooth oscillator representation of $\widetilde{\mathrm{Sp}}(W)$ associated to ψ. For a smooth irreducible genuine representation π of $\widetilde{G}$, we have the full theta lift $\Theta(\pi)$ of π, which is a smooth representation of $\widetilde{G'}$ of finite length. For readers' convenience, we have included a brief review of the theory of dual pairs and local theta correspondence in Section 2. The question we wish to address in the current article is the behavior of these invariants under local theta correspondence, more specifically the relationship of the nilpotent invariants of $\Theta(\pi)$ with those of π.

In the first part of this article (Section 3), we focus on the behavior of the generalized Whittaker models under local theta correspondence. We shall summarize the result as follows. Let $\mathcal{O}$ be a nilpotent orbit in the Lie algebra $\mathfrak{g}$ of G. Let $\gamma = \{H, X, Y\}$ be an $\mathfrak{sl}_2$-triple with $X \in \mathcal{O}$ and $P = MN$ the associated parabolic subgroup of G. Then we have an oscillator representation or a 1-dimensional character $\mathscr{S}_\gamma$ of $\widetilde{M_X} \ltimes N$, where $\widetilde{M_X}$ is a double cover of the stabilizer group M_X of X in M. For a reductive subgroup R of M_X and a genuine representation τ of $\widetilde{R}$, we consider its model

$$\mathrm{Wh}_{\mathcal{O},\tau}(\pi) = \mathrm{Hom}_{\widetilde{R} \ltimes N}(\mathscr{V}, \mathscr{V}_\tau \widehat{\otimes} \mathscr{S}_\gamma).$$

(When R is the trivial group, we will simply write $\mathrm{Wh}_{\mathcal{O}}(\pi)$.)

To any nilpotent orbit $\mathcal{O}' \subset \mathfrak{g}'$ in the image of the moment map, one may associate a nilpotent orbit $\mathcal{O} \subset \mathfrak{g}$ via the moment maps, called the generalized descent of $\mathcal{O}'$. The associated stabilizer groups M_X and $M'_{X'}$ are of the form

$$M_X \supset M_{X,X'} \times L, \qquad M'_{X'} = M_{X,X'} \times L',$$

for some reductive dual pair (L, L') of the same type as (G, G'). Then the main result (which generalizes [**15**], in the so-called descent case where L is trivial and

$M_X = M_{X,X'}$) says that

$$\mathrm{Wh}_{\mathcal{O}',\tau'}(\Theta(\pi)) \simeq \mathrm{Wh}_{\mathcal{O},\Theta(\tau')^\vee}(\pi^\vee)$$

as $\widetilde{M_{X,X'}}$-modules. Here $\Theta(\tau')$ is the full theta lift of τ' with respect to the dual pair (L, L'), and the symbol $\vee$ indicates the dual in the category of Casselman-Wallach representations. We also prove that if $\mathcal{O}'$ is not in the image of the moment map, then

$$\mathrm{Wh}_{\mathcal{O}'}(\Theta(\pi)) = 0.$$

For precise statements and unexplained notations, see Sections 3.2 and 3.3.

In the second part of this article (Section 4), we focus on the behavior of the associated character under local theta correspondence (over $\mathbb{R}$). We shall also summarize the result. Let $\mathcal{O}$ be a nilpotent orbit in $\mathfrak{g}$, now the complexified Lie algebra of G. To any $(\mathfrak{g}, \widetilde{K})$ module Π of finite length whose complex associated variety is contained in the closure of $\mathcal{O}$, Vogan associates an element (called the associated character)

$$\mathrm{Ch}_{\mathcal{O}}(\Pi) \in \mathcal{K}_{\mathcal{O}}(\widetilde{\mathbf{K}}) := \bigoplus \mathcal{K}_{\mathscr{O}}(\widetilde{\mathbf{K}})$$

where $\mathscr{O}$ runs over $\mathbf{K}$-orbits in $\mathcal{O} \cap \mathfrak{p}$, and $\mathcal{K}_{\mathscr{O}}(\widetilde{\mathbf{K}})$ is the Grothendieck group of $\widetilde{\mathbf{K}}$-equivariant algebraic vector bundles on $\mathscr{O}$.

Suppose that nilpotent orbits $\mathcal{O} \subset \mathfrak{g}$ and $\mathcal{O}' \subset \mathfrak{g}'$ are the images of an element of maximal rank under the moment maps. Then the result of [**35**] gives an upper bound

$$\mathrm{Ch}_{\mathcal{O}'}(\Theta(\pi^\vee)) \preceq \vartheta_{\mathcal{O},\mathcal{O}'}(\mathrm{Ch}_{\mathcal{O}}(\pi)),$$

where $\vartheta_{\mathcal{O},\mathcal{O}'}$ is a certain sum of maps between the Grothendieck groups of algebraic vector bundles and is also defined via the moment maps. Moreover, if (π, G') is in the so-called convergent range (see Definition 4.7), then this inequality becomes an equality with $\Theta(\pi^\vee)$ replaced by a certain quotient $\bar{\Theta}(\pi)$ of $\Theta(\pi^\vee)$. We refer the reader to Sections 4.4 and 4.5 for precise statements and unexplained notations.

Our results have the following immediate consequences: if $\mathrm{Wh}_{\mathcal{O},\Theta(\tau')^\vee}(\pi^\vee) \neq 0$ for some $\mathcal{O}$ and τ', then $\Theta(\pi) \neq 0$. For example if (G, G') is in the so-called stable range with G the smaller member [**29**], one may find a nilpotent orbit $\mathcal{O}'$ in $\mathfrak{g}'$ which descents to the zero orbit in $\mathfrak{g}$ (such an $\mathcal{O}'$ is called quadratic nilpotent). This immediately implies that $\Theta(\pi) \neq 0$ for any smooth irreducible genuine representation π of $\widetilde{G}$. Similarly, if (π, G') is in the convergent range, and if $\vartheta_{\mathcal{O},\mathcal{O}'}(\mathrm{Ch}_{\mathcal{O}}(\pi)) \neq 0$ for some $\mathcal{O} \subset \mathfrak{g}$ and $\mathcal{O}' \subset \mathfrak{g}'$ in the images of an element of maximal rank under the moment maps, then $\bar{\Theta}(\pi) \neq 0$ and consequently $\Theta(\pi^\vee) \neq 0$.

Each of the above provides an effective criterion for the nonvanishing of $\Theta(\pi)$. Both are internal in the sense that one does not invoke other dual pairs or other representations. Of course conservation relations (conjectured by Kudla and Rallis [**27**] and established by B. Sun and the author in its full generality [**50**]) provide one other way to determine the vanishing/nonvanishing, which is external. (We refer the reader to [**51**] for a easy read on the conservation relations.) Note that vanish/non-vanishing of $\Theta(\pi)$ amounts to an inequality on the first occurrence index of π in a (generalized) Witt tower, and so will yield information on the non-vanishing/vanishing in another (generalized) Witt tower. The two results on the correspondence of nilpotent invariants surveyed in this article, when combined with conservation relations, will thus provide additional information on the questions of vanishing and nonvanishing in local theta correspondence.

As a concluding remark, the author wishes to highlight in this article the special role of the double fibration of moment maps in questions on the local theta correspondence. Important earlier work on this double fibration of moment maps include those of Kraft-Procesi [**24**] and Prezbinda [**44**]. More recent work connected to the theme of this article include [**15, 33, 35, 41, 42**].

Acknowledgements: The author would like to thank the anonymous referee for his thoughtful comments and suggestions, which helped to substantially improve the exposition of this article. The author would also like to thank Kyo Nishiyama whose collaboration a decade ago has helped to shape the author's thinking on nilpotent invariants of smooth representations.

2. Dual pairs and local theta correspondence: a brief review

We review the basic set-up of the theory of dual pairs [**18**].

Let F be a local field of characteristic zero, and let D be one of the following division algebras over F: the field F itself, a quadratic field extension of F or the central quaternion division algebra over F.

Let $\epsilon = \pm 1$, and V be an ϵ-Hermitian space, namely a finite dimensional right D-vector space, equipped with the ϵ-Hermitian form $(\cdot,\cdot)_V$. The isometry group $G = \mathrm{G}(V)$ is a type I classical group, as in the following table.

D	F	quadratic extension	quaternion algebra
$\epsilon=1$	orthogonal group	unitary group	quaternionic symplectic group
$\epsilon=-1$	symplectic group	unitary group	quaternionic orthogonal group

Let V' be an ϵ'-Hermitian right D-vector space, equipped with the ϵ'-Hermitian form $(\cdot,\cdot)_{V'}$, where $\epsilon\epsilon' = -1$. Denote by $G' = \mathrm{G}(V')$ the corresponding isometry group. Let $W =: \mathrm{Hom}_{\mathrm{D}}(V, V')$, and we define a symplectic form $\langle\cdot,\cdot\rangle_W$ on W by setting

$$\langle T, S\rangle_W := \mathrm{Tr}_{\mathrm{D}/\mathrm{F}}(T^*S), \qquad T,\ S \in \mathrm{Hom}_{\mathrm{D}}(V, V'),$$

where $\mathrm{Tr}_{\mathrm{D}/\mathrm{F}}(T^*S)$ is the trace of T^*S as an F-linear transformation, and $T^* \in \mathrm{Hom}_D(V', V)$ is defined by

$$(Tv, v')_{V'} = (v, T^*v')_V, \qquad v \in V,\ v' \in V'.$$

Let $\mathrm{Sp}(W)$ be the isometry group of $\langle\cdot,\cdot\rangle_W$.

There is a natural homomorphism: $G \times G' \longrightarrow \mathrm{Sp}(W)$ given by

$$(g, g')\cdot T = g'Tg^{-1} \qquad \text{for } T \in \mathrm{Hom}_{\mathrm{D}}(V, V'),\ g \in G,\ g' \in G'.$$

Fix a non-trivial unitary character ψ of F, and let $(\omega, \mathscr{Y})$ be the smooth oscillator representation of $\widetilde{\mathrm{Sp}}(W)$ associated to ψ. Here $\widetilde{\mathrm{Sp}}(W)$ is the metaplectic cover of $\mathrm{Sp}(W)$, namely the unique topological central extension of the symplectic group $\mathrm{Sp}(W)$ by $\{\pm 1\}$ which does not split unless $W = 0$ or $\mathrm{F} = \mathbb{C}$. For a subgroup E of $\mathrm{Sp}(W)$, let $\widetilde{E}$ be the inverse image of E in $\widetilde{\mathrm{Sp}}(W)$. By restriction, $(\omega, \mathscr{Y})$ yields a representation $\omega|_{\widetilde{G}\times\widetilde{G'}}$, and the basic problem is to understand this restriction.

The main assertion of the theory of local theta correspondence is the Howe duality conjecture [**18**]: the condition $\mathrm{Hom}_{\widetilde{G}\times\widetilde{G'}}(\omega, \pi\widehat{\otimes}\pi') \neq 0$ [1] determines a one

[1] Here $\widehat{\otimes}$ denotes algebraic tensor product for F non-Archimedean, and completed projective tensor product for F Archimedean. See [**1**] on the issue of algebraic versus smooth correspondences for F Archimedean.

to one correspondence

$$\{\pi \in \mathrm{Irr}(\widetilde{G}) \mid \mathrm{Hom}_{\widetilde{G}}(\omega, \pi) \neq 0\} \longleftrightarrow \{\pi' \in \mathrm{Irr}(\widetilde{G'}) \mid \mathrm{Hom}_{\widetilde{G'}}(\omega, \pi') \neq 0\}.$$

Here and as usual, $\mathrm{Irr}(R)$ denotes the set of equivalent classes of irreducible smooth representations of a reductive group R.

The Howe duality conjecture is now a theorem in its full generality. It was established by Howe when F is Archimedean [**20**], by Waldspurger when F is p-adic (for $p \neq 2$) [**56**], and completed by Minguez [**36**], Gan-Takeda [**13**] and Gan-Sun [**12**], without any restriction on the residue characteristics.

It is convenient to introduce the following version of the Howe duality theorem: Let $(\pi, \mathscr{V})$ be a smooth irreducible genuine representation of $\widetilde{G}$, and consider the maximal π-isotypic quotient of $\mathscr{Y}$, called the Howe quotient of π. It is of the form $\mathscr{V}\widehat{\otimes}\Theta(\mathscr{V})$, where $\Theta(\mathscr{V})$ carries a smooth representation $\Theta(\pi)$ of $\widetilde{G'}$. The representation $\Theta(\pi)$ is often referred to as the full theta lift (or colloquially the big theta lift) of π. The fundamental results of Howe, Waldspurger, Minguez, Gan-Takeda, and Gan-Sun say that $\Theta(\pi)$ is an admissible representation of finite length, moreover, when $\Theta(\pi)$ is non-zero, it has a unique irreducible quotient $\theta(\pi)$, called the (local) theta lift of π.

3. Correspondence of generalized Whittaker models

In this section we will discuss a refinement, due to Gomez and the author, of a previous work [**15**], which computes the generalized Whittaker models of the full theta lift in a general setting. Related earlier works on transition of models include those of Furusawa [**9**], Mœglin [**37**] and Ginzburg-Jiang-Soudry [**14**]. See also Gan's article in this volume [**10**].

3.1. Generalized Whittaker models. We recall some basics of generalized Whittaker models [**39**, **57**, **63**]. Let G be a reductive group over F, $\mathfrak{g}$ its Lie algebra, on which we fix an $\mathrm{Ad}\, G$-invariant non-degenerate bilinear form κ. Let $\mathcal{O} \subset \mathfrak{g}$ be a nilpotent orbit, and $X \in \mathcal{O}$. We complete X to an $\mathfrak{sl}_2$-triple $\gamma = \{X, H, Y\} \subset \mathfrak{g}$, namely

$$[H, X] = 2X, \qquad [H, Y] = -2Y, \qquad [X, Y] = H.$$

Set $\mathfrak{g}_j = \{Z \in \mathfrak{g} \mid \mathrm{ad}(H)Z = jZ\}$, for $j \in \mathbb{Z}$. Then, from standard $\mathfrak{sl}_2$-theory, we have a finite direct sum $\mathfrak{g} = \oplus_{j \in \mathbb{Z}} \mathfrak{g}_j$. Define the Lie subalgebras $\mathfrak{u} = \oplus_{j \leq -2} \mathfrak{g}_j$, $\mathfrak{n} = \oplus_{j \leq -1} \mathfrak{g}_j$, $\mathfrak{p} = \oplus_{j \leq 0} \mathfrak{g}_j$ and $\mathfrak{m} = \mathfrak{g}_0$. Let U, N, P, and M be the corresponding subgroups of G. Thus $U = \exp \mathfrak{u}$, $N = \exp \mathfrak{n}$, $P = \{p \in G \mid \mathrm{Ad}(p)\mathfrak{p} \subset \mathfrak{p}\}$ and $M = \{m \subset G \mid \mathrm{Ad}(m)H = H\}$. Define the (non degenerate) character χ_γ of U by

$$\chi_\gamma(\exp Z) = \psi(\kappa(X, Z)), \qquad \forall\ Z \in \mathfrak{u}. \tag{1}$$

Let $M_X = \{m \in M \mid \mathrm{Ad}(m)X = X\}$, the stabilizer group of X in M. Then it is well-known [**6**, Section 3.4] that

$$M_X = G_\gamma := \{g \in G \mid \mathrm{Ad}(g)X = X,\ \mathrm{Ad}(g)Y = Y,\ \mathrm{Ad}(g)H = H\}\ .$$

In particular M_X is reductive. For the moment assume that $\mathfrak{g}_{-1} \neq 0$. In this case $\mathrm{ad}(X)|_{\mathfrak{g}_{-1}} : \mathfrak{g}_{-1} \longrightarrow \mathfrak{g}_1$ is an isomorphism, and we may define a symplectic structure on $\mathfrak{g}_{-1}$ by setting

$$\kappa_{-1}(S, T) = \kappa(\mathrm{ad}(X)S, T) = \kappa(X, [S, T]), \qquad \text{for } S,\ T \in \mathfrak{g}_{-1}. \tag{2}$$

We may exhibit a canonical surjective group homomorphism from N to the associated Heisenberg group $\mathrm{H}_{\mathfrak{g}_{-1}}$ which maps $\exp Z$ to $\kappa(X, Z)$ in the center of $\mathrm{H}_{\mathfrak{g}_{-1}}$, for $Z \in \mathfrak{u}$. Then, according to the Stone-von Neumann theorem, there exists a unique, up to equivalence, smooth irreducible (unitarizable) representation $\mathscr{S}_\gamma$ of N such that U acts by the character χ_γ. Since M_X preserves γ and thus the symplectic form κ_{-1}, it is well-known [**61**] that the representation $\mathscr{S}_\gamma$ extends to the semi-direct product $\widetilde{M_X} \ltimes N$, where $\widetilde{M_X}$ is the inverse image of M_X in $\widetilde{Sp}(\mathfrak{g}_{-1})$. If $\mathfrak{g}_{-1} = 0$, we will use the same notation $\mathscr{S}_\gamma$ to denote the 1-dimensional representation of $N = U$ given by the character χ_γ, and $\widetilde{M_X} := M_X$ acts on it trivially.

Definition 3.1. (a) Let $(\pi, \mathscr{V})$ be a smooth representation of G. We define the space of generalized Whittaker models of π associated to the nilpotent orbit $\mathscr{O}$ by

$$\mathrm{Wh}_{\mathscr{O}}(\pi) = \mathrm{Hom}_N(\mathscr{V}, \mathscr{S}_\gamma).$$

$\mathrm{Wh}_{\mathscr{O}}(\pi)$ is naturally an $\widetilde{M_X}$-module.
(b) Let R be a reductive subgroup of M_X. Also define

$$\mathrm{Wh}_{\mathscr{O},\tau}(\pi) = \mathrm{Hom}_{\widetilde{R}N}(\mathscr{V}, \mathscr{V}_\tau \widehat{\otimes} \mathscr{S}_\gamma),$$

where $(\tau, \mathscr{V}_\tau)$ is a smooth genuine representation of $\widetilde{R}$.

Remark 3.2. In the representation theory literature, elements of $\mathrm{Wh}_{\mathscr{O},\tau}(\pi)$ are referred to as (local) models, which are generalizations of Bessel and Fourier-Jacobi models. (Typically one takes R to be a relatively big (e.g. spherical) subgroup of M_X.) We refer the reader to [**11**] for more information on these special models such as uniqueness. See also [**21**] which discusses their roles in a general theory of periods of automorphic forms.

We now assume that G is a type I classical group defined by an ϵ-Hermitian space (V, B), where B is the ϵ-Hermitian form. Set $\mathfrak{g}_\gamma = \mathrm{Span}_{\mathrm{F}}\{X, H, Y\} \subset \mathfrak{g}$. Under the action of $\mathfrak{g}_\gamma$, we have the isotypic decomposition:

$$V = \bigoplus_{j=1}^{l} V^{\gamma, t_j},$$

where V^{γ,t_j} is a direct sum of irreducible t_j-dimensional $\mathfrak{g}_\gamma$-modules, and $t_1 > t_2 > \ldots > t_l > 0$.

We need one more notation. Denote by (ρ_m, F^m) the irreducible representation of $\mathfrak{g}_\gamma \simeq \mathfrak{sl}_2$ of dimension m. There is a non-degenerate invariant bilinear form on F^m, which is unique up to scalar and is $(-1)^{m-1}$-symmetric. We fix one such form $(\cdot,\cdot)_m$, and denote the corresponded formed space by $(\mathrm{F}^m, (\cdot,\cdot)_m)$.

Let $V^{\gamma,t_j}_{t_j-1}$ be the space of highest weight vectors in V^{γ,t_j} (the subscript refers to the H-weight), and let $i_j = \dim V^{\gamma,t_j}_{t_j-1}$, which is the multiplicity of $(\rho_{t_j}, \mathrm{F}^{t_j})$ in V. The isotypic decomposition gives rise to a partition or equivalently a Young diagram $\mathbf{d}^\gamma = [t_1^{i_1}, \ldots, t_l^{i_l}]$ of size $\dim V$. Using $\mathfrak{sl}_2$ theory and the ϵ-Hermitian form B on V, one may define a non-degenerate ϵ_j-Hermitian form $B^{\gamma,t_j}_{t_j-1}$ on $V^{\gamma,t_j}_{t_j-1}$, where $\epsilon_j = (-1)^{t_j-1}\epsilon$. As ϵ-Hermitian spaces, we have

$$(3) \qquad (V, B) \simeq \bigoplus_j (V^{\gamma,t_j}_{t_j-1}, B^{\gamma,t_j}_{t_j-1}) \otimes (\mathrm{F}^{t_j}, (\cdot,\cdot)_{t_j}).$$

We may also describe the stabilizer group M_X as follows. Since M_X acts on $V_{t_j-1}^{\gamma,t_j}$ preserving $B_{t_j-1}^{\gamma,t_j}$, there is an embedding of $G(V_{t_j-1}^{\gamma,t_j})$ into M_X, and we identify $G(V_{t_j-1}^{\gamma,t_j})$ with its image in M_X. Then we have

$$M_X \cong \prod_{j=1}^{l} G(V_{t_j-1}^{\gamma,t_j}). \tag{4}$$

It is well-known [**6**] that the pair, consisting of

(i) the Young diagram $\mathbf{d}^\gamma = [t_1^{i_1}, \ldots, t_l^{i_l}]$, and the collection of
(ii) the formed spaces $(V_\gamma(j), B_\gamma(j)) := (V_{t_j-1}^{\gamma,t_j}, B_{t_j-1}^{\gamma,t_j})$ of dimension t_j for $1 \leq j \leq l$,

uniquely determines the $\mathfrak{sl}_2$-triple γ up to the Adjoint action of G. This is an example of an admissible ϵ-Hermitian Young tableaux (or admissible Young tableaux in short; admissibility refers to the requirement in (3)). With an obvious notion of equivalence, nilpotent orbits in $\mathfrak{g}$ are then parameterized by equivalence classes of admissible ϵ-Hermitian Young tableaux. See [**15**, Section 3.2]).

3.2. Generalized descent of $\mathfrak{sl}_2$-triples. We are back in the setting of Section 2, namely we are given $(G, G') = (G(V), G(V')) \subset \mathrm{Sp}(W)$, a type I reductive dual pair.

To ease notation, we will skip D in $\mathrm{Hom}_D(V, V')$ from now on. Recall the moment maps: [**7, 24**]

$$\begin{array}{ccc} & \mathrm{Hom}(V, V') & \\ \swarrow^{\varphi} & & \searrow^{\varphi'} \\ \mathfrak{g} & & \mathfrak{g}' \end{array}$$

where $\varphi(T) = T^*T$ and $\varphi'(T) = TT^*$.

Define

$$\mathrm{Gen\,Hom}(V, V') = \{T \in \mathrm{Hom}(V, V') \mid \mathrm{Ker}(T) \text{ is non-degenerate}\}. \tag{5}$$

For a given $\mathfrak{sl}_2$-triple $\gamma = \{X, H, Y\} \subset \mathfrak{g}$, set $V_k = \{v \in V \mid Hv = kv\}$, for $k \in \mathbb{Z}$. We have by standard $\mathfrak{sl}_2$-theory,

$$V = \bigoplus_{k \in \mathbb{Z}} V_k.$$

Similar notations apply for an $\mathfrak{sl}_2$-triple $\gamma' = \{X', H', Y'\} \subset \mathfrak{g}'$.

DEFINITION 3.3. Let $\gamma \subset \mathfrak{g}$, $\gamma' \subset \mathfrak{g}'$ be two $\mathfrak{sl}_2$-triples, and $T \subset \mathrm{Hom}(V, V')$. We say that T lifts γ to γ' if

(1) $T \in \mathrm{Gen\,Hom}(V, V')$.
(2) $\varphi(T) = X$ and $\varphi'(T) = X'$.
(3) $T(V_k) \subset V'_{k+1}$ for all k.

We set

$$\mathcal{O}_{\gamma,\gamma'} = \{T \in \mathrm{Hom}(V, V') \mid T \text{ lifts } \gamma \text{ to } \gamma'\}. \tag{6}$$

LEMMA 3.4. *Let $\gamma' = \{X', H', Y'\} \subset \mathfrak{g}'$ be an $\mathfrak{sl}_2$-triple. If X' is in the image of φ', then there is a unique conjugate class of $\mathfrak{sl}_2$-triple $\gamma = \{X, H, Y\} \subset \mathfrak{g}$, such that $\mathcal{O}_{\gamma,\gamma'}$ is non-empty. Furthermore $\mathcal{O}_{\gamma,\gamma'}$ is a single $M_X \times M'_{X'}$-orbit.*

PROOF. We will only show the existence part, namely there is an $\mathfrak{sl}_2$-triple $\gamma = \{X, H, Y\} \subset \mathfrak{g}$ such that $\mathcal{O}_{\gamma,\gamma'}$ is non-empty.

Decompose

$$V' = \bigoplus_{j=1}^{l} (V')^{\gamma', t_j},$$

into $\mathfrak{sl}_2$-isotypic components, where $(V')^{\gamma', t_j}$ is a direct sum of irreducible t_j-dimensional $\mathfrak{sl}_2$-modules, and $t_1 > t_2 > \ldots > t_l > 0$. We fix one such irreducible module F^{r+1} of dimension $r+1$, as well as a basis $\{v'_{-r}, v'_{-(r-2)}, ..., v'_{r-2}, v'_r\}$ consisting of H'-weight vectors.

Write $X' = \varphi'(S)$, for some $S \in \mathrm{Hom}(V, V')$. Define the following set of vectors in V:

$$v_{-(r-1)} = S^* v'_{-r}, v_{-(r-3)} = S^* v'_{-(r-2)}, ..., v_{r-3} = S^* v'_{r-4}, v_{r-1} = S^* v'_{r-2}.$$

It is easy to see that $\{v_{-(r-1)}, v_{-(r-3)}, ..., v_{r-3}, v_{r-1}\}$ span a non-degenerate subspace of V, of dimension r.

Now define a new element $T \in \mathrm{Hom}(V, V')$ by setting

$$T^* = \begin{cases} S^*, & \text{on } v'_{-r}, v'_{-(r-2)}, ..., v'_{r-4}, v'_{r-2}, \\ 0, & \text{on } v'_r. \end{cases}$$

Clearly the image of T^* will be non-degenerate, and so is $\mathrm{Ker}(T)$. Putting together the above construction for all $(V')^{\gamma', t_j}$'s, the newly defined T and $X := \varphi(T)$ will have all the required properties.

The rest of proof follows the same line of argument as the proof of Proposition 5.3 and Lemma 5.7 of [**15**], which is the special case when $\mathrm{Ker}(T) = 0$. □

DEFINITION 3.5. We are in the setting of Lemma 3.4. We will say that γ (resp. $\mathcal{O}$) is the generalized descent of γ' (resp. $\mathcal{O}'$), and we write

$$\mathcal{O} = \nabla^{\mathrm{gen}}_{V',V}(\mathcal{O}').$$

When $T \in \mathcal{O}_{\gamma,\gamma'}$ is injective, we will refer to it as the descent case, and we call γ (resp. $\mathcal{O}$) the descent of γ' (resp. $\mathcal{O}'$). We also write

$$\mathcal{O} = \nabla_{V',V}(\mathcal{O}').$$

We will describe the generalized descent $\mathcal{O}' \mapsto \mathcal{O}$ using the parametrization of nilpotent orbits by admissible Young tableaux.

Let the nilpotent orbit $\mathcal{O}' \subset \mathfrak{g}'$ correspond to the admissible ϵ'-Hermitian Young tableaux $(\mathbf{d}^{\gamma'}, (V'_{\gamma'}(j), B_{\gamma'}(j)))$, with

$$\mathbf{d}^{\gamma'} = [(t_1+1)^{i_1}, \ldots, (t_l+1)^{i_l}, 2^a, 1^s], \quad t_l \geq 2,$$

and a (resp. s) is the dimension of the ϵ-Hermitian space $((V')_1^{\gamma',2}, B_1^{\gamma',2})$ (resp. ϵ'-Hermitian space $((V')_0^{\gamma',1}, B_0^{\gamma',1})$.

If $\mathcal{O}'$ is in the image of the moment map, then we have an embedding of ϵ-Hermitian spaces (via T^*):

$$\{\oplus_{1\leq j\leq l}((V')_{t_j}^{\gamma',t_j+1}, B_{t_j}^{\gamma',t_j+1})\otimes(\mathrm{F}^{t_j}, (\cdot,\cdot)_{t_j})\}\oplus((V')_1^{\gamma',2}, B_1^{\gamma',2})\otimes(\mathrm{F}^1, (\cdot,\cdot)_1)\hookrightarrow(V, B).$$

Denote by (U, B_U) be the orthogonal complement of $\{\oplus_{1\le j\le l}((V')^{\gamma',t_j+1}_{t_j}, B^{\gamma',t_j+1}_{t_j})\otimes(\mathrm{F}^{t_j}, (\cdot,\cdot)_{t_j})\}$ in V with respect to the form B. Then (U, B_U) contains

$$(U_1, B_{U_1}) := ((V')^{\gamma',2}_1, B^{\gamma',2}_1)\otimes(\mathrm{F}^1, (\cdot,\cdot)_1) \simeq ((V')^{\gamma',2}_1, B^{\gamma',2}_1)$$

as a non-degenerate ϵ-Hermitian subspace.

The admissible ϵ-Hermitian Young tableau $(\mathbf{d}^\gamma, (V_\gamma(j), B_\gamma(j))$ corresponding to $\mathcal{O}\subset\mathfrak{g}$ is as follows.

- $\mathbf{d}^\gamma = [t_1^{i_1}, \ldots, t_l^{i_l}, 1^{a+b}]$, where $\dim U = a+b$.
- $(V^{\gamma,t_j}_{t_j-1}, B^{\gamma,t_j}_{t_j-1}) \cong ((V')^{\gamma',t_j+1}_{t_j}, B^{\gamma',t_j+1}_{t_j})$, for all $j = 1,\ldots,l$;

Or in descriptive words:

- The Young diagram of $\mathcal{O}$ is obtained by erasing the first column off the Young diagram of $\mathcal{O}'$, and then adding some additional boxes (this is the b) in the first column.
- After erasing the first column off the Young diagram of $\mathcal{O}'$, one keeps the forms on the spaces of multiplicities unchanged.

In the setting of the generalized descent, the two stabilizer groups M_X and $M'_{X'}$ are closely related. We proceed to describe their relationship.

DEFINITION 3.6. Let γ, γ' be as before, and $T\in\mathcal{O}_{\gamma,\gamma'}$. Define the following non-degenerate subspaces of V and V':

$$V_{\gamma,\gamma'} := \mathrm{Ker}(T) \qquad \text{and} \quad V'_{\gamma,\gamma'} := (V')^{\gamma',1}_0 \quad \text{(the space of } \gamma'\text{-invariants)}.$$

Let L and L' be the isometry groups of $V_{\gamma,\gamma'}$ and $V'_{\gamma,\gamma'}$, respectively.

We define a group homomorphism

$$\alpha = \alpha_T : M'_{X'} \longrightarrow M_X \tag{7}$$

by setting $\alpha_T(m')$ to be the identity map on $\mathrm{Ker}(T)$, and equal to $T^{-1}m'T$ on $(\mathrm{Ker}T)^\perp$, for $m'\in M'_{X'}$.

We have a direct product

$$M'_{X'} = M_{X,X'}\times L', \quad \text{where} \quad M_{X,X'} = \prod_{j=1}^{l} G((V')^{\gamma',t_j+1}_{t_j})\times G(U_1). \tag{8}$$

Similarly we have

$$M_X = \prod_{j=1}^{l} G(V^{\gamma,t_j}_{t_j-1})\times G(U) \cong \prod_{j=1}^{l} G((V')^{\gamma',t_j+1}_{t_j})\times G(U). \tag{9}$$

Note that the latter contains $M_{X,X'}\times L$ as a symmetric subgroup, due to the fact that $U = U_1\oplus\mathrm{Ker}(T)$ (orthogonal direct sum) and so $G(U)\supset G(U_1)\times L$ as a symmetric subgroup. Under the above identification, $\alpha : M'_{X'}\to M_X$ is the identity homomorphism on the first factor of (8), and the trivial homomorphism on the second factor of (8).

3.3. Transition of generalized Whittaker models. We are in the setting of Section 2: we are given $(G, G') \subset \mathrm{Sp}(W)$, a type I reductive dual pair, and the smooth oscillator representation $(\omega, \mathscr{Y})$ of $\widetilde{\mathrm{Sp}}(W)$ associated to a fixed ψ. We are interested in knowing all generalized Whittaker models of $\Theta(\pi)$, where $\pi \in \mathrm{Irr}(\widetilde{G})$.

We state our result on the transition of generalized Whittaker models. This is an extension of the main result of [**15**], which is the descent case (i.e., $\mathrm{Ker}(T) = 0$, in which case L is trivial and $M_X = M_{X,X'}$) in the terminology of this article.

THEOREM 3.7. ([Gomez and Zhu]) *Let* $\pi \in \mathrm{Irr}(\widetilde{G})$.
(a) If $\mathcal{O}'$ *is not in the image of the moment map* φ', *then*

$$\mathrm{Wh}_{\mathcal{O}'}(\Theta(\pi)) = 0.$$

(b) If $\mathcal{O}'$ *is in the image of the moment map* φ', *let* $\mathcal{O} = \nabla^{\mathrm{gen}}_{V',V}(\mathcal{O}')$ *be its generalized descent. We have (as in* (8) *and* (9)*)*

$$M_X \supset M_{X,X'} \times L,$$
$$M'_{X'} = M_{X,X'} \times L',$$

and (L, L') *forms a dual pair of the same type as* (G, G'). *Then for any* $\tau' \in \mathrm{Irr}(\widetilde{L'})$, *we have*

$$\mathrm{Wh}_{\mathcal{O}',\tau'}(\Theta(\pi)) \simeq \mathrm{Wh}_{\mathcal{O},\Theta(\tau')^\vee}(\pi^\vee)$$

as $\widetilde{M_{X,X'}}$*-modules. Here* $\Theta(\tau')$ *is the full theta lift of* τ' *with respect to the dual pair* (L, L'), *and the symbol* $\vee$ *indicates the dual in the category of Casselman-Wallach representations.*

Let $\gamma \subset \mathfrak{g}$ and $\gamma' \subset \mathfrak{g}'$ be two $\mathfrak{sl}_2$-triples of type $\mathcal{O}$ and $\mathcal{O}'$, where $\mathcal{O} = \nabla^{\mathrm{gen}}_{V',V}(\mathcal{O}')$. We may assume that

$$V = \bigoplus_{k=-r}^{r} V_k \qquad \text{and} \qquad V' = \bigoplus_{k=-r-1}^{r+1} V'_k, \tag{10}$$

for some r. Write $T = \oplus_{k=-r}^{r} T_k$, where $T_k = T|_{V_k}$. The following diagram is instructive:

$$\begin{array}{c} V_{-r} \oplus V_{-r+1} \oplus \cdots \oplus V_{r-1} \oplus V_r \\ \searrow^{T_{-r}} \quad \searrow^{T_{-r+1}} \quad \cdots \quad \searrow^{T_{r-1}} \searrow^{T_r} \\ V'_{-r-1} \oplus V'_{-r} \oplus V'_{-r+1} \oplus \cdots \oplus V'_{r-1} \oplus V'_r \oplus V'_{r+1}. \end{array} \tag{11}$$

We have $\mathrm{Ker}(T) = \mathrm{Ker}(T_0) \subseteq V_0$, and T_k is injective for $k \neq 0$, by our non-degeneracy assumption on $\mathrm{Ker}(T)$.

Recall that $(G, G') \subseteq \mathrm{Sp}(W)$ is a type I reductive dual pair, and $(\omega, \mathscr{Y})$ is the smooth oscillator representation of $\widetilde{\mathrm{Sp}}(W)$ associated to a non-trivial unitary character ψ of F, which we fix. Denote by $\mathscr{Y}_{U',\chi_{\gamma'}}$ the space of $(U', \chi_{\gamma'})$-covariants of the smooth oscillator representation $(\omega, \mathscr{Y})$.

The key technical result to derive Theorem 3.7 is the following

PROPOSITION 3.8. *Let* $\gamma \subset \mathfrak{g}$ *and* $\gamma' \subset \mathfrak{g}'$ *be two* $\mathfrak{sl}_2$*-triples of type* $\mathcal{O}$ *and* $\mathcal{O}'$, *where* $\mathcal{O} = \nabla^{\mathrm{gen}}_{V',V}(\mathcal{O}')$. *Then, given any* $T \in \mathcal{O}_{\gamma,\gamma'}$, *there exists a* $\widetilde{G} \times \widetilde{M'_{X'}}N'$*-intertwining isomorphism*

$$\Psi_T : \mathscr{Y}_{U',\chi_{\gamma'}} \longrightarrow \mathscr{C}(\widetilde{L}N\backslash\widetilde{G}; \mathscr{S}_{\tilde{\gamma}} \otimes \mathscr{S}_{\gamma'} \otimes \mathscr{Y}_{\gamma,\gamma'}),$$

where $\check{\gamma} = \{-X, H, -Y\}$, *and* $\mathscr{Y}_{\gamma,\gamma'}$ *is the smooth oscillator representation associated to the symplectic subspace* $W_{\gamma,\gamma'} := \mathrm{Hom}(V_{\gamma,\gamma'}, V'_{\gamma,\gamma'})$ *of* W.

REMARK 3.9. In the above proposition and in the sequel, the unexplained notation $\mathscr{C}$ denotes certain space of rapidly decreasing functions defined by an appropriate collection of semi-norms. See [**15**, Section 4.3]. We also refer to [**15**, Proposition 6.3] for other unexplained actions such as the action of $\widetilde{M'_{X'}}N'$ on the right hand of the isomorphism Ψ_T.

In the descent case, $V_{\gamma,\gamma'} = 0$, L is the trivial group, $M_X = M_{X,X'}$, and the above proposition is just [**15**, Proposition 6.5]. The proof in the general case is along the same line, and it is by induction on r. We will outline the main ingredients of the inductive step in the next subsection.

3.4. The inductive step. To facilitate the discussion, we introduce some notations. For $l \leq r$ and $m \leq r+1$, set

$$V_{(l)} = \bigoplus_{k=-l}^{l} V_k, \qquad \text{and} \qquad {V'}_{(m)} = \bigoplus_{k=-m}^{m} {V'}_k.$$

Clearly both $V_{(l)}$ and ${V'}_{(m)}$ are non-degenerate. Let $G_{(l)} := G(V_{(l)})$ (the isometry group of $V_{(l)}$), which we view as a subgroup of G in the obvious way. We define ${G'}_{(m)}$ similarly. Let $(\omega_{(l),(m)}, \mathscr{Y}_{(l),(m)})$ be the smooth oscillator representation associated to the dual pair $(G_{(l)}, {G'}_{(m)})$ and the character ψ of F. Let the symbol $\mathscr{S}$ stand for the Schwartz space (of a vector space). Set $\mathscr{S}_{(l),m} = \mathscr{S}(\mathrm{Hom}(V_{(l)}, {V'}_m))$, $\mathscr{S}_{l,(m)} = \mathscr{S}(\mathrm{Hom}(V_l, {V'}_{(m)}))$ and $\mathscr{S}_{l,m} = \mathscr{S}(\mathrm{Hom}(V_l, {V'}_m))$. Note that $\mathscr{Y}_{(l),(m)}$ can be identified with the Schwartz space of a Lagrangian subspace of $\mathrm{Hom}(V_{(l)}, {V'}_{(m)})$, through the Schrodinger model.

Under these notations, we have

$$V = V_{(r)}, \qquad \text{and} \qquad V' = {V'}_{(r+1)};$$
$$G = G_{(r)}, \qquad \text{and} \qquad G' = G'_{(r+1)};$$
$$(\omega, \mathscr{Y}) = (\omega_{(r),(r+1)}, \mathscr{Y}_{(r),(r+1)}).$$

Let ${P'}_m$ be the stabilizer of ${V'}_m$ in ${G'}_{(m)}$. We have ${P'}_m = {M'}_{(m)}{N'}_m$, where ${N'}_m$ is the unipotent radical of ${P'}_m$, ${M'}_{(m)} = {M'}_m \times {G'}_{(m-1)}$, and ${M'}_m \cong \mathrm{GL}({V'}_m)$. Set

$$N'_{(m)} = N' \cap G_{(m)}, \qquad U'_m = U' \cap N'_m, \qquad \text{and} \qquad U'_{(m)} = U' \cap N'_{(m)}.$$

Recall that $\chi' =: \chi_{\gamma'}$ is the character of U' associated to the $\mathfrak{sl}_2$-triple γ', as in (1). We set

$$\chi'_m = \chi'|_{U'_m}, \quad \text{and} \quad \chi'_{(m)} = \chi'|_{U_{(m)}}.$$

Using these notations, we then have

$$U' = U'_{(r+1)} = U'_{(r)}U'_{r+1}, \quad \text{and} \quad \chi' = \chi'_{(r+1)} = \chi'_{(r)}\chi'_{r+1}. \tag{12}$$

Instead of working with the covariant space $\mathscr{Y}_{U',\chi'}$, it is more convenient to work dually with the space of quasi-invariant distributions $(\mathscr{Y}')^{U',\chi'}$. It is also convenient to adopt the notion of $\mathscr{V}$-distributions on a manifold, where $\mathscr{V}$ is a Fréchet space. We refer the reader to [**25**] for a general introduction on such distributions.

To carry out the induction, we will apply the decomposition

$${V'}_{(r+1)} = {V'}_{r+1} \oplus {V'}_{-r-1} \oplus {V'}_{(r)}, \tag{13}$$

followed by the decomposition

(14) $$V_{(r)} = V_{-r} \oplus V_r \oplus V_{(r-1)}.$$

This gives rise to the decomposition

$$\begin{aligned}\mathrm{Hom}(V_{(r)}, {V'}_{(r+1)}) &= \mathrm{Hom}(V_{(r)}, {V'}_{r+1}) \oplus \mathrm{Hom}(V_{(r)}, {V'}_{-r-1}) \oplus \mathrm{Hom}(V_{(r)}, {V'}_{(r)})\\ &= (\mathrm{Hom}(V_{(r)}, {V'}_{r+1}) \oplus \mathrm{Hom}(V_{-r}, {V'}_{(r)})) \oplus (\mathrm{Hom}(V_{(r)}, {V'}_{-r-1}) \oplus \mathrm{Hom}(V_r, {V'}_{(r)}))\\ &\quad \oplus \mathrm{Hom}(V_{(r-1)}, {V'}_{(r)}).\end{aligned}$$

Note that $\mathrm{Hom}(V_{(r)}, {V'}_{r+1}) \oplus \mathrm{Hom}(V_{-r}, {V'}_{(r)})$, $\mathrm{Hom}(V_{(r)}, {V'}_{-r-1}) \oplus \mathrm{Hom}(V_r, {V'}_{(r)})$ are totally isotropic, complementary subspaces, and thus via the mixed models, we will have the identification

(15) $$\begin{aligned}\mathscr{Y}_{(r),(r+1)} &\cong \mathscr{S}_{(r),r+1} \otimes \mathscr{Y}_{(r),(r)}\\ &\cong [\mathscr{S}_{(r),r+1} \otimes \mathscr{S}_{-r,(r)}] \otimes \mathscr{Y}_{(r-1),(r)}, \qquad r > 0.\end{aligned}$$

This identification allows us to do the induction with respect to r.

There are two steps in each induction, corresponding to the two isomorphisms in (15).

Our goal is to understand the space $(\mathscr{Y}'_{(r),(r+1)})^{{U'}_{(r+1)}, {\mathfrak{X}'}_{(r+1)}}$. In view of (12), we start with a (tempered) distribution $\lambda \in (\mathscr{Y}'_{(r),(r+1)})^{{U'}_{r+1}, {\mathfrak{X}'}_{r+1}}$.

Step 1: We use the explicit realization of $\mathscr{Y}_{(r),(r+1)}$ given by the first isomorphism in (15). We identify λ with a $\mathscr{Y}'_{(r),(r)}$-valued distribution on $\mathrm{Hom}(V_{(r)}, {V'}_{r+1})$, as in [**25**]. By using (derived) action of elements in the center of $\mathfrak{u}'_{r+1}$ and the fact that $\mathrm{Im}(X')$ contains V'_{r+1}, one shows that λ may be identified with a $\mathscr{Y}'_{(r),(r)}$-valued distribution that *lives on* $\mathrm{Hom}_{\mathrm{GNM}}(V_{(r)}, {V'}_{r+1})$ (i.e., having transverse order zero at all points of $\mathrm{Hom}_{\mathrm{GNM}}(V_{(r)}, {V'}_{r+1})$), where

$$\begin{aligned}&\mathrm{Hom}_{\mathrm{GNM}}(V_{(r)}, {V'}_{r+1}) =\\ &\{T \in \mathrm{Hom}(V_{(r)}, {V'}_{r+1}) \,|\, TT^* = 0 \text{ and } T \text{ has maximal rank } \dim {V'}_{r+1}\}.\end{aligned}$$

(Here the subscript GNM stands for generic null mappings.)

(A1) Assume that $\mathrm{Hom}_{\mathrm{GNM}}(V_{(r)}, {V'}_{r+1})$ is non-empty.

Then $\mathrm{Hom}_{\mathrm{GNM}}(V_{(r)}, {V'}_{r+1})$ is a single orbit under the action of $G = G_{(r)}$. We pick one representative of $\mathrm{Hom}_{\mathrm{GNM}}(V_{(r)}, {V'}_{r+1})$ called T_r, and by permuting under the G-action if necessary, we may pick a T_r with the following property: there exists a totally isotropic subspace V_r of $V = V(r)$ having the same dimension as ${V'}_{r+1}$, and

(16) $$T_r|_{V_{(r-1)} \oplus V_{-r}} = 0, \quad \text{and} \quad T_r : V_r \to V'_{r+1} \text{ is a linear isomorphism.}$$

Here V_{-r} is the isotropic subspace dual to V_r, and $V_{(r-1)} = (V_r \oplus V_{-r})^{\perp}$.

REMARK 3.10. Note that in Part (a) of Theorem 3.7, we are only given the nilpotent orbit $\mathcal{O}'$, and not $\mathcal{O}$. Thus we have the decomposition (13), but not the decomposition (14). Nevertheless, the just concluded discussion allows us, under the hypothesis (A1), to find V_r, V_{-r}, $V_{(r-1)}$ so that the decomposition (14) holds. Of course these spaces (V_r, V_{-r}, $V_{(r-1)}$) no longer carry any meanings in terms of eigenvalues of the neutral element of an $\mathfrak{sl}_2$-triple. With the decomposition (14) in place, we then continue the induction.

Let P_r be the stabilizer of V_{-r} in G. Then $P_r = M_{(r)}N_r$, where N_r is the unipotent radical of P_r, $M_{(r)} = M_r \times G_{(r-1)}$, and $M_r \cong \mathrm{GL}(V_{-r})$. Note that $N_r G_{(r-1)}$ is the stabilizer of T_r.

Step 2: Since we have the decomposition $V = V_{-r} \oplus V_{(r-1)} \oplus V_r$, we could now use the explicit realization of $\mathscr{Y}_{(r),(r+1)}$ given by the second isomorphism in (15). After evaluating at T_r, and by using (derived) action of other elements of $\mathfrak{u}'_{r+1}$, one finds that λ may be identified with a distribution that *lives on* $G \times \mathscr{A}$, where $\mathscr{A}$ is the affine space given by the linear system:

$$\{S \in \mathrm{Hom}(V_{-r}, V'_{(r)}) \mid \mathrm{Tr}(R'T_rS^*) = \mathrm{Tr}(R'X'),\ \forall R' \in \mathrm{Hom}(V'_{r+1}, V'_{(r-1)} \oplus V'_{-r})\}.$$

REMARK 3.11. The above affine space $\mathscr{A}$ appears incorrectly in line 18, page 842 of [**15**]: Tr should be inserted on both sides.

$$\text{(A2)} \qquad\qquad \text{Assume that } \mathscr{A} \text{ is non-empty.}$$

Observe that the corresponding homogeneous system of $\mathscr{A}$ has $\mathrm{Hom}(V_{-r}, V'_{-r})$ as the general solution. Due to the way we have arranged T_r, we may choose a particular solution T_{-r} or equivalently T^*_{-r} so that

$$\text{(17)} \qquad T^*_{-r} \in \mathrm{Hom}(V'_{r-1}, V_r) \text{ and } T_rT^*_{-r} = X' \text{ on } V'_{r-1}.$$

Figuratively, we have completed a triangle:

$$\begin{array}{ccc} & V_r & \\ T^*_{-r} \nearrow & & \searrow T_r \\ V'_{r-1} & \xrightarrow{X'} & V'_{r+1} \end{array}$$

The full solution of the linear system is then $T_{-r} + \mathrm{Hom}(V_{-r}, V'_{-r})$, and λ is identified as distribution that *lives on* $G \times [T_{-r} + \mathrm{Hom}(V_{-r}, V'_{-r})]$.

Given a function $f \in \mathscr{Y}_{(r),(r+1)} \cong [\mathscr{S}_{(r),r+1} \otimes \mathscr{S}_{-r,(r)}] \otimes \mathscr{Y}_{(r-1),(r)}$, define a new function $f_r \in C^\infty(G; \mathscr{S}_{-r,-r} \otimes \mathscr{Y}_{(r-1),(r)})$ by

$$f_r(g)(S) = [\omega_{(r),(r+1)}(g)f](T_r, T_{-r} + S), \qquad g \in G, S \in \mathrm{Hom}(V_{-r}, V'_{-r}).$$

The final result of the inductive step, which will relate covariants of $\mathscr{Y}_{(r),(r+1)}$ with those of $\mathscr{Y}_{(r-1),(r)}$, is the following

LEMMA 3.12. *(a) For $(\mathscr{Y}_{(r),(r+1)})_{U'_{r+1},\mathcal{X}'_{r+1}}$ to be non-zero, both (A1) and (A2) must be satisfied.*
(b) Assume both (A1) and (A2). For $r > 0$, the map $f \mapsto f_r$ induces a G-intertwining isomorphism

$$\Psi_r : (\mathscr{Y}_{(r),(r+1)})_{U'_{r+1},\mathcal{X}'_{r+1}} \longrightarrow \mathscr{C}(N_rG_{(r-1)}\backslash G; \mathscr{S}_{-r,-r} \otimes \mathscr{Y}_{(r-1),(r)}),$$

where the action of $N_rG_{(r-1)}$ on $\mathscr{S}_{-r,-r} \otimes \mathscr{Y}_{(r-1),(r)}$ is given by equations (6.37)-(6.39) in [**15**].

3.5. Sketch of proof of Theorem 3.7. Continuing the induction provided by Lemma 3.12, Part (a) of Theorem 3.7 follows. This is because the nonvanishing of $(\mathscr{Y}_{(r),(r+1)})_{U'_{(r+1)},\mathcal{X}'_{(r+1)}}$ requires both (A1) and (A2) to be satisfied for all r. We may thus find T_r and T_{-r} as in equations (16) and (17). It is then clear that the moment map φ' will send $T =: \oplus_{k=-r}^{r} T_k$ to X', and so $\mathcal{O}'$ is in the image of the moment map φ'.

For Part (b) of Theorem 3.7, we proceed with the induction as in the descent case. For details, see Sections 6.4 and 6.5 of [**15**]. At the end of the induction, we will need the following lemma, which is a replacement of [**15**, Lemma 5.8] in the descent case. Together with the inductive steps, it implies Proposition 3.8, which in turn implies Part (b) of Theorem 3.7.

Recall $\mathfrak{g}_{-1}$ is a symplectic space whose symplectic structure is defined through an $\operatorname{Ad} G$-invariant non-degenerate F-bilinear form κ on $\mathfrak{g}$, as in (2). We now fix the bilinear form κ as $\frac{1}{2}\operatorname{Tr}(T^*S)$ (for $T, S \in \mathfrak{g}$), and likewise for κ'.

LEMMA 3.13. *Let W_0 be the symplectic subspace of W defined by*

$$W_0 = \bigoplus_{k=-r}^{r} \operatorname{Hom}(V_k, V_k') \subset \operatorname{Hom}(V, V').$$

Given $T \in \mathcal{O}_{\gamma,\gamma'}$, the map

$$\begin{array}{rrcl} J_T: & -\mathfrak{g}_{-1} \oplus \mathfrak{g}'_{-1} & \longrightarrow & W_0, \\ & (R, R') & \mapsto & TR + R'T \end{array}$$

is a symplectic imbedding, with $(\operatorname{Im} J_T)^\perp = W_{\gamma,\gamma'}$.

REMARK 3.14. The smooth oscillator representation of $\widetilde{\operatorname{Sp}}(W_0)$ is realized in the tensor product space $\mathscr{S}_{-r,-r} \otimes \mathscr{S}_{-r+1,-r+1} \cdots \otimes \mathscr{S}_{-1,-1} \otimes \mathscr{Y}_{(0),(0)}$. This representation naturally appears when we carry out the induction. See Lemma 3.12.

PROOF. A straightforward computation shows that

$$\langle J_T(R_1, R'_1), J_T(R_2, R'_2)\rangle = -\kappa_{-1}(R_1, R_2) + \kappa'_{-1}(R'_1, R'_2),$$

for all R_1, $R_2 \in \mathfrak{g}_{-1}$, R'_1, $R'_2 \in \mathfrak{g}'_{-1}$. Since κ_{-1} and κ'_{-1} are non-degenerate, we see that J_T is injective.

It is also straightforward to show that

$$\langle S, J_T(R, R')\rangle = 0, \qquad \text{for } S \in W_0,\ R \in \mathfrak{g}_{-1} \text{ and } R' \in \mathfrak{g}'_{-1}.$$

It remains to show that the dimensions match. Observe that there is a linear isomorphism between $\oplus_{k \leq 0} \operatorname{Hom}(V_k, V_{k-1})$ and $\mathfrak{g}_{-1}$. From this we conclude that

$$\begin{aligned} \dim \mathfrak{g}_{-1} &= \sum_{k=0}^{-r+1} \dim V_k \cdot \dim V_{k-1} \\ &= \sum_{k=0}^{-r+1} \dim V'_{k-1} \cdot \dim V_{k-1} + \dim(\operatorname{Ker} T) \cdot \dim V_{-1} \\ &= \sum_{k=-1}^{-r} \dim V'_k \cdot \dim V_k + \dim(\operatorname{Ker} T) \cdot \dim V_{-1}. \end{aligned}$$

Here we have used the easily checked fact that $\dim V_0 = \dim V'_{-1} + \dim(\mathrm{Ker} T)$ and $\dim V_k = \dim V'_{k-1}$ for $k \leq 1$. C.f. the diagram in (11). Similarly,

$$\begin{aligned}
\dim \mathfrak{g}'_{-1} &= \sum_{k=0}^{-r} \dim V'_k \cdot \dim V'_{k-1} \\
&= \sum_{k=0}^{-r} \dim V'_k \cdot \dim V_k - \dim V'_0 \cdot \dim(\mathrm{Ker} T) \\
&= \sum_{k=0}^{r} \dim V'_k \cdot \dim V_k - \dim V'_0 \cdot \dim(\mathrm{Ker} T).
\end{aligned}$$

Here we have the fact that $\dim V_k = \dim V_{-k}$. Therefore we have

$$\begin{aligned}
\dim \mathfrak{g}_{-1} + \dim \mathfrak{g}'_{-1} &= \sum_{k=-r}^{r} \dim V'_k \cdot \dim V_k - \dim(\mathrm{Ker} T) \cdot [\dim V'_0 - \dim V_{-1}] \\
&= \dim(W_0) - \dim(\mathrm{Ker} T) \cdot [\dim V'_0 - \dim V'_{-2}] \\
&= \dim(W_0) - \dim(\mathrm{Ker} T) \cdot \dim((V')_0^{\gamma',1}).
\end{aligned}$$

□

4. Correspondence of associated characters

In this section, the local field F will be $\mathbb{R}$. We will discuss a recent result of Ma, Sun and the author [**35**], which computes the associated character of the local theta lift, in the so-called convergent range. Related earlier works on transition of associated cycles include [**33, 41, 42**].

4.1. The associated character map. We first recall some basics of Vogan's theory of associated varieties [**54**]. We will need a minor extension of the theory to allow covering groups which appear in the theory of dual pairs.

Let $\mathbf{G}$ be a complex reductive Lie group and $\mathfrak{g}$ is its Lie algebra. Denote by $\mathrm{Nil}_{\mathbf{G}}(\mathfrak{g})$ the set of nilpotent $\mathbf{G}$-orbits in $\mathfrak{g}$. We will be concerned with genuine Casselman-Wallach representations of $\widetilde{G}$ and their Harish-Chandra modules. Here $(G, G') \subseteq \mathrm{Sp}(W)$ is a type I reductive dual pair as in Section 2. In this context, $\mathbf{G}$ is the complexification of G, which is a complex classical group.

Suppose $\mathcal{O} \in \mathrm{Nil}_{\mathbf{G}}(\mathfrak{g})$. We say that a finite length $(\mathfrak{g}, \widetilde{K})$-module Π is $\mathcal{O}$-*bounded* (or *bounded by* $\mathcal{O}$) if the associated variety of the annihilator ideal $\mathrm{Ann}(\Pi)$ is contained in $\overline{\mathcal{O}}$ (the closure of $\mathcal{O}$). Let

$$\mathfrak{g} = \mathfrak{k} \oplus \mathfrak{p}$$

be the complexified Cartan decomposition fixed by our choice of the maximal compact subgroup K of G. It follows from [**54**, Theorem 8.4] that Π is $\mathcal{O}$-bounded if and only if its associated variety $\mathrm{AV}(\Pi)$ is contained in $\overline{\mathcal{O}} \cap \mathfrak{p}$. Let $\mathcal{M}_{\mathcal{O}}(\mathfrak{g}, \widetilde{K})$ denote the category of genuine $\mathcal{O}$-bounded finite length $(\mathfrak{g}, \widetilde{K})$-modules, and write $\mathcal{K}_{\mathcal{O}}(\mathfrak{g}, \widetilde{K})$ for its Grothendieck group.

Under the adjoint action of $\mathbf{K}$, the complex variety $\mathcal{O} \cap \mathfrak{p}$ is a union of finitely many orbits, each of dimension $\frac{\dim_{\mathbb{C}} \mathcal{O}}{2}$. For any $\mathbf{K}$-orbit $\mathscr{O} \subset \mathcal{O} \cap \mathfrak{p}$, let $\mathcal{K}_{\mathscr{O}}(\widetilde{\mathbf{K}})$ denote the Grothendieck group of $\widetilde{\mathbf{K}}$-equivariant coherent sheaves (or the Grothedieck group of the category of $\widetilde{\mathbf{K}}$-equivariant algebraic vector bundles) on $\mathscr{O}$. Taking the isotropy representation at a point $X \in \mathscr{O}$ yields an identification

$$\mathcal{K}_{\mathscr{O}}(\widetilde{\mathbf{K}}) = \mathcal{R}(\widetilde{\mathbf{K}}_X), \tag{18}$$

where the right hand side denotes the Grothendieck group of the category of algebraic representations of the stabilizer group $\widetilde{\mathbf{K}}_X$.

Put

$$\mathcal{K}_{\mathcal{O}}(\widetilde{\mathbf{K}}) := \bigoplus_{\mathscr{O} \text{ is a } \mathbf{K}\text{-orbit in } \mathcal{O} \cap \mathfrak{p}} \mathcal{K}_{\mathscr{O}}(\widetilde{\mathbf{K}}). \tag{19}$$

According to Vogan [**54**, Theorem 2.13], we have a canonical homomorphism, called the associated character map:

$$\mathrm{Ch}_{\mathcal{O}} : \mathcal{K}_{\mathcal{O}}(\mathfrak{g}, \widetilde{K}) \to \mathcal{K}_{\mathcal{O}}(\widetilde{\mathbf{K}}). \tag{20}$$

Briefly, starting from a good filtration of a finite length $(\mathfrak{g}, \widetilde{K})$-module Π, one obtains a finitely generated $(S(\mathfrak{g}), \widetilde{K})$-module M by taking the graded module. Vogan shows that there is a finite filtration $\{M_j\}$ of M by $(S(\mathfrak{g}), \widetilde{K})$-submodules with the property that every subquotient is generically reduced along every minimal prime in the associated variety of M. The virtual representation Ch_X of $\widetilde{\mathbf{K}}_X$ that Vogan attaches to $X \in \mathscr{O}$ is then

$$\mathrm{Ch}_X = \sum_j M_j / (\mathfrak{m}(X) M_j + M_{j-1}),$$

where $\mathfrak{m}(X)$ is the maximal ideal in $S(\mathfrak{g})$ corresponding to X. It is independent of all the choices that have been made. See [**54**, Section 2] for details.

For a Casselman-Wallach representation of $\widetilde{G}$, we define its associated character by using its Harish-Chandra module.

4.2. Algebraic theta lifting. We are back in the setting of Section 1, namely we are given $(G, G') = (G(V), G(V')) \subset \mathrm{Sp}(W)$, a type I reductive dual pair, and the smooth oscillator representation $(\omega, \mathscr{Y})$ of $\widetilde{\mathrm{Sp}}(W)$ associated to a fixed ψ. Let $\Omega_{V,V'}$ be the Harish-Chandra module of $(\omega, \mathscr{Y})$, which is naturally a $(\mathfrak{g} \times \mathfrak{g}', \widetilde{K} \times \widetilde{K'})$-module.

DEFINITION 4.1. For a genuine $(\mathfrak{g}, \widetilde{K})$-module Π of finite length, define

$$\check{\Theta}_{V,V'}(\Pi) := (\Omega_{V,V'} \otimes \Pi)_{\mathfrak{g}, \widetilde{K}}, \qquad \text{(the coinvariant space)}.$$

The $(\mathfrak{g}', \widetilde{K}')$-module $\check{\Theta}_{V,V'}(\Pi)$, or $\check{\Theta}(\Pi)$ in short, is genuine and of finite length [**20**].

REMARK 4.2. If π is a smooth irreducible genuine representation of $\widetilde{G}$, and let Π denote the Harish-Chandra module of π. Then

$$\check{\Theta}(\Pi) \simeq \Theta(\pi^\vee)^{\mathrm{al}},$$

where $\pi^\vee$ is the contragredient representation of π, and $\Theta(\pi^\vee)$ is the full theta lift of $\pi^\vee$ (in the smooth category), and $\Theta(\pi^\vee)^{\mathrm{al}}$ is the Harish-Chandra module of $\Theta(\pi^\vee)$.

For the moment, we let $(\mathbf{G}, \mathbf{G}') \subseteq \mathrm{Sp}(\mathbf{W})$ be a complex reductive dual pair [**18**]. Recall we have the moment maps: [**7, 24**]

$$\begin{array}{ccc} & \mathbf{W} & \\ {}^{\mathbf{M}}\swarrow & & \searrow^{\mathbf{M}'} \\ \mathfrak{g} & & \mathfrak{g}' \end{array}$$

Let us describe the moment maps explicitly for a type I complex dual pair. Let $\mathbf{V}$ be an ϵ-symmetric bilinear space and $\mathbf{V}'$ be an ϵ'-symmetric bilinear space, where $\epsilon, \epsilon' \in \{\pm 1\}$ with $\epsilon\epsilon' = -1$. We have the complex symplectic space $\mathbf{W} := \mathrm{Hom}(\mathbf{V}, \mathbf{V}')$, with the symplectic form:

$$< T_1, T_2 >_{\mathbf{W}} := \mathrm{Tr}(T_1^\star T_2), \qquad T_1, T_2 \in \mathbf{W}.$$

Here $\star \colon \mathrm{Hom}(\mathbf{V}, \mathbf{V}') \to \mathrm{Hom}(\mathbf{V}', \mathbf{V})$ is the adjoint map induced by the form $(\cdot, \cdot)_{\mathbf{V}}$ on $\mathbf{V}$ and the form $(\cdot, \cdot)_{\mathbf{V}'}$ on $\mathbf{V}'$:

$$< Tv, v' >_{\mathbf{V}'} = < v, T^\star v' >_{\mathbf{V}}, \qquad v \in \mathbf{V}, v' \in \mathbf{V}', T \in \mathrm{Hom}(\mathbf{V}, \mathbf{V}').$$

Then $(\mathbf{G}, \mathbf{G}') := (\mathbf{G}_{\mathbf{V}}, \mathbf{G}_{\mathbf{V}'}) \subseteq \mathrm{Sp}(\mathbf{W})$. The moment maps are then given by

$$\mathbf{M}(T) = T^\star T, \quad \text{and} \quad \mathbf{M}'(T) = TT^\star.$$

We also recall the notion of theta lift of nilpotent orbits for a complex dual pair $(\mathbf{G}, \mathbf{G}') \subseteq \mathrm{Sp}(\mathbf{W})$. By [**7**, Theorem 1.1], for any $\mathcal{O} \in \mathrm{Nil}_{\mathbf{G}}(\mathfrak{g})$, $\mathbf{M}'(\mathbf{M}^{-1}(\overline{\mathcal{O}}))$ equals the closure of a unique nilpotent orbit $\mathcal{O}' \in \mathrm{Nil}_{\mathbf{G}'}(\mathfrak{g}')$. We call $\mathcal{O}'$ the *theta lift* of $\mathcal{O}$, and we write

$$\mathcal{O}' = \boldsymbol{\theta}_{\mathbf{V},\mathbf{V}'}(\mathcal{O}). \tag{21}$$

Now we fix compatible Cartan forms L on $\mathbf{V}$, L' on $\mathbf{V}'$ and $L_{\mathbf{W}}$ on $\mathbf{W}$ (as in [**35**, Section 2]), and this leads us to a real reductive dual pair in $(G, G') \subseteq \mathrm{Sp}(W)$ (also called a rational dual pair). Denote $\mathbf{K} = \mathbf{G}^L$ and $\mathbf{K}' = (\mathbf{G}')^{L'}$.

Let $\mathbf{i}$ denote a fixed $\sqrt{-1}$. We decompose

$$\mathbf{W} = \mathcal{X} \oplus \mathcal{Y}$$

where $\mathcal{X}$ and $\mathcal{Y}$ are $+\mathbf{i}$ and $-\mathbf{i}$ eigenspaces of $L_{\mathbf{W}}$, respectively. Restriction on $\mathcal{X}$ induces a pair of maps:

$$\begin{array}{ccc} & \mathcal{X} & \\ {}^{M := \mathbf{M}|_{\mathcal{X}}}\swarrow & & \searrow^{M' := \mathbf{M}'|_{\mathcal{X}}} \\ \mathfrak{p} & & \mathfrak{p}' \end{array}$$

which are also called moment maps (with respect to the rational dual pair).

We have the following estimate of the size of $\check{\Theta}(\pi)$.

PROPOSITION 4.3 ([**33**, Theorem B and Corollary E]). *For any genuine $(\mathfrak{g}, \widetilde{K})$-module Π of finite length, we have*

$$\mathrm{AV}(\check{\Theta}(\Pi)) \subset M'(M^{-1}(\mathrm{AV}(\Pi))).$$

Consequently, if Π is $\mathcal{O}$-bounded for a nilpotent orbit $\mathcal{O} \in \mathrm{Nil}_{\mathbf{G}}(\mathfrak{g})$, then $\check{\Theta}(\Pi)$ is $\boldsymbol{\theta}_{\mathbf{V},\mathbf{V}'}(\mathcal{O})$-bounded.

4.3. Descent and lift of nilpotent orbits. Let

$$\mathbf{W}^\circ := \{T \in \mathbf{W} \mid T \text{ is an injective map from } \mathbf{V} \text{ to } \mathbf{V}'\}.$$

Clearly $\mathbf{W}^\circ \neq \emptyset$ only if $\dim \mathbf{V} \leq \dim \mathbf{V}'$.

Suppose $\mathbf{e} \in \mathcal{O} \in \mathrm{Nil}_{\mathbf{G}}(\mathfrak{g})$ and $\mathbf{e}' \in \mathcal{O}' \in \mathrm{Nil}_{\mathbf{G}'}(\mathfrak{g}')$. We call $\mathbf{e}$ (resp. $\mathcal{O}$) a descent of $\mathbf{e}'$ (resp. $\mathcal{O}'$), if there exists $T \in \mathbf{W}^\circ$ such that

$$\mathbf{M}(T) = \mathbf{e} \quad \text{and} \quad \mathbf{M}'(T) = \mathbf{e}'.$$

For a rational dual pair $(G, G') \subseteq \mathrm{Sp}(W)$, put

$$\mathcal{X}^\circ := \mathbf{W}^\circ \cap \mathcal{X}.$$

Suppose $X \in \mathscr{O} \in \mathrm{Nil}_{\mathbf{K}}(\mathfrak{p})$ and $X' \in \mathscr{O}' \in \mathrm{Nil}_{\mathbf{K}'}(\mathfrak{p}')$. We call X (resp. $\mathscr{O}$) a descent of X' (resp. $\mathscr{O}'$), if there exists $T \in \mathcal{X}^\circ$ such that

$$M(T) = X \quad \text{and} \quad M'(T) = X'.$$

In both cases, we will say that T realizes the descent, and $\mathcal{O}'$ (resp. $\mathscr{O}'$) is the lift of $\mathcal{O}$ (resp. $\mathscr{O}$). We will write

$$\mathcal{O} = \boldsymbol{\nabla}(\mathcal{O}') = \boldsymbol{\nabla}_{\mathbf{V}',\mathbf{V}}(\mathcal{O}') \quad \text{and} \quad \mathscr{O} = \nabla(\mathscr{O}') = \nabla_{\mathbf{V}',\mathbf{V}}(\mathscr{O}'). \tag{22}$$

By using explicit formulas in [**7**, **24**] (for complex dual pairs) and [**43**, Lemma 14] (for rational dual pairs), we have the following key property:

$$\mathbf{M}(\mathbf{M}'^{-1}(\overline{\mathcal{O}})) = \overline{\mathcal{O}'} \quad \text{and} \quad M(M'^{-1}(\overline{\mathscr{O}})) = \overline{\mathscr{O}'},$$

where "$\overline{}$" means taking Zariski closure. (The notion of lift is thus stronger than the notion of theta lift for a complex nilpotent orbit.)

Given $T \in \mathcal{X}^\circ$ which realizes the descent from $X' = M'(T) \in \mathscr{O}' \in \mathrm{Nil}_{\mathbf{K}'}(\mathfrak{p}')$ to $X = M(T) \in \mathscr{O} \in \mathrm{Nil}_{\mathbf{K}}(\mathfrak{p})$, we denote the respective isotropy subgroups by

$$\mathbf{S}_T := \mathrm{Stab}_{\mathbf{K}\times\mathbf{K}'}(T), \quad \mathbf{K}_X := \mathrm{Stab}_{\mathbf{K}}(X) \quad \text{and} \quad \mathbf{K}'_{X'} := \mathrm{Stab}_{\mathbf{K}'}(X').$$

Then there is a unique homomorphism

$$\alpha = \alpha_T \colon \mathbf{K}'_{X'} \to \mathbf{K}_X \tag{23}$$

such that $\mathbf{S}_T$ is the graph of α:

$$\mathbf{S}_T = \{(\alpha(k'), k') \in \mathbf{K}_X \times \mathbf{K}'_{X'} \mid k' \in \mathbf{K}'_{X'}\}.$$

The homomorphism α is uniquely determined by the requirement that

$$T(\alpha(k')(v)) = k'(T(v)) \quad \text{for all } v \in \mathbf{V},\ k' \in \mathbf{K}'_{X'}.$$

4.4. Lift of algebraic vector bundles. Attached to a dual pair $(G(V), G(V'))$, there is a distinguished genuine character $\varsigma_{V,V'}$ of $\widetilde{\mathbf{K}} \times \widetilde{\mathbf{K}'}$ arising from the oscillator module $\Omega_{V,V'}$, as in [**35**, Section 2.7].

Recall the notations from Section 4.1. Thus $\mathscr{O} \in \mathcal{O} \cap \mathfrak{p}$ is a $\mathbf{K}$-orbit, and we identify $\mathcal{K}_{\mathscr{O}}(\widetilde{\mathbf{K}})$ with $\mathcal{R}(\widetilde{\mathbf{K}}_X)$, the Grothendieck group of the category of genuine algebraic representations of the stabilizer group $\widetilde{\mathbf{K}}_X$.

Let ρ be a genuine algebraic representation of $\widetilde{\mathbf{K}}_X$. Then the representation $\varsigma_{V,V'}|_{\widetilde{\mathbf{K}}_X} \otimes \rho$ of $\widetilde{\mathbf{K}}_X$ descends to a representation of $\mathbf{K}_X$. Define

$$\vartheta_T(\rho) := \varsigma_{V,V'}|_{\widetilde{\mathbf{K}'}_{X'}} \otimes (\varsigma_{V,V'}|_{\widetilde{\mathbf{K}}_X} \otimes \rho) \circ \alpha,$$

which is a genuine algebraic representation of $\widetilde{\mathbf{K}'}_{X'}$.

Clearly ϑ_T induces a homomorphism from $\mathcal{R}(\widetilde{\mathbf{K}}_X)$ to $\mathcal{R}(\widetilde{\mathbf{K}}'_{X'})$. In view of (18), we thus have a homomorphism

$$\vartheta_{\mathscr{O},\mathscr{O}'} : \mathcal{K}_{\mathscr{O}}(\widetilde{\mathbf{K}}) \longrightarrow \mathcal{K}_{\mathscr{O}'}(\widetilde{\mathbf{K}'}).$$

This is independent of the choice of T.

Suppose $\mathcal{O} \in \mathrm{Nil}_{\mathbf{G}}(\mathfrak{g})$, $\mathcal{O}' \in \mathrm{Nil}_{\mathbf{G}'}(\mathfrak{g}')$ and $\mathcal{O} = \boldsymbol{\nabla}(\mathcal{O}')$. Using the decomposition in (19), we define a homomorphism

$$\vartheta_{\mathcal{O},\mathcal{O}'} := \sum_{\substack{\mathscr{O}'\subset\mathcal{O}'\cap\mathfrak{p}' \\ \mathscr{O}=\nabla(\mathscr{O}')\subset\mathfrak{p}}} \vartheta_{\mathscr{O},\mathscr{O}'} : \mathcal{K}_{\mathcal{O}}(\widetilde{\mathbf{K}}) \longrightarrow \mathcal{K}_{\mathcal{O}'}(\widetilde{\mathbf{K}'}) \tag{24}$$

where the summation is over all pairs $(\mathscr{O},\mathscr{O}')$ such that $\mathscr{O} \subset \mathfrak{p}$ is the descent of $\mathscr{O}' \subset \mathcal{O}' \cap \mathfrak{p}'$.

4.5. An upper bound and an equality of associated characters. We have an upper bound of associated characters in the descent situation.

Recall the natural partial order $\succeq$ on $\mathcal{K}_{\mathcal{O}}(\widetilde{\mathbf{K}})$ and $\mathcal{K}_{\mathscr{O}}(\widetilde{\mathbf{K}})$ defined as follows. We say $c_1 \succeq c_2$ (or $c_2 \preceq c_1$) if $c_1 - c_2$ is represented by a $\widetilde{\mathbf{K}}$-equivariant coherent sheaf.

PROPOSITION 4.4 ([**35**, Theorem 4.3]). *Let $\mathcal{O} \in \mathrm{Nil}_{\mathbf{G}}(\mathfrak{g})$, $\mathcal{O}' \in \mathrm{Nil}_{\mathbf{G}'}(\mathfrak{g}')$ and $\mathcal{O}$ is a descent of $\mathcal{O}'$. Then for every genuine $\mathcal{O}$-bounded $(\mathfrak{g},\widetilde{K})$-module Π of finite length, $\check{\Theta}_{V,V'}(\Pi)$ is $\mathcal{O}'$-bounded and*

$$\mathrm{Ch}_{\mathcal{O}'}(\check{\Theta}_{V,V'}(\Pi)) \preceq \vartheta_{\mathcal{O},\mathcal{O}'}(\mathrm{Ch}_{\mathcal{O}}(\Pi)).$$

REMARK 4.5. We may also define the notion of generalized descent in the setting of complex or rational dual pairs [**35**, Section 2.6]. Proposition 4.4 extends to a more general situation, where $\mathcal{O}'$ is "good for generalized descent". See [**35**, Section 4.1].

The proofs of Propositions 4.3 and 4.4 are similar and are largely geometric. There are natural good filtrations on Π and $\check{\Theta}_{V,V'}(\Pi)$, which are generated by the minimal degree $\widetilde{K}$-types and $\widetilde{K}'$-types, respectively. After twisting by $\varsigma_{V,V'}$, their graded spaces may be viewed as a $\mathbf{K}$-equivariant coherent sheaf on $\mathfrak{p}$ and $\mathbf{K}'$-equivariant coherent sheaf on $\mathfrak{p}'$, respectively. Thus it boils down to understanding this lifting of equivariant coherent sheaves induced by the double fibration of moment maps, and this can be readily analyzed via algebraic-geometric techniques. See [**35**, Section 4.1] for details.

The final theorem we will discuss is more precise and it is concerned with the associate character of a quotient of $\check{\Theta}_{V,V'}(\pi)$, where π is a certain genuine irreducible Casselman-Wallach representation of $\widetilde{G}$. Note that by Proposition 4.4, we will have an upper bound on the full maximal quotient $\check{\Theta}_{V,V'}(\pi)$.

We will review the notion of convergent range in the setting of local theta correspondence ([**35**, Sections 3.1 and 3.2]). First recall that a function on a real reductive group G is called ν-bounded if it is bounded by the ν-th power of the Harish-Chandra's Ξ-function Ξ_G on G, up to a function of logarithmic growth. From the well-known properties of Ξ_G, we see that every ν-bounded continuous function on G is integrable, for $\nu > 2$. One may also define the notion of ν-boundedness for a Casselman-Wallach representation in the obvious way, by using its matrix coefficients. Under this terminology, tempered representations are then 1-bounded.

Fix a type I dual pair $(G, G') = (G(V), G(V'))$ as in Section 2, where V is an ϵ-Hermitian space and V' is an ϵ'-Hermitian space. Define

$$\dim^\circ_{\mathrm{F}} V = \dim_{\mathrm{F}} V - 2\frac{d_1}{d}, \tag{25}$$

where $d = \dim_{\mathrm{F}} D$, and $d_1 = \dim_{\mathrm{F}}\{t \in D | \bar{t} = \epsilon t\}$. Here the superscript bar denotes the standard involutive anti-automorphism of D.

We have the following estimate of matrix coefficients. See [**35**, Lemma 3.8] or [**29**, Theorem 3.2].

LEMMA 4.6. *The representation* $\omega_{V,V'}|_{\widetilde{G}}$ *is* $\frac{\dim_{\mathrm{F}} V'}{\dim^\circ_{\mathrm{F}} V}$*-bounded.*

We thus make the following

DEFINITION 4.7. Assume that $\dim^\circ_{\mathrm{F}} V > 0$. Let π be a genuine Casselman-Wallach representation of $\widetilde{G}$. The pair (π, V') (or (π, G')) is said to be in the *convergent range* if π is ν_π-bounded for some $\nu_\pi > 2 - \frac{\dim_{\mathrm{F}} V'}{\dim^\circ_{\mathrm{F}} V}$.

Suppose that (π, V') is in the convergent range. This ensures that the following integral is convergent:

$$\begin{aligned} (\pi\widehat{\otimes}\omega_{V,V'}) \times (\pi^\vee\widehat{\otimes}\omega^\vee_{V,V'}) &\longrightarrow \mathbb{C}, \\ (u, v) &\longmapsto \int_{\widetilde{G'}} < g\cdot u, v > \mathrm{d}g. \end{aligned} \tag{26}$$

Define

$$\bar{\Theta}_{V,V'}(\pi) := \frac{\pi\widehat{\otimes}\omega_{V,V'}}{\text{the left kernel of (26)}}.$$

This is a Casselman-Wallach representation of $\widetilde{G'}$, since it is a quotient of $(\pi\widehat{\otimes}\omega_{V,V'})_G$ (the Hausdorff coinvariant space, which is the full theta lift of $\pi^\vee$).

As usual we parameterize a nilpotent $\mathbf{G}'$-orbit $\mathcal{O}'$ in $\mathfrak{g}'$ by its Young diagram $\mathbf{D}(\mathcal{O}')$, but labelled by its column partition $[c_0, c_1, \cdots, c_k]$, where $c_0 \geq c_1 \geq \cdots \geq c_k > 0$, and $\sum_{l=0}^k c_l = \dim \mathbf{V}'$ (and with some additional constraints on c_l's depending on $\mathfrak{g}'$).

THEOREM 4.8 ([**35**, Theorem 4.4]). *Let* $\mathcal{O} \in \mathrm{Nil}_{\mathbf{G}}(\mathfrak{g})$ *and* $\mathcal{O}' \in \mathrm{Nil}_{\mathbf{G}'}(\mathfrak{g}')$ *such that* $\mathcal{O}$ *is the descent of* $\mathcal{O}'$*. Write* $\mathbf{D}(\mathcal{O}') = [c_0, c_1, \cdots, c_k]$ $(k \geq 1)$*. Assume that* $c_0 > c_1$ *when* G *is a real symplectic group. Then for every* $\mathcal{O}$*-bounded irreducible Casselman-Wallach representation* π *of* $\widetilde{G}$ *such that* (π, V') *is in the convergent range,* $\bar{\Theta}_{V,V'}(\pi)$ *is* $\mathcal{O}'$*-bounded and*

$$\mathrm{Ch}_{\mathcal{O}'}(\bar{\Theta}_{V,V'}(\pi)) = \vartheta_{\mathcal{O},\mathcal{O}'}(\mathrm{Ch}_{\mathcal{O}}(\pi)).$$

REMARK 4.9. (a). Theorem 4.8 is the most crucial technical ingredient of [**35**] on the construction of unipotent representations of real classical groups. When (V, V') is in the stable range, the result was proved for unitary representations π in [**33**]. (b) Note that in [**35**] the authors only consider the cases of real (or quaternionic) orthogonal and symplectic groups. The case of unitary groups is proved by a similar method.

We outline the basic ideas in the proof of Theorem 4.8 for the cases of real (or quaternionic) orthogonal and symplectic groups and offer brief remarks for the case of unitary groups.

First observe that the Harish-Chandra module of $\bar{\Theta}_{V,V'}(\pi)$ is isomorphic to a quotient of $\check{\Theta}(\pi^{\mathrm{al}})$, where π^{al} denotes the Harish-Chandra module of π. Thus Proposition 4.4 implies that $\bar{\Theta}_{V,V'}(\pi)$ is $\mathcal{O}'$-bounded and

$$\mathrm{Ch}_{\mathcal{O}'}(\bar{\Theta}_{V,V'}(\pi)) \preceq \vartheta_{\mathcal{O},\mathcal{O}'}(\mathrm{Ch}_{\mathcal{O}}(\pi)). \tag{27}$$

In order to show that the equality in (27) is actually achieved, we will apply another (appropriate) theta lifting by integration. This idea is due to He [**17**], who used it to show the nonvanishing of a certain theta lift. We choose a split space U, of the same type as V, which is roughly twice the size of V'. The reader is referred to [**35**, Section 3.5] for the precise description of U. The space V is then realized as a non-degenerate subspace of U and we write $V^{\perp}$ for the orthogonal complement of V in U. We will consider the two-step theta lifting by integration $\bar{\Theta}_{V',V^{\perp}}(\bar{\Theta}_{V,V'}(\pi))$.

The Rallis quotient (the maximal quotient of the trivial representation of $G(V')$) with respect to the dual pair (U, V') will now sit inside a degenerate principal series representation of $\widetilde{G(U)}$ induced from a certain character of its Siegel parabolic subgroup. Because of our choice of U, the corresponding degenerate principal series is on the unitary axis when G is a real orthogonal or a real symplectic group, and is induced from $\det^{\frac{1}{2}}$ (resp. $\det^{-\frac{1}{2}}$) when G is a quaternionic orthogonal group (resp. a quaternionic symplectic group). Here det denotes the reduced norm of an appropriate general linear group in the quaternionic case. Note that the structure of afore-mentioned degenerate principal series of $\widetilde{G(U)}$ relative to Rallis quotients is well-known by the work of Kudla-Rallis [**26**], Lee-Zhu [**31**,**32**] (for real orthogonal and real symplectic groups) and Yamana [**62**] (for quaternionic orthogonal and quaternionic symplectic groups). (For unitary groups, the corresponding results are in [**30**].)

One shows that $(\bar{\Theta}_{V,V'}(\pi), V^{\perp})$ is in the convergent range. The decisive observation ([**35**, Lemma 4.5] and [**35**, Proposition 3.16]) which enables us to conclude the desired equality of associated characters is as follows: via a variant of the well-known doubling method (c.f. [**38**,**45**]), the two-step theta lifting $\bar{\Theta}_{V',V^{\perp}}(\bar{\Theta}_{V,V'}(\pi))$ is inside a certain degenerate principal series representation of $\widetilde{G(V^{\perp})}$ which is intimately related to the aforementioned degenerate principal series of $\widetilde{G(U)}$ and in a completely explicit way. A result of Barbasch [**2**] then allows us to determine the associate cycle of $\bar{\Theta}_{V',V^{\perp}}(\bar{\Theta}_{V,V'}(\pi))$.

On the other hand, there is an upper bound of the associate character of $\bar{\Theta}_{V',V^{\perp}}(\bar{\Theta}_{V,V'}(\pi))$ (by applying Proposition 4.4 twice). By comparing the associated cycles, one concludes that there is no other choice except to have the desired equality in (27). This comparison is also the step where one needs to do a case-by-case check for real orthogonal and real symplectic groups, and for quaternionic orthogonal and quaternionic symplectic groups, and similarly for unitary groups.

References

[1] Y. Bao and B. Sun, *Coincidence of algebraic and smooth theta correspondences*, Represent. Theory **21** (2017), 458–466, DOI 10.1090/ert/508. MR3718456

[2] D. Barbasch, *Orbital integrals of nilpotent orbits*, The mathematical legacy of Harish-Chandra (Baltimore, MD, 1998), Proc. Sympos. Pure Math., vol. 68, Amer. Math. Soc., Providence, RI, 2000, pp. 97–110, DOI 10.1090/pspum/068/1767894. MR1767894

[3] D. Barbasch and D. A. Vogan Jr., *The local structure of characters*, J. Funct. Anal. **37** (1980), no. 1, 27–55, DOI 10.1016/0022-1236(80)90026-9. MR576644

[4] W. Casselman, *Canonical extensions of Harish-Chandra modules to representations of G*, Canad. J. Math. **41** (1989), no. 3, 385–438, DOI 10.4153/CJM-1989-019-5. MR1013462

[5] F. du Cloux, *Sur les représentations différentiables des groupes de Lie algébriques* (French), Ann. Sci. École Norm. Sup. (4) **24** (1991), no. 3, 257–318. MR1100992

[6] D. H. Collingwood and W. M. McGovern, *Nilpotent orbits in semisimple Lie algebras*, Van Nostrand Reinhold Mathematics Series, Van Nostrand Reinhold Co., New York, 1993. MR1251060

[7] A. Daszkiewicz, W. Kraśkiewicz, and T. Przebinda, *Nilpotent orbits and complex dual pairs*, J. Algebra **190** (1997), no. 2, 518–539, DOI 10.1006/jabr.1996.6910. MR1441961

[8] A. Daszkiewicz, W. Kraśkiewicz, and T. Przebinda, *Dual pairs and Kostant-Sekiguchi correspondence. II. Classification of nilpotent elements*, Cent. Eur. J. Math. **3** (2005), no. 3, 430–474, DOI 10.2478/BF02475917. MR2152470

[9] M. Furusawa, *On the theta lift from SO_{2n+1} to $\widetilde{\mathrm{Sp}}_n$*, J. Reine Angew. Math. **466** (1995), 87–110, DOI 10.1515/crll.1995.466.87. MR1353315

[10] W. T. Gan, *Periods and theta correspondence*, this volume.

[11] W. T. Gan, B. H. Gross, and D. Prasad, *Restrictions of representations of classical groups: examples* (English, with English and French summaries), Astérisque **346** (2012), 111–170. Sur les conjectures de Gross et Prasad. I. MR3202557

[12] W. T. Gan and B. Sun, *The Howe duality conjecture: quaternionic case*, Representation theory, number theory, and invariant theory, Progr. Math., vol. 323, Birkhäuser/Springer, Cham, 2017, pp. 175–192. MR3753911

[13] W. T. Gan and S. Takeda, *A proof of the Howe duality conjecture*, J. Amer. Math. Soc. **29** (2016), no. 2, 473–493, DOI 10.1090/jams/839. MR3454380

[14] D. Ginzburg, D. Jiang, and D. Soudry, *Poles of L-functions and theta liftings for orthogonal groups, II*, On certain L-functions, Clay Math. Proc., vol. 13, Amer. Math. Soc., Providence, RI, 2011, pp. 141–158. MR2767515

[15] R. Gomez and C.-B. Zhu, *Local theta lifting of generalized Whittaker models associated to nilpotent orbits*, Geom. Funct. Anal. **24** (2014), no. 3, 796–853, DOI 10.1007/s00039-014-0276-5. MR3213830

[16] Harish-Chandra, *Admissible invariant distributions on reductive p-adic groups*, Lie theories and their applications (Proc. Ann. Sem. Canad. Math. Congr., Queen's Univ., Kingston, Ont., 1977), Queen's Univ., Kingston, Ont., 1978, pp. 281–347. Queen's Papers in Pure Appl. Math., No. 48. MR0579175

[17] H. He, *Unipotent representations and quantum induction*, preprint (2002), arXiv:math/0210372.

[18] R. Howe, *θ-series and invariant theory*, Automorphic forms, representations and L-functions (Proc. Sympos. Pure Math., Oregon State Univ., Corvallis, Ore., 1977), Proc. Sympos. Pure Math., XXXIII, Amer. Math. Soc., Providence, R.I., 1979, pp. 275–285. MR546602

[19] R. Howe, *Wave front sets of representations of Lie groups*, Automorphic forms, representation theory and arithmetic (Bombay, 1979), Tata Inst. Fund. Res. Studies in Math., vol. 10, Tata Inst. Fundamental Res., Bombay, 1981, pp. 117–140. MR633659

[20] R. Howe, *Transcending classical invariant theory*, J. Amer. Math. Soc. **2** (1989), no. 3, 535–552, DOI 10.2307/1990942. MR985172

[21] D. Jiang, *Periods of automorphic forms*, Proceedings of the International Conference on Complex Geometry and Related Fields, AMS/IP Stud. Adv. Math., vol. 39, Amer. Math. Soc., Providence, RI, 2007, pp. 125–148. MR2338623

[22] N. Kawanaka, *Generalized Gel′fand-Graev representations and Ennola duality*, Algebraic groups and related topics (Kyoto/Nagoya, 1983), Adv. Stud. Pure Math., vol. 6, North-Holland, Amsterdam, 1985, pp. 175–206. MR803335

[23] B. Kostant, *On Whittaker vectors and representation theory*, Invent. Math. **48** (1978), no. 2, 101–184, DOI 10.1007/BF01390249. MR507800

[24] H. Kraft and C. Procesi, *On the geometry of conjugacy classes in classical groups*, Comment. Math. Helv. **57** (1982), no. 4, 539–602, DOI 10.1007/BF02565876. MR694606

[25] J. A. C. Kolk and V. S. Varadarajan, *On the transverse symbol of vectorial distributions and some applications to harmonic analysis*, Indag. Math. (N.S.) **7** (1996), no. 1, 67–96, DOI 10.1016/0019-3577(96)88657-5. MR1621372

[26] S. S. Kudla and S. Rallis, *Degenerate principal series and invariant distributions*, Israel J. Math. **69** (1990), no. 1, 25–45, DOI 10.1007/BF02764727. MR1046171

[27] S. S. Kudla and S. Rallis, *On first occurrence in the local theta correspondence*, Automorphic representations, L-functions and applications: progress and prospects, Ohio State Univ. Math. Res. Inst. Publ., vol. 11, de Gruyter, Berlin, 2005, pp. 273–308, DOI 10.1515/9783110892703.273. MR2192827

[28] R. P. Langlands, *On the classification of irreducible representations of real algebraic groups*, Representation theory and harmonic analysis on semisimple Lie groups, Math. Surveys Monogr., vol. 31, Amer. Math. Soc., Providence, RI, 1989, pp. 101–170, DOI 10.1090/surv/031/03. MR1011897

[29] J.-S. Li, *Singular unitary representations of classical groups*, Invent. Math. **97** (1989), no. 2, 237–255, DOI 10.1007/BF01389041. MR1001840

[30] S. T. Lee and C.-B. Zhu, *Degenerate principal series and local theta correspondence*, Trans. Amer. Math. Soc. **350** (1998), no. 12, 5017–5046, DOI 10.1090/S0002-9947-98-02036-4. MR1443883

[31] S. T. Lee and C.-B. Zhu, *Degenerate principal series and local theta correspondence. II*, Israel J. Math. **100** (1997), 29–59, DOI 10.1007/BF02773634. MR1469104

[32] S. T. Lee and C.-B. Zhu, *Degenerate principal series of metaplectic groups and Howe correspondence*, Automorphic representations and L-functions, Tata Inst. Fundam. Res. Stud. Math., vol. 22, Tata Inst. Fund. Res., Mumbai, 2013, pp. 379–408. MR3156858

[33] H. Y. Loke and J. Ma, *Invariants and K-spectrums of local theta lifts*, Compos. Math. **151** (2015), no. 1, 179–206, DOI 10.1112/S0010437X14007520. MR3305312

[34] H. Matumoto, *$C^{-\infty}$-Whittaker vectors corresponding to a principal nilpotent orbit of a real reductive linear Lie group, and wave front sets*, Compositio Math. **82** (1992), no. 2, 189–244. MR1157939

[35] J.-J. Ma, B. Sun and C.-B. Zhu, *Unipotent representations of real classical groups*, preprint (2017), arXiv:1712.05552.

[36] A. Mínguez, *Correspondance de Howe explicite: paires duales de type II* (French, with English and French summaries), Ann. Sci. Éc. Norm. Supér. (4) **41** (2008), no. 5, 717–741, DOI 10.24033/asens.2080. MR2504432

[37] C. Mœglin, *Correspondance de Howe et front d'onde* (French), Adv. Math. **133** (1998), no. 2, 224–285, DOI 10.1006/aima.1997.1689. MR1604742

[38] C. Mœglin, M.-F. Vignéras, and J.-L. Waldspurger, *Correspondances de Howe sur un corps p-adique* (French), Lecture Notes in Mathematics, vol. 1291, Springer-Verlag, Berlin, 1987. MR1041060

[39] C. Mœglin and J.-L. Waldspurger, *Modèles de Whittaker dégénérés pour des groupes p-adiques* (French), Math. Z. **196** (1987), no. 3, 427–452, DOI 10.1007/BF01200363. MR913667

[40] M. E. Novodvorskiĭ and I. I. Pjateckiĭ-Šapiro, *Generalized Bessel models for the symplectic group of rank* 2 (Russian), Mat. Sb. (N.S.) **90(132)** (1973), 246–256, 326. MR0338279

[41] K. Nishiyama, H. Ochiai, K. Taniguchi, H. Yamashita, and S. Kato, *Nilpotent orbits, associated cycles and Whittaker models for highest weight representations*, Société Mathématique de France, Paris, 2001. Astérisque No. 273 (2001) (2001). MR1845713

[42] K. Nishiyama and C.-B. Zhu, *Theta lifting of unitary lowest weight modules and their associated cycles*, Duke Math. J. **125** (2004), no. 3, 415–465, DOI 10.1215/S0012-7094-04-12531-X. MR2166751

[43] T. Ohta, *The closures of nilpotent orbits in the classical symmetric pairs and their singularities*, Tohoku Math. J. (2) **43** (1991), no. 2, 161–211, DOI 10.2748/tmj/1178227492. MR1104427

[44] T. Przebinda, *Characters, dual pairs, and unitary representations*, Duke Math. J. **69** (1993), no. 3, 547–592, DOI 10.1215/S0012-7094-93-06923-2. MR1208811

[45] S. Rallis, *On the Howe duality conjecture*, Compositio Math. **51** (1984), no. 3, 333–399. MR743016

[46] W. Rossmann, *Picard-Lefschetz theory for the coadjoint quotient of a semisimple Lie algebra*, Invent. Math. **121** (1995), no. 3, 531–578, DOI 10.1007/BF01884311. MR1353308

[47] J. Sekiguchi, *Remarks on real nilpotent orbits of a symmetric pair*, J. Math. Soc. Japan **39** (1987), no. 1, 127–138, DOI 10.2969/jmsj/03910127. MR867991

[48] J. A. Shalika, *The multiplicity one theorem for* GL_n, Ann. of Math. (2) **100** (1974), 171–193, DOI 10.2307/1971071. MR0348047

[49] W. Schmid and K. Vilonen, *Characteristic cycles and wave front cycles of representations of reductive Lie groups*, Ann. of Math. (2) **151** (2000), no. 3, 1071–1118, DOI 10.2307/121129. MR1779564

[50] B. Sun and C.-B. Zhu, *Conservation relations for local theta correspondence*, J. Amer. Math. Soc. **28** (2015), no. 4, 939–983, DOI 10.1090/S0894-0347-2014-00817-1. MR3369906

[51] B. Sun and C.-B. Zhu, *On the conservation conjectures of Kudla and Rallis*, Representation theory, number theory, and invariant theory, Progr. Math., vol. 323, Birkhäuser/Springer, Cham, 2017, pp. 587–602. MR3753924

[52] D. A. Vogan Jr., *Gel′fand-Kirillov dimension for Harish-Chandra modules*, Invent. Math. **48** (1978), no. 1, 75–98, DOI 10.1007/BF01390063. MR0506503

[53] D. A. Vogan Jr., *The algebraic structure of the representation of semisimple Lie groups. I*, Ann. of Math. (2) **109** (1979), no. 1, 1–60, DOI 10.2307/1971266. MR519352

[54] D. A. Vogan Jr., *Associated varieties and unipotent representations*, Harmonic analysis on reductive groups (Brunswick, ME, 1989), Progr. Math., vol. 101, Birkhäuser Boston, Boston, MA, 1991, pp. 315–388. MR1168491

[55] D. A. Vogan Jr., *The method of coadjoint orbits for real reductive groups*, Representation theory of Lie groups (Park City, UT, 1998), IAS/Park City Math. Ser., vol. 8, Amer. Math. Soc., Providence, RI, 2000, pp. 179–238. MR1737729

[56] J.-L. Waldspurger, *Démonstration d'une conjecture de dualité de Howe dans le cas p-adique,* $p \neq 2$ (French), Festschrift in honor of I. I. Piatetski-Shapiro on the occasion of his sixtieth birthday, Part I (Ramat Aviv, 1989), Israel Math. Conf. Proc., vol. 2, Weizmann, Jerusalem, 1990, pp. 267–324. MR1159105

[57] N. R. Wallach, *Lie algebra cohomology and holomorphic continuation of generalized Jacquet integrals*, Representations of Lie groups, Kyoto, Hiroshima, 1986, Adv. Stud. Pure Math., vol. 14, Academic Press, Boston, MA, 1988, pp. 123–151. MR1039836

[58] N. R. Wallach, *Real reductive groups. I*, Pure and Applied Mathematics, vol. 132, Academic Press, Inc., Boston, MA, 1988. MR929683

[59] N. R. Wallach, *Real reductive groups. II*, Pure and Applied Mathematics, vol. 132, Academic Press, Inc., Boston, MA, 1992. MR1170566

[60] H. Weyl, *The Classical Groups. Their Invariants and Representations*, Princeton University Press, Princeton, N.J., 1939. MR0000255

[61] A. Weil, *Sur certains groupes d'opérateurs unitaires* (French), Acta Math. **111** (1964), 143–211, DOI 10.1007/BF02391012. MR0165033

[62] S. Yamana, *Degenerate principal series representations for quaternionic unitary groups*, Israel J. Math. **185** (2011), 77–124, DOI 10.1007/s11856-011-0102-9. MR2837129

[63] H. Yamashita, *On Whittaker vectors for generalized Gel′fand-Graev representations of semisimple Lie groups*, J. Math. Kyoto Univ. **26** (1986), no. 2, 263–298, DOI 10.1215/kjm/1250520922. MR849220

[64] H. Yamashita, *Cayley transform and generalized Whittaker models for irreducible highest weight modules* (English, with English and French summaries), Astérisque **273** (2001), 81–137. Nilpotent orbits, associated cycles and Whittaker models for highest weight representations. MR1845715

Department of Mathematics, National University of Singapore, Block S17, 10 Lower Kent Ridge Road, Singapore 119076

Email address: matzhucb@nus.edu.sg

Published Titles in This Series

101 **Avraham Aizenbud, Dmitry Gourevitch, David Kazhdan, and Erez M. Lapid, Editors,** Representations of Reductive Groups, 2019

100 **Chiu-Chu Melissa Liu and Motohico Mulase, Editors,** Topological Recursion and its Influence in Analysis, Geometry, and Topology, 2018

99 **Vicente Muñoz, Ivan Smith, and Richard P. Thomas, Editors,** Modern Geometry, 2018

98 **Amir-Kian Kashani-Poor, Ruben Minasian, Nikita Nekrasov, and Boris Pioline, Editors,** String-Math 2016, 2018

97 **Tommaso de Fernex, Brendan Hassett, Mircea Mustaţă, Martin Olsson, Mihnea Popa, and Richard Thomas, Editors,** Algebraic Geometry: Salt Lake City 2015 (Parts 1 and 2), 2018

96 **Si Li, Bong H. Lian, Wei Song, and Shing-Tung Yau, Editors,** String-Math 2015, 2017

95 **Izzet Coskun, Tommaso de Fernex, and Angela Gibney, Editors,** Surveys on Recent Developments in Algebraic Geometry, 2017

94 **Mahir Bilen Can, Editor,** Algebraic Groups: Structure and Actions, 2017

93 **Vincent Bouchard, Charles Doran, Stefan Méndez-Diez, and Callum Quigley, Editors,** String-Math 2014, 2016

92 **Kailash C. Misra, Daniel K. Nakano, and Brian J. Parshall, Editors,** Lie Algebras, Lie Superalgebras, Vertex Algebras and Related Topics, 2016

91 **V. Sidoravicius and S. Smirnov, Editors,** Probability and Statistical Physics in St. Petersburg, 2016

90 **Ron Donagi, Sheldon Katz, Albrecht Klemm, and David R. Morrison, Editors,** String-Math 2012, 2015

89 **D. Dolgopyat, Y. Pesin, M. Pollicott, and L. Stoyanov, Editors,** Hyperbolic Dynamics, Fluctuations and Large Deviations, 2015

88 **Ron Donagi, Michael R. Douglas, Ljudmila Kamenova, and Martin Rocek, Editors,** String-Math 2013, 2014

87 **Helge Holden, Barry Simon, and Gerald Teschl, Editors,** Spectral Analysis, Differential Equations and Mathematical Physics: A Festschrift in Honor of Fritz Gesztesy's 60th Birthday, 2013

86 **Kailash C. Misra, Daniel K. Nakano, and Brian J. Parshall, Editors,** Recent Developments in Lie Algebras, Groups and Representation Theory, 2012

85 **Jonathan Block, Jacques Distler, Ron Donagi, and Eric Sharpe, Editors,** String-Math 2011, 2012

84 **Alex H. Barnett, Carolyn S. Gordon, Peter A. Perry, and Alejandro Uribe, Editors,** Spectral Geometry, 2012

83 **Hisham Sati and Urs Schreiber, Editors,** Mathematical Foundations of Quantum Field Theory and Perturbative String Theory, 2011

82 **Michael Usher, Editor,** Low-dimensional and Symplectic Topology, 2011

81 **Robert S. Doran, Greg Friedman, and Jonathan Rosenberg, Editors,** Superstrings, Geometry, Topology, and C^*-algebras, 2010

80 **D. Abramovich, A. Bertram, L. Katzarkov, R. Pandharipande, and M. Thaddeus, Editors,** Algebraic Geometry, 2009

79 **Dorina Mitrea and Marius Mitrea, Editors,** Perspectives in Partial Differential Equations, Harmonic Analysis and Applications, 2008

78 **Ron Y. Donagi and Katrin Wendland, Editors,** From Hodge Theory to Integrability and TQFT, 2008

77 **Pavel Exner, Jonathan P. Keating, Peter Kuchment, Toshikazu Sunada, and Alexander Teplyaev, Editors,** Analysis on Graphs and Its Applications, 2008

76 **Fritz Gesztesy, Percy Deift, Cherie Galvez, Peter Perry, and Wilhelm Schlag, Editors,** Spectral Theory and Mathematical Physics: A Festschrift in Honor of Barry Simon's 60th Birthday, 2007

ISBN 978-1-4704-4284-2